Extremal Paths in Graphs

Mathematical Topics

Volume 10

Extremal Paths in Graphs

Foundations, Search Strategies, and Related Topics

Ulrich Huckenbeck

Akademie Verlag

Author:

Dr. Ulrich Huckenbeck, Ernst-Moritz-Arndt University, Greifswald, Germany

1st edition
With 28 figures
Editorial Director: Dipl.-Math. Gesine Reiher

Die Deutsche Bibliothek – CIP-Einheitsaufnahme

Huckenbeck, Ulrich:
Extremal paths in graphs : foundations, search strategies, and related
topics / Ulrich Huckenbeck. – 1. ed. – Berlin : Akad. Verl., 1997
　(Mathematical topics ; Vol. 10)
　ISBN 3-05-501658-0 Gb.

ISSN 0946-3844

© Akademie Verlag GmbH, Berlin 1997
Akademie Verlag is a member of VCH – a Wiley company.

Printed on non-acid paper.
The paper used corresponds to both the U.S. standard ANSI Z.39.48 – 1984
and the European standard ISO TC 46.

Printing: GAM Media GmbH, Berlin
Bookbinding: Verlagsbuchbinderei Mikolai GmbH, Berlin

Printed in the Federal Republic of Germany

Akademie Verlag GmbH
Mühlenstraße 33–34
D-13187 Berlin
Federal Republic of Germany

Preface

The search for optimal paths in graphs is a classical problem of Graph Theory. This problem appears in many practical situations. The simplest example is the search for a shortest route in a network of streets or railways. A more abstract situation is the search for an optimal sequence of decisions; this situation can often be reduced to the search for an optimal path in a graph. These and other examples are described in Section 1.1.

Originally, the search for optimal paths in graphs belongs to *Graph Theory* and to *Combinatorial Optimization*. As mentioned above, the problem to find an optimal sequence of decisions can often be reduced to an optimal path problem; therefore, many applications of extremal paths in graphs are settled in *Operations Research* and in *Artificial Intelligence*.

The close relationship of the optimal path problem to other fields of Mathematics and Computer Science is one of the reasons for the large collection of literature about paths in graphs. Many of these sources will be considered later in this book. Here, we only mention several sources of literature which are very important for this book. Probably, the first algorithms to find optimal paths were developed by Dijkstra [Dijk59], Dantzig [Dan60], and by Ford [Ford46] and Bellman [Bell58]. There exist many results about very short or very long paths in graphs without describing a search strategy for these paths; several of these results are collected in [Dir52] and [BVMS90]. General surveys about optimization of paths in graphs are given by Deo and Pang [DeoP84] and by Mahr [Mahr80].

This book is an English translation and a revised version of [Huck92], and it has been strongly influenced by the papers [DePe85], [HuRu90], [LT91], [LTh91] about generalized optimal path algorithms.

The main purpose of this book is to give general results about the following aspects of paths in graphs:

- *Paths with non-additive cost measures*

  A cost measure is additive if the cost of a path is the sum of the costs of the arcs or edges used by P. A cost measure is non-additive if the cost of a path cannot be computed in this way.

  The practical relevance of non-additive cost measures will be described in

Section 1.1. The sections 2.3 and 2.4 are about structural investigations of non-additive cost measures. Sections 4.2, 4.3, 4.4, and 4.9 describe the search for optimal or almost optimal paths if the underlying cost measure is not additive. Section 4.2 is strongly influenced by Lengauer and Theune [LT91], [LTh91] and by Dechter and Pearl [DePe85].

- *Relationships to continuous objects*

  Two observations show that paths in graphs and continuous curves are similar objects: The first is that the approximation of a curve by a piecewise linear function looks like a path in a graph (see Figure 26 on page 403). The second observation is that cost measures can be defined for continuous curves or functions as well as for paths in graphs. Finding cost minimal curves is the central problem of *Variational Calculus*.

  In Chapter 5, we give a detailed description of relationships between paths in graphs and continuous objects. We translate many graph theoretic definitions and optimization problems into the continuous setting.

Also, this book gives an overview over the literature about extremal paths in graphs and their applications in other settings.

This book is organized as follows:

Simple examples of path problems are given in *Chapter* 1. Basic terminology and elementary mathematical results are presented in the second part of this chapter.

Chapter 2 is about cost or utility measures for paths in graphs. Each path P is assigned a cost or utility value $H(P)$, which is very often a real number. Thus, we can distinguish between good and bad paths, and we can define the optimality of a path. We consider structural properties of H like additivity, order preservation, and Bellman conditions, and we present modifications and generalizations of these properties.

The chapter about cost and utility measures appears very early in this book. The reason is that these measures must be defined before paths can be compared with each other and optimal candidates can be searched for. That is the reason why we consider structure of cost or utility functions earlier than other aspects of optimal paths in graphs.

Chapter 3 describes, compares, and classifies many combinatorial results about (almost) extremal paths in graphs. (The search for these paths, however, will be described in Chapter 4.) The cost or utility of each path P is measured by the length of P (i.e. the number of arcs of edges used by P). Most results in this chapter are of the following form: Given a graph with a particular property and a class of paths in this graph (for example the class of all paths visiting no node twice) then the length of the longest path in this class must be greater than a given number B, and the length of the shortest candidate must be smaller than a given number B.

The properties of (almost) extremal paths are considered earlier than the search for these paths. The reason is that it is often sensible to study the properties of extremal paths before searching for them.

Chapter 4 is about algorithms to find optimal or almost optimal paths in graphs. Section 4.1 gives an introduction to optimal path problems. In particular, we shall formulate a simple search problem for optimal paths, and we shall introduce Dijkstra's algorithm, which solves this problem. In Sections 4.2, 4.3, and 4.4, we shall formulate generalizations of standard search problems for optimal paths. Moreover, we study generalized versions of the algorithms of Dijkstra, of Ford and Bellman, and of Floyd.

The other sections of Chapter 4 are about modifications of standard optimization problems. An example of such a modification is the Traveling Salesman Problem, which means to search for an optimal path with side constraints.

Chapter 5 compares paths in directed graphs to continuous curves in the plane.

Each chapter is almost *independent* of the other ones. For example, Chapter 4 can be read without studying Chapters 1 – 3; it is sufficient to consult these chapters when a term or a notation in Chapter 4 is not clear.

We make several technical remarks. Five independent counters are used in this book. The first counter is used for theorems, lemmas, definitions, remarks, etc.; the second is used for equations; the third is used for problems; the fourth is used for algorithms; the fifth is used for Figures. The first and the second counter depend on the current chapter, the last three counters do not. For example, equation (i, j) is the jth equation of Chapter i; Problem j is the jth numbered problem of the whole book.

The following marks are used up to few exceptions: Assumptions of theorems are marked with (i), (ii), (iii), etc.; an example is Theorem 4.34. Parts of definitions or of remarks and assertions of theorems are marked with a), b), c), etc.; for example, Theorem 4.111 consists of the assertions 4.111 a) – 4.111 f). In contrast to this, the marks (a),(b),(c), etc. are used if a theorem claims the logical relationship between the marked assertions; for example, Theorem 2.19 does not say that (a) or (b) is true, but it does say that (a) and (b) are equivalent.

The symbol "□" is used to mark the end of a definition, of a remark or of a proof.

Acknowlededements:

I thank all persons who supported my project to write a book about external paths in graphs. In particular, I thank Prof. Widmayer, whose enthusiastic report helped to convince the Akademie Verlag to publish my book.

I am indebted to my colleages Prof. Asser, Prof. Bär, Dr. Cieslik, Dr. Gaßner, Prof. Hemmerling, Mrs. Köhler, Prof. Schreiber, Dr. Tews, and Prof. Voelkel at Greifswald University; they appreciated my decision to write this book, and they always understood my problems when I was preparing the manuscript.

I thank all persons who gave me advice about English language and about the typesetting system LaTeX. Also, I thank all persons who have patiently waited for my book.

At last, I wish to thank Mrs. Reiher (Akademie Verlag), who was always ready to help me when I was writing this book.

Contents

Chapter 1

Introduction

1.1 Examples of extremal paths in graphs

Many practical problems can be reduced to the search for extremal paths in graphs. Here, we give several examples.

The most obvious situation is a network of rails or roads; the problem is to find a shortest route in this network. For example, Figure 1 shows the situation of a traveller who wants to go by train from A to G. If the fare is proportional to the distance the traveller will use the shortest route $[A, D, E, F, G]$, which has the length $212 + 206 + 200 + 206 = 824$.

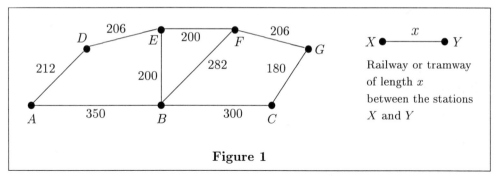

Figure 1

A more abstract situation is described in [Nils82]. An $n \times n$-puzzle must be transformed from a start configuration into an end configuration by executing a minimal number of moves. Each sequence of moves is represented by a path in a directed graph.

The next example is a modification of the first. The original cost measure in Figure 1 is additive, i.e., the cost (= length) of a route $[v_0, \ldots, v_k]$ is computed by adding the costs of the pieces $[v_\kappa, v_{\kappa+1}]$, $\kappa = 0, \ldots, k - 1$. Now we use another cost measure. We interpret Figure 1 as the network of tram lines. A short trip with at most one intermediate stop costs \$ 1,–, and a longer trip costs \$ 2,–. Consequently, the trip $[A, D, E, B, F]$ costs \$ 2,–. Adding the costs of the edges, however, yields another price; each of the four trips $[A, D]$, $[D, E]$, $[E, B]$, and $[B, F]$ costs \$ 1,–, and the sum of these fares is \$ 4,–. That means that this cost measure is not additive. So, searching for an optimal route means to search for a cost minimal path which a non-additive cost measure. In this special case, however, we can reduce this search problem to the search for a path with the minimal number of edges, and the number of edges of a path is an additive cost measure.

The following example shows that the maximum cost during a procedure may be more relevant than the cost at the end of that procedure. We assume that a chemical experiment occurs in a capsule. A sequence of chemical reactions $X_1, \ldots, X_k$ is interpreted as a path $[X_1, \ldots, X_k]$ in a graph. We define the cost of each path $[X_1, \ldots, X_\kappa]$ ($\kappa \leq k$) as the temperature in the capsule immediately after the reaction X_κ. The fact that the maximum cost of the path $[X_1, \ldots, X_k]$ is too high means that one of the reactions $X_1, \ldots, X_k$ causes a too high temperature, and the capsule is destroyed. Therefore, it is reasonable to search for a path $[X_1, \ldots, X_k]$ of low maximum cost.

Another situation is given in [HuRu90], Example 1.2, where a person buys and sells houses; each sequence of purchases and sales is represented as a path in a directed graph. The owner of the real estate intends to maximize his profit or to minimize his debt after the last transaction; that means that he is searching for a path of minimal cost or of maximal profit. His bank, however, tries to minimize its customer's intermediate debts; that means that the bank is searching for a path whose maximum cost is minimal.

Sometimes, the worst path is more important than the optimal one. This is the case if a complicated project is described with the help of a digraph. More details are given in Example 3.51 and in Theorem 3.52 of this book and in [Neu75].

Another example for the practical relevance of paths in graphs is a cleaning process described in Remark 4.126. Further examples can be found in [LT91] and in [LTh91].

1.2 Basic Terminology

Mathematical terminology:

- *Sets:*
$\mathrm{I\!N} := \{1, 2, 3, \ldots\}$ is the set of all natural numbers, and $\mathrm{I\!R}$ is set of all real numbers. Open bounds of intervals are symbolized by round brackets, and closed bounds are symbolized by square brackets. For example, $(5, 18] := \{x \in \mathrm{I\!R} \mid 5 < x \leq 18\}$.
$|X| :=$ cardinality of the set X.
$X \subseteq Y$ means that X is a subset of Y; X may be equal to Y.

- *Functions and other relations:*
Given two sets X, Y, a relation $R \subseteq X \times Y$, and $x \in X$.
Then we define $R[x] := \{y \in Y \mid (x, y) \in R\}$. Let $\mathrm{def}(R) := \{x \in X \mid R[x] \neq \emptyset\}$; $\mathrm{def}(R)$ is the *domain of definition* of R.
$R : X \dashrightarrow Y$ means that R is a partial function, i.e., for each $x \in X$, there exists at most one $y \in Y$ with $(x, y) \in R$; in this case, we write $R : X \dashrightarrow Y$.
$R : X \to Y$ means that R is a (total) function, i.e., for each $x \in X$, there exists exactly one $y \in Y$ with $(x, y) \in R$.
Let $Y^X :=$ set of all total functions $R : X \to Y$.
We shall sometimes write $x \mapsto R(x)$ if R is a partial or even a total function.

Graph theoretic terminology:

• *Directed Graphs (Digraphs):*

A *directed graph* or *digraph* is a pair $(\mathcal{V}, \mathcal{R})$ with $\mathcal{V} \neq \emptyset$, $\mathcal{R} \subseteq \mathcal{V} \times \mathcal{V}$, and $\mathcal{V} \cap \mathcal{R} = \emptyset$.[1]
A *node* or *vertex* is an element of $\mathcal{V}$; an *arc* is an element of $\mathcal{R}$.
If $r = (v, w) \in \mathcal{R}$ then $\alpha(r) := v$ (*start node of r*), and $\omega(r) := w$ (*end node of r*).[2]
We define $\alpha(v) := \omega(v) := v$ for all $v \in \mathcal{V}$.

• *Subgraphs and supergraphs:*

Let $\mathcal{G}_i = (\mathcal{V}_i, \mathcal{R}_i)$, $i = 1, 2$.
$\mathcal{G}_1$ is a *subgraph* of $\mathcal{G}_2$ if $\mathcal{V}_1 \subseteq \mathcal{V}_2$ and $\mathcal{R}_1 \subseteq \mathcal{R}_2$.
Let $\mathcal{G}_1$ be a subgraph of $\mathcal{G}_2$. Then $\mathcal{G}_1$ is *induced by $\mathcal{V}_1$* if $\mathcal{R}_1 = \mathcal{R}_2 \cap (\mathcal{V}_1 \times \mathcal{V}_1)$, and $\mathcal{G}_1$ is *induced by $\mathcal{R}_1$* if $\mathcal{V}_1 = \mathcal{V}_2$.
$\mathcal{G}_2$ is a *supergraph* of $\mathcal{G}_1$ if and only if $\mathcal{G}_1$ is a subgraph of $\mathcal{G}_2$.

• *Local properties of digraphs:*

Given a digraph $\mathcal{G} = (\mathcal{V}, \mathcal{R})$; let $v, w \in \mathcal{V}$ and $r, r' \in \mathcal{R}$. Then we define
$$\mathcal{R}^+(v) := \{\widetilde{r} \in \mathcal{R} \mid \alpha(\widetilde{r}) = v\}; \quad \deg^+(v) := |\mathcal{R}^+(v)| \; (\textit{outdegree of } v);$$
$$\mathcal{R}^-(v) := \{\widetilde{r} \in \mathcal{R} \mid \omega(\widetilde{r}) = v\}; \quad \deg^-(v) := |\mathcal{R}^-(v)| \; (\textit{indegree of } v);$$
We define $\deg(v) := \deg^-(v) + \deg^+(v)$ (*degree of v*).
The node w is a *successor* of v if $(v, w) \in \mathcal{R}$; w is a *predecessor* of v if $(w, v) \in \mathcal{R}$.
Let $\mathcal{N}^+(v) :=$ set of all successors of v and $\mathcal{N}^-(v) :=$ set of all predecessors of v;
r is a *loop* if $r = (x, x)$ for some $x \in \mathcal{V}$.
r and r' are *inverse to each other* if two nodes x, y exist with $r = (x, y)$, $r' = (y, x)$.

• *Further properties of digraphs:*

Given a digraph $\mathcal{G} = (\mathcal{V}, \mathcal{R})$. The graph $\mathcal{G}$ is *finite* if $\mathcal{V}$ is a finite set; then $\mathcal{R} \subseteq \mathcal{V} \times \mathcal{V}$ is finite, too. $\mathcal{G}$ is *locally finite* if every node of $\mathcal{G}$ has a finite outdegree.
The graph $\mathcal{G}$ is *loopfree* or *simple* if $\mathcal{G}$ has no loops[3]. $\mathcal{G}$ is *oriented* if $\mathcal{G}$ is simple and has no pair of inverse arcs. $\mathcal{G}$ is *acyclic* if $\mathcal{G}$ has no sequence $[v_0, v_1 \ldots, v_k]$ of nodes such that $k > 0$, $v_k = v_0$, and $(v_\kappa, v_{\kappa+1}) \in \mathcal{R}$ for all κ.
The graph $\mathcal{G}$ is *complete* if $\mathcal{R} = \mathcal{V} \times \mathcal{V}$ or $\mathcal{R} = (\mathcal{V} \times \mathcal{V}) \backslash \{(v, v) \mid v \in \mathcal{V}\}$.

• *Paths in digraphs:*

Given a digraph $\mathcal{G} = (\mathcal{V}, \mathcal{R})$.
A sequence $P = [v_0, \ldots, v_k]$ is a (*finite*) *path* if $(v_\kappa, v_{\kappa+1}) \in \mathcal{R}$ for all κ. If P is given we define $\alpha(P) := v_0$ (*start node of P*), $\omega(P) := v_k$ (*end node of P*), and $\ell(P) := k$ (*length of P*).
In particular, if $k = 0$ then $P = [v_0]$ is a path of length 0. If $k = 1$ then the path $P = [v_0, v_1]$ is different from the arc $r := (v_0, v_1)$; nevertheless, we shall sometimes identify P and r. — We call P a *substantial path* if $k > 0$.
We write $v \in P$ and $r \in P$, respectively, if v is a node of P and r is an arc of P.

---

[1]This property is not trivial. E.g., let $\mathcal{V} = \{1, 2, (1, 2)\}$. Then $(1, 2) \in \mathcal{V} \cap (\mathcal{V} \times \mathcal{V})$.
[2]Our definition of arcs as ordered pairs of nodes implies that $r = r'$ if $\alpha(r) = \alpha(r')$ and $\omega(r) = \omega(r')$. There exist, however, other definitions of arcs in graphs such that the following situation may occur: $r \neq r'$, $\alpha(r) = \alpha(r')$, $\omega(r')$; then r and r' are *parallel* or *multiple arcs*; a *parallel free* digraph has no parallel arcs.
[3]Usually, a graph is simple if it has no loops and no parallel arcs; but $\mathcal{G} = (\mathcal{V}, \mathcal{R})$ cannot have parallel arcs; so, $\mathcal{G}$ is automatically simple if it is loopfree.

P is a *cycle* or a *closed path* if $\alpha(P) = \omega(P)$.

(So, the graph $\mathcal{G}$ is acyclic if and only if it has no cycle of length > 0.)

P is *simple* or *arc injective* if it does not use any arc twice.

P is *elementary* or *node injective* if one of the following assertions is true:

- $v_0, \ldots, v_k$ are pairwise distinct;
- $v_0, \ldots, v_{k-1}$ are pairwise distinct, and $v_k = v_0$ (then P is a node injective cycle).

P is *acyclic* if P does not visit any node more than once; consequently, the only acyclic cycles are of the form $[v]$ where $v \in \mathcal{V}$.

A *Eulerian path* in $\mathcal{G}$ is an arc injective path that uses all arcs of $\mathcal{G}$ at least once (and in consequence exactly once). A *Hamiltonian path* is a node injective path that visits each node of $\mathcal{G}$ at least once.

A *Eulerian [Hamiltonian] cycle* is a Eulerian [Hamiltonian] path P with $\alpha(P) = \omega(P)$. A sequence $P = [v_0, v_1, v_2, \ldots)$ is an *infinite path* if $(v_\kappa, v_{\kappa+1}) \in \mathcal{R}$ for all κ. Many properties like "*node injective*", "*acyclic*", and "*arc injective*" are also defined for infinite paths. A path P is always meant to be finite unless the contrary is stated.

- *Sets of paths:*

Let $\mathcal{G} = (\mathcal{V}, \mathcal{R})$ be a digraph, $v, w \in \mathcal{R}$, $\mathcal{V}', \mathcal{V}'' \subseteq \mathcal{V}$. Then we define

$\mathcal{P}(\mathcal{G}) :=$ set of all paths in $\mathcal{G}$, $\qquad \mathcal{P}^\circ(\mathcal{G}) :=$ set of all cycles in $\mathcal{G}$,

$\mathcal{P}(v) := \{P \in \mathcal{P}(\mathcal{G}) \,|\, \alpha(P) = v\}$, $\quad \mathcal{P}(v, w) := \{P \in \mathcal{P}(\mathcal{G}) \,|\, \alpha(P) = v, \ \omega(P) = w\}$,

$\mathcal{P}(v, \mathcal{V}'') := \{P \in \mathcal{P}(\mathcal{G}) \,|\, \alpha(P) = v, \omega(P) \in \mathcal{V}''\}$,

$\mathcal{P}(\mathcal{V}', \mathcal{V}'') := \{P \in \mathcal{P}(\mathcal{G}) \,|\, \alpha(P) \in \mathcal{V}', \omega(P) \in \mathcal{V}''\}$.

Each element of $\mathcal{P}(v, w)$ [of $\mathcal{P}(v, \mathcal{V}'')$, of $\mathcal{P}(\mathcal{V}', \mathcal{V}'')$, resp.] is called a *$v$-$w$-path* [a *$v$-$\mathcal{V}''$-path*, a *$\mathcal{V}'$-$\mathcal{V}''$-path*, resp.].

- *Concatenation of paths and arcs:*

Let P, Q be two paths with $\omega(P) = \alpha(Q)$; then the *concatenation* $P \oplus Q$ is defined as the path generated by appending Q to P; also, the operands of $'\oplus'$ may be arcs. For example, if $X = [v_0, \ldots, v_5]$, we may write X as follows:

$$
\begin{aligned}
X \;&= [v_0, v_1, v_2] \oplus [v_2, v_3, v_4, v_4] \;= [v_0, \ldots, v_4] \oplus [v_4, v_5] \;= [v_0, \ldots, v_4] \oplus (v_4, v_5) \\
&= [v_0, v_1, v_2] \oplus (v_2, v_3) \oplus [v_3, v_4, v_5] \oplus (v_2, v_3) \;= (v_0, v_1) \oplus (v_1, v_2) \oplus \ldots \oplus (v_4, v_5) \ \text{ etc.}
\end{aligned}
$$

If P is a cycle then P^ν is the ν-fold concatenation of P; for example, $P^0 = [\alpha(P)]$, $P^1 = P$, $P^2 = P \oplus P$, and so on.

- *Subpaths:*

Given a path $P = [v_0, \ldots, v_k]$.

Any path $Q = [v_i, \ldots, v_j]$ with $0 \leq i \leq j \leq k$ is called a *subpath* or an *infix* of P; this is abbreviated by $Q \subseteq P$. If P and Q have the same start node (i.e. $i = 0$) then Q is called a *prefix* of P; this is abbreviated by $Q \leq P$. If Q and P have the same end node (i.e. $j = k$) then Q is called a *suffix* of P; this is abbreviated by $Q \preceq P$. In particular, P is a subpath, a prefix and a suffix of itself.

Q is called a *proper infix* of P if $Q \subseteq P$ and $Q \neq P$. Proper prefixes and suffixes of P are defined analogically.

Prefixes of infinite paths are defined in the same way as prefixes of finite paths.

A set $\mathcal{P}$ of paths is *prefix closed* if $(P \in \mathcal{P} \land Q \leq P) \Rightarrow Q \in \mathcal{P}$ for all paths P, Q; e.g., $\mathcal{P} = \mathcal{P}(v)$ is prefix closed for all nodes v.

- *Graph homomorphisms:*

Given two digraphs $\mathcal{G}_i = (\mathcal{V}_i, \mathcal{R}_i)$ $(i = 1, 2)$. A function $\Psi : \mathcal{V}_1 \cup \mathcal{R}_1 \to \mathcal{V}_2 \cup \mathcal{R}_2$ is a *(graph) homomorphism* if Ψ is surjective and the following is true:

$$(\forall\, x \in \mathcal{V}_1 \cup \mathcal{R}_1)\ \big(\alpha(\Psi(x)) = \Psi(\alpha(x)) \quad \text{and} \quad \omega(\Psi(x)) = \Psi(\omega(x))\big). \tag{1.1}$$

It is well-known that $\Psi(\mathcal{V}_1) = \mathcal{V}_2$.

Ψ is an *isomorphism* if Ψ is bijective. If, in addition, $\mathcal{G}_1 = \mathcal{G}_2$ then Ψ is an *automorphism*.

Any graph homomorphism can be extended to $\mathcal{P}(\mathcal{G}_1)$; if $P = [v_0, v_1, \ldots, v_k]$ is a path in $\mathcal{G}_1$, then we define $\Psi(P)$ as follows: $\Psi(P) := [\Psi(v_0)]$ if $k = 0$; $\Psi(P) := [\Psi(v_0, v_1)]$ if $k = 1$ and $\Psi(v_0, v_1) \in \mathcal{V}_2$; $\Psi(P) := [\Psi(v_0), \Psi(v_1)]$ if $k = 1$ and $\Psi(v_0, v_1) \in \mathcal{R}_2$; $\Psi(P) := \Psi([v_0, v_1]) \oplus \Psi([v_1, v_2]) \oplus \ldots \oplus \Psi([v_{k-1}, v_k])$ if $k > 1$.

Then (1.1) is also true if x is a path.

Extended homomorphisms are not always surjective as shown in the following example: Let $\mathcal{V}_1 := \{x_1, x_2, \ldots, x_6\}$, $\mathcal{R}_1 := \{(v_i, v_{i+1}) \,|\, 1 \leq i \leq 5\}$, and let $\mathcal{V}_2 := \{y_1, y_3, y_4, y_6, \overline{y}\}$, $\mathcal{R}_2 := \{(y_1, \overline{y}), (\overline{y}, y_3), (y_3, y_4), (y_4, \overline{y}), (\overline{y}, y_6)\}$. Let $\Psi(x_2) := \Psi(x_5) := \overline{y}$, $\Psi(x_i) := y_i$ for all other x_i, and let $\Psi(v, w) := (\Psi(v), \Psi(w))$ for all arcs $(v, w) \in \mathcal{R}_1$. Then there exists no path $P \in \mathcal{P}(\mathcal{G}_1)$ with $\Psi(P) = [y_1, \overline{y}, y_6]$.

- *Expanded version of a digraph:*

Given a digraph $\mathcal{G} = (\mathcal{V}, \mathcal{R})$ and a node s; we assume that an s-v-path exists for each node $v \in \mathcal{V}$. Then the *s-expansion of $\mathcal{G}$* is defined as $\mathcal{G}^s = (\mathcal{V}^s, \mathcal{R}^s)$ with $\mathcal{V}^s := \mathcal{P}(s)$ and $\mathcal{R}^s := \{(P, P \oplus r) \,|\, P \in \mathcal{P}(s)\,,\, r \in \mathcal{R}\,,\, \alpha(r) = \omega(P)\}$.

$\mathcal{G}^s$ has the following property: If $P = [s, v_1, \ldots, v_k] \in \mathcal{P}(s)$ and $P_i = [s, v_1, \ldots, v_i]$, $i = 0, \ldots, k$, then $[P_0, P_1, \ldots, P_j] \in \mathcal{P}(\mathcal{G}^s)$ for each j; vice versa, if $[[s], Q_1, \ldots, Q_j] \in \mathcal{P}(\mathcal{G}^s)$ then there exists j nodes $w_1, \ldots, w_j \in \mathcal{V}$ such that $Q_i = [s, w_1, \ldots, w_i]$ for all $i = 0, \ldots, j$.

Moreover, $\mathcal{G}$ is a homomorphic image of $\mathcal{G}^s$; let $\Psi(P) := \omega(P)$ for all $P \in \mathcal{V}^s$, and $\Psi(P, P \oplus r) := r$ for all $(P, P \oplus r) \in \mathcal{R}^s$; then Ψ is a homomorphism; in particular, Ψ is surjective because an s-v-path exists for each $v \in \mathcal{V}$.

We extend Ψ to $\mathcal{P}(\mathcal{G}^s)$. Let $[P, P \oplus (v, w)] \in \mathcal{P}(\mathcal{G}^s)$. Then

$$\Psi\big([P, P \oplus (v, w)]\big) = [v, w] \tag{1.2}$$

because $\Psi\big(P, P \oplus (v, w)\big) = (v, w) \in \mathcal{R}$. Moreover,

$$\Psi(P') = \omega(P') \tag{1.3}$$

for all $\mathcal{G}^s$-paths $P' \in \mathcal{P}([s])$. To see this we assume that $P_0 = [s]$ and $P' = [P_0, P_1, \ldots, P_k]$. If $k = 0$ then $P' = [P_0]$, and (1.3) follows from $\Psi(P') = P_0 = \omega(P')$. If $k > 0$, then k nodes $v_1, \ldots, v_k \in \mathcal{V}$ exist such that $P_i = [s, v_1, \ldots, v_i]$ for all $i = 0, \ldots, k$. Then

$$\Psi(P') = \Psi([P_0, P_1]) \oplus \Psi([P_1, P_2]) \oplus \ldots \oplus \Psi([P_{k-1}, P_k]) =$$
$$\Psi([P_0, P_0 \oplus (s, v_1)]) \oplus \Psi([P_1, P_1 \oplus (v_1, v_2)]) \oplus \ldots \oplus \Psi([P_{k-1}, P_{k-1} \oplus (v_{k-1}, v_k)])$$
$$\overset{(1.2)}{=} (s, v_1) \oplus \ldots \oplus (v_{k-1}, v_k) = P_k = \omega(P').$$

Each path $P = [s, v_1, \ldots, v_k] \in \mathcal{P}(s)$ has an inverse image under Ψ; let $P_i := [s, v_1, \ldots, v_i]$, $0 \leq i \leq k$, and let $P' := [P_0, P_1, \ldots, P_k]$; then $\Psi(P') \overset{(1.3)}{=} \omega(P') = P_k = P$.

- *Copying graphs into several levels:*

Given a digraph $\mathcal{G} = (\mathcal{V}, \mathcal{R})$ and a fixed constant $K \geq 2$. We copy the nodes of $\mathcal{G}$ into the levels $\kappa = 0, \ldots, K - 1$, and we use the levels to record the lengths of paths. (This idea is called "timestamp method".) More precisely, we construct the digraph $\mathcal{G}^{[K]}$ with the following properties:

a) $\mathcal{G}$ is a homomorphic image of $\mathcal{G}^{[K]}$.

b) $\mathcal{G}^{[K]}$ is *acyclic*; moreover, each $\mathcal{G}$-path of a length $\leq K - 1$ has a homomorphic inverse in $\mathcal{G}^{[K]}$.

We define $\mathcal{G}^{[K]} := (\mathcal{V}^{[K]}, \mathcal{R}^{[K]})$ as follows: Let $\mathcal{V}^{[K]} := \mathcal{V} \times \{0, \ldots, K - 1\}$ and
$$\mathcal{R}^{[K]} := \{((v, i), (w, i + 1)) \mid (v, w) \in \mathcal{R}, \ i = 0, \ldots, K - 2\}.$$
Let $\Psi(v, i) := v$ for all $(v, i) \in \mathcal{V}^{[K]}$, and let $\Psi((v, i), (w, j)) := (v, w)$ for all $((v, i), (w, j)) \in \mathcal{R}^{[K]}$. Then Ψ is a homomorphism from $\mathcal{G}^{[K]}$ onto $\mathcal{G}$.

Each path $P = [v_0, \ldots, v_k] \in \mathcal{P}(\mathcal{G})$ with $k \leq K - 1$ is represented by the path $P' = [(v_0, 0), (v_1, 1), \ldots, (v_k, k)]$ in $\mathcal{G}^{[K]}$. Moreover, $\Psi(P') = P$.

The digraph $\mathcal{G}^{[K]}$ is useful when searching an optimal path of a length $< K$ in $\mathcal{G}$. When solving the same problem in $\mathcal{G}^{[K]}$ we can make use of the fact that $\mathcal{G}^{[K]}$ is acyclic. Examples for replacing $\mathcal{G}$ by $\mathcal{G}^{[K]}$ can be found in 4.60 and in 4.117.

- *Undirected Graphs:*

An *undirected graph* is a pair $(\mathcal{V}, \mathcal{R})$ with $\mathcal{V} \neq \emptyset$, $\mathcal{R} \subseteq \{\{x, y\} \mid x, y \in \mathcal{V}\}$, and $\mathcal{V} \cap \mathcal{R} = \emptyset$.[4]

A *node* or *vertex* is an element of $\mathcal{V}$; an *edge* is an element of $\mathcal{R}$.

If $r = \{v, w\} \in \mathcal{R}$ then v is an *endpoint* of r, and v and w are *neighbours* of each other.[5] If $v = w$ then $r = \{v\}$ is a *loop*. Let $\mathcal{N}(v)$ be the set of all neighbours of v.

The *degree* or *valence* of a node v is defined as follows: Let $X := \{\{v, w\} \in \mathcal{R} \mid w \neq v\}$; then $\deg(v) := |X|$ if $\{v\} \notin \mathcal{R}$, and $\deg(v) := |X| + 2$ if $\{v\} \in \mathcal{R}$; that means that each loop $\{v\}$ is counted twice.

Finite, *loopfree* and *simple* undirected graphs are defined in the same way as in the directed case.

The graph $\mathcal{G} = (\mathcal{V}, \mathcal{R})$ is *complete* if $\mathcal{R} = \{\{v, w\} \mid v \neq w\}$;[6] in this case, we define $\mathcal{K}_\mathcal{V} := \mathcal{G}$. Let $n := |\mathcal{V}|$, and let $\widetilde{\mathcal{G}}$ be isomorphic to $\mathcal{G}$; then $\widetilde{\mathcal{G}}$ is called a $\mathcal{K}_n$.

The graph $\mathcal{G} = (\mathcal{V}, \mathcal{R})$ is called *complete bipartite* if there exist two disjoint nonempty sets $\mathcal{V}', \mathcal{V}'' \subseteq \mathcal{V}$ of nodes such that $\mathcal{V}' \cup \mathcal{V}'' = \mathcal{V}$ and $\mathcal{R}$ is the set of all edges $\{v', v''\}$ with $v' \in \mathcal{V}', v'' \in \mathcal{V}''$; this graph is written as $\mathcal{K}_{\mathcal{V}', \mathcal{V}''}$. Let $n' := |\mathcal{V}'|$, $n'' := |\mathcal{V}''|$, and let $\widetilde{\mathcal{G}}$ be isomorphic to $\mathcal{G}$; then $\widetilde{\mathcal{G}}$ is called a $\mathcal{K}_{n', n''}$.

A *directed version of* $\mathcal{G}$ is a digraph $\vec{\mathcal{G}} = (\mathcal{V}, \vec{\mathcal{R}})$ constructed by orienting each edge of the graph $\mathcal{G}$. For example, let $\mathcal{V} := \{x_1, x_2, x_3\}$, $\mathcal{R} := \{\{x_1, x_2\} \{x_2, x_3\}\}$, and $\vec{\mathcal{R}} := \{(x_1, x_2), (x_3, x_2)\}$; then $\vec{\mathcal{G}} = (\mathcal{V}, \vec{\mathcal{R}})$ is a directed version of $\mathcal{G} = (\mathcal{V}, \mathcal{R})$.

---

[4]This condition is not trivial. E.g., let $\mathcal{V} = \{1, 2, \{1, 2\}\}$ and $\mathcal{R} = \{\{1, 2\}\}$; then $\{1, 2\} \in \mathcal{V} \cap \mathcal{R}$.

[5]Our definition of edges implies that $r = r'$ if r and r' have the same endpoints. There exist, however, other definitions of graphs such that the following situation may occur: $r \neq r'$ although r and r' have the same endpoints; then r and r' are *parallel* or *multiple edges*; a *parallel free* graph has no parallel edges.

[6]We do not define complete undirected graphs *with* loops because they do not appear in this book.

- *Paths in undirected graphs:*

A *path* is a finite sequence $P = [v_0, v_1, \ldots, v_k]$ such that $\{v_\kappa, v_{\kappa+1}\} \in \mathcal{R}$ for all κ. We define $\alpha(P) := v_0$ (*start node*), $\omega(P) := v_k$ (*end node*), $\ell(P) := k$ (*length of P*). The terms "*cycle*", "*simple path*", "*elementary path*", "*Eulerian path*", "*Hamiltonian path*", etc. are analogous to the directed case; in particular, a simple path in an undirected graph must not use any edge twice, not even in different directions; such a path is called *edge injective* instead of *arc injective*.

The concatenation operator $''\oplus''$ for paths is defined in the same way as in the directed case; we admit the use of $''\oplus''$ to describe mixed concatenations of paths and edges, for example $P \oplus r \oplus Q$ with $r \in \mathcal{R}$; the notation P^ν means the ν-fold concatenation of a cycle P.

Generalized order relations

- *Definition of new relation with the help of a given one:*

Let $\mathbf{R}$ be a nonempty set, and let $\Re \subseteq \mathbf{R} \times \mathbf{R}$ be transitive and reflexive. We define the following relations. Let $\varrho_1, \varrho_2 \in \mathbf{R}$; then

$$\varrho_1 \, \Re^+ \, \varrho_2 \quad :\Longleftrightarrow \quad \varrho_1 \, \Re \, \varrho_2 \wedge \neg \, \varrho_2 \, \Re \, \varrho_1, \qquad \varrho_1 \, \Re^0 \, \varrho_2 \quad :\Longleftrightarrow \quad \varrho_1 \, \Re \, \varrho_2 \wedge \varrho_2 \, \Re \, \varrho_1,$$

$$\varrho_1 \, \overline{\Re} \, \varrho_2 \quad :\Longleftrightarrow \quad \varrho_2 \, \Re \, \varrho_1, \qquad \varrho_1 \, \overline{\Re}^+ \, \varrho_2 \quad :\Longleftrightarrow \quad \varrho_2 \, \Re \, \varrho_1 \wedge \neg \, \varrho_1 \, \Re \, \varrho_2.$$

We define $\mathbf{R}/\Re^0$ as the set of all equivalence classes with regard to the equivalence relation $\Re^0$.

Given the following (possibly subjective) assumption:

$$\varrho_1 \, \Re^+ \, \varrho_2 \text{ means that } \varrho_1 \text{ is "smaller" than } \varrho_2. \tag{1.4}$$

Then we write $\varrho_1 \preceq \varrho_2$ instead of $\varrho_1 \, \Re \, \varrho_2$, and we write $\varrho_1 \prec \varrho_2$ if $\varrho_1 \, \Re^+ \, \varrho_2$, $\varrho_1 \equiv \varrho_2$ if $\varrho_1 \, \Re^0 \, \varrho_2$, $\varrho_1 \succeq \varrho_2$ if $\varrho_1 \, \overline{\Re} \, \varrho_2$, and $\varrho_1 \succ \varrho_2$ if $\varrho_1 \, \overline{\Re}^+ \, \varrho_2$. $\mathbf{R}/\equiv$ is the set of all equivalence classes with regard to $\equiv$.

For example, assumption (1.4) is given if $\varrho_1 \Re^+ \varrho_2$ means that $\varrho_1 < \varrho_2$ for $\varrho_1, \varrho_2 \in \mathbb{R}$ or that the set ϱ_1 is a proper subset of ϱ_2. A counterexample is that $\Re$ equals $\geq$; therefore, we shall never write $\varrho_1 \preceq \varrho_2$ instead of $\varrho_1 \geq \varrho_2$.

Given the following (possibly subjective) assumption:

$$\varrho_1 \, \Re^+ \, \varrho_2 \text{ means that } \varrho_1 \text{ is "greater" than } \varrho_2. \tag{1.5}$$

Then we write $\varrho_1 \succeq \varrho_2$ instead of $\varrho_1 \, \Re \, \varrho_2$, and we write $\varrho_1 \succ \varrho_2$ if $\varrho_1 \, \Re^+ \, \varrho_2$, $\varrho_1 \equiv \varrho_2$ if $\varrho_1 \, \Re^0 \, \varrho_2$, $\varrho_1 \preceq \varrho_2$ if $\varrho_1 \, \overline{\Re} \, \varrho_2$, and $\varrho_1 \prec \varrho_2$ if $\varrho_1 \, \overline{\Re}^+ \, \varrho_2$. $\mathbf{R}/\equiv$ is the set of all equivalence classes with regard to $\equiv$.

For example, assumption (1.5) is given if $\Re$ equals $'' \geq ''$.

- *Structural properties of relations:*

Let $\Re \subseteq \mathbf{R} \times \mathbf{R}$ be transitive and reflexive.

$\Re$ is *total* if $(\varrho_1 \, \Re \, \varrho_2 \vee \varrho_2 \, \Re \, \varrho_1)$ for all $\varrho_1, \varrho_2 \in \mathbf{R}$; of course, $\Re$ is total if and only if $\overline{\Re}$ is total.

$\Re$ is *identitive* if $(\varrho_1 \, \Re \, \varrho_2 \wedge \varrho_2 \, \Re \, \varrho_1) \Rightarrow \varrho_1 = \varrho_2$ for all $\varrho_1, \varrho_2 \in \mathbf{R}$; of course, $\Re$ is identitive if and only if $\overline{\Re}$ is identitive.

For example, the relation $\varrho_1 \leq \varrho_2$ ($\varrho_1, \varrho_2 \in \mathbb{R}$) is total and identitive. — The relation $X \subseteq Y$ between two sets X, Y is identitive but not total. — Let $X \, \Re_1 \, Y :\Leftrightarrow |X| \leq |Y|$ for all subsets $X, Y \subseteq \{1, \ldots, 5\}$; then $\Re_1$ is total; but $\Re_1$ is not identitive because

$\{1\}\, \Re_1\, \{2\}$ and $\{2\}\, \Re_1\, \{1\}$. — Let $X\, \Re_2\, Y\ :\Longleftrightarrow\ (X \cap \{1,3,5\}) \subseteq (Y \cap \{1,3,5\})$ for all subsets $X, Y \subseteq \{1,\ldots,5\}$. Then $\Re_2$ is not total because neither $X\Re_2 Y$ nor $Y\Re_2 X$ for $X := \{1,3\}$ and $Y := \{3,5\}$; moreover, $\Re_2$ is not identitive because $X\Re_2 Y$ and $Y\Re_2 X$ for $X := \{1,2\}$ and $Y := \{1,4\}$.

- *Chains, antichains:*

Let $\Re \subseteq \mathbf{R} \times \mathbf{R}$ be transitive and reflexive, and let $\mathbf{R}' \subseteq \mathbf{R}$. We call $\mathbf{R}'$ a a *chain* if $(\varrho_1 \Re \varrho_2 \vee \varrho_2 \Re \varrho_1)$ for all $\varrho_1, \varrho_2 \in \mathbf{R}'$, and we call $\mathbf{R}'$ call an *antichain* if no elements $\varrho_1, \varrho_2 \in \mathbf{R}'$ exist with $\varrho_1 \Re \varrho_2$ and $\varrho_1 \neq \varrho_2$. The *width* of $\Re$ is defined as the maximal cardinality of all antichains $\mathbf{R}'$; the width of $\Re$ may be infinite.

- *Extremal elements:*

Let $\Re \subseteq \mathbf{R} \times \mathbf{R}$ be transitive and reflexive. Let $X \subseteq \mathbf{R}$ and $\varrho^* \in X$.
ϱ^* is an *extremal element of X* (*with respect to $\Re$*) if $\varrho \Re \varrho^* \Rightarrow \varrho^* \Re \varrho$ for all $\varrho \in X$; let $EXT_\Re(X) := EXT(X) := $ set of all extremal elements of X.
Let (1.4) or (1.5) be true; then ϱ^* is a *minimal [maximal]* element of X if ϱ^* is extremal with respect to $"\preceq"$ [to $"\succeq"$]. Let $MIN(X) := EXT_\preceq(X)$, and let $MAX(X) := EXT_\succeq(X)$.
For example, let $\mathbf{R} = \mathbb{R}^2$, and let

$$(\forall\, (x_1, y_1),\, (x_2, y_2) \in \mathbb{R}^2)\ (x_1, y_1) \preceq (x_2, y_2) \iff (x_1 \le x_2 \vee y_1 \le y_2).$$

Then $MIN(X)$ is called the *Pareto*-set of X.
If $\Re$ is total and identitive then $EXT(X)$ has at most one element; if $\Re$ is not total then $EXT(X)$ may have two incomparable elements $\varrho_1^* \neq \varrho_2^*$; if $\Re$ is not identitive then $EXT(X)$ may have two elements $\varrho_1^* \neq \varrho_2^*$ with $\varrho_1^* \equiv \varrho_2^*$.

Let ϱ^* be the only extremal, minimal, and maximal element of X, respectively; then ϱ^* is called the *extremum, minimum,* and *maximum* of X, and we write $ext(X) := ext_\Re(X) := \varrho^*$, $\min(X) := \varrho^*$, and $\max(X) := \varrho^*$, respectively.

- *Monotonicity and injectivity of functions:*

Given a set $\mathbf{R} \neq \emptyset$ and a transitive, reflexive relation $\preceq$ between elements of $\mathbf{R}$. Let $X \subseteq \mathbf{R}$ and $f : X \to \mathbf{R}$.
f is *monotonically increasing* if $\varrho_1 \preceq \varrho_2 \Rightarrow f(\varrho_1) \preceq f(\varrho_2)$ for all $\varrho_1, \varrho_2 \in X$. f is *monotonically decreasing* if $\varrho_1 \preceq \varrho_2 \Rightarrow f(\varrho_1) \succeq f(\varrho_2)$ for all $\varrho_1, \varrho_2 \in X$.
f is called *strictly increasing [decreasing]* if $\varrho_1 \prec \varrho_2$ implies that $f(\varrho_1) \prec f(\varrho_2)$ $[f(\varrho_1) \succ f(\varrho_2)]$.
f is called $\equiv$-*injective* if $\varrho_1 \not\equiv \varrho_2 \Rightarrow f(\varrho_1) \not\equiv f(\varrho_2)$ for all $\varrho_1, \varrho_2 \in X$.
W a r n i n g : Strict monotonicity does not always imply monotonicity. It is possible that f is strictly increasing and that there exist $\varrho_1, \varrho_2 \in X$ such that $\varrho_1 \equiv \varrho_2$, $\neg(f(\varrho_1) \preceq f(\varrho_2))$, and $\neg(f(\varrho_2) \npreceq f(\varrho_1))$; then f is not monotonically increasing. — If, however, $\preceq$ is total then the strict monotonicity of f implies that f is monotone.
If f is injective then f need not be $\equiv$-injective because the following situation may arise: There are $\varrho_1, \varrho_2 \in X$ with $\varrho_1 \not\equiv \varrho_2$, $f(\varrho_1) \neq f(\varrho_2)$ but $f(\varrho_1) \equiv f(\varrho_2)$; then f is not injective.
If f is $\equiv$-injective then f need not be injective because the following situation may arise: There exist ϱ_1, ϱ_2 with $\varrho_1 \neq \varrho_2$, $\varrho_1 \equiv \varrho_2$, and $f(\varrho_1) = f(\varrho_2)$ so that f is not injective. Hence, there is no general logical relationship between usual injectivity and $\equiv$-injectivity.

Chapter 2

Path functions

Before searching for optimal paths we must have a measure for the cost or the utility of a path in a graph. The simplest method is assigning a real number $H(P)$ to each path P; in most cases, we consider a path as good if its H-value is small.

Such a function $H : \mathcal{P}(\mathcal{G}) \to \mathbb{R}$ is an example of a path function. In this chapter, we shall introduce more general path functions; they are of the form $H : \mathcal{P}(\mathcal{G}) \to \mathbf{R}$ where $\mathbf{R}$ is an arbitrary set with an generalized order relation $\Re$. We shall investigate structural properties of path functions; roughly speaking, we shall compare $H(P)$ and $H(Q)$ where P is a subpath of Q.

2.1 Introduction of path functions and of related terms

We give the exact definition of path functions, and we shall define node, arc and edge functions.

Definition/Remark 2.1 Given a directed or undirected graph $\mathcal{G} = (\mathcal{V}, \mathcal{R})$. Let $\mathcal{P} \subseteq \mathcal{P}(\mathcal{G})$.

A function $H : \mathcal{P} \to \mathbf{R}$ is called a *path function*.

We define $\mathcal{B}(\mathcal{G}, \mathbf{R})$ as the set of all path functions with values in $\mathbf{R}$. More formally,
$$\mathcal{B}(\mathcal{G}, \mathbf{R}) = \{H : \mathcal{P} \to \mathbf{R} \mid \mathcal{P} \subseteq \mathcal{P}(\mathcal{G})\} \,.$$

Any function $p : \mathcal{V} \to \mathbf{R}$ is called a *node function*.

Any function $h : \mathcal{R} \to \mathbf{R}$ is called an *arc function* if $\mathcal{G}$ is directed and an *edge function* if $\mathcal{G}$ is undirected.

Let $H : \mathcal{P} \to \mathbb{R}$, and let $p : \mathcal{V} \to \mathbb{R}$. Then we define the path function $(H+p) : \mathcal{P} \to \mathbb{R}$ as follows: $(H + p)(P) := H(P) + p(\omega(P))$ for all paths $P \in \mathcal{P}$.

Many optimal path algorithms use a path function $(C + p)$ where C is the given cost function and p is a so-called heuristic function (see Subsection 2.5.2). $\qquad\square$

Next, we introduce several terms in the context of extremal paths.

Definition 2.2 Given the terminology of Definition 2.1. Let $\Re \subseteq \mathbf{R} \times \mathbf{R}$ be a transitive and reflexive relation; we do not assume that $\Re$ is total or identitive. Given a path function $H : \mathcal{P} \to \mathbf{R}$ and a set $\mathcal{P}' \subseteq \mathcal{P}$.

Then a path $P \in \mathcal{P}'$ is called H-extremal in $\mathcal{P}'$ (with respect to $\Re$) if $H(P) \in EXT_\Re\{H(Q) \mid Q \in \mathcal{P}'\}$; this is equivalent to the fact that $H(P') \, \Re \, H(P) \Rightarrow H(P) \, \Re \, H(P')$ for all $P' \in \mathcal{P}'$.

Let P be H-extremal; then P is called H-minimal and H-maximal, respectively, if $\Re$ equals $\preceq$ and $\succeq$. $\qquad\qquad\qquad\qquad\qquad\qquad\qquad\qquad\qquad\qquad\qquad\qquad$ $\Box$

Next, we define "H-extremal connections" between two sets $\mathcal{P}_1$, $\mathcal{P}_2$ of paths. Roughly speaking, we connect the endnodes of the paths $P_1 \in \mathcal{P}_1$ with the startnodes of the paths $P_2 \in \mathcal{P}_2$ such that the resulting path is extremal.

Definition 2.3 Given the terminology of Definition 2.1.
Let $\Re \subseteq \mathbf{R} \times \mathbf{R}$ be transitive and reflexive; we do not assume that $\Re$ is total or identitive.
Moreover, given a path function $H : \mathcal{P} \to \mathbf{R}$, and let $\mathcal{P}_1$, $\mathcal{P}_2 \subseteq \mathcal{P}(\mathcal{G})$.

A path $P \in \mathcal{P}(\mathcal{G})$ is called an H-extremal connection of $\mathcal{P}_1$ and $\mathcal{P}_2$ (with respect to $\Re$) if there exist two paths $P_1 \in \mathcal{P}_1$ and $P_2 \in \mathcal{P}_2$ with the following property: The value $H(P_1 \oplus P \oplus P_2)$ is extremal with respect to $\Re$ among all values $H(P_1' \oplus P' \oplus P_2')$ with $P_1' \in \mathcal{P}_1$, $P_2' \in \mathcal{P}_2$, and $P_1' \oplus P' \oplus P_2' \in \mathcal{P}$. Recalling Definition 2.2, we may say that P is an H-extremal connection of $\mathcal{P}_1$ and $\mathcal{P}_2$ if $P_1 \oplus P \oplus P_2$ is an H-extremal element of $\mathcal{P}' := \{P_1' \oplus P' \oplus P_2' \mid P_1' \in \mathcal{P}_1, P_2' \in \mathcal{P}_2, \; P_1' \oplus P \oplus P_2' \in \mathcal{P}\}$.

If (1.4) or (1.5) is true then H-extremal connections with regard to $\preceq$ and to $\succeq$, respectively, are called H-minimal and H-maximal connections.

We write the set of all H-extremal connections P with respect to $\Re$ as $\mathcal{P}_{Ext}(\mathcal{P}_1, \mathcal{P}_2)$ or as $\mathcal{P}_{Ext_\Re}(\mathcal{P}_1, \mathcal{P}_2)$.

We simplify the notation $\mathcal{P}_{Ext}(\mathcal{P}_1, \mathcal{P}_2)$ if the following cases are given:

    (i) All paths of $\mathcal{P}_i$ have length 0 for $i = 1$ or $i = 2$.
    (ii) $\mathcal{P}_1$ or $\mathcal{P}_2$ has exactly one element.

If (i) is true then there is a set $\Gamma_i \subseteq \mathcal{V}$ with $\mathcal{P}_i = \{[v] \mid v \in \Gamma_i\}$. Then we replace $\mathcal{P}_i$ by Γ_i and write $\mathcal{P}_{Ext}(\Gamma_1, \mathcal{P}_2)$, $\mathcal{P}_{Ext}(\mathcal{P}_1, \Gamma_2)$ or even $\mathcal{P}_{Ext}(\Gamma_1, \Gamma_2)$.
If (ii) is true then we replace $\mathcal{P}_i$ by its single element X_i and, and we write $\mathcal{P}_{Ext}(X_1, \mathcal{P}_2)$, $\mathcal{P}_{Ext}(\mathcal{P}_1, X_2)$ or $\mathcal{P}_{Ext}(X_1, X_2)$.
If both (i) and (ii) are given then $\mathcal{P}_i$, $i = 1, 2$, consists of a single path $[v_i]$; then we write $\mathcal{P}_{Ext}(v_1, v_2)$.
For example, let $\mathcal{P} = \{Q_1\}$ and $\mathcal{P}_2 = \{[v_2]\}$. Then $\mathcal{P}_{Ext}(Q_1, v_2) = \mathcal{P}_{Ext}(\{Q_1\}, \{[v_2]\})$ is the set of all H-extremal connections P from the path Q_1 to the node v_2; that means that $H(Q_1 \oplus P) = H(Q_1 \oplus P \oplus [v_2])$ is extremal among all values $H(Q_1 \oplus P')$ where P' is a path from $\omega(Q_1)$ to v_2.

If (1.4) or (1.5) is true then write the set of all H-minimal and of all H-maximal connections with the subscript 'Min' and 'Max', respectively. For example, $\mathcal{P}_{Min}(\mathcal{P}_1, \mathcal{P}_2)$

denotes the set of all H-minimal connections from $\mathcal{P}_1$ to $\mathcal{P}_2$. $\mathcal{P}_{Max}(v_1, v_2)$ denotes the set of all H-maximal connections from v_1 to v_2. $\square$

The next remark says that the search for an extremal connection between two sets of nodes is often equivalent to the search for an extremal connection between two nodes.

Remark 2.4 Let $\mathcal{G} = (\mathcal{V}, \mathcal{R})$ be a digraph, let $H : \mathcal{P} \to \mathbf{R}$, and let $\Re \subseteq \mathbf{R} \times \mathbf{R}$ be transitive and reflexive. Let $\Gamma_1, \Gamma_2 \subseteq \mathcal{V}$.

The search for an H-extremal connection from Γ_1 to Γ_2 can often be reduced to the search for an extremal connection between two nodes v_1, v_2.

If Γ_1 or Γ_2 has more than one element then add $\widetilde{v}_1$ and $\widetilde{v}_2$ to the digraph $\mathcal{G}$. Generate new arcs from $\widetilde{v}_1$ to all nodes of Γ_1 and from all nodes of Γ_2 to $\widetilde{v}_2$. For each path $\widetilde{P} = \widetilde{r}_1 \oplus P \oplus \widetilde{r}_2$ in the new graph with $\alpha(\widetilde{r}_1) = \widetilde{v}_1$, $\omega(\widetilde{r}_1) \in \Gamma_1$, $\alpha(\widetilde{r}_2) \in \Gamma_2$, $\omega(\widetilde{r}_2) = \widetilde{v}_2$, we define $H(\widetilde{P}) := H(P)$. So, searching for an element of $\mathcal{P}_{Ext}(\Gamma_1, \Gamma_2)$ is equivalent to searching for an element of $\mathcal{P}_{Ext}(\widetilde{v}_1, \widetilde{v}_2)$. $\square$

Next, we consider the following situation: Given a path Q; then there exists an extremal path Q^* such that $H(Q)$ and $H(Q^*)$ are comparable with regard to $\Re$.

Definition/Remark 2.5 Let $\mathcal{G} = (\mathcal{V}, \mathcal{R})$ be a digraph, let $H : \mathcal{P} \to \mathbf{R}$ be a path function, and let $\Re \subseteq \mathbf{R} \times \mathbf{R}$ be transitive and reflexive. Let $\mathcal{P}_1, \mathcal{P}_2 \subseteq \mathcal{P}(\mathcal{G})$.

H has the $\mathcal{P}_1$-$\mathcal{P}_2$-*extremality property* (*with respect to* $\Re$) if for each path of the form $Q = P_1 \oplus P \oplus P_2$ with $P_1 \in \mathcal{P}_1$, $P_2 \in \mathcal{P}_2$, there exists Q^* with the following properties:

  • Q^* is an H-extremal element of $\mathcal{P}' := \{ P_i \oplus P \oplus P_2 \in \mathcal{P} \mid P_i \in \mathcal{P}_1, \ P_2 \in \mathcal{P}_2 \}$.
  • $H(Q^*) \, \Re \, H(Q)$.

Each set $\mathcal{P}_i$ of paths, $i = 1, 2$, may be replaced by

  • the set $\Gamma_i := \{ v \mid [v] \in \mathcal{P}_i \}$ of nodes if all paths in $\mathcal{P}_i$ are of the length 0;
  • the only element of $\mathcal{P}_i$ and of Γ_i, respectively, if $|\mathcal{P}_i| = 1$ and if $|\Gamma_i| = 1$.

For example, P_1-v_2-extremality property means that for each path $Q = P_1 \oplus P$ with $\omega(Q) = v_2$, there exists a path $Q^* = P_1 \oplus P^*$ such that $P^* \in \mathcal{P}_{Ext}(P_1, v_2)$ and $H(Q^*) \, \Re \, H(Q)$.

The terms "$\mathcal{P}_1$-$\mathcal{P}_2$-*minimality property*" and "$\mathcal{P}_1$-$\mathcal{P}_2$-*maximality property*" mean the $\mathcal{P}_1$-$\mathcal{P}_2$-extremality property with respect to $\preceq$ and to $\succeq$, respectively.

Let H have the X_1-X_2-extremality property for all paths $X_1, X_2 \in \mathcal{P}(\mathcal{G})$. Then the following are true:

  • H has the v_1-X_2-extremality property for all $v_1 \in \mathcal{V}$ and all $X_2 \in \mathcal{P}(\mathcal{G})$;
    this follows by considering the special case $X_1 = [v_1]$,
  • H has the X_1-v_2-extremality property for all $X_1 \in \mathcal{P}(\mathcal{G})$ and $v_2 \in \mathcal{V}$;
    this follows by considering the special case $X_2 = [v_2]$.
  • H has the v_1-v_2-extremality property for all $v_1, v_2 \in \mathcal{V}$.

Next, we see that the $\mathcal{P}_1$-$\mathcal{P}_2$-extremality property does not always imply the P_1-P_2-extremality property for all $P_1 \in \mathcal{P}_1$, $P_2 \in \mathcal{P}_2$ and that the reversed implication may be false, too. For example, let $v_1 \in \mathcal{V}$ and $\Gamma_2 \subseteq \mathcal{V}$. Let $\mathcal{P}_1 = \{[v_1]\}$

and $\mathcal{P}_2 = \{[w] \mid w \in \Gamma_2\}$. Then there is no logical relationship between the following assertions:

> H has v_1-w-extremality property for all nodes $w \in \Gamma_2$.
>
> (This is equivalent to the P_1-P_2-extremality property for all paths $\qquad$ (2.1)
> $P_1 = [v_1] \in \mathcal{P}_1$ and for all paths $P_2 = [w] \in \mathcal{P}_2$).

> H has the v_1-Γ_2-extremality property.
>
> (This is equivalent to the $\mathcal{P}_1$-$\mathcal{P}_2$-extremality property.) $\qquad$ (2.2)

If (2.1) is true then there exists an H-extremal v_1-w-path Q_w for each $w \in \Gamma_2$. If Γ is infinite and the set $\{H(Q_w), \mid w \in \Gamma_2\}$ has no extremal element then (2.2) is false although (2.1) is true.

A counterexample to $(2.2) \Rightarrow (2.1)$ is the following situation: Let Q be a path with $\alpha(Q) = v_1$, $w^* := \omega(Q) \in \Gamma_2$ and $H(Q) \, \Re \, H(P)$ for all paths P from v_1 to a node $w \in \Gamma_2$. Then (2.2) is true. But (2.1) is false if H does not have the v_1-w-extremality property for some $w \in \Gamma_2 \backslash \{w^*\}$. $\qquad\square$

Definition 2.6 The notation $H : \mathcal{P} \to (\mathbf{R}, \Re)$ means that all definitions in connection with H-extremal paths automatically refer to $\Re$; for example, the term "H-extremal connection" means "H-extremal connection with respect to $\Re$" (and not with respect to another relation).

In particular, $H : \mathcal{P} \to (\mathbf{R}, \preceq)$ means that H-extremal paths are automatically H-minimal paths and that the $\mathcal{P}_1$-$\mathcal{P}_2$-extremality property of two sets $\mathcal{P}_1, \mathcal{P}_2$ is equivalent to the $\mathcal{P}_1$-$\mathcal{P}_2$-minimality property; the notation $H : \mathcal{P} \to (\mathbf{R}, \succeq)$ means that H-extremal paths are automatically H-maximal paths and that the $\mathcal{P}_1$-$\mathcal{P}_2$-extremality property of two sets $\mathcal{P}_1, \mathcal{P}_2$ is equivalent to the $\mathcal{P}_1$-$\mathcal{P}_2$-maximality property. $\qquad\square$

Next, we define cost and utility functions.

Definition 2.7 Given a path function $H : \mathcal{P} \to (\mathbf{R}, \Re)$ where $\Re \in \{\preceq, \succeq\}$.

We shall call H a *cost function* or a *cost measure* if we think that H-minimal paths are good, and we shall call H a *utility function* if we think that H-maximal paths are good.

H-optimal paths are defined as H-minimal paths if H is a cost function, and they are defined as H-maximal paths if H is a utility function. $\qquad\square$

Remark 2.8 Definition 2.7 has the following consequences:

a) If $H : \mathcal{P} \to (\mathbf{R}, \preceq)$ is a cost function then all H-extremal paths are H-minimal, and these paths are good.

b) If $H : \mathcal{P} \to (\mathbf{R}, \preceq)$ is a utility function then all H-extremal paths are H-minimal, and these paths are bad.

c) If $H : \mathcal{P} \to (\mathbf{R}, \succeq)$ is a cost function then all H-extremal paths are H-maximal, and these paths are bad.

d) If $H : \mathcal{P} \to (\mathbf{R}, \succeq)$ is a utility function then all H-extremal paths are H-maximal, and these paths are good. $\qquad\square$

Remark 2.9 Let $H : \mathcal{P} \to (\mathbf{R}, \Re)$ be a path function with $\Re \in \{\preceq, \succeq\}$. Then we can formulate two questions:

1) Are H-extremal paths H-minimal? (Or are H-extremal paths H-maximal?)
2) Are H-extremal paths good? (Or are they bad?)

The answers to these two questions are independent of each other. This has been shown in Remark 2.8. For example, the answer 'yes' to the first question and 'no' to the second is realized in Remark 2.8 b).

Moreover, we can formulate the following question:

3) Are we searching for H-minimal paths?
   (Or are we searching for H-maximal candidates?)

The answer to the questions 1) – 3) will be 'yes' in almost every situation described in this book. That means that extremal paths are minimal, that H is a cost function and that we are searching for paths of minimal costs.

Another situation, however, is given in Theorem 3.52; let $H : \mathcal{P}(\mathcal{G}) \to (\mathbb{R}, \geq)$ be defined as $H := \ell$. Then H-extremal paths are H-maximal and bad, and we are interested in paths of maximal length to describe the longest duration of the project. That means that all questions 1) – 3) are answered with 'no'. In particular, we are searching for bad paths and not for optimal ones in the situation of 3.52. □

Next, we define several equivalence relations between path functions.

Definition/Remark 2.10 Given a directed or undirected graph $\mathcal{G} = (\mathcal{V}, \mathcal{R})$. Let $\mathcal{P} \subseteq \mathcal{P}(\mathcal{G})$.
Given the nonempty sets $\mathbf{R}_1, \mathbf{R}_2$ and the transitive, reflexive relations $\preceq_1 \subseteq \mathbf{R}_1 \times \mathbf{R}_1$ and $\preceq_2 \subseteq \mathbf{R}_2 \times \mathbf{R}_2$; we do not assume that $\preceq_1$ or $\preceq_2$ is total or identitive. Given two path frunctions $H_i : \mathcal{P} \to (\mathbf{R}_i, \preceq_i)$, $i = 1, 2$.

H_1 and H_2 are called *order equivalent* ($H_1 \cong H_2$) if $H_1(P) \preceq_1 H_1(Q) \iff H_2(P) \preceq_2 H_2(Q)$ for all paths $P, Q \in \mathcal{P}$.

H_1 and H_2 are called *equivalent* ($H_1 \equiv H_2$) if $(\mathbf{R}_1, \preceq_1) = (\mathbf{R}_2, \preceq_2)$ and $H_1(P) \equiv H_2(P)$ for all paths $P \in \mathcal{P}$.

Obviously, $H_1 \equiv H_2 \Rightarrow H_1 \cong H_2$ for all path functions. The inverse conclusion is not always correct even if $(\mathbf{R}_1, \preceq_1) = (\mathbf{R}_2, \preceq_2)$. For example, let $H_1, H_2 : \mathcal{P} \to (\mathbb{R}, \leq)$ with $(\forall P \in \mathcal{P})\ H_2(P) = 2 \cdot H_1(P)$. Then H_1 and H_2 are order equivalent but not equivalent. □

The following definition of cost structures is almost the same as Definition 1 of [LT91] and of [LTh91].

Definition/Remark 2.11 A *cost structure* is a quadruple $(\mathbf{R}, \odot, \varrho_0, \preceq)$ with the following components. $\mathbf{R}$ is a nonempty set. $\odot : \mathbf{R} \times \mathbf{R} \to \mathbf{R}$ is binary operation. ϱ_0 is a left neutral element for $\odot$; the term "*left neutral*" means that $\varrho_0 \odot \varrho = \varrho$ for all $\varrho \in \mathbf{R}$. $\preceq \subseteq \mathbf{R} \times \mathbf{R}$ is a transitive, reflexive relation; we do not assume that $\preceq$ is total or identitive.

Lengauer and Theune consider path functions $\lambda : \mathcal{P}(s) \to (\mathbf{R}, \preceq)$ with $\lambda([s]) = \varrho_0$. Each arc r has a weight $\lambda(r)$. If P is a path then $\lambda(P)$ is computed by successively applying $"\odot"$ to the weights $\lambda(r)$ of the arcs $r \in P$. This process of computing will be exactly described in Definition 2.21 b).

We leave it to the reader to create an adequate name for a quadruple $(\mathbf{R}, \odot, \varrho_0, \Re)$ in the case that assumption (1.4) is not given for $\Re$ and the path function λ is not considered to be a cost measure. $\qquad\qquad\qquad\qquad\qquad\qquad\qquad\qquad\qquad\qquad\square$

The next result says the following: If two arbitrary path functions F and G are given, we can find path functions λ_0 and λ_1 with the following properties: λ_0 is order equivalent to F, λ_1 is order equivalent to G; in addition, there is a close relationship between λ_0 and λ_1 even if F and G have nothing to do with each other.

Theorem 2.12 *Let $F, G : \mathcal{P} \to (\mathbf{R}', \preceq')$ be two path functions.*
Then a pair $(\mathbf{R}, \preceq)$ and two path functions $\lambda_0, \lambda_1 : \mathcal{P} \to (\mathbf{R}, \preceq)$ exist such that the following are true: λ_0 is order equivalent to F; λ_1 is order equivalent to G; there exists a function $g : V \times \mathbf{R} \to \mathbf{R}$ such that $\lambda_0(P) = g(\omega(P), \lambda_1(P))$ for all $P \in \mathcal{P}$.

P r o o f : We define $\mathbf{R} := \mathbf{R}' \times \mathbf{R}'$ and $\lambda_1(P) := (F(P), G(P))$ for all $P \in \mathcal{P}$. Let $g(v, (a, b)) := (b, a)$ for all $v \in V$, $a, b \in \mathbf{R}'$. (g is independent of v.) Let $\lambda_0(P) := g(\omega(P), \lambda_1(P))$ $(P \in \mathcal{P})$, and let $(a, b) \preceq (\widetilde{a}, \widetilde{b})$ $:\Leftrightarrow$ $b \preceq' \widetilde{b}$ for all $a, b, \widetilde{a}, \widetilde{b} \in \mathbf{R}'$.
Then $\lambda_1(P) \preceq \lambda_1(\widetilde{P})$ $\Longleftrightarrow$ $G(P) \preceq G(\widetilde{P})$ for all paths $P, \widetilde{P} \in \mathcal{P}$. The equation $\lambda_0(P) = g(\omega(P), (F(P), G(P))) = (G(P), F(P))$ and the analogous equation about $\widetilde{P}$ imply that $\lambda_0(P) \preceq \lambda_0(\widetilde{P})$ $\Longleftrightarrow$ $F(P) \preceq' F(\widetilde{P})$.
Consequently, λ_0 is order equivalent to F, and λ_1 is order equivalent G. The assertion that $\lambda_0(P) = g(\omega(P), \lambda_1(P))$ for all paths P follows from the definition of λ_0. $\quad\square$

Remark/Definition 2.13 Given the situation as described in Theorem 2.12.
The function g can easily be replaced by a function $f : V \times \mathbf{R} \to \mathbf{R}_\equiv$ where $\mathbf{R}_\equiv$ is a system of representatives of the equivalence relation $\equiv$.
To see this we define $\pi_\equiv : \mathbf{R} \to \mathbf{R}_\equiv$ as the projection from $\mathbf{R}$ onto $\mathbf{R}_\equiv$; that means, if $\varrho \in \mathbf{R}$ then $\pi_\equiv(\varrho)$ is defined as the unique element $\varrho' \in \mathbf{R}_\equiv$ with $\varrho' \equiv \varrho$. Then we define $f(v, \varrho) := \pi_\equiv(g(v, \varrho))$ for all $v \in V$ and $\varrho \in \mathbf{R}$. The resulting path function $\lambda_0(P) := f(v, \lambda_1(P))$ is order equivalent to F, and the values of $\lambda_0(P)$ are in $\mathbf{R}_\equiv$.
We shall call each function $f : V \times \mathbf{R} \to \mathbf{R}_\equiv$ an *evaluation function*. This definition is similar to the one in [LT91] and in [LTh91] where evaluation functions are of the form $f : V \times \mathbf{R} \to \mathbf{R}/\equiv$. (The set $\mathbf{R}/\equiv$ has been defined on page 19.) The difference, however, between both types of evaluation functions is only formal and not substantial. $\qquad\qquad\qquad\qquad\qquad\qquad\qquad\qquad\qquad\qquad\qquad\qquad\qquad\square$

2.2   Simple methods to define path functions

Here, we describe how to define particular path functions with the help of node functions, arc functions and other functions. The most well-known example is that the cost of a path P is defined as the sum of the costs of the arcs used by P.

Definition/Remark 2.14 Given a digraph $\mathcal{G} = (\mathcal{V}, \mathcal{R})$ and a path function $H : \mathcal{P} \to \mathbf{R}$.

H is called a *potential* if a node function $p : \mathcal{V} \to \mathbf{R}$ exists such that $H(P) = p(\omega(P))$ for all paths $P \in \mathcal{P}$. If $\mathbf{R} = \mathbb{R}$ and H is a potential then $H(P)$ may be considered as the "height" (the "potential") of the end node $\omega(P)$.

A non-trivial example of a potential is the following: Let $\mathcal{G}' := \mathcal{G}^s$ be the s-expansion of a digraph $\mathcal{G} = (\mathcal{V}, \mathcal{R})$. Let Ψ be the canonical homomorphism from $\mathcal{G}' = \mathcal{G}^s$ to $\mathcal{G}$.

Let $H : \mathcal{P}(s) \to \mathbf{R}$ be a path function in $\mathcal{G}$. Then $\Psi(P') \overset{(1.3)}{=} \omega(P') \in \mathcal{P}(s)$ for all $\mathcal{G}'$-paths $P' \in \mathcal{P}([s])$. Hence, the following path function $H' : \mathcal{P}([s]) \to \mathbf{R}$ is well-defined: $H'(P') := H(\Psi(P')) = H(\omega(P'))$ for all $P' \in \mathcal{P}([s])$. Then H' is a potential for $\mathcal{G}'$ because H is a node function for $\mathcal{G}'$.

H is a *potential difference* if $\mathbf{R} = \mathbb{R}$ and there exists a node function $p : \mathcal{V} \to \mathbb{R}$ such that $H(P) = p(\omega(P)) - p(\alpha(P))$ for all paths $P \in \mathcal{P}$.

Let $\mathbf{R} = \mathbb{R}$. H is called *additive* if there is an arc function $h : \mathcal{R} \to \mathbb{R}$ with the following properties: If $P = [v_0]$ then $H(P) = 0$; if $P = [v_0, v_1, \ldots, v_{k-1}, v_k]$ with $k > 0$ then $H(P) = \sum_{i=0}^{k-1} h(v_i, v_{i+1})$. This relationship between H and h is written as $H = SUM_h$.

For example, the length function $\ell(P)$ is additive because $\ell = SUM_h$ with $h(v, w) = 1$ for all $(v, w) \in \mathcal{R}$. Moreover, if H is a potential difference then $H = SUM_h$ with $h(v, w) := p(w) - p(v)$, $(v, w) \in \mathcal{R}$.

H is called *multiplicative* if there is an arc function h with $H([v_0]) = 1$ and $H(P) = \prod_{i=0}^{k-1} h(v_i, v_{i+1})$ for $P = [v_0, v_1, \ldots, v_{k-1}, v_k]$ with $k > 0$.

Let $\mathbf{R} = \mathbb{R}$. H is called *semi-additive* if there are a node function $p : \mathcal{V} \to \mathbb{R}$ and an arc function $h : \mathcal{R} \to \mathbb{R}$ with the following properties: If $P = [v_0]$ then $H(P) = p(v_0)$; if $P = [v_0, v_1, \ldots, v_k]$ with $k > 0$ then $H(P) = p(v_0) + \sum_{i=0}^{k-1} h(v_i, v_{i+1})$.

For example, all additive path functions are semi-additive.

Moreover, all path functions of the form $H = H_0 + p$ are semi-additive if $H_0 = SUM_{h_0}$ is additive and $p : \mathcal{V} \to \mathbb{R}$. To see this we define $h(v, w) := h_0(v, w) + p(w) - p(v)$ for all $(v, w) \in \mathcal{R}$; then H can be computed with the help of h and p. Note that H is not additive if $H([v]) = p(v) \neq 0$ for some $v \in \mathcal{V}$. — It follows that each potential is semiadditive.

It is easy to see that the sum $H(P) := H_1(P) + H_2(P)$ of two semi-additive [additive] path functions H_1 and H_2 is always semi-additive [additive]. $\qquad\square$

Next, we introduce further methods to define path functions.

Definition 2.15 Let $\mathcal{G} = (\mathcal{V}, \mathcal{R})$ be a digraph.

Given a partial function $\phi_r : \mathbf{R} \dashrightarrow \mathbf{R}$ for each arc $r \in \mathcal{R}$. Moreover, given a node function $p : \mathcal{V} \to \mathbf{R}$. Let $\mathcal{P} = \mathcal{P}(\mathcal{G})$ or $\mathcal{P} = \mathcal{P}(s)$ for some node s. Then we define the path function $H = \phi \circ\!\circ p : \mathcal{P} \to \mathbf{R}$ such that its values are computed by recursively evaluating the functions ϕ_r; more precisely, $H(P) := p(v_0)$ if $P = [v_0]$, and

$$H(P) := (\phi \circ\circ p)(P) := \phi_{(v_{k-1},v_k)} \left(\phi_{(v_{k-2},v_{k-1})} \left(\cdots \left(\phi_{(v_0,v_1)} \big(p(v_0)\big) \right) \cdots \right) \right)$$

if $P = [v_0, v_1, \ldots, v_k]$ with $k > 0$.

The function H is only correctly defined if each function ϕ_r is exclusively applied to arguments in $\mathrm{def}(\phi_r)$. For this reason, we make the following assumption:

$$\phi_{(v_{k-2},v_{k-1})} \left(\phi_{(v_{k-3},v_{k-2})} \left(\cdots \phi_{(v_0,v_1)} \big(p(v_0)\big) \cdots \right) \right) \in \mathrm{def}\left(\phi_{(v_{k-1},v_k)} \right) \qquad (2.3)$$
for all paths $[v_0, v_1, \ldots, v_k] \in \mathcal{P}$ with $k > 0$.

A consequence of the definition of H is the recursion formula $H(Q \oplus r) = \phi_r\big(H(Q)\big)$ for all paths Q and all arcs r.

For example, let H be semi-additive with $p : \mathcal{V} \to \mathbb{R}$ and $h : \mathcal{R} \to \mathbb{R}$. Define $\phi_r(x) := x + h(r)$ for all arcs r and all $x \in \mathbb{R}$. Then $H = \phi \circ\circ p$.

The idea of computing costs of paths by recursively applying functions ϕ_r appears frequently in literature. Two examples are [KaHe67] and [Mor82]. □

Definition 2.16 Given two digraphs $\mathcal{G}_i = (\mathcal{V}_i, \mathcal{R}_i)$ $(i = 1, 2)$, a graph homomorphism $\Psi : \mathcal{V}_1 \cup \mathcal{R}_1 \to \mathcal{V}_2 \cup \mathcal{R}_2$ and a path function $H : \mathcal{P} \to (\mathbf{R}, \preceq)$ with $\mathcal{P} \subseteq \mathcal{P}(\mathcal{G}_2)$. Then we define $H_\Psi(P) := H(\Psi(P))$ for all $P \in \mathcal{P}(\mathcal{G}_1)$ with $\Psi(P) \in \mathcal{P}$; consequently, $\mathrm{def}(H_\Psi) = \Psi^{-1}(\mathcal{P})$. We say that H_Ψ is *induced* by Ψ. □

Definition 2.17 Let $\mathcal{G} = (\mathcal{V}, \mathcal{R})$ be a digraph, and let $H : \mathcal{P} \to (\mathbf{R}, \preceq)$ where $\mathcal{P} \subseteq \mathcal{P}(\mathcal{G})$ is prefix closed and $\preceq$ is total and identitive.(E.g., $\mathbf{R}$ may be equal to $\mathbb{R}$, and "$\preceq$" may be equal to "$\leq$".) Let $P \in \mathcal{P}$.

Then we define $H_{\max}(P) := \max\{H(Q) \,|\, Q \text{ is a prefix of } P\}$.
$H_{\min}(P)$ is defined analogically.

For example, let $P = [v_0, v_1, v_2, v_3]$ and let $H([v_0]) = 0$, $H([v_0, v_1]) = 12$, $H([v_0, v_1, v_2]) = -5$, $H([v_0, v_1, v_2, v_3]) = 7$. Then $H_{\max}(P) = 12 = H([v_0, v_1])$, and $H_{\min}(P) = -5 = H([v_0, v_1, v_2])$. □

Next, we show a simple relationship between path functions constructed in Definition 2.16 and path functions constructed in Definition 2.17.

Theorem 2.18 *Given two digraphs* $\mathcal{G}_i = (\mathcal{V}_i, \mathcal{R}_i)$, $i = 1, 2$. *Let* Ψ *be a graph homomorphism from* $\mathcal{G}_1$ *to* $\mathcal{G}_2$. *Let* $s_2 \in \mathcal{V}_2$, *let* $\mathcal{P}_2 := \mathcal{P}(s_2)$. *Let* $H_2 : \mathcal{P}_2 \to (\mathbf{R}, \preceq)$ *be a path function where* $\preceq$ *is total and identitive. (E.g.,* $\mathbf{R}$ *may be equal to* $\mathbb{R}$, *and* "$\preceq$" *may be equal to* "$\leq$".*)*

   a) *The set* $\mathcal{P}_1 := \Psi^{-1}(\mathcal{P}(s_2))$ *is prefix closed.*
   b) *Let* $H_1 : \mathcal{P}_1 \to (\mathbf{R}, \preceq)$ *be defined as* $H_1 := (H_2)_\Psi = H_2 \circ \Psi$.
      Then $(H_1)_{\max} = (H_2)_{\max} \circ \Psi$.

P r o o f   o f   a) : Let $P_1 \in \mathcal{P}_1$ and $X \preceq P_1$. Then $\Psi(P_1) \in \mathcal{P}_2 = \mathcal{P}(s_2)$. That is the reason why relationship $(*)$ of the following equation is valid: $\alpha(\Psi(X)) =$

$$\Psi(\alpha(X)) \overset{X \leq P_1}{=} \Psi(\alpha(P_1)) = \alpha(\Psi(P_1)) \overset{(*)}{=} s_2. \text{ Hence, } \Psi(X) \in P_2 \text{ so that } X \in P_1.$$

P r o o f   o f   b) : Let $P_1 \in \mathcal{P}_1$ and $P_2 := \Psi(P_1)$. Then the following is true:

$$\{\Psi(X) \,|\, X \leq P_1\} = \{Y \,|\, Y \leq P_2\}. \tag{2.4}$$

Relationship " $\subseteq$" follows from the fact that $\Psi(X) \leq \Psi(P_1) = P_2$ for all $X \leq P_1$. Relationship " $\supseteq$" can be seen as follows: Let $l := \ell(P_1)$, and let X_λ be the prefix of P_1 with $\ell(X_\lambda) = \lambda$ $(\lambda = 0, \ldots, l)$. Then $\ell(\Psi(X_{\lambda+1})) \leq \ell(\Psi(X_\lambda)) + 1$ for all λ. That means that the sequence $\Psi(X_0), \ldots, \Psi(X_l)$ does not leave out any prefix of P_2. Then the following equation implies fact b):

$$(H_1)_{\max}(P_1) = \max\{H_1(X) \,|\, X \leq P_1\} = \max\{H_2(\Psi(X)) \,|\, X \leq P_1\}$$
$$\overset{(2.4)}{=} \max\{H_2(Y) \,|\, Y \leq P_2\} = (H_2)_{\max}(P_2) = (H_2)_{\max}(\Psi(P_1)). \qquad \square$$

Next, we characterize the path functions of the form $\phi \circ\!\circ p$ with total functions ϕ_r.

Theorem 2.19 *Let $\mathcal{G} = (\mathcal{V}, \mathcal{R})$ be a digraph and $s \in \mathcal{V}$. Let $\mathcal{P} = \mathcal{P}(\mathcal{G})$ or $\mathcal{P} = \mathcal{P}(s)$, and let $H : \mathcal{P} \to \mathbf{R}$ be a path function. Then the following assertions are equivalent:*

(a) *H preserves equality in the following sense:*
$$H(P_1) = H(P_2) \;\Rightarrow\; H(P_1 \oplus Q) = H(P_2 \oplus Q)$$
for all paths P_1, P_2, Q with $\omega(P_1) = \omega(P_2) = \alpha(Q)$.
(b) *There exist a function $p : \mathcal{V} \to \mathbf{R}$ and a total function $\phi_r : \mathbf{R} \to \mathbf{R}$ for each $r \in \mathcal{R}$ such that $H = \phi \circ\!\circ p$.*

P r o o f   o f   (a)$\Rightarrow$(b) : First, we define p. If $\mathcal{P} = \mathcal{P}(\mathcal{G})$ then $p(v) := H([v])$ for all v. If $\mathcal{P} = \mathcal{P}(s)$ then $p(v) := H([s])$ for all v.
Next, we define the functions ϕ_r. Let $r \in \mathcal{R}$ and $z \in \mathbf{R}$. Then $\phi_r(z) := H(P \oplus r)$ if a path $P \in \mathcal{P}$ exists with $H(P) = z$ and $\omega(P) = \alpha(r)$; if such a path does not exist, we define $\phi_r(z)$ as an arbitrary element of $\mathbf{R}$. Then ϕ_r is well-defined; to see this we assume that $P_1, P_2 \in \mathcal{P}$ with $z = H(P_1) = H(P_2)$. Then $H(P_1 \oplus r) = H(P_2 \oplus r)$ by (a); therefore, the definitions $\phi_r(z) := H(P_1 \oplus r)$ and $\phi_r(z) := H(P_2 \oplus r)$ are not in conflict with each other.
It is easy to see that $H = \phi \circ\!\circ p$.

P r o o f   o f   (b)$\Rightarrow$(a) : We assume that (b) is true. The proof of (a) is an induction on the length of Q. If $\ell(Q) = 0$ then nothing is to be shown. Let fact (a) be true for all paths of a length l, and let $\ell(Q) = l + 1$. Then Q can be written as $Q = Q' \oplus r$ where $\ell(Q') = l$. We obtain the following equation, in which (*) follows from the induction hypothesis.

$$H(P_1 \oplus Q) = H(P_1 \oplus Q' \oplus r) \overset{(b)}{=} \phi_r(H(P_1 \oplus Q')) \overset{(*)}{=}$$
$$\phi_r(H(P_2 \oplus Q')) \overset{(b)}{=} H(P_2 \oplus Q' \oplus r) = H(P_2 \oplus Q). \qquad \square$$

Remark 2.20 At first sight, condition (a) in Theorem 2.19 is satisfied by all practically relevant path functions. A counterexample, however, is the average weight function H, which is defined as follows: Let $h : \mathcal{R} \to \mathbb{R}$ be an arc function and let $H^+ := SUM_h : \mathcal{P}(\mathcal{G}) \to \mathbb{R}$. We define $H(P) := 0$ if $\ell(P) = 0$, and we define

$H(P) := H^+(P)/\ell(P)$ if $\ell(P) > 0$.

Then condition (a) of 2.19 is not satisfied in the following example: Let $\mathcal{G} = (\mathcal{V}, \mathcal{R})$ with $\mathcal{V} := \{x_0, x_1, x_2, x_3\}$ and $\mathcal{R} := \{(x_0, x_1), (x_0, x_2), (x_1, x_2), (x_2, x_3)\}$; let $h(x_2, x_3) := 9$ and $h(r) := 3$ for all other arcs. Let $P_1 := [x_0, x_1, x_2]$, $P_2 := [x_0, x_2]$, and $Q := [x_2, x_3]$. Then $H(P_1) = \frac{6}{2} = 3 = \frac{3}{1} = H(P_2)$ but $H(P_1 \oplus Q) = \frac{15}{3} = 5 \neq 6 = \frac{12}{2} = H(P_2 \oplus Q)$.

Another simple path function without the above property condition (a) is given in Remark 4.208.   □

Next, we introduce a class of path functions that can be easily computed with the help of node and arc functions; this class contains all semi-additive path functions.

Definition/Remark 2.21 Given a digraph $\mathcal{G} = (\mathcal{V}, \mathcal{R})$, a set $\mathcal{P} \subseteq \mathcal{P}(\mathcal{G})$ of paths and a path function $H : \mathcal{P} \to \mathbf{R}$.

We say that H is *easily computable of type A* if there exist a node function $p : \mathcal{V} \to \mathbf{R}$, an arc function $h : \mathcal{R} \to \mathbf{R}$, and a binary operation $\odot : \mathbf{R} \times \mathbf{R} \to \mathbf{R}$ such that the following is true for all paths $P = [v_0, \ldots, v_k] \in \mathcal{P}$: If $k = 0$ then $H(P) = p(v_0)$; if $k > 0$ then

$$H(P) = \left[\left(\cdots \left(\left(p(v_0) \odot h(v_0, v_1) \right) \odot h(v_1, v_2) \right) \odot \cdots \right) \odot h(v_{k-1}, v_k) \right]. \quad (2.5)$$

(The brackets are necessary because $\odot$ need not be associative.)

For example, let H be a semi-additive path functions basing on $p : \mathcal{V} \to \mathbb{R}$ and $h : \mathcal{R} \to \mathbb{R}$; then H is easily computable of type A with the help of p, h and the operation $\varrho \odot \varrho' := \varrho + \varrho'$ $(\varrho, \varrho' \in \mathbf{R} = \mathbb{R})$.

If H is easily computable of type A then H is of the form $\phi \circ \circ p$; to see this we define $\phi_r(\zeta) := \zeta \odot h(r)$ for all $r \in \mathcal{R}$ and all $\zeta \in \mathcal{R}$.

This and Theorem 2.19 imply that the function H in Remark 2.20 is not easily computable of type A.

We say that H is *easily computable of type B* if H is easily computable of type A and p has the following property: $(\forall\, v)\; p(v) = \varrho_0$ where $\varrho_0 \odot \varrho = \varrho$ for all ϱ; then the term $p(v_0)$ in (2.5) may be omitted; this version of (2.5) is equivalent to the the representation of path functions in Definition 2 of [LT91] and of [LTh91].

We say that H is *easily computable of type C* if the following situation is given: There exists a set $\widetilde{\mathbf{R}}$, a path function $\widetilde{H} : \mathcal{P} \to \widetilde{\mathbf{R}}$ that is easily computable of type A and a function $f : \widetilde{\mathbf{R}} \to \mathbf{R}$ such that $H(P) = f(\widetilde{H}(P))$ for all paths $P \in \mathcal{P}$.

For example, the path function $H(P) = H^+(P)/\ell(P)$ in Remark 2.20 is not easily computable of type A, but it is easily computable of type C. To see this we define $\widetilde{\mathbf{R}} := \mathbb{R}^2$, and we define $\widetilde{H} : \mathcal{P} \to \widetilde{\mathbf{R}}$ as $\widetilde{H}(P) := (H^+(P), \ell(P))$ for all paths P. Then $\widetilde{H}$ is easily computable of type A with $\widetilde{p}(v) := (0,0)$ for all $v \in \mathcal{V}$, $\widetilde{h}(v, w) := (h(v, w), 1)$ for all $(v, w) \in \mathcal{R}$, and $(a, b) \widetilde{\odot} (c, d) := (a + c, b + d)$.

We define $f : \widetilde{\mathbf{R}} \to \mathbf{R}$ as $f(a, b) := a/b$ if $b \neq 0$ and as $f(a, b) := 0$ if $b = 0$. Then $H = f \circ \widetilde{H}$ where $\widetilde{H}$ is easily computable of type C.   □

Remark 2.22 It is not difficult to create new types of of easy computability by changing the order of the brackets. For example, we can replace (2.5) by the following definition:

$$H(P) := p(v_0) \odot \left[\left(\cdots \left(\left(h(v_0,v_1) \odot h(v_1,v_2) \right) \odot h(v_2,v_3) \right) \odot \cdots \right) \odot h(v_{k-1},v_k) \right].$$

Further modifications of (2.5) are given in Definition 2.4 and Remark 2.5 of [Huck92] and perhaps in a second edition of this book.   ☐

Next, we prove a concatenation formula for easily computable path functions.

Theorem 2.23 (*see also* 4.160)

Let $H : \mathcal{P} \to \mathbf{R}$ easily computable of type B. We assume that the underlying operation $\odot$ is associative and that the element $\varrho_0 \in \mathbf{R}$ is neutral. (That means that $\varrho_0 \odot \varrho = \varrho \odot \varrho_0 = \varrho$ for all $\varrho \in \mathbf{R}$.)

Then $H(P \oplus Q) = H(P) \odot H(Q)$ for all paths $P, Q \in \mathcal{P}$ with $P \oplus Q \in \mathcal{P}$.

P r o o f : Let $\ell(Q) = 0$; then the assertion follows from the fact that $H(Q)$ is equal to the neutral element ϱ_0. The other case is that $k := \ell(Q) > 0$ and that $Q = r_1 \oplus \ldots \oplus r_k$; then we obtain an equality where $(*),(* * *)$ follow from the easy computability of H and where $(**)$ follows from the associativity of $\odot$.

$$H(P \oplus Q) \overset{(*)}{=} [\ldots [H(P) \odot h(r_1)] \odot \ldots] \odot h(r_k) \overset{(**)}{=} H(P) \odot [h(r_1) \odot \ldots \odot h(r_k)]$$
$$\overset{(* * *)}{=} H(P) \odot H(Q).$$
                                                                                              ☐

2.3   Order preserving path functions

This section mainly is about a structural property of path functions, the so-called order preservation. This property means that the order of $H(P_1)$ and $H(P_2)$ is preserved if a common continuation Q is appended. More formally, a real-valued path function H is called order preserving if $H(P_1) \leq H(P_2) \Rightarrow H(P_1 \oplus Q) \leq H(P_2) \oplus Q$ for all paths with $\alpha(P_1) = \alpha(P_2)$ and $\omega(P_1) = \omega(P_2) = \alpha(Q)$.

Additive path functions are always order preserving but not each order preserving function is additive. For example, if H is a multiplicative path function basing on an arc function $h : \mathcal{R} \to (0, \infty)$ then H is order preserving but not additive; further examples are the cleaning processes described in in Remark 4.126 of this book and the financial transactions described in Example 1.2 of [HuRu90].

It turns out that order preservation is almost equivalent to the so-called monotonicity of a path function. Monotonicity means that the path function can be written as $\phi \circ\circ p$ (see Definition 2.14) where all functions ϕ_r are monotonically increasing.

Order preserving and monotone path functions are widely studied in literature. Examples are given in [DePe85], page 506, in [Mor82], in [LT91], [LTh91] (property A3left), and in [HuRu90], page 43; the preference structures in [EiMS85], page 78, have a remote relationship to order preservation.

Order preserving path functions are very important because they make the search for optimal paths easier. For example, Theorems 4.24, 4.34, and 4.121 describe the behaviour of search strategies that work correctly because of the order preservation of path functions.

It is easy to formulate conditions that are similar to the original order preservation. These modified principles of order preservation yield many problems and results about the structure of path functions. Typical examples are Theorems 2.19, 2.95, and 4.158. Also, we shall use modified principles of order preservation when proving Theorems 4.118, 4.122, 4.153 and other results about the behaviour of optimal path algorithms.

This section is structured as follows: First, we define order preservation and monotonicity and give several examples. Then we show that these properties are almost equivalent. After this, we describe methods to generate order preserving path functions. At last, we consider modifications and generalizations of order preservation.

2.3.1   Definition of order preservation and monotonicity

Here, we give an exact definition of order preservation and of monotonicity of a path function; moreover, we define $\prec$-preservation and $=$-preservation of a path function.

Definition 2.24 Given a digraph $\mathcal{G} = (\mathcal{V}, \mathcal{R})$ and a path function $H : \mathcal{P} \rightarrow (\mathbf{R}, \preceq)$. We assume that $\mathcal{P} = \mathcal{P}(\mathcal{G})$ or $\mathcal{P} = \mathcal{P}(s)$ for some $s \in \mathcal{V}$. Let $\widehat{\Re} \in \{\preceq, \prec, \equiv, =\}$.

H is called *order preserving* or $\preceq$-*preserving* if the following is true for all nodes v, v', for all paths $P_1, P_2 \in \mathcal{P}$ from v to v', and for all paths Q with start node v':
$$H(P_1) \preceq H(P_2) \quad \Rightarrow \quad H(P_1 \oplus Q) \preceq H(P_2 \oplus Q). \tag{2.6}$$

In general, H is called $\widehat{\Re}$-*preserving* if the following is true for all nodes v, v', for all paths $P_1, P_2 \in \mathcal{P}$ from v to v', and for all paths Q with start node v':
$$H(P_1) \, \widehat{\Re} \, H(P_2) \Rightarrow H(P_1 \oplus Q) \, \widehat{\Re} \, H(P_2 \oplus Q). \tag{2.7}$$

That means that we have defined $\prec$-*preserving*, $\equiv$-*preserving*, and $=$-*preserving* path functions.[1]

H is called *monotonic* if there exist a node function $p : \mathcal{V} \rightarrow \mathbf{R}$ and monotonically increasing functions $\phi_r : \mathbf{R} \dashrightarrow \mathbf{R}$ $(r \in \mathcal{R})$ such that $H = \phi \circ \circ p$. □

Example 2.25 Let $H : \mathcal{P}(\mathcal{G}) \rightarrow \mathrm{I\!R}$ be a additive. Then H is $\widehat{\Re}$-preserving for all $\widehat{\Re} \in \{\leq, <, =\}$; to see this we assume that $H(P_1)\widehat{\Re}H(P_2)$; then $H(P_1 \oplus Q) = H(P_1) + H(Q) \, \widehat{\Re} \, H(P_2) + H(Q) = H(P_2 \oplus Q)$.

Let H be semi-additive; then H is likewise $\widehat{\Re}$-preserving for all $\widehat{\Re} \in \{\leq, <, =\}$. Moreover, H is monotone. To see this we assume that H is defined with the help of

---

[1] Each additive path function is $=$-preserving but the path function in 4.208 is not.

$p : V \to \mathbb{R}$ and $h : \mathcal{R} \to \mathbb{R}$; then $H = \phi \circ\circ\, p$ where $\phi_r(x) := x + h(r)$ for all $x \in \mathbb{R}$ and $r \in \mathcal{R}$; the function ϕ_r is monotonically increasing for all r.

All potentials $H : \mathcal{P}(\mathcal{G}) \to \mathbf{R}$ are $\widehat{\Re}$-preserving for all $\widehat{\Re} \in \{\preceq, \prec, =\}$. $\qquad\square$

Remark 2.26 (Elementary properties of order preserving path functions)
Let $\mathcal{G} = (V, \mathcal{R})$ be a digraph, and let $\mathcal{P} = \mathcal{P}(\mathcal{G})$ or $\mathcal{P} = \mathcal{P}(s)$ where $s \in V$.

a) Let $H : \mathcal{P} \to (\mathbf{R}, \preceq)$ be order preserving. Then the following is true for all paths P_1, P_2, Q with $\alpha(P_1) = \alpha(P_2)$, $\omega(P_1) = \omega(P_2) = \alpha(Q)$, and P_1, $P_1 \oplus Q$, P_2, $P_2 \oplus Q \in \mathcal{P}$:
$$H(P_1) \equiv H(P_2) \implies H(P_1 \oplus Q) \equiv H(P_2 \oplus Q).$$
P r o o f : Let $H(P_1) \equiv H(P_2)$. Then $H(P_1) \preceq H(P_2)$, and $H(P_2) \preceq H(P_1)$. Then $H(P_1 \oplus Q) \preceq H(P_2 \oplus Q)$ and $H(P_2 \oplus Q) \preceq H(P_1 \oplus Q)$ by order preservation. Hence, $H(P_1 \oplus Q) \equiv H(P_2 \oplus Q)$.

b) Let $H_1 : \mathcal{P} \to (\mathbf{R}_1, \preceq_1)$ be order equivalent to an order preserving function $H_0 : \mathcal{P} \to (\mathbf{R}_0, \preceq_0)$. Then H_1 is also order preserving.
(In particular, H_0 is order preserving if it is equivalent to H_0 because equivalence implies order equivalence as seen in 2.10.)
P r o o f : Let $H_1(P_1) \preceq_1 H_1(P_2)$. Then the order equivalence of H_0 and H_1 implies $H_0(P_1) \preceq_0 H_0(P_2)$. Then $H_0(P_1 \oplus Q) \preceq_0 H_0(P_2 \oplus Q)$ because H_0 is order preserving. The order equivalence implies $H_1(P_1 \oplus Q) \preceq_1 H_1(P_2 \oplus Q)$.
If $\mathbf{R} = \mathbb{R}$ we can use this result to construct order preserving functions with "small" domains of values. Let $H_0 : \mathcal{P} \to \mathbb{R}$ be an order preserving path function. Let $\psi : \mathbb{R} \to X$ be a strictly increasing real function with $X \subseteq \mathbb{R}$. We define $H_1(P) := \psi(H_0(P))$ for all $P \in \mathcal{P}$. Then H_1 is order equivalent to H_0 and consequently order preserving. Moreover, all values of H_1 lie in $X \subseteq \mathbb{R}$. For example, let $H_1(P) := \arctan(H_0(P))$ then $X = (-\pi, +\pi)$.

c) Order preservation is inherited by subgraphs. More precisely, let $\mathcal{G}'$ be a subgraph of $\mathcal{G}$ and let $\mathcal{P}' = \mathcal{P} \cap \mathcal{P}(\mathcal{G}')$; then order preservation of H implies that the restriction $H|_{\mathcal{P}'}$ has the same property. $\qquad\square$

Next, we prove a simple result about $=$-preserving path functions.

Lemma 2.27 *Given a path function $H : \mathcal{P}(\mathcal{G}) \to (\mathbf{R}, \preceq)$. We assume that the following assertions are true:*

(i) H *is $=$-preserving.*

(ii) H *is $=$-preserving in the following sense:*
$$(\forall\, Q', P_1, P_2,\ \omega(Q') = \alpha(P_1) = \alpha(P_2),\ \omega(P_1) = \omega(P_2))$$
$$H(P_1) = H(P_2) \implies H(Q' \oplus P_1) = H(Q' \oplus P_2).^2$$
Let X_1, X_2, Y_1, Y_2 be four paths with the following properties:
(iii) $\alpha(X_1) = \alpha(X_2)$, $\omega(X_1) = \omega(X_2) = \alpha(Y_1) = \alpha(Y_2)$, $\omega(Y_1) = \omega(Y_2)$.
(iv) $H(X_1) = H(X_2)$, *and* $H(Y_1) = H(Y_2)$.

Then $H(X_1 \oplus Y_1) = H(X_2 \oplus Y_2)$.

---
^{2}That means that H is $=$-preserving with regard to $\mathbb{L}'_1 \cap \mathbb{L}''_2 \cap \mathbb{L}_1$ (see Example 2.47).

P r o o f : Assumption (iv) says that $H(X_1) = H(X_2)$. Then (iii) and (i) imply that $H(X_1 \oplus Y_1) = H(X_2 \oplus Y_1)$. Moreover, assumption (iv) says that $H(Y_1) = H(Y_2)$. Then (iii) and (ii) imply that $H(X_2 \oplus Y_1) = H(X_2 \oplus Y_2)$. Consequently, $H(X_1 \oplus Y_1) = H(X_2 \oplus Y_1) = H(X_2 \oplus Y_2)$.   □

2.3.2 Structural properties of order preserving path functions

Next, we study the structure of order preserving functons, and we describe methods to generate such functions.

The following result characterizes order preserving path functions that are defined on a set $\mathcal{P}(s)$. Roughly speaking, H is order preserving if and only if H is "almost equal" to a monotone path function H'.

Theorem 2.28 *Given a finite digraph $\mathcal{G} = (\mathcal{V}, \mathcal{R})$, a node $s \in \mathcal{V}$, and a path function $H : \mathcal{P}(s) \to (\mathbf{R}, \preceq)$. Then the following assertions are equivalent:*

(a) *H is order preserving.*
(b) *There exists a monotone path function $H' : \mathcal{P}(s) \to (\mathbf{R}, \preceq)$ such that the following is true for all paths $P \in \mathcal{P}(s)$:*
    $H(P) = H'(P)$ *if* $P = [s]$, *and* $H(P) \equiv H'(P)$ *if* $\ell(P) > 0$.

Before proving this result we make the following remarks:

1. Assertion (a) does not always imply that $H' = H$. A counterexample will be given in Remark 2.31.
2. Let H be semi-additive with $p : \mathcal{V} \to \mathbb{R}$ and $h : \mathcal{R} \to \mathbb{R}$. Define $\phi_r(x) := x + h(r)$ for all arcs r and all $x \in \mathbb{R}$. Then the path function $H' := \phi \circ \circ p$ is monotone because all functions ϕ_r are monotonically increasing. Consequently, $H = H'$ in this case.
3. The direction (b)⇒(a) has been proven in [Mor82] and in [ChMi82] in the special case that $(\mathbf{R}, \preceq)$ is equal to $(\mathbb{R}, \leq)$.

P r o o f   o f (a)⇒(b) : Let H be order preserving. We must define a function H' with the help of monotonically increasing functions ϕ_r, $r \in \mathcal{R}$. We shall define the ϕ_r's on the sets $\mathbf{R}_v \subseteq \mathbf{R}$, which are now exactly described. Let $v \in \mathcal{V}$. Then we define $\mathbf{R}_v \subseteq \mathbf{R}$ as the set of all H-values that can be generated by some s-v-path; more formally, let $\mathbf{R}_v := \{H(P) \mid P \in \mathcal{P}(s,v)\}$ for all $v \in \mathcal{V}$.

Next, we define the functions ϕ_r. Let $r \in \mathcal{R}$ with $x := \alpha(r)$, and let $\varrho \in \mathbf{R}_x$. Then there exists a path $P_\varrho \in \mathcal{P}(s,x)$ with $H(P_\varrho) = \varrho$. We define $\phi_r : \mathbf{R}_x \to \mathbf{R}$ such that $\phi_r(\varrho) := H(P_\varrho \oplus r)$ for all $\varrho \in \mathbf{R}_x$.

Note that $P_\varrho \oplus r$ is a path from s to $\omega(r)$. Hence, $\phi_r(\varrho) \in \mathbf{R}_{\omega(r)}$. This implies

$$\phi_r : \mathbf{R}_{\alpha(r)} \to \mathbf{R}_{\omega(r)} \quad \text{for all } r \in \mathcal{R}. \tag{2.8}$$

Next, we define $H' := \phi \circ \circ p$ where $p(v) := H([v])$ for all nodes v. It must be shown that H' is well-defined. For this purpose we prove fact (2.3), which means the following in our current situation:

Let $P = [v_0 = s, v_1, \ldots, v_k]$ and let $r_\kappa := (v_{\kappa-1}, v_\kappa)$, $(\kappa = 1, \ldots, k-1)$.
Then $\phi_{r_{k-1}} \left(\phi_{r_{k-2}} \left(\ldots \phi_{r_1} (H'([s])) \ldots \right) \right) \in \mathbf{R}_{v_{k-1}} = \mathrm{def}\,(\phi_{r_k})$ (2.9)

The proof is an induction on k. If $k = 1$ then $\phi_{r_{k-1}} \left(\phi_{r_{k-2}} \left(\ldots \phi_{r_1} (H'([s])) \ldots \right) \right) = H'([s]) = H([s]) \in \mathbf{R}_s = \mathbf{R}_{v_{k-1}}$. (The relation $H([s]) \in \mathbf{R}_s$ in this equation follows from the fact that $H([s])$ is the H-value of the s-s-path $[s]$.)

We assume that assertion (2.9) is true for k. We have to prove (2.9) for every path $P = [v_0 = s, \ldots, v_k, v_{k+1}]$.
The induction hypothesis says that $\phi_{r_{k-1}} \left(\phi_{r_{k-2}} \left(\ldots \phi_{r_1} (H'([s])) \ldots \right) \right) \in \mathbf{R}_{v_{k-1}} = \mathrm{def}\,(\phi_{r_k})$. Moreover, fact (2.8) implies that $\phi_{r_k} : \mathbf{R}_{v_{k-1}} \to \mathbf{R}_{v_k}$ and that $\mathbf{R}_{v_k} = \mathrm{def}(\phi_{r_{k+1}})$. Consequently, $\phi_{r_k} \left(\phi_{r_{k-1}} \left(\ldots \phi_{r_1} (H'([s])) \ldots \right) \right) \in \phi_{r_k} \left(\mathbf{R}_{v_{k-1}} \right) \subseteq \mathbf{R}_{v_k} = \mathrm{def}\,(\phi_{r_{k+1}})$.

We have shown fact (2.9); hence, H' is well-defined.

We show that ϕ_r is monotonically increasing. Let $\varrho_1, \varrho_2 \in \mathbf{R}_{\alpha(r)}$ with $\varrho_1 \preceq \varrho_2$. Then $H(P_{\varrho_1}) = \varrho_1 \preceq \varrho_2 = H(P_{\varrho_2})$. The order preservation of H implies that

$$\phi_r(\varrho_1) = H(P_{\varrho_1} \oplus r) \preceq H(P_{\varrho_2} \oplus r) = \phi_r(\varrho_2).$$

At last, we prove that $H(P) \equiv H'(P)$ for all paths $P = [v_0 = s, v_1, \ldots, v_k]$. If $k = 0$ then $H'(P) = H'([s]) = H([s]) = H(P)$ by the definition of H'.
Let $k > 0$, and let $r_\kappa := (v_{\kappa-1}, v_\kappa)$, $\kappa = 1, \ldots, k$. We assume that our assertion is true for all paths of length $k-1$. Let $P_* := [v_0, v_1, \ldots, v_{k-1}]$; then $P = P_* \oplus r_k$. The induction hypothesis says that

$$\varrho := H'(P_*) \equiv H(P_*). \tag{2.10}$$

Moreover, $\varrho \in \mathbf{R}_{v_{k-1}}$ by fact (2.9). Consequently, there exists a path $P_\varrho \in \mathcal{P}(s, v_{k-1})$ with

$$H(P_\varrho) = \varrho \overset{(2.10)}{\equiv} H'(P_*) \overset{(2.10)}{\equiv} H(P_*). \tag{2.11}$$

The order preservation of H and 2.26 a) imply that

$$H(P_\varrho \oplus r_k) \overset{(2.11)}{\equiv} H(P_* \oplus r_k) = H(P). \tag{2.12}$$

Moreover, $H(P_\varrho \oplus r_k) = \phi_{r_k}(\varrho)$ because of $H(P_\varrho) = \varrho$ and the definition of ϕ_{r_k}. Hence, the following is true:

$$H(P_\varrho \oplus r_k) = \phi_{r_k}(\varrho) = \phi_{r_k}(H'(P_*)) \overset{(2.11)}{\underset{\text{Def. of } H'}{=}} H'(P_* \oplus r_k) = H'(P). \tag{2.13}$$

Hence, $H(P) \overset{(2.12)}{\equiv} H(P_\varrho \oplus r_k) \overset{(2.13)}{=} H'(P)$ so that indeed $H(P) \equiv H'(P)$.

P r o o f o f (b)$\Rightarrow$(a) : First, we show that the path function H' itself is order preserving. Given the paths P_1, P_2, Q with $\alpha(P_1) = \alpha(P_2) = s$ and $\omega(P_1) = \omega(P_2) = \alpha(Q) =: v'$. We assume that

$$H'(P_1) \preceq H'(P_2). \tag{2.14}$$

If $Q = [v']$ then $H'(P_1 \oplus Q) = H'(P_1) \overset{(2.14)}{\preceq} H'(P_2) = H'(P_2 \oplus Q)$. The other case is that $Q = \hat{r}_1 \oplus \ldots \oplus \hat{r}_j$. Then assumption (b) says that these arcs are marked with monotonically increasing functions $\phi_{\widehat{r}_1}, \ldots, \phi_{\widehat{r}_j}$. Then $\widehat{\phi} := \phi_{\widehat{r}_j} \circ \ldots \circ \phi_{\widehat{r}_1}$ is monotonically increasing, too, and $H'(P_i \oplus Q) = \widehat{\phi}(H(P_i))$, $i = 1, 2$. Hence,

$$H'(P_1 \oplus Q) = \widehat{\phi}(H'(P_1)) \overset{(2.14)}{\preceq} \widehat{\phi}(H'(P_2)) = H'(P_2 \oplus Q).$$

We have seen that H' is order preserving. Moreover, $H \equiv H'$ by assumption (b). Remark 2.26 b) implies that H is order preserving, too. $\qquad\square$

Remark 2.29 Of course, direction (b)$\Rightarrow$(a) of Theorem 2.28 is true if the monotone functions ϕ_r are defined on the entire set $\mathbf{R}$. This observation yields a simple method to construct order preserving path functions: Choose an arbitrary value $H([s]) \in \mathbf{R}$; attach each arc r with a monotnically increasing function $\phi_r : \mathbf{R} \to \mathbf{R}$. Then define $H(P) := \phi_{r_k}(\phi_{r_{k-1}}(\ldots(H([s]))\ldots)$ for all paths $P = r_1 \oplus \ldots \oplus r_k$ with $\alpha(P) = s$. Moreover, let $H' := \phi \circ\circ p$ where $p(v) := H([v])$, $v \in V$.
Then H' is even equal to H. Consequently, H is order preserving by (b)$\Rightarrow$(a) of Theorem 2.28.
This construction of order preserving functions has been described in [Mor82] and [ChMi82]. $\qquad\square$

Remark 2.30 The question arises whether the conclusion in Remark 2.29 can be reversed. More precisely, let H be order preserving, and let $p(v) := H([v])$, $v \in V$; can we find a monotone function $H' = \phi \circ\circ p$ with $H' \equiv H$ such that each monotone function ϕ_r is defined on the entire set $\mathbf{R}$ (that is, $\phi_r : \mathbf{R} \to \mathbf{R}$ for all arcs r)?
The answer is NO, and we give the following counterexample: Let $\mathcal{G} = (V, \mathcal{R})$ where $V := \{s, v^+, w^+, v^-, w^-, y, z\}$ and

$$\mathcal{R} := \{(s, v^+), (s, v^-), (v^+, w^+), (w^+, v^+), (v^-, w^-), (w^-, v^-), (v^+, y), (v^-, y), (y, z)\}.$$

Let $\mathbf{R} := \mathbb{R}\backslash\{1\}$, and let $\preceq$ be the $\leq$-relation.
We define $p(v) := 0$ for all $v \in V$, and we define the following functions ϕ_r, $r \in \mathcal{R}$:

$$\phi_r(\zeta) := \begin{cases} 0.9\,\zeta + 0.1 & \text{if } r = (s, v^+), \ \zeta \in \mathbf{R}, \\ 1.1\,\zeta - 0.1 & \text{if } r = (s, v^-), \ \zeta \in \mathbf{R}, \\ \zeta^3 & \text{if } r \in \{(v^+, w^+), (v^-, w^-)\}, \ \zeta \in \mathbf{R}, \\ \zeta & \text{if } r \in \{(w^+, v^+), (w^-, v^-), (v^+, y), (v^-, y)\}, \ \zeta \in \mathbf{R}, \\ \zeta^3 + 1 & \text{if } r = (y, z), \ \zeta \in \mathbf{R}\backslash\{0\} = \mathbb{R}\backslash\{0, 1\}. \end{cases}$$

Then it is easy to see that $\phi_r \in \mathbf{R}$ for all $r \in \mathcal{R}$ and all $\zeta \in \text{def}(\phi_r)$.
We define $H : \mathcal{P}(s) \to \mathbf{R}$ as $H := \phi \circ\circ p$. Then H is well-defined. To see this we assume that $P = [v_0, \ldots, v_k]$ with $v_0 = s$. Then (2.3) is almost trivial if $(v_{k-1}, v_k) \neq (y, z)$. So, we must still show that (2.3) is true if $(v_{k-1}, v_k) = (y, z)$. For this purpose, we observe that the current H-value is cubed when moving along the lower or the upper circle. Hence, the set $\mathbf{R}_y := \{H(P) \mid P \in \mathcal{P}(s, y)\}$ is equal to $\mathbf{R}^+ \cup \mathbf{R}^-$ where

$$\mathbf{R}^+ := \{(0.1)^{3^j} \mid j = 0, 1, 2, \ldots\} \quad \text{and} \quad \mathbf{R}^- := \{-(0.1)^{3^j} \mid j = 0, 1, 2, \ldots\}.$$

In particular, $\mathbf{R}_y \subseteq \mathbf{R}\backslash\{0\} = \text{def}(\phi_{(y,z)})$ so that $\phi_{(y,z)}(H([v_0, \ldots, v_{k-1}]))$ is well-defined if $(v_{k-1}, v_k) = (y, z)$.
The path function H is order preserving. This follows from the monotonicity of the functions ϕ_r.

Next, we try to extend $\phi_{(y,z)}$ to a monotonic function on the entire set $\mathbf{R}$. That means that we must find a value $\phi_{(y,z)}(0)$. The monotonicity of the extended function $\phi_{(y,z)}$ implies that

$$\underbrace{\left[-0.1^{3^j}\right]^3}_{\in\mathbf{R}^-} + 1 = \phi_{(y,z)}\left(-0.1^{3^j}\right) \leq \phi_{(y,z)}(0) \leq \phi_{(y,z)}\left(0.1^{3^j}\right) \leq \underbrace{\left[0.1^{3^j}\right]^3}_{\in\mathbf{R}^+} + 1$$

for all $j \in \mathbb{N}$. The left and the right side of this inequality tend to 1 for $j \to \infty$. Hence, the only way to define $\phi_{(y,z)}(0)$ is $\phi_{(y,z)}(0) := 1$, but this value is not an element of $\mathbf{R}$. Consequently, $\phi_{(y,z)}$ cannot be extended to a function defined on the entire set $\mathbf{R}$.

If we replace $\mathbf{R}$ by $\mathbf{R}' := \mathbb{R}$ then all ϕ_r's in our example can be defined on the entire set $\mathbf{R}'$. This set $\mathbf{R}'$ is the closed hull of $\mathbf{R}$. In the general case, however, it is not clear whether a set $\mathbf{R}$ can be replaced by another set $\mathbf{R}'$ such that the ϕ_r's can be defined as total, monotone functions $\phi_r : \mathbf{R}' \to \mathbf{R}'$. $\qquad\square$

Remark 2.31 Let H be order preserving. Then we cannot guarantee that $H(P) = H'(P)$ in assertion (b) of Theorem 2.28. It is possible that there exist two paths P_1, P_2 and an arc r with the following properties:

$$H(P_1) = H(P_2), \quad H(P_1 \oplus r) \equiv H(P_2 \oplus r), \quad \text{but } H(P_1 \oplus r) \neq H(P_2 \oplus r).$$

If H were equal to H' then H itself were monotone, and there were a function ϕ_r with $H(P_1 \oplus r) = \phi_r(H(P_1)) = \phi_r(H(P_2)) = H(P_2 \oplus r)$, where the second equality follows from $H(P_1) = H(P_2)$. The equation $H(P_1 \oplus r) = H(P_2 \oplus r)$, however, is a contradiction to the assumption that $H(P_1 \oplus r) \neq H(P_2 \oplus r)$. $\qquad\square$

The next theorem says that all order preserving functions $H : \mathcal{P}(\mathcal{G}) \to \mathbf{R}$ can be generated with order preserving functions of the form $H_v : \mathcal{P}(v) \to \mathbf{R}$.

Theorem 2.32 Let $H : \mathcal{P}(\mathcal{G}) \to (\mathbf{R}, \preceq)$ be a path function, and let $H_v := H|_{\mathcal{P}_v}$ for all nodes $v \in \mathcal{V}$.

Then the following assertions are equivalent:
(a) H is order preserving.
(b) All restrictions H_v, $v \in \mathcal{V}$, are order preserving.[3]

P r o o f   o f   (a)$\Rightarrow$(b) : Let H be order preserving, and let $v \in \mathcal{V}$. We show that H_v is order preserving, too.
Let $P_1, P_2 \in \mathcal{P}(v)$ and $Q \in \mathcal{P}(\mathcal{G})$ such that $\omega(P_1) = \omega(P_2) = \alpha(Q)$. Then $P_1, P_2, P_1 \oplus Q, P_2 \oplus Q \in \mathcal{P}(v) = \mathrm{def}(H_v)$. Consequently, if $H_v(P_1) \preceq H_v(P_2)$ then $H(P_1) \preceq H(P_2)$ and $H_v(P_1 \oplus Q) = H(P_1 \oplus Q) \overset{(a)}{\preceq} H(P_2 \oplus Q) = H_v(P_2 \oplus Q)$.

P r o o f   o f   (b)$\Rightarrow$(a) : Let $P_1, P_2, Q \in \mathcal{P}(\mathcal{G})$ such that $v := \alpha(P_1) = \alpha(P_2)$ and $\omega(P_1) = \omega(P_2) = \alpha(Q)$. Then $P_1, P_2, P_1 \oplus Q, P_2 \oplus Q \in \mathcal{P}(v) = \mathrm{def}(H_v)$. Consequently, if $H(P_1) \preceq H(P_2)$ then $H_v(P_1) \preceq H_v(P_2)$ and $H_v(P_1 \oplus Q) = H_v(P_1 \oplus Q) \overset{(b)}{\preceq} H_v(P_2 \oplus Q) = H(P_2 \oplus Q)$. $\qquad\square$

---

[3]This result yields the following method to construct an order preserving function H: First generate an order preserving function $H_v : \mathcal{P}(v) \to \mathbf{R}$ for each $v \in \mathcal{V}$. (This is described in 2.29.) Then define $H(P) := H_{\alpha(P)}(P)$ for all paths $P \in \mathcal{P}$. The assertion (b)$\Rightarrow$(a) of the current theorem says that H is order preserving.

2.3.3   Further methods to construct order preserving functions

The next results show how several given order preserving functions can be used to construct a new one.

The simplest idea is to add two given real-valued order preserving functions H' and H''. But this method does not work in general because the following situation may occur:

$$H'(P_1) = 0, \quad H'(P_2) = 2, \quad H'(P_1 \oplus Q) = 0, \quad H'(P_2 \oplus Q) = 2,$$
$$H''(P_1) = 1, \quad H''(P_2) = 0, \quad H''(P_1 \oplus Q) = 4, \quad H''(P_2 \oplus Q) = 0.$$

Then $H'(P_1) + H''(P_1) = 1 < 2 = H'(P_2) + H''(P_2)$ but $H'(P_1 \oplus Q) + H''(P_1 \oplus Q) = 4 > 2 = H'(P_2 \oplus Q) + H''(P_2 \oplus Q)$.

This next theorem gives sufficient conditions for the order preservation of $H' + H''$.

Theorem 2.33 *Given the order preserving path functions $H', H'' : \mathcal{P} \to (\mathbb{R}, \leq)$ where $\mathcal{P} = \mathcal{P}(s)$ or $\mathcal{P} = \mathcal{P}(\mathcal{G})$.*
Moreover, we assume H' is $<$-preserving; that means that H' and H'' have property (2.7) where $\widehat{\mathfrak{R}}$ must be replaced by " $<$".
In addition, we assume that there exists a $\Lambda > 0$ with the following properties:
    (i) $(\forall P_1, P_2 \in \mathcal{P}) \; H'(P_1) \neq H'(P_2) \implies |H'(P_1) - H'(P_2)| > 2\Lambda,$
    (ii) $(\forall P \in \mathcal{P}) \; |H''(P)| \leq \Lambda.$
Then $H := H' + H''$ is order preserving, too.

P r o o f : The proof is given in [Huck92], page 36 – 37; one of the ideas is to distinguish between the cases $H'(P_1) = H'(P_2)$ and $H'(P_1) \neq H'(P_2)$ if $H(P_1) \leq H(P_2)$.   □

The next result describes how an order preserving path function H for a graph $\mathcal{G}$ can be extended to an order preserving path function H' for a supergraph $\mathcal{G}'$ of $\mathcal{G}$. The idea is to assign a path a very high H'-value if this path uses an arc in $\mathcal{G}'\backslash\mathcal{G}$.

Theorem 2.34 *Given the digraphs $\mathcal{G} = (V, \mathcal{R})$ and $\mathcal{G}' = (V, \mathcal{R}')$ with $\mathcal{R}' \supseteq \mathcal{R}$. Let $\mathcal{P} = \mathcal{P}(\mathcal{G})$ and $\mathcal{P}' := \mathcal{P}(\mathcal{G}')$. Given an order preserving path function $H : \mathcal{P} \to \mathbb{R}$ and a number $\Lambda \in \mathbb{R}$ such that $H(P) < \Lambda$ for all $P \in \mathcal{P}$.*

Then there exists an order preserving path function $H' : \mathcal{P}' \to \mathbb{R}$ such that $H = H'|_{\mathcal{P}}$.

P r o o f : We define $H'(P) := H(P)$ if $P \in \mathcal{P}$ and $H'(P) := \Lambda$ if $P \in \mathcal{P}'\backslash\mathcal{P}$. Then $H = H'|_{\mathcal{P}}$, and we must show that H' is order preserving.

Let $v, w \in V$, and let $P_1, P_2 \in \mathcal{P}'$ two v-w-paths with $H'(P_1) \leq H'(P_2)$. Let $Q \in \mathcal{P}(\mathcal{G}')$ with $\alpha(Q) = w$.

If $P_2 \oplus Q \notin \mathcal{P}$ then $H'(P_1 \oplus Q) \leq \Lambda = H'(P_2 \oplus Q)$, which must be proven.

The other case is that $P_2 \oplus Q \in \mathcal{P}$. Then $P_2 \in \mathcal{P}$. Consequently, $H'(P_1) \leq H'(P_2) = H(P_2) < \Lambda$ so that $P_1 \in \mathcal{P}$. Moreover, $Q \in \mathcal{P}$ because $P_2 \oplus Q \in \mathcal{P}$. Consequently, $P_1 \oplus Q \in \mathcal{P}$.

We have seen that $P \in \mathcal{P}$ for all $P \in \{P_1, P_2, P_1 \oplus Q, P_2 \oplus Q\}$ so that $H'(P) = H(P)$ for all these paths. Therefore, $H'(P_1) \leq H'(P_2)$ implies that $H(P_1) \leq H(P_2)$, and this and the order preservation of H imply that $H'(P_1 \oplus Q) = H(P_1 \oplus Q) \leq H(P_2 \oplus Q) = H'(P_2 \oplus Q)$.   □

Remark 2.35 Given an order preserving path function $H : \mathcal{P}(\mathcal{G}) \to \mathbb{R}$. Then we may often assume that $\mathcal{G}$ is equal to the complete graph $\mathcal{G}^* := (\mathcal{V}, \mathcal{V} \times \mathcal{V})$. If $\mathcal{G} \neq \mathcal{G}^*$, we carry out the following construction: We define $\tilde{H} := \arctan \circ H$ as proposed in Remark 2.26 b); let $\Lambda := \pi$; then $\tilde{H}$ is order preserving, and $\tilde{H}(P) < \Lambda = \pi$ for all $P \in \mathcal{P}(\mathcal{G})$. Then we apply 2.34 to $\tilde{H}$ and $\mathcal{G}' := \mathcal{G}^*$. Thus, we obtain an order preserving path function $H' : \mathcal{P}(\mathcal{G}^*) \to \mathbb{R}$ such that $H'|_{\mathcal{P}(\mathcal{G})} = \tilde{H}$ is order equivalent to H.   □

Next, we describe the construction of order preserving functions with the help of graph homomorphisms.

Theorem 2.36 *Given two digraphs $\mathcal{G} = (\mathcal{V}, \mathcal{R})$ and $\tilde{\mathcal{G}} = (\tilde{\mathcal{V}}, \tilde{\mathcal{R}})$. Let Ψ be a homomorphism from $\tilde{\mathcal{G}}$ to $\mathcal{G}$. Let $\mathcal{P} := \mathcal{P}(\mathcal{G})$ or $\mathcal{P} := \mathcal{P}(s)$ where $s \in \mathcal{V}$. Let $H : \mathcal{P} \to (\mathbb{R}, \preceq)$ be a path function in $\mathcal{G}$, and let Let $H_\Psi : \Psi^{-1}(\mathcal{P}) \to (\mathbb{R}, \preceq)$ be the path function induced by Ψ.*

Then the following assertions are true:

a) If H is order preserving then H_Ψ is order preserving.

b) If H_Ψ is order preserving then H is not always order preserving.

P r o o f   o f   a) :   Given three paths $\tilde{P}_1, \tilde{P}_2, \tilde{Q}$ in $\mathcal{P}(\tilde{\mathcal{G}})$ with the following properties:

$$\alpha(\tilde{P}_1) = \alpha(\tilde{P}_2), \quad \omega(\tilde{P}_1) = \tilde{\omega}(\tilde{P}_2), \quad \tilde{\omega}(\tilde{P}_1) = \tilde{\alpha}(\tilde{Q}), \tag{2.15}$$

$$(\alpha) \ \tilde{P}_1, \tilde{P}_2 \in \Psi^{-1}(\mathcal{P}), \qquad (\beta) \ H_\Psi(\tilde{P}_1) \preceq H_\Psi(\tilde{P}_2). \tag{2.16}$$

We define the following paths in $\mathcal{G}$: $P_1 := \Psi(\tilde{P}_1)$, $P_2 := \Psi(\tilde{P}_2)$, $Q := \Psi(\tilde{Q})$. Then

$$P_i = \Psi(\tilde{P}_i) \overset{(2.16)(\alpha)}{\in} \mathcal{P}, \quad i = 1, 2. \tag{2.17}$$

Moreover, the functional equations for Ψ and the definition of H_Ψ imply the following equations:

$$\alpha(P_1) = \alpha(\Psi(\tilde{P}_1)) = \Psi(\alpha(\tilde{P}_1)) \overset{(2.15)}{=} \Psi(\alpha(\tilde{P}_2)) = \alpha(\Psi(\tilde{P}_2)) = \alpha(P_2),$$

$$\omega(P_1) = \omega(\Psi(\tilde{P}_1)) = \Psi(\omega(\tilde{P}_1)) \overset{(2.15)}{=} \Psi(\omega(\tilde{P}_2)) = \omega(\Psi(\tilde{P}_2)) = \omega(P_2),$$

$$\omega(P_1) = \Psi(\omega(\tilde{P}_1)) \overset{(2.15)}{=} \Psi(\omega(\tilde{Q})) = \alpha(\Psi(\tilde{Q})) = \alpha(Q),$$

$$H(P_1) = H(\Psi(\tilde{P}_1)) = H_\Psi(\tilde{P}_1) \overset{(2.16)(\beta)}{\preceq} H_\Psi(\tilde{P}_2) = H(\Psi(\tilde{P}_2)) = H(P_2).$$

The fact that $\alpha(P_1) = \alpha(P_2)$ and the order preservation of H imply that

$$P_1 \oplus Q, P_2 \oplus Q \overset{(2.17)}{\in} \mathcal{P} \quad \text{and} \quad H(P_1 \oplus Q) \preceq H(P_2 \oplus Q). \tag{2.18}$$

Moreover, (2.15) and the functional equations about the homomorphism Ψ imply that

$$P_i \oplus Q = \Psi(\tilde{P}_i) \oplus \Psi(\tilde{Q}) = \Psi(\tilde{P}_i \oplus \tilde{Q}) \ (i = 1, 2). \tag{2.19}$$

This and (2.18) imply that $\tilde{P}_i \oplus \tilde{Q} \in \Psi^{-1}(\mathcal{P})$ $(i = 1, 2)$ so that $H_\Psi(\tilde{P}_1)$ and $H_\Psi(\tilde{P}_2)$ are well-defined. Moreover, we obtain the following assertion:

$$H_\Psi(\widetilde{P}_1 \oplus \widetilde{Q}) = H(\Psi(\widetilde{P}_1 \oplus \widetilde{Q})) \overset{(2.19)}{=} H(P_1 \oplus Q) \overset{(2.18)}{\preceq} H(P_2 \oplus Q) \overset{(2.19)}{=} H(\Psi(\widetilde{P}_2 \oplus \widetilde{Q})) = H_\Psi(\widetilde{P}_2 \oplus \widetilde{Q}).$$

Hence, H_Ψ is order preserving.

P r o o f o f b) : We construct the following counterexample: Let $s \in V$ such that all other nodes of G can be reached from s. Let $H : \mathcal{P}(s) \to \mathbf{R}$ be a path function that is not order preserving. Let $\widetilde{G} = G^s$. Then 2.14 says that H_Ψ is a potential; hence, H_Ψ is order preserving although this is not true for H. $\square$

The previous results were about constructing an order preserving function with the help of other order preserving functions. The next results are about constructing an order preserving function H' with the help of a function H that are not necessarily order preserving.

If H and H' are closely related we can use the order preservation of H' to find H-minimal paths.

We give an example of such a search strategy; in this example, we define H' as a path function induced by a graph homomorphism.

1. Given a digraph G, a node s in G and a path function $H : \mathcal{P}(s) \to \mathbf{R}$ that is not order preserving.
   Construct a digraph G' and a homomorphism Ψ from G' into G such that
   - $\Psi\left(\Psi^{-1}(\mathcal{P}(s))\right) = \mathcal{P}$. (That means that Ψ is surjective on $\Psi^{-1}(\mathcal{P}(s))$.)
   - H_Ψ is order preserving.

   An example is the canonical homomorphism Ψ from the expanded digraph $G' := G^s$ to G (see the proof to 2.36 b)).

2. Search for an H_Ψ-optimal path P' in G'.
   (The order preservation of H_Ψ makes the search easy.)

3. Let $P^* := \Psi(P')$.

The assumption that $\Psi(\Psi^{-1}(\mathcal{P}(s))) = \mathcal{P}(s)$ guarantees that no path in $\mathcal{P}$ is omitted when searching the set $\Psi^{-1}(\mathcal{P}(s))$.

The simplest homomorphic inverse of $G = (V, \mathcal{R})$ is $G' = G^s$. Next, we construct other homomorphic inverses G' that have much fewer nodes and arcs than G^s. The set of nodes of G' is a Cartesian product $V' = \mathcal{M} \times V$ or $V' = V \times X$ where $\mathcal{M}$ or X is a nonempty set. The construction is similar to the one of $G^{[K]}$ on page 18. Each node $(\mu, v) \in \mathcal{M}$ and $(v, \xi) \in V \times X$ may be interpreted as a copy of v in the μth and in the ξth level, respectively. The component μ or ξ is used to record information about the path function H in G. This information need not be recorded any more in the values of the new path function H' in G'. Therefore, we can define H' as an order preserving function.

Further digraphs with Cartesian products as sets of nodes are constructed in [Snie86], page 171 – 173 and in [Rom88]. In the first paper, a special path function H is transformed in an order preserving one. The second paper uses the Cartesian product to record two things: a node of a given digraph and a state of a finite automaton.

Theorem 2.37 *Given a digraph $G = (V, \mathcal{R})$. Let s be a node in G such that all nodes $v \in V$ can be reached from s (by an s-v-path).*
Let $H : \mathcal{P}(s) \to (\mathbb{R}, \leq)$ be a path function.

We assume that $H = \phi \circ p$ where $p : V \to \mathbb{R}$ and $\phi_r : \mathbb{R} \to \mathbb{R}$, $r \in \mathcal{R}$.

Moreover, given $m \geq 1$ pairwise disjoint sets $I_1, \ldots, I_m \subseteq \mathbb{R}$. Let $I := I_1 \cup \ldots \cup I_m$. We assume that $p(s) \in I$ and that each ϕ_r ($r \in \mathcal{R}$) maps each I_μ monotonically into a set $I_{\mu'}$. More precisely, we assume that there exists a function $\pi_r : \{1, \ldots, m\} \to \{1, \ldots, m\}$ for each arc $r \in R$ such that

(i) $(\forall r \in \mathcal{R}, m \in \{1, \ldots, m\})$ $\phi_r(I_\mu) \subseteq I_{\pi_r(\mu)}$.

(ii) $(\forall r \in \mathcal{R}, m \in \{1, \ldots, m\})$ $\phi_r|_{I_\mu}$ *is monotonically increasing or decreasing.*

Then we can construct

- *a digraph $\mathcal{G}' = (\mathcal{V}', \mathcal{R}')$ with a node $s' \in \mathcal{V}'$,*
- *a graph homomorphism $\Psi : (\mathcal{V}' \cup \mathcal{R}' \cup \mathcal{P}(s')) \to (\mathcal{V} \cup \mathcal{R} \cup \mathcal{P}(s))$,*
- *a node function $p' : \mathcal{V}' \to \mathbb{R}$,*
- *monotonically increasing or decreasing partial functions $\phi'_{r'} : \mathbb{R} \dashrightarrow \mathbb{R}$ ($r' \in \mathcal{R}'$),*
- *the path function $H' : \mathcal{P}(s') \to (\mathbf{R}, \preceq)$ of the form $H' = \phi' \circ\circ p'$*

such that the following are true:

a) $\Psi(\mathcal{V}') = \mathcal{V}$, $\quad \Psi(\mathcal{R}') = \mathcal{R}$, $\quad \Psi(\mathcal{P}(s')) = \mathcal{P}(s)$.

b) *H' is well-defined, and H' is induced by Ψ, i.e., $H' = H_\Psi = H \circ \Psi$.*

c) *If all restrictions $\phi_r|_{I_\mu}$ are monotonically increasing then H' is order preserving.*

P r o o f : First, we outline the idea. We shall define the set of nodes in the new digraph as $\mathcal{V}' := \{1, \ldots, m\} \times \mathcal{V}$. Given a $\mathcal{G}$-path $P = [s = v_0, v_1, \ldots, v_k] \in \mathcal{P}(s)$ with $H([v_0, \ldots, v_\kappa]) \in I_{\mu_\kappa}$ for all $\kappa = 0, \ldots, k$. Then its inverse under Ψ is a $\mathcal{G}'$-path of the form $P' = [(\mu_0, v_0), \ldots, (\mu_k, v_k)]$. So, the components μ_κ indicate that the original H-value is an element of I_{μ_κ}. Therefore, the transition function $\phi'_{((\mu_\kappa, v_\kappa),(\mu_{\kappa+1}, v_{\kappa+1}))}$ must only represent the behaviour of $\phi_{(v_\kappa, v_{\kappa+1})}$ on the interval I_{μ_κ}; more precisely, we may define $\phi_{(v_\kappa, v_{\kappa+1})}$ as the restriction of $\phi'_{((\mu_\kappa, v_\kappa),(\mu_{\kappa+1}, v_{\kappa+1}))}$ on the interval I_{μ_κ}, and this restriction is monotonically increasing or decreasing.

Next, we describe the details of the construction: Let $M := \{1, \ldots, m\}$. For all $x \in I$, we define $pr(x) \in M$ as the unique number for which $x \in I_{pr(x)}$. Then we define the digraph $\mathcal{G}' = (\mathcal{V}', \mathcal{R}')$ as follows:

$$\mathcal{V}' := M \times \mathcal{V}, \text{ and } \mathcal{R}' := \{ ((\mu, v), (\tilde{\mu}, \tilde{v})) \mid (v, \tilde{v}) \in \mathcal{R}, \tilde{\mu} = \pi_{(v, \tilde{v})}(\mu) \}.$$

Let $\mu_0 := pr(H[(s)])$; then $H[s] = p(s) \in I_{\mu_0}$. Let $s' := (\mu_0, s)$; then $s' \in \mathcal{V}'$. We define the homomorphism Ψ as follows: Let $\Psi((\mu, v)) := v$ for all nodes $(\mu, v) \in \mathcal{V}'$, and let $\Psi(((\mu, v), (\tilde{\mu}, \tilde{v}))) := (v, \tilde{v})$ for all arcs $((\mu, v), (\tilde{\mu}, \tilde{v})) \in \mathcal{R}'$. We define $p'(v') := H([s])$ for all nodes $v' \in \mathcal{V}'$. For each arc $r' = ((\mu, v), (\tilde{\mu}, \tilde{v})) \in \mathcal{R}'$, we define $\phi'_{r'} := \phi_{(v, \tilde{v})}|_{I_\mu}$. Moreover, we define $H' : \mathcal{P}(s') \to \mathbb{R}$ as $H' := \phi' \circ\circ p'$.

Next, we prove the assertions a), b), and c).

P r o o f   o f   a) : We show the surjectivity of Ψ by constructing Ψ-inverses. If $v \in \mathcal{V}$ then $\Psi((\mu_0, v)) = v$. If $r := (v, \tilde{v}) \in \mathcal{R}$ then $\Psi(((\mu_0, v), (\pi_r(\mu_0), \tilde{v}))) = (v, \tilde{v}) = r$. Let $P = [v_0 = s, v_1, \ldots, v_k] \in \mathcal{P}(s)$. If $k = 0$ then $\Psi([s']) = [s] = P$ by the definition of Ψ. The other case is that $k > 0$. Then let $r_\kappa := (v_{\kappa-1}, v_\kappa)$ and $\mu_\kappa := pr(H([v_0, \ldots, v_\kappa]))$, $\kappa = 1, \ldots, k$. It is $\mu_{\kappa+1} = \pi_{r_\kappa}(\mu_\kappa)$ for all $\kappa = 0, \ldots, k-1$.

Hence, the path $P' := [(\mu_0, s), (\mu_1, v_1), \ldots, (\mu_k, v_k)]$ is an element of $\mathcal{P}(s')$, and it is $\Psi(P') = P$.

P r o o f o f b) : To prove that H' is well-defined we show that the following is true for all $k \geq 0$:

> $H'(P')$ is well-defined for each path $P' \in \mathcal{P}(s')$ of length k.
> Moreover, if $\omega(P') = (\mu, v)$ then $H'(P') \in I_\mu$.
> $\hfill(2.20)$

If $k = 0$ then $H'(P')$ is well-defined. Moreover, the definition of H' says that $H'(P') = H'([s']) = p'(s') = H([s])$, and $H([s]) \in I_{\mu_0}$ because μ_0 has been defined as $pr(H([s]))$. Hence, $H'(P') \in I_{\mu_0}$, and this is the right set I_μ because $\omega(P') = s' = (\mu_0, s)$.

We assume that (2.20) is true for all paths of length $k - 1$. Let P' have k arcs, and let Q' consist of the first $(k - 1)$ arcs of P'. Let $(\overline{\mu}, \overline{v})$ be the last node of Q', and let (μ, v) be the last node of P'. Then $r' := ((\overline{\mu}, \overline{v}), (\mu, v))$ is the last arc of P', and $P' = Q' \oplus r'$.

The induction hypothesis says that $H'(Q')$ is well-defined and $H'(Q') \in I_{\overline{\mu}}$, and this set is equal to $\mathrm{def}\left(\phi'_{r'}\right)$ by the definition of $\phi'_{r'}$. That means that $H'(P')$ is well-defined, too.

We must still show that $H'(P') \in I_\mu$. For this end, we recall the definition of $\mathcal{R}'$. The fact that $r' := ((\overline{\mu}, \overline{v}), (\mu, v)) \in \mathcal{R}'$ implies that $\mu = \pi_{(\overline{v}, v)}(\overline{\mu})$; hence, $\phi_{(\overline{v}, v)}$ maps $I_{\overline{\mu}}$ into I_μ. This and the assumption that $H'(Q') \in I_{\overline{\mu}}$ imply relation $(*)$ in the following assertion:

$$H'(P') = \phi'_{r'}(H'(Q')) = \left(\phi_{(\overline{v}, v)}\big|_{I_{\overline{\mu}}}\right)\left(H'(Q')\right) \overset{(*)}{\in} I_\mu.$$

Next, we prove that $H' = H \circ \Psi$. For this end, we show the following assertion by an induction on k: $\qquad H'(P') = H(\Psi(P'))$ for all paths P' of length k. $\hfill(2.21)$

Let $k = 0$. Then $P = [s']$ and $H'(P') = H'([s']) = p'(s') = H([s]) = H(\Psi([s']))$.

We assume that (2.21) is true for all paths of length $k - 1$. Let P' be a path of length P', and let Q', r', etc. be the same objects as in the proof of (2.20). Then we obtain an assertion, in which fact $(*)$ follows from the definition of $\phi'_{r'}$, fact $(**)$ follows from the induction hypothesis, and fact $(* * *)$ follows from the recursion formula for H.

$$H'(P') = \phi'_{r'}\big(H'(Q')\big) \overset{(*)}{=} \phi_{(\overline{v}, v)}\big(H'(Q')\big) \overset{(**)}{=} \phi_{(\overline{v}, v)}\big(H(\Psi(Q'))\big) \overset{(* * *)}{=}$$
$$H(\Psi(Q') \oplus (\overline{v}, v)) = H(\Psi(Q') \oplus \Psi(r')) = H(\Psi(Q' \oplus r')) = H(\Psi(P')).$$

Hence, $H'(P') = H(\Psi(P'))$. That means that fact (2.21) and assertion b) have been proven.

P r o o f o f c) : This claim is an immediate consequence of Theorem 2.28. $\hfill\square$

Remark 2.38 The digraph $\mathcal{G}'$ and the path function H' in 2.37 are often useful to find optimal paths in $\mathcal{G}$. Now we describe the search problem and the strategy in detail.

Suppose that we are searching for an H-optimal path $P_* \in \mathcal{P}(s, \Gamma)$ where $\Gamma \subseteq V \backslash \{s\}$ is a given set of goal nodes in $\mathcal{G}$. Then the following is true:

P_* may be chosen as $\Psi(P'_*)$ where P'_* is an H'-optimal $\mathcal{G}'$-path among all candidates that start with s' and end with a node $v' \in \Gamma' := \{1, \ldots, m\} \times \Gamma$. (2.22)

Fact (2.22) is shown as follows: Assertion 2.37 b) implies that $H(P_*)$ is optimal among all values $H(\Psi(P'))$, $P' \in \mathcal{P}(s', \Gamma')$; moreover, this optimization process considers all candidates $P \in \mathcal{P}(s, \Gamma)$ because $\Psi(\mathcal{P}(s', \Gamma')) = \mathcal{P}(s, \Gamma)$ by 2.37 a). So, assertion (2.22) is true.

Hence, an H-optimal s-Γ-path P_* can be found by searching for an optimal $\mathcal{G}'$-path P'_* from s' to Γ'. Here, we can make use of the order preservation of H' if all restrictions $\phi_r|_{I_\mu}$ are monotonically increasing.

In the same way we can find an $H_{\max}$-optimal s-Γ-path $P_\bullet$ in $\mathcal{G}$: Search for a $\mathcal{G}'$-path $P'_\bullet$ with minimal $H'_{\max}$-value among all candidates from s' to Γ'; then let $P_\bullet := \Psi(P'_\bullet)$. The correctness of this method follows from the surjectivity of Ψ (see 2.37 a)) and from the fact that $H'_{\max}(P') = H_{\max}(\Psi(P'))$ for all $P' \in \mathcal{P}(s')$; this fact has been shown in Theorem 2.18.

Further details about the search for $H_{\max}$-optimal paths will be given in Theorem 4.47. □

The next theorem says the following: Given a digraph $\mathcal{G}$ and a path function $H = \phi \circ p$ we assume that each function ϕ_r is monotonically increasing or decreasing. Then we can construct a digraph $\widetilde{\mathcal{G}}$ and a path function $\widetilde{H} = \widetilde{\phi} \circ \widetilde{p}$ with the following properties: $\widetilde{H}$ is closely related to H, and all functions $\widetilde{\phi}$ are monotonically increasing; in consequence, $\widetilde{H}$ is order preserving.

For example, the digraph $\mathcal{G}'$ constructed in Theorem 2.37 has the property that all arcs are marked with monotone functions. So, we may apply the next result to $\mathcal{G}'$.

A second example is a graph with a multiplicative path function H. H can be written as $H = \phi \circ p$ where $p(v) = 1$ for all nodes v; the functions ϕ_r, $r \in \mathcal{R}$, are of the form $x \mapsto h(r) \cdot x$ where h is an arc function. Then every function ϕ_r is monotone; ϕ_r is increasing if $h(r) \geq 0$, and ϕ_r is decreasing if $h(r) < 0$.

Theorem 2.39 is a generalization of a result in [Snie86]; Sniedovich has transformed multiplicative path functions in order preserving ones.

Theorem 2.39 *Given a digraph* $\mathcal{G} = (\mathcal{V}, \mathcal{R})$. *We assume that s is a node of $\mathcal{G}$ such that an s-v-path exists for each node $v \in \mathcal{V}$.*
Let $X(v) \subseteq \mathbb{R}$ be a nonempty set, and let $\sigma_v : X(v) \to X(v)$ for all nodes v. Moreover, let each arc r be marked with a function ϕ_r. We assume that the following are true:

(i) $(\forall\ r \in \mathcal{R})$ $\phi_r : X(\alpha(r)) \to X(\omega(r))$ *is monotonically increasing or decreasing.*
(ii) $\sigma_v : X(v) \to X(v)$ *is bijective and strictly decreasing.*[4]
(iii) $H([s]) \in X(s)$.

---

[4]For example, let $X(v) = (a, b)$ or $X(v) = [a, b]$; then we can choose $\sigma_v(x) := a + b - x$; if $X(v) = (a, \infty)$ choose $\sigma_v(x) := \frac{1}{x-a}$. If, however, $X(v) = [a, \infty)$ then no σ_v with property (ii) exists because we cannot define the value $\sigma_v(a)$.
When choosing the sets $X(v)$ such that (ii) is true one must take respect of condition (i), which is also related to the sets $X(v)$.

T h e n   there exist

- *a digraph $\widetilde{\mathcal{G}} = (\widetilde{\mathcal{V}}, \widetilde{\mathcal{R}})$ whose set of nodes is $\widetilde{\mathcal{V}} = \mathcal{V} \times \{-1, +1\}$,*
- *a graph homomorphism $\Lambda : (\widetilde{\mathcal{V}} \cup \widetilde{\mathcal{R}} \cup \mathcal{P}(\widetilde{s})) \rightarrow (\mathcal{V} \cup \mathcal{R} \cup \mathcal{P}(s))$,*
- *a node function $\widetilde{p} : \widetilde{\mathcal{V}} \rightarrow \mathbf{R}$,*
- *a function $\widetilde{\phi}_{\widetilde{r}} : X(v) \rightarrow X(w)$ for each arc $\widetilde{r} = ((v, i), (w, j)) \in \widetilde{\mathcal{R}}$,*
- *a path function $\widetilde{H} : \mathcal{P}(\widetilde{s}) \rightarrow \mathbb{R}$*

such that the following are true (with $\widetilde{s} := (s, 1)$):

a) *$\Lambda(\widetilde{\mathcal{V}}) = \mathcal{V}$, $\Lambda(\widetilde{\mathcal{R}}) = \mathcal{R}$ and $\Lambda(\mathcal{P}(\widetilde{s})) = \mathcal{P}(s)$.*

b) *The path function $\widetilde{H} := \widetilde{\phi} \circ \circ \widetilde{p}$, and $\widetilde{H}$ is well-defined.*
   Moreover, there is the following relationship between H and $\widetilde{H}$.
   Let $\widetilde{P} \in \mathcal{P}(\widetilde{s})$, let $P = \Lambda(\widetilde{P})$, and let $v = \omega(P)$; then

$$\widetilde{H}(\widetilde{P}) = \begin{cases} H(P) & = & H(\Lambda(\widetilde{P})) & \text{if } \omega(\widetilde{P}) \in \mathcal{V} \times \{+1\}, \\ \sigma_v(H(P)) & = & \left(\sigma_{\omega(\Lambda(\widetilde{P}))}\right)(H(\Lambda(\widetilde{P}))) & \text{if } \omega(\widetilde{P}) \in \mathcal{V} \times \{-1\}. \end{cases}$$

c) *$\widetilde{H}$ is order preserving.*

P r o o f : First, we define the set of arcs in $\widetilde{\mathcal{G}}$. For this we note that the set $\mathcal{R}$ of all arcs in $\mathcal{G}$ can be partitioned into

$$\mathcal{R}^+ := \{r \in \mathcal{R} \,|\, \phi_r \text{ is monotonically increasing (possibly constant)}\},$$

$$\mathcal{R}^- := \{r \in \mathcal{R} \,|\, \phi_r \text{ is monotonically decreasing and not constant}\}.$$

The new set $\widetilde{\mathcal{R}}$ represents all $r = (v, w) \in \mathcal{R}$ with the help of new arcs $((v, i), (w, j))$ where $i = j$ if and only if $r \in \mathcal{R}^+$ and where $i = -j$ if and only if $r \in \mathcal{R}^-$. More precisely,

$$\begin{aligned} \widetilde{\mathcal{R}} := \quad &\{((v, +1), (w, +1)) \,|\, (v, w) \in \mathcal{R}^+\} \cup \{((v, -1), (w, -1)) \,|\, (v, w) \in \mathcal{R}^+\} \cup \\ &\{((v, +1), (w, -1)) \,|\, (v, w) \in \mathcal{R}^-\} \cup \{((v, -1), (w, +1)) \,|\, (v, w) \in \mathcal{R}^-\}. \end{aligned}$$

We now define the homomorphism Λ. Let $\Lambda((v, i)) := v$ for all $(v, i) \in \widetilde{\mathcal{V}}$, and let $\Lambda(((v, i), (w, j))) := (v, w)$ for all $((v, i), (w, j)) \in \widetilde{\mathcal{R}}$.

Next, we define $\widetilde{p} : \widetilde{\mathcal{V}} \rightarrow \mathbf{R}$ such that $\widetilde{p}(\widetilde{v}) := H([s])$ for all $\widetilde{v} \in \widetilde{\mathcal{V}}$.

Next, we define the functions $\widetilde{\phi}_{\widetilde{r}} : X(v) \rightarrow X(w)$ for all $\widetilde{\mathcal{G}}$-arcs $\widetilde{r} = ((v, i), (w, j))$.

$$\widetilde{\phi}_{\widetilde{r}} := \begin{cases} \phi_{(v,w)}, & \text{if } i = 1, \quad j = 1, \\ \sigma_w \circ \phi_{(v,w)}, & \text{if } i = 1, \quad j = -1, \\ \phi_{(v,w)} \circ \sigma_v^{-1}, & \text{if } i = -1, \quad j = 1, \\ \sigma_w \circ \phi_{(v,w)} \circ \sigma_v^{-1}, & \text{if } i = -1, \quad j = -1. \end{cases}$$

Then we define $\widetilde{H} : \mathcal{P}(\widetilde{s}) \rightarrow \mathbb{R}$ as $\widetilde{H} := \widetilde{\phi} \circ \circ \widetilde{p}$.

P r o o f   o f   a) : If $v \in \mathcal{V}$ then $(v, 1)$ is an inverse of v under Λ. Let $r \in \mathcal{R}$. If $r = (v, w) \in \mathcal{R}^+$ and $r = (v, w) \in \mathcal{R}^-$, respectively, then $((v, 1), (w, 1)) \in \widetilde{\mathcal{R}}$ and $((v, 1), (w, -1)) \in \widetilde{\mathcal{R}}$ is a Λ-inverse of r.
We construct the inverse of a given path $P = [s = v_0, v_1, \ldots, v_k] \in \mathcal{P}(s)$ under Λ. If $k = 0$ then $\Lambda([\widetilde{s}]) = P$. If $k > 0$ then let $r_\kappa := (v_{\kappa-1}, v_\kappa)$, $\kappa = 1, \ldots, k$. For each κ, we

define an arc $\tilde{r}_\kappa = ((v_{\kappa-1}, i_{\kappa-1}), (v_\kappa, i_\kappa))$ by recursively defining the components i_κ:

$$i_0 := 1, \quad (\forall\, \kappa = 1, \ldots, k) \quad i_\kappa := \begin{cases} +i_{\kappa-1}, & \text{if } r_\kappa \in \mathcal{R}^+, \\ -i_{\kappa-1}, & \text{if } r_\kappa \in \mathcal{R}^-. \end{cases}$$

Then $\tilde{r}_1, \ldots, \tilde{r}_k \in \tilde{\mathcal{R}}$; moreover, $\alpha(\tilde{r}_\kappa) = \omega(\tilde{r}_{\kappa-1})$ for all κ. Hence, the arcs $\tilde{r}_1, \ldots, \tilde{r}_k$ form a path $\tilde{P} \in \mathcal{P}((s,1))$, and it is $\Lambda(\tilde{P})) = P$.

P r o o f of b) : First, we see that $\tilde{H}$ is well-defined; that means that each function $\tilde{\phi}_{\tilde{r}}$ is only applied to arguments $x \in \mathrm{def}\left(\tilde{\phi}_{\tilde{r}}\right)$; we show that

$$\tilde{H}(\tilde{P}) \in \mathrm{def}\left(\tilde{\phi}_{\tilde{r}}\right) \text{ for all } \tilde{P} \in \mathcal{P}(\tilde{s}) \text{ and } \tilde{r} \in \tilde{\mathcal{R}} \text{ with } \omega(\tilde{P}) = \alpha(\tilde{r}). \qquad (2.23)$$

The proof of this fact is an induction on the length of $\tilde{P}$. If $\tilde{P} = [\tilde{s}]$ and $\alpha(\tilde{r}) = \tilde{s} = (s,1)$ then $\tilde{H}(\tilde{P}) = \tilde{H}([\tilde{s}]) = \tilde{p}(\tilde{s}) = H([s]) \overset{\text{(iii)}}{\in} X(s) = \mathrm{def}\left(\tilde{\phi}_{\tilde{r}}\right)$.

Otherwise, the path $\tilde{P}$ can be written as $\tilde{Q} \oplus \tilde{r}$ where $\tilde{r}$ is of the form $\tilde{r} = ((u,i),(v,j))$. The induction hypothesis implies that $\tilde{H}(\tilde{Q}) \in \mathrm{def}\left(\tilde{\phi}_{\tilde{r}}\right) = X(u)$. Then we obtain an assertion, which does not depend on i and j; relationship $(*)$ follows from $(v,j) = \alpha(\tilde{r})$:

$$\tilde{H}(\tilde{P}) = \tilde{\phi}_{\tilde{r}}(\tilde{H}(\tilde{Q})) \in \phi_{(u,v)}(X(u)) \subseteq X(v) \overset{(*)}{=} \mathrm{def}\left(\tilde{\phi}_{\tilde{r}}\right).$$

Consequently, (2.23) is true, and $\tilde{H}$ is well-defined.

Next, we show the second part of b), i.e., the relationship between $\tilde{H}$ and H. This is proven by an induction on the length of the path $\tilde{P}$. If $\tilde{P} = [\tilde{s}]$ then our assertion follows from $\omega(\tilde{P}) = (s,1) \in \mathcal{V} \times \{1\}$ and $\tilde{H}(\tilde{P}) = \tilde{p}(\tilde{s}) = H([s]) = H(P)$.

Let $\ell(\tilde{P}) > 0$. Suppose that the equality at the end of assertion b) is valid for all paths shorter than $\tilde{P}$. We write $\tilde{P} = \tilde{Q} \oplus \tilde{r}$ where $\tilde{Q}$ is the longest proper prefix of $\tilde{P}$ and $\tilde{r}$ is the last arc of $\tilde{P}$. Let $\tilde{r} = ((u,i),(v,j))$. We define $Q := \Lambda(\tilde{Q})$ and $P := \Lambda(\tilde{P})$ so that $P = Q \oplus (u,v)$. We now distinguish between two cases:

Case 1: $i = 1$. Then $\tilde{H}(\tilde{Q}) = H(Q)$ by the induction hypothesis. The definition of $\tilde{\phi}_{\tilde{r}}$ implies that

$$\tilde{H}(\tilde{P}) = \tilde{\phi}_{\tilde{r}}(\tilde{H}(\tilde{Q})) = \tilde{\phi}_{\tilde{r}}(H(Q)) = \begin{cases} \phi_{(u,v)}(H(Q)) = H(P) & (j = 1), \\ \sigma_v\left(\phi_{(u,v)}(H(Q))\right) = \sigma_v(H(P)) & (j = -1). \end{cases}$$

Case 2: $i = -1$. Then $\tilde{H}(\tilde{Q}) = \sigma_u(H(Q))$ by the induction hypothesis. The definition of $\tilde{\phi}_{\tilde{r}}$ implies that

$$\tilde{H}(\tilde{P}) = \tilde{\phi}_{\tilde{r}}(\tilde{H}(\tilde{Q})) = \tilde{\phi}_{\tilde{r}}(\sigma_u(H(Q))) = \begin{cases} \phi_{(u,v)}(\sigma_u^{-1}(\sigma_u(H(Q))))=H(P) & (j = 1), \\ \sigma_v\left(\phi_{(u,v)}(\sigma_u^{-1}(\sigma_u(H(Q))))\right)=\sigma_v(H(P)) & (j = -1). \end{cases}$$

Consequently, the following is true for *all* $i = -1, +1$: If $\omega(\tilde{P}) \in \mathcal{V} \times \{1\}$ then $j = 1$ and $\tilde{H}(\tilde{P}) = H(\Lambda(\tilde{P}))$; if, however, $\omega(\tilde{P}) \in \mathcal{V} \times \{-1\}$ then $j = -1$ and $\tilde{H}(\tilde{P}) = \sigma_v(H(P)) = \sigma_{\omega(\Lambda(\tilde{P}))}\left(H(\Lambda(\tilde{P}))\right)$.

P r o o f o f c) : The definition of the functions $\tilde{\phi}_{\tilde{r}}$ implies that each of them is monotonically increasing. Hence, $\tilde{H}$ is order preserving.   □

Example 2.40 We apply Theorem 2.39 to a multiplicative path function $H': P(s) \to \mathbb{R}$. We assume that an arc function $h : \mathcal{R} \to \mathbb{R}$ is given such that

$$H(P) = h(r_1) \cdot \ldots \cdot h(r_k) \text{ for all paths } P = r_1 \oplus \ldots \oplus r_k.$$

For each node $v \in \mathcal{G}$, let $X(v) := \mathbb{R}$; we define $\sigma_v(x) := -x$ for all $x \in \mathbb{R} = X(v)$.

Let now $\phi_r(\xi) = h(r) \cdot \xi$ for some arc $r = (u, v)$.

If $h(r) \geq 0$ then r is represented by two arcs $((u, 1), (v, 1))$ and $((u, -1), (v, -1))$, and each of them is marked with the function $\xi \mapsto h(r) \cdot \xi$.

If $h(r) < 0$ then r is represented by two arcs $((u, 1), (v, -1))$ and $((u, -1), (v, 1))$, and each of them is marked with the function $\xi \mapsto (-h(r)) \cdot \xi$.

That means that a multiplication with $h(r)$ is replaced by a multiplication with $|h(r)|$. Consequently, $\widetilde{H}(\widetilde{P}) = |H(\Lambda(\widetilde{P}))| = |H(P)|$ for all paths $\widetilde{P} \in \mathcal{P}(\widetilde{s})$ and $P \in \mathcal{P}(s)$ with $P = \Lambda(\widetilde{P})$. The sign of $H(P)$ is recorded in the second component of $\omega(\widetilde{P})$; if $\omega(\widetilde{P}) = (\omega(P), i)$ then $H(P) = i \cdot |H(P)| = i \cdot \widetilde{H}(\widetilde{P})$.

The digraphs constructed in 2.39 and in [Snie86], Example 6.1, are almost the same. The main difference is that Sniedovich's digraph has $\mathcal{V} \times \{-1, 0, 1\}$ as its set of nodes and not $\mathcal{V} \times \{-1, 1\}$. Then the nodes $(v, 0)$ in [Snie86] are reached after a multiplication with $h(r) = 0$; in our construction in 2.39, however, a multiplication with $h(r) = 0$ is treated like the case $h(r) > 0$. $\square$

Remark 2.41 (Finding minimal s-γ-paths)

Given the situation as described in Theorem 2.39. Let $\gamma \in \mathcal{V} \backslash \{s\}$. The surjectivity of $\Lambda : \mathcal{P}(s) \to \mathcal{P}(\widetilde{s})$ implies that $\mathcal{P}(s, \gamma) = \{\Lambda(\widetilde{P}) \,|\, \widetilde{P} \in \mathcal{P}(\widetilde{s}, (\gamma, 1)) \cup \mathcal{P}(\widetilde{s}, (\gamma, -1))\}$. Consequently,

$$\min\{H(P) \,|\, P \in \mathcal{P}(s, \gamma)\} =$$

$$\min\left(\min\{\widetilde{H}(\widetilde{P}) \,|\, \widetilde{P} \in \mathcal{P}(\widetilde{s}, (\gamma, 1))\}, \ \min\{\sigma_\gamma^{-1}(\widetilde{H}(\widetilde{P})) \,|\, \widetilde{P} \in \mathcal{P}(\widetilde{s}, (\gamma, -1))\}\right) =$$

$$\min\left(\min\{\widetilde{H}(\widetilde{P}) \,|\, \widetilde{P} \in \mathcal{P}(\widetilde{s}, (\gamma, 1))\}, \ \sigma_\gamma^{-1}\left(\max\{\widetilde{H}(\widetilde{P}) \,|\, \widetilde{P} \in \mathcal{P}(\widetilde{s}, (\gamma, -1))\}\right)\right).$$

So, the following search strategy will output an H-minimal s-γ-path P^* in $\mathcal{G}$:

Apply a search strategy for extremal paths to the digraph $\widetilde{\mathcal{G}}$ with the order preserving path function $\widetilde{H}$. Find an $\widetilde{H}$-minimal path $\widetilde{P}^+$ from $\widetilde{s}$ to $(v, 1)$ and an $\widetilde{H}$-maximal path $\widetilde{P}^-$ from $\widetilde{s}$ to $(v, -1)$. If $\widetilde{H}(\widetilde{P}^+) \leq \sigma_\gamma^{-1}(\widetilde{H}(\widetilde{P}^-))$ then output $P^* := \Lambda(\widetilde{P}^+)$, and if $\widetilde{H}(\widetilde{P}^+) > \sigma_\gamma^{-1}(\widetilde{H}(\widetilde{P}^-))$ then output $P^* := \Lambda(\widetilde{P}^-)$. This path P^* is H-minimal.

On the other hand, the digraph $\widetilde{\mathcal{G}}$ cannot always be used to find an $H_{\max}$-minimal path in $\mathcal{G}$. The reason is that there is no general relationship between $\widetilde{H}_{\max}(\widetilde{P})$ and $H_{\max}(\Lambda(\widetilde{P}))$; such a relationship does not exist because $H_{\max}(\Lambda(\widetilde{P}))$ is computed by comparing values of the function H whereas $\widetilde{H}_{\max}(\widetilde{P})$ is computed by comparing values of the functions H and $(\sigma_v \circ H)$.

For example, let $\mathcal{G} = (\mathcal{V}, \mathcal{R})$ with $\mathcal{V} := \{s, x_0, x_1, x_2, x_3, x_4, \gamma\}$ and $\mathcal{R} := \{(s, x_i), (x_i, \gamma) \,|\, i = 0, 1, 2, 3, 4\}$. Let H be the multiplicative path function with the following arc function $h : \mathcal{R} \to \mathbb{R}$:

$$h(s,x_0) := -5, \quad h(s,x_1) := 4, \quad h(s,x_2) := 4, \quad h(s,x_3) := -8, \quad h(s,x_4) := 8,$$
$$h(x_0,\gamma) := -0.4, \quad h(x_1,\gamma) := 0.5, \quad h(x_2,\gamma) := -0.5, \quad h(x_3,\gamma) := -0.5, \quad h(x_4,\gamma) := -0.5.$$

Let $X_i := [s,x_i]$ and $Y_i := [s,x_i,\gamma]$ for all $i = 0,\ldots,4$; we assume that X_i and Y_i, respectively, are represented by the paths $\widetilde{X}_i$ and $\widetilde{Y}_i$ in $\widetilde{\mathcal{G}}$; then $\widetilde{H}(\widetilde{X}_i) = |H(X_i)|$, and $\widetilde{H}(\widetilde{Y}_i) = |H(Y_i)|$ for all $i = 0,\ldots,4$. Consequently, the following are true:

$H([s]) = 1,$	$H([s]) = 1,$	$H([s]) = 1,$	$H([s]) = 1,$	$H([s]) = 1,$
$H(X_0) = -5,$	$H(X_1) = 4,$	$H(X_2) = 4,$	$H(X_3) = -8,$	$H(X_4) = 8,$
$H(Y_0) = 2,$	$H(Y_1) = 2,$	$H(Y_2) = -2,$	$H(Y_3) = 4,$	$H(Y_4) = -4,$
$H_{\max}(Y_0) = 2,$	$H_{\max}(Y_1) = 4,$	$H_{\max}(Y_2) = 4,$	$H_{\max}(Y_3) = 4,$	$H_{\max}(Y_4) = 8,$
$\widetilde{H}([\widetilde{s}]) = 1,$	$\widetilde{H}([\widetilde{s}]) = 1,$	$\widetilde{H}([\widetilde{s}]) = 1,$	$\widetilde{H}([\widetilde{s}]) = 1,$	$\widetilde{H}([\widetilde{s}]) = 1,$
$\widetilde{H}(\widetilde{X}_0) = 5,$	$\widetilde{H}(\widetilde{X}_1) = 4,$	$\widetilde{H}(\widetilde{X}_2) = 4,$	$\widetilde{H}(\widetilde{X}_3) = 8,$	$\widetilde{H}(\widetilde{X}_4) = 8,$
$\widetilde{H}(\widetilde{Y}_0) = 2,$	$\widetilde{H}(\widetilde{Y}_1) = 2,$	$\widetilde{H}(\widetilde{Y}_2) = 2,$	$\widetilde{H}(\widetilde{Y}_3) = 4,$	$\widetilde{H}(\widetilde{Y}_4) = 4,$
$\widetilde{H}_{\max}(\widetilde{Y}_0) = 5,$	$\widetilde{H}_{\max}(\widetilde{Y}_1) = 4,$	$\widetilde{H}_{\max}(\widetilde{Y}_2) = 4,$	$\widetilde{H}_{\max}(\widetilde{Y}_3) = 8,$	$\widetilde{H}_{\max}(\widetilde{Y}_4) = 8.$

Next, we apply the following algorithm $\mathcal{A}$ to $\widetilde{\mathcal{G}}$:

1) Search for an $\widetilde{H}_{\max}$-minimal $\widetilde{s}$-$(\gamma,1)$-path $\widetilde{U}_1$; let $U_1 := \Lambda(\widetilde{U}_1)$.
2) Search for an $\widetilde{H}_{\max}$-minimal $\widetilde{s}$-$(\gamma,-1)$-path $\widetilde{U}_2$; let $U_2 := \Lambda(\widetilde{U}_2)$.
3) Search for an $\widetilde{H}_{\max}$-maximal $\widetilde{s}$-$(\gamma,1)$-path $\widetilde{U}_3$; let $U_3 := \Lambda(\widetilde{U}_3)$.
4) Search for an $\widetilde{H}_{\max}$-maximal $\widetilde{s}$-$(\gamma,-1)$-path $\widetilde{U}_4$; let $U_4 := \Lambda(\widetilde{U}_4)$.
5) Select $P^{**} \in \{U_1, U_2, U_3, U_4\}$ such that $H_{\max}(P^{**}) = \min\{H_{\max}(U_i) \mid i = 1,2,3,4\}$.

This algorithm finds the paths $\widetilde{U}_1 = \widetilde{Y}_i$ and consequently, $U_i = Y_i$, $i = 1,2,3,4$. That means that $\mathcal{A}$ will output a path $P^{**} \in \{P_1, P_2, P_3, P_4\}$; so, $\mathcal{A}$ does not find the path P_0, which is actually $H_{\max}$-minimal. That means that even algorithm $\mathcal{A}$, which searches very carefully, does not always find $H_{\max}$-minimal paths. □

Next, we describe the combined application of Theorems 2.37 and 2.39.

Remark 2.42 Given the situation in Theorem 2.37. We assume that for each $\mu = 1,\ldots,m$, a function δ_μ exists with the following property:

$$\delta_\mu : I_\mu \to I_\mu \text{ is strictly decreasing and bijective.} \tag{2.24}$$

Then we can first transform the graph $\mathcal{G}$ in 2.37 into a graph $\mathcal{G}'$ whose functions $\phi'_{r'}$ are monotone. In particular, Theorem 2.37 yields a graph homomorphis Ψ from $\mathcal{G}'$ to $\mathcal{G}$ and the path function $H' := H \circ \Psi$.

The next step is to apply Theorem 2.39 to the digraph $\mathcal{G}'$; thus, we construct a digraph $\widetilde{\mathcal{G}'}$ whose arcs are marked with monotonically increasing functions $\phi'_{\widetilde{r'}}$. Theorem 2.39 yields a graph homomorphism Λ from $\widetilde{\mathcal{G}'}$ to $\mathcal{G}'$ and an order preserving path function $\widetilde{H'}$.

We give more details of this construction: For each node $(\mu,v) \in \mathcal{V}'$, we define the set $X(\mu,v) := I_\mu$ and the function $\sigma_{(\mu,v)} := \delta_\mu$. We first verify condition (i) in 2.39. For this we consider an arbitrary arc $r' := ((\mu,v),(\widehat{\mu},w)) \in \mathcal{R}'$; then $\phi'_{r'} : I_\mu \to I_{\widehat{\mu}}$ is also a function from $X(\mu,v)) = X(\alpha(r'))$ to $X(\widehat{\mu},w)) = X(\omega(r'))$; the monotonicity of $\phi'_{r'}$ is one of the results of 2.37. Condition (ii) of 2.39 follows from (2.24) and the definitions of $X(\mu, v)$ and of $\sigma_{(\mu,v)}$. We still must prove that condition (iii) of 2.39 is also satisfied. For this purpose, recall the definition of the node s' and the number μ_0 in the proof of 2.37; an immediate consequence of that definition is that $H([s]) \in I_{\mu_0}$; consequently, $H'([s']) = H'([(\mu_0, s)]) = H([s]) \in I_{\mu_0} = X(s')$. We just have seen the following: If (2.24) is satisfied then the digraph $\mathcal{G}'$ constructed in 2.37 satisfies the conditions (i),(ii),(iii) of 2.39 so that this theorem can be applied to $\mathcal{G}'$. The nodes of the resulting digraph are of the form $((\mu, v), i)$ with $\mu \in \{1, \ldots, m\}$, $v \in V$ and $i \in \{\pm 1\}$; it is more convenient to write each node $((\mu, v), i)$ as a triple (μ, v, i).

Moreover, the order preserving path function $\widetilde{H}'$ in $\widetilde{\mathcal{G}}'$ has the following property: Let $\widetilde{P}' \in \mathcal{P}(\widetilde{s}')$, and let $\omega(\widetilde{P}') = (\mu, v, i)$; then

$$
\begin{aligned}
\widetilde{H}'(\widetilde{P}') &\overset{2.39b)}{=} H'(\Lambda(\widetilde{P}')) \overset{2.37b)}{=} H(\Psi(\Lambda(\widetilde{P}'))) && \text{if} \quad i = 1, \\
\widetilde{H}'(\widetilde{P}') &\overset{2.39b)}{=} \sigma_{(\mu,v)}\big(H'(\Lambda(\widetilde{P}'))\big) \overset{2.37b)}{=} \delta_\mu\big(H(\Psi(\Lambda(\widetilde{P}')))\big) && \text{if} \quad i = -1
\end{aligned}
\tag{2.25}
$$

The search for an H-minimal s-γ-path P^* in the original graph $\mathcal{G}$ consists of the following steps:

1) For each $\mu \in \{1, \ldots, m\}$, search for an $\widetilde{H}'$-minimal $\widetilde{\mathcal{G}}'$-path P_μ^+ from $\widetilde{s}' = (\mu_0, s, 1)$ to $(\mu, \gamma, 1)$ and an $\widetilde{H}'$-maximal path P_μ^- from $\widetilde{s}'$ to $(\mu, \gamma, -1)$.

2) As seen in 2.41, the path $P'_\mu = \Lambda(P_\mu^+)$ or the path $P'_\mu := \Lambda(P_\mu^-)$ is an H'-minimal path in $\mathcal{G}'$ among all candidates that start with $s' = (\mu_0, s)$ and end with (μ, γ).

3) Remark 2.38 says that the desired $\mathcal{G}$-path P^* can be found by minimizing the H-value of all candidates $\Psi(P'_1), \ldots, \Psi(P'_m)$. $\square$

Example 2.43 Given the same situation as in Example 2.40; in addition let $h(r) \neq 0$ for all r. Then $H = \phi \circ o \circ p$ where $p(v) = 1$ for all v and $\phi_r(\zeta) = h(r) \cdot \zeta$ for all $r \in \mathcal{R}$, $\zeta \in \mathbb{R}$.

Let $I_1 := (-\infty, 0)$ and $I_2 := (0, \infty)$. Let $\delta_\mu(x) := 1/x$ for all $\mu = 1, 2$ and all $x \in I_\mu$; then (2.24) is true.

Using Theorem 2.37, we obtain a digraph $\mathcal{G}'$ whose nodes are of the form (μ, v) with $\nu \in \{1, 2\}$ and $v \in V$. In particular, $s' = (2, s)$ because $H([s]) = 1 \in I_2$. Moreover, we obtain a graph homomorphism Ψ from $\mathcal{G}'$ to $\mathcal{G}$ and the path function $H' = H \circ \Psi$. The following is true for all paths $P \in \mathcal{P}(s)$: Let $w := \omega(P)$, and let P be represented by P' in $\mathcal{G}'$; then $\omega(P') = (1, w)$ if $H(P) < 0$ (i.e., $H(P) \in I_1$), and $\omega(P') = (2, w)$ if $H(P) > 0$ (i.e., $H(P) \in I_2$). Moreover, $H'(P') = H(P)$.

The digraph $\widetilde{\mathcal{G}}'$ constructed in Remark 2.42 consists of nodes (μ, v, i) where $\mu \in \{1, 2\}$, $v \in V$, and $i \in \{-1, +1\}$. The start node is $\widetilde{s}' := (2, s, 1)$. Theorem 2.39 yields a homomorphism Λ from $\widetilde{\mathcal{G}}'$ to $\mathcal{G}'$. Let $\widetilde{H}'$ be the path function described in Remark 2.42. Let $P \in \mathcal{P}(s)$ and $w := \omega(P)$. Then P is represented by a path $\widetilde{P}'$ in $\widetilde{\mathcal{G}}'$, i.e., $\Psi(\Lambda(\widetilde{P}')) = P$. Equation (2.25) implies that the following is true:

$$
\begin{aligned}
&\text{If } H(P) > 0 \text{ then } \widetilde{P}' \text{ ends with } (2, w, 1), \text{ and } \widetilde{H}'(\widetilde{P}') = H(P). \\
&\text{If } H(P) < 0 \text{ then } \widetilde{P}' \text{ ends with } (1, w, -1), \text{ and } \widetilde{H}'(\widetilde{P}') = \tfrac{1}{H(P)}.
\end{aligned}
\tag{2.26}
$$

The graph $\widetilde{\mathcal{G}}'$ is more complicated than the graph $\widetilde{\mathcal{G}}$ in Example 2.40.[5] The advantage of $\widetilde{\mathcal{G}}'$, however, is that it can be used to find H_{max}-minimal paths[6]; this will be shown in Remark 4.48. □

The next remark is about group theoretic path functions.

Remark 2.44 A similar construction as in 2.37 and in 2.39 can also be applied in the following situation:

Given a finite digraph $\mathcal{G} = (\mathcal{V}, \mathcal{R})$. Let $\mathbf{G}$ be a group, and let $g_1 = 1$ be the neutral element of $\mathbf{G}$. Let $h : \mathcal{R} \to \mathbf{G}$ be an arc function. We define $H : \mathcal{P}(s) \to \mathbf{G}$ as follows: $H[s] := g_1$ and $H(P) := h(r_1) \cdot \ldots \cdot h(r_k)$ for all paths $P = r_1 \oplus \ldots \oplus r_k$.

Let now $\mathbf{U} \leq \mathbf{G}$ be a subgroup of $\mathbf{G}$. We generate a digraph $\mathcal{G}'$ and a path function H' with the following properties:

- All values $H'(P')$ are elements of $\mathbf{U}$.
- The values $H(P)$ of the original paths $P \in \mathcal{P}$ can be reconstructed from P' and from $H(P')$.

For this purpose, we partition $\mathbf{G}$ into the right cosets $\mathbf{U} \cdot g_1, \ldots, \mathbf{U} \cdot g_m$ of $\mathbf{U}$. Then we define $\mathcal{G}' := (\mathcal{V}', \mathcal{R}')$ with $\mathcal{V}' := \{g_1, \ldots, g_m\} \times \mathcal{V}$ and

$$\mathcal{R}' := \left\{ \left((g_\mu, v), (g_{\widetilde{\mu}}, \widetilde{v}) \right) \mid (v, \widetilde{v}) = r \in \mathcal{R} \wedge g_\mu \cdot h(r) \in \mathbf{U} \cdot g_{\widetilde{\mu}} \right\}.$$

Moreover, we define $s' := (g_1, s)$.

For all nodes $v' = (g_\mu, v) \in \mathcal{V}'$ and for all arcs $r' = ((g_\mu, v), (g_{\widetilde{\mu}}, \widetilde{v})) \in \mathcal{R}'$, we define $\Psi(v') := v$ and $\Psi(r') := (v, \widetilde{v})$; moreover, we define $h'(r') := g_\mu \cdot h(\Psi(r')) \cdot g_{\widetilde{\mu}}^{-1}$.

Then $h'(r') \in \mathbf{U}$ by the definition of $\mathcal{R}'$. Hence, $H'(P') \in \mathbf{U}$ for all paths P' where $H' : \mathcal{P}(s') \to \mathbf{G}$ is defined as follows: $H([s']) := g_1$, and $H'(P') := h(r'_1) \cdot \ldots \cdot h(r'_k)$ for all paths $P' \in \mathcal{P}(s')$ with the arcs $r'_1, \ldots, r'_k$.

Next, we show how to reconstruct a path $P \in \mathcal{P}(s)$ and its value $H(P)$ from P' and $H'(P')$. For this purpose, we first prove the following assertions:

$$P \text{ has an inverse } P' \in \mathcal{P}(s') \text{ under } \Psi. \tag{2.27}$$

$$\text{If } \omega(P') = (g_\mu, v) \text{ for this inverse } P' \text{ then } H(P) = H'(P') \cdot g_\mu. \tag{2.28}$$

We prove (2.27) and (2.28) simultaneously by an induction on the length of P. If $P = [s]$ let $P' := [s'] = [(g_1, s)]$; then $\Psi(P') = P$ and $H(P') = g_1 = H'([s']) \cdot g_1$.

Suppose that (2.27) and (2.28) are true for all paths of length k. Let then P be path of length $k + 1$. Then P can be written as $P = Q \oplus r$ where Q is the prefix of P of length k and $r = (\overline{v}, v)$ is the last arc of P. Then the induction hypothesis yields a $\mathcal{G}'$-path $Q' \in \mathcal{P}(s')$ with $\Psi(Q') = Q$. Moreover, let $\omega(Q') = (g_{\overline{\mu}}, \overline{v})$. Then a further consequence of the induction hypothesis is that $H(Q) = H'(Q') \cdot g_{\overline{\mu}}$.

---

[5] We have seen that each path $\widetilde{P}' \in \mathcal{P}(\widetilde{s}')$ ends with a node $(2, w, 1)$ or $(1, w, -1)$; $\widetilde{P}'$ never ends with a node $(2, w, -1)$ or $(1, w, 1)$. Therefore, we can simplify $\widetilde{\mathcal{G}}'$ as follows: We replace each node $(2, w, 1)$ by a pair $(w, +)$ and each node $(1, w, -1)$ by a pair $(w, -)$.

[6] This is probably not possible in $\widetilde{\mathcal{G}}$ as shown in 2.41.

Next, we choose μ such that $g_{\overline{\mu}} \cdot h(r) \in \mathbf{U} \cdot g_{\mu}$. Then the definition of $\mathcal{R}'$ implies that
$r' := ((g_{\overline{\mu}}, \overline{v}), (g_{\mu}, v)) \in \mathcal{R}'$. Let $P' := Q' \oplus r'$. Then $\Psi(P') = \Psi(Q') \oplus \Psi(r') \overset{(*)}{=}$ (*)
$Q \oplus r = P$, where (*) follows from $\Psi(Q') = Q$. Consequently, (2.27) is true for P.

Moreover, the following assertion is true, in which (*) follows from the above equation
$H(Q) = H'(Q') \cdot g_{\overline{\mu}}$.

$$H'(P') \cdot g_{\mu} = H'(Q' \oplus r') \cdot g_{\mu} = H'(Q') \cdot h'(r') \cdot g_{\mu} =$$

$$H'(Q') \cdot [g_{\overline{\mu}} \cdot h(r) \cdot g_{\mu}^{-1}] \cdot g_{\mu} = [H'(Q') \cdot g_{\overline{\mu}}] \cdot h(r) \overset{(*)}{=} H(Q) \cdot h(r) = H(P).$$

This proves assertion (2.28).
                                                                                                    $\square$

Definition/Remark 2.45 Another method to generate order preservation is presented in Section 6 of [LT91] and of [LTh91]. The idea is to modify the order relation
$\preceq$. We now discuss Lengauer's and Theune's construction in detail.
Given an arbitrary path function $H : \mathcal{P}(s) \to (\mathbf{R}, \preceq)$. We assume that H is easily
computable of type B where $\odot : \mathbf{R} \times \mathbf{R} \to \mathbf{R}$ is the given binary operation. A relation
$\preceq_m \subseteq \preceq$ is called a *monotone reduction of* $\preceq$ if the following is true:

$$(\forall \varrho_1, \varrho_2, \varrho \in \mathbf{R}) \quad \varrho_1 \preceq_m \varrho_2 \Rightarrow \varrho_1 \odot \varrho \preceq_m \varrho_2 \odot \varrho. \tag{2.29}$$

If $\preceq_m$ is a monotone reduction of $\preceq$ then the following is true:

$$\begin{aligned} H(P_1) \preceq_m H(P_2) &\implies H(P_1 \oplus Q) \preceq_m H(P_2 \oplus Q) \\ &\text{for all paths } P_1, P_2 \in \mathcal{P}(s) \text{ and } Q \in \mathcal{P}(\mathcal{G}) \text{ with } \omega(P_1) = \omega(P_2) = \alpha(Q). \end{aligned} \tag{2.30}$$

We call this property "$\preceq_m$-preservation".

Fact (2.30) is proven as follows: Mark each arc r with the following function function:
$\phi_r : \mathbf{R} \to \mathbf{R}, \; \varrho \mapsto \varrho \odot h(r)$. Then every ϕ_r is monotonically increasing with respect to
$\preceq_m$. Let $p(v) := \varrho_0$ for all nodes v. Then $H = \phi \circ \circ p$ is order preserving with respect
to $\preceq_m$.

An example of a monotone reduction is given in [LT91], Definition 9: For all pairs
$(\varrho_1, \varrho_2) \in \mathbf{R} \times \mathbf{R}$, let $\varrho_1 \preceq_{\max} \varrho_2 \; :\Longleftrightarrow \; (\varrho_1 \preceq \varrho_2 \text{ and } (\forall \rho \in \mathbf{R}) \; (\varrho_1 \odot \varrho \preceq \varrho_2 \odot \varrho))$.

Lemma 3 in [LT91] says that $\preceq_{\max}$ is a monotone reduction of $\preceq$; moreover, $\preceq_{\max}$ is
maximal, that is, $\preceq_m \subseteq \preceq_{\max}$ for all monotone reductions $\preceq_m$.

Next, we discuss the advantages and disadvantages of Lengauer's and Theune's construction.
Its first advantage is its generality; the relation $\preceq_{\max}$ can be defined for a large class
of path functions H (namely the type-B-computable ones) and for all relations $\preceq$.
The second advantage of Lengauer's and Theune's construction is its simplicity; the
two authors do not define a new digraph or a new path function, and the definition of
$\preceq_{\max}$ is easy to understand.

The disadvantages of this construction are the following:
newline The first is that the question "$\varrho_1 \preceq_{\max} \varrho_2$?" is difficult to answer because the
relationship $\varrho_1 \odot \varrho \preceq \varrho_2 \odot \varrho$ must be verified for all ϱ.
The second disadvantage is that $\preceq_{\max}$ may collapse to the relation "$=$". As an example, we consider a strongly connected digraph $\mathcal{G}$. Let $H : \mathcal{P}(s) \to \mathbb{R}$ be a multiplicative

path function basing on the arc function $h : \mathcal{R} \to \mathbb{R}$. Then we may choose $\odot$ as the multiplication of reals numbers and $\preceq$ as the relation $\leq$. Then the following is true:

$$(\forall \varrho_1, \varrho_2 \in \mathbf{R} = \mathbb{R}) \quad \varrho_1 \preceq_{\max} \varrho_2 \quad \Longleftrightarrow \quad \varrho_1 = \varrho_2 .$$

To prove $" \Rightarrow "$ we assume that $\varrho_1 \preceq_{\max} \varrho_2$. Then $\varrho_1 \leq \varrho_2$ and $\varrho_1 \cdot \varrho \leq \varrho_2 \cdot \varrho$ for all $\varrho \in \mathbb{R}$. For $\varrho := -1$, we obtain that $-\varrho_1 \leq -\varrho_2$. This and $\varrho_1 \leq \varrho_2$ imply that $\varrho_1 = \varrho_2$. The reverse direction $" \Leftarrow "$ is trivial.

If $"\preceq_{\max}"$ collapses to $" = "$ then the $" \preceq_{\max}$-preservation is equivalent to the $=$-preservation of H. But this is a trivial property, which immediately follows from the computability of H with the help of $\odot$. □

2.3.4 Modified versions of order preservation

Here, we describe several variants of order preservation. Moreover, we introduce a global concept that makes it easy to formulate modified order preserving properties or Bellman principles, which are studied in Section 2.4.

2.3.4.1 A general method to formulate principles of order preservation

Let us outline the underlying idea of the next definition. The usual order preservation says:

$$\left(\begin{array}{ccc} \text{I f} & H(P_1) & \preceq & H(P_2) \\ \text{t h e n} & H(P_1 \oplus Q) & \preceq & H(P_2 \oplus Q) \end{array} \right)$$

w h e r e the paths P_1, P_2, and Q satisfy the following condition:

$$\alpha(P_1) = \alpha(P_2) \text{ and } \omega(P_1) = \omega(P_2) = \alpha(Q).$$

The general formulation of order preservation says the following:

$$\left(\begin{array}{cccc} \text{I f} & H(X_{1,1} \oplus X_{1,2} \oplus X_{1,3}) & \preceq & H(X_{2,1} \oplus X_{2,2} \oplus X_{2,3}) \\ \text{t h e n} & H(X_{3,1} \oplus X_{3,2} \oplus X_{3,3}) & \preceq & H(X_{4,1} \oplus X_{4,2} \oplus X_{4,3}) \end{array} \right)$$

w h e r e the paths $X_{i,j}$ satisfy the following condition:

$$\text{The matrix } \mathbf{X} := \begin{pmatrix} X_{1,1} & X_{1,2} & X_{1,3} \\ X_{2,1} & X_{2,2} & X_{2,3} \\ X_{3,1} & X_{3,2} & X_{3,3} \\ X_{4,1} & X_{4,2} & X_{4,3} \end{pmatrix} \text{ is an element of a given set } L.$$

The representation of the paths $X_{i,j}$ as a matrix makes it easy to memorize the structure of the generalized order preservation: First, concatenate the paths in each line in order to get the paths $X_i := X_{i,1} \oplus X_{i,2} \oplus X_{i,3}$ for $i = 1, 2, 3, 4$. Then formulate the following condition about the paths $X_1, \ldots, X_4$: If $H(X_1) \preceq H(X_2)$ then $H(X_3) \preceq H(X_4)$.

Next, we describe the general concept of order preservation in detail; this concept also includes $\prec$- and $=$-preservation, which were defined in 2.24.

Definition 2.46 Given a digraph $\mathcal{G} = (\mathcal{V}, \mathcal{R})$, a set $\mathcal{P} \subseteq \mathcal{P}(\mathcal{G})$, and a path function $H : \mathcal{P} \to \mathbf{R}$.

a) First, we define $\mathcal{L}$ as the set of all path matrices allowing a concatenation of the paths in each line.

$$\mathcal{L} := \left\{ \mathbf{X} = \begin{pmatrix} X_{1,1} & X_{1,2} & X_{1,3} \\ X_{2,1} & X_{2,2} & X_{2,3} \\ X_{3,1} & X_{3,2} & X_{3,3} \\ X_{4,1} & X_{4,2} & X_{4,3} \end{pmatrix} \middle| \begin{array}{l} X_{i,j} \in \mathcal{P}(\mathcal{G}), \, \omega(X_{i,1}) = \alpha(X_{i,2}), \\ \text{and } \omega(X_{i,2}) = \alpha(X_{i,3}) \text{ for all } i, j. \end{array} \right\}.$$

If the elements of a matrix in $\mathcal{L}$ are written with two subscripts i, j then the single index i automatically means the concatenation of the three paths in the ith line. For example, if the matrices $\mathbf{X} = (X_{i,j})$, $\mathbf{Y}' = (Y'_{i,j})$, and $\tilde{\mathbf{P}} = (\tilde{P}_{i,j})$ are given then $X_i := X_{i,1} \oplus X_{i,2} \oplus X_{i,3}$, $Y'_i := Y'_{i,1} \oplus Y'_{i,2} \oplus Y'_{i,3}$, and $\tilde{P}_i := \tilde{P}_{i,1} \oplus \tilde{P}_{i,2} \oplus \tilde{P}_{i,3}$.

b) Let $L \subseteq \mathcal{L}$. Then H is called L-*order preserving* or *order preserving with respect to* L if the following is true for all matrices $\mathbf{X} = (X_{i,j}) \in \mathcal{L}$:

$$\begin{aligned} &\left(X_1 \in \mathcal{P} \wedge X_2 \in \mathcal{P} \wedge H(X_1) \preceq H(X_2) \right) \implies \\ &\left(X_3 \in \mathcal{P} \wedge X_4 \in \mathcal{P} \wedge H(X_3) \preceq H(X_4) \right). \end{aligned} \quad (2.31)$$

For example, standard order preservation is equivalent to the $\mathbb{L}_0$-order preservation with the following set of matrices

$$\mathbb{L}_0 := \left\{ \mathbf{X} := \begin{pmatrix} [v] & P_1 & [v'] \\ [v] & P_2 & [v'] \\ [v] & P_1 & Q \\ [v] & P_2 & Q \end{pmatrix} \in \mathcal{L} \middle| v, v' \in \mathcal{V} \right\}.$$

(Recall the conditions in (2.31) about $\mathcal{P}$. If $\mathcal{P} = \operatorname{def}(H) = \mathcal{P}(s)$ then $(X_1 = P_1 \in \mathcal{P} \wedge X_2 = P_2 \in \mathcal{P})$ implies that $v = s$; this implies that $(X_3 = P_1 \oplus Q \in \mathcal{P} \wedge X_4 = P_2 \oplus Q \in \mathcal{P})$.)

c) Let $\Re_1, \Re_2 \in \{\preceq, \prec, \succeq, \succ, \equiv, =\}$, and let $L \subseteq \mathcal{L}$. We say that H is $\Re_1$-$\Re_2$-*preserving* (*with respect to* L), if the following is true for all matrices $\mathbf{X} = (X_{i,j}) \in L$:

$$\begin{aligned} &\left(X_1 \in \mathcal{P} \wedge X_2 \in \mathcal{P} \wedge H(X_1) \, \Re_1 \, H(X_2) \right) \implies \\ &\left(X_3 \in \mathcal{P} \wedge X_4 \in \mathcal{P} \wedge H(X_3) \, \Re_2 \, H(X_4) \right). \end{aligned} \quad (2.32)$$

Let $\Re_1 = \Re_2 = \widehat{\Re}$; then H is $\widehat{\Re}$-*preserving* (*with respect to* L) if H is $\Re_1$-$\Re_2$-preserving (with respect to L); for example, $\preceq$-preservation, $\prec$-preservation, $\equiv$-preservation, and $=$-preservation mean that $\Re_1 = \Re_2 = \widehat{\Re}$ is equal to to "$\preceq$", "$\prec$", "$\equiv$", and "$=$", respectively. □

The next example describes several natural conditions about the paths appearing in modified principles of order preservation; such principles have been formulated in Theorem 2.19 where P_1 and P_2 need not have the same start node and in Theorem 2.27 where P_1 and P_2 are extended by a common prefix. We now formulate such properties of a path function with the help of Definition 2.46.

Example 2.47 Let $H : \mathcal{P} \to (\mathbf{R}, \preceq)$ where $\mathcal{P} \subseteq \mathcal{P}(\mathcal{G})$. A natural generalization of usual order preservation is the following property of H:

$$\left(\begin{array}{ccc} \text{If} & H(P_1) & \preceq & H(P_2) \\ \text{then} & H(Q_1' \oplus P_1 \oplus Q_1'') & \preceq & H(Q_2' \oplus P_2 \oplus Q_2'') \end{array} \right)$$

where Q_i', P_i, Q_i'' $(i = 1, 2)$ possibly satisfy several additional conditions.

For example, in the case of the usual order preservation, the six paths must satisfy the three conditions $Q_1' = Q_2'$, $Q_1' = [\alpha(P_1)]$, and $Q_1'' = Q_2''$.

If no additional conditions are given then the above order preservation is equivalent to $\mathbb{L}^*$-order preservation where $\mathbb{L}^*$ is defined as follows:

$$\mathbb{L}^* := \left\{ \mathbf{X} = \begin{pmatrix} [\alpha(P_1)] & P_1 & [\omega(P_1)] \\ [\alpha(P_2)] & P_2 & [\omega(P_2)] \\ Q_1' & P_1 & Q_1'' \\ Q_2' & P_2 & Q_2'' \end{pmatrix} \in \mathcal{L} \right\}.$$

Next, we define several subsets of $\mathbb{L}^*$ by formulating natural conditions about the paths Q_i', P_i, Q_i'', $i = 1, 2$.

The prefixes Q_1', Q_2' are the same: $\mathbb{L}_1' := \{\mathbf{X} \in \mathbb{L}^* \mid Q_1' = Q_2' =: Q'\}$.

The suffixes Q_1'', Q_2'' are the same: $\mathbb{L}_1'' := \{\mathbf{X} \in \mathbb{L}^* \mid Q_1'' = Q_2'' =: Q''\}$.

The prefixes Q_1', Q_2' are of length 0 : $\mathbb{L}_2' := \{\mathbf{X} \in \mathbb{L}^* \mid Q_1' = [\alpha(P_1)], \ Q_2' = [\alpha(P_2)]\}$.

The suffixes Q_1'', Q_2'' are of length 0 : $\mathbb{L}_2'' := \{\mathbf{X} \in \mathbb{L}^* \mid Q_1'' = [\omega(P_1)], \ Q_2'' = [\omega(P_2)]\}$.

The prefixes Q_1', Q_2' are of length 1 : $\mathbb{L}_3' := \{\mathbf{X} \in \mathbb{L}^* \mid \ell(Q_1') = \ell(Q_2') = 1\}$.

The suffixes Q_1'', Q_2'' are of length 1 : $\mathbb{L}_3'' := \{\mathbf{X} \in \mathbb{L}^* \mid \ell(Q_1'') = \ell(Q_2'') = 1\}$.

For example, the set $\mathbb{L}_0$, which induces standard order preservation, is the intersection of $\mathbb{L}_1'$, $\mathbb{L}_2'$, and $\mathbb{L}_1''$. If only two of these three sets are intersected we obtain the following principles of L-order preservation:

- *The case* $L = \mathbb{L}_2' \cap \mathbb{L}_1''$:
    Then $H(P_1) \preceq H(P_2) \Rightarrow H(P_1 \oplus Q'') \preceq H(P_2 \oplus Q'')$
    for all paths P_1, P_2, Q'.
    The paths P_1 and P_2 need not have the same start node.
- *The case* $L = \mathbb{L}_1' \cap \mathbb{L}_1''$:
    Then $H(P_1) \preceq H(P_2) \Rightarrow H(Q' \oplus P_1 \oplus Q'') \preceq H(Q' \oplus P_2 \oplus Q'')$
    for all paths Q', P_1, P_2, Q''.
    The path Q' may visit more than one node.
- *The case* $L = \mathbb{L}_1' \cap \mathbb{L}_2'$:
    Then $H(P_1) \preceq H(P_2) \Rightarrow H(P_1 \oplus Q_1'') \preceq H(P_2' \oplus Q_2'')$
    for all paths P_1, P_2 and Q_1'', Q_2'' with $\alpha(P_1) = \alpha(P_2)$.
    The end points of P_1, P_2 and the paths Q_1'' and Q_2'' may be different.

E.g., assertion (a) in 2.19 is equivalent to $=$-preservation with respect to $\mathbb{L}_2' \cap \mathbb{L}_1''$. □

The next remark describes several relationships between set theoretic operations for sets of matrices and the corresponding principles of order preservation.

Remark 2.48   a) Let $L_1 \subseteq L_2$; then $\Re_1$-$\Re_2$-preservation with respect to L_1 implies the same property with respect to L_2. This follows from Definition 2.46 b).
For example, let $H : \mathcal{P}(\mathcal{G}) \to \mathbf{R}$; then the equality $\mathbb{L}_0 = \mathbb{L}_1' \cap \mathbb{L}_2' \cap \mathbb{L}_1''$ in 2.47 has the following consequence: If H is order preserving with respect to $(\mathbb{L}_2' \cap \mathbb{L}_1'')$ or with respect to $(\mathbb{L}_1' \cap \mathbb{L}_1'')$ or with respect to $(\mathbb{L}_1' \cap \mathbb{L}_2')$ then H is order preserving in the original sense.

b) Let $L_1, L_2 \subseteq \mathcal{L}$. Then the following are true:
  b1) H is $(L_1 \cup L_2)$-order preserving if and only if H is both L_1- and L_2-order preserving.
  b2) If H is L_1- or L_2-order preserving then H is $(L_1 \cap L_2)$-order preserving. The reverse conclusion is not always correct.   ☐

In the following remark, we compare $\preceq$-preservation with $\prec$-preservation.

Remark 2.49 There is no logical relationship between the following properties of H:
  • the usual order preservation, i.e., the $\preceq$-preservation with respect to $\mathbb{L}_0$,
  • the $\prec$-preservation with respect to $\mathbb{L}_0$.
To see this we consider the formula: $\big(H(P_1) \; \Re' \; H(P_2) \; \wedge \; H(P_1 \oplus Q) \; \Re'' \; H(P_2 \oplus Q)\big)$.
If H is $\preceq$-preserving it may occur that $\Re'$ equals "$\prec$" and $\Re''$ equals "$\equiv$"; in this case, H is not $\prec$-preserving.
If H is $\prec$-preserving it may occur that $\Re'$ equals "$\equiv$" and $\Re''$ equals "$\succ$"; in this case, H is not $\preceq$-preserving.   ☐

The following remark says that several previous results about order preserving functions remain true for modified order preservation.

Remark 2.50   a) Recall the set $L := \mathbb{L}_2' \cap \mathbb{L}_1''$, which has been described in 2.47. (L-order preservation means that the paths P_1 and P_2 need not have the same start node.) Then the following results are also true for L-order preservation:
  • *Theorem* 2.28: It can easily be proven that the following assertions are equivalent:
    (a) $H : \mathcal{P}(\mathcal{G}) \to \mathbf{R}$ is L-order preserving.
    (b) There exists a monotone path function $H' : \mathcal{P}(\mathcal{G}) \to \mathbf{R}$ such that
       $H'([v]) = H([v])$ for all $v \in V$ and $H'(P) \equiv H(P)$ for all $P \in \mathcal{P}(\mathcal{G})$ with $\ell(P) > 0$.
The main idea to replace all paths $P \in \mathcal{P}(s)$ in the proof of 2.28 by paths with arbitrary start nodes; in particular, define $\mathbf{R}_v := \{H(P) \,|\, \omega(P) = v\}$.
  • *Theorem* 2.33 – *Theorem* 2.36: We can apply these results to the problem of constructing a new L-order preserving path function if other L-order preserving ones are given. The condition about $<$-preservation in 2.33, however, must be replaced by the $<$-preservation with respect to $\mathbb{L}_1''$.

b) The following assertion is true, and it is analogous to (b)$\Rightarrow$(a) of Theorem 2.28. Let each arc r is marked with a strictly increasing function ϕ_r, and let $p : V \to \mathbf{R}$. Moreover, let $H, H' : \mathcal{P} \to \mathbf{R}$ where $\mathcal{P} = \mathcal{P}(\mathcal{G})$ or $\mathcal{P} = \mathcal{P}(s)$; we assume that

$H' = \phi \circ o \circ p$ and that $H \equiv H'$. Then H is $\prec$-preserving with respect to $\mathbb{L}_0$. We can construct many examples of $\prec$-preserving functions in this way. But it is not possible to generate all these functions because the following case may occur: H is $\prec$-preserving, and there exist three paths P_1, P_2, Q with $H(P_1) = H(P_2)$ and $H(P_1 \oplus Q) \prec H(P_2 \oplus Q)$. In this case there exist no functions p and ϕ_r and no path function H' such that $H \equiv H'$ and $H' = \phi \circ o \circ p$.

c) Theorem 2.32 is also true for $\prec$-preservation.

   We leave it to the reader to decide which of the results 2.33 – 2.36 are true for $\prec$-preservation instead of $\preceq$-preservation.

d) Also, we leave it to the reader to prove or disprove the following conjecture: $H : \mathcal{P}(\mathcal{G}) \to \mathbf{R}$ is order preserving with respect to $\mathbb{L}'_1 \cap \mathbb{L}''_1$ if and only if for each path $Q' \in \mathcal{P}(\mathcal{G})$, the path function $H_{Q'}(P) := H(Q' \oplus P)$ $(P \in \mathcal{P}(\mathcal{G}))$ is order preserving in the original sense. $\square$

2.3.4.2 Minimum preserving path functions

The main difference between order and minimum preservation is the following: Order preservation says that one inequality implies another inequality. Minimum preservation says that the minimality of one path implies the minimality of another path.

We start with a principle of minimum preservation that is similar to (2.6).

Definition 2.51 Let $\mathcal{P} := \mathcal{P}(\mathcal{G})$ [$\mathcal{P} := \mathcal{P}(s)$, resp.]. A path function $H : \mathcal{P} \to (\mathbf{R}, \preceq)$ is called *minimum preserving* if for all nodes v, v' [with $v = s$] and for all paths $P_1 \in \mathcal{P}(v, v')$ and $Q \in \mathcal{P}(v')$, the following is true:

$$H(P_1) \ \in \ MIN \{H(P_2) \,|\, P_2 \in \mathcal{P}(v, v')\} \quad \Longrightarrow \qquad (2.33)$$

$$H(P_1 \oplus Q) \ \in \ MIN \{H(P_2 \oplus Q) \,|\, P_2 \in \mathcal{P}(v, v')\} \,. \qquad (2.34) \qquad \square$$

The next result describes a relationship between order preservation and minimum preservation.

Theorem 2.52 Let $\mathcal{G} = (\mathcal{V}, \mathcal{R})$ be a digraph and $s \in \mathcal{V}$. Let $\mathcal{P} := \mathcal{P}(\mathcal{G})$ [$\mathcal{P} := \mathcal{P}(s)$, resp.], and let $H : \mathcal{P} \to (\mathbf{R}, \preceq)$ be an order preserving path function. We assume that $\preceq$ be total.
Then H is minimum preserving.

P r o o f : Let $v, v' \in \mathcal{V}$ [with $v = s$]. Let P_1 be a v-v'-path with property (2.33), and let Q be a path with start node v'. Then (2.33) and the totality of $\preceq$ imply that $H(P_1) \preceq H(P_2)$ for all $P_2 \in \mathcal{P}(v, v')$. This and the order preservation of H imply that $H(P_1 \oplus Q) \preceq H(P_2 \oplus Q)$ for all $P_2 \oplus \mathcal{P}(v, v')$. That means that (2.34) is true. $\square$

Next, we generalize the definition of minimum preservation in the case that $def(H) = \mathcal{P}(\mathcal{G})$; we leave it to the reader to do the same the case that $def(H)$ is a proper subset of $\mathcal{P}(\mathcal{G})$.

Definition 2.53 A path function $H : \mathcal{P}(\mathcal{G}) \to \mathbf{R}$ is called *L-minimum preserving* if the following is true for all paths $X_{1,1}, X_{1,2}, X_{1,3}, X_{3,1}, X_{3,2}, X_{3,3} \in \mathcal{P}(\mathcal{G})$ with $\alpha(X_{1,2}) = \omega(X_{1,1})$, $\alpha(X_{1,3}) = \omega(X_{1,2})$, $\alpha(X_{3,2}) = \omega(X_{3,1})$, $\alpha(X_{3,3}) = \omega(X_{3,2})$:

$$\text{I f} \quad H(X_1) \in MIN\{H(X_2) \,|\, X_2 \in S_2\} \tag{2.35}$$

$$\text{t h e n} \quad H(X_3) \in MIN\{H(X_4) \,|\, X_4 \in S_4\} \tag{2.36}$$

The set $S_2 = S_2(L, X_{1,1}, X_{1,2}, X_{1,3}, X_{3,1}, X_{3,2}, X_{3,3})$ and the set $S_4 = S_4(L, X_{1,1}, X_{1,2}, X_{1,3}, X_{3,1}, X_{3,2}, X_{3,3})$ are defined as follows:

$$S_2 := \left\{ X_2 \in \mathcal{P}(\mathcal{G}) \;\middle|\; \begin{array}{l} \text{There exist } X_{2,1}, X_{2,2}, X_{2,3}, \ X_{4,1}, X_{4,2}, X_{4,3} \in \mathcal{P}(\mathcal{G}) \text{ such that} \\[4pt] X_2 = X_{2,1} \oplus X_{2,2} \oplus X_{2,3} \text{ and } \mathbf{Y} := \begin{pmatrix} X_{1,1} & X_{1,2} & X_{1,3} \\ X_{2,1} & X_{2,2} & X_{2,3} \\ X_{3,1} & X_{3,2} & X_{3,3} \\ X_{4,1} & X_{4,2} & X_{4,3} \end{pmatrix} \in L \end{array} \right\},$$

$$S_4 := \left\{ X_4 \in \mathcal{P}(\mathcal{G}) \;\middle|\; \begin{array}{l} \text{There exist } X_{2,1}, X_{2,2}, X_{2,3}, \ X_{4,1}, X_{4,2}, X_{4,3} \in \mathcal{P}(\mathcal{G}) \text{ such that} \\[4pt] X_4 = X_{4,1} \oplus X_{4,2} \oplus X_{4,3} \text{ and } \mathbf{Y} := \begin{pmatrix} X_{1,1} & X_{1,2} & X_{1,3} \\ X_{2,1} & X_{2,2} & X_{2,3} \\ X_{3,1} & X_{3,2} & X_{3,3} \\ X_{4,1} & X_{4,2} & X_{4,3} \end{pmatrix} \in L \end{array} \right\}.$$

By the way, we define $S_2 := S_4 := \emptyset$ if $\alpha(X_{1,2}) \neq \omega(X_{1,1}) \vee \alpha(X_{1,3}) \neq \omega(X_{1,2}) \vee \alpha(X_{3,2}) \neq \omega(X_{3,1}) \vee \alpha(X_{3,3}) \neq \omega(X_{3,2})$. □

In our next example, we give an explicit description of L-minimum preservation for $L = \mathbb{L}_0$. Recall that $\mathbb{L}_0$-preservation is equivalent to standard order preservation. It turns out that if $\mathrm{def}(H) = \mathcal{P}(\mathcal{G})$ then $\mathbb{L}_0$-minimum preservation is equivalent to minimum preservation as defined in 2.51 on page 55.

Example 2.54 Let $H : \mathcal{P}(\mathcal{G}) \to \mathbf{R}$. Given the set $\mathbb{L}_0$. Then the sets S_2 and S_4 in Definition 2.53 depend on the paths $[v]$, P_1, $[v']$, Q where $v = \alpha(P_1)$ and $v' = \omega(P_1) = \alpha(Q)$; it is $S_2 = \{ P_2 \,|\, P_2 \in \mathcal{P}(v, v') \} = \mathcal{P}(v, v')$ and $S_4 = \{ P_2 \oplus Q \,|\, P_2 \in \mathcal{P}(v, v') \}$. Hence, we can formulate the $\mathbb{L}_0$-minimum preservation in Definition 2.53 as follows: Let $v, v' \in V$, and let $P_1 \in \mathcal{P}(v, v')$, $Q \in \mathcal{P}(v')$; then $H(P_1) \in MIN\{H(P_2) \,|\, P_2 \in \mathcal{P}(v, v')\}$ implies that $H(P_1 \oplus Q) \in MIN\{H(P_2 \oplus Q) \,|\, P_2 \in \mathcal{P}(v, v')\}$. That means that $\mathbb{L}_0$-minimum preservation is equivalent to minimum preservation as defined in 2.51 on page 55. □

If $\mathrm{def}(H) = \mathcal{P}(\mathcal{G})$, we can generalize Theorem 2.52 as follows: We shall show that L-order preservation implies L-minimum preservation if L satisfies a particular condition; this condition is formulated with the help of the following definition.

Definition 2.55 Given a matrix $\mathbf{X} \in \mathcal{L}$. Then the matrix $\vartheta(\mathbf{X})$ is generated by copying the first line of $\mathbf{X}$ into the second line and the third line of $\mathbf{X}$ into the fourth line. More formally,

$$\text{if } \mathbf{X} = \begin{pmatrix} X_{1,1} & X_{1,2} & X_{1,3} \\ X_{2,1} & X_{2,2} & X_{2,3} \\ X_{3,1} & X_{3,2} & X_{3,3} \\ X_{4,1} & X_{4,2} & X_{4,3} \end{pmatrix} \quad \text{then } \vartheta(\mathbf{X}) := \begin{pmatrix} X_{1,1} & X_{1,2} & X_{1,3} \\ X_{1,1} & X_{1,2} & X_{1,3} \\ X_{3,1} & X_{3,2} & X_{3,3} \\ X_{3,1} & X_{3,2} & X_{3,3} \end{pmatrix}.$$

A set $L \subseteq \mathcal{L}$ is called ϑ-closed if $\vartheta(\mathbf{X}) \in L$ for all $\mathbf{X} \in L$.
For example, $\mathbb{L}_0$ is ϑ-closed. $\qquad\square$

Theorem 2.56 *Let $H : \mathcal{P}(\mathcal{G}) \to (\mathbf{R}, \preceq)$ where $\preceq$ be total. Let $L \subseteq \mathcal{L}$ be ϑ-closed. Then L-order preservation implies L-minimum preservation.*

P r o o f : Given the paths $X_1 = X_{1,1} \oplus X_{1,2} \oplus X_{1,3}$ and $X_3 = X_{3,1} \oplus X_{3,2} \oplus X_{3,3}$. We assume that the assertion in (2.35) be true. We must show the assertion formulated in (2.36).
Let $X_4 \in S_4$. By the definition of S_4, there are six paths $X_{i,j}$, $i = 2,4$, $j = 1,2,3$, such that $X_4 = X_{4,1} \oplus X_{4,2} \oplus X_{4,3}$ and $\mathbf{X} := (X_{i,j}) \in L$. In particular, $X_2 \in S_2$. This fact together with assertion (2.35) and the totality of $\preceq$ imply that $H(X_1) \preceq H(X_2)$; consequently, $H(X_3) \preceq H(X_4)$ because H is L-order preserving.
We have shown that $H(X_3) \preceq H(X_4)$ for all $X_4 \in S_4$; it remains to be shown that $H(X_3)$ is an element of $\{H(X_4) \mid X_4 \in S_4\}$. For this purpose, we prove that $X_3 \in S_4$. Assumption (2.35) implies that $\{H(X_2) \mid X_2 \in S_2\} \neq \emptyset$ so that $S_2 \neq \emptyset$. Consequently, there exists a path $X_2 \in S_2$. By the definition of this set, there exist six paths $X_{i,j}$, $i = 2,4$, $j = 1,2,3$, such that $X_2 = X_{2,1} \oplus X_{2,2} \oplus X_{2,3}$ and $\mathbf{X} := (X_{i,j}) \in L$. Then

$$\vartheta(\mathbf{X}) := \begin{pmatrix} X_{1,1} & X_{1,2} & X_{1,3} \\ X_{1,1} & X_{1,2} & X_{1,3} \\ X_{3,1} & X_{3,2} & X_{3,3} \\ X_{3,1} & X_{3,2} & X_{3,3} \end{pmatrix} \in L \text{ because } L \text{ is } \vartheta\text{-closed. Considering the second}$$

and the fourth line of $\vartheta(\mathbf{X})$, we see that $X_3 \in S_4$. $\qquad\square$

Remark 2.57 Let $\mathrm{def}(H) = \mathcal{P} = \mathcal{P}(\mathcal{G})$. Then the assertion of Theorem 2.52 follows from Theorem 2.56 because $\mathbb{L}_0$ is ϑ-closed. $\qquad\square$

2.3.4.3   Order preservation with respect to minimal continuations

The transformation of order preservation into minimum preservation shows that modifications of order preservation can be formulated with the help of minimal paths. Next, we introduce another version of this idea. We replace the path Q appearing in the usual order preservation by two paths Q_i ($i = 1,2$) that are minimal continuations of P_i to an end node v''.

Definition/Remark 2.58 Let $H : \mathcal{P} \to (\mathbf{R}, \preceq)$ where $\mathcal{P} = \mathcal{P}(\mathcal{G})$ [$\mathcal{P} = \mathcal{P}(s)$, resp.]. We consider the following order preserving principle:
For all nodes v, v', v'' [with $v = s$], for all v-v'-paths P_1, P_2, for all paths $Q_1 \in \mathcal{P}_{\mathrm{Min}}(P_1, v'')$, and for all paths $Q_2 \in \mathcal{P}_{\mathrm{Min}}(P_2, v'')$, the following is true:
$$H(P_1) \preceq H(P_2) \implies H(P_1 \oplus Q_1) \preceq H(P_2 \oplus Q_2).$$

This property of H is the same as the order preservation with respect to the set $\mathbb{L}'_{min}$, which is defined as follows:

$$\mathbb{L}'_{min} := \left\{ \left(\begin{array}{ccc} [v] & P_1 & [v'] \\ [v] & P_2 & [v'] \\ [v] & P_1 & Q_1 \\ [v] & P_2 & Q_2 \end{array} \right) \in \mathcal{L} \;\middle|\; \begin{array}{l} v, v' \in \mathcal{V},\; \omega(Q_1) = \omega(Q_2) =: v'', \\ Q_1 \in \mathcal{P}_{Min}(P_1, v''), \\ Q_2 \in \mathcal{P}_{Min}(P_2, v'') \end{array} \right\}.$$

(Recall the conditions in (2.31) about $\mathcal{P}$. If $\mathcal{P} = \mathrm{def}(H) = \mathcal{P}(s)$ then $(X_1 = P_1 \in \mathcal{P} \wedge X_2 = P_2 \in \mathcal{P})$ implies that $v = s$; this implies that $(X_3 = P_1 \oplus Q_1 \in \mathcal{P} \wedge X_4 = P_2 \oplus Q_2 \in \mathcal{P})$.) $\qquad\square$

Next, we show that standard order preservation implies order preservation with respect to $\mathbb{L}'_{min}$.

Theorem 2.59 Let $\mathcal{G} = (\mathcal{V}, \mathcal{R})$ be a digraph. Let $\mathcal{P} = \mathcal{P}(\mathcal{G})$ [$\mathcal{P} = \mathcal{P}(s)$, resp.], and let $H : \mathcal{P} \to (\mathbf{R}, \preceq)$ where $\preceq$ is total. Let $\Re \in \{\preceq, \prec\}$.
Then $\Re$-preservation of H with respect to $\mathbb{L}_0$ implies the same property with respect to $\mathbb{L}'_{min}$.

P r o o f : We only consider the case that $\Re$ equals $\prec$; the proof in the case that $\Re$ equals $"\preceq"$ can be generated by replacing $"\prec"$ by the symbol $"\preceq"$.

Let H be $\prec$-preserving. Given $v, v' \in \mathcal{V}$ [with $v = s$]. Let $P_1, P_2 \in \mathcal{P}(v, v')$ with $H(P_1) \prec H(P_2)$. Moreover, let $v'' \in \mathcal{V}$ and $Q_i \in \mathcal{P}_{Min}(P_i, v'')$, $i = 1, 2$. Then we obtain the following inequality; fact $(*)$ follows from the totality of $\preceq$ and from $Q_1 \in \mathcal{P}_{Min}(P_1, v'')$, fact $(**)$ follows from $H(P_1) \prec H(P_2)$ and the $\prec$-preservation of H.

$$H(P_1 \oplus Q_1) \overset{(*)}{\preceq} H(P_1 \oplus Q_2) \overset{(**)}{\prec} H(P_2 \oplus Q_2).$$

Consequently, $H(P_1 \oplus Q_1) \prec H(P_2 \oplus Q_2)$. $\qquad\square$

Let $\mathrm{def}(H) = \mathcal{P}(\mathcal{G})$. Then we show the Theorem 2.62, which is a generalization Theorem 2.59 in the case that $\mathrm{def}(H) = \mathcal{P}(\mathcal{G})$. The idea is the following: If L is a set of matrices we shall define another set $\Theta_H(L) \supseteq L$. We shall show in Theorem 2.62 that $\Re$-preservation with respect to L is equivalent to $\Re$-preservation with respect to $\Theta_H(L)$ where $\Re \in \{\preceq, \prec\}$. Moreover, we shall see that $\mathbb{L}'_{min} \subseteq \Theta_H(\mathbb{L}_0)$. So, $\Re$-preservation with respect to $\Theta_H(\mathbb{L}_0)$ (which is equivalent to $\Re$-preservation with respect to $\mathbb{L}_0$) implies the same property with respect to $\mathbb{L}'_{min}$.
The matrices in $\Theta_H(L)$ are constructed as follows: Given a matrix

$$\mathbf{Y} = \left(\begin{array}{ccc} X_{1,1} & X_{1,2} & X_{1,3} \\ X_{2,1} & X_{2,2} & X_{2,3} \\ Y_{3,1} & Y_{3,2} & Y_{3,3} \\ Y_{4,1} & Y_{4,2} & Y_{4,3} \end{array} \right) \in L.$$

Then choose six paths $X_{i,j}$, $i = 3, 4$, $j = 1, 2, 3$, such that

$$H(X_3) = H(X_{3,1} \oplus X_{3,2} \oplus X_{3,3}) \preceq H(Y_{3,1} \oplus Y_{3,2} \oplus Y_{3,3}) = H(Y_3)$$
$$\text{and} \qquad H(Y_4) = H(Y_{4,1} \oplus Y_{4,2} \oplus Y_{4,3}) \preceq H(X_{4,1} \oplus X_{4,2} \oplus X_{4,3}) = H(X_4).$$

Then the matrix $(X_{i,j})$ with the $X_{3,j}$'s in the third line and the $X_{4,j}$'s in the fourth line is an element of $\Theta_H(L)$.
Now we give the formal definition of $\Theta_H(L)$.

Definition 2.60 Given a set $L \subseteq \mathcal{L}$ of matrices and a path function $H : \mathcal{P}(\mathcal{G}) \to (\mathbf{R}, \preceq)$ where $\preceq$ is total. Then we define the following subset of $\mathcal{L}$:

$$
\Theta_H(L) := \left\{
\begin{pmatrix}
X_{1,1} & X_{1,2} & X_{1,3} \\
X_{2,1} & X_{2,2} & X_{2,3} \\
X_{3,1} & X_{3,2} & X_{3,3} \\
X_{4,1} & X_{4,2} & X_{4,3}
\end{pmatrix}
\in \mathcal{L}
\left|
\begin{array}{l}
\text{There are } Y_{3,1}, \; Y_{3,2}, \; Y_{3,3}, \; Y_{4,1}, \; Y_{4,2}, \; Y_{4,3} \\[4pt]
\text{with } \mathbf{Y} :=
\begin{pmatrix}
X_{1,1} & X_{1,2} & X_{1,3} \\
X_{2,1} & X_{2,2} & X_{2,3} \\
Y_{3,1} & Y_{3,2} & Y_{3,3} \\
Y_{4,1} & Y_{4,2} & Y_{4,3}
\end{pmatrix}
\in L \\[4pt]
\text{and } H(X_3) \preceq H(Y_3), \quad H(Y_4) \preceq H(X_4)
\end{array}
\right.
\right\}.
$$

$\square$

Next, we give an explicit description of $\Theta_H(\mathbb{L}_0)$.

Example 2.61 Let $H : \mathcal{P}(\mathcal{G}) \to (\mathbf{R}, \preceq)$. The following equation describes the set $\Theta_H(\mathbb{L}_0)$.

$$
\Theta_H(\mathbb{L}_0) = \left\{
\begin{pmatrix}
[v] & P_1 & [v'] \\
[v] & P_2 & [v'] \\
X_{3,1} & X_{3,2} & X_{3,3} \\
X_{4,1} & X_{4,2} & X_{4,3}
\end{pmatrix}
\in \mathcal{L}
\left|
\begin{array}{l}
v, v' \in \mathcal{V}, \text{ and there exists a path } Q \in \mathcal{P}(v') \\
\text{with } H(X_3) \preceq H(P_1 \oplus Q) \\
\text{and } H(P_2 \oplus Q) \preceq H(X_4)
\end{array}
\right.
\right\}.
$$

The matrix $\mathbf{Y}$ mentioned in 2.60 has the form $\mathbf{Y} = \begin{pmatrix} [v] & P_1 & [v'] \\ [v] & P_2 & [v'] \\ [v] & P_1 & Q \\ [v] & P_2 & Q \end{pmatrix}$ $(\in \mathbb{L}_0)$.

Then $\Theta_H(\mathbb{L}_0)$-order preservation means that $H(P_1) \preceq H(P_2) \Rightarrow H(X_3) \preceq H(X_4)$ for all paths P_1, P_2, X_3, X_4 with the following properties: $v := \alpha(P_1) = \alpha(P_2)$ $v' := \omega(P_1) = \omega(P_2)$, and there exists path $Q \in \mathcal{P}(v')$ with $H(X_3) \preceq H(P_1 \oplus Q)$ and $H(P_2 \oplus Q) \preceq H(X_4)$.

$\square$

Theorem 2.62 *Given the situation of Definition 2.60. Let $\Re \in \{\preceq, \prec\}$.*
Then H is $\Re$-preserving with respect to L if and only if H is $\Re$-preserving with respect to $\Theta_H(L)$.

P r o o f : We show the result in the case that $\Re$ equals $\prec$; the proof is analogic if $\Re$ equals $\preceq$.

Suppose that H be $\prec$-preserving with respect to L. We show that H is $\prec$-preserving with respect to $\Theta_H(L)$. Let $(X_{i,j}) \in \Theta_H(L)$, and let $H(X_1) \prec H(X_2)$. Let $Y_{i,j}$ $(i = 3, 4, \; j = 1, 2, 3)$ and $\mathbf{Y} \in L$ be the same objects as in Definition 2.60. Then

$$H(X_3) \preceq H(Y_3) \quad \text{and} \quad H(Y_4) \preceq H(X_4). \tag{2.37}$$

Moreover, $H(Y_3) \prec H(Y_4)$ because $H(X_1) \prec H(X_2)$, $\mathbf{Y} \in L$, and H is $\prec$-preserving. Consequently, $H(X_3) \overset{(2.37)}{\preceq} H(Y_3) \prec H(Y_4) \overset{(2.37)}{\preceq} H(X_4)$.

So, we have shown that $H(X_1) \prec H(X_2) \Rightarrow H(X_3) \prec H(X_4)$ for all matrices $(X_{i,j}) \in \Theta_H(L)$; that means that H is $\prec$-preserving with respect to $\Theta_H(L)$.

Next, we show the reversed direction, i.e., $\prec$-preservation with respect to $\Theta(L)$ implies $\prec$-preservation with respect to L. This implication is correct because $L \subseteq \Theta_H(L)$. $\square$

Remark 2.63 Let $\mathrm{def}(H) = \mathcal{P}(\mathcal{G})$. Then we show Theorem 2.59 with the help of Theorem 2.62.

First, we show that $\mathbb{L}'_{\min} \subseteq \Theta_H(\mathbb{L}_0)$ if $\preceq$ is total. For this purpose, let

$$
\mathbf{X} = \begin{pmatrix} [v] & P_1 & [v'] \\ [v] & P_2 & [v'] \\ [v] & P_1 & Q_1 \\ [v] & P_2 & Q_2 \end{pmatrix} \in \mathbb{L}'_{\min} \, .
$$

The optimality of Q_1 and the totality of $\preceq$ imply that $Q := Q_2$ has the following properties:

$$ H(P_1 \oplus Q_1) \preceq H(P_1 \oplus Q) \quad \text{and} \quad H(P_2 \oplus Q) \preceq H(P_2 \oplus Q_2) \, . $$

Hence, the matrix $\mathbf{Y}$ described in Example 2.61 (with $Q := Q_2$) can be used to show that $\mathbf{X} \in \Theta_H(\mathbb{L}_0)$.

Let now $\Re \in \{\preceq, \prec\}$, and let H be $\Re$-preserving with respect to $\mathbb{L}_0$. Then H is $\Re$-preserving with respect to $\Theta_H(\mathbb{L}_0)$ by Theorem 2.62 and with respect to $\mathbb{L}'_{\min}$ because $\mathbb{L}'_{\min} \subseteq \Theta_H(\mathbb{L}_0)$. $\square$

2.3.4.4 Minimum preservation with respect to minimal continuations

Here, we combine the principle of minimum preservation and the idea of minimal continuations of paths. In particular, we shall give an explicit formulation of $\mathbb{L}'_{\min}$-minimum preservation.

Definition 2.64 Let $\mathcal{P} = \mathcal{P}(\mathcal{G})$ [$\mathcal{P} = \mathcal{P}(s)$, resp.]. Let $H : \mathcal{P} \to (\mathbf{R}, \preceq)$ where $\preceq$ is total. Then we call H *minimum preserving with respect to minimal continuations* if and only if the following is true for all for all nodes v, v', v'' [with $v = s$], for all paths P_1 from v to v', and for all paths $Q_1 \in \mathcal{P}_{Min}(P_1, v)$:

$$ H(P_1) \in MIN\,\{H(P_2)\,|\,P_2 \in \mathcal{P}(v,v')\} \quad \Longrightarrow \quad \text{(2.38)} $$
$$ H(P_1 \oplus Q_1) \in MIN\,\{H(P_2 \oplus Q_2)\,|\,P_2 \in \mathcal{P}(v,v') \wedge Q_2 \in \mathcal{P}_{Min}(P_2,v'')\}\,. \quad \text{(2.39)} \qquad \square $$

Theorem 2.65 *Given the situation in Definition 2.64. In addition let $\mathcal{P} = \mathcal{P}(\mathcal{G})$, and let H have the $\mathcal{P}(X,w)$-minimality property for all $X \in \mathcal{P}$ and all $w \in V$. The H is minimum preserving with respect to minimal continuations if and only if H is $\mathbb{L}'_{\min}$-minimum preserving.*

P r o o f : We describe the sets S_2, S_4 from Definition 2.53. Let $v, v', v'' \in V$ and $P_1 \in \mathcal{P}(v,v')$, $Q_1 \in \mathcal{P}(v',v'')$. If $Q_1 \notin \mathcal{P}_{Min}(P_1, v'')$ then $S_2 = S_4 = \emptyset$. The other case is that $Q_1 \in \mathcal{P}_{Min}(v,v')$. Then $S_2 = \{P_2\,|\,P_2 \in \mathcal{P}(v,v')\} = \mathcal{P}(v,v')$ for the following reason: If $P_2 \in \mathcal{P}(v,v')$ then the P_2-v''-minimality property implies that a path $Q_2 \in \mathcal{P}_{Min}(P_2, v'')$ exists; then $\mathbf{Y} := \begin{pmatrix} [v] & P_1 & [v'] \\ [v] & P_2 & [v'] \\ [v] & P_1 & Q_1 \\ [v] & P_2 & Q_2 \end{pmatrix} \in \mathbb{L}'_{min}$.

Moreover, it is easy to see that $S_4 = \{P_2 \oplus Q_2 \mid P_2 \in \mathcal{P}(v, v'),\ Q_2 \in \mathcal{P}_{Min}(P_2, v'')\}$.

Consequently, H is $\mathbb{L}'_{min}$-minimum preserving if and only if the premise in (2.38) implies assertion (2.39). □

Next, we show that both $\mathbb{L}'_{min}$-order preservation and $\mathbb{L}_0$-minimum preservation imply minimum preservation with respect to minimal continuations.

Theorem 2.66 *Given a digreaph $\mathcal{G} = (\mathcal{V}, \mathcal{R})$ and a node $s \in \mathcal{V}$. Let $\mathcal{P} = \mathcal{P}(\mathcal{G})$ or $\mathcal{P} = \mathcal{P}(s)$. Let $H : \mathcal{P} \to (\mathbf{R}, \preceq)$ where $\preceq$ is total. Then the following assertions are true:*

a) *If H is $\mathbb{L}'_{min}$-order preserving then H is minimum preserving with respect to minimal continuations.*

b) *Let even $\mathcal{P} = \mathcal{P}(\mathcal{G})$, and let H have the P-w-minimality property for all $P \in \mathcal{P}(\mathcal{G})$ and all $w \in \mathcal{V}$.*
   If H is minimum preserving then H is minimum preserving with respect to minimal continuations.

P r o o f   o f   a) :   Let H be $\mathbb{L}'_{min}$-order preserving. Given two paths $P_1 \in \mathcal{P}(v, v')$ and $Q_1 \in \mathcal{P}_{Min}(v, v')$. Let (2.38) be true. This and the totality of $\preceq$ imply that $H(P_1) \preceq H(P_2)$ for all $P_2 \in \mathcal{P}(v, v')$. Using the $\mathbb{L}'_{min}$-order preservation, we see that $H(P_1 \oplus Q_1) \preceq H(P_2 \oplus Q_2)$ for all $P_2 \in \mathcal{P}(v, v')$ and all $Q_2 \in \mathcal{P}_{Min}(P_2, v'')$. That means that (2.39) is true.

We can give a more elegant proof if $\mathcal{P} = \mathcal{P}(\mathcal{G})$ and if H has the P-w-minimality property for all $P \in \mathcal{P}(\mathcal{G})$ and all $w \in \mathcal{V}$. In this case, we observe that $\mathbb{L}'_{min}$ is ϑ-closed. So, Theorem 2.56 implies that H is $\mathbb{L}'_{min}$-minimum preserving. Theorem 2.65 says that this property of H is equivalent to the minimum preservation with respect to minimal continuations.

P r o o f   o f   b) :   Let H be minimum preserving. Let (2.38) (= (2.33)) be true. Moreover, let $Q_1 \in \mathcal{P}_{Min}(v', v'')$.
Then relation $(*)$ of the next inequality follows from $Q_1 \in \mathcal{P}_{Min}(v', v'')$ and the totality of $\preceq$; relation $(**)$ follows from (2.33) and the minimum preservation of H.

$$(\forall P_2 \in \mathcal{P}(v, v'),\ Q_2 \in \mathcal{P}_{Min}(P_2, v''))\ H(P_1 \oplus Q_1) \overset{(*)}{\preceq} H(P_1 \oplus Q_2) \overset{(**)}{\preceq} H(P_2 \oplus Q_2).$$

That means that (2.39) is true. □

2.3.4.5   Transitive inference methods

Let $H : \mathcal{P} \to (\mathbf{R}, \preceq)$ where $\mathcal{P} = \mathcal{P}(\mathcal{G})$ or $\mathcal{P} = \mathcal{P}(s)$. Given the following local order preserving principle, which is only valid if a single arc is appended to P_1 and P_2.

$H(P_1) \preceq H(P_2) \implies H(P_1 \oplus r) \preceq H(P_2 \oplus r)$
for all paths P_1, P_2 and all arcs r with $\alpha(P_1) = \alpha(P_2)$ and $\omega(P_1) = \omega(P_2) = \alpha(r)$.

This is order preservation with respect to $L := \mathbb{L}_1' \cap \mathbb{L}_2' \cap \mathbb{L}_3''$; these sets were introduced in 2.47.

If H is order preserving then H is locally order preserving. The reason is that the usual order preservation includes the case that the appended path Q is of length 1.

But the reverse conclusion is also true. Let H be locally order preserving, and let $Q = r_1 \oplus r_2 \oplus \ldots \oplus r_k$. Then $H(P_1) \preceq H(P_2)$ implies $H(P_1 \oplus r_1) \preceq H(P_2 \oplus r_1)$, and this implies $H(P_1 \oplus r_1 \oplus r_2) \preceq H(P_2 \oplus r_1 \oplus r_2)$, and so on; in the k-th step, we obtain $H(P_1 \oplus Q) \preceq H(P_2 \oplus Q)$.

That means that local order preservation is equivalent to the usual one. Similar conclusions can be drawn in the case of order preservation with respect to general sets $L \in \mathcal{L}$. More details can be found in Definition 2.34 – Corollary 2.38 of [Huck92].

2.3.5   Further remarks and definitions concerning modified order preserving principles

All modified order preserving principles are assertions about a single path function. Next, we see that some of our definitions can be used to describe a relationship of two path functions.

Remark 2.67   a) Several sets $L \subseteq \mathcal{L}$ of matrices depend on a path function H, i.e. $L = L^{(H)}$; an example is $L^{(H)} = \Theta_H(L)$; another example is $L^{(H)} = \mathbb{L}_{\min}'$ because the sets $\mathcal{P}_{\min}(P_1, v'')$, $\mathcal{P}_{\min}(P_2, v'')$ in the definition of $\mathbb{L}_{\min}'$ are defined with the help of H. We have considered the order $L^{(H)}$-order preservation of $H' := H$; we leave it to the reader to investigate the $L^{(H)}$-order preservation of the path function $H' \neq H$.

b) Let $L \subseteq \mathcal{L}$ and $H, H' : \mathcal{P}(\mathcal{G}) \to (\mathbf{R}, \preceq)$ where $\preceq$ is total. We write $H \trianglelefteq_L H'$ if the following modified principle of L-order preservation is true:
$$(\forall (X_{i,j}) \in L) \quad H(X_1) \preceq H(X_2) \implies H'(X_3) \preceq H'(X_4).$$
For example, let $L := \{(X_{i,j}) \in \mathcal{L} \mid (\forall j = 1, 2, 3) (X_{1,j} = X_{3,j} \wedge X_{2,j} = X_{4,j})\}$. Then H and H' are order equivalent if and only if $(H \trianglelefteq_L H' \wedge H' \trianglelefteq_L H)$.   □

The next remark is about the question whether order or minimum preservation remains true for subgraphs.

Remark 2.68 Let $H : \mathcal{P}(\mathcal{G}) \to (\mathbf{R}, \preceq)$. All properties of order preservation in Example 2.47 are inherited by subgraphs (see 2.26 c)).
The $\mathbb{L}_{\min}'$-order preservation, however, is not always inherited by subgraphs. The following situation is a counterexample: H is $\mathbb{L}_{\min}'$-order preserving in $\mathcal{G}$, and $H(P_1) \preceq H(P_2)$; moreover, $Q_i \in \mathcal{P}_{Min}(P_i, v'')$, $i = 1, 2$; $\widetilde{\mathcal{G}}$ is a subgraph of $\mathcal{G}$ in which Q_1 is interrupted and Q_2 is preserved. Let $\widetilde{Q}_1 \in \mathcal{P}_{Min}(P_1, v'')$ in the subgraph $\widetilde{\mathcal{G}}$; then it may occur that $H(P_1 \oplus \widetilde{Q}_1) \succ H(P_2 \oplus Q_2)$. In this case, H is not $\mathbb{L}_{\min}'$-order preserving in the subgraph $\widetilde{\mathcal{G}}$.
In general, the following question is still open: Can we formulate a simple property of

the set $L \subseteq \mathcal{L}$ which implies that L-order preservation or L-minimum preservation is always inherited by subgraphs? ☐

Remark/Definition 2.69 We can express many variants of order preservation as L-order preservation. Nevertheless, there are principles of order preservation that probably cannot be formulated as order preservation with respect to an appropriate set $L \subseteq \mathcal{L}$. The reason is that all assertions in Definition 2.46 b) are of the form $\big((\forall \mathbf{X} = (X_{i,j}) \in L) \; \mathcal{A}(H, X_1, X_2) \Rightarrow \mathcal{A}(H, X_3, X_4)\big)$ where $\mathcal{A}(H, P, Q)$ denotes the formula $\big(P \in \mathrm{def}(H) \wedge Q \in \mathrm{def}(H) \wedge H(P) \preceq H(Q)\big)$.

The following two principles of order preservation have another structure; therefore, it is doubtful whether we can formulate these principles as L-order preservation.

The first of them is about six instead of four paths, and the second version of order preservation contains an existential quantifier.

We say that H is *order preserving with respect to infixes* if for all nodes $v, w, v', w' \in V$, for all paths P_1, P_2 from v to w, for all paths P_1', P_2' from v' to w', and for all inserted infixes Q from w to v' the following is true:

$$\big(H(P_1) \preceq H(P_2) \wedge H(P_1') \preceq H(P_2')\big) \;\Rightarrow\; \big(H(P_1 \oplus Q \oplus P_1') \preceq H(P_2 \oplus Q \oplus P_2')\big).$$

We say that H is *existentially order preserving* if for all nodes v, v', v'' and for all paths P_1, P_2 from v to v' the following is true:

> If $H(P_1) \preceq H(P_2)$ then for each path $Q_2 \in \mathcal{P}(v', v'')$, there exists a path $Q_1 \in \mathcal{P}(v', v'')$ such that $H(P_1 \oplus Q_1) \preceq H(P_2 \oplus Q_2)$.

This order preserving principle implies $\mathbb{L}'_{\min}$-order preservation. To see this we assume that $H(P_1) \preceq H(P_2)$ and $\widetilde{Q}_i \in \mathcal{P}_{Min}(P_i, v'')$, $i = 1, 2$. If H is extentially order preserving then a path $Q_1 \in \mathcal{P}(v', v'')$ exists such that $H(P_1 \oplus Q_1) \preceq H(P_2 \oplus \widetilde{Q}_2)$. This and $\widetilde{Q}_1 \in \mathcal{P}_{Min}(P_1, v'')$ imply that $H(P_1 \oplus \widetilde{Q}_1) \preceq H(P_1 \oplus Q_1) \preceq H(P_2 \oplus \widetilde{Q}_2)$. ☐

The next definition is motivated by comparing order preservation with Bellman principles (see Section 2.4). The standard order preservation is a conclusion from the paths P_1, P_2 to their extensions $P_1 \oplus Q$, $P_2 \oplus Q$; the same is true for standard minimum preservation. The usual Bellman principles, however, mean to draw a conclusion from paths to their subpaths.

We now define several properties of sets $L \subseteq \mathcal{L}$ These properties make it possible to predict that the order or minimum preservation with respect to L is a conclusion from given paths to their extensions or their subpaths.

Definition/Remark 2.70 We call a set $L \subseteq \mathcal{L}$ *(weakly) expanding*, if $X_1 \subseteq X_3$ for all matrices $(X_{i,j}) \in L$; if each matrix $(X_{i,j}) \in L$ has the additional property that $X_2 \subseteq X_4$ then we say that L is *strongly expanding*. For example the set $\mathbb{L}_0$ is strongly expanding.

We call a set L *(weakly) contracting* if $X_3 \subseteq X_1$ for all matrices $(X_{i,j}) \in L$; if every matrix $(X_{i,j}) \in L$ has the additional property that $X_4 \subseteq X_2$ then we say that L is *strongly contracting*.

Let L be strongly expanding. Then L-order preservation is a conclusion from the property of "short" paths X_1, X_2 to a property of "longer" paths $X_3 \supseteq X_1$ and $X_4 \supseteq X_2$. An example is the usual order preservation; it is equivalent to the $\mathbb{L}_0$-preservation where $\mathbb{L}_0$ is strongly contracting.

If L is strongly contracting then L-order preservation is a conclusion from a property of "long" paths X_1, X_2 to a property of "shorter" paths $X_3 \subseteq X_1$ and $X_4 \subseteq X_2$.

If L is weakly expanding [contracting] then minimum preservation is a conclusion from a path X_1 to a "longer" path $X_3 \supseteq X_1$ [to a "shorter" path $X_3 \subseteq X_1$].   □

2.4   Bellman principles for paths in graphs

One of the most well-known examples of Bellman's principle of optimality is the following: Let a and b two points in the Euclidean plane. The optimal, i.e. the shortest curve between a and b is the line segment $\mathcal{O} = \overline{a, b}$. Then each line segment $\mathcal{O}' \subseteq \mathcal{O}$ is also an optimal connection between its two endpoints.

In general, a Bellman principle is a assertion of the following form:

> If an object $\mathcal{O}$ is optimal then particular parts $\mathcal{O}'$ of $\mathcal{O}$ are optimal, too.

For paths in directed or undirected graphs, we obtain the following formulation:

> If a path P is optimal then particular infixes P' are optimal, too.

The simplest example of a Bellman principle for paths in graphs arises from counting their arcs. If P has minimal length among all s-v-paths then each subpath $P' \subseteq P$ has minimal length among all paths from $\alpha(P')$ to $\omega(P')$. On the other hand, Bellman properties are not trivial; Remark 4.207 shows a simple path function that does not have a particular Bellman property.

Bellman's principle of optimality appears in many further situations of Mathematics and Computer Science. Bellman himself has given several examples in [Bell67], in [BeDr62] in [BeKa65], and in many other sources. In particular, Bellman has considered optimal curves whose parts are also optimal, and he has derived the Euler-Lagrange differential equation of Variational Calculus; in our above examples, these optimal curves are line segments.

Also, Bellman principles are often valid for optimal trees. A typical example is a search tree with minimal expected time of searching for $x \in \mathbb{R}$ where x is randomly chosen. The Bellman principle says that all subtrees of an optimal search tree are optimal, too. More details about optimal search trees with Bellman properties are given in Volume 3 of [Knu82], in [Aign88], in [Nolt82], in [Helm86], and in many other sources.

Bellman principles also appear in many models of decision processes. These models can often be described with paths in digraphs. Two examples of such decision models are presented in [Mor82] and in [Snie86], which are mentioned in Section 2.4.8 of this book.

Bellman's principle of optimality is very important for the correct behaviour of dynamic programming and many other search strategies for optimal objects.

The algorithm of Ford and Bellman is mainly a dynamic programming method applied to paths in digraphs.

Bellman's principle of optimality makes it easy to record and to find optimal paths. For example, let P be an optimal path from a node s to a node γ, and let $P' \subseteq P$ be a subpath from v to w. If the Bellman principle is given then P' is also optimal. That means that P automatically contains an optimal $v\text{-}v'$-path. Consequently, if P is given we do not need to compute and to record a new optimal $v\text{-}v'$-path so that we save time and space. If, however, P is not given, then we can find this path very fast. We must only consider candidates of the form $P = P'_1 \oplus \ldots \oplus P'_k$ where all subpaths P'_κ themselves are optimal. So, we have only few candidates for an optimal $s\text{-}\gamma$-path, and we save time by discarding irrelevant competitors.

Next, we give exact definitions of Bellman principles of path functions. We first formulate strong Bellman conditions, and then we consider the corresponding weak conditions.

2.4.1 Introduction to strong Bellman principles for path functions

In this section, we first formulate a generalized strong Bellman principle, and then we define several special Bellman conditions. The general Bellman principle has a simpler structure than the generalized order or minimum preservation because we avoid matrices. In Section 2.4.5, however, we give a unifying approach to L-minimum preservation and Bellman principles.

An important object in the context of generalized Bellman principle is the relation of optimal connections (ROC). It describes the situation that a path P connects two paths P_1, P_2 in an optimal way.

Definition 2.71 Given a digraph $\mathcal{G} = (\mathcal{V}, \mathcal{R})$ and a cost function $H : \mathcal{P} \to (\mathbf{R}, \preceq)$. The *relation of optimal connections* (ROC) is defined as follows:
$$ROC := \{(P_1, P, P_2) \in \mathcal{P}(\mathcal{G})^3 \mid P \in \mathcal{P}_{Min}(P_1, P_2)\} \ .$$
E.g., let $s, \gamma \in \mathcal{V}$, and let P be a minimal $s\text{-}\gamma$-path. Then $([s], P, [\gamma]) \in ROC$.
We say that P is *an optimal connection from P_1 to P_2* if $(P_1, P, P_2) \in ROC$; that means that $H(P_1 \oplus P \oplus P_2)$ is minimal among all values $H(P_1 \oplus \widetilde{P} \oplus P_2)$. $\qquad\square$

Our generalized Bellman principles about paths in graphs have the following form:

> For particular paths P_1, P, P_2, Q_1, Q, Q_2, the following is true:
>> If P is optimal among all connections from P_1 to P_2
>> then the subpath Q of P optimal among all connections from Q_1 to Q_2.
>
> More formally, if $(P_1, P, P_2) \in ROC$ then $(Q_1, Q, Q_2) \in ROC$ for particular 6-tuples $(P_1, P, P_2, Q_1, Q, Q_2)$.

These 6-tuples form a relation B. We shall use this relation to define Bellman-principles. A formal description of the relations B and the corresponding Bellman principles is given in the following two definitions.

Definition 2.72 Given a cost function $H : \mathcal{P} \to (\mathbf{R}, \preceq)$.
We define the following fixed relation $\mathcal{B}$, which depends on $\mathcal{P} = \mathrm{def}(H)$:

$$\mathcal{B} := \left\{ (P_1, P, P_2, Q_1, Q, Q_2) \in (\mathcal{P}(\mathcal{G}))^6 \;\middle|\; \begin{array}{l} \alpha(P){=}\omega(P_1),\, \alpha(P_2){=}\omega(P), \\ \alpha(Q){=}\omega(Q_1),\, \alpha(Q_2){=}\omega(Q), \\ P_1 \oplus P \oplus P_2 {\in} \mathrm{def}(H),\, Q_1 \oplus Q \oplus Q_2 {\in} \mathrm{def}(H) \end{array} \right\}.$$

A relation $B \subseteq (\mathcal{P}(\mathcal{G}))^6$ is called a $\mathcal{B}$-*relation* if $B \subseteq \mathcal{B} \subseteq (\mathcal{P}(\mathcal{G}))^6$. □

Definition 2.73 (General Bellman principles for path functions)
Given a digraph $\mathcal{G}$ with a cost function $H : \mathcal{P} \to \mathbf{R}$. Let B be a $\mathcal{B}$-relation, i.e., $B \subseteq \mathcal{B}$.

H satisfies *the strong Bellman B-condition* if the following is true for all 6-tuples $(P_1, P, P_2, Q_1, Q, Q_2) \in B$:

$$\text{If } (P_1, P, P_2) \in ROC \text{ then } (Q_1, Q, Q_2) \in ROC.$$

An equivalent formulation is the following: H satisfies the strong Bellman B-condition if for all triples $(P_1, P, P_2) \in ROC$, the following is true:

$$(P_1, P, P_2, Q_1, Q, Q_2) \in B \implies (Q_1, Q, Q_2) \in ROC. \qquad (2.40)$$

□

Next, we define several concrete strong Bellman principles for paths in digraphs; we see that they can be formulated as B-conditions with appropriate sets $B \subseteq \mathcal{B}$. Note that every additive path function satisfies all strong Bellman conditions of type $0-6$, which are now defined.

Definition 2.74 Let $\mathcal{G}$ be a digraph, and let $H : \mathcal{P} \to (\mathbf{R}, \preceq)$ be a cost function.

H satisfies the *strong prefix oriented [infix oriented, suffix oriented, resp.] Bellman condition* if the following is true for all nodes $s, z \in V$, all H-minimal paths P from s to z, and for all prefixes [infixes, suffixes, resp.] Q of P:

$$Q \text{ is an } H\text{-minimal path from } \alpha(Q) \text{ to } \omega(Q).$$

This property is also called the strong prefix oriented [infix oriented, suffix oriented, resp.] Bellman property *of type 0*.

The strong prefix oriented [infix oriented, suffix oriented, resp.] Bellman property of type 0 is equivalent to the strong $B^{(0)}_{\leq}$-condition [$B^{(0)}_{\subseteq}$-condition, $B^{(0)}_{\preceq}$-condition, resp.] where the relations $B^{(0)}_{\leq}, B^{(0)}_{\subseteq}, B^{(0)}_{\preceq}$ are defined as follows:

$$B^{(0)}_{\leq} := \left\{ (P_1, P, P_2, Q_1, Q, Q_2) \in \mathcal{B} \;\middle|\; \begin{array}{l} P_1{=}[\alpha(P)],\, P_2{=}[\omega(P)], \\ Q_1{=}[\alpha(P)],\, Q_2{=}[\omega(Q)],\ \ Q \leq P \end{array} \right\},$$

$$B^{(0)}_{\subseteq} := \left\{ (P_1, P, P_2, Q_1, Q, Q_2) \in \mathcal{B} \;\middle|\; \begin{array}{l} P_1{=}[\alpha(P)],\, P_2{=}[\omega(P)], \\ Q_1{=}[\alpha(Q)],\, Q_2{=}[\omega(Q)],\ \ Q \subseteq P \end{array} \right\},$$

$$B^{(0)}_{\preceq} := \left\{ (P_1, P, P_2, Q_1, Q, Q_2) \in \mathcal{B} \;\middle|\; \begin{array}{l} P_1{=}[\alpha(P)],\, P_2{=}[\omega(P)], \\ Q_1{=}[\alpha(Q)],\, Q_2{=}[\omega(P)],\ \ Q \preceq P \end{array} \right\}.$$

□

Definition/Remark 2.75 Given a cost function $H : \mathcal{P} \to (\mathbf{R}, \preceq)$. Let v and w be two nodes, and let P be a v-w-path.

We say that P is a *Bellman path* if each prefix $Q \leq P$ has the minimal H-value among all paths from v to $\omega(Q)$.

Consequently, every Bellman path P is an H-minimal connection from $\alpha(P)$ to $\omega(P)$; this follows from the fact that $Q := P$ is a prefix of P.

The strong prefix oriented Bellman principle of type 0 can be formulated as follows: If P is an H-minimal path from $\alpha(P)$ to $\omega(P)$ then P is a Bellman path. ☐

The next Bellman condition says that the conclusion from the optimality of P to the optimality of Q may only be drawn if Q is "almost as long as P".

Definition 2.76 Let Q be a subpath of P. We say that Q is *similar to* P if one of the following assertions 1) – 4) is true:

  1) Q is equal to P.
  2) Q is generated by deleting the first arc of P.
  3) Q is generated by deleting the last arc of P.
  4) Q is generated by deleting the first and the last arc of P. ☐

Definition 2.77 Let $\mathcal{G}$ be a digraph, and let $H : \mathcal{P} \to (\mathbf{R}, \preceq)$ be a cost function.

H satisfies the *strong prefix oriented [infix oriented, suffix oriented, resp.] Bellman condition of type 1* if the following is true for all nodes $s, z \in V$, for all H-minimal paths P from s to z, and for all prefixes [infixes] [suffixes] that are similar to P:

Q is an H-minimal path from $\alpha(Q)$ to $\omega(Q)$.

The strong prefix oriented [infix oriented, suffix oriented, resp.] Bellman property of type 1 is equivalent to the strong $B_{\preceq}^{(1)}$-condition [$B_{\subseteq}^{(1)}$-condition, $B_{\preceq}^{(1)}$-condition, resp.] where the relations $B_{\preceq}^{(1)}, B_{\subseteq}^{(1)}, B_{\preceq}^{(1)}$ are defined as follows:

$$B_{\preceq}^{(1)} := \left\{ (P_1, P, P_2, Q_1, Q, Q_2) \in \mathcal{B} \;\middle|\; \begin{array}{l} P_1 = [\alpha(P)], P_2 = [\omega(P)], \\ Q_1 = [\alpha(P)], Q_2 = [\omega(Q)], \\ (\exists\, r'' \in \mathcal{R})\, (P = Q \oplus r'' \vee P = Q) \end{array} \right\},$$

$$B_{\subseteq}^{(1)} := \left\{ (P_1, P, P_2, Q_1, Q, Q_2) \in \mathcal{B} \;\middle|\; \begin{array}{l} P_1 = [\alpha(P)], P_2 = [\omega(P)], \\ Q_1 = [\alpha(Q)], Q_2 = [\omega(Q)], \\ (\exists\, r', r'' \in \mathcal{R}) \left(\begin{array}{l} P = r' \oplus Q \oplus r'' \vee P = Q \oplus r'' \\ \vee\, P = r' \oplus Q \vee P = Q \end{array} \right) \end{array} \right\},$$

$$B_{\preceq}^{(1)} := \left\{ (P_1, P, P_2, Q_1, Q, Q_2) \in \mathcal{B} \;\middle|\; \begin{array}{l} P_1 = [\alpha(P)], P_2 = [\omega(P)], \\ Q_1 = [\alpha(Q)], Q_2 = [\omega(P)], \\ (\exists\, r' \in \mathcal{R})\, (P = r' \oplus Q \vee P = Q) \end{array} \right\}. \qquad ☐$$

Next, we define the Bellman properties of type 2 and 3. They are stronger than the properties of type 0 and 1, respectively; this will be shown in Theorem 2.91. The Bellman principles of type 2 and 3 are precise formulations of an idea in the section "Principle of Optimality" of [Snie86].

Definition 2.78 Let G be a digraph, and let $H : \mathcal{P} \to (\mathbf{R}, \preceq)$ be a cost function.

We say that H satisfies the *strong prefix oriented [infix oriented, suffix oriented, resp.] Bellman condition of type 2* if the following is true for all triples (P_1, P, P_2):

If P is an optimal connection from P_1 to P_2 and if P is a prefix [infix, suffix, resp.] of $P_1 \oplus P \oplus P_2$ then $Q := P$ is an optimal connection from $\alpha(P)$ to $\omega(P)$.

The strong prefix oriented [infix oriented, suffix oriented, resp.] Bellman property of type 2 is equivalent to the strong Bellman $B_{\preceq}^{(2)}$-condition [$B_{\subseteq}^{(2)}$-condition, $B_{\preceq}^{(2)}$-condition, resp.] where the relations $B_{\preceq}^{(2)}, B_{\subseteq}^{(2)}, B_{\preceq}^{(2)}$ are defined as follows:

$$B_{\preceq}^{(2)} := \left\{ (P_1, P, P_2, Q_1, Q, Q_2) \in \mathcal{B} \;\middle|\; \begin{array}{l} P_1 = [\alpha(P)], \\ Q_1 = [\alpha(P)], \; Q = P, \; Q_2 = [\omega(P)] \end{array} \right\},$$

$$B_{\subseteq}^{(2)} := \left\{ (P_1, P, P_2, Q_1, Q, Q_2) \in \mathcal{B} \;\middle|\; Q_1 = [\alpha(P)], \; Q = P, \; Q_2 = [\omega(P)] \right\},$$

$$B_{\preceq}^{(2)} := \left\{ (P_1, P, P_2, Q_1, Q, Q_2) \in \mathcal{B} \;\middle|\; \begin{array}{l} P_2 = [\omega(P)], \\ Q_1 = [\alpha(P)], \; Q = P, \; Q_2 = [\omega(P)] \end{array} \right\}. \qquad \square$$

Remark 2.79 The strong Bellman properties of type 2 are equivalent to the following conditions:

- *in the prefix oriented case* :
$$(\forall\, v_1 \in \mathcal{V}, \; P_1 \in \mathcal{P}(G)) \;\; \mathcal{P}_{Min}(v_1, P_2) \subseteq \mathcal{P}_{Min}(v_1, \alpha(P_2));$$
- *in the infix oriented case* :
$$(\forall\, P_1, P_2 \in \mathcal{P}(G)) \;\; \mathcal{P}_{Min}(P_1, P_2) \subseteq \mathcal{P}_{Min}(\omega(P_1), \alpha(P_2)); \qquad (2.41)$$
- *in the suffix oriented case* :
$$(\forall\, P_1 \in \mathcal{P}(G), \; v_2 \in \mathcal{V}) \;\; \mathcal{P}_{Min}(P_1, v_2) \subseteq \mathcal{P}_{Min}(\omega(P_1), v_2).$$

For example, let H be additive. Then the suffix oriented Bellman property of type 2 is an exact description of the following observation: An optimal extension P of a path P_1 to a node v_2 can be found by constructing an optimal path from $\omega(P_1)$ to v_2.

We can formulate new properties of H by replacing the relations $"\subseteq"$ in (2.41) by the relations $"\supseteq"$ or $"="$. It is an open question what these new properties actually mean and whether they may be considered as B-conditions for some $B \subseteq \mathcal{B}$. $\qquad \square$

The next Bellman property has the same structure as the type-2-principle. In analogy to type 1, we only consider subpaths $Q := P$ that are similar to $P_1 \oplus P \oplus P_2$.

Definition 2.80 Let $\mathcal{G}$ be a digraph, and let $H : \mathcal{P} \to (\mathbf{R}, \preceq)$ be a cost function.

We say that H satisfies the *strong prefix oriented [infix oriented, suffix oriented, resp.] Bellman condition of type 3* if the following is true for all triples (P_1, P, P_2):

If P is an optimal connection from P_1 to P_2 and if P is a prefix [infix] [suffix] of $P_1 \oplus P \oplus P_2$ and if $\ell(P_1), \ell(P_2) \leq 1$ then $Q := P$ is an optimal connection from $\alpha(P)$ to $\omega(P)$.

The strong prefix oriented [infix oriented, suffix oriented, resp.] Bellman property of type 3 is equivalent to the strong $B_{\preceq}^{(3)}$-condition [$B_{\subseteq}^{(3)}$-condition, $B_{\preceq}^{(3)}$-condition, resp.] where the relations $B_{\preceq}^{(3)}, B_{\subseteq}^{(3)}, B_{\preceq}^{(3)}$ are defined as follows:

$$B_{\preceq}^{(3)} := \left\{ (P_1, P, P_2, Q_1, Q, Q_2) \in \mathcal{B} \;\middle|\; \begin{array}{l} P_1 = [\alpha(P)], \ \ell(P_2) \leq 1, \\ Q_1 = [\alpha(P)], \ Q = P, \ Q_2 = [\omega(P)] \end{array} \right\},$$

$$B_{\subseteq}^{(3)} := \left\{ (P_1, P, P_2, Q_1, Q, Q_2) \in \mathcal{B} \;\middle|\; \begin{array}{l} \ell(P_1) \leq 1, \ \ell(P_2) \leq 1, \\ Q_1 = [\alpha(P)], \ Q = P, \ Q_2 = [\omega(P)] \end{array} \right\},$$

$$B_{\preceq}^{(3)} := \left\{ (P_1, P, P_2, Q_1, Q, Q_2) \in \mathcal{B} \;\middle|\; \begin{array}{l} \ell(P_1) \leq 1, \ P_2 = [\omega(P)], \\ Q_1 = [\alpha(P)], \ Q = P, \ Q_2 = [\omega(P)] \end{array} \right\}. \qquad \square$$

Many further Bellman conditions can be defined with the concept of B-conditions. Three examples are given in the next definition.

Definition 2.81 Let $\mathcal{G} = (\mathcal{V}, \mathcal{R})$ be a digraph. Let $\mathcal{P} = \mathcal{P}(\mathcal{G})$ or $\mathcal{P} = \mathcal{P}(s)$ where $s \in \mathcal{V}$. Let $H : \mathcal{P} \to (\mathbf{R}, \preceq)$.

We say that H satisfies the *strong Bellman condition of type 4* if for all paths P_1, P, Q and all nodes v' the following is true:
If P is an optimal connection from P_1 to v' and Q is a prefix of P then Q is an optimal connection from $\alpha(Q) = \alpha(P) = \omega(P_1)$ to $\omega(Q)$.
The Bellman property of type 4 is equivalent to the strong Bellman B-condition with the following set $B = B^{(4)} \subseteq \mathcal{B}$; this can be seen by recalling that $v' = \omega(P)$.

$$B^{(4)} := \{ (P_1, P, P_2, Q_1, Q, Q_2) \in \mathcal{B} \mid P_2 = [\omega(P)], \ Q_1 = [\alpha(P)], \ Q_2 = [\omega(Q)], \ Q \leq P \} .$$

We say that H satisfies the *strong Bellman condition of type 5* if the following is true for all paths P_1, P, Q and for all nodes v':
If P is an optimal connection from P_1 to v' and if Q is a prefix of P then Q is an optimal connection from $\alpha(Q)$ to the remaining part Q_2 from $\omega(Q)$ to $\omega(P)$.
This Bellman property is equivalent to the strong Bellman B-condition with the following set $B = B^{(5)} \subseteq \mathcal{B}$; this can be seen by recalling that $v' = \omega(P)$.

$$B^{(5)} := \{ (P_1, P, P_2, Q_1, Q, Q_2) \in \mathcal{B} \mid P_2 = [\omega(P)], \ Q_1 = [\alpha(P)], \ Q \oplus Q_2 = P \} .$$

We say that H satisfies the *strong Bellman condition of type 6* if the following is true for all paths $P_1, P, P_2, Q_1', Q, Q_2'$:
If P is an optimal connection from P_1 to P_2 and if Q is a subpath of P with $P = Q_1' \oplus Q \oplus Q_2'$ then Q is an optimal connection between the paths $P_1 \oplus Q_1'$ and $Q_2' \oplus P_2$, which are generated by removing Q from $P_1 \oplus P \oplus P_2$.

This Bellman property is equivalent to the strong Bellman B-condition with the following set $B = B^{(6)} \subseteq \mathcal{B}$:

$$B^{(6)} := \left\{ (P_1, P, P_2, Q_1, Q, Q_2) \in \mathcal{B} \,\middle|\, \begin{matrix} \text{There are a prefix } Q_1' \leq P \text{ and a suffix } Q_2' \preceq P \text{ with} \\ P = Q_1' \oplus Q \oplus Q_2', \; Q_1 = P_1 \oplus Q_1', \; Q_2 = Q_2' \oplus P_2 \end{matrix} \right\}. \qquad \Box$$

The next remark is a warning about a plausible but false conclusion

Remark 2.82 All B-properties have the following structure:

If $P_1 \oplus P \oplus P_2$ is an H-minimal path then $Q_1 \oplus Q \oplus Q_2$ is an H-minimal path; this is true for all $(P_1, P, P_2, Q_1, Q, Q_2) \in B$.

We replace 'minimal' by 'maximal' and obtain the following condition:

If $P_1 \oplus P \oplus P_2$ is an H-maximal path then $Q_1 \oplus Q \oplus Q_2$ is an H-maximal path; this is true for all $(P_1, P, P_2, Q_1, Q, Q_2) \in B$.

At first sight, these two assertions are equivalent. But in reality, the B-condition about H-minimal paths has no logical relationships to the B-condition about H-minimal paths. The reason is that the properties of H-minimal paths have nothing to do with the properties of H-maximal paths.

For example, if H has the prefix oriented Bellman property of type 0 then each prefix Q of a minimal path P is minimal as well; but that does not automatically mean that all prefixes Q' of an H-maximal path P' must be maximal. $\qquad \Box$

2.4.2 Introduction to weak Bellman principles for path functions

Weak Bellman principles are existential versions of strong ones. For example, let v, w be two nodes of a digraph. The strong prefix oriented Bellman principle of type 0 says that each optimal v-w-path P is a Bellman path. At first sight, the corresponding weak Bellman principle should say that there exists a Bellman path from v to w. But then the following situation is possible: H has the strong prefix oriented Bellman property of type 0, and there exists no Bellman path from v to w so that H would not have the weak prefix oriented Bellman property of type 0. That is in in conflict with the expectation that a strong Bellman principle alway implies a weak one. The following definition avoids this conflict.

Definition 2.83 Given a digraph $\mathcal{G} = (\mathcal{V}, \mathcal{R})$ and a cost function $H : \mathcal{P} \to \mathbf{R}$.

H satisfies *the weak Bellman B-condition* if the following assertion is true for all triples $(P_1, P', P_2) \in ROC$: There exists a path P with $H(P_1 \oplus P \oplus P_2) \equiv H(P_1 \oplus P' \oplus P_2)$ such that for all triples (Q_1, Q, Q_2),

$$(P_1, P, P_2, Q_1, Q, Q_2) \in B \implies (Q_1, Q, Q_2) \in ROC. \qquad (2.42) \qquad \Box$$

We now compare the strong B-conditions with the weak ones.

Remark 2.84 The weak Bellman B-principle with the additional condition $P = P'$ is equivalent to the strong Bellman B-principle. Consequently, any strong Bellman principle is at least as strong as the weak version.

The strong and the weak B-condition of Bellman are even equivalent if for all paths P_1, P_2, exactly one path P' exists with $(P_1, P', P_2) \in ROC$. This situation occurs if $\preceq$ is total and the restriction $H|_{\mathcal{P}(v,w)}$ is $\equiv$-injective for all nodes $v, w \in \mathcal{V}$. In this case, all values $H(P_1 \oplus P' \oplus P_2)$ and $H(P_1 \oplus P'' \oplus P_2)$ can be compared with each other; if both of them are H-minimal then $H(P_1 \oplus P' \oplus P_2) \equiv H(P_1 \oplus P'' \oplus P_2)$; this and the $\equiv$-injectivity imply that $P' = P''$. □

We now give an exact definition of the weak Bellman property of type i where $i = 0, \ldots, 3$.

Definition 2.85 Given a digraph $\mathcal{G}$ and a cost function $H : \mathcal{P} \to (\mathbf{R}, \preceq)$. Let $i \in \{0, 1, 2, 3\}$.

We say that H satisfies the *weak prefix oriented, infix oriented*, and *suffix oriented Bellman condition of type* i, respectively, if H satisfies the weak Bellman $B^{(i)}_{\preceq}$-, $B^{(i)}_{\subseteq}$-, and $B^{(i)}_{\preceq}$-condition. □

We now give explicit formulations of the weak Bellman conditions of type 0 and 2. It turns out that the weak infix oriented Bellman condition of type 0 is equivalent to condition 2 and condition 3 formulated in the beginning of this subsection.

Theorem 2.86
a) *Given a cost function H. Then the following assertions are equivalent:*
  (a1) *H satisfies the weak prefix, infix, and suffix oriented Bellman condition of type 0, respectively.*
  (a2) *The following is true for all nodes v, w:*
     If there exists an H-minimal v-w path P' then there exists a v-w-path P with $H(P) \equiv H(P')$ such that all prefixes [infixes, suffixes, resp.] of P are H-minimal connections from $\alpha(Q)$ to $\omega(Q)$.
b) *The following assertions are equivalent:*
  (b1) *H satisfies the weak prefix, infix, and suffix oriented Bellman condition of type 2, respectively.*
  (b2) *The following is true for all paths P_1, P', P_2:*
     If P' is an H-minimal connection of P_1 and P_2 and if P' is the prefix [infix, suffix, resp.] of $P_1 \oplus P' \oplus P_2$ then there exists a path P with $H(P_1 \oplus P \oplus P_2) \equiv H(P_1 \oplus P' \oplus P_2)$ such that P is also an optimal connection between the nodes $\alpha(P) = \alpha(P')$ and $\omega(P) = \omega(P')$.

P r o o f   o f   a) :   First, we show (a1)⇒(a2) and assume that assertion (a1) be true. Let P' be an optimal connection from v to w. Then $(P_1 \oplus P' \oplus P_2) \in ROC$ for $P_1 = [\alpha(P')] = [v]$ and $P_2 = [\omega(P')] = [w]$. Assertion (a1) yields an optimal connection P' from P_1 to P_2 such that (2.42) is true; it follows from the definition of the relations $B_{\leq}^{(0)}$, $B_{\subseteq}^{(0)}$ and $B_{\preceq}^{(0)}$ that P has the property formulated in (a2).

Next, we show (a2)$\Rightarrow$(a1) and assume that (a2) be true for one of the sets $B \in \left\{ B_{\leq}^{(0)}, B_{\subseteq}^{(0)}, B_{\preceq}^{(0)} \right\}$. Let $(P_1, P', P_2) \in ROC$. We must show that there exists an optimal path P from P_1 to P_2 such that assertion (2.42) is true for all triples (Q_1, Q, Q_2). First, we consider the case that not $\left(P_1 = [\alpha(P')] \wedge P_2 = [\omega(P')] \right)$. Then (2.42) is trivially true for $P = P'$ because the definition of B implies that the premise of (2.42) false. The other case is that $P_1 = [\alpha(P')]$ and $P_2 = [\omega(P')]$. Then $(P_1, P', P_2) \in ROC$ implies that P' is an H-minimal path from $v := \alpha(P')$ to $w := \omega(P')$. The existence of the desired path P follows immediately from (a2).

P r o o f   o f   b) : First, we prove (b1)$\Rightarrow$(b2) and assume that (b1) be true. If P' is an H-minimal connection from P_1 to P_2 then $(P_1, P', P_2) \in ROC$. Assertion (b1) yields a path P with $H(P_1 \oplus P \oplus P_2) \equiv H(P_1 \oplus P' \oplus P_2)$ and with property (2.42). Then the definition of the relations $B_{\leq}^{(2)}$, $B_{\subseteq}^{(2)}$, and $B_{\preceq}^{(2)}$ imply that P has the properties described in (b2).

Next, we prove the direction (b2)$\Rightarrow$(b1). Let (b2) be true for one of the sets $B \in \left\{ B_{\leq}^{(2)}, B_{\subseteq}^{(2)}, B_{\preceq}^{(2)} \right\}$. Moreover, let $(P_1, P', P_2) \in ROC$. Then P' is an optimal connection from P_1 to P_2. The definition of $B = B_{\leq}^{(2)}$, $B = B_{\subseteq}^{(2)}$, and $B = B_{\preceq}^{(2)}$ implies that P' is a prefix, infix, and a suffix of $P_1 \oplus P \oplus P_2$, respectively. Let P be the path described in (b2). Then $H(P_1 \oplus P \oplus P_2) \equiv H(P_1 \oplus P' \oplus P_2)$. It remains to be proven that for all triples (Q_1, Q, Q_2) the following is true:

If $(P_1, P, P_2, Q_1, Q, Q_2) \in B$ then $(Q_1, Q, Q_2) \in ROC$.

Let the premise of this assertion be true. Then the definition of B implies that Q_1 consists of the start node of $Q = P$ and that Q_2 consists of the end node of $Q = P$. This and (b2) imply that $Q = P$ is an optimal condition from Q_1 to Q_2 so that $(Q_1, Q, Q_2) \in ROC$. $\square$

By the way, the path function in Remark 4.207 does not have the weak prefix oriented Bellman property of type 0.

2.4.3   Comparing Bellman principles with each other

In this subsection, we investigate whether or not a given Bellman condition is stronger than another one. We mainly focus on the strong B-conditions. We start with several obvious observations.

Theorem 2.87 *Given the 6-ary relations $B, \widetilde{B}$ with $B \subseteq \widetilde{B}$. Then the strong [weak] $\widetilde{B}$-condition implies the strong [weak] B-condition.*

P r o o f : To show the strong or the weak B-property we must prove that for particular 6-tuples $\tau := (P_1 \widehat{P}, P_2, Q_1, \widehat{Q}, Q_2) \in B$, the path $\widehat{Q}$ is optimal; this can be done by considering τ as an element of $\widetilde{B}$ and using the strong and the weak Bellman $\widetilde{B}$-condition, respectively. $\square$

This theorem implies many relationships between B- and $\tilde{B}$-conditions where B and $\tilde{B}$ are sets of the form $B_\leq^{(i)}$, $B_\subseteq^{(i)}$, $B_\preceq^{(i)}$ $(i = 0, 1, 2, 3)$ or of the form $B^{(i)}$ $(i = 4, 5, 6)$.

Corollary 2.88

a) If $i = 0, 1, 2, 3$ then the strong [weak] Bellman $B_\subseteq^{(i)}$-property implies the strong [weak] Bellman $B_\leq^{(i)}$-property.

b) If $i = 0, 1, 2, 3$ then the strong [weak] Bellman $B_\subseteq^{(i)}$-property implies the strong [weak] Bellman $B_\preceq^{(i)}$-property.

c) If $\rho \in \{\leq, \subseteq, \preceq\}$ then the strong [weak] $B_\rho^{(0)}$-property implies the strong [weak] $B_\rho^{(1)}$-property.

d) If $\rho \in \{\leq, \subseteq, \preceq\}$ then the strong [weak] $B_\rho^{(2)}$-property implies the strong [weak] $B_\rho^{(3)}$-property.

e) For all $i = 4, 5$, the following is true: The strong [weak] $B^{(i)}$-condition implies the strong [weak] $B_\preceq^{(2)}$ condition.

P r o o f   o f   a) : Let $B := B_\leq^{(i)}$ and $\tilde{B} := B_\subseteq^{(i)}$. Then $B \subseteq \tilde{B}$ because a prefix of a path is a special version of an infix. Our assertion follows from Theorem 2.87.

P r o o f   o f   b) : Let $B := B_\preceq^{(i)}$ and $\tilde{B} := B_\subseteq^{(i)}$. Then $B \subseteq \tilde{B}$ because a suffix of a path is a special version of an infix. Our assertion follows from Theorem 2.87.

P r o o f   o f   c) : Let $\rho \in \{\leq, \subseteq, \preceq\}$; then let $B := B_\rho^{(1)}$ and $\tilde{B} := B_\rho^{(0)}$. Then $B \subseteq \tilde{B}$ because for all tuples $(P_1, P, P_2, Q_1, Q, Q_2) \in B_\rho^{(1)}$, the subpath $Q \rho P$ has the additional property that Q is obtained by removing at most one arc at the beginning of P and at most one arc at the end of P. Since $B \subseteq \tilde{B}$ we may apply Theorem 2.87 to obtain the desired assertions.

P r o o f   o f   d) : Let $\rho \in \{\leq, \subseteq, \preceq\}$; then let $B := B_\rho^{(2)}$ and $\tilde{B} := B_\rho^{(3)}$. Then $B \subseteq \tilde{B}$ because the definition of $B_\rho^{(3)}$ requires that $\ell(P_1) \leq 1$ or that $\ell(P_2) \leq 1$; this condition does not appear in the definition of $B_\rho^{(2)}$. As $B \subseteq \tilde{B}$ we may apply Theorem 2.87 to show our assertions.

P r o o f   o f   e) : Let $i \in \{4, 5\}$. Let $B := B_\preceq^{(2)}$ and $\tilde{B} := B^{(i)}$. Then $B \subseteq \tilde{B}$.
This relationship is obvious if $i = 4$ because the the conditions $Q_2 = [\omega(P)]$ and $Q = P$ in the definition of $B_\preceq^{(2)}$ imply the condition $Q_2 = [\omega(Q)]$ and $Q \leq P$ in the definition of $B^{(4)}$.
If $i = 5$ we observe that the conditions $Q_2 = [\omega(P)]$ and $Q = P$ in the definition of $B_\preceq^{(2)}$ imply that $Q \oplus Q_2 = P$; hence, $B = B_\preceq^{(2)} \subseteq B^{(5)} = B$
Hence, $B \subseteq \tilde{B} = B^{(i)}$ for each $i = 4, 5$. Therefore, we may apply Theorem 2.87. □

Next, we show that the strong prefix oriented Bellman property of type 0 is sometimes weaker than the corresponding infix oriented condition.

Theorem 2.89 *In general, the strong $B_\leq^{(0)}$-property implies neither the strong $B_\subseteq^{(0)}$-property nor the strong $B_\preceq^{(0)}$-property.*

P r o o f :   We show the claim with the help of the graph $\mathcal{G} = (\mathcal{V}, \mathcal{R})$ with $\mathcal{V} := \{s, x_1, x_2, y, z\}$ and the following arcs:

$$r_1 := (s, y), \ r_1' := (s, x_1), \ r_1'' := (x_1, y), \quad r_2 := (y, z), \ r_2' := (y, x_2), \ r_2'' := (x_2, z).$$

We define $p(v) := 10$ for all nodes v, and we define

$$\phi_{r_1}(\zeta) := \zeta - 6, \quad \phi_{r_1'}(\zeta) := \zeta - 3, \quad \phi_{r_1''}(\zeta) := \zeta - 2,$$
$$\phi_{r_2}(\zeta) := 2 \cdot \zeta, \quad \phi_{r_2'}(\zeta) := \zeta, \quad \phi_{r_2''}(\zeta) := \zeta + 7.$$

We define $H : \mathcal{P}(\mathcal{G}) \to \mathbb{R}$ as $H := \phi \circ p$. Then we have the situation as described in Remark 2.50 b), so that H is $<$-preserving. Theorem 2.95 says that H has the strong prefix oriented Bellman property of type 0.

Next, we show that H does not satisfy the corresponding suffix oriented condition. For this purpose, we compute the H-values of all s-z-paths and of all y-z-paths.

$$
\begin{array}{rclcl}
H(r_1 \oplus r_2) & = & 2 \cdot (\mathbf{10} - 6) & = & 8, \\
H(r_1 \oplus r_2' \oplus r_2'') & = & (\mathbf{10} - 6) + 7 & = & 11, \\
H(r_1' \oplus r_1'' \oplus r_2) & = & 2 \cdot ((\mathbf{10} - 3) - 2) & = & 10, \\
H(r_1' \oplus r_1'' \oplus r_2' \oplus r_2'') & = & ((\mathbf{10} - 3) - 2) + 7 & = & 12, \\
H(r_2) & = & 2 \cdot 10 & = & 20, \\
H(r_2' \oplus r_2'') & = & 10 + 7 & = & 17.
\end{array}
$$

It follows that $r_1 \oplus r_2$ is the optimal connection from s to z; the optimal y-z-path, however, is not the suffix $r_2 \preceq r_1 \oplus r_2$ but the path $r_2' \oplus r_2''$. Hence, H does not have the strong suffix oriented Bellman property of type 0.

This and Corollary 2.88 a) imply that H does not satisfy the corresponding infix oriented condition. $\square$

The next result about the structure of ROC is useful for our further comparisons of L-conditions.

Lemma 2.90 (ABCDE-Lemma) *Given a path of the form* $A \oplus B \oplus C \oplus D \oplus E$. *Then* $(A, B \oplus C \oplus D, E) \in ROC \Rightarrow (A \oplus B, C, D \oplus E) \in ROC$.

This result is illustrated in the following sketch:

P r o o f : We show the following implication:

$$(A \oplus B, C, D \oplus E) \notin ROC \Rightarrow (A, B \oplus C \oplus D, E) \notin ROC.$$

Let $(A \oplus B, C, D \oplus E) \notin ROC$. Then there exists a path C' with

$$H((A \oplus B) \oplus C' \oplus (D \oplus E)) \prec H((A \oplus B) \oplus C \oplus (D \oplus E)).$$

This and the associativity of $\oplus$ yield the assertion $H(A \oplus (B \oplus C' \oplus D) \oplus E) \prec H(A \oplus (B \oplus C \oplus D) \oplus E)$. Consequently, $B \oplus C' \oplus D$ is a better connection from A to E than $B \oplus C \oplus D$. Hence, $(A, B \oplus C \oplus D, E) \notin ROC$. $\square$

Next, we show that the strong prefix, infix and suffix oriented Bellman property of type 2 is stronger than the corresponding property of type 0; moreover, the strong Bellman conditions of type 3 are stronger than the corresponding conditions of type 1. We use the ABCDE-Lemma in the proof of these assertions.

Theorem 2.91

a) *For all cost functions H and all $\rho \in \{\leq, \subseteq, \preceq\}$, the following is true:*
If H has the strong Bellman $B_\rho^{(2)}$-property then H has the strong Bellman $B_\rho^{(0)}$-property.

b) *For all cost functions H and all $\rho \in \{\leq, \subseteq, \preceq\}$, the following is true:*
If H has the strong Bellman $B_\rho^{(3)}$-property then H has the strong Bellman $B_\rho^{(1)}$-property.

c) *None of the conclusions in a) and b) may be reversed. More precisely, let $\rho \in \{\leq, \subseteq, \preceq\}$ and $i \in \{0,1\}$. Then the strong $B_\rho^{(i)}$-condition does not always imply the strong $B_\rho^{(i+2)}$-condition.*

P r o o f   o f   a) : Let H satisfy the strong $B_\rho^{(2)}$-condition. We show that H has the strong Bellman $B_\rho^{(0)}$-property.

Let P_1, P, P_2, Q_1, Q, Q_2 be six paths with the following properties: $(P_1, P, P_2) \in ROC$ and $(P_1, P, P_2, Q_1, Q, Q_2) \in B_\rho^{(0)}$. We must derive that $(Q_1, Q, Q_2) \in ROC$.

The definition of $B_\rho^{(0)}$ implies that for all $\rho \in \{\leq, \subseteq, \preceq\}$, the following is true:
$$P_1 = [\alpha(P)], \quad P_2 = [\omega(P)], \quad Q_1 = [\alpha(Q)], \quad Q_2 = [\omega(Q)]. \tag{2.43}$$

Next, we define the paths $\widehat{P} := \widehat{Q} := Q$, and we choose the paths $\widehat{P}_1$, $\widehat{P}_2$ such that $\widehat{P}_1 \oplus \widehat{P} \oplus \widehat{P}_2 = P$. The following figure shows the position of these paths in the infix oriented case; that means that ρ equals $\subseteq$. (If ρ is equal to $\leq$ then $\widehat{P}_1 = P_1 = Q_1 = [\alpha(P)]$, and if ρ is equal to $\preceq$ then $\widehat{P}_2 = P_2 = Q_2 = [\omega(P)]$.)

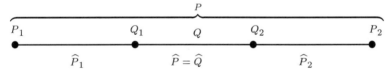

Now we define $A := P_1$, $B := \widehat{P}_1$, $C := \widehat{P} = \widehat{Q} = Q$, $D := \widehat{P}_2$, $E := P_2$.

Then $B \oplus C \oplus D = P$ and our assumption implies that $(A, B \oplus C \oplus D, E) = (P_1, P, P_2) \in ROC$. It follows from the ABCDE-Lemma that $(A \oplus B, C, D \oplus E) \in ROC$; that means that $(\widehat{P}_1, \widehat{P}, \widehat{P}_2) \in ROC$.

Moreover, the facts $\widehat{Q} = \widehat{P} = Q$, $Q_1 \overset{(2.43)}{=} [\alpha(Q)] = [\alpha(\widehat{P})]$ and $Q_2 \overset{(2.43)}{=} [\omega(Q)] = [\omega(\widehat{P})]$ imply the following relationship: $(\widehat{P}_1, \widehat{P}, \widehat{P}_2, Q_1, \widehat{Q}, Q_2) \in B_\rho^{(2)}$.

We have assumed the strong $B_\rho^{(2)}$-property of H, and we have seen that $(\widehat{P}_1, \widehat{P}, \widehat{P}_2) \in ROC$ and that $(\widehat{P}_1, \widehat{P}, \widehat{P}_2, Q_1, \widehat{Q}, Q_2) \in B_\rho^{(2)}$. So, we may conclude that $(Q_1, \widehat{Q}, Q_2)$ lies in ROC. Then the desired relationship $(Q_1, Q, Q_2) \in ROC$ follows from $\widehat{Q} = Q$.

Next, we give a less formal proof of assertion a). We use the paths defined above, and we focus on the infix oriented case. The strong $B_\subseteq^{(2)}$-property says:

$\widehat{P} = \widehat{Q}$ is an optimal connection from $\alpha(\widehat{P})$ to $\omega(\widehat{P})$ if $\widehat{P} = \widehat{Q}$ is an optimal connection from $\widehat{P}_1$ to $\widehat{P}_2$ where $\widehat{P}_1$ and $\widehat{P}_2$ are arbitrary paths.

The strong $B_{\subseteq}^{(0)}$-property can be formulated as follows:

$\widehat{P} = \widehat{Q}$ is an optimal connection from $\alpha(Q)$ to $\omega(Q)$ if $\widehat{P} = \widehat{Q}$ is an optimal connection from $\widehat{P}_1$ to $\widehat{P}_2$ where $\widehat{P}_1$ is a prefix and $\widehat{P}_2$ is a suffix of an optimal connection P from $\alpha(P)$ to $\omega(P)$.

Roughly speaking, the $B_{\subseteq}^{(2)}$-condition means the optimality of $\widehat{P} = \widehat{Q}$ if $\widehat{P}_1$ and $\widehat{P}_2$ are arbitrary paths; the $B_{\subseteq}^{(2)}$-condition means the optimality of $\widehat{P} = \widehat{Q}$ only if $\widehat{P}_1$ and $\widehat{P}_2$ are parts of an optimal path. Consequently, if H has the strong $B_{\subseteq}^{(2)}$-property, then H has also the strong $B_{\subseteq}^{(0)}$-property.
This is mainly the same argumentation as in [Snie86], page 166.

P r o o f o f b) : Our argumentation is analogous to the one in part a); the main difference is now the paths $\widehat{P}_1$ and $\widehat{P}_2$ in part a) have at most one arc.

We assume that H satisfies the strong $B_\rho^{(3)}$-condition. We show that H has the strong Bellman $B_\rho^{(1)}$-property.
Let P_1, P, P_2, Q_1, Q, Q_2 be six paths with the following properties: $(P_1, P, P_2) \in ROC$ and $(P_1, P, P_2, Q_1, Q, Q_2) \in B_\rho^{(1)}$. We must derive that $(Q_1, Q, Q_2) \in ROC$.

The definition of $B_\rho^{(1)}$ implies that for all $\rho \in \{\leq, \subseteq, \preceq\}$, the following is true:

$$P_1 = [\alpha(P)], \quad P_2 = \omega[(P)], \quad Q_1 = [\alpha(Q)], \quad Q_2 = \omega[(Q)]. \tag{2.44}$$

Let $\widehat{P}_1$, $\widehat{P}$ and $\widehat{P}_2$ be the same paths as in part a). Then the definition of $B_\rho^{(1)}$ implies that $\ell(\widehat{P}_1) \leq 1$ and $\ell(\widehat{P}_2) \leq 1$.

We apply the ABCDE-Lemma in the same way as in part a) and see that $(\widehat{P}_1, \widehat{P}, \widehat{P}_2) \in ROC$.

Moreover, the equations $\widehat{Q} = \widehat{P} = Q$, $Q_1 \overset{(2.44)}{=} [\alpha(Q)] = [\alpha(\widehat{P})]$, and $Q_2 \overset{(2.44)}{=} [\omega(Q)] = [\omega(\widehat{P})]$ imply that $(\widehat{P}_1, \widehat{P}, \widehat{P}_2, Q_1, \widehat{Q}, Q_2) \in B_\rho^{(2)}$; this and $\ell(\widehat{P}_1) \leq 1$, $\ell(\widehat{P}_2) \leq 1$ imply that even $(\widehat{P}_1, \widehat{P}, \widehat{P}_2, Q_1, \widehat{Q}, Q_2) \in B_\rho^{(3)}$.

We have seen that $(\widehat{P}_1, \widehat{P}, \widehat{P}_2) \in ROC$, and that $(\widehat{P}_1, \widehat{P}, \widehat{P}_2, Q_1, \widehat{Q}, Q_2) \in B_\rho^{(3)}$. Moreover, we have assumed that H have the strong $B^{(3)}$-property. So, we may conclude that $(Q_1, \widehat{Q}, Q_2)$ lies in ROC. This implies the claim $(Q_1, Q, Q_2) \in ROC$ because $\widehat{Q} = Q$.

P r o o f o f c) : The following example will show that neither the implication in a) nor the one in b) can be reversed. We only consider the prefix and the infix oriented Bellman properties because the suffix oriented case can be generated by reversing the directions of all arcs.

Let $\mathcal{G} = (\mathcal{V}, \mathcal{R})$ be the same graph as in the proof of 2.89. We use the same names for the arcs of $\mathcal{G}$ but we define other paths and another path function.
Let $X := [s, y]$, and let $X' := [s, x_1, y]$. We define the cost function $H : \mathcal{P}(\mathcal{G}) \to \mathbb{R}$ as follows:

$$H(X) := 3, \quad H(X \oplus r_2) := 60, \quad H(X \oplus r_2') = 10, \quad H(X \oplus r_2' \oplus r_2'') := 30,$$
$$H(X') := 4, \quad H(X' \oplus r_2) := 40, \quad H(X' \oplus r_2') = 20, \quad H(X' \oplus r_2' \oplus r_2'') := 50,$$
$$H(P) := 0 \quad \text{for all other paths.}$$

Next, we show the following assertions are true:

H has the strong infix oriented Bellman property of type 0.

(Consequently, H has the corresponding prefix oriented property.)

(2.45)

H does not have the strong prefix oriented Bellman property of type 3.

(Consequently, H does not have the corresponding infix oriented property.)

(2.46)

We prove (2.45). Let u and u' be two nodes of $\mathcal{G}$, let P be an optimal u-u'-path, and let Q be an infix of P. We must show that Q is optimal as well.
This is true if $H(Q) = 0$ so that we must only consider the case $H(Q) > 0$. Moreover, if $Q = P$ then Q is optimal as well as P. So, the only critical case is that $H(Q) > 0$ and Q is a proper subpath of P. Then Q cannot be an s-z-path because Q is a proper subpath of P and no s-z-path Q' is a proper prefix of any path P'. Therefore, we must only consider the paths $Q \in \{X, X \oplus r_2', X', X' \oplus r_2'\}$ because all other paths have H-value 0 or are s-z-paths.
If $Q = X$ then Q is an optimal s-y-connection, and if $Q = X \oplus r_2'$ then Q is an optimal s-x_2-connection.
The cases $Q = X'$ and $Q = X' \oplus r_2'$, however, do not occur because they are in conflict with the optimality of P; this is seen as follows: If $Q = X'$ then P is one of the paths $X' \oplus r_2$, $X' \oplus r_2'$, and $X' \oplus r_2' \oplus r_2''$. If $Q = X' \oplus r_2'$ then $P = X' \oplus r_2' \oplus r_2''$. None of these paths P is an optimal connection from its start to its end node because $H(X' \oplus r_2) = 40 > 30 = H(X \oplus r_2' \oplus r_2'')$, $H(X' \oplus r_2') = 20 > 10 = H(X \oplus r_2')$, and $H(X' \oplus r_2' \oplus r_2'') = 50 > 30 = H(X \oplus r_2' \oplus r_2'')$.
We have seen the following: If Q is a subpath of an optimal path P then $H(Q) = 0$, $Q = P$ or $Q \in \{X, X \oplus r_2'\}$. In all these cases, Q is an optimal connection from $\alpha(Q)$ to $\omega(Q)$. Hence, H has the strong infix oriented Bellman property of type 0.

Now, we prove fact (2.46). For this end, we define the following paths:

$$U_1 := [s], \quad U := X', \quad U_2 := [y, z], \quad V_1 := [s], \quad V := U = X', \quad V_2 := [y].$$

It is obvious that $(U_1, U, U_2, V_1, V, V_2) \in B_{\leq}^{(3)}$. Moreover, the following is true:

$$H(U_1 \oplus U \oplus U_2) = H(X' \oplus r_2) = 40 < 60 = H(X \oplus r_2) = H(U_1 \oplus X \oplus U_2).$$

Hence, the path $U = X'$ (and not the path X) is the optimal connection from $U_1 = [s]$ to $U_2 = [y, z]$. Consequently, $(U_1, U, U_2) \in ROC$.
On the other hand, the path X (and not the path $V = X'$) is the optimal connection from $\alpha(V) = s$ to $\omega(V) = y$ so that $(V_1, V, V_2) \notin ROC$. That means that H does not have the strong prefix oriented Bellman property of type 3.

Now it is easy to show that the implication in assertion a) must not be reversed. The cost function H in our example has the strong prefix and infix oriented Bellman properties of type 0; this follows from fact (2.45). But H does not have the prefix or the infix oriented Bellman property of type 2. This is a consequence of fact (2.46) and Corollary 2.88 d).

The reversed implication in assertion b) is also false. H has the strong prefix and the infix oriented Bellman property of type 1 by fact (2.45) and 2.88 c); fact (2.46), however, says that H does not have the corresponding properties of type 3. □

Next, we compare the "global" Bellman properties of type 0 and of type 2 with the "local" properties of type 1 and of type 3, respectively. Similar investigations were made in Subsection 2.3.4.5. Our proofs were based on the following idea: We assumed that appending a single arc to the paths P_1, P_2 does not change the order of $H(P_1)$ and $H(P_2)$; then we concluded that we can append several arcs (and consequently a whole path Q) to P_1 and P_2 without changing the order of $H(P_1)$ and $H(P_2)$.

A similar argumentation can be used in the first part but not in the second part of the following result.

Theorem 2.92 *Let $\mathcal{G}$ be a digraph, and let $H : \mathcal{P}(\mathcal{G}) \to \mathbf{R}$. Let $\rho \in \{\leq, \subseteq, \preceq\}$. We define $B^{(i)} := B^{(i)}_{\leq}$ for all $i = 0, 1, 2, 3$.*

Then the following are true:

  a) *The strong $B^{(1)}$-property of H is equivalent to the strong $B^{(0)}$-property of H.*

  b) *The $B^{(3)}$-property of H is weaker than the $B^{(2)}$-property of H.*

P r o o f : Let $i = 0, 1$. Corollary 2.88 c) and d) say that the Bellman properties of type i are at least as strong as the corresponding properties of type $(i + 2)$. So, we must only investigate whether or not type $(i + 2)$ implies type i.

P r o o f o f a) : We assume that H has the strong $B^{(1)}$-property.
Let $(X_1, X, X_2, Y_1, Y, Y_2)$ be an arbitrary 6-tuple in $B^{(0)}$ with $(X_1, X, X_2) \in ROC$. We must show that then $(Y_1, Y, Y_2) \in ROC$.

For this end, we remove one arc after the other from X until we have the path $Y \rho X$. More formally, let $\bar{\rho}$ be the inverse relation to ρ, i.e.,

$$(\forall A, B \in \mathcal{P}(\mathcal{G})) \quad A \bar{\rho} B \quad :\Longleftrightarrow \quad B \rho A .$$

We choose a sequence $U^{(0)}, \ldots, U^{(k)}$ of paths that satisfies the following conditions:

$$X \;=\; U^{(0)} \,\bar{\rho}\, U^{(1)} \,\bar{\rho}\, \ldots \,\bar{\rho}\, U^{(k)} \;=\; Y . \tag{2.47}$$

For all $\kappa = 1, \ldots, k$, the following is true: Either $U^{(\kappa)} = U^{(\kappa-1)}$, or $U^{(\kappa)}$ is generated by removing exactly one arc from $U^{(\kappa-1)}$. (For example, the case $X = U^{(0)} = \ldots = U^{(k)} = Y$ is possible.) $\tag{2.48}$

Moreover, we define the paths $U_1^{(\kappa)} := [\alpha(U^{(\kappa)})]$ and $U_2^{(\kappa)} := [\omega(U^{(\kappa)})]$ for all $\kappa = 0, \ldots, k$. The assumption $(X_1, X, X_2, Y_1, Y, Y_2) \in B^{(0)}$ and fact (2.47) imply that

$$
\begin{array}{llll}
U_1^{(0)} \stackrel{(2.47)}{=} [\alpha(X)] = X_1, & U_2^{(0)} \stackrel{(2.47)}{=} [\omega(X)] = X_2, & \\
U_1^{(k)} \stackrel{(2.47)}{=} [\alpha(Y)] = Y_1, & U_2^{(k)} \stackrel{(2.47)}{=} [\omega(Y)] = Y_2.
\end{array}
\tag{2.49}
$$

The following fact is a consequence of (2.47) and (2.48):

$$(\forall \kappa = 1, \ldots, k) \quad \left(U_1^{(\kappa-1)}, U^{(\kappa-1)}, U_2^{(\kappa-1)}, U_1^{(\kappa)}, U^{(\kappa)}, U_2^{(\kappa)} \right) \in B^{(1)} . \tag{2.50}$$

Now we show that $\left(U_1^{(\kappa)}, U^{(\kappa)}, U_2^{(\kappa)}\right) \in ROC$ for all $\kappa = 0, \ldots, k$.

Let $\kappa = 0$; then $\left(U_1^{(\kappa)}, U^{(\kappa)}, U_2^{(\kappa)}\right) = \left(U_1^{(0)}, U^{(0)}, U_2^{(0)}\right) \in ROC$ because $U_1^{(0)} \overset{(2.49)}{=} X_1$, $U^{(0)} \overset{(2.47)}{=} X$, $U_2^{(0)} \overset{(2.49)}{=} X_2$, and because we have assumed that $(X_1, X, X_2) \in ROC$.

We now conclude from $\kappa - 1$ to κ. Let $\left(U_1^{(\kappa-1)}, U^{(\kappa-1)}, U_2^{(\kappa-1)}\right) \in ROC$. Then fact (2.50) and the assumed $B^{(1)}$-property of H imply that $\left(U_1^{(\kappa)}, U^{(\kappa)}, U_2^{(\kappa)}\right) \in ROC$.

We have shown that $\left(U_1^{(\kappa)}, U^{(\kappa)}, U_2^{(\kappa)}\right) \in ROC$ for all κ. This implies that $(Y_1, Y, Y_2) \overset{(2.47),(2.49)}{=} \left(U_1^{(k)}, U^{(k)}, U_2^{(k)}\right) \in ROC$.

P r o o f   o f   b) :   We only consider the prefix and the infix oriented case; the suffix oriented assertion can be obtained by inverting the directions of all arcs in the following example and by translating the proof for the prefix oriented case into the new situation.

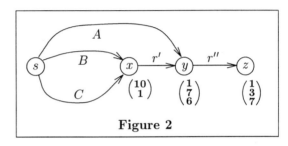

Figure 2

Let $\mathcal{G}$ be the digraph in Figure 2. The objects A, B, and C are paths in $\mathcal{G}$. We define the cost function $H : \mathcal{P}(\mathcal{G}) \to \mathbb{R}$ as follows:

$$H(A) := 1, \quad H(A \oplus r'') = 1,$$
$$H(B) := 10, \quad H(B \oplus r') := 7, \quad H(B \oplus r' \oplus r'') := 3,$$
$$H(C) := 1, \quad H(C \oplus r') := 6, \quad H(C \oplus r' \oplus r'') := 7,$$
$$H(P) := 0 \text{ for all other paths.}$$

The nodes x, y, and z in Figure 2 are marked with the columns $\begin{pmatrix} H(B) \\ H(C) \end{pmatrix}$, $\begin{pmatrix} H(A) \\ H(B \oplus r') \\ H(C \oplus r') \end{pmatrix}$, and $\begin{pmatrix} H(A \oplus r'') \\ H(B \oplus r' \oplus r'') \\ H(C \oplus r' \oplus r'') \end{pmatrix}$, respectively.

Next, we show the following fact:

H has the strong infix oriented Bellman property of type 3. $\quad$ (2.51)

(Consequently, H has the corresponding prefix oriented property.)

Let X, P_1, P, P_2 be four paths with $X := P_1 \oplus P \oplus P_2$, $\ell(P_1)$, $\ell(P_2) \leq 1$, and $(P_1, P, P_2) \in ROC$. We must show that P is an optimal connection from its start node to its end node.

If $\alpha(P) \neq s$, then $H(P) = 0$ so that P is an optimal path from $\alpha(P)$ to $\omega(P)$.

So, we must only consider the case $\alpha(P) = s$. Observing that s has no predecessor in $\mathcal{G}$, we obtain

$$P_1 = [s], \text{ and consequently, } P \leq X \text{ and } X = P \oplus P_2. \quad (2.52)$$

If the endnode of X is not in $\{x, y, z\}$ then $\omega(X)$ is situated on one of the paths A, B or C; hence, X is a proper prefix of A, B or C, and the same is true for $P \leq X$. Then $H(P) = 0$, and P is optimal.

Next, we consider the remaining critical cases.

Case 1: $\omega(X) = x$, $X = B$.
We show that P is a proper prefix of $X = B$. Fact (2.52) says that P is a prefix of X so that we must only refute the case $P = X$.
If $P = X$ then $P_2 = [\omega(X)] = x$ because $P_1 \oplus P \oplus P_2 = X$. Moreover, (2.52) says that $[s] = P_1$. Consequently, $(P_1, P, P_2) = ([s], B, [x])$. This and the assumption $(P_1, P, P_2) \in ROC$ would imply that $([s], B, [x]) \in ROC$. But this is not true because $H(C) < H(B)$.
Hence, P is a proper prefix of $X = B$; then $H(P) = 0$ so that P is indeed an optimal path from $\alpha(P)$ to $\omega(P)$.

Case 2: $\omega(X) = x$, $X = C$.
If $P = X$ then $P = X = C$ an optimal path from $s = \alpha(P)$ to $x = \omega(P)$. If, however, P is a proper prefix of $X = C$ then $H(P) = 0$, and P is also optimal.

Case 3: $\omega(X) = y$, $\ell(P_2) = 0$.
Then $P_2 = [\omega(X)] = [y]$ and $(P_1, P, P_2) \overset{(2.52)}{=} ([s], P, [y])$. This and the assumption $(P_1, P, P_2) \in ROC$ imply that P is an optimal s-y-connection.

Case 4: $\omega(X) = y$, $\ell(P_2) = 1$.
Then P_2 consists of r' or of the last arc of A.
We first consider the case $P_2 = [x, y]$. The assumption $([s], P, [x, y]) \in ROC$ is only satisfied for the path $P = C$ because there are exactly two candidates $P = B$ and $P = C$, and they have the property that $H([s] \oplus B \oplus r') = 7 > 6 = H([s] \oplus C \oplus r')$. Then the path $P = C$ is also an optimal connection from $\alpha(P) = s$ to $\omega(P) = x$.

If P_2 consists of the last arc of A then $X \overset{(2.52)}{=} P \oplus P_2$ implies that P is a proper prefix of $X = A$. Hence, $H(P) = 0$ so that P is an optimal path from $\alpha(P)$ to $\omega(P)$.

Case 5: $\omega(X) = z$, $\ell(P_2) = 0$.
Then $P_2 = [\omega(X)] = [z]$. Hence, $(P_1, P, P_2) = ([s], P, [z])$. This and the assumption $(P_1, P, P_2) \in ROC$ imply that P is an optimal path from $s = \alpha(P)$ to $z = \omega(P)$.

Case 6: $\omega(X) = z$, $\ell(P_2) = 1$.
Then P_2 consists of r'', i.e., $P_2 = [y, z]$. The assumption $([s], P, [y, z]) = (P_1, P, P_2) \in ROC$ is only true for $P = A$. To see this we consider the following list of H-values:

     If $P = A$ then $H(P_1 \oplus P \oplus P_2) = H(A \oplus r'') = 1$,
     If $P = B \oplus r'$ then $H(P_1 \oplus P \oplus P_2) = H(B \oplus r' \oplus r'') = 3$,
     If $P = C \oplus r'$ then $H(P_1 \oplus P \oplus P_2) = H(C \oplus r' \oplus r'') = 7$.

This list shows that $H(P_1 \oplus P \oplus P_2)$ is minimal if and only $P = A$.

This path $P = A$ is also the optimal connection from $s = \alpha(P)$ to $y = \omega(P)$.

We have shown that H has the strong infix (and consequently prefix) oriented Bellman property of type 3.

On the other hand, H does not have the strong prefix or even the infix oriented Bellman property of type 2. To see this we consider the paths $P_1 := [s]$, $\quad P := B$, $\quad P_2 :=$ $[x, y, z]$. Then P the best connection from P_1 to P_2 because the only other competitor C has the property that $H(P_1 \oplus C \oplus P_2) = 7 > 3 = H(P_1 \oplus P \oplus P_2)$. But $P = B$ is not the best connection from $s = \alpha(P)$ to $x = \omega(P)$ because the competitor C with $H(C) = 1 < 10 = H(B) = H(P)$ is better.

Hence, H does not have the $B_{\leq}^{(2)}$-property. $\qquad\qquad\qquad\qquad\qquad\qquad\qquad$ $\square$

Remark 2.93 The cost function in the proof to 2.92 b) is neither order preserving nor $<$-preserving because $H(C \oplus r') = 6 < 7 = H(B \oplus r')$ and $H(C \oplus r' \oplus r'') = 7 > 3 = H(B \oplus r' \oplus r'')$.

It is an interesting open question whether the strong $B_\rho^{(2)}$- and the strong $B_\rho^{(3)}$-property are equivalent if H is order preserving or $\prec$-preserving. The answer to this question would fit in Subsection 2.4.4 about relationships between order preservation and Bellman principles. $\qquad\qquad\qquad\qquad\qquad\qquad\qquad\qquad\qquad$ $\square$

The next theorem describes a situation in which weak and strong Bellman properties are equivalent.

Theorem 2.94 *Given a digraph $\mathcal{G}$ and a node s in $\mathcal{G}$. Let $H : \mathcal{P}(s) \to (\mathbf{R}, \preceq)$. We assume that the restriction $H|_{\mathcal{P}(s,v')}$ is $\equiv$-injective for each $v' \in V$.*
Then the strong and the weak prefix oriented Bellman properties of type 0 are equivalent for H.
(Analogic results are valid for the infix and the suffix oriented Bellman properties.)

P r o o f : We must show that the weak Bellman property of H implies the strong one. For this purpose, we assume that H has the weak prefix oriented Bellman property of type 0. Let $v \in V$, and let P^* be an H-minimal s-v-path. By the weak Bellman property of H, there exists a Bellman path P^{**} from s to v such that $H(P^{**}) = H(P^*)$. This and the $\equiv$-injectivity of H imply that $P^* = P^{**}$. Consequently, every H-optimal s-v-path P^* is a Bellman path. $\qquad\qquad\qquad\qquad\qquad\qquad\qquad\qquad$ $\square$

2.4.4 Relationships between Bellman properties and principles of order preservation

This subsection is about assertions (A)$\Rightarrow$(B) and (B)$\Rightarrow$(A) where (A) and (B) are assertions of the following form:

    (A) The path function H meets a particular principle of order preservation.
    (B) The path function H meets a particular Bellman principle.

Similar results are given in [Snie86], page 170, in [Mor82] and in other papers, and it is possible that some of our following results can be shown with the help of earlier sources.

The following results are arranged in the following way: The later the theorem appears, the weaker is the Bellman principle formulated in the claim. In particular, we shall prove the following results:

1) If a path function H is $\prec$-preserving with respect to $\mathbb{L}_0$ then H has the strong prefix oriented Bellman property of type 2.

2) If H is $\prec$-preserving with respect to $\mathbb{L}'_{\min}$ then H has the strong prefix oriented Bellman property of type 0.

3) If H is order preserving in the usual sense and if the underlying digraph $\mathcal{G}$ is acyclic then H has the weak prefix oriented Bellman property of type 0.

Theorem 2.95 *Let $\mathcal{G}$ be a digraph, and let s be a node in $\mathcal{G}$. Let $\mathcal{P} \in \{\mathcal{P}(\mathcal{G}), \mathcal{P}(s)\}$. Given a $\prec$-preserving cost function $H : \mathcal{P} \to (\mathbf{R}, \preceq)$.*
Then H has the strong prefix oriented Bellman property of type 2 (and consequently the corresponding property of type 0).[7]

P r o o f : Let P be an optimal prefix of $P \oplus P_2$. We must show that P is an optimal connection from $\alpha(P)$ to $\omega(P)$.

Otherwise, there exists a better path P'; this path has the properties $\alpha(P') = \alpha(P)$, $\omega(P') = \omega(P)$ and $H(P') \prec H(P)$. But then the $\prec$-preservation of H implies that $H(P' \oplus P_2) \prec H(P \oplus P_2)$; hence, P is not an optimal prefix of $P \oplus P_2$, which is a contradiction. □

Our next result is similar to Theorem 1 in [Mor82].

Corollary 2.96 *Let $H = \phi \circ\circ p$ where $p : V \to \mathbf{R}$ and where all functions ϕ_r, $r \in \mathcal{R}$, are strictly increasing. Then H has the strong prefix oriented Bellman properties of type 2 and of type 0.*

P r o o f : Remark 2.50 b) says that H is $\prec$-preserving. The strong prefix oriented Bellman property of type 2 follows from 2.95, and the Bellman property of type 2 implies the one of type 0. □

In our next result, we assume a weaker principle of order preservation than in 2.95. We describe a situation where this weak version of order preservation implies the prefix oriented Bellman property of type 0.

Theorem 2.97 *Given a digraph $\mathcal{G}$ and a cost function $H : \mathcal{P}(\mathcal{G}) \to \mathbf{R}$.*
Let $\preceq$ be total and identitive. (E.g., $\mathbf{R}$ may be equal to $\mathbb{R}$, and "$\preceq$" may be equal to "$\leq$".) Let H have the U-u-minimality property for all paths U and all nodes u (see Definition 2.5). Moreover, we assume that H is $\prec$-preserving with respect to $\mathbb{L}'_{\min}$.[8]
Then H has the strong prefix oriented Bellman property of type 0.

P r o o f : Let $v, v'' \in V$. Let P be an H-minimal v-v''-path. Let Q be a prefix of P with $v' := \omega(Q)$, and let Y be the suffix of P with the property $P = Q \oplus Y$.

---

[7]The proof of 2.89 shows that $\prec$-preservation does not always imply the strong suffix or even infix oriented Bellman property of type 2 or 0.

[8]Theorem 2.59 says that this property is at most as strong as the $\prec$-preservation with respect to $\mathbb{L}_0$.

Suppose that Q is not an optimal connection from v to v'. Then there exists a path $X \in \mathcal{P}(v, v')$ with $H(X) \prec H(Q)$. The assumed extremality property of H implies that a path $Z \in \mathcal{P}_{Min}(X, v'')$ exists. Moreover, $Y \in \mathcal{P}_{Min}(Q, v'')$ because $Q \oplus Y = P$ is an H-minimal v-v''-path. We have assumed that H is $\prec$-preserving with respect to $\mathbb{L}_{min}$. So, $H(X) \prec H(Q)$ implies that $H(X \oplus Z) \prec H(Q \oplus Y) = H(P)$, and that is a contradiction to the optimality of P. $\qquad\square$

Next, we investigate logical relationships between the standard order preservation and Bellman properties. The next example shows that we must not expect very strong results.

Remark 2.98 The standard order preservation does not always imply the strong prefix oriented Bellman property of type 0. For example, let $\mathcal{G} := (\mathcal{V}, \mathcal{R})$, with $\mathcal{V} := \{s, w_1, w_2, x, z\}$ and $\mathcal{R} := \{(s, w_1), (s, w_2), (w_1, x), (w_2, x), (x, z)\}$

Let $H([s, w_i, x]) := i$ and $H([s, w_i, x, z]) := 8$ for all $i = 1, 2$; let $H(P) := 0$ for all other paths. Then H is order preserving. Moreover, $P := [s, w_2, x, z]$ is an H-minimal s-z-path but its prefix $Q := [s, w_2, x]$ is not an optimal s-x-path. Hence, H does not have the strong prefix oriented Bellman property of type 0. $\qquad\square$

On the other hand, we can prove the following positive result:

Theorem 2.99 *Given a finite and acyclic digraph $\mathcal{G}$. Let $H : \mathcal{P}(\mathcal{G}) \to (\mathbf{R}, \preceq)$ be order preserving.*

Then H has the weak prefix oriented Bellman property of type 0.

P r o o f : First, we define the quantity $K(X)$ for optimal paths X. Let X be an optimal connection between two nodes v, w. Let $l := \ell(X)$, and let X_λ be the prefix of X of length $l - \lambda$ ($\lambda = 0, \dots, l$); for example, $X_0 = X$ and $X_l = [v]$. Then $K(X)$ is the maximum of all numbers k with the following property:

All prefixes $X_0 = X$, $X_1, \dots, X_k$ are optimal connections from v to their endpoints.

For example, $K(X) = l$ means that all prefixes of X (namely $X_0, X_1, \dots, [v] = X_l$) are optimal.

(The number $K(X)$ can be computed as follows: Remove the arcs at the end of X one by one until the resulting prefix Y is of length 0 or Y is not any more an optimal path from v to $\omega(Y)$. If Y is of length 0 then $K(X) = l$. If Y is not optimal then $K(X) = \ell(X) - \ell(Y) - 1$.)

The proof of our assertion is illustrated in Figure 3. Let $P \in \mathcal{P}(v, w)$ be an optimal connection from v to w. We must find an optimal v-w-connection P' with the additional property that all prefixes of P' are optimal, too.

Since $\mathcal{G}$ is finite and acyclic there exist only finitely many paths $\widetilde{P} \in \mathcal{P}(v, w)$ with $H(\widetilde{P}) \equiv H(P)$. Among these candidates we choose a path P' of maximal K-value. We assume that P' visits the nodes $v = u_l, u_{l-1}, u_{l-2}, \dots, u_0 = w$ in this order. For all $\lambda = 0, \dots, l$, let P'_λ be the prefix of P' from $v = u_l$ to u_λ.

We show that all prefixes $P'_\lambda \preceq P'$ are optimal paths from v to $\omega(P'_\lambda)$. For this end, we prove that $K(P') = l$.

If not then $i := K(P') < l$. Then the prefix P'_{i+1} is not an optimal path from $v = u_l$ to $\omega(P'_{i+1}) = u_{i+1}$. As $\mathcal{G}$ is finite and acyclic there are only finitely many u_l-u_{i+1}-paths. Hence, there exists an H-minimal path P''_{i+1} from $v = u_l$ to u_{i+1}.

The assumption that P'_{i+1} is not optimal implies that $H(P''_{i+1}) \prec H(P'_{i+1})$.

For all $\lambda = i + 1, i, i - 1, \ldots, 0$, let U_λ denote the infix of P' that visits exactly the nodes $u_{i+1}, u_i, u_{i-1}, \ldots, u_\lambda$.

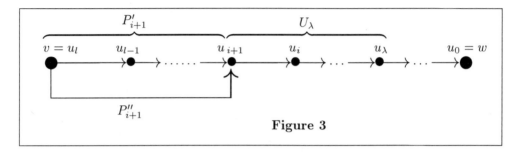

Figure 3

The assumption that $H(P''_{i+1}) \prec H(P'_{i+1})$ and the order preservation of H imply that

$$(\forall \lambda = i + 1, i, i - 1, \ldots, 0) \ \ H(P''_{i+1} \oplus U_\lambda) \preceq H(P'_{i+1} \oplus U_\lambda) = H(P'_\lambda). \quad (2.53)$$

We now show the following fact:

All paths $P''_{i+1} \oplus U_\lambda$ with $\lambda = i + 1, i \ldots, 0$ are H-minimal v-u_λ-paths. $\quad (2.54)$

This is true for $\lambda = i + 1$ because P''_{i-1} has been assumed to be an H-minimal v-u_{i-1}-path. Let $\lambda \in \{i, i - 1, \ldots, 0\}$. Then P'_λ is optimal because $\lambda \leq i = K(P')$; the H-minimality of P'_λ and fact (2.53) imply that $P''_{i+1} \oplus U_\lambda$ is optimal, too.

An immediate consequence of fact (2.54) is that $K(P''_{i+1} \oplus U_0) \geq i + 1 > i = K(P')$. That is a contradiction to the maximality of $K(P')$. $\qquad\qquad \square$

In 2.99, we have assumed that $\mathcal{G}$ is finite and that $\mathcal{G}$ is acyclic. We next show that none of these assumptions may be omitted.

Theorem 2.100 *We formulate the following assertions:*

    (A) *The digraph $\mathcal{G}$ is finite but not acyclic.*
    (B) *The digraph $\mathcal{G}$ is acyclic but not finite.*
    (C) *The cost function H has the X-u-extremality property for all paths X and for all nodes u of the digraph $\mathcal{G}$.*
    (D) *The cost function H has the weak prefix oriented Bellman property.*

Then each premise $((A)\wedge(C))$ and $((B)\wedge(C))$ does not always imply assertion (D).

P r o o f : The following example shows that (D) may be false although (A) and (C) are true. Let $\mathcal{G} = (\mathcal{V}, \mathcal{R})$ with $\mathcal{V} := \{x, y, z\}$ and with the arcs $r_0 := (x, y)$, $r_1 := (y, z)$, $r_2 := (z, x)$. We define $H : \mathcal{P}(\mathcal{G}) \to (\mathbb{R}, \leq)$ as follows:

$$H(P) := \begin{cases} 10, & \text{if} & \alpha(P) = x \wedge \ell(P) \leq 1, \\ 5, & \text{if} & \alpha(P) = x \wedge \ell(P) \geq 2, \\ 0, & \text{if} & \alpha(P) \neq x. \end{cases}$$

Let $p(x) := 10$ and $p(y) := p(z) := 0$. To show that H is order preserving we write H as $H = \phi \circ \circ p$ where each arc r is marked with a monotonically increasing function ϕ_r. For this purpose we define $\phi_{r_0} : \xi \mapsto \min(\xi, 10)$ and $\phi_{r_1} := \phi_{r_2} : \xi \mapsto \min(\xi, 5)$; then indeed $H = \phi \circ \circ p$.

Moreover, H has the X-u'-minimality property for all paths X and all nodes u'. Let P be a path such that $X \oplus P$ ends with u'. We must find a path Q_{Min} from $w(X)$ to u' such that $Q^{Min} := X \oplus Q_{Min}$ is optimal among all candidates $X \oplus Q$ with endpoint u'. If $H(X \oplus P)$ is equal to 0 or to 5 then $Q_{Min} := P$ has the desired properties. Let $H(X \oplus P) = 10$. Then $X \oplus P = [x]$ (with $u' = x$) or $X \oplus P = [x, y]$ (with $u' = y$). If $X \oplus P = [x]$ let $Q_{Min} := r_0 \oplus r_1 \oplus r_2$; if $X \oplus P = [x, y]$ let $Q_{Min} := P \oplus r_1 \oplus r_2 \oplus r_0$; in any case, $Q^{Min} = X \oplus Q_{Min}$ has the desired properties.

Nevertheless, H does not have the weak Bellman property of type 0 because all paths from x to y (for example $[x, y, z, x]$) have the prefix $Q := [x, y]$, which is not an optimal x-y-path.

The next example shows that (D) may be false although (B) and (C) are true. Let $\mathcal{G} = (\mathcal{V}, \mathcal{R})$ with $\mathcal{V} := \{s, z, w_1, w_2, w_3, \ldots\}$ and with the arcs $r_i := (s, w_i)$ $(i = 1, 2, 3, \ldots)$, $r'_0 := (w_1, z)$, and $r'_i := (w_{i+1}, w_i)$ $(i = 1, 2, 3, \ldots)$.
We define the following cost function $H : \mathcal{P}(\mathcal{G}) \to (\mathbb{R}, \leq)$:

$$H(P) := \begin{cases} 0, & \text{if} & P \text{ uses one of the arcs } r'_0, r'_1, r'_2, \ldots, \\ 1, & \text{if} & P \text{ does not use one the arcs } r'_0, r'_1, r'_2, \ldots. \end{cases}$$

The H is order preserving because $H = \phi \circ \circ p$ with $p(v) = 1$ $(v \in \mathcal{V})$ and with the monotonically increasing functions $\phi_{r_i} : \xi \mapsto \min(\xi, 1)$ and $\phi_{r'_{i-1}} : \xi \mapsto \min(\xi, 0)$, $i = 1, 2, 3, 4, \ldots$.
Moreover, H has the X-u'-minimality property for all paths X and all nodes u'. Let P be a path such that $X \oplus P$ ends with u'. We must find a path Q_{Min} from $w(X)$ to u' such that $Q^{Min} = X \oplus Q_{Min}$ is optimal among all candidates $X \oplus Q$ with endpoint u'. If $H(X \oplus P) = 0$ then $Q_{Min} := P$ has the desired properties. If $\ell(P) = 0$ then $w(X) = \alpha(P) = w(P) = u'$; this and the fact that $\mathcal{G}$ is acyclic imply that $Q_{Min} := P$ is the only (and consequently optimal) connection from X to u'. The last case is that $H(X \oplus P) = 1$ and $\ell(P) > 0$; then $X = [s]$ and $P = [s, w_i]$ for some i; choose $Q_{Min} := [s, w_{i+1}, w_i]$ in this case; then $H(Q^{Min}) = 0$, and Q^{Min} is optimal.

Nevertheless, H does not have the weak Bellman property of type 0. The reason is that all paths (and consequently all optimal paths) from v to w begin with the prefix $[v, w_i]$ with $H([v, w_i]) = 1$. This prefix is not an optimal v-w_i-path because $H([v, w_{i+1}, w_i]) = 0$. That means: If P is an optimal v-w_i-path then there is not optimal v-w_i-path P' with the additional property that all prefixes of P' are optimal. Hence, H does not have the weak Bellman property of type 0. □

2.4.5   Relationships between Bellman properties and L-minimum preservation

All Bellman conditions are essentially conclusions of the following form: If a path P is optimal then a path Q is optimal. The same type of conclusion is drawn in the definition of minimum preservation (see Subsection 2.3.5). From this observation, the conjecture arises that several Bellman properties and principles of minimum preservation are logically equivalent.

In fact, $\mathbb{L}_0$-minimum preservation is equivalent to the strong prefix oriented Bellman property of type 2. This is shown in 2.103. Moreover, we prove that all B-properties in 2.73 may be considered as (possibly modified) principles of minimum preservation.

We start with a general result, which will be used to show the equivalence of $\mathbb{L}_0$-minimum preservation and the Bellman property of type 2.

Theorem 2.101 *Given a digraph G and a cost function $H : P(G) \to (\mathbf{R}, \preceq)$. Let $\preceq$ be total. Moreover, let H have the X-Y-minimality property for all paths X, Y. We assume that all restrictions $H|_{P(v,\widehat{v})}$ are $\equiv$-injective. Let $B \subseteq \mathcal{B}$ be a set with the following properties:*

(i) For all $(U_1, U, U_2, V_1, V, V_2) \in B$ and all paths $U' \in P(\omega(U_1), \alpha(U_2))$, the following is true: If $U' \neq U$, then a path $V' \neq V$ exists with $(U_1, U', U_2, V_1, V', V_2) \in B$.

(ii) H satisfies the strong Bellman B-condition.

Then H has the strong Bellman property with respect to

$$B^{-1} := \{ (V_1, V, V_2, U_1, U, U_2) \mid (U_1, U, U_2, V_1, V, V_2) \in B \} .$$

P r o o f : Let $(U_1, U, U_2, V_1, V, V_2) \in B$ and let $(V_1, V, V_2) \in ROC$. We must show that $(U_1, U, U_2) \in ROC$.

We assume that this is not the case. Then the X-Y-minimality property yields a path U' with $(U_1, U', U_2) \in ROC$.

This and the assumption that $(U_1, U, U_2) \notin ROC$ imply that $U' \neq U$. Applying assumption (i) we obtain a path V' with

$$(\alpha) \ \ (U_1, U', U_2, V_1, V', V_2) \in B , \qquad (\beta) \ \ V' \neq V . \tag{2.55}$$

We have assumed that $(U_1, U', U_2) \in ROC$, and we have seen that fact $(2.55)(\alpha)$ is true; using assumption (ii), we see that $(V_1, V', V_2) \in ROC$. This and the fact that $\preceq$ is total imply $H(V_1 \oplus V' \oplus V_2) \preceq H(V_1 \oplus V \oplus V_2)$. Hence, $H(V_1 \oplus V' \oplus V_2) \prec H(V_1 \oplus V \oplus V_2)$ because $V' \neq V$ and $H|_{P(v,\widehat{v})}$ is $\equiv$-injective for $v := \alpha(V_1)$ and $\widehat{v} := \omega(V_2)$. The fact that $H(V_1 \oplus V' \oplus V_2) \prec H(V_1 \oplus V \oplus V_2)$ implies that $(V_1, V, V_2) \notin ROC$, and that is a contradiction to the assumption that $(V_1, V, V_2) \in ROC$. $\qquad\qquad \square$

The following remark will be useful in the proof of 2.103.

Remark 2.102   a) Let $\mathcal{P}, \mathcal{Q} \subseteq P(G)$ be two sets of paths with the following property: $U_1 \in \mathcal{P}$ and $V_1 \in \mathcal{Q}$ for all tuples $(U_1, U, U_2, V_1, V, V_2) \in B$.

Then the general X-Y-minimality property in 2.101 may be replaced by the weaker assumption that H has the X-Y-minimality property for all paths $X \in \mathcal{P}$ and $Y \in \mathcal{Q}$.

b) Given the following situation: Firstly, condition (i) in 2.101 is true for B and for B^{-1}; secondly, $\mathcal{P}$, $\mathcal{Q} \subseteq \mathcal{P}(\mathcal{G})$ are two sets of paths such that $U_1 \in \mathcal{P}$ and $V_1 \in \mathcal{Q}$ for all tuples $(U_1, U, U_2, V_1, V, V_2) \in B \cup B^{-1}$; thirdly, H has the X-Y-minimality property for all paths $X \in \mathcal{P}$ and $Y \in \mathcal{Q}$. Then the strong Bellman B- and B^{-1}-property are equivalent. This follows from applying Theorem 2.101 and part a) of this remark to B and to B^{-1}. $\qquad\square$

Corollary 2.103 *Let $\preceq$ be total. Let $H : \mathcal{P}(\mathcal{G}) \to (\mathbf{R}, \preceq)$ have the v-Y-minimality property for all nodes v and all paths Y. We assume that all restrictions $H\big|_{\mathcal{P}(v, \widehat{v})}$ are $\equiv$-injective.*

Then the following assertions are equivalent:
(a) *H is $\mathbb{L}_0$-minimum preserving.*
(b) *H has the strong prefix oriented Bellman property of type 2.*

P r o o f : First, we formulate condition (a) as follows:
$$(\forall\, U, V_2)\ (U \in \mathcal{P}_{Min}([\alpha(U)], [\omega(U)]) \implies U \in \mathcal{P}_{Min}([\alpha(U)], V_2)).$$
Hence, condition (a) is equivalent to the strong Bellman B-condition with the following set B:
$$B := \{(U_1, U, U_2, V_1, V, V_2) \in \mathcal{B} \,|\, U_1 = [\alpha(U)],\ U_2 = [\omega(U)],\ V_1 = [\alpha(V)],\ V = U\}.$$
Then the following is true:
$$B^{-1} = \{(U_1, U, U_2, V_1, V, V_2) \in \mathcal{B} \,|\, U_1 = [\alpha(U)],\ V_1 = [\alpha(V)],\ V_2 = [\omega(V)],\ V = U\}.$$

We observe that $B^{-1} = B^{(2)}_{\preceq}$. Therefore, it is sufficient to show the equivalence of the strong B- and the strong B^{-1}-condition.

For this purpose, we check whether the conditions in 2.101 and in 2.102 b) are satisfied. Assumption (i) in 2.101 is true for each of the sets B and B^{-1}; this can be seen by choosing $V' := U' (\neq U = V)$.

Moreover, if $(U_1, U, U_2, V_1, V, V_2) \in B \cup B^{-1}$ then U_1 is of length 0 so that $U_1 \in \mathcal{P} := \{P \in \mathcal{P}(\mathcal{G}) \,|\, \ell(P) = 0\}$ and $U_2 \in \mathcal{Q} := \mathcal{P}(\mathcal{G})$.

In the formulation of our current result, we have assumed that H has the X-Y-minimality property for all paths $X = [u] \in \mathcal{P}$ and all paths $Y \in \mathcal{Q}$.

Consequently, we may apply Remark 2.102 b); thus, we see that the strong Bellman $B-$ and B^{-1}-conditions are equivalent. $\qquad\square$

We have studied an example of an equivalence between a Bellman B-condition and a principle of minimum preservation.

We shall see that each strong B-condition can be formulated as a principle of minimum preservation with regard to a particular set L_B of matrices; this is the assertion of Theorem 2.106; this result, however, will not imply 2.103 because $\mathbb{L}_0 \neq L_B$ for $B = B^{(2)}_{\preceq}$ (see Example 2.107).

Moreover, we shall formulate the weak B-condition as a so-called weak principle of minimum preservation with regard to the set L_B.

It will turn out that the usual Bellman principles are equivalent to L_B-minimum preservation with a weakly contracting set L_B (see Definition/Remark 2.70); the reason for this property of L_B is that all typical Bellman principle draw a conclusion from the optimality of a path A to the optimality of a subpath $B \subseteq A$. Hence, the following are true:

> A typical order preserving property is a principle of order preservation with respect to a strongly expanding set L.
> A typical Bellman property is a principle of minimum preservation with respect to a weakly contracting set L.

Next we define the set L_B of matrices if $B \subseteq \mathcal{B}$ is given.

Definition 2.104 Let $B \subseteq \mathcal{B}$. Then we define the following set $L_B \subseteq L$ of matrices:

$$
L_B := \left\{ \begin{pmatrix} U_1 & U & U_2 \\ U_1 & U' & U_2 \\ V_1 & V & V_2 \\ V_1 & V' & V_2 \end{pmatrix} \in L \;\middle|\; (U_1, U, U_2, V_1, V, V_2) \in B \right\}. \qquad \square
$$

The next observations are useful for the proof of the equivalence between B-conditions and L_B-minimum preservation.

Introductory Remark 2.105 Let $H : \mathcal{P}(\mathcal{G}) \to (\mathbf{R}, \preceq)$.

a) In Definition 2.53, two sets S_2 and S_4 of paths were defined. Let $B \subseteq \mathcal{B}$. We show that the following is true for all paths U_1, U, U_2, V_1, V, V_2:

$$S_2 = S_2(L_B, U_1, U, U_2, V_1, V, V_2) =$$
$$\begin{cases} S_2' := \{U_1 \oplus U' \oplus U_2 \,|\, U' \in \mathcal{P}(\omega(U_1), \alpha(U_2))\} & \text{if } (U_1, U, U_2, V_1, V, V_2) \in B, \\ \emptyset & \text{if } (U_1, U, U_2, V_1, V, V_2) \notin B. \end{cases}$$

To see this we observe that all matrices in the definition of the set $S_2 =$

$$S_2(L_B, U_1, U, U_2, V_1, V, V_2) \text{ are of the form } \mathbf{Y} = \begin{pmatrix} U_1 & U & U_2 \\ X_{2,1} & X_{2,2} & X_{2,3} \\ V_1 & V & V_2 \\ X_{4,1} & X_{4,2} & X_{4,3} \end{pmatrix}.$$

First, we consider the case that $(U_1, U, U_2, V_1, V, V_2) \in B$.
Then $(U_1, U, U_2, V_1, V, V_2) \in B \subseteq \mathcal{B}$. This and the definition of $\mathcal{B}$ imply that $\alpha(U) = \omega(U_1)$, $\alpha(U_2) = \omega(U)$, $\alpha(V) = \omega(V_1)$, $\alpha(V_2) = \omega(V)$.
Moreover, $S_2' \subseteq S_2$ because each path $U_1 \oplus U' \oplus U_2 \in S_2'$ is an element of S_2; this can be seen by choosing the free parameters $X_{2,j}$ and $X_{4,j}$ as follows:

$$X_{2,1} := U_1, \; X_{2,2} := U', \; X_{2,3} := U_2, \; X_{4,1} := V_1, \; X_{4,2} := V, \; X_{4,3} := V_2;$$

these paths $X_{2,j}$ and $X_{4,j}$ yield a matrix $\mathbf{Y} \in L_B$.
Moreover, $S_2 \subseteq S_2'$ for the following reasons: Let $X_2 = X_{2,1} \oplus X_{2,2} \oplus X_{2,3} \in S_2$. The definition of S_2 yields three paths $X_{4,1}, X_{4,2}, X_{4,3}$ such that the matrix $\mathbf{Y}$ lies in L_B. Consequently, $X_{2,1} = U_1$ and $X_{2,3} = U_2$, and this implies that $X_2 \in S_2'$.

We now consider the case that $(U_1, U, U_2, V_1, V, V_2) \notin B$. Then the matrix $\mathbf{Y}$ is not in L_B, and this is independent of the choice of the paths $X_{2,1}$, $X_{2,2}$, $X_{2,3}$, $X_{4,1}$, $X_{4,2}$, $X_{4,3}$. Hence, S_2 is empty.

An analogous argumentation shows that

$$S_4 = S_4(L_B, U_1, U, U_2, V_1, V, V_2) =$$

$$\begin{cases} S_4' := \{V_1 \oplus V' \oplus V_2 \mid V' \in \mathcal{P}(\omega(V_1), \alpha(V_2))\} & \text{if } (U_1, U, U_2, V_1, V, V_2) \in B, \\ \emptyset, & \text{if } (U_1, U, U_2, V_1, V, V_2) \notin B. \end{cases}$$

b) For all paths $U_1, U, U_2, V_1, V, V_2 \in \mathcal{P}(\mathcal{G})$ with $\alpha(U) = \omega(U_1)$, $\alpha(U_2) = \omega(U)$, $\alpha(V) = \omega(V_1)$, $\alpha(V_2) = \omega(V)$, the following assertions (α) and (β) are equivalent:

(α) $\quad H(U_1 \oplus U \oplus U_2) \in MIN\{H(W) \mid W \in S_2(L_B, U_1, U, U_2, V_1, V, V_2)\}$

(β) $\quad (U_1, U, U_2, V_1, V, V_2) \in B \wedge (U_1, U, U_2) \in ROC$.

First, we show $(\alpha) \Rightarrow (\beta)$. If (α) is true then $S_2 \neq \emptyset$.[9] Then part a) implies that $(U_1, U, U_2, V_1, V, V_2) \in B$ and $S_2 = S_2'$. Hence, S_2 consists of all paths $W = U_1 \oplus U' \oplus U_2$; so, the assertion $(U_1, U, U_2) \in ROC$ is a further consequence of (α).

Now we show $(\beta) \Rightarrow (\alpha)$. Let (β) be true. Then $(U_1, U, U_2, V_1, V, V_2) \in B$. Consequently, $S_2 = S_2'$ by part a). The assertions $S_2 = S_2'$ and $(U_1, U, U_2) \overset{(\beta)}{\in} ROC$ imply assertion (α).

An analogous argumentation shows that the following assertions (γ) and (δ) are equivalent:

(γ) $\quad H(V_1 \oplus V \oplus V_2) \in MIN\{H(W) \mid W \in S_4(L_B, U_1, U, U_2, V_1, V, V_2)\}$

(δ) $\quad (U_1, U, U_2, V_1, V, V_2) \in B \wedge (V_1, V, V_2) \in ROC$. $\qquad\square$

Remark 2.105 is useful in the next result, which shows the relations between generalized strong Bellman properties and principles of minimum preservation.

Theorem 2.106 *Let $H : \mathcal{P}(\mathcal{G}) \to (\mathbf{R}, \preceq)$. Given $B \subseteq \mathcal{B}$. Then the strong Bellman B-condition is equivalent to the minimum preservation with respect to L_B.*

P r o o f : We formulate the strong Bellman B-condition as follows: For all paths U_1, U, U_2, V_1, V, V_2 with $\alpha(U) = \omega(U_1)$, $\alpha(U_2) = \omega(U)$, $\alpha(V) = \omega(V_1)$, $\alpha(V_2) = \omega(V)$,

$$\left[(U_1, U, U_2, V_1, V, V_2) \in B \wedge (U_1, U, U_2) \in ROC\right] \implies (V_1, V, V_2) \in ROC.$$

That means that for all paths an assertion of the form $[a_1 \wedge a_2] \Rightarrow a_3$ is true, and this assertion is equivalent to $[a_1 \wedge a_2] \Rightarrow [a_1 \wedge a_3]$. Therefore, we may write the strong Bellman principle in the following way:

$$[(U_1, U, U_2, V_1, V, V_2) \in B \wedge (U_1, U, U_2) \in ROC] \implies$$
$$[(U_1, U, U_2, V_1, V, V_2) \in B \wedge (V_1, V, V_2) \in ROC]$$

---

[9] Note that $H(U_1 \oplus U \oplus U_2) \in MIN\{H(W) \mid W \in S_2(L_B, U_1, U, U_2, V_1, V, V_2)\} \neq \emptyset$ by condition (α).

for all paths U_1, U, U_2, V_1, V, V_2 with $\alpha(U) = \omega(U_1)$, $\alpha(U_2) = \omega(U)$, $\alpha(V) = \omega(V_1)$, $\alpha(V_2) = \omega(V)$.

We use 2.105 b) to formulate a further condition, which is equivalent to the strong B-property: For all paths

$$X_{1,1} = U_1, \quad X_{1,2} = U, \quad X_{1,3} = U_2,$$
$$X_{3,1} = V_1, \quad X_{3,2} = V, \quad X_{3,3} = V_2$$

with $\alpha(U) = \omega(U_1)$, $\alpha(U_2) = \omega(U)$, $\alpha(V) = \omega(V_1)$, $\alpha(V_2) = \omega(V)$, the following is true:

$$H(X_{1,1} \oplus X_{1,2} \oplus X_{1,3}) \in MIN\{H(X_2) \mid X_2 \in S_2(L_B, X_{1,1}, X_{1,2}, X_{1,3}, X_{3,1}, X_{3,2}, X_{3,3})\} \Rightarrow$$

$$H(X_{3,1} \oplus X_{3,2} \oplus X_{3,3}) \in MIN\{H(X_4) \mid X_4 \in S_4(L_B, X_{1,1}, X_{1,2}, X_{1,3}, X_{3,1}, X_{3,2}, X_{3,3})\}.$$

This condition is exactly the minimum preservation with respect to L_B. □

Example 2.107 Let $B = B_{\leq}^{(2)}$; then the following are true:

$$L_B = \left\{ \begin{pmatrix} [\alpha(P)] & P & P_2 \\ [\alpha(P)] & P' & P_2 \\ [\alpha(P)] & P & [\omega(P)] \\ [\alpha(P)] & P'' & [\omega(P)] \end{pmatrix} \in \mathcal{L} \right\}, \quad \mathbb{L}_0 = \left\{ \begin{pmatrix} [\alpha(P)] & P & [\omega(P)] \\ [\alpha(P)] & P' & [\omega(P)] \\ [\alpha(P)] & P & P_2 \\ [\alpha(P)] & P' & P_2 \end{pmatrix} \in \mathcal{L} \right\}.$$

Consequently, the matrices in $\mathbb{L}_0$ can be generated from the ones in L_B by exchanging the upper half with the lower half and replacing P'' by P'.

By Theorem 2.106 and Corollary 2.103, the L_B-minimum preservation for $B = B_{\leq}^{(2)}$, the strong Bellman $B_{\leq}^{(2)}$-condition, and the $\mathbb{L}_0$-minimum preservation are equivalent to each other.

Theorem 2.106 itself does not imply Corollary 2.103 because $\mathbb{L}_0$ does not equal the set L_B for $B = B_{\leq}^{(2)}$. □

Theorem 2.106 says that any strong B-condition is equivalent to minimum preservation with respect to L_B. A similar result says that any weak B-condition is equivalent to the so-called weak minimum preservation with respect to L_B. This term is now introduced.

Definition 2.108 Let $L \subseteq \mathcal{L}$. Given a digraph $\mathcal{G} = (\mathcal{V}, \mathcal{R})$. Let $H : \mathcal{P}(\mathcal{G}) \to (\mathbf{R}, \preceq)$. We say that H is *weakly minimum preserving with respect to L* if the following is true:

  For all paths $X_{1,1}, X_{1,2}, X_{1,3}$, there exists a path $Y_{1,2}$ with
  $H(X_{1,1} \oplus X_{1,2} \oplus X_{1,3}) \equiv H(X_{1,1} \oplus Y_{1,2} \oplus X_{1,3})$ such that for
  $X_1' := X_{1,1} \oplus Y_{1,2} \oplus X_{1,3}$ and for all paths $X_{3,1}, X_{3,2}, X_{3,3}$,
  the following implication is true:

$$H(X_1') = MIN\{H(X_2) \mid X_2 \in S_2(L, X_{1,1}, Y_{1,2}, X_{1,3}, X_{3,1}, X_{3,2}, X_{3,3})\} \Longrightarrow$$
$$H(X_3) = MIN\{H(X_4) \mid X_4 \in S_4(L, X_{1,1}, Y_{1,2}, X_{1,3}, X_{3,1}, X_{3,2}, X_{3,3})\}.$$ □

Remark 2.109 If H is minimum preserving with respect to L then H is also weakly minimum preserving with respect to L; this can be seen by choosing $Y_{1,2} := X_{1,2}$. □

Theorem 2.110 *Given a digraph* $\mathcal{G} = (\mathcal{V}, \mathcal{R})$, *a node* s *in* $\mathcal{G}$ *and a path function* $H : \mathcal{P}(\mathcal{G}) \to (\mathbf{R}, \preceq)$. *Let* $B \subseteq \mathcal{B}$.

Then the weak Bellman B-condition is equivalent to the weak minimum preservation with respect to L_B; *the set* L_B *of matrices was defined in 2.104.*

P r o o f : The proof is given in Remark 2.63 and in Theorems 2.64, 2.65 of [Huck92].[10] The main idea is to reformulate the weak Bellman condition with respect to L_B. $\quad\square$

We have seen that each Bellman B-condition can be formulated as a principle of minimum preservation. The question arises, whether the converse problem can be solved: Given a set $L \subseteq \mathcal{L}$ of matrices, find a relation $B_L \subseteq \mathcal{B}$ such that the strong Bellman B_L-condition is equivalent to the L-minimum preservation.

We conjecture that this problem cannot always be solved. To make this conjecture more obvious we next introduce the Bellman conditions of type A, B1 and B2. In many cases, we can transform these Bellman properties into principles of L-minimum preservation with $L \subseteq \mathcal{L}$. But it is difficult and probably impossible to formulate any of these Bellman properties as a B-condition. The reason is that the Bellman principles of type A, B1 and B2 have another structure than any B-condition; the Bellman principle of type A is about candidate paths with various start and end nodes; the principles of type B1, B2 are about candidate paths satisfying a side constraint.

We introduce the new Bellman principles in Definition/Remark 2.111. In Remark 2.112, we represent the three Bellman properties as principles of L-minimum preservation, and we show why it is so difficult to express them as B-conditions.

Definition/Remark 2.111 Given a digraph $\mathcal{G}$, a node s in $\mathcal{G}$, and two path functions $H, H_* : \mathcal{P}(\mathcal{G}) \to (\mathbf{R}, \preceq)$.

B e l l m a n   p r i n c i p l e   o f   t y p e   A :
Given the sets $\mathcal{V}_1, \ldots, \mathcal{V}_N \subseteq \mathcal{V}$. We call a path P an H-*minimal connection from* $\mathcal{V}_i$ *to* $\mathcal{V}_j$ if $\alpha(P) \in \mathcal{V}_i$, $\omega(P) \in \mathcal{V}_j$ and $H(P) \in MIN\{H(\widetilde{P}) \,|\, \alpha(\widetilde{P}) \in \mathcal{V}_i , \omega(\widetilde{P}) \in \mathcal{V}_j \}$.

Then we say that H satisfies the *strong prefix [infix] [suffix] oriented Bellman condition of type A* if for all sets $\mathcal{V}_i, \mathcal{V}_j, \mathcal{V}_{i'}, \mathcal{V}_{j'}$ and for all H-minimal paths P from $\mathcal{V}_i$ to $\mathcal{V}_j$, the following is true:

If Q *is a prefix [infix] [suffix] of* P *and if* $\alpha(Q) \in \mathcal{V}_{i'}$ *and if* $\omega(Q) \in \mathcal{V}_{j'}$
then Q *is an optimal connection from* $\mathcal{V}_{i'}$ *to* $\mathcal{V}_{j'}$.

This principle is equivalent to the strong prefix [infix] [suffix] oriented Bellman property of type 0 if the following special case if given: All sets $\mathcal{V}_i$ have one element and are pairwise disjoint; moreover, $\bigcup \mathcal{V}_i = \mathcal{V}$.

---

[10]The symbol "=" must be replaced by "≡" in the equation $H(X_{1,1} \oplus X_{1,2} \oplus X_{1,3}) = H(X_{1,1} \oplus Y_{1,2} \oplus X_{1,3})$ in Definition 2.62 of [Huck92]; also, the symbol "=" must be replaced by "≡" in equations $(*)$, $(***)$, $(+++)$ on page 84 of [Huck92].

B e l l m a n   p r i n c i p l e   o f   t y p e   B1 :

Let $B \subseteq \mathcal{B}$. We say that H satisfies the *strong Bellman B-condition of type B1 with respect to H_** if the following is true for all 6-tuples $(U_1, U, U_2, V_1, V, V_2) \in B$:

If $U_1 \oplus U \oplus U_2$ is H-minimal among all paths $U_1 \oplus U' \oplus U_2$ with $H_*(U_1 \oplus U' \oplus U_2) = H_*(U_1 \oplus U \oplus U_2)$ then $V_1 \oplus V \oplus V_2$ is H-minimal among all paths $V_1 \oplus V' \oplus V_2$ with $H_*(V_1 \oplus V' \oplus V_2) = H_*(V_1 \oplus V \oplus V_2)$.

B e l l m a n   p r i n c i p l e   o f   t y p e   B2 :

We say that H has the *prefix oriented Bellman property of type B2* if for all nodes v, v', v'' and for all paths $P \in \mathcal{P}(v, v')$, $Q \in \mathcal{P}(v', v'')$, the following is true:

If $(P \oplus Q)$ is an H-minimal path among all candidates $\widetilde{P} \oplus \widetilde{Q} \in \mathcal{P}(v, v'')$ with $\omega(\widetilde{P}) = \omega(P) = v'$ then $P \in \mathcal{P}_{Min}(v, v')$.

The prefix oriented Bellman property of type B2 implies the one of type 0. To see this we assume that $X \in \mathcal{P}_{Min}(v, v'')$ and that P be a prefix of X with end node v'; moreover, let Q be the suffix of X for that $X = P \oplus Q$. The optimality of $X = P \oplus Q$ among all v-v''-paths implies the optimality of X among all v-v''-paths $\widetilde{P} \oplus \widetilde{Q}$ visiting the node v'. So, the Bellman principle of type B2 may be applied to X, and we obtain the desired assertion $P \in \mathcal{P}_{Min}(v, v')$. ◻

Remark 2.112 Here, we show how the Bellman properties of type A, B1, B2 can be interpreted principles of minimum preservation.[11] Moreover, we give reasons why it is difficult and probably impossible to express the three modified Bellman properties as strong B-principles.

First, we consider the *infix oriented Bellman principle of type A*. We assume that each $v \in \mathcal{V}$ lies in exactly one set $\mathcal{V}_\nu$; then there exists a function $pr : \mathcal{V} \to \{1, \dots, N\}$ such that $v \in V_{pr(v)}$ for all $v \in \mathcal{V}$. We define

$$L := \left\{ \left(\begin{array}{ccc} [\alpha(P)] & P & [\omega(P)] \\ [\alpha(\widetilde{P})] & \widetilde{P} & [\omega(\widetilde{P})] \\ [\alpha(Q)] & Q & [\omega(Q)] \\ [\alpha(\widetilde{Q})] & \widetilde{Q} & [\omega(\widetilde{Q})] \end{array} \right) \left| \begin{array}{l} Q \subseteq P, \\ \left(\begin{array}{ll} pr(\alpha(P))=pr(\alpha(\widetilde{P})):=i, & pr(\omega(P))=pr(\omega(\widetilde{P})):=j, \\ pr(\alpha(Q))=pr(\alpha(\widetilde{Q})):=i', & pr(\omega(Q))=pr(\omega(\widetilde{Q}))=:j' \end{array} \right) \end{array} \right. \right\} .$$

Then the infix oriented Bellman principle of type A is equivalent to the principle of L-minimum preservation. To see this we assume that $Q \subseteq P$. Then

$$S_2(L, [\alpha(P)], P, [\omega(P)], [\alpha(Q)], Q, [\omega(Q)]) = \{\widetilde{P} \in \mathcal{P}(\mathcal{G}) \,|\, pr(\alpha(\widetilde{P})) = pr(\alpha(P)), \; pr(\omega(\widetilde{P})) = pr(\omega(P))\}$$

because $\mathbf{Y} := \left(\begin{array}{ccc} [\alpha(P)] & P & [\omega(P)] \\ [\alpha(\widetilde{P})] & \widetilde{P} & [\omega(\widetilde{P})] \\ [\alpha(Q)] & Q & [\omega(Q)] \\ [\alpha(Q)] & Q & [\omega(Q)] \end{array} \right) \in L$ if and only if $pr(\alpha(\widetilde{P})) = pr(\alpha(P))$

and $pr(\omega(\widetilde{P})) = pr(\omega(P))$. Moreover,

---

[11] The sets S_2, S_4 and the matrices $\mathbf{Y}$ appearing in Remark 2.112 are the same as in 2.53.

$$S_4(L\,,\,[\alpha(P)]\,,P,[\omega(P)]\,,\,[\alpha(Q)],Q,[\omega(Q)]) =$$
$$\{\widetilde{Q} \in \mathcal{P}(\mathcal{G})\,|\,pr(\alpha(\widetilde{Q})) = pr(\alpha(Q))\,,\ pr(\omega(\widetilde{Q})) = pr(\omega(Q))\}$$

because $\mathbf{Y} := \begin{pmatrix} [\alpha(P)] & P & [\omega(P)] \\ [\alpha(P)] & P & [\omega(P)] \\ [\alpha(Q)] & Q & [\omega(Q)] \\ [\alpha(\widetilde{Q})] & \widetilde{Q} & [\omega(\widetilde{Q})] \end{pmatrix} \in L$ if and only if $pr(\alpha(\widetilde{Q})) = pr(\alpha(Q))$

and $pr(\omega(\widetilde{Q})) = pr(\omega(Q))$.

Moreover, $S_i(L\,,\,X_{1,1},X_{1,2},X_{1,3}\,,\,X_{3,1},X_{3,2},X_{3,3}) = \emptyset$ $(i = 2,4)$ whenever the paths $X_{i,j}$ does not satisfy each of the following conditions: $X_{1,1} = [\alpha(X_{1,2})]$, $X_{1,3} = [\omega(X_{1,2})]$, $X_{3,1} = [\alpha(X_{3,2})]$, $X_{3,3} = [\omega(X_{3,2})]$, and $X_{3,2} \subseteq X_{1,2}$.

If the $\mathcal{V}_\nu$'s are not pairwise disjoint then the following situation may occur: There exists a path P with $\alpha(P) \in \mathcal{V}_1 \cap \mathcal{V}_2$ and $\omega(P) \in \mathcal{V}_3$ such that P is an H-minimal $\mathcal{V}_1$-$\mathcal{V}_3$-path but not an H-minimal $\mathcal{V}_2$-$\mathcal{V}_3$-path. It is not clear how to define a set L such that these minimality properties of P can be expressed as conditions of the form $H(P) \in MIN\{X_2|X_2 \in S_2\}$, $H(P) \notin MIN\{X_2|X_2 \in S_2\}$, $H(P) \in MIN\{X_4|X_4 \in S_4\}$ or $H(P) \notin MIN\{X_4|X_4 \in S_4\}$ (see (2.35) and (2.36)).

Next, we consider the *Bellman B-principle of type B1*. Let $L_B \subseteq \mathcal{L}$ be the same set as in Definition 2.104. Then we define

$$L := \left\{ \begin{pmatrix} U_1 & U & U_2 \\ U_1 & U' & U_2 \\ V_1 & V & V_2 \\ V_1 & V' & V_2 \end{pmatrix} \in L_B \,\middle|\, \begin{matrix} H_*(U_1 \oplus U \oplus U_2) = \\ H_*(U_1 \oplus U' \oplus U_2)\,, \\ H_*(V_1 \oplus V \oplus V_2) = \\ H_*(V_1 \oplus V' \oplus V_2) \end{matrix} \right\}.$$

Then the Bellman B-principle of type B1 is equivalent to the L-minimum preservation. To see this we assume that $(U_1, U, U_2, V_1, V, V_2) \in B$. Then

$$S_2(L\,,\,U_1, U, U_2\,,\,V_1, V, V_2) =$$
$$\{U_1 \oplus U' \oplus U_2\,|\,H_*(U_1 \oplus U' \oplus U_2) = H_*(U_1 \oplus U \oplus U_2)\}$$

because $\mathbf{Y} := \begin{pmatrix} U_1 & U & U_2 \\ U_1 & U' & U_2 \\ V_1 & V & V_2 \\ V_1 & V & V_2 \end{pmatrix} \in L$ if and only if $H_*(U_1 \oplus U' \oplus U_2) = H_*(U_1 \oplus U \oplus U_2)$.

Moreover,

$$S_4(L\,,\,U_1, U, U_2\,,\,V_1, V, V_2) =$$
$$\{V_1 \oplus V' \oplus V_2\,|\,H_*(V_1 \oplus V' \oplus V_2) = H_*(V_1 \oplus V \oplus V_2)\}$$

because $\mathbf{Y} := \begin{pmatrix} U_1 & U & U_2 \\ U_1 & U & U_2 \\ V_1 & V & V_2 \\ V_1 & V' & V_2 \end{pmatrix} \in L$ if and only if $H_*(V_1 \oplus V' \oplus V_2) = H_*(V_1 \oplus V \oplus V_2)$.

If, however, $(U_1, U, U_2, V_1, V, V_2) \notin B$ then $S_2 = S_4 = \emptyset$.

Next, we consider the *Bellman principle of type B2* and define

$$
L \; := \; \left\{ \begin{pmatrix} [v] & P & Q \\ [v] & \widetilde{P} & \widetilde{Q} \\ [v] & P & [v'] \\ [v] & \widetilde{P} & [v'] \end{pmatrix} \in \mathcal{L} \; \middle| \; \begin{array}{l} v, v' \in \mathcal{V}, \\ w(\widetilde{Q}) = w(Q) =: v'' \end{array} \right\}.
$$

Then the Bellman B-principle of type B2 is equivalent to the L-minimum preservation. To see this we assume that $v, v' \in \mathcal{V}$ and that $P \in \mathcal{P}(v, v')$ and $Q \in \mathcal{P}(v')$. Then

$$
S_2(L, [v], P, Q, [v], P, [v']) = \{\widetilde{P} \oplus \widetilde{Q} \,|\, \alpha(\widetilde{P}) = v, \; w(\widetilde{P}) = v', \; w(\widetilde{Q}) = w(Q)\}
$$

because $\mathbf{Y} := \begin{pmatrix} [v] & P & Q \\ [v] & \widetilde{P} & \widetilde{Q} \\ [v] & P & [v'] \\ [v] & \widetilde{P} & [v'] \end{pmatrix} \in L$ if and only $\alpha(\widetilde{P}) = v, \; w(\widetilde{P}) = v'$

and $w(\widetilde{Q}) = w(Q)$. Moreover,

$$
S_4(L, [v], P, Q, [v], P, [v']) = \{\widetilde{P} \,|\, \alpha(\widetilde{P}) = v, \; w(\widetilde{P}) = v'\}
$$

because $\mathbf{Y} := \begin{pmatrix} [v] & P & Q \\ [v] & \widetilde{P} & Q \\ [v] & P & [v'] \\ [v] & \widetilde{P} & [v'] \end{pmatrix} \in L$ if and only $\alpha(\widetilde{P}) = v$ and $w(\widetilde{P}) = v'$.

Moreover, $S_i(L, X_{1,1}, X_{1,2}, X_{1,3}, X_{3,1}, X_{3,2}, X_{3,3}) = \emptyset$ $(i = 2, 4)$ whenever the paths $X_{i,j}$ does not satisfy each of the following conditions: $X_{1,1} = [\alpha(X_{1,2})]$, $X_{1,3} = [w(X_{1,2})]$, and $X_{3,1} = [\alpha(X_{3,2})]$, $X_{3,2} = X_{1,2}$, $X_{3,3} = [w(X_{3,2})]$.

The Bellman principles $A, B1, B2$ can probably not be written as Bellman B-conditions. To see this we compare the structures of the new and the previous Bellman properties.

The essential part of each B-property is an implication of the form

$$
(U_1, U, U_2) \in ROC \implies (V_1, V, V_2) \in ROC.
$$

If $U := \{U_1 \oplus U' \oplus U_2 \,|\, \alpha(U') = w(U_1), \; w(U') = \alpha(U_2)\}$ and $V := \{V_1 \oplus V' \oplus V_2 \,|\, \alpha(V') = w(V_1), \; w(V') = \alpha(V_2)\}$, the above implication can be formulated as follows:

If $U_1 \oplus U \oplus U_2$ is an H-minimal path in U then $V_1 \oplus V \oplus V_2$ is an H-minimal path in V.

The three modified Bellman principles are similar to this condition but they use other candidate sets U and V for their processes of minimization.

The Bellman property of type A uses two sets U, V whose elements may have different start nodes and different end nodes. The principles of type B1 and B2 are based on subsets $U \subseteq \{U_1 \oplus U' \oplus U_2 \,|\, \alpha(U') = w(U_1), \; w(U') = \alpha(U_1)\}$ and $V \subseteq \{V_1 \oplus V' \oplus V_2 \,|\, \alpha(V') = w(V_1), \; w(V') = \alpha(V_1)\}$ with the property that U and V are described with the help of side-constraints about $U_1 \oplus U' \oplus U_2$ and $V_1 \oplus V' \oplus V_2$.

Consequently, the Bellman conditions of type A, B1, and B2 can probably not be transformed into a B-property. $\Box$

2.4.6   Structural properties of optimal paths if Bellman principles are given

At the beginning of this section we gave two reasons why Bellman conditions are useful.

The first is that many optimal paths can easily be recorded as subpaths of other optimal ones. One of the simplest realization of this idea is to represent optimal paths as a rooted tree. Theorem 2.113 and Remark 2.114 give answers to the question when optimal paths can be arranged in this way and when not.

The second advantage of a Bellman principle is that optimal paths can be composed of optimal subpaths. This will be demonstrated in Theorem 2.118.

Theorem 2.113 *Given a digraph $\mathcal{G}$ and a node s in $\mathcal{G}$. Let $H : \mathcal{P}(s) \rightarrow (\mathbf{R}, \preceq)$ where $\preceq$ is total. Let $\mathcal{V}'$ be the set of all nodes v' in $\mathcal{G}$ for which an H-minimal s-v'-path exists, and let $s \in \mathcal{V}'$. We assume that each restriction $H|_{\mathcal{P}(s,v')}$ $(v' \in \mathcal{V}')$ is $\equiv$-injective and that H has the weak prefix oriented Bellman property of type 0.*

Then there exists a rooted tree $\mathcal{T}$ with the following properties:
s is the root of $\mathcal{T}$, and $\mathcal{V}'$ is the set of all nodes in $\mathcal{T}$; moreover, $\mathcal{T}$ is a subgraph of $\mathcal{G}$;
all $\mathcal{T}$-paths $Q \in \mathcal{P}(s,v)$ are optimal among the $\mathcal{G}$-paths $Q' \in \mathcal{P}(s,v)$.

P r o o f :   By the definition of $\mathcal{V}'$, there exists an optimal s-v-path $P(v)$ in $\mathcal{G}$ for each $v \in \mathcal{V}'$. Moreover, Theorem 2.94 implies that H has the strong prefix oriented Bellman property of type 0. Consequently, the following is true for all $v \in \mathcal{V}'$:

> If $v \in \mathcal{V}'$ then all prefixes $U \leq P(v)$ are optimal paths from $s = \alpha(U)$ to $\omega(U)$.   (2.56)

Let $\mathcal{R}'$ be the set of all arcs appearing in some path $P(v)$. We claim that $\mathcal{R}'$ forms a rooted tree $\mathcal{T}$ with root s. To see this we prove three assertions:

> No arc of $\mathcal{R}'$ ends with s.   (2.57)

> For each other node $v \in \mathcal{V}'$, there exists exactly one arc in $\mathcal{R}'$ ending with v.   (2.58)

> For each node $v \in \mathcal{V}'$, there exists exactly one path $Q(v)$ in $\mathcal{T}$ from s to v; this path $Q(v)$ is optimal among all s-v-paths in $\mathcal{G}$, i.e., $Q(v) = P(v)$.   (2.59)

To show (2.57) we assume that r is an arc ending with s. Then r belongs to a path $P(v)$. Hence, this path must have a substantial cycle X with $\alpha(X) = \omega(X) = s$. Consequently, $P(v)$ can be written as $P(v) = X \oplus Q$. Then (2.56) implies that both prefixes $[s]$ and X are H-minimal s-s-paths. This and the fact that $\preceq$ is total imply that $H(X) \equiv H([s])$ although $X \neq [s]$. That is a contradiction to the $\equiv$-injectivity of H.

Next, we show fact (2.58). If $v \neq s$ then $P(v)$ has one or more arcs, and the last of them ends with v. We must still show that at most one arc has v as its endnode. We assume that this is not the case for some node w. Let r be the last arc of $P(w)$; then there exists another arc $r' \in \mathcal{R}'$ with $\omega(r') = w$. This arc appears in a path $P(w')$. (It is possible that $w' = w$.) Let P' be the prefix of $P(w')$ that ends with r'. Then (2.56) implies that P' is an optimal connection from s to $\omega(P') = w$.

Hence, $H(P') \equiv H(P(w))$ because $\preceq$ is total. On the other hand, $P' \neq P(w)$ because P' ends with r' and $P(w)$ ends with r. This situation means a contradiction to the $\equiv$-injectivity of H on the set $\mathcal{P}(s, w)$.

We now prove fact (2.59). The existence of $Q(v)$ follows from the definition of $\mathcal{R}'$, which implies that $P(v)$ is a path in $\mathcal{T}'$ with the desired properties. Let $Q(v) := P(v)$. We must still show the uniqueness of $Q(v)$. If $v = s$ then (2.57) implies that $Q(s) = [s]$ is the only s-s-path in $\mathcal{T}$. Let $v \neq s$, and let $Q'(v)$ be an arbitrary s-v-path in $\mathcal{T}$. We assume that $Q(v) = r_k \oplus r_{k-1} \oplus \ldots \oplus r_1$ and that $Q'(v) = r'_l \oplus r'_{l-1} \oplus \ldots \oplus r'_1$. Then (2.58) implies that

$$r_\kappa = r'_\kappa \text{ for all } \kappa = 1, \ldots, m := \min\{k, l\}. \tag{2.60}$$

If $m = k$ then $\alpha(r_m) = \alpha(Q(v)) = s$, and if $m = l$ then $\alpha(r_m) \overset{(2.60)}{=} \alpha(r'_m) = \alpha(Q'(v)) = s$ so that $\alpha(r_m) = s$ in any case. This and (2.57) imply that the cases $m < k$ and $m < l$ are impossible so that $l = m = k$. Consequently, $Q(v) = Q'(v)$ by (2.60).

The existence and uniqueness of the paths $Q(v)$ imply that $\mathcal{T}$ is a rooted tree with root s. The optimality of $Q(v)$ in $\mathcal{G}$ follows from the equation $Q(v) = P(v)$. $\qquad\square$

Remark 2.114   a) The $\equiv$-injectivity of H in 2.113 must not be omitted, even if we replace the weak Bellman property of H by the strong one. For example, let $\mathcal{G} = (\mathcal{V}, \mathcal{R})$ with $\mathcal{V} = \{s, x_1, x_2, y, z_1, z_2\}$, $\mathcal{R} := \{(s, x_i), (x_i, y), (y, z_i) \mid i = 1, 2\}$, and the following path function $H : \mathcal{P}(s) \to \mathbb{R}$:

$H([s]) := 0,$ $H([s, x_1]) := 0,$ $H([s, x_1, y]) := 5,$ $H([s, x_1, y, z_1]) := 1,$ $H([s, x_1, y, z_2]) := 6,$
              $H([s, x_2]) := 0,$ $H([s, x_2, y]) := 5,$ $H([s, x_2, y, z_1]) := 4,$ $H([s, x_2, y, z_2]) := 2.$

Moreover, we define $P_i := [s, x_i, y]$ and $Q_i := [s, x_i, y, z_i]$ for $i = 1, 2$.

Then the restriction $H|_{\mathcal{P}(s, y)}$ is not $\equiv$-injective because $H(P_1) = 5 = H(P_2)$.

H has the strong Bellman property of type 0. The only critical case is that an optimal path Y has a proper prefix X with $H(X) > 0$. Therefore, it is sufficient to consider the paths $Y = Q_i$, $i = 1, 2$, and their prefixes $X = P_i$. This prefix P_i is an optimal s-y-path for all $i = 1, 2$. Consequently, H has indeed the desired Bellman property. Each node z_i, $i = 1, 2$, can be reached by the optimal s-z_i-path $[s, x_i, y, z_i]$; so, z_i is an element of the set $\mathcal{V}'$ defined in 2.113.

It is easy to see that there is no rooted tree $\mathcal{T}$ with an optimal s-v-path $P(v)$ for each node $v \in \mathcal{V}'$. The reason is that no tree can contain the only H-minimal s-z_1-path Q_1 and the only H-minimal s-z_2-path Q_2 at the same time.

     b) The condition that $\preceq$ is total must not be omitted, even if H has the strong Bellman property of type 0.

For example, let $\mathcal{G} = (\mathcal{V}, \mathcal{R})$ and P_1, P_2, P_1, P_2 be the same objects as in part a). We define $\mathbf{R} := \mathbb{R}^2$ and the following relation $\preceq$:

$$(\forall\, (a, b),\, (c, d) \in \mathbb{R}^2)\ \ (a, b) \preceq (c, d)\ :\Longleftrightarrow\ (a \leq c \wedge b = d).$$

Then $\preceq$ is not total because we can only compare pairs with the same second component.

We define $H : \mathcal{P}(s) \to (\mathbf{R}, \preceq)$ as follows: Let $H([s]) := (0, 0)$ and

$H([s, x_1]) := (0, 0)$ $H([s, x_1, y]) := (2, 3),$ $H([s, x_1, y, z_1]) := (3, 3),$ $H([s, x_1, y, z_2]) := (9, 5),$
$H([s, x_2]) := (0, 0)$ $H([s, x_2, y]) := (1, 5),$ $H([s, x_2, y, z_1]) := (4, 3),$ $H([s, x_2, y, z_2]) := (4, 5).$

Then H has the strong Bellman property of type 0. The only critical case is that an optimal path Y has a proper prefix X with $H(X) \neq (0,0)$. Therefore, it is sufficient to consider the paths $Y = Q_i$, $i = 1, 2$, and their prefixes $X = P_i$. This prefix is an optimal s-y-path for each $i = 1, 2$ because neither $H(P_1) \preceq H(P_2)$ nor $H(P_2) \preceq H(P_1)$. Hence, H has indeed the desired Bellman property.

The restriction $H|_{\mathcal{P}(s,v)}$ is $\equiv$-injective for each node $v \in \mathcal{V}$. This can easily be checked for $v \in \{u, z_1, z_2\}$; for all other nodes v, the set $\mathcal{P}(s,v)$ has exactly one element.

Each node z_i, $i = 1, 2$ can be reached by the optimal s-z_i-path $P_i \oplus r_i$; therefore, z_i is an element of the set $\mathcal{V}'$ defined in 2.113.

So, there exists no rooted tree $\mathcal{T}$ with an optimal s-v-path $P(v)$ for each node $v \in \mathcal{V}'$. The reason is that no tree can contain the only H-minimal s-z_1-path Q_1 and the only H-minimal s-z_2-path Q_2 at the same time.

c) If H is order preserving then we cannot always construct a rooted tree of optimal paths. It is possible that each optimal s-v-connection has a cycle as its infix. An example is given in Remark 4.124. □

The next results are about replacing optimal s-v-connections by optimal and acyclic s-v-paths. In contrast to 2.113, we do not claim that we can form a tree of optimal paths.

Lemma 2.115 *Given a digraph $\mathcal{G}$ and two nodes s, v of $\mathcal{G}$. Let $H : \mathcal{P}(s) \to (\mathbf{R}, \preceq)$ where $\preceq$ is total and identitive. (E.g., $\mathbf{R}$ may be equal to $\mathbb{R}$, and "$\preceq$" may be equal to "$\leq$".) We assume that H is $=$-preserving.*
Let P^+ be a Bellman path from s to v. Let $P^+ = P' \oplus X \oplus P''$ where X is a cycle.
Then $P' \oplus P''$, too, is a Bellman path from s to v.

P r o o f : Let $\widetilde{P}$ be a prefix of $P' \oplus P''$. If $\widetilde{P}$ is even a prefix of P' then $\widetilde{P} \leq P' \leq P^+$ so that $\widetilde{P}$ is H-minimal. If, however, $\widetilde{P} \leq P'$ then there exists a path $Z \leq P''$ with $\widetilde{P} = P' \oplus Z$. Then $P' \oplus X \oplus Z$ is a prefix of P^+.

The two paths P' and $P' \oplus X$ are prefixes of P' and have the same endpoint. Our assumption about P implies that that both P' and $P \oplus X$ are H-minimal. This and the assumption that $\preceq$ is total and identitive imply that $H(P') = H(P' \oplus X)$. This and the $=$-preservation of H have the consequence that $H(\widetilde{P}) = H(P' \oplus Z) = H(P' \oplus X \oplus Z)$. Therefore, $\widetilde{P}$ is an H-minimal path from s to $\omega(\widetilde{P}) = \omega(Z)$ because the same is true for the prefix $P' \oplus X \oplus Z$ of P^+. □

Theorem 2.116 *Given a digraph $\mathcal{G}$ and two nodes s, v of $\mathcal{G}$. Let $H : \mathcal{P}(s) \to (\mathbf{R}, \preceq)$ where $\preceq$ is total and identitive. (E.g., $\mathbf{R}$ may be equal to $\mathbb{R}$, and "$\preceq$" may be equal to "$\leq$".) We assume that H is $=$-preserving and has the weak prefix oriented Bellman property of type 0. Moreover, we assume that an H-minimal s-v-path P exists.*
Then there exists an a c y c l i c Bellman path Q from s to v.

P r o o f : We have assumed the existence of an H-minimal s-v-path P. The weak Bellman property of H implies that there exists an H-minimal Bellman path P^* from s to v.

Now, we generate a sequence $P_0, P_1, \ldots, P_k$ of Bellman paths from s to v. We define

$P_0 := P^*$. If P_κ is given and has a substantial cycle X then we construct $P_{\kappa+1}$ by removing this cycle. At last, we obtain an acyclic path P_k.

Lemma 2.115 implies that all paths P_κ are Bellman paths from s to v. In particular, this is true for the path $Q := P_k$, and this path is acyclic. $\qquad\square$

Remark 2.117 The assertions in 2.115 and in 2.116 may be false if H is not $=$-preserving; in this case, we cannot even guarantee the existence of a Bellman path of a length $\leq (|\mathcal{V}| - 1)$. An example is the digraph $\mathcal{G} = (\mathcal{V}, \mathcal{R})$ with $\mathcal{V} := \{v^{(i)}|i = 1,\ldots,5\}$, $s := v^{(1)}$, and $\mathcal{R} := \{(s, v^{(2)}), (v^{(2)}, v^{(3)}), (v^{(3)}, v^{(4)}), (v^{(4)}, v^{(2)}), (v^{(2)}, v^{(5)})\}$. Let $X := [v^{(2)}, v^{(3)}, v^{(4)}, v^{(2)}]$. We define $H : \mathcal{P}(s) \to \mathbb{R}$ as follows:

$$H([s]) := H((s, v^{(2)}) \oplus X^2 \oplus (v^{(2)}, v^{(5)})) := 0, \quad H(P) := 1 \text{ for all other paths } P.$$

Then all paths ending with a node $v \neq v^{(5)}$ have the same H-values; consequently, each s-v-path is H-minimal.

The function H has even the strong prefix oriented Bellman property of type 0. To see this we assume that P is an H-optimal path and that P' is a proper prefix of P. Then P' does not end with $v^{(5)}$ and is consequently an H-minimal path from s to $\omega(P')$.

Also, H is not $=$-preserving because $H((s, v^{(2)}) \oplus X) = 1 = H((s, v^{(2)}) \oplus X^2)$ but $H((s, v^{(2)}) \oplus X \oplus (v^{(2)}, v^{(5)})) = 1 \neq 0 = H((s, v^{(2)}) \oplus X^2 \oplus (v^{(2)}, v^{(5)}))$.

The graph $\mathcal{G}$ and the path function H do not have any of the properties asserted in Lemma 2.115 and in Theorem 2.116. To see this we define $P' := [s, v^{(2)}] \oplus X$ and $P'' := [v^{(2)}, v^{(5)}]$. Then $P' \oplus X \oplus P''$ is a Bellman path, but $P' \oplus P''$ is not a Bellman path because $P' \oplus P''$ itself is not H-minimal; this situation is in conflict with the assertion of Lemma 2.115. Moreover, there exists no H-optimal and acyclic path Q from s to $v^{(5)}$ because the only acyclic path $[s, v^{(2)}, v^{(5)}]$ is not H-minimal. There does not even exist any H-minimal s-$v^{(5)}$-path of a length $\leq n - 1 = 4$ because $(s, v^{(2)}) \oplus X^2 \oplus (v^{(2)}, v^{(5)})$ is the only H-minimal s-$v^{(5)}$-path, and the length of this path is 8. $\qquad\square$

The next result shows how Bellman principles can be used to construct optimal paths with the help of optimal subpaths.

Theorem 2.118 *Given a digraph $\mathcal{G}$ and a node s of $\mathcal{G}$. Let $H : \mathcal{P}(s) \to (\mathbf{R}, \preceq)$ be a cost function with the strong prefix oriented Bellman property of type 0.*

Then for all nodes $v \neq s$ and for all optimal s-v-paths P the following is true: P consists of a prefix $P' \leq P$ and an arc r, and the prefix P' itself is optimal. More formally,

$$(\forall v \neq s) \quad \mathcal{P}_{Min}(s, v) \subseteq \{P' \oplus r \mid P' \in \mathcal{P}_{Min}(s, \omega(P')), \, \omega(r) = v\} := A(v).^{12}$$

---

[12]This result has the following consequence: If $\mathcal{P}_{Min}(s, v')$ is already known for all predecessors of v then $\mathcal{P}_{Min}(s, v)$ can be easily constructed by searching the H-minimal paths among all candidates $P' \oplus r$ with $P' \in \mathcal{P}_{Min}(s, v')$, $\alpha(r) = v$, $\omega(r) = v$. This is the basic idea of the Ford-Bellman algorithm for optimal paths; Section 4.3 gives a detailed description of this search strategy.

P r o o f : Let $P \in \mathcal{P}_{Min}(s, v)$. Let r be the last arc of P and let P' be the prefix of P, for that $P = P' \oplus r$. The strong Bellman property implies that $P' \in \mathcal{P}_{Min}(s, \omega(P'))$. Hence, P is in $A(v)$.

$\qquad\qquad\qquad\qquad\qquad\qquad\qquad\qquad\qquad\qquad\qquad\qquad\qquad\qquad\qquad$ □

Remark 2.119 Let $H : \mathcal{P}(\mathcal{G}) \to (\mathbf{R}, \preceq)$ have the strong suffix oriented Bellman property of type 0. Let γ be a fixed node of $\mathcal{G}$.

Then we can formulate a "symmetric" version of Theorem 2.118 by exchanging the roles of prefixes and suffixes. Thus we obtain the following assertion about all nodes $v \neq \gamma$:

$$\mathcal{P}_{Min}(v, \gamma) \subseteq \{ r \oplus P', | \alpha(r) = v , P' \in \mathcal{P}_{Min}(\omega(P'), \gamma) \} := A^-(v).$$

Hence, all paths in $\mathcal{P}_{Min}(v, \gamma)$ can be found by searching for the H-minimal paths among the candidates $r \oplus P'$ where $r \in \mathcal{R}$ and P' is an optimal connection from $\omega(r) = \alpha(P')$ to γ.

$\qquad\qquad\qquad\qquad\qquad\qquad\qquad\qquad\qquad\qquad\qquad\qquad\qquad\qquad\qquad$ □

Remark 2.120 We have seen that the strong prefix and suffix oriented Bellman principles of type 0 make it possible to construct optimal paths with the help of optimal prefixes and suffixes, respectively.

We conjecture that one can formulate other Bellman principles implying the following assertion; this assertion says that particular optimal paths can be constructed with optimal prefixes and suffixes.

> Let $\mathcal{P}(v, v', v'')$ be the set of all v-v''-paths that visit v'. Let $P \in \mathcal{P}(v, v', v'')$ be a path with minimal H-value among all paths $Q \in \mathcal{P}(v, v', v'')$.
> Then there exist optimal paths $P' \in \mathcal{P}_{Min}(v, v')$ and $P'' \in \mathcal{P}_{Min}(v', v'')$ such that $P = P' \oplus P''$.

A similar idea is described in Remark 4.161; an almost optimal path $P \in \mathcal{P}(v, v', v'')$ is found by concatenating two almost optimal paths $P' \in \mathcal{P}(v, v')$ and $P'' \in \mathcal{P}(v', v'')$; this method yields a fast algorithm to find optimal v-v''-paths for all nodes v, v''. □

2.4.7 Functional equations and similar structural properties of cost functions with Bellman properties

In this subsection, we study a generalization of the following well-known result, which is of great importance in Dynamic Programming:

Let H be an additive cost function with arc function $h : \mathcal{R} \to \mathbb{R}$. For each node v, let $f(v)$ be the minimal H-value of any s-v-connection.

Then the following functional equation is true for all $v \neq s$:

$$f(v) = \min\{f(v') + h(r) \mid r = (v', v) \in \mathcal{R}\}. \tag{2.61}$$

We shall prove Theorem 2.121, which is a general version of (2.61). Roughly speaking, Theorem 2.121 says the following:

Let H be a cost function with the weak prefix oriented Bellman property of type 0. Then all H-values of optimal s-v-paths can be computed with the help of the values

$H(P' \oplus r)$ *where*
  - P' *is an H-optimal path from s to some predecessor $v' \in \mathcal{N}^-(v)$,*
  - $r = (v', v) \in \mathcal{R}$.

This assertion is similar to Theorem 2.118, which may be interpreted as follows:

Let H be a cost function with the strong prefix oriented Bellman property of type 0.

Then all optimal s-v-paths can be constructed with the help of the paths $P' \oplus r$ where
  - P' *is an H-optimal path from s to some predecessor $v' \in \mathcal{N}^-(v)$,*
  - $r = (v', v) \in \mathcal{R}$.

Roughly speaking, Theorem 2.121 is better than 2.118 because it requires only the weak and not the strong Bellman property. On the other hand, Theorem 2.121 is worse than 2.118 because it only yields the H-values of optimal s-v-connections and not the optimal s-v-paths themselves.

Theorem 2.121 *Given a digraph $\mathcal{G} = (\mathcal{V}, \mathcal{R})$ and a node $s \in \mathcal{V}$. Let $\preceq$ be total and identitive[13]. (E.g., $\mathbf{R}$ may be equal to $\mathbb{R}$, and "$\preceq$" may be equal to "$\leq$".) Let $H : \mathcal{P}(s) \to (\mathbf{R}, \preceq)$ be a path function of the form $H = \phi \circ p$. We assume that H has the weak prefix oriented Bellman property of type 0, and we assume that an H-minimal s-v-path $P^+(v)$ exists for each node $v \in \mathcal{V}$.*

Let $f(v) := H(P^+(v))$, $v \in \mathcal{V}$.

Then the following is true for all nodes $v \neq s$:

$$ f(v) = \min \underbrace{\{\phi_r(f(v')) \mid r = (v', v) \in \mathcal{R}\}}_{=: \mathcal{M}(v)} . \tag{2.62} $$

(If H is additive then (2.62) implies (2.61). This can be seen by defining $\phi_r(\xi) = \xi + h(r)$, $\xi \in \mathbb{R}$; then $\phi_r(f(v')) = f(v') + h(r)$.)

P r o o f : Let $v \neq s$. The Bellman property of H implies that there is a Bellman path $P^*(v) \in \mathcal{P}(s, v)$ with $H(P^*(v)) \equiv H(P^+(v))$; then $H(P^*(v)) = H(P^+(v)) = f(v)$ because $\preceq$ is identitive.

The path $P^*(v)$ has at least one arc because $v \neq s$. Let (v', v) be the last arc of $P^*(v)$, and let Q be the prefix of $P^*(v)$ with $P^*(v) = Q \oplus (v', v)$. Then $H(Q) = f(v')$ because Q is a Bellman path. Consequently, $f(v) = H(P^*(v)) = H(Q \oplus (v', v)) = \phi_{(v',v)}(f(v'))$ so that $f(v) \in \mathcal{M}(v)$.

Next, we see that $f(v) \preceq \varrho$ for all $\varrho \in \mathcal{M}(v)$. Let $\varrho = \phi_{(v'',v)}(f(v'')) \in \mathcal{M}(v)$. Then $H(P^+(v'')) = f(v'')$, and the following equation is true where $(*)$ results from the minimality of $P^+(v)$:

$$ f(v) = H(P^+(v)) \overset{(*)}{\preceq} H(P^+(v'') \oplus (v'', v)) = \phi_{(v'',v)}(H(P^+(v''))) = \phi_{(v'',v)}(f(v'')) = \varrho . $$

We have seen that $f(v) \in \mathcal{M}(v)$ and that $f(v) \preceq \varrho$ for all $\varrho \in \mathcal{M}(v)$. This implies fact (2.62).   □

---

[13] This implies that $MIN(X)$ has at most one element for all sets $X \subseteq \mathbf{R}$; therefore, we may write $\min(X)$. If $\preceq$ is not total then $MIN(X)$ may have two incomparable elements, and if $\preceq$ is not identitive then $MIN(X)$ may have two elements $x \neq y$ with $x \equiv y$.

Remark 2.122   a) We can formulate the following symmetric version of Theorem 2.121: Let all arc r be marked with a function ψ_r, and let $H(r \oplus P) = \psi_r(H(P))$ for arcs r and all paths P. Let γ be a fixed node and let $f^-(v)$ be the minimal H-value of all v-γ-paths. If H has the weak suffix oriented Bellman property of type 0 then the following is true for all $v \neq \gamma$:
$$f^-(v) \;=\; \min\{\psi_{(v,v')}(f^-(v')) \mid (v, v') \in \mathcal{R}\}\ .$$
For example, if H is additive we define $\psi_r(\zeta) := \zeta + h(r)$.

    b) A general version of 2.121 has been proven in Theorem 2.113 of [Huck92]. The path function H in that result need only have the weak prefix oriented Bellman property of type 0; it is not required that H is of the form $\phi \circ \circ p$. Also, $\preceq$ need not to be total or identitive.   □

2.4.8 Decision models with Bellman properties

Many sources of literature give formal descriptions of economic decision processes and their costs. Typical examples are [BeDr62] [BeKa65], [Bell67], [KaHe67], [Mor82], [Helm86], [Snie86], [Hck92], [Hck93], [MaCh93], and [deMo94]. Many search strategies for optimal decisions use Bellman properties of cost functions; in particular, this is the case if Dynamic Programming is used to find optimal sequences of decisions.

Probably, all optimization problems in these economic models can be formulated as search problems for optimal paths in graphs; each path in a graph represents a sequence of decisions. If the cost function of decision processes has a Bellman property then the corresponding path function often has a Bellman property, too.

The author of this book has translated the decision processes of [Mor82] and of [Snie86] into the world of paths in graphs. The details can be found in Subsection 2.3.6 of [Huck92] and perhaps in a second edition of this book.

2.5 Further definitions and results related to path functions

In addition to order preservation and Bellman properties, there are many further terms that have something to do with path functions. Several of them are now presented.

2.5.1 Prefix, infix, and suffix monotone path functions

Definition/Remark 2.123 Given a prefix closed set $\mathcal{P}$ of paths and a path function $H : \mathcal{P} \to (\mathbf{R}, \preceq)$. We say that H is *prefix monotone* if $H(Q) \preceq H(P)$ for all paths $P \in \mathcal{P}$ and all prefixes $Q \leq P$.

For example, any additive function $H = SUM_h$ is prefix monotone if and only if $h(r) \geq 0$ for all $r \in \mathcal{R}$; the same is true for all semi-additive functions H.

Prefix monotonicity must not be confused with monotonicity, which has been defined in 2.24. Any function $H = \phi \circ \circ p$ is monotone if and only if all functions ϕ_r are monotonically increasing. In contrast to this, $H = \phi \circ \circ p$ is prefix monotone if $\phi_r(\zeta) \succeq \zeta$ for all r and ζ.   □

Next, we construct a special prefix monotone function; this function describes the cost of the cheapest extension of a path P into a set Γ of goal nodes.

Remark 2.124 Given a digraph $\mathcal{G} = (\mathcal{V}, \mathcal{R})$. Let $\Gamma \subseteq \mathcal{V}$ be a set of goal nodes, and let $s \in \mathcal{V}$ be a start node. Moreover, let $\mathcal{P} \subseteq \mathcal{P}(\mathcal{G})$ be prefix closed.
Let $H : \mathcal{P} \to (\mathbf{R}, \preceq)$ be a path function where $\preceq$ is total and identitive. (E.g., $\mathbf{R}$ may be equal to $\mathbb{R}$, and "$\preceq$" may be equal to "$\leq$".) We assume that H have the X-Γ-minimality property for all paths $X \in \mathcal{P}(s)$.
Then we define the path function $H^* : \mathcal{P} \to \mathbf{R}$ as follows: For all $P \in \mathcal{P}$, let
$$H^*(P) := \min\{H(P \oplus Q) \,|\, \omega(Q) \in \Gamma\}\,.$$
The function H^* is prefix monotone; this is a consequence of the following inequality, which is valid for all paths P and for all arcs r:
$$H^*(P \oplus r) = \min\{H(P \oplus r \oplus Q) \,|\, \omega(Q) \in \Gamma\} \succeq \min\{H(P \oplus Q') \,|\, \omega(Q') \in \Gamma\}\,;$$
the relationship $\succeq$ follows from the fact that each path $P \oplus r \oplus Q$ with $\omega(Q) \in \Gamma$ is also a path of the form $P \oplus Q'$ with $\omega(Q') \in \Gamma$.

We consider the example that $H : \mathcal{P}(\mathcal{G}) \to (\mathbb{R}, \leq)$ is additive. Let
$$(\forall\, v \in \mathcal{V})\ \ p^*(v) := \min\{H(Q) \,|\, \alpha(Q) = v\,,\ \omega(Q) \in \Gamma\}\,.$$
Then $H^*(P) = H(P) + p^*(\omega(P))$ for all paths P.   □

The next result describes a situation in which only finitely many acyclic paths have an H-value not greater than a fixed bound; the path function H is assumed to be prefix monotone.

Lemma 2.125 *Given a locally finite digraph* $\mathcal{G} = (\mathcal{V}, \mathcal{R})$ *and a node* $s \in \mathcal{V}$. *Let* $H : \mathcal{P}(s) \to (\mathbf{R}, \preceq)$ *be a prefix monotone path function where* $\preceq$ *is total and identitive.* *(E.g.,* $\mathbf{R}$ *may be equal to* $\mathbb{R}$, *and "*$\preceq$*" may be equal to "*$\leq$*".) Let* $M \in \mathbf{R}$. *We assume that the sequence* $\big(H([s, v_1, v_2, \dots, v_i])\big)_{i \in \mathbb{N}}$ *has no upper bound for each acyclic infinite path* $[s, v_1, v_2, v_3, \dots)$.[14]
Then there exist only finitely many acyclic paths $Q \in \mathcal{P}(s)$ *with* $H(Q) \preceq M$.

P r o o f : Our idea is similar to the Infinity Lemma (see [Knu82], Vol. 1). Suppose that there exist infinitely many node injective paths $Q \in \mathcal{P}(s)$ with $H(Q) \preceq M$. We generate an infinite path $P = [s, v_1, v_2, \dots)$ such that the H-values of its finite prefixes are bounded.

For this purpose, we recursively construct the paths $P_i = [s, v_1, \dots, v_i]$ such that the following is true for all i:

P_i is a prefix of infinitely many node injective paths Q with $H(Q) \preceq M$.   (2.63)

It is clear that $P_0 := [s]$ has this property. Given P_i with property (2.63). Since $\mathcal{G}$ is locally finite there are only finitely many arcs $r_1, \dots, r_k$ with start node v_i. On the other hand, there exist infinitely many node injective extensions Q of P_i with

---

[14]Note that $\preceq$ is total so that this condition can be formulated as follows: For each $\varrho \in \mathbf{R}$, there exists an i such that $C_0 \left([v_0, \dots, v_i]\right) \succ \varrho$.

the property $H(Q) \preceq M$; this follows from (2.63). Hence, there exists an arc r_κ that is used by infinitely many of these paths Q. Let $v_{i+1} := \omega(r_\kappa)$. Then the path $P_{i+1} := [s, v_1, \ldots, v_i, v_{i+1}]$ has property (2.63) as well.

Then each path P_i is a prefix of infinitely many acyclic paths Q with $H(Q) \preceq M$. Let Q_i be one of these paths Q. Then P_i is acyclic as a prefix of the acyclic path Q_i. Moreover, $H(P_i) \preceq H(Q_i) \preceq M$ because H is prefix monotone. The fact that $H(P_i) \preceq M$ for all i is a contradiction to the assumption that the values $H(P_1)$, $i \in \mathbb{N}$, are unbounded. □

Remark 2.126 If H is not prefix monotone then 2.125 is not always true. A counterexample is the digraph $\mathcal{G} = (\mathcal{V}, \mathcal{R})$ with $\mathcal{V} := \{a_i, b_i \mid i = 0, 1, 2, 3, \ldots\}$ and $\mathcal{R} := \{(a_i, a_i + 1), (a_i, b_i), (b_{i+1}, b_i) \mid i = 0, 1, 2, 3, \ldots\}$. Let $s := a_0$. We define H such that $H(P) = i$ if P ends with a_i or b_i, $i = 0, 1, 2, 3, \ldots$.

Then $H([s, v_1, \ldots, v_i])$ is not bounded for all infinite paths $P = [s, v_1, v_2, \ldots)$. The reason is that the only infinite path with start node $s = a_0$ is the path $P = [a_0, a_1, a_2, a_3, \ldots)$.

Nevertheless there are infinitely many node injective paths Q with $H(Q) \leq 1$. Each path $Q = [s = a_0, a_1, \ldots, a_i, b_i, b_{i-1}, \ldots, b_0]$ has this property. □

2.5.2 Introduction to heuristic functions

Heuristic functions are node functions used to estimate the minimal cost of all paths from a given node to a set of goal nodes. In the following remark, we give a more precise description of heuristic functions.

Remark/Definition 2.127 Given the situation in 2.124. Obviously, the path function H^* is very useful when searching for H-minimal paths from s to Γ; if two paths $P, Q \in \mathcal{P}(s)$ are given then comparing $H^*(P)$ with $H^*(Q)$ answers the question whether P or Q should be continued to Γ.

But unfortunately, H^* itself is not always available. In this case it is often sensible to approximate H^* by a path function H'. If H is additive we shall choose a node function $p : \mathcal{V} \to \mathbb{R}$ to estimate p^*, and we shall use the path function $H' := H + p$ to approximate H^*. (This notation was introduced in Definition 2.1.)

Any node function p that is used as an extimate for p^* is called a *heuristic function*. Several important search strategies for H-optimal s-Γ-paths use heuristic functions; two of the earliest examples are the algorithms A and A*, which are described in [Nils82] and in paragraph 4.2.2.1.6 of this book.

An extended version of heuristic functions is given in [BaMa83]. Every function $\bar{p} : \mathcal{V} \times \mathbb{R} \to \mathbb{R}$ is called a *path dependent heuristic*. If the given path function H is additive then Bagchi and Mahanti approximate H^* with the path function $H''(P) := H(P) + \bar{p}(\omega(P), H(P))$, $P \in \mathcal{P}(\mathcal{G})$. □

2.5.3   Negative cycles

First, we define negative cycles.

Definition 2.128 Given a digraph $\mathcal{G}$ and a path function $H : \mathcal{P} \to (\mathbf{R}, \preceq)$. A cycle X is called a *negative cycle* if there exist paths X_1, X_2 with $H(X_1 \oplus X \oplus X_2) \prec H(X_1 \oplus X_2)$.
We also say that X is a *H-negative cycle* or that *H causes the negative cycle X*. □

Next, we give several results about path functions that do not cause negative cycles.

Lemma 2.129 *If H is prefix monotone and order preserving then $\mathcal{G}$ has no negative cycles.*

P r o o f : If X is a cycle then $H(X_1) \preceq H(X_1 \oplus X)$ by the prefix monotonicity and $H(X_1 \oplus X_2) \preceq H(X_1 \oplus X \oplus X_2)$ by the order preservation. □

The next result is similar to 2.129.

Lemma 2.130 *Given a digraph $\mathcal{G} = (\mathcal{V}, \mathcal{R})$ with $s \in \mathcal{V}$. Let $H : \mathcal{P}(s) \to (\mathbf{R}, \preceq)$ where $\preceq$ is total and identitive. (E.g., $\mathbf{R}$ may be equal to $\mathbb{R}$, and "$\preceq$" may be equal to "$\leq$".) We make the following assumptions:*
  (i) *H is order preserving.*
  (ii) *The sequence $\big(H([s, v_1, v_2, \dots, v_i])\big)_{i \in \mathbb{N}}$ has no upper bound for any infinite path $[s, v_1, v_2, v_3, \dots)$.*[15]
Then $\mathcal{G}$ has no H-negative cycle.

P r o o f : Suppose that X is a negative cycle. Then there exist $P_1 \in \mathcal{P}(s)$ and $P_2 \in \mathcal{P}(\mathcal{G})$ such that $H(P_1 \oplus X \oplus P_2) \prec H(P_1 \oplus P_2)$. This, assumption (i), and the totality of $\preceq$ imply that $H(P_1 \oplus X) \prec H(P_1)$ because otherwise, $H(P_1) \preceq H(P_1 \oplus X)$ and assumption (i) had the consequence that $H(P_1 \oplus P_2) \preceq H(P_1 \oplus X \oplus P_2)$.
The fact that $H(P_1 \oplus X) \prec H(P_1)$ and the order preservation of H imply that $H(P_1 \oplus X^k \oplus U) = H((P_1 \oplus X) \oplus X^{k-1} \oplus U) \preceq H(P_1 \oplus X^{k-1} \oplus U)$ for all $k \in \mathbb{N}$ and all prefixes $U \leq X$. Hence,

$$H(P_1 \oplus X^k \oplus U) \preceq H(P_1 \oplus X^{k-1} \oplus U) \preceq H(P_1 \oplus X^{k-2} \oplus U) \preceq \dots \preceq H(P_1 \oplus U) \quad (2.64)$$

for all $k \in \mathbb{N}$ and all $U \leq X$.
We define the infinite path $P := P_1 \oplus X \oplus X \oplus X \dots$. Then each finite prefix $Q \leq P$ is of the form $Q \leq P_1$ or of the form $Q \leq P_1 \oplus X^k \oplus U$ with $k \in \mathbb{N} \cup \{0\}$ and $U \leq X$; this and (2.64) imply that $H(Q) \preceq \max\{H(Q') \,|\, Q' \leq P_1 \text{ or } Q' = P_1 \oplus U, U \leq X\}$. Hence, the values $H(Q)$ of all proper prefixes $Q \leq P$ are bounded; that is a contradiction to assumption (ii). □

The next result has the following consequence: If H does not cause negative cycles then $H_{\max}$ does not cause negative cycles, too.

---

[15] Note that $\preceq$ is total so that this condition can be formulated as follows: For each $\varrho \in \mathbf{R}$, there exists an i such that $H([s, v_1, \dots, v_i]) \succ \varrho$.

Lemma 2.131 *Let $\mathcal{G} = (\mathcal{V}, \mathcal{R})$ be a digraph, and let $s \in \mathcal{V}$. Let $\widetilde{H}, \widetilde{\widetilde{H}} : \mathcal{P}(s) \to (\mathbf{R}, \preceq)$ where $\preceq$ is total. (E.g., $\mathbf{R}$ may be equal to $\mathbb{R}$, and "$\preceq$" may be equal to "$\leq$".) We assume that $\widetilde{\widetilde{H}}(P) \in MAX\{\widetilde{H}(Q) \mid Q \leq P\}$ for all $P \in \mathcal{P}(s)$. (That means that $\widetilde{\widetilde{H}} = \widetilde{H}_{\max}$ if $\preceq$ is total and identitive.) Also, we assume that $\mathcal{G}$ has no $\widetilde{H}$-negative cycles.*

Then $\mathcal{G}$ has no $\widetilde{\widetilde{H}}$-negative cycles.

P r o o f : The above relationship between $\widetilde{H}$ and $\widetilde{\widetilde{H}}$ and the totality of $\preceq$ have the following consequence:

$$(\forall P \in \mathcal{P}(s))\,(\forall Q \leq P)\quad \widetilde{H}(Q) \preceq \widetilde{\widetilde{H}}(P). \tag{2.65}$$

Let now $X_1 \oplus X \oplus X_2 \in \mathcal{P}(s)$ where X is a cycle. We show that $\widetilde{\widetilde{H}}(X_1 \oplus X_2) \preceq \widetilde{\widetilde{H}}(X_1 \oplus X \oplus X_2)$. For this purpose, we choose a prefix $Q \leq X_1 \oplus X_2$ such that $\widetilde{H}(Q) = \widetilde{\widetilde{H}}(X_1 \oplus X_2)$. If even $Q \leq X_1$ then $Q \leq X_1 \oplus X \oplus X_2$; this and fact (2.65) imply that $\widetilde{\widetilde{H}}(X_1 \oplus X_2) = \widetilde{H}(Q) \preceq \widetilde{\widetilde{H}}(X_1 \oplus X \oplus X_2)$. If not $Q \leq X_1$ there exists a prefix $Q' \leq X_2$ such that $Q = X_1 \oplus Q'$. Then we obtain the following inequality; relationship $(*)$ follows from the assumptions that $\preceq$ is total and X is not a $\widetilde{H}$-negative cycle; relationship $(**)$ follows from $X_1 \oplus X \oplus Q' \leq X_1 \oplus X \oplus X_2$ by applying (2.65).

$$\widetilde{\widetilde{H}}(X_1 \oplus X_2) = \widetilde{H}(Q) = \widetilde{H}(X_1 \oplus Q') \overset{(*)}{\preceq} \widetilde{H}(X_1 \oplus X \oplus Q') \overset{(**)}{\preceq} \widetilde{\widetilde{H}}(X_1 \oplus X \oplus X_2). \qquad \square$$

The next result says: If H is order preserving and does not cause negative cycles then each path P can be replaced by an acyclic path Q with $H(Q) \preceq H(P)$.

Lemma 2.132 *Given a locally finite, digraph $\mathcal{G} = (\mathcal{V}, \mathcal{R})$. Let $s \in \mathcal{V}$. Let $H : \mathcal{P}(s) \to (\mathbf{R}, \preceq)$ be a path function where $\preceq$ is total. (E.g., $\mathbf{R}$ may be equal to $\mathbb{R}$, and "$\preceq$" may be equal to "$\leq$".) We assume that $\mathcal{G}$ has no H-negative cycles. Let $v \in \mathcal{V}$ and $P \in \mathcal{P}(s, v)$. Then an acyclic path $Q \in \mathcal{P}(s, v)$ exists such that $H(Q) \preceq H(P)$.*

P r o o f: We recursively remove cycles from P. Thus, we obtain a sequence $P = P_0, P_1, \ldots, P_k = Q$ of s-v-paths such that the following are true for all $\kappa = 1, \ldots, k$:

(α) If $\kappa < k$ then $P_{\kappa+1}$ is shorter than P_κ and $H(P_{\kappa+1}) \preceq H(P_\kappa)$.
(β) If $\kappa = k$ then P_κ is acyclic. $\tag{2.66}$

First, we define $P_0 := P$.
Suppose that the paths $P = P_0, \ldots, P_j$ are given and that (2.66) is true for all $\kappa = 0, \ldots, j-1$. If P_j is already acyclic then let $k := j$, and (2.66) is also satisfied for $\kappa := j = k$. If, however, P_j is not yet acyclic we write P_j as $P_j = U \oplus X \oplus V$ where X is a substantial cycle. Let $P_{j+1} := U \oplus V$. Then P_{j+1} is an s-v-path, which is shorter than P_j. Moreover, the relationship $H(U \oplus X \oplus V) \prec H(U \oplus V)$ is not valid because $\mathcal{G}$ does not contain negative cycles. This and the totality of $\preceq$ imply that $H(U \oplus V) \preceq H(U \oplus X \oplus V)$ so that $H(P_{j+1}) = H(U \oplus V) \preceq H(U \oplus X \oplus V) = H(P_j)$. Consequently, assertion (2.66) is true for $\kappa := j$. (Note that $\kappa = j < j+1 \leq k$.)
The construction of the paths $P_0, P_1, \ldots$ terminates after finitely many iterations.

This follows from the fact that the paths P_κ become shorter and shorter. That means that a node injective s-v-path $Q = P_k$ can be found after finitely many steps. Then part (β) of (2.66) implies that $H(Q) = H(P_k) \preceq H(P_0) = H(P)$. $\qquad \square$

The next result gives sufficient conditions for the existence of optimal path from a start node s to a set Γ of goal nodes; one of these conditions is that $\mathcal{G}$ has no negative cycles.

Lemma 2.133 *Given a locally finite digraph $\mathcal{G} = (\mathcal{V}, \mathcal{R})$, a node $s \in \mathcal{V}$, and a set $\Gamma \subseteq \mathcal{V}$ of goal nodes. Let $H : \mathcal{P}(s) \to (\mathbf{R}, \preceq)$ be a path function where $\preceq$ is total and identitive. (E.g., $\mathbf{R}$ may be equal to $\mathbb{R}$, and "$\preceq$" may be equal to "$\leq$".) We make the following assumptions:*
  (i) *There exists an s-Γ-path P^+.*
  (ii) *The sequence $\big(H([s, v_1, v_2, \ldots, v_i])\big)_{i \in \mathbb{N}}$ has no upper bound for any node injective infinite path $[s, v_1, v_2, v_3, \ldots)$.[16]*
  (iii) *H is prefix monotone.*
  (iv) *$\mathcal{G}$ has no H-negative cycles.[17]*
Then there exists an H-minimal goal path P^* such that P^* is acyclic.

P r o o f : Let $M := H(P^+)$. Lemma 2.125 implies that there are only finitely many node injective s-Γ-paths $P_1, \ldots, P_k$ with $H(P_\kappa) \preceq M$. Without loss of generality, $H(P_1)$ is minimal among all values $H(P_\kappa)$. Hence, $P^* := P_1$ has the following property:

$$H(P^*) \text{ is minimal among all values } H(P) \text{ where } P \text{ is an acyclic } s\text{-}\Gamma\text{-path.} \qquad (2.67)$$

Moreover, $H(P^*) \preceq H(Q)$ for all s-Γ-paths Q that are not node injective. This follows from (iv) and from result 2.132, which yields an acyclic s-Γ-paths P with $H(P) \preceq H(Q)$; this implies that $H(P^*) \overset{(2.67)}{\preceq} H(P) \preceq H(Q)$.

Hence, P^* is an H-minimal s-Γ-path. Moreover, P^* is acyclic by the definition of P^*. $\qquad \square$

Remark 2.134 Condition (ii) in 2.133 must not be omitted. The digraph in 2.126 is a counterexample. We define $\Gamma := \{b_0\}$, and we define the additive path function $H := SUM_h$ whith the help of the following arc function h:

$$(\forall i = 0, 1, 2, \ldots) \quad h(a_i, a_{i+1}) := \frac{9}{10^{i+1}}, \quad h(a_i, b_i) := \frac{2}{10^i}, \quad h(b_{i+1}, b_i) := 0.$$

Then H is prefix monotone and even order preserving. The graph $\mathcal{G}$ has no H-negative cycles because $\mathcal{G}$ is acyclic. Moreover, the values $H([v_0, v_1, \ldots, v_i]) = 1 - \frac{1}{10^i}$ are bounded by 1 so that condition (ii) in 2.133 is not satisfied. All s-Γ-paths are of the form $Q_i = [a_0, a_1, \ldots, a_i, b_i, b_{i-1}, \ldots, b_0]$. (Recall that $a_0 = s$). Moreover, $H(Q_i) = 1 + \frac{1}{10^i}$. Hence, there exists no C-minimal goal path. $\qquad \square$

---
[16]Note that $\preceq$ is total so that this condition can be formulated as follows: For each $\varrho \in \mathbf{R}$, there exists an i such that $H([s, v_1, \ldots, v_i]) \succ \varrho$.
[17]This assumption may be replaced by the order preservation of H; this follows from Lemma 2.129.

2.5.4 Consistent functions

Consistent functions are studied in several sources of literature. We define two versions
of the term "consistent" in 2.135 and in 2.136. We shall show in 2.137 and in 2.138
that consistent functions can be used to construct prefix monotone path functions. A
motivation of using consistent heuristic functions is given in Remark 2.140.

Definition/Remark 2.135 (*see* [DePe85], *page* 522)
Given an additive cost function $H : \mathcal{P}(\mathcal{G}) \to (\mathbb{R}, \leq)$ and a node function $p : V \to \mathbb{R}$.
We assume that H has the v-v'-minimality property for all nodes v, v'. Moreover, we
define
$$d_H(v, v') := \min\{H(Q) \,|\, Q \in \mathcal{P}(v, v')\}$$
for all nodes v, v'. (This quantity may be interpreted as the distance from v to v'.)
Then p is called *consistent* if $p(v) \leq d_H(v, v') + p(v')$ for all $v, v' \in V$.

For example, let p^* be the node function in Remark 2.124; then $H^*(P) :=
H(P) + p^*(\omega(P))$ is prefix monotone, and 2.137 implies that p^* is consistent.   □

The following definition is closely related to condition E3 in [LT91] and [LTh91].

Definition 2.136 Given a digraph $\mathcal{G} = (V, \mathcal{R})$ and path function $H : \mathcal{P} \to (\mathbb{R}, \preceq)$.
Given a function $f : V \times \mathbb{R} \to \mathbb{R}$ (e.g. an evaluation function $f : V \times \mathbb{R} \to \mathbb{R}_{\equiv}$, see
Remark/Definition 2.13). Then f is called *consistent* if $f(\alpha(r), \varrho) \preceq f(\omega(r), \varrho \odot
h(r))$ for all $r \in \mathcal{R}$ and $\varrho \in \mathbb{R}$.   □

The next result describes a relationship between consistent node functions and prefix
monotone path functions.

Theorem 2.137 *Given the terminology of Definition 2.135. Then the following assertions are equivalent:*

(a) *p is consistent (see Definition 2.135).*   (b) *$H' := H + p$ is prefix monotone.*

P r o o f o f (a)$\Rightarrow$(b) : Let $P = Q \oplus Q' \in \mathcal{P}(\mathcal{G})$. We show that $H'(Q) \leq H'(P)$. Let
$v := \omega(Q)$ and $w := \omega(Q')$. Then the following assertion is true, in which (∗) follows
from the minimality of $d_H(v, w)$:
$$H'(P) = H'(Q \oplus Q') = H(Q \oplus Q') + p(w) = H(Q) + H(Q') + p(w)$$
$$\overset{(*)}{\geq} H(Q) + d_H(v, w) + p(w) \overset{(a)}{\geq} H(Q) + p(v) = H'(Q).$$
Hence, $H'(P) \geq H'(Q)$ if $Q \leq P$.

P r o o f o f (b)$\Rightarrow$(a) : Let $u, v, w \in V$, and let $P \in \mathcal{P}(u, v)$, $Q \in \mathcal{P}_{Min}(v, w)$. Then
$H(Q) = d_H(v, w)$, and consequently, $H(P) + d_H(v, w) = H(P) + H(Q) = H(P \oplus Q)$.
This and assumption (b) yield the following assertions:
$$H(P) + d_H(v, w) \;=\; H(P \oplus Q),$$
$$H'(P) \;\leq\; H'(P \oplus Q).$$
Subtracting the upper equality from the lower inequality, we obtain:
$$p(\omega(P)) - d_H(v, w) = H'(P) - (H(P) + d_H(v, w)) \leq H'(P \oplus Q) - H(P \oplus Q) = p(\omega(Q)).$$

This and the equations $v = \omega(P)$, $w = \omega(Q)$ imply that p is consistent. $\qquad\square$

The next result describes a relationship between consistent functions $f : V \times \mathbf{R} \to \mathbf{R}$ and prefix monotone path functions.

Theorem 2.138 *Given a prefix closed set $\mathcal{P}$ of paths. Let $H : \mathcal{P} \to (\mathbf{R}, \preceq)$ be easily computable of type B (This property has been exactly described in Definition 2.21.) Let $p : V \to \mathbb{R}$ be the node function, let $h : \mathcal{R} \to \mathbf{R}$ be the arc function, and let $\odot : \mathbf{R} \times \mathbf{R} \to \mathbf{R}$ be the binary operation making H easily computable of type B. Moreover, given a function $f : V \times \mathbf{R} \to \mathbf{R}$. We define a further path function $H' : \mathcal{P} \to \mathbf{R}$ as follows: $H'(P) := f(\omega(P), H(P))$ for all paths $P \in \mathcal{P}$. Then the implication (a)$\Rightarrow$(b) concerning the following assertions (a), (b) is is valid.*

    (a) *f is consistent (see Definition 2.136).*   (b) *H' is prefix monotone.*

P r o o f : It suffices to prove that $H'(P \oplus r) \succeq H'(P)$ for all paths P and all arcs r. This is shown with the help of the following inequality.

$$\begin{aligned} H'(P \oplus r) &= f(\omega(P \oplus r), H(P \oplus r)) = f(\omega(r), H(P) \odot h(r)) \overset{(a)}{\succeq} \\ f(\alpha(r), H(P)) &= f(\omega(P), H(P)) = H(P). \end{aligned}$$
$\qquad\square$

2.5.5 Admissibility of path functions, of heuristic functions, and of evaluation functions

We define the term "admissible" and describe its relevance.

Definition 2.139   a) Given an additive path function $H : \mathcal{P}(\mathcal{G}) \to \mathbb{R}$, a set Γ of goal nodes, and a node function $p : V \to \mathbb{R}$. We say that p is *admissible (with respect to H)* if $p(v) \leq H(Q)$ for all nodes v and all v-Γ-paths Q. The admissibility of p means that for each node v, the value $p(v)$ is an optimistic assessment of the best path from v to Γ.

    b) Given a path function $H : \mathcal{P}(\mathcal{G}) \to (\mathbf{R}, \preceq)$ and an evaluation function $f : V \times \mathbf{R} \to \mathbf{R}_{=}$ (see 2.13). We say that f is *admissible (with respect to H)* if $f(\omega(P'), H(P')) \preceq H(P'')$ for all goal paths P'' and for all prefixes $P' \leq P''$. A relationship between the admissibility of f and the admissibility of heuristic functions will be given in 2.141. $\qquad\square$

Remark 2.140 Admissible heuristic functions are often used when searching for H-minimal paths. An example is Algorithm A, which is described in [Nils82]. This search strategy measures the quality of every path P by the value $H'(P) = H(P) + p(\omega(P))$. It is shown in [Nils82] and in paragraph 4.2.2.1.6 of this book that Algorithm A will find an H-minimal path if the heuristic function p is admissible; in this case, Algorithm A is called A*. Admissibility of p means that $H'(P) \preceq H^*(P)$ where H^* is the path function defined in 2.124; the same remark says that H^* is prefix monotone. Then the approximating path function H' should be prefix monotone, too. If not then it may occur that

$H'(P) \succ H(P \oplus r)$ so that $H^*(P \oplus r) \succeq H^*(P) \succeq H'(P) \succ H'(P \oplus r)$. That means that $H'(P \oplus r)$ is a worse approximation of $H^*(P \oplus r)$ than $H'(P)$; this situation can be avoided if H' is chosen as a prefix monotone function. By 2.137, this is equivalent to the fact that p is consistent. $\qquad\square$

The next theorem describes a relationship between admissible heuristic functions and admissible evaluation functions.

Theorem 2.141 *Let $\mathcal{G} = (\mathcal{V}, \mathcal{R})$ be a digraph, and let $s \in \mathcal{V}$ and $\Gamma \subseteq \mathcal{V}$.*
Given the cost structure $(\mathbb{R}, +, 0, \leq)$. Let h be an arc function. (Then the given cost structure yields the additive cost function $H := \lambda := SUM_h$.) Let p be a heuristic function, and let $f(v, \varrho) := \varrho + p(v)$ for all $\varrho \in \mathbb{R}$ and $v \in \mathcal{V}$. Then the following assertions are equivalent:

(a) *p is admissible.*   (b) *f is admissible.*

P r o o f   o f   (a)$\Rightarrow$(b) : Let P'' be a goal path and let P', Q' be two paths with $P'' = P' \oplus Q'$. Then $p(\omega(P')) \leq H(Q')$ by the admissibility of p. Consequently,

$$f\big(\omega(P'), H(P')\big) = H(P') + p\big(\omega(P')\big) \leq H(P') + H(Q') = H(P'').$$

Hence, f is admissible.

P r o o f   o f   (b)$\Rightarrow$(a) : Let P'' be a goal path, and let P', Q' be two paths with $P'' = P' \oplus Q'$. Let $v := \omega(P') = \alpha(Q')$. Then the following inequality is true where $(*)$ follows from the admissibility of f:

$$p(v) = H(P') + p(\omega(P')) - H(P') = f(\omega(P'), H(P')) - H(P') \overset{(*)}{\leq} H(P'') - H(P') = H(Q').$$

Consequently, p is admissible. $\qquad\square$

2.5.6   Triangle inequalities

A triangle inequality is usually an assertion of the form $d(x, x'') \leq d(x, x') + d(x', x'')$ where d denotes the distance in a metric space. Here, we discuss triangle inequalities related to path functions. There exist various assertions that are called 'triangle inequality', and it is somewhat difficult to embed them into a uniform concept.

2.5.6.1   Examples of triangle inequalities

We now give several examples of triangle inequalities.

Example 2.142   a) Given the situation as described in Definition 2.135. In 2.145, we shall show the following triangle for all nodes $v, v', v'' \in \mathcal{V}$:

$$d_H(v, v'') \leq d_H(v, v') + d_H(v', v''). \tag{2.68}$$

b) Let $\mathcal{G} = K_{\mathcal{V}}$ be a complete undirected graph, and let $h : \mathcal{R} \to \mathbb{N} \cup \{0\}$ be an edge function. If $h(\{x, y\})$ is interpreted as the geometric distance between two cities x, y then the following triangle inequality is true for all nodes u, v, w:

$$h(\{u, w\}) \leq h(\{u, v\}) + h(\{v, w\}) \tag{2.69}$$

This triangle inequality plays an important part in Problem 61 on page 383.

c) In [DePe85] and in Definition 2.135, a node function p is called consistent if the following is true for all nodes $v, v' \in \mathcal{V}$:
$$p(v) \leq d_H(v, v') + p(v') \tag{2.70}$$
This formula is called 'triangle inequality' in Dechter's and Pearl's definition of consistent node functions.

d) Given a cost structure $(\mathbf{R}, \odot, \varrho_0, \preceq)$ (see Definition 2.11); let H be easily computable of type B with the help of $\odot$ and an arc function $h : \mathcal{R} \to \mathbf{R}$.
In Definition 7 of [LT91] and of [LT91], the following assertion about $\varrho, \varrho' \in \mathbf{R}$ and $(v, w) \in \mathcal{R}$ is called a triangle inequality:
$$\varrho \preceq \varrho' \odot h(v, w). \tag{2.71}$$
This inequality, too, appears in Lengauer's and Theune's generalized Ford-Bellman algorithm (see [LT91], step (13) in Figure 3.2 or [LTh91], step (9) in Figure 2).

e) Let $\preceq$ be total. Then the following inequality is the negation of (2.71):
$$\varrho' \odot h(v, w) \prec \varrho. \tag{2.72}$$
This inequality is likewise checked in Lengauer's and Theune's generalized Ford-Bellman algorithm (see [LT91], step (14) in Figure 3.2 or [LTh91], step (4) in Figure 2).

In particular, given the cost structure $(\mathbb{R}, +, 0, \leq)$, and let $H = SUM_h$. Moreover, let $P(v)$ and $P(w)$ be two paths with the following properties: $\alpha(P(v)) = \alpha(P(w))$, $\omega(P(v)) = v$, $\omega(P(w)) = w$, $H(P(v)) = \varrho'$, and $H(P(w)) = \varrho$. Then $H(P(v) \oplus (v, w)) = \varrho + h(v, w)$ so that (2.72) is equivalent to
$$H(P(v) \oplus (v, w)) < H(P(w)). \tag{2.73}$$
This inequality is checked in step 6c of Dijkstra's algorithm on page 174 (with $H := C$, $v := \bar{v}$, $w := v'$); also, this inequality is is checked in step 6c of Algorithm BF** on page 185 (with $H := F^{(2)}$, $v := \bar{v}$, $w := v'$). □

2.5.6.2 General triangle inequalities

We define two general types of triangle inequalities; we shall show in 2.145 – 2.148 that many graph theoretic triangle inequalities are special cases of these two types. In several cases, we explain why a particular formula is called a triangle inequality.

Definition 2.143 (*see Figure 4*)
Given the path functions H, H_1, H_2 : $\mathcal{P} \to (\mathbf{R}, \preceq)$ and a binary operation $\odot : \mathbf{R} \times \mathbf{R} \to \mathbf{R}$.

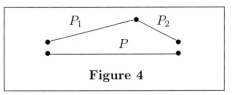
Figure 4

a) We call the following formula a *triangle inequality of type 1*.
$$\left(P, P_1 \oplus P_2 \in \mathcal{P}\right) \wedge \left(H(P) \preceq H(P_1 \oplus P_2)\right). \tag{2.74}$$

b) We call the following formula a *triangle inequality of type 2*.

$$\big(P, P_1, P_2 \in \mathcal{P}\big) \quad \wedge \quad \big(H(P) \preceq H_1(P_1) \odot H_2(P_2)\big). \qquad (2.75)$$

$\square$

Next, we show that the triangle inequality of type 2 is often more general than the one of type 1.

Remark 2.144 A triangle inequality of type 1 can often be expressed as a triangle inequality of type 2.

More precisely, let $\mathcal{P} = \mathcal{P}(\mathcal{G})$, and let $H : \mathcal{P} \to (\mathbf{R}, \preceq)$.

If H is additive then (2.74) is equivalent to $H(P) \leq H(P_1) + H(P_2)$, and this is a special case of (2.75) with $H_1 := H_2 := H$.

A more general situation is that H is easily computable of type B and $H(P \oplus Q) = H(P) \odot H(Q)$ for all paths P, Q. (This equation is shown in 2.23 and in 4.160.) Then (2.74) is equivalent to $H(P) \preceq H(P_1) \odot H(P_2)$, and this is a special case of (2.75) with $H_1 := H_2 := H$.

The following situation is yet more general: There exist two functions $H_1, H_2 : \mathcal{P}(\mathcal{G}) \to (\mathbf{R}, \preceq)$ and a binary operation $\odot$ such that the following is true for all P_1, P_2:

$$H(P_1 \oplus P_2) = H_1(P_1) \odot H_2(P_2). \qquad (2.76)$$

Then (2.74) is trivially equivalent to (2.75).

Now we show how to make (2.76) true. We may assume that that $\mathbf{R}$ has at least the cardinality of $\mathcal{P}(\mathcal{G})$. (If $\mathbf{R}$ does not have this property, we replace $\mathbf{R}$ with $\mathbf{R}' := \mathbf{R} \cup \mathcal{P}(\mathcal{G})$.) Then we choose $H_1, H_2 : \mathcal{P}(\mathcal{G}) \to \mathbf{R}$ as two arbitrary injective path functions; we may choose H_2 such that $H_1 = H_2$. We define the binary operation $\odot : \mathbf{R} \times \mathbf{R} \to \mathbf{R}$ as follows: Let $x, y \in \mathbf{R}$. Then $x \odot y := H\big(H_1^{-1}(x) \oplus H_2^{-1}(y)\big)$ if $x \in H_1\big(\mathcal{P}(\mathcal{G})\big)$, $y \in H_2\big(\mathcal{P}(\mathcal{G})\big)$, and $\omega\big(H_1^{-1}(x)\big) = \alpha\big(H_2^{-1}(y)\big)$. (Note that $H_1^{-1}(x)$ and $H_2^{-1}(y)$ are paths in $\mathcal{G}$.) In all other cases, we choose $x \oplus y$ as an arbitrary element of $\mathbf{R}$.

To see that (2.76) is true we consider two paths P_1, P_2 with $\omega(P_1) = \alpha(P_2)$, and we define $x := H_1(P_1)$, $y := H_2(P_2)$; then $x \in H_1\big(\mathcal{P}(\mathcal{G})\big)$, $y \in H_2\big(\mathcal{P}(\mathcal{G})\big)$, and $\omega\big(H_1^{-1}(x)\big) = \alpha\big(H_2^{-1}(y)\big)$; hence,

$$H_1(P_1) \odot H_2(P_2) = x \odot y = H\big(H_1^{-1}(x) \oplus H_2^{-1}(y)\big) = H(P_1 \oplus P_2).$$

Consequently, H_1, H_2 and $\odot$ have indeed property (2.76).

The advantage of the above construction is that it works in almost every case. The disadvantage is that H_1, H_2, and $\odot$ may have a very complicated structure; for example, if $\mathbf{R} = \mathbb{R}$ it is sometimes difficult to express $f(x, y) := x \odot y$ in terms of the operations $+, -, *, /$. In this situation, the simple triangle inequality (2.74) itself is better than its complicated formulation as a type 2 inequality.

Another consequence of the sometimes complicated structure of H_1, H_2, and $\odot$ is the requirement that a "reasonable" triangle inequality of type 2 should be a formula about *simply structured* functions H_1, H_2.

$\square$

2.5.6.3   Several results about triangle inequalities

We show several simple assertions about triangle inequalities. The main purpose of the results 2.145 – 2.148 is to formulate the graph theoretic triangle inequalities in 2.142 as inequalities of type 1 or type 2.

Remark 2.145 Let H and d_H be the same functions as in 2.135. Let $P^*(v, w)$ be an H-minimal v-w-path for each pair (v, w) of nodes. Then (2.68) is equivalent to $H(P^*(v, v'')) \leq H(P^*(v, v')) + H(P^*(v', v''))$ so that (2.68) may be considered as a triangle inequality of type 2.

This inequality is true for all nodes v, v', v''. To see this we observe that $H(P^*(v, v'')) \leq H(P^*(v, v') \oplus P^*(v', v''))$ by the H-minimality of $P^*(v, v'')$ in $\mathcal{P}(v, v'')$. Moreover, $H(P^*(v, v') \oplus P^*(v', v'')) = H(P^*(v, v')) + H(P^*(v', v''))$ because H is additive. Consequently, $d_H(v, v'') = H(P^*(v, v'')) \leq H(P^*(v, v') \oplus P^*(v', v'')) = H(P^*(v, v')) + H(P^*(v', v'')) = d_H(v, v') + d_H(v', v'')$. □

Remark 2.146 Given the situation as described in 2.142 b). Let $H := SUM_h$. Then (2.69) is equivalent to the formula $H([v, v'']) \leq H([v, v']) + H([v', v''])$, which is a triangle inequality of type 2. □

Next, we show that (2.70) is equivalent to a triangle inequality of type 2.

Remark 2.147 Let $H : \mathcal{P}(\mathcal{G}) \to \mathbb{R}$ be an additive path function. We assume that H has the v-v'-minimality property for all nodes v, v'.
Moreover, given a node function $p : \mathcal{V} \to \mathbb{R}$.
We define the path functions $\tilde{H}$, $\tilde{H}_1$, $\tilde{H}_2 : \mathcal{P}(\mathcal{G}) \to \mathbb{R}$ as follows: For each path P, let $\tilde{H}(P) := p(\alpha(P))$, $\tilde{H}_1(P) := H(P)$ (i.e., $\tilde{H}_1 = H$), and $\tilde{H}_2(P) := p(\alpha(P))$ (i.e., $\tilde{H}_2 = \tilde{H}$).
Let $P^*(x, y)$ be an H-minimal x-y-path for each pair (x, y) of nodes; let $P(x)$ be an arbitrary path starting with x. Then for all nodes v, v', assertion (2.70) is equivalent to the following triangle inequality of type 2:

$$\tilde{H}(P(v)) \leq \tilde{H}_1(P^*(v, v')) + \tilde{H}_2(P(v')). \tag{2.77}$$

This equivalence follows from the equations $\tilde{H}(P(v)) = p(\alpha(P(v))) = p(v)$, $\tilde{H}_1(P^*(v, v')) = H(P^*(v, v')) = d_H(v, v')$, and $\tilde{H}(P(v')) = p(\alpha(P(v'))) = p(v')$.

An immediate consequence of the equivalence of (2.70) and (2.77) is the following fact: p is consistent if and only if $\tilde{H}(P) \leq \tilde{H}_1(P^*(v, v')) + \tilde{H}_2(P(v'))$ for all nodes v, v' and for all paths $P \in \mathcal{P}(v)$, $P^*(v, v') \in \mathcal{P}_{Min}(v, v')$, and $P(v') \in \mathcal{P}(v')$. □

Now we discuss triangle inequality (2.71).

Remark 2.148 Given the situation as described in 2.142 d). Let $P_1 \in \mathcal{P}(s, v)$, $P_2 := [v, w]$, and $P := P_1 \oplus P_2$. Moreover, let $\varrho := H(P)$ and $\varrho' := H(P_1)$. Then $\varrho' \odot h(v, w) = H(P_1) \odot h(v, w) = H(P_1 \oplus (v, w)) = H(P_1 \oplus P_2)$, and $\varrho' \odot h(v, w) = H(P_1) \odot H(P_2)$. Consequently, (2.71) is equivalent to each of the triangle inequalities $(2.78)(\alpha)$ and (β), which are of type 1 and of type 2, respectively.

$$(\alpha) \ H(P_1) \preceq H(P_1 \oplus P_2) \qquad (\beta) \ H(P_1) \preceq H(P_1) \odot H(P_2) \qquad (2.78)$$ $\qquad \square$

Next, we prove several triangle inequalities.

Theorem 2.149 *Given a path function $H : \mathcal{P}(\mathcal{G}) \to (\mathbf{R}, \preceq)$; we assume that $\preceq$ is total and identitive. (E.g., $\mathbf{R}$ may be equal to $\mathbb{R}$, and "$\preceq$" may be equal to "$\leq$".)*
Let v, v' be two nodes, and let Γ be a set of nodes.
We assume that H has the v-Γ-minimality property.
Then all paths $P \in \mathcal{P}_{Min}(v, \Gamma)$, $P_1 \in \mathcal{P}(v, v')$, and $P_2 \in \mathcal{P}(v', \Gamma)$ meet the following triangle inequality of type 1:

$$H(P) \ \preceq \ H(P_1 \oplus P_2).$$

P r o o f : The path $P_1 \oplus P_2$ is one of the candidates when searching for an H-minimal v-Γ-path. This and $P \in \mathcal{P}_{Min}(v, \Gamma)$ imply that $H(P) \preceq H(P_1 \oplus P_2)$. $\qquad \square$

Remark 2.150 Theorem 2.149 is more general than the result in Remark 2.145 about the validity of (2.68). To see this we define $P := P^*(v, v'')$, $P_1 := P^*(v, v')$, and $P_2 := P^*(v', v'')$. Then $d_H(v, v'') = H(P)$, $d_H(v, v') = H(P_1)$, and $d_H(v', v'') = H(P_2)$; consequently, $d_H(v, v'') = H(P) \overset{\text{Theorem 2.149}}{\leq} H(P_1 \oplus P_2) = H(P_1) + H(P_2) = d_H(v, v') + d_H(v', v'')$.
That means that (2.68) is a consequence of Theorem 2.149. $\qquad \square$

The following result is a generalization of 2.149; roughly speaking, each of the paths P and $P_1 \oplus P_2$ is appended to a prefix X.

Theorem 2.151 *Given a path function $H : \mathcal{P}(\mathcal{G}) \to (\mathbf{R}, \preceq)$; we assume that $\preceq$ is total and indentitive. (E.g., $\mathbf{R}$ may be equal to $\mathbb{R}$, and "$\preceq$" may be equal to "$\leq$".)*
Let s, v, v' be three nodes, and let Γ be a set of nodes. Let X be an s-v-path.
Suppose that H have the X-Γ-minimality property.
Then the following is true for all paths $P \in \mathcal{P}_{Min}(X, \Gamma)$, $P_1 \in \mathcal{P}(v, v')$, and $P_2 \in \mathcal{P}(v', \Gamma)$:

$$H(X \oplus P) \ \preceq \ H(X \oplus P_1 \oplus P_2). \qquad (2.79)$$

P r o o f : The path $X \oplus P_1 \oplus P_2$ is one of the candidates when searching for an H-minimal path $X \oplus P'$ with $\omega(P') \in \Gamma$. This and $P \in \mathcal{P}_{Min}(X, \Gamma)$ imply that $H(X \oplus P) \preceq H(X \oplus P_1 \oplus P_2)$. $\qquad \square$

Remark 2.152 Theorem 2.151 is somewhat more general than 2.149; this can be seen by considering the special case $s = v$ and $X = [s] = [v]$; then the current result implies Theorem 2.149.
Note that (2.79) is a triangle inequality of type 1. To see this we define $Q := X \oplus P$, $Q_1 := X \oplus Q_1$, and $Q_2 := P_2$; then (2.79) is equivalent to $H(Q) \preceq H(Q_1 \oplus Q_2)$. $\square$

2.5.7 Nonmisleading node functions, proper node functions and nonmisleading path functions

We now define what it means that node function is nonmisleading or proper; both properties are defined in [BaMa83]; moreover, we define the property "nonmisleading" of path functions.

Definition/Remark 2.153 Given a digraph $\mathcal{G} = (\mathcal{V}, \mathcal{R})$, a node $s \in \mathcal{V}$ and a set $\Gamma \subseteq \mathcal{V}$. Let $\mathcal{P} = \mathcal{P}(\mathcal{G})$ or $\mathcal{P} = \mathcal{P}(s)$. Let $H : \mathcal{P} \to \mathbb{R}$ be additive. Moreover, let Γ be a set of nodes. We assume that H have the X-Γ-minimality property for all paths $X \in \mathcal{P}(\mathcal{G})$. Let $p^* : \mathcal{V} \to \mathbb{R}$ be the node function defined in 2.124. Moreover, let $p : \mathcal{V} \to \mathbb{R}$ be a further node function.

Then p is called *nonmisleading* if the assertion

$$H(P_1) + p(\omega(P_1)) < H(P_2) + p(\omega(P_2)) \ \Rightarrow \ H(P_1) + p^*(\omega(P_1)) \leq H(P_2) + p^*(\omega(P_2)) \quad (2.80)$$

is true for all paths $P_1, P_2 \in \mathcal{P}$ with the following properties: $\alpha(P_1) = \alpha(P_2)$, $\omega(P_1) \neq \alpha(P_1)$, and $\omega(P_2) \neq \alpha(P_2)$.

Let p satisfy condition (2.80); then p actually does not mislead anybody who searches for an optimal s-Γ-path; if $H(P_1)+p(\omega(P_1)) \neq H(P_2)+p(\omega(P_2))$ then comparing these values yields the correct $\leq$-relationship between the (unknown) values $H(P_1) + p^*(\omega(P_1))$ and $H(P_2) + p^*(\omega(P_2))$; thus we can easily decide which of the paths P_1 and P_2 should preferably be extended to Γ.

We say that p is *proper* if assertion (2.80) is true for all paths $P_1, P_2 \in \mathcal{P}$ with the following properties: $\alpha(P_1) = \alpha(P_2)$, $\omega(P_1) \neq \alpha(P_1)$, $\omega(P_2) \neq \alpha(P_2)$, P_1 is not a prefix of P_2, and P_2 is not a prefix of P_1. — Of course, any nonmisleading node function p is proper. □

Next, we define the property "nonmisleading" of path functions. The definitions of "nonmisleading" will be compared in Remark 2.155.

Definition 2.154 Given a digraph $\mathcal{G} = (\mathcal{V}, \mathcal{R})$, a node $s \in \mathcal{V}$ and a set $\Gamma \subseteq \mathcal{V}$. Let $\mathcal{P} = \mathcal{P}(s)$ or $\mathcal{P} = \mathcal{P}(\mathcal{G})$.
Given the path functions $H, H' : \mathcal{P} \to (\mathbf{R}, \preceq)$ where $\preceq$ is total and identitive. (E.g., $\mathbf{R}$ may be equal to $\mathbb{R}$, and "$\preceq$" may be equal to "$\leq$".) We assume that H has the X-Γ-minimality property for all paths $X \in \mathcal{P}$.

Then we say that the path function H' is H-*nonmisleading* if

$$H'(P_1) \prec H'(P_2) \ \Rightarrow \ H(P_1 \oplus Q_1) \preceq H(P_2 \oplus Q_2)$$

for all paths $P_1, P_2 \in \mathcal{P}$ with $\alpha(P_1) = \alpha(P_2)$, $\omega(P_1) \neq \alpha(P_1)$, $\omega(P_2) \neq \alpha(P_2)$ and for all paths $Q_1, Q_2 \in \mathcal{P}_{Min}^{(H)}(P_i, \Gamma)$.[18]

The path function H is called *nonmisleading* if H is H-nonmisleading; that is, H is nonmisleading with respect to itself. □

---

[18] We write $\mathcal{P}_{Min}^{(H)}(X, \Gamma)$ in order to emphasize that $\mathcal{P}_{Min}(X, \Gamma)$ is defined by comparing H-values (and not H'-values).

Remark 2.155 The definition of "nonmisleading" in 2.153 is a special version of the definition in 2.154. This can be seen as follows:

Let H, p, p^* be the same functions as in 2.153. Let $P_1, P_2 \in \mathcal{P}$ with $\alpha(P_1) = \alpha(P_2)$ and $\omega(P_i) \neq \alpha(P_i)$ $(i = 1, 2)$; let $Q_i \in \mathcal{P}_{Min}^{(H)}(P_i, \Gamma)$, $i = 1, 2$. Then $H(Q_i) = p^*(\omega(P_i))$ so that

$$H(P_i \oplus Q_i) = H(P_i) + H(Q_i) = H(P_i) + p^*(\omega(P_i)) \quad (i = 1, 2). \tag{2.81}$$

Let now $\widehat{H} := H + p$. Then (2.81) implies that condition (2.80) in 2.153 is equivalent to the following assertion:

$$\widehat{H}(P_1) < \widehat{H}(P_2) \implies H(P_1 \oplus Q_1) \leq H(P_2 \oplus Q_2).$$

Consequently, the node function p is nonmisleading in the sense of 2.153 if and only if the path function $\widehat{H} := H + p$ is H-nonmisleading in the sense of 2.154. □

Remark 2.156 Any path function $H : \mathcal{P}(s) \to (\mathbf{R}, \preceq)$ is nonmisleading if and only if the following is true for all paths $P_1, P_2 \in \mathcal{P}(s)$ with $\omega(P_1) \neq \alpha(P_1)$, $\omega(P_2) \neq \alpha(P_2)$ and for all paths $Q_1 \in \mathcal{P}_{Min}(P_1, \Gamma)$, $Q_2 \in \mathcal{P}_{Min}(P_2, \Gamma)$:

$$H(P_1) \prec H(P_2) \implies H(P_1 \oplus Q_1) \preceq H(P_2 \oplus Q_2).$$

Consequently, H is nonmisleading if the equality $H(P \oplus Q) = H(P)$ is valid for all paths $P \in \mathcal{P}(\mathcal{G})$ and for all optimal extensions $Q \in \mathcal{P}_{Min}(P, \Gamma)$.

For example, let H be an additive function with an arc function h, and let $h(r) \geq 0$ for all arcs. We assume that for each node v, there exists a path $Q' \in \mathcal{P}(v)$ with $\omega(Q') \in \Gamma$ and $H(Q') = 0$. Then $H(Q) = 0$ for all optimal paths from v to Γ, and $H(P \oplus Q) = H(P) + H(Q) = H(P)$. □

Remark 2.157 Let $\mathcal{P} := \mathcal{P}(\mathcal{G})$ or $\mathcal{P} := \mathcal{P}(s)$. Suppose that $H : \mathcal{P} \to (\mathbf{R}, \preceq)$ has the X-Γ-minimality property for all paths $X \in \mathcal{P}$. Then the fact that H is nonmisleading can be formulated as an L-condition with $L \subseteq \mathcal{L}$. More precisely, H-nonmisleading if and only if H is $\prec$-$\preceq$-preserving with respect tthe following set $\mathbb{L}'_{\min}(H, \Gamma)$:

$$\mathbb{L}'_{\min}(H, \Gamma) := \left\{ \begin{pmatrix} [v] & P_1 & [v_1'] \\ [v] & P_2 & [v_2'] \\ [v] & P_1 & Q_1 \\ [v] & P_2 & Q_2 \end{pmatrix} \in \mathcal{L} \,\middle|\, \begin{array}{l} v, v_1', v_2' \in \mathcal{V}, \; v_1' \neq v, \; v_2' \neq v \\ \omega(Q_1), \omega(Q_2) \in \Gamma, \\ Q_i \in \mathcal{P}_{Min}^{(H)}(P_i, \Gamma), \; i = 1, 2 \end{array} \right\}.$$

(Recall the conditions about $\mathcal{P}$ in (2.32). If $\mathcal{P} = \text{def}(H) = \mathcal{P}(s)$ then $(X_1 = P_1 \in \mathcal{P} \wedge X_2 = P_2 \in \mathcal{P})$ implies that $v = s$; this implies that $(X_3 = P_1 \oplus Q_1 \in \mathcal{P} \wedge X_4 = P_2 \oplus Q_2 \in \mathcal{P})$.)

By the way, we have intentionally chosen the name "$\mathbb{L}'_{\min}(H, \Gamma)$" such that it is similar to the name "$\mathbb{L}'_{\min}$" in Definition/Remark 2.58 because both sets have a similar structure. □

2.5.8 Lexicographic path functions

Here, we define and study lexicographic path functions. The idea is to mark each arc r with a symbol $h(r)$ and to assign each path $r_1 \oplus \ldots \oplus r_k$ the string $h(r_1) \ldots h(r_k)$; we use a lexicographic order to compare these strings with each other.

Search algorithms for lexicographically minimal paths are described in [AnMa87].

We start with the definition of lexicographic orderings for strings.

Definition 2.158 Let Σ be a finite alphabet with an order relation $<_\Sigma$. Let Σ^* be the set of all finite strings over Σ, and let ε be the empty word. If $w \in \Sigma^*$, the length $\ell(w)$ of w is the number of symbols in w. For all k, let Σ_k be the set of all words in Σ^* of a length $\leq k$.

The *lexicographic order* $\leq_{ab,z}$ is defined in analogy to the usual alphabetic order (for example in telephone books). Let $v, w \in \Sigma^*$. Then $v \leq_{ab,z} w$ if one of the following assertions (i), (ii) is true:

  (i) There exist $u, v', w' \in \Sigma^*$, $\alpha, \beta \in \Sigma$ with $v = u\alpha v'$, $w = u\beta w'$, and $\alpha <_\Sigma \beta$.
  (ii) There exists $v' \in \Sigma^*$ such that $w = vv'$.

In particular, $ab \leq_{ab,z} z$; that is the reason why using the subscript $"ab, z"$.

The *lexicographic order* $\leq_{z,ab}$ is defined such that shorter words are smaller than longer ones. Let $v, w \in \Sigma^*$. Then $v \leq_{z,ab} w$ if one of the following assertions (i), (ii) is true:

  (i) $\ell(v) < \ell(w)$.   (ii) $\ell(v) = \ell(w)$ and $v \leq_{ab,z} w$.

In particular, $z \leq_{z,ab} ab$; that is the reason why using the subscript $"z, ab"$. ☐

Next, we define a path functions that translates the arcs of any path into a string; two paths are compared with each other by comparing their strings lexicographically.

Definition 2.159 Given a digraph $\mathcal{G} = (\mathcal{V}, \mathcal{R})$, a finite alphabet Σ, and an arc function $h : \mathcal{R} \to \Sigma$. Let $P = [v_0, \ldots, v_k]$.
Then we define $\mathcal{W}_h(P) := \varepsilon$ if $k = 0$, and $\mathcal{W}_h(P) := h(v_0, v_1)h(v_1, v_2), \ldots, h(v_{k-1}, v_k)$ if $k > 0$. (For example, $\mathcal{W}_h([v_0, v_1, v_2, v_3]) = axb$ if $h(v_0, v_1) = "a"$, $h(v_1, v_2) = "x"$, and $h(v_2, v_3) = "b"$.)

The path function $\mathcal{W}_h$ is called *lexicographic* if we a use the lexicographic order $\leq_{ab,z}$ or $\leq_{z,ab}$ to compare paths; in this case, we write $\mathcal{W}_h : \mathcal{P} \to (\Sigma^*, \leq_{ab,z})$ and $\mathcal{W}_h : \mathcal{P} \to (\Sigma^*, \leq_{z,ab})$, respectively. ☐

Next, we define two real-valued path functions that are closely related to the lexicographic ones.

Definition 2.160 Given a digraph $\mathcal{G} = (\mathcal{V}, \mathcal{R})$, a natural number $N > 1$, and an arc function $h : \mathcal{R} \to \{1, \ldots, N-1\}$. Let $P = [v_0, \ldots, v_k]$
We define $\mathbf{n}_h(P)$ as the number $0.h(v_1, v_2)h(v_2, v_3) \ldots h(v_{k-1}, v_k)$ in the N-nary number system. More formally, $\mathbf{n}_h(P) := 0$ if $k = 0$, and $\mathbf{n}_h(P) := \sum_{i=1}^{k} h(v_{i-1}, v_i) \cdot N^{-i}$ if $k > 0$.
We define $\mathbf{N}_h(P)$ as the number $h(v_1, v_2)h(v_2, v_3) \ldots h(v_{k-1}, v_k)$ in the N-nary number system. More formally, $\mathbf{N}_h(P) := 0$ if $k = 0$, and $\mathbf{N}_h(P) := N^k \cdot \mathbf{N}_h(P) = \sum_{i=1}^{k} h(v_{i-1}, v_i) \cdot N^{k-i}$ if $k > 0$.
For example, let $N = 10$, $P = [v_0, v_1, v_2, v_3]$, $h(v_0, v_1) = 6$, $h(v_1, v_2) = 9$, $h(v_2, v_3) = 6$; then $\mathbf{n}_h(P) = 0.696$, and $\mathbf{N}_h(P) = 696$. ☐

Remark 2.161   The lexicographic functions $\mathcal{W}_h : \mathcal{P}(\mathcal{G}) \to (\Sigma^*, \leq_{ab,z})$ and $\mathcal{W}_h : \mathcal{P}(\mathcal{G}) \to (\Sigma^*, \leq_{z,ab})$ and the functions $\mathbf{n}_h$, $\mathbf{N}_h$ are prefix monotone.   □

Next, we describe the close relationship between the path functions in Definition 2.159 and the ones in Definition 2.160.

Theorem 2.162 *Let* $N > 1$ *and let* $\Sigma = \{a_1, \ldots, a_{N-1}\}$ *with* $a_1 <_\Sigma a_2 <_\Sigma \ldots <_\Sigma a_{N-1}$. *Let* $h : \mathcal{R} \to \{1, \ldots, N-1\}$, *and let* $\widehat{h}(r) := a_{h(r)}$ *for all* $r \in \mathcal{R}$.
Then $\mathbf{n}_h : \mathcal{P}(\mathcal{G}) \to (\mathbb{R}, \leq)$ *is order equivalent to* $\mathcal{W}_{\widehat{h}} : \mathcal{P}(\mathcal{G}) \to (\Sigma^*, \leq_{ab,z})$, *and* $\mathbf{N}_h : \mathcal{P}(\mathcal{G}) \to (\mathbb{R}, \leq)$ *is order equivalent to* $\mathcal{W}_{\widehat{h}} : \mathcal{P}(\mathcal{G}) \to (\Sigma^*, \leq_{z,ab})$.

P r o o f :   The following three assertions are equivalent for all paths P:

$$\mathbf{n}_h(P) = 0 \,.\, x_1 \ldots x_l . \qquad \mathbf{N}_h(P) = x_1 \ldots x_l . \qquad \mathcal{W}_{\widehat{h}}(P) = a_{x_1} \ldots a_{x_l} .$$

This observation is helpful to show that the following are true for all paths $P, Q \in \mathcal{P}(\mathcal{G})$:

$$\mathbf{n}_h(P) \leq \mathbf{n}_h(Q) \iff \mathcal{W}_{\widehat{h}}(P) \leq_{ab,z} \mathcal{W}_{\widehat{h}}(Q) \quad \text{and}$$

$$\mathbf{N}_h(P) \leq \mathbf{N}_h(Q) \iff \mathcal{W}_{\widehat{h}}(P) \leq_{z,ab} \mathcal{W}_{\widehat{h}}(Q).$$

We leave the proof of these equivalences to the reader.   □

Next, we describe structural properties of $\mathbf{n}_h$ and $\mathbf{N}_h$.

Theorem 2.163 *If the function* h *in 2.160 is injective then* $\mathbf{n}_h$ *is nonmisleading.*

P r o o f :   Let $P_1, P_2 \in \mathcal{P}(\mathcal{G})$ with $\alpha(P_1) = \alpha(P_2)$ and with $\mathbf{n}_h(P_1) < \mathbf{n}_h(P_2)$; moreover, let $Q_i \in \mathcal{P}_{Min}(P_i, \Gamma)$, $i = 1, 2$. We must show that $\mathbf{n}_h(P_1 \oplus Q_1) \leq \mathbf{n}_h(P_2 \oplus Q_2)$.
We consider two cases. The first is that P_1 is a prefix of P_2. Then there exists a path Q with $P_1 \oplus Q = P_2$. Then $Q \oplus Q_2$ is a connection from $\omega(P_1)$ to Γ. This and the optimality of Q_1 imply that $\mathbf{n}_h(P_1 \oplus Q_1) \leq \mathbf{n}_h(P_1 \oplus Q \oplus Q_2) = \mathbf{n}_h(P_2 \oplus Q_2)$.
The other case is that P_1 is not a prefix of P_2. Then P_2 is not a prefix of P_1 for the following reasons: If $P_2 \leq P_1$ then $\mathbf{n}_h(P_2) \leq \mathbf{n}_h(P_1)$ by 2.161 although we have assumed that $\mathbf{n}_h(P_1) < \mathbf{n}_h(P_2)$.
Hence, P_1 and P_2 are no prefixes of each other. Consequently, P_1 is of the form $P_1 = [v_0, \ldots, v_j, v_0', \ldots v_k']$, and P_2 is of the form $P_2 = [v_0, \ldots, v_j, v_0'', \ldots, v_l'']$ where $j \geq 0$, $k, l > 0$, and $v_0' \neq v_0''$. Then

$$\mathbf{n}_h(P_1) = 0 \,.\, h(v_0, v_1) \ldots h(v_{j-1}, v_j) h(v_j, v_0') h(v_0', v_1') \ldots h(v_{k-1}', v_k'),$$
$$\mathbf{n}_h(P_1) = 0 \,.\, h(v_0, v_1) \ldots h(v_{j-1}, v_j) h(v_j, v_0'') h(v_0'', v_1'') \ldots h(v_{l-1}'', v_l'').$$

The injectivity of h implies that $h(v_j, v_0') \neq h(v_j, v_0'')$; this and the assumption $\mathbf{n}_h(P_1) < \mathbf{n}_h(P_2)$ imply $h(v_j, v_0') < h(v_j, v_0'')$. The values $\mathbf{n}_h(P_i \oplus Q_i)$, $i = 1, 2$, are obtained by appending further digits at the end of $\mathbf{n}_h(P_i)$ (if $\ell(Q_i) > 0$). Hence, $\mathbf{n}_h(P_1 \oplus Q_1)$ and $\mathbf{n}_h(P_2 \oplus Q_2)$ are of the following form:

$$\mathbf{n}_h(P_1 \oplus Q_1) = 0 \,.\, h(v_0, v_1) \ldots h(v_{j-1}, v_j) h(v_j, v_0') h(v_0', v_1') \ldots h(v_{k-1}', v_k') a_1 \ldots a_{m'},$$
$$\mathbf{n}_h(P_2 \oplus Q_2) = 0 \,.\, h(v_0, v_1) \ldots h(v_{j-1}, v_j) h(v_j, v_0'') h(v_0'', v_1'') \ldots h(v_{l-1}'', v_l'') b_1 \ldots b_{m''}.$$

The leftmost different digits are again $h(v_j, v_0')$ and $h(v_j, v_0'')$. Consequently, $\mathbf{n}_h(P_1 \oplus Q_1) < \mathbf{n}_h(P_2 \oplus Q_2)$.   □

Remark 2.164 The injectivity of h in 2.163 must not be omitted. For example, let $\mathcal{G} = (\mathcal{V}, \mathcal{R})$ with $\mathcal{V} := \{s, x, y, z\}$, $\Gamma := \{z\}$, and $\mathcal{R} := \{(s, x), (s, y), (y, x), (x, z)\}$. Let $N = 10$, and let $h(s, x) := 1$, $h(s, y) := 1$, $h(y, x) := 3$, and $h(x, z) := 9$. Then h is not injective.

We define $P_1 := [s, x]$, $P_2 := [s, y, x]$, and $Q_1 := Q_2 := [x, z]$. Then for each $i = 1, 2$, the path Q_i is the only (and in consequence, the optimal) continuation of P_i to Γ. It is $\mathbf{n}_h(P_1) = 0.1 < 0.13 = \mathbf{n}_h(P_2)$ but $\mathbf{n}_h(P_1 \oplus Q_1) = 0.19 > 0.139 = \mathbf{n}_h(P_2 \oplus Q_2)$. That means that $\mathbf{n}_h$ is not nonmisleading. $\qquad\square$

Theorem 2.165 *If* $h : \mathcal{R} \to \{1, \ldots, N - 1\}$ *is injective and* $\mathcal{G}$ *is acyclic then* $\mathbf{n}_h$ *is $\le$- and $<$-preserving.*

P r o o f : Let P_1, P_2, Q be three paths with $\alpha(P_1) = \alpha(P_2)$ and $\alpha(Q) = \omega(P_1) = \omega(P_2)$.

First, we consider the case that $\mathbf{n}_h(P_1) < \mathbf{n}_h(P_2)$. Then P_1 is not a prefix of P_2 and P_2 is not a prefix of P_1. Otherwise, there existed a path X with $P_2 = P_1 \oplus X$ or $P_1 = P_2 \oplus X$. In the first case, $\alpha(X) = \omega(P_1) = \omega(P_2) = \omega(X)$, and in the second case, $\alpha(X) = \omega(P_2) = \omega(P_1) = \omega(X)$. Hence, X is a cycle. The fact that $\mathcal{G}$ is acyclic implies that $\ell(X) = 0$, and consequently, $P_1 = P_2$; that is a contradiction to $\mathbf{n}_h(P_1) < \mathbf{n}_h(P_2)$.

The further argumentation follows the proof of 2.163. The values $\mathbf{n}_h(P_i \oplus Q)$ are generated by appending digits to $\mathbf{n}_h(P_i)$, $i = 1, 2$. As seen in 2.163, this does not influence the $<$-order of the $\mathbf{n}_h$-values. Consequently, $\mathbf{n}_h(P_1 \oplus Q) < \mathbf{n}_h(P_2 \oplus Q)$.

(For example, let $N = 10$ and $\mathbf{n}_h(P_1) = 0.135$, $\mathbf{n}_h(P_2) = 0.1675$. If $\mathbf{N}_h(Q) = 892$ then $\mathbf{n}_h(P_1 \oplus Q) = 0.135892 < 0.1675892 = \mathbf{n}_h(P_2 \oplus Q)$.)

Now we consider the case that $\mathbf{n}_h(P_1) = \mathbf{n}_h(P_2)$. Appending Q to P_1 and P_2 means that the N-ary digits effected by Q are appended to $\mathbf{n}_h(P_1)$ and $\mathbf{n}_h(P_2)$. Hence, $\mathbf{n}_h(P_1 \oplus Q) = \mathbf{n}_h(P_2 \oplus Q)$.

We have seen that $\mathbf{n}_h(P_1) < \mathbf{n}_h(P_2)$ implies $\mathbf{n}_h(P_1 \oplus Q) < \mathbf{n}_h(P_2 \oplus Q)$ and that $\mathbf{n}_h(P_1) = \mathbf{n}_h(P_2)$ implies $\mathbf{n}_h(P_1 \oplus Q) = \mathbf{n}_h(P_2 \oplus Q)$. Consequently, $\mathbf{n}_h$ is $<$- and $\le$-preserving. $\qquad\square$

Remark 2.166 The assertion in 2.165 may be false if $\mathcal{G}$ is not acyclic. For example, let $\mathcal{G} = (\mathcal{V}, \mathcal{R})$ with $\mathcal{V} := \{s, x, y, z\}$ and $\mathcal{R} := \{(s, x), (x, y), (y, x), (x, z)\}$. Let $N = 10$, and let $h(s, x) := 1$, $h(x, y) := 2$, $h(y, x) := 3$, and $h(x, z) := 9$. Then h is injective but $\mathcal{G}$ is not acyclic.

We define $P_1 := [s, x]$, $P_2 := [s, x, y, x]$, and $Q := [x, z]$. Then $\mathbf{n}_h(P_1) = 0.1 < 0.123 = \mathbf{n}_h(P_2)$ but $\mathbf{n}_h(P_1 \oplus Q) = 0.19 > 0.1239 = \mathbf{n}_h(P_2 \oplus Q)$. That means that $\mathbf{n}_h$ is not $<$-preserving and not order preserving.

Also, the assertion in 2.165 may be false if h is not injective. To see this we recall the example in Remark 2.164. The graph $\mathcal{G}$ is acyclic but h is not injective. Let $Q := [x, z]$. Then $\mathbf{n}_h(P_1) = 0.1 < 0.13 = \mathbf{n}_h(P_2)$ but $\mathbf{n}_h(P_1 \oplus Q) = 0.19 > 0.139 = \mathbf{n}_h(P_2 \oplus Q)$. That means that $\mathbf{n}_h$ is not $<$-preserving and not order preserving. $\qquad\square$

Theorem 2.167 *The function* $\mathbf{N}_h$ *is in any case order preserving and* $<$*-preserving.*

P r o o f : Let $p(v) := 0$ for all nodes, and let $\phi_r(\xi) := N \cdot \xi + h(r)$ for all $r \in \mathcal{R}$ and all $\xi \in \mathbb{R}$. Then each function ϕ_r is strictly increasing, and $\mathbf{N}_h = \phi \circ op$. Consequently, $\mathbf{N}_h$ is order preserving and $<$-preserving. $\qquad\qquad\square$

Next, we describe several relationships between path functions $\mathcal{W}_{\widehat{h}}$, $\mathbf{n}_h$, $\mathbf{N}_h$ and the strategies Breadth-First-Search (BFS) and Depth-First-Search (DFS) in trees.

Remark 2.168 Given a finite alphabet $\Sigma = \{a_1, \ldots, a_{N-1}\}$; let $a_1 <_\Sigma a_2 <_\Sigma \cdots <_\Sigma a_{N-1}$. Let $k \geq 0$.
We define $\mathcal{T}(N, k) = (\mathcal{V}, \mathcal{R})$ as the rooted tree with the following properties: All leaves are of height k, and each internal node v has exactly $(N-1)$ sons.[19] Let $n := |\mathcal{V}|$, let s be the root of $\mathcal{T}(J, k)$, and let $P(v)$ be the unique path from s to $v \in \mathcal{V}$.
We assume that an arc function $h : \mathcal{R} \to \{1, \ldots, N-1\}$ is given such that h maps each set $\mathcal{R}^+(v)$ bijectively onto $\{1, \ldots, N-1\}$. Moreover, we define $\widehat{h}(r) := a_{h(r)}$ for all arcs $r \in \mathcal{R}$.
We assume that Breadth-First-Seach (BFS) and Depth-First-Search (DFS) are applied to the nodes of $\mathcal{T}(N, k)$ and that BFS and DFS mark each node when visiting it the first time. It often occurs that there are several candidates $u_1, \ldots, u_j$ that may be marked in the next step of BFS of of DFS; this situation can only occur if all $u_1, \ldots, u_j$ have the same father u. We assume that in this case, BFS and DFS prefer the node u_i with the property $h(u, u_i) = \min\{h(u, u_1), \ldots, h(u, u_j)\}$; this node has the additional properties that $\mathbf{n}_h(P(u_i)) = \min\{\mathbf{n}_h(P(u_1)), \ldots, \mathbf{n}_h(P(u_j))\}$ and $\mathbf{N}_h(P(u_i)) = \min\{\mathbf{N}_h(P(u_1)), \ldots, \mathbf{N}_h(P(u_j))\}$.
It is well-known that then BFS and DFS behave as follows:
If BFS marks the nodes $x_1, \ldots, x_n$ of $\mathcal{V}$ in this order then $\mathbf{N}_h(P(x_1)) < \mathbf{N}_h(P(x_2)) < \cdots < \mathbf{N}_h(P(x_n))$, and $\mathcal{W}_{\widehat{h}}(P(x_1)) <_{z,ab} \mathcal{W}_{\widehat{h}}(P(x_2)) <_{z,ab} \cdots <_{z,ab} \mathcal{W}_{\widehat{h}}(P(x_n))$.
If DFS marks the nodes $y_1, \ldots, y_n$ of $\mathcal{V}$ in this order then $\mathbf{n}_h(P(y_1)) < \mathbf{n}_h(P(y_2)) < \cdots < \mathbf{n}_h(P(y_n))$, and $\mathcal{W}_{\widehat{h}}(P(y_1)) <_{ab,z} \mathcal{W}_{\widehat{h}}(P(y_2)) <_{ab,z} \cdots <_{ab,z} \mathcal{W}_{\widehat{h}}(P(y_n))$. $\quad\square$

2.5.9 Dioid properties of paths in graphs

Here we define dioids, and we describe their relationships to order preservation.

Definition 2.169 A *weak dioid* is a quadruple $(X, Y, \boxplus, \boxtimes)$ with the following properties: X and Y are two nonempty sets; the components $\boxplus : X \times X \to X$ and $\boxtimes : X \times Y \dashrightarrow X$ are two binary operations; the operation $\boxplus$ is associative; that means that $(x \boxplus x') \boxplus x'' = x \boxplus (x' \boxplus x'')$ for all $x, x', x'' \in X$.
A *dioid* (*path algebra, semiring*) is a weak dioid $(X, Y, \boxplus, \boxtimes)$ with the following properties:

---

[19] Recall that $\mathcal{R}^+(v)$ is defined as the set of all arcs with start node v.

- $X = Y$, and $\text{def}(\boxtimes) = X \times Y = X \times X$.
- $x_1 \boxplus x_2 = x_2 \boxplus x_1$, and $(x_1 \boxtimes x_2) \boxtimes x_3 = x_1 \boxtimes (x_2 \boxtimes x_3)$ for all $x_1, x_2, x_3 \in X$.
- There exist two neutral elements $\mathbf{0}, \mathbf{1} \in X$ such that $x \boxplus \mathbf{0} = \mathbf{0} \boxplus x = x$,
  $x \boxtimes \mathbf{1} = \mathbf{1} \boxtimes x = x$, and $x \boxtimes \mathbf{0} = \mathbf{0} \boxtimes x = \mathbf{0}$ for all $x \in X$.
- $x_1 \boxtimes (x_2 \boxplus x_3) = (x_1 \boxtimes x_2) \boxplus (x_1 \boxtimes x_3)$, and $(x_1 \boxplus x_2) \boxtimes x_3 = (x_1 \boxtimes x_3) \boxplus (x_2 \boxtimes x_3)$
  for all $x_1, x_2, x_3 \in X$.

Dioids are defined in [PanR87], [PanR89], [Rote90], [LenT91], and [Ju94].   □

In the definition of weak dioids, we have not required any distributive laws for $\boxplus$ and $\boxtimes$. Next, we show that there are close relationships between order preservation of path functions and distributive laws for weak dioids.

Theorem 2.170 *Let* $H : \mathcal{P}(\mathcal{G}) \to (\mathbb{R}, \leq)$ *be a path function. Let* $(\mathcal{P}(\mathcal{G}), \mathcal{P}(\mathcal{G}), \boxplus, \boxtimes)$ *be the weak dioid with the following operations* $\boxplus$ *and* $\boxtimes$:

$$(\forall\, P', P'', P, Q)\ P' \boxplus P'' := \left\{ \begin{array}{ll} P' & \text{if} \quad H(P') \leq H(P'') \\ P'' & \text{if} \quad H(P') > H(P'') \end{array} \right\}, \quad P \boxtimes Q := P \oplus Q.$$

(*Then* $\boxplus$ *is associative. Applying* $\boxplus$ *to* k *paths* $P_1, \ldots, P_k$ *always yields the first path* P_κ *in the sequence* $(P_1, P_2, \ldots, P_k)$ *that has the following property:* $H(P_\kappa) = \min\{H(P_1), \ldots, H(P_k)\}$; *this outcome is independent of the position of the brackets.*)

Then the following assertions are equivalent:

(a) *H is $<$- and $=$-preserving with respect to the set* $\mathbb{L}'_2 \cap \mathbb{L}''_1$ *of matrices.* [20]

(b) *For all paths* P', P'', Q *with* $\omega(P') = \omega(P'') = \alpha(Q)$, *the following distributive law is valid:* $(P' \boxplus P'') \boxtimes Q = (P' \boxtimes Q) \boxplus (P'' \boxtimes Q)$.

P r o o f   o f   (a)⇒(b) : Given three paths P', P'', Q with $\omega(P') = \omega(P'') = \alpha(Q)$. The first case is that $P' \boxplus P'' = P'$. Then $H(P') \leq H(P'')$, and (a) implies that $H(P' \oplus Q) \leq H(P'' \oplus Q)$. Hence, $(P' \boxtimes Q) \boxplus (P'' \boxtimes Q) = (P' \oplus Q) \boxplus (P'' \oplus Q) = P' \oplus Q = (P' \boxplus P'') \boxtimes Q$.
The second case is that $P' \boxplus P'' = P''$. Then $H(P'') < H(P')$, and (a) implies that $H(P'' \oplus Q) < H(P' \oplus Q)$. Hence, $(P' \boxtimes Q) \boxplus (P'' \boxtimes Q) = (P' \oplus Q) \boxplus (P'' \oplus Q) = P'' \oplus Q = (P' \boxplus P'') \boxtimes Q$.

P r o o f   o f   (b)⇒(a) : Let (b) be true, and let P', P'', Q be three paths with $\omega(P') = \omega(P'') = \alpha(Q)$.
First, we consider the case that $H(P') < H(P'')$. Then $P'' \boxplus P' = P'$ by the definition of $\boxplus$.[21] Hence, $(P'' \boxplus P') \boxtimes Q = P' \boxtimes Q$. This is used in $(*)$ of the following equation:

$$(P'' \oplus Q) \boxplus (P' \oplus Q) = (P'' \boxtimes Q) \boxplus (P' \boxtimes Q) \overset{(b)}{=} (P'' \boxplus P') \boxtimes Q \overset{(*)}{=} P'' \boxtimes Q = P'' \oplus Q.$$

---

[20] The sets $\mathbb{L}'_2$ and $\mathbb{L}''_1$ were defined in Example 2.47.
Assertion (a) means that for all paths P', P'', Q with $\omega(P') = \omega(P'') = \alpha(Q)$ (but possibly $\alpha(P') \neq \alpha(P'')$), the following are true:
$H(P') < H(P'') \Rightarrow H(P' \oplus Q) < H(P'' \oplus Q)$, and $H(P') = H(P'') \Rightarrow H(P' \oplus Q) = H(P'' \oplus Q)$.
By the way, $<$- and $=$-preservation imply $\leq$-preservation.

[21] We have intentially concluded that $P'' \boxplus P' = P'$ and not that $P' \boxplus P'' = P'$; the latter conclusion would not yield equation (2.82) but it would yield the equation $(P' \oplus Q) \boxplus (P'' \oplus Q) = P' \oplus Q$; this assertion is not strong enough to conclude that $H(P' \oplus Q) < H(P'' \oplus Q)$.

It follows that
$$(P'' \oplus Q) \boxplus (P' \oplus Q) = P'' \oplus Q \qquad (2.82)$$
This is only possible if $H(P' \oplus Q) < H(P'' \oplus Q)$.

Next, we consider the case that $H(P') = H(P'')$; we will show that $H(P' \oplus Q) = H(P'' \oplus Q)$. This is trivial if $P' = P''$. Now let $P' \neq P''$. Then $P' \boxplus P'' = P' \neq P'' = P'' \boxplus P'$. This and the definition of $\boxtimes$ imply that $(P' \boxplus P'') \boxtimes Q = P' \oplus Q \neq P'' \oplus Q = (P'' \boxplus P') \boxtimes Q$. Applying (b) and the definition of $\boxtimes$, we obtain
$$(P' \oplus Q) \boxplus (P'' \oplus Q) \neq (P'' \oplus Q) \boxplus (P' \oplus Q). \qquad (2.83)$$
Now suppose that $H(P' \oplus Q) \neq H(P'' \oplus Q)$; then $H(P' \oplus Q) < H(P'' \oplus Q)$ or $H(P'' \oplus Q) < H(P'' \oplus Q)$; if $H(P' \oplus Q) < H(P'' \oplus Q)$ then the definition of $\boxplus$ implies that $(P' \oplus Q) \boxplus (P'' \oplus Q) = P' \oplus Q = (P'' \oplus Q) \boxplus (P' \oplus Q)$, which is in conflict with (2.83); if $H(P'' \oplus Q) < H(P' \oplus Q)$ then the definition of $\boxplus$ implies that $(P' \oplus Q) \boxplus (P'' \oplus Q) = P'' \oplus Q = (P'' \oplus Q) \boxplus (P' \oplus Q)$, which is in conflict with (2.83) as well. Hence, $H(P' \oplus Q) = H(P'' \oplus Q)$.

We have seen that $H(P') < H(P'')$ implies $H(P' \oplus Q) < H(P'' \oplus Q)$ and that $H(P') = H(P'')$ implies $H(P' \oplus Q) = H(P'' \oplus Q)$. Consequently, (a) is true. $\qquad \square$

Theorem 2.171 *Let H be order preserving with respect to $\mathbb{L}'_2 \cap \mathbb{L}''_1$. Let $\boxplus$ and $\boxtimes$ be the same operations as in 2.170. Then for all paths P', P'', Q with $\omega(P') = \omega(P'') = \alpha(Q)$, the following modified distributive law is true: $H((P' \boxplus P'') \boxtimes Q) = H((P' \boxtimes Q) \boxplus (P'' \boxtimes Q))$.*

P r o o f: Let $P = P' \boxplus P''$. Then $H(P) = \min(H(P'), H(P''))$. This and the order preservation of H imply that
$$H((P' \boxplus P'') \boxtimes Q) = H(P \oplus Q) = \min\big(H(P' \oplus Q), H(P'' \oplus Q)\big).$$
Moreover, the definition of $\boxplus$ and $\boxtimes$ implies that $H((P' \boxtimes Q) \boxplus (P'' \boxtimes Q)) = H((P' \oplus Q) \boxplus (P'' \oplus Q)) = \min(H(P' \oplus Q), H(P'' \oplus Q))$.
So, we may conclude that $H((P' \boxplus P'') \boxtimes Q) = \min\big(H(P' \oplus Q), H(P'' \oplus Q)\big) = H((P' \boxtimes Q) \boxplus (P'' \boxtimes Q))$. $\qquad \square$

Theorem 2.172 *Let $H : \mathcal{P}(\mathcal{G}) \to \mathbf{R}$ be easily computable of type B with an arc function $h : \mathcal{R} \to \mathbb{R}$ and a binary operation $\odot$. Let $x \boxplus y := \min(x, y)$ and $x \boxtimes y := x \odot y$ for all real numbers x, y. Then the following assertions are true:*

a) *If $\boxtimes = \odot$ is associative and has a neutral element ϱ_0, and if H is order preserving with respect to $\mathbb{L}'_2 \cap \mathbb{L}''_1$ then the following distributive law is valid for all paths P', P'', Q with $\omega(P') = \omega(P'') = \alpha(Q)$:*
$$\big(H(P') \boxplus H(P'')\big) \boxtimes H(Q) = \big(H(P') \boxtimes H(Q)\big) \boxplus \big(H(P'') \boxtimes H(Q)\big).$$

b) *Suppose that the following distributive law is valid for all real numbers x, y, z (and not only for values of H): $(x \boxplus y) \boxtimes z = (x \boxtimes z) \boxplus (y \boxtimes z)$. Then H is order preserving with regard to $\mathbb{L}'_2 \cap \mathbb{L}''_1$.*

P r o o f   o f   a) :   We define the operations $\boxplus$ and $\boxtimes = \oplus$ in the same way as in 2.171. Then relationship (1) in the next equation follows from $\boxplus = min$ and from

$\boxtimes$ $= \odot$; assertion (2) follows from the definition of $\boxplus$; assertion (3) follows from Theorem 2.23 and from $\oplus = \boxtimes$; assertion (4) follows from 2.171; relation (5) follows from the definition of $\boxplus$ and from $\min = \boxplus$; assertion (6) follows from $\boxtimes = \oplus$; relation (7) follows from Theorem 2.23, and relation (8) follows from $\odot = \boxtimes$.

$$\left(H(P')\,\boxplus\,H(P'') \right) \boxtimes H(Q) \overset{(1)}{=} \left(\min(H(P'), H(P'')) \right) \odot H(Q)$$
$$\overset{(2)}{=} H(P' \boxplus P'') \odot H(Q) \overset{(3)}{=} H((P' \boxplus P'') \boxtimes Q) \overset{(4)}{=} H((P' \boxtimes Q) \boxplus (P'' \boxtimes Q))$$
$$\overset{(5)}{=} H(P' \boxtimes Q)\,\boxplus\,H(P'' \boxtimes Q) \overset{(6)}{=} H(P' \oplus Q)\,\boxplus\,H(P'' \oplus Q)$$
$$\overset{(7)}{=} \left(H(P') \odot H(Q) \right) \boxplus \left(H(P'') \odot H(Q) \right)$$
$$\overset{(8)}{=} \left(H(P')\,\boxtimes\,H(Q) \right) \boxplus \left(H(P'')\,\boxtimes\,H(Q) \right).$$

This equation implies the desired distributive law.

P r o o f   o f   b) : Let $f_z(u) := u \odot z = u\,\boxtimes\,z$ for all $u, z \in \mathbb{R}$. Then f_z is monotonically increasing; to see this we assume that $x \le y$. Then $\min(x, y) = x$ so that $x\,\boxplus\,y = x$. Then the following fact is true where $(*)$ follows from the distributive law.

$$\min(f_z(x), f_z(y)) = f_z(x)\,\boxplus\,f_z(y) = (x\,\boxtimes\,z)\,\boxplus\,(y\,\boxtimes\,z) \overset{(*)}{=}$$
$$(x\,\boxplus\,y)\,\boxtimes\,z = x\,\boxtimes\,z = f_z(x).$$

Hence, $f_z(x) \le f_z(y)$ if $x \le y$, and f_z has been shown to be monotonically increasing. We recall that H is easily computable of type B. Hence, for all paths P and all arcs r, the following recursion formula is valid: $H(P \oplus r) = H(P) \odot h(r) = f_{h(r)}(H(P))$. Let $p(v) := H([v])$, $v \in V$, and let $\phi_r := f_{h(r)}$, $r \in \mathcal{R}$. The recursion formula implies that $H(P \oplus r) = \phi_r(P)$ for all paths P and all arcs r. Consequently, $H = \phi \circ\circ p$ where all functions ϕ_r are monotonically increasing. Our assertion follows from the modified version of Theorem 2.28 that has been described in Remark 2.50. $\qquad\square$

Chapter 3

Combinatorial results about paths in graphs

This chapter is mainly about the properties of very long or very short paths in graphs. Most results are obtained with combinatorial methods. Many recent sources of literature are quoted in this chapter; this shows the great interest of graph theorists for combinatorial questions about paths in graphs.

We focus on the length of extremal paths, but we also consider their intersections, i.e. common nodes, edges or arcs of two extremal paths. Many results in this chapter say essentially the following:

If a graph satisfies a particular condition then there exists a cycle
or an acyclic path of a length $\leq l$ or $\geq L$.

We may consider this type of assertion as a result about extremal paths because it implies that the shortest cycle or acyclic path has a length $\leq l$ and the longest cycle or acyclic path has a length $\geq L$.

The current part of the book is written in another style than the previous one. In Chapter 2, we presented new concepts and results in the context of path functions; most of our theorems were carefully proven. This chapter, however, is an overview over the literature. Most of the the concepts and results we describe are quoted from earlier sources. We show theorems only if their proofs are short and simple; otherwise we omit the proofs, trusting that the quoted articles or the reports in Mathematical Reviews are without mistakes. Instead of proving theorems, we classify and compare many results. In consequence, the reader finds an ordered and commented survey of the literature about combinatorial results on paths in graphs.

A further difference between the chapters 2 and 3 is the following: The previous part of the book was about generalized and sometimes complicated path functions. In this chapter, however, we mainly consider the length function, which describes the number of arcs or edges used by a path; other additive path functions or even non-additive ones are very seldom mentioned in this chapter.

Moreover, this chapter is the only one in this book in which undirected graphs play a more important rôle than digraphs.

In this chapter, all cycles and other paths are node injective if we do not say the contrary. The abbreviation "MR" means Mathematical Reviews; for example, MR 85a:05067 is the abstract no. 05067 in Volume 85a of the Mathematical Reviews.

The current chapter is structured according to the following principles:
- First, we consider very short paths, then paths of arbitrary lengths, then very long paths. (Roughly speaking, we consider longer and longer paths.)
- We first consider digraphs and then undirected graphs.
- We first consider cycles, and then we give results about other paths.

3.1 Basic definitions and results

Here, we give several graph theoretic definitions and results that are essential in this chapter. Many terms have almost the same meaning in directed and in undirected graphs; we define these terms only for one type of graphs and shall use them in the undirected and in the directed case.

Definition 3.1 Let $\mathcal{G}$ be a finite, strongly connected digraph. We define the following minimum and maximum degrees:

$$\delta_1^+(\mathcal{G}) := \min\{\deg^+(v) \,|\, v \in \mathcal{V}\}, \qquad \Delta_1^+(\mathcal{G}) := \max\{\deg^+(v) \,|\, v \in \mathcal{V}\},$$

$$\delta_1^-(\mathcal{G}) := \min\{\deg^-(v) \,|\, v \in \mathcal{V}\}, \qquad \Delta_1^-(\mathcal{G}) := \max\{\deg^-(v) \,|\, v \in \mathcal{V}\},$$

$$\delta_1^{\min}(\mathcal{G}) := \min\{\delta_1^+, \delta_1^-\}, \quad \delta_1^{\max}(\mathcal{G}) := \max\{\delta_1^+, \delta_1^-\}, \quad \Delta_1^{\max}(\mathcal{G}) := \max\{\Delta_1^+, \Delta_1^-\},$$

$$\delta_1(\mathcal{G}) := \min\{\deg(v) \,|\, v \in \mathcal{V}\}, \qquad \delta_2(\mathcal{G}) := \min\{\deg^+(u) + \deg^-(v) \,|\, (u,v) \notin \mathcal{R}\}.$$

We shall often use the abbreviations $\delta_1^+, \Delta_1^+, \delta_1^-$, etc. $\qquad\qquad\square$

Next, we define several terms in the context of extremal paths.

Definition 3.2 Let $\mathcal{G}$ be a directed or undirected graph.
The *girth* $g(\mathcal{G})$ of $\mathcal{G}$ is the minimal length of any substantial node injective cycle. (A path is substantial if it has a length > 0.)
The *circumference* $c(\mathcal{G})$ of $\mathcal{G}$ is the maximal length of any node injective cycle.
The *distance* $d(x, y)$ of two nodes x and y is the minimal length of any x-y-path; we define $d(x, y) := \infty$ if such a path does not exist. If $\mathcal{G}$ is directed then $d(x, y)$ may be unequal to $d(y, x)$ for some nodes x, y; if, however, $\mathcal{G}$ is an undirected graph then $d(x, y) = d(y, x)$ for all nodes x, y.
The *diameter* of $\mathcal{G}$ is defined as $d(\mathcal{G}) := \max\{d(x, y) \,|\, x, y \in \mathcal{V}\}$. The diameter is closely related to shortest paths: $d(\mathcal{G})$ is the maximal length of all shortest x-y-paths where x and y are nodes in $\mathcal{G}$. $\qquad\qquad\square$

Remark 3.3 The diameter of a digraph is important for broadcasting problems: Suppose that a message must be transferred from a single processor v_0 to all other processors of a network $\mathcal{G}$; we assume that in each time unit, all processors that already have got the message forward their information to all successors. Then the complete broadcasting process may take $d(\mathcal{G})$ time units in the worst case; this situation occurs if we choose v_0 such that there exists a node w with $d(v_0, w) = d(\mathcal{G})$. □

Next, we define centres and radii of graphs.

Definition 3.4 Let $\mathcal{G}$ be an undirected graph.
For each node v, we define $d_{\max}(v)$ as as the maximal distance between v and any node $w \in \mathcal{G}$; more formally, $d_{\max}(v) := \max\{d(v, w) \,|\, w \in \mathcal{V}\}$.
The *radius* of $\mathcal{G}$ is defined as $r(\mathcal{G}) := \min\{d_{\max}(v) \,|\, v \in \mathcal{V}\}$.
(In contrast to this, the diameter of $\mathcal{G}$ has the property $d(\mathcal{G}) = \max\{d_{\max}(v) \,|\, v \in \mathcal{V}\}$.)

Let $\mathcal{G}$ be a digraph.
For each node v, let $d_{\max}^+(v) := \max\{d(v, w) \,|\, w \in \mathcal{V}\}$ and $d_{\max}^-(v) := \max\{d(w, v) \,|\, w \in \mathcal{V}\}$. Preferring $d_{\max}^+$ to $d_{\max}^-$, we define the *radius of a digraph* as $r(\mathcal{G}) := \min\{d_{\max}^+(v) \,|\, v \in \mathcal{V}\}$.
A *centre of a digraph* is a node v with $d_{\max}^+(v) = r(\mathcal{G})$. □

Next, we define several further terms in the context of undirected graphs.

Definition 3.5 Let $\mathcal{G}$ be a connected and simple undirected graph with n nodes. (A simple undirected graph has no loops and no parallel edges.)
A set $\mathcal{V}' \subseteq \mathcal{V}$ of nodes is called *independent* if no vertices $x, y \in \mathcal{V}'$ are connected by an edge; also, we say that the elements of $\mathcal{V}'$ are independent.
The maximal cardinality $\alpha(\mathcal{G})$ of any independent set is called the *independence number* of $\mathcal{G}$; we shall sometimes write α instead of $\alpha(\mathcal{G})$. □

Definition 3.6 Let $\mathcal{G}$ be a connected and simple undirected graph with n nodes.
The *node connectivity number* $\kappa(\mathcal{G})$ is the minimum number of nodes that must be removed to generate a disconnected induced subgraph. $\mathcal{G}$ is called *k-node connected* if $\kappa(\mathcal{G}) \geq k$.
If $\mathcal{G}$ is equal to the complete graph $\mathcal{K}_\mathcal{V}$ then $\kappa(\mathcal{G}) = \infty$; the reason for this definition is that removing any set of nodes from $\mathcal{K}_\mathcal{V}$ never yields an induced subgraph with more than one connected component.
Let $\kappa(\mathcal{G}) = 1$. A *cutpoint* is a node $v \in \mathcal{V}$ with the property that the subgraph induced by $\mathcal{V}\backslash\{v\}$ is not connected.
The *edge connectivity number* $\lambda(\mathcal{G})$ is the minimal number of edges that must be removed to generate a disconnected induced subgraph. $\mathcal{G}$ is called *k-edge connected* if $\lambda(\mathcal{G}) \geq k$.
For example, if $\mathcal{G} = \mathcal{K}_\mathcal{V}$ then $\lambda(\mathcal{G}) \leq |\mathcal{V}| - 1$ because choosing node $v \in \mathcal{V}$ and deleting all edges $\{v, w\}$, $w \in \mathcal{V}\backslash\{v\}$, yields a subgraph in which v is separated from all other nodes.
A *bridge* is an edge r with the property that removing r destroys the connectivity

of $\mathcal{G}$. The graph $\mathcal{G}$ is called *bridgeless* if $\mathcal{G}$ is 2-edge connected.

We shall use the abbreviations κ for $\kappa(\mathcal{G})$ and λ for $\lambda(\mathcal{G})$.   □

Definition/Remark 3.7 (*See* [Chva73] *and* [BVMS90], *page* 59)
Let $\mathcal{G}$ be a simple, connected, undirected graph with n nodes.
$\mathcal{G}$ is called *t-tough* if the following is true for any subset $\mathcal{V}' \subseteq \mathcal{V}$:

> If $\mathcal{V} \setminus \mathcal{V}'$ induces a disconnected subgraph of $\mathcal{G}$ then this subgraph
> has not more than $|\mathcal{V}'|/t$ connected components.   (3.1)

The *toughness* $\tau = \tau(\mathcal{G})$ of the graph $\mathcal{G}$ is the greatest number t such that $\mathcal{G}$ is t-tough.

If $\mathcal{G} = \mathcal{K}_\mathcal{V}$ then $\tau(\mathcal{G}) := \infty$. This definition is reasonable because the premise of (3.1) is always false and (3.1) is consequently always true; therefore, $\mathcal{G} = \mathcal{K}_\mathcal{V}$ is t-tough even if t is arbitrarily great.

Let $\mathcal{G}$ be 1-tough. Then $\mathcal{G}$ is 2-node connected. To see this we suppose that there existed a node v_0 such that the graph $\mathcal{G}_0$ induced by $\mathcal{V} \setminus \{v_0\}$ were not connected. It follows from (3.1) that $\mathcal{G}_0$ would have at most $|\{v_0\}|/t = |\{v_0\}|/1 = 1$ connected components, which is a contradiction to the assumption that $\mathcal{G}_0$ is not connected.   □

Definition 3.8 Let $\mathcal{G}$ be a simple and connected undirected graph. Let $k \geq 1$. We define the following quantities:

$$\delta_k(\mathcal{G}) := \min \left\{ \sum_{i=1}^k \deg(v_k) \,\middle|\, v_1, \ldots, v_k \text{ are pairwise distinct and independent} \right\},$$

$$\overline{\delta}_k(\mathcal{G}) := \min \left\{ \left| \bigcup_{i=1}^k \mathcal{N}(v_i) \right| \,\middle|\, v_1, \ldots, v_k \text{ are pairwise distinct and independent} \right\},$$

$$\overline{\overline{\delta}}_k(\mathcal{G}) := \min \left\{ \left| \bigcup_{i=1}^k \mathcal{N}(v_i) \right| \,\middle|\, v_1, \ldots, v_k \text{ are pairwise distinct} \right\}.$$

Then $\overline{\overline{\delta}}_1(\mathcal{G}) = \overline{\delta}_1(\mathcal{G}) = \delta_1(\mathcal{G})$ and $\overline{\overline{\delta}}_k \leq \overline{\delta}_k(\mathcal{G}) \leq \delta_k(\mathcal{G})$ for all $k = 2, 3, \ldots, n$.

The number $\delta_1(\mathcal{G})$ is called the *minimum degree* of $\mathcal{G}$.

We define $\Delta_1(\mathcal{G}) := \max\{\deg(v) \,|\, v \in \mathcal{V}\}$; this number is called the *maximum degree* of $\mathcal{G}$.

We abbreviate these notations by δ_k, $\overline{\delta}_k$, $\overline{\overline{\delta}}_k$ and Δ_1, respectively.   □

Next, we define the transmission of a graph.

Definition 3.9 Let $\mathcal{G}$ be an undirected or a directed graph.
The *transmission* of $\mathcal{G}$ is defined as $\sigma := \sigma(\mathcal{G}) := \sum_{(u,v) \in \mathcal{V} \times \mathcal{V}} d(u, v)$.
This is an addition over all ordered pairs (u, v) although $d(u, v) = d(v, u)$ in the undirected case.   □

Relationships between graph theoretic parameters are studied in [Xu91]. The author of that paper considers the minimum degree δ_1, the node connectivity κ, the girth g, the circumference c and many other parameters. Modified versions of connectivity are described in [Fau93]. The next result describes relationships between node and edge connectivity.

Theorem 3.10

   a) *If $\mathcal{G}$ is a finite and simple undirected graph $\mathcal{G}$ then $\kappa(\mathcal{G}) \leq \lambda(\mathcal{G}) \leq \delta_1(\mathcal{G})$.*

   b) *For each $\delta \geq 2$, there exist a finite and simple undirected graph $\mathcal{G}$ with $\kappa(\mathcal{G}) = 1$ and $\lambda(\mathcal{G}) = \delta_1(\mathcal{G}) = \delta$. (That means that we must not conclude from δ-edge-connectivity to a high node connectivity.)*

P r o o f   o f   a) : (*See* [Ju94], *Theorem* 11.6.1 *and* [Whit32].) Let v be a node of $\mathcal{G}$ with $\deg(v) = \delta_1$. Removing all edges with endpoint v yields a disconnected graph. Therefore, $\lambda(\mathcal{G}) \leq \delta_1(\mathcal{G})$.

We now show that $\kappa(\mathcal{G}) \leq \lambda(\mathcal{G})$. If $\lambda(\mathcal{G}) = 1$ then there exists a bridge $r = \{u, v\}$; then removing the node u destroys the connectivity as well. Let $k := \lambda(\mathcal{G}) \geq 2$. Then there exist k edges $r_1, \ldots, r_k$ such that removing $r_2, \ldots, r_k$ generates a graph with a bridge $r_1 = \{u, v\}$. Another way to destroy all connections $r_2, \ldots, r_k$ is the following: For each $\kappa = 2, \ldots, k$, delete an endpoint w_κ of r_κ with $w_\kappa \notin \{u, v\}$; the w_κ's need not be pairwise distinct. If the resulting graph $\mathcal{G}'$ is not connected any more then removing the nodes $w_2, \ldots, w_k$ has destroyed the connectivity; if, however, $\mathcal{G}'$ is still connected then deleting u or v destroys the connectivity. In any case, we can generate a disconnected graph by deleting k or fewer nodes. Therefore, $\kappa(\mathcal{G}) \leq k = \lambda(\mathcal{G})$.

P r o o f   o f   b) : An example is the undirected graph $\mathcal{G}$ in 3.68. The node v_0 is an cutpoint of $\mathcal{G}$ so that $\kappa(\mathcal{G}) = 1$. On the other hand, $\lambda(\mathcal{G}) = \delta_1(\mathcal{G}) = \delta$. ☐

Next, we quote Menger's Theorem and three variants of it. We shall later see that several results about weighted paths can be interpreted as a modifications of Menger's Theorem.

All results 3.11 to 3.12, 3.14 to 3.15 have been shown in [Hara69].

Theorem 3.11 *Let $\mathcal{G} = (\mathcal{V}, \mathcal{R})$ be a connected and simple undirected graph with n nodes. For all nodes x, y, we define the following quantities $a(x, y)$ and $b(x, y)$:*

$a(x, y)$ *is defined as the minimum cardinality of any subset $\mathcal{V}' \subseteq \mathcal{V} \setminus \{x, y\}$ of nodes with the following property: x and y lie in different connected components of the subgraph of $\mathcal{G}$ induced by $\mathcal{V} \setminus \mathcal{V}'$.*

$b(x, y)$ *is defined as the maximum of all numbers k with the following property: There exist k node injective x-y-paths $P_1, \ldots, P_k$ such that all paths $P_i \neq P_j$ have only the nodes x and y in common. (Then we say that $P_1, \ldots, P_k$ are internally node disjoint.)*

Then $a(x, y) = b(x, y)$ for all nodes x, y. ☐

The idea behind the next theorem is to replace the word "node" in 3.11 with the word "edge".

Theorem 3.12 *Let $\mathcal{G} = (\mathcal{V}, \mathcal{R})$ be a connected and simple undirected graph with n nodes. For all nodes $x \neq y$, we define the following quantities $a(x, y)$ and $b(x, y)$:*

$a(x, y)$ *is defined as the minimal cardinality of any subset $\mathcal{R}' \subseteq \mathcal{R}$ of edges with the following property: x and y lie in different connected components of the subgraph of $\mathcal{G}$ induced by $\mathcal{R} \setminus \mathcal{R}'$.*

$b(x,y)$ *is defined as the maximum of all numbers k with the following property:
There exist k edge injective x-y-paths $P_1, \ldots, P_k$ such that all paths $P_i \neq P_j$ have no edge in common. (That is, there exist k pairwise edge disjoint x-y-paths.)*

T h e n $a(x,y) = b(x,y)$ for all nodes x,y. □

Remark 3.13 The following kind of result has a similar structure as 3.11 and 3.12. Therefore, we call these assertions modified Menger theorems.

Let $\mathcal{G}$ be a connected and simple undirected graph with n nodes. For all $x, y \in V$, we define the following quantities $a(x,y)$ and $b(x,y)$

$a(x,y)$ *is defined as the minimum cardinality of a set S of nodes [of edges] with the property that deleting all elements of S destroys all x-y-paths belonging to a given set $\mathcal{P}_{x,y}$ of x-y-paths.*

$b(x,y)$ *is defined as the maximum of all numbers k with the following property: There exist k node injective [edge injective] x-y-paths $P_1, \ldots, P_k$ in a given set $\mathcal{Q}_{x,y} \subseteq \mathcal{P}(x,y)$ such that $P_1, \ldots, P_k$ are pairwise node disjoint [edge disjoint].*

T h e n there is a particular relationship between $a(x,y)$ and $b(x,y)$. □

The next two variants of the Menger theorem follow from 3.11 and 3.12, respectively.

Theorem 3.14 *A simple and finite undirected graph is j-node connected if and only if for all nodes $x \neq y$, there exist at least j pairwise internally node disjoint x-y-paths.* □

Theorem 3.15 *A simple and finite undirected graph is j-edge connected if and only if for all nodes $x \neq y$ there exist at least j pairwise edge disjoint x-y-paths.* □

Remark 3.16 With 3.14 and 3.15 in mind, we consider any statement of the form (a)⇒(b), (b)⇒(a), or (a)⇔(b) as a modified Menger theorem if the assertions (a) and (b) have the following structure:

(a) *For a particular number k, for particular sets S of nodes [edges, resp.] and for particular sets $\mathcal{P}$ of paths, the following is true: If k arbitrary nodes [edges, resp.] are removed from S then the induced subgraph has at least one path in $\mathcal{P}$.*

(b) *For a particular number l and for particular sets $\mathcal{Q}_1, \ldots, \mathcal{Q}_l$ of paths, the following is true: There exist l pairwise node disjoint [edge disjoint, resp.] paths $P_1 \in \mathcal{Q}_1, \ldots, P_l \in \mathcal{Q}_l$.*

For example, Theorem 3.14 is a statement of the form (a)⇒(b) with the following parameters: $k = j-1$; $S = V\backslash\{x,y\}$ where x and y are two arbitrary nodes; $\mathcal{P} = \mathcal{P}(x,y)$; $l = k+1 = j$; $\mathcal{Q} = \mathcal{P}(x,y)$ where x and y are two arbitrary nodes. □

3.2 Short cycles and other short paths in directed and half directed graphs

We start this section with the consideration of short *cycles*. A survey about this topic is given in [BeTh81].

The following result, which has been shown in [Nish86], estimates the length of the short node injective cycle.

Theorem 3.17 *The girth $g(\mathcal{G})$ of any digraph $\mathcal{G}$ with n nodes has the following property:* $g(\mathcal{G}) \leq \lceil n/\delta_1^+ \rceil + 304.$ $\qquad\square$

Remark 3.18   a) Inverting all arcs in a digraph does not influence its girth. Therefore, the estimation in 3.17 is also true for δ_1^- instead of δ_1^+ so that $g(\mathcal{G}) \leq \lceil n/\delta_1^- \rceil + 304.$ Consequently, $g(\mathcal{G}) \leq \min\left(\lceil n/\delta_1^+ \rceil, \lceil n/\delta_1^- \rceil\right) + 304$ $= \lceil n/\delta_1^{\max} \rceil + 304.$

b) The following conjecture is mentioned in MR 90a:05119: $g(\mathcal{G}) \leq \lceil n/\delta_1^+ \rceil.$

c) The same abstract MR 90a:05119 describes a special case in which the inequality in part b) is true.
Let $\mathcal{V} = G$ be a finite group with the neutral element $\mathbf{1}$; let A be a subset of $G\backslash\{\mathbf{1}\}$. The set of all arcs of $\mathcal{G}$ is defined as $\mathcal{R} := \{(x, xa) \,|\, x \in G, \, a \in A\}$. Then it is obvious that $\delta_1^+ = |A|$ because the digraph $\mathcal{G}$ is regular. Let $A^i := \{a_1 \cdot \ldots \cdot a_i \,|\, a_1, \ldots, a_i \in A\}$ for all $i \in \mathbb{N} \cup \{0\}$. A theorem mentioned in MR 90a:05119 says that $\mathbf{1} \in A \cup A^2 \cup \ldots \cup A^k$ for $k := \lceil |G|/|A| \rceil = \lceil n/\delta_1^+ \rceil$ Hence, there exists a number $j \leq k = \lceil n/\delta_1^+ \rceil$ and elements $\tilde{a}_1, \ldots, \tilde{a}_j \in A$ such that $\tilde{a}_1 \cdot \ldots \cdot \tilde{a}_j = \mathbf{1}$. Hence, the path
$$X := [\mathbf{1}, \tilde{a}_1, \, \tilde{a}_1 \cdot \tilde{a}_2, \, \tilde{a}_1 \cdot \tilde{a}_2 \cdot \tilde{a}_3, \, \ldots\ldots, \, \tilde{a}_1 \cdot \ldots \cdot \tilde{a}_j = \mathbf{1}]$$
is a substantial cycle. (Note that $\tilde{a}_1 \neq \mathbf{1}$ because $\tilde{a}_1 \in A \subseteq \mathcal{G}\backslash\{\mathbf{1}\}$.) So, X contains a node injective substantial cycle $X' \subseteq X$ (possibly $X' = X$). Then $g(\mathcal{G}) \leq \ell(X') \leq \ell(X) = j \leq \lceil n/\delta_1^+ \rceil.$ $\qquad\square$

The next result is about shortest cycles through a fixed set of points.

Theorem 3.19 *Given a parallel free digraph $\mathcal{G}$ with n nodes. If $\delta_2 \geq n$ then for all subsets $S \subseteq \mathcal{V}$ of nodes, the following is true: There exists a cycle that visits all nodes of S, and the shortest of these cycles has a length $\leq \min(n, 2s)$.*

This result is proven in [Frai86]. $\qquad\square$

The articles [Behz73] and [Knya90] are about the girths of digraphs with $\deg^-(v) = \deg^+(v) = 2$ for all nodes v; in particular, $g(\mathcal{G}) \leq \lceil n/2 \rceil$.

A further article about shortest cycles in digraphs is [Dong88]. The paper is about the existence of digraphs with a given girth and a given exponent; the exponent of a digraph has been defined in [Dong88].

The authors of [BeaP90] study the behaviour of the maximal cycle length if a so-called linear mapping L is applied to a digraph $\mathcal{G}$; more precisely, Beasley and Pullman characterize the functions L for which $\mathcal{G}$ and $L(\mathcal{G})$ have the same girths.

Next, we describe several results about short paths in digraphs; these paths need not be cycles.

[Šolt86] and [McCa88] are about the following problem: Given an undirected graph; orientate all edges such that the diameter or the radius of the resulting digraph is minimal.

An implicit lower bound of the diameter is shown in [Xu92]:

Theorem 3.20 *Given a strongly connected digraph $\mathcal{G} = (\mathcal{V}, \mathcal{R})$. Let κ be the node connectivity number of $\mathcal{G}$. (This is the minimum number of nodes that must be removed to destroy the strong connectivity of $\mathcal{G}$ or to produce a digraph with a single node.)*
Then the diameter $d = d(\mathcal{G})$ has the following property:

$$\kappa \cdot \frac{(\Delta_1^{\max})^d - \Delta_1^{\max}}{\Delta_1^{\max} - 1} + \Delta_1^{\max} \geq n - 1.$$

There is a similar result about the so-called arc connectivity number of $\mathcal{G}$.   □

The article [MiFr88] investigates the diameter of regular digraphs. The article [MiFr93] is about the existence of regular digraphs with a given diameter.

One of the chapters in [SBvL87] describes how the diameter of a digraph is increased when deleting arcs.

Kulkarni [Kulk86] develops methods to compute the probability distribution of the lengths of H-minimal paths in digraphs; Kulkarni considers additive path functions $H = SUM_h$ with the property that the values $h(r)$, $r \in \mathcal{R}$, are independent and exponentially distributed random variables.

All sources [Šolt86], [McCa88], [MiFr88], [SBvL87], [Kulk86] give solutions of very special problems. A more general question is the following: How can the diameter of an arbitrary digraph be estimated if other parameters of this graph are given? (The relationship between the diameter and shortest paths has been described in Definition 3.2.) Perhaps, there exist two fixed functions F_1 and F_2 such that for all digraphs $\mathcal{G}$ with n nodes and m arcs,

$$F_1(n, m, \delta_1^+(\mathcal{G}), \delta_1^-(\mathcal{G}), \delta_1(\mathcal{G}), \delta_2(\mathcal{G})) \leq d(\mathcal{G}) \leq F_2(n, m, \delta_1^+(\mathcal{G}), \delta_1^-(\mathcal{G}), \delta_1(\mathcal{G}), \delta_2(\mathcal{G})) . \quad (3.2)$$

For example, if F_1 were identical to zero and $F_2(n, m, \delta_1^+(\mathcal{G}), \delta_1^-(\mathcal{G}), \delta_1(\mathcal{G}), \delta_2(\mathcal{G})) = \mathbf{a} \cdot \lceil n/\delta_1^+(\mathcal{G}) \rceil + \mathbf{b}$ with fixed constants $\mathbf{a}$, $\mathbf{b}$, we would obtain the assertion $d(\mathcal{G}) \leq \mathbf{a} \cdot \lceil n/\delta_1^+ \rceil + \mathbf{b}$, which is analogous to Theorem 3.17.

The author of this book, however, does not know anything about a source yielding a result like (3.2).

New measures of distance are defined and studied in [ChTi90] and in [Tian90]. More precisely, let $\mathcal{G} = (\mathcal{V}, \mathcal{R})$ be a strongly connected digraph, and let $u, v \in \mathcal{V}$. The *maximum distance* is defined as $md(u, v) := \max(d(u, v), d(v, u))$, and the *sum distance* is defined as $sd(u, v) := d(u, v) + d(v, u)$. The paper [ChTi90] is about the maximum distance, and [Tian90] is about the sum distance. A further source about distances in digraphs is [CJT92].

The transmission of a digraph has been studied in [Pl84]. Many results are shown in that paper. The next assertion follows from Theorems 1b and 7b in [Pl84].

Theorem 3.21 *Let $\mathcal{G}$ be a strongly connected digraph with n nodes and m arcs. Then*
$$2n(n-1) - m \;\leq\; \sigma(\mathcal{G}) \;\leq\; \tfrac{1}{2}n^2(n-1).$$
$\square$

By the way, it is shown in Section 5 of [Pl84] that the mean distance (= average distance) $\mu(\mathcal{G}) := \dfrac{\sigma(\mathcal{G})}{n\,(n-1)}$ is independent of the radius and the diameter of a digraph. That means that there are no substantial relationships between $\mu(\mathcal{G})$ and the quantities $d(\mathcal{G})$ and $r(\mathcal{G})$.

Shortest paths in half directed graphs are studied in [Hic82]. More precisely, the paper is about geodetic graphs. This kind of graph is will be defined in Definition/Remark 3.34 in the undirected case; geodetic graphs are closely related to shortest paths.

3.3 Short cycles and other short paths in undirected graphs

The author of this book has found much literature about short paths in undirected graphs and only few papers about the directed case. That is the reason why the current section 3.3 is divided into subsections and Section 3.2 is not.

3.3.1 Short cycles in undirected graphs

Many results about this topic describe the length of shortest *node injective* cycles, and this quantity is the girth of the given digraph. To estimate the girth in a similar way as in Theorem 3.17 we show an auxiliary result about directed versions of undirected graphs. (Directed versions of undirected graphs were defined on page 18.)

Lemma 3.22 *Let $\mathcal{G} = (\mathcal{V}, \mathcal{R})$ be a connected and simple undirected graph with n nodes. For each directed version $\vec{G}$ of $\mathcal{G}$ and all nodes v, let $\deg^+(\vec{G}, v)$ and $\deg^-(\vec{G}, v)$ denote the outdegree and the indegree of v in $\vec{G}$, respectively. Then the following assertions are true:*

a) *There exists a directed version $\vec{G}^* = (\mathcal{V}, \mathcal{R}^*)$ of $\mathcal{G}$ such that for all nodes v, the outdegree $\deg^+(\vec{G}^*, v)$ is roughly the half of the degree $\deg(v)$ in $\mathcal{G}$; more precisely,*
$$\deg^+(\vec{G}^*, v) \;=\; \lfloor \deg(v)/2 \rfloor \quad \text{for all } v \in \mathcal{V}.$$

b) *Let $\vec{\delta}_1^+ = \vec{\delta}_1^+(\mathcal{G})$ be the maximum of all minimal outdegrees $\delta_1^+(\vec{\mathcal{G}})$ where $\vec{\mathcal{G}}$ is a directed version of $\mathcal{G}$. Then $\vec{\delta}_1^+ \geq \lfloor \delta_1/2 \rfloor$.*

c) *The bound in part b) is tight for all cardinalities $n = |\mathcal{V}|$.*

P r o o f   o f   a) : We consider two cases.

• *Case 1: All nodes of $\mathcal{G}$ have even degree.*
Then we can partition the set $\mathcal{R}$ of edges into arc disjoint cycles. More precisely, we can construct one or more edge injective cycles $X_1, \ldots, X_j$ such that each edge $r \in \mathcal{R}$ appears in exactly one cycle X_i. The well-known construction is the following: We first generate X_1. For this purpose, we construct a sequence $v_0, v_1, v_2, \ldots$ of nodes such that all pairs (v_i, v_{i+1}) are connected by an edge and all edges $\{v_i, v_{i+1}\}$ are pairwise distinct. We do not arrive at a dead end because $\deg(v)$ is even for all nodes v. Hence, we shall run into a cycle, i.e., we find a node v_k with $v_k = v_j$ for some $j < k$. When the first node v_k with this property has been found we define $X_1 := [v_j, v_{j+1}, \ldots, v_{k-1}, v_k]$. Then we remove all edges of X_1. The remaining graph still has the property that all degrees are even so that we can construct X_2 in the same way as X_1. In general, each cycle X_i, $i > 1$, is constructed like X_1 after removing $X_1, \ldots, X_{i-1}$.
The edges of $\mathcal{G}$ are orientated as follows: If $i = 1, \ldots, j$ and $X_i = [w_0, w_1, \ldots, w_l = w_0]$ then $\{w_\lambda, w_{\lambda+1}\}$ is replaced with $(w_\lambda, w_{\lambda+1})$. Let $\vec{\mathcal{G}}^*$ be the digraph resulting from these orientations.
Then for all nodes v the following is true: v is visited by exactly $\deg(v)/2$ cycles; each cycle has one arc ending in v and one arc starting from v. Hence, $\deg^+(\vec{\mathcal{G}}^*, v) = \deg(v)/2 = \lfloor \deg(v)/2 \rfloor$.

• *Case 2: $\mathcal{G}$ has a node with odd degree.*
Let $\mathcal{V}_e$ be the set of all nodes with even degree and $\mathcal{V}_o$ the set of all nodes with odd degree. Then $|\mathcal{V}_o|$ is an even number because $2 \cdot |\mathcal{V}| = \sum_{v \in \mathcal{V}} \deg(v) = \sum_{v \in \mathcal{V}_e} \deg(v) + \sum_{v \in \mathcal{V}_o} \deg(v)$ so that $2 \cdot |\mathcal{V}| - \sum_{v \in \mathcal{V}_e} \deg(v) = \sum_{v \in \mathcal{V}_o} \deg(v)$ is an even number.
Let $\mathcal{V}_o = \{u_1, \ldots, u_j\}$. We construct a graph $\mathcal{G}_1 = (\mathcal{V}_1, \mathcal{R}_1)$ by adding a new edge to each node of $\mathcal{V}_o$. More precisely, we generate new nodes $w_1, \ldots, w_{j/2}$ and define the graph $\mathcal{G}_1 := (\mathcal{V}_1, \mathcal{R}_1)$ with the set $\mathcal{V}_1 := \mathcal{V} \cup \{w_1, \ldots, w_{j/2}\}$ of nodes and the set $\mathcal{R}_1 := \mathcal{R} \cup \{\{u_i, w_i\}, \{w_i, u_{i+j/2}\} \mid i = 1, \ldots, j/2\}$ of edges. Let $\deg(\mathcal{G}_1, v)$ be the degree of v in the new graph $\mathcal{G}_1$; then $\deg(\mathcal{G}_1, v)$ is even for all nodes $v \in \mathcal{V}_1$. Hence, there exists a directed version $\vec{\mathcal{G}}_1^*$ of $\mathcal{G}_1$ such that

$$(\forall\, v \in \mathcal{V}) \quad \deg^+(\vec{\mathcal{G}}_1^*, v) \geq \deg(\mathcal{G}_1, v)/2. \tag{3.3}$$

We define $\vec{\mathcal{G}}^* = (\mathcal{V}, \mathcal{R}^*)$ as the subgraph of $\vec{\mathcal{G}}_1^*$ induced by $\mathcal{V}$; roughly speaking, we remove all nodes and arcs from $\vec{\mathcal{G}}_1^*$ that do not have their origins in $\mathcal{G}$. We consider the degrees in $\vec{\mathcal{G}}^*$.
If $v \in \mathcal{V}_e$ then no edge $\{v, w_i\}$, $i = 1, \ldots, j/2$, has been generated when transforming $\mathcal{G}$ into $\mathcal{G}_1$; moreover, no arc (v, w_i) or (w_i, v), $i = 1, \ldots, j/2$, has been removed when transforming $\vec{\mathcal{G}}_1^*$ into $\vec{\mathcal{G}}^*$. Consequently,

$$\deg^+(\vec{\mathcal{G}}^*, v) = \deg^+(\vec{\mathcal{G}}_1^*, v) \overset{(3.3)}{\geq} \deg(v, \mathcal{G}_1)/2 = \deg(v)/2 = \lfloor \deg(v)/2 \rfloor .$$

If $v \in V_o$ then exactly one edge $\{v, w_i\}$, $i = 1, \ldots, j/2$, has been generated when transforming $\mathcal{G}$ into $\mathcal{G}_1$; moreover, one of the arcs (v, w_i) or (w_i, v), $i = 1, \ldots, j/2$, has been removed when transforming $\vec{\mathcal{G}}_1^*$ into $\vec{\mathcal{G}}^*$. Consequently,

$$\deg^+(\vec{\mathcal{G}}^*, v) \geq \deg^+(\vec{\mathcal{G}}_1^*, v) - 1 \overset{(3.3)}{\geq} \deg(\mathcal{G}_1, v)/2 - 1 = (\deg(v) + 1)/2 - 1 = \lfloor \deg(v)/2 \rfloor .$$

Hence, $\deg^+(\vec{\mathcal{G}}^*, v) \geq \lfloor \deg(v)/2 \rfloor$ for all nodes $v \in V$.

P r o o f   o f   b) : Let $\vec{\mathcal{G}}^*$ be the digraph mentioned in part a). Then $\deg^+(\vec{\mathcal{G}}^*, v) \overset{a)}{\geq} \lfloor \deg(v)/2 \rfloor$ for all nodes $v \in V$. Consequently, the following inequality is true where $(*)$ follows from the monotonicity of the function $\xi \mapsto \lfloor \xi/2 \rfloor$.

$$\delta_1^+(\vec{\mathcal{G}}^*) = \min\{\deg^+(\vec{\mathcal{G}}^*, v) \,|\, v \in V\} \geq \min\{\lfloor \deg(v)/2 \rfloor \,|\, v \in V\} \overset{(*)}{=}$$
$$\lfloor (\min\{\deg(v) \,|\, v \in V\})/2 \rfloor = \lfloor \delta_1(\mathcal{G})/2 \rfloor .$$

Consequently, $\vec{\delta}^+ \geq \delta_1^+(\vec{\mathcal{G}}^*) \geq \lfloor \delta_1(\mathcal{G})/2 \rfloor$.

P r o o f   o f   c) : We show this assertion by defining a graph $\mathcal{G}_0$ with $\vec{\delta}_1^+(\mathcal{G}_0) = \lfloor \delta_1(\mathcal{G}_0)/2 \rfloor$. Let $\mathcal{G}_0 := \mathcal{K}_V$, and let $\mathcal{R}_0$ be the set of all edges in $\mathcal{G}_0$. Let $\vec{\mathcal{G}}_0 = (V, \vec{\mathcal{R}}_0)$ be an arbitrary directed version of $\mathcal{G}_0$. We obtain the following inequality, in which relationship $(*)$ results from the fact that $\delta_1^+(\vec{\mathcal{G}}_0)$ is the minimal outdegree in $\vec{\mathcal{G}}_0$.

$$n \cdot \delta_1^+(\vec{\mathcal{G}}_0) \overset{(*)}{\leq} \sum_{v \in V} \deg^+(\vec{\mathcal{G}}_0, v) = |\mathcal{R}_0| = \frac{n \cdot (n-1)}{2} .$$

Hence, $\delta_1^+(\vec{\mathcal{G}}_0) \leq \frac{n-1}{2}$. Moreover, $\delta_1(\mathcal{G}_0) = n - 1$ because all nodes of this graph have degree $n - 1$. This and $\delta_1^+(\vec{\mathcal{G}}_0) \leq \frac{n-1}{2}$ imply that $\delta_1^+(\vec{\mathcal{G}}_0) \leq \frac{\delta_1(\mathcal{G}_0)}{2}$ so that even $\delta_1^+(\vec{\mathcal{G}}_0) \leq \left\lfloor \frac{\delta_1(\mathcal{G}_0)}{2} \right\rfloor$. This inequality is valid for all directed versions $\vec{\mathcal{G}}_0$, and consequently, it is also true for the maximal value $\vec{\delta}_1^+(\mathcal{G}_0)$. Hence, $\vec{\delta}_1^+(\mathcal{G}_0) \leq \left\lfloor \frac{\delta_1(\mathcal{G}_0)}{2} \right\rfloor$. $\square$

The next assertion gives upper bounds of the girth.

Corollary 3.23 *Let $\mathcal{G}$ be a connected and simple undirected graph with n nodes and m edges. Let $\vec{\delta}_1^+$ be the same quantity as in 3.22. Then the following are true:*

a) $g(\mathcal{G}) \leq \lceil n/\vec{\delta}_1^+ \rceil + 304.$ b) $g(\mathcal{G}) \leq \lceil 2\,n/(\delta_1 - 1) \rceil + 304.$

P r o o f   o f   a) : The assertion is an immediate consequence of Theorem 3.17 and the definition of $\vec{\delta}_1^+$.

P r o o f   o f   b) : Theorem 3.22 b) implies that $\vec{\delta}_1^+ \geq \lfloor \delta_1(\mathcal{G})/2 \rfloor \geq (\delta_1(\mathcal{G}) - 1)/2$. This and part a) yield the current assertion. $\square$

It is shown in [LZh94] that for all natural numbers k, j with $j \geq 3$, there exists an undirected graph $\mathcal{G}$ with $\kappa(\mathcal{G}) = k$ and $g(\mathcal{G}) = j$. The article [TeoK92] investigates the number of cycles of length $g(\mathcal{G})$. The so-called girth pair is investigated in [HWMG94].

Next, we quote several papers about the girth of special undirected graphs.

One of the results in [Bra91] is the following: Given a fixed number $c \in (0, 1]$. Let $\mathcal{G}$ be a cubic[1], 2-edge connected undirected graph with n nodes and m edges; suppose that $\mathcal{G}$ has at most $\frac{n}{1+c}$ nodes of degree 2. Then the girth of $\mathcal{G}$ is sublinear. More precisely, there exists a function $F : \mathbb{N} \to \mathbb{N}$ with $\lim_{n \to \infty} F(n)/n = 0$ such that $g(\mathcal{G}) \le F(n)$ for all above graphs $\mathcal{G}$.

In contrast to this, Corollary 3.23 a) and b) only yield a linear upper bound; this follows from the fact that $\vec{\delta}_1^+(\mathcal{G}) \le \delta_1(\mathcal{G}) \le 3$.

The abstract MR 86c:05982 about [Wei84] says that any cubic graph has the property that $g(\mathcal{G}) \le 2 \cdot \log_2(n)$. For this reason, it is probably interesting to compare [Bra91] with [Wei84].

The author of [Entr81] defines graphs with an annular symmetry and investigates their girths.

Theorem 1.5.3 in [Ju94] implies that $g(\mathcal{G}) \le \frac{2m}{m-n+2}$ for all connected planar graphs $\mathcal{G}$ with n nodes and with m edges. (Note that $m - n + 2 > 0$ because $\mathcal{G}$ is connected.) An immediate consequence of this result is that the inequality $g(\mathcal{G}) > \frac{2m}{m-n+2}$ can be used to show that $\mathcal{G}$ is not planar.

The article [Se93] gives an upper bound for the girth of transitive, infinite graphs. A further result is that there exist infinite graphs $\mathcal{G}$ with particular properties, and one of these properties is that $g(\mathcal{G}) \ge l$ where $l \ge 3$ is a given number.

Another question about short cycles is studied in [TaNS80]: How long is the shortest cycle visiting all nodes of $\mathcal{G}$? The cycles in that paper need not be node injective.

3.3.2   Short acyclic paths in undirected graphs

We now study combinatorial results about short acyclic paths in undirected graphs. Several terms like "distance", "diameter" and "radius" are closely related to shortest paths. An overview over these parameters is given in [ChOe89], and a special survey about diameters is given in [BeBo81].

We now quote several results about the diameter. There exist two estimations of this quantity.

Theorem 3.24 *Let $\mathcal{G}$ be an connected and simple undirected graph with n nodes, m edges and the minimal degree δ_1.*

    a) (Upper bound for $d = d(\mathcal{G})$ depending on n and m):

        If $\mathcal{G} = \mathcal{K}_\mathcal{V}$ then $d = 1$. Otherwise, $2 \le d \le n - \left\lceil \frac{1}{2} \left(\sqrt{8(m-n)+17} - 1 \right) \right\rceil$.

    b) (Upper bound for $d = d(\mathcal{G})$ depending on n and δ_1):

        If $d \ge 3$ then $d \le 3n/\delta_1$.

---

[1] A graph is *cubic* if all nodes have a degree ≤ 3.

P r o o f   o f   a) :   The proof of assertion a) is given in [Smy87].

P r o o f   o f   b) :   It is shown in [GoMF81] that

$$n \ge \frac{(3+d)(\delta_1 + 1) + a \cdot (2 - \delta_1)}{3} \tag{3.4}$$

where $a \in \{0, 1, 2\}$ is the remainder of the division of d by 3. Then $(2-a)\delta_1 \ge 0 \ge -2a - 2$, and consequently, $a \cdot (2 - \delta_1) \ge -2(\delta_1 + 1)$. This fact is used in $(*)$ of the inequality $n \overset{(3.4)}{\ge} \frac{(3+d)(\delta_1 + 1) + a \cdot (2 - \delta_1)}{3} \overset{(*)}{\ge} \frac{(1+d)(\delta_1 + 1)}{3} \ge \frac{d \cdot \delta_1}{3}$, and this fact yields the desired estimation of d.   □

In [Broe93], a relation between the toughness $\tau(\mathcal{G})$ and the diameter $d(\mathcal{G})$ is proven.

The paper [PaSu78] is about the minimal number of edges in an undirected graph if the fixed diameter 2 is given and $\delta_1 = c \cdot n$ with $0 < c < 1$. The papers [Znam90] and [Znam92] describe the asymptotic behaviour of the minimum number of edges of undirected graphs that have diameter 2 and a given maximum degree.

The article [Jor92] is about the diameters of cubic graphs. It is shown that $n \le 3 \cdot 2^{d(\mathcal{G})} - 4$ if $\mathcal{G}$ is a cubic graph of diameter $d(\mathcal{G}) \ge 4$.

The articles [SBvL87] and [Chung92] investigate the behaviour of the diameter when removing edges from the given graph. For example, one of the results in [SBvL87] says that the diameter after deleting k edges is at most $(k+1)$ times as great as it was before deleting the edges. More precisely, the following result is valid:

Theorem 3.25 *Let $f(k, d)$ be the maximum diameter that can be obtained by deleting k edges from a simple, connected graph with n nodes and diameter d. Then*

$$\begin{array}{rcccll} (k+1) \cdot d - k & \le & f(k, d) & \le & (k+1) \cdot d & \text{if } d \text{ is even,} \\ (k+1) \cdot d - 2k + 2 & \le & f(k, d) & \le & (k+1) \cdot d & \text{if } d \text{ is odd.} \end{array}$$ □

The article [Fur92] describes graphs with diameter 2 and with the property that the diameter increases if an arbitrary edge is removed.

The diameter of a graph after removing nodes is often called *fault diameter*. If the given graph is a network of processors and several processors are faulty then the fault diameter is the longest of all shortest u-v-paths that do not visit any defective processor. The fault diameter is studied in [RSr93].

The diameters of undirected Cayley graphs and of other graphs arising from Group Theory are investigated in [AnB93] and [ChHM93].

The papers [BeDQ82] and [DiHa94] are about the following problem: Given Δ and d, find a graph $\mathcal{G}$ with a large number of nodes such that the maximum degree $\Delta_1(\mathcal{G})$ equals Δ and diameter $d(\mathcal{G})$ equals d.

The paper [BHBA91] determines circulant graphs with minimum diameter and minimum transmission; a *circulant graph* is defined as $\mathcal{G} = (\mathcal{V}, \mathcal{R})$ with $\mathcal{V} := \{1, \dots, n\}$ and

$\mathcal{R} := \left\{ \{i\,,\,i + \rho_j\,(\mathrm{mod}\,n)\} \,\middle|\, i = 1,\dots,n,\; j = 1,\dots,k \right\}$ where $\rho_1,\dots,\rho_k$ are fixed numbers.

The diameter of so-called diagonal mesh networks is investigated in [TaPa92].

The authors of [Ched93] introduce modified hypercubes with small diameters; these graphs are called general twisted hypercubes. (Standard hypercubes are defined in Definition 3.97.)

The diameters of random graphs are investigated in [HaZe85], in [BCDJ93], and in [KoSW93].

The next result gives an upper bound for the radius of a graph; the assertion follows from Theorems 1 and 2 in [HaWa81].

Theorem 3.26 *Let $k \geq 1$, and let $\mathcal{G}$ be a simple and $(2k-1)$-node connected undirected graph with n nodes. Then $r(\mathcal{G}) \leq \frac{1}{2k}n + O(\log n)$.* $\qquad\Box$

The paper [Va92] is likewise about shortest paths in undirected graphs; an *open Hamiltonian walk* is defined as a path of minimal length among all candidates visiting all nodes.

The results 3.22 – 3.26 were about the girth, the diameter, and the radius of undirected graph; each of these quantities may be considered as the length of a single path; for example, the girth is the length of the shortest edge injective cycle. In contrast to this, we next consider objects depending on several shortest paths or their lengths. An example of such an object is the set of all shortest u-v-paths. A further example is the transmission $\sigma(\mathcal{G})$, which is computed from all distances $d(u,v)$, $u,v \in V$.

Now we consider the transmission. We shall later describe other mathematical constructions depending on shortest paths in an undirected graph.

The following theorem is analogous to 3.21; a lower and an upper bound of the transmission of an undirected graph is given; the result follows from Theorems 1 and 4 in [Pl84].

Theorem 3.27 *Given a connected and simple undirected graph $\mathcal{G}$ with n nodes. Then $2n(n-1) \leq \sigma(\mathcal{G}) \leq \frac{1}{3}n(n-1)(n+1)$.* $\qquad\Box$

Another paper about the transmission is [CoMi80]. The problem in that paper is easy to understand: Given a an undirected graph $\mathcal{G}$ that only consists of a single node injective path; that is, $\mathcal{G}$ has exactly the edges $\{v^{(0)}, v^{(1)}\}, \dots, \{v^{(n-1)}, v^{(n)}\}$. Let $\mathcal{G}(e)$ be the graph generated by adding the new edge $e = \{v^{(i)}, v^{(j)}\}$ to $\mathcal{G}$. Which edge e^* must be added to make the transmission $\sigma(\mathcal{G}(e^*))$ minimal among all transmissions $\sigma(\mathcal{G}(e))$?

The authors of [GYCh94] show the following: For all even numbers J with $J \notin \{4, 10\}$, there exists a graph $\mathcal{G}$ without loops such that $\sigma(\mathcal{G}) = J$.

The paper [BuSu81] is about the mean distance (= average distance) in undirected graphs; the author investigates graphs for which the this quantity is equal to the radius. Further results about mean distances are given in [Pl84], [Alth90], [AltS92], [Zhou92], and in [Shi94].

It is not clear whether the average or mean distance in all these sources is the same quantity. For example, Plésnik defines the average distance in an undirected graphs as follows: $\mu(\mathcal{G}) := \sum_{(u,v)\in\mathcal{V}\times\mathcal{V}} d(u,v)/(n \cdot (n-1))$; a similar definition is given in the paper [Alth90]. In [BuSu81], however, the average distance is defined as $\sum_{u,v\in\mathcal{V}} d(u,v)/\binom{n}{2}$; a similar definition is given in [Shi94]. Then $\sum_{u,v\in\mathcal{V}} d(u,v)/\binom{n}{2}$ is equal to $2 \cdot \mu(\mathcal{G})$ if $\sum_{u,v\in\mathcal{V}} d(u,v)$ means the sum over all ordered pairs (u,v); if, however, $\sum_{u,v\in\mathcal{V}} d(u,v)/\binom{n}{2}$ means the sum over all unordered pairs $\{u,v\}$ then $\sum_{u,v\in\mathcal{V}} d(u,v)/\binom{n}{2}$ is equal to $\mu(\mathcal{G})$.

Next, we consider another object based on all distances in an undirected graph.

Definition/Remark 3.28 Given a connected and simple undirected graph $\mathcal{G}$ with n nodes.

The *sequence of distances* (= *distance distribution*) is defined as the d-tuple $D(\mathcal{G}) := (D_1, \ldots, D_d)$ with the following properties: $d = d(\mathcal{G})$, and each component D_i equals the number of all ordered pairs (u,v) with distance i.

If $D(\mathcal{G})$ is given the transmission can be computed as follows: $\sigma(\mathcal{G}) \quad = \quad \sum_{i=1}^{d(\mathcal{G})} i \cdot D_i$.

That means that $D(\mathcal{G})$ gives more detailed information about the distances in $\mathcal{G}$ than $\sigma(\mathcal{G})$ does.

We say that $\mathcal{G}$ has a *uniform distance distribution* if $D_1 = \cdots = D_d$. $\qquad\square$

The following result about graphs with uniform distance distribution has been shown in [BuSu81].

Theorem 3.29 *Given a number n such that $\binom{n}{2}$ is a multiple of 3. Then the following assertions are equivalent:*
  (a) *There exists a connected and simple undirected graph $\mathcal{G} \neq \mathcal{K}_\mathcal{V}$ with n nodes and with a uniform distance distribution.*
  (b) $\binom{n}{2}$ *can be written as $\binom{n}{2} = 3\,pq$ with $pq \geq 6$ and $p \geq q \geq 2$.*

(For example, let $n = 7$. Then $\mathcal{K}_\mathcal{V}$ is the only graph with uniform distance distribution. The reason is that there are no $p, q \geq 2$ with $\binom{7}{2} = 21 = 3 \cdot 7 = 3\,pq$ with $p \geq q \geq 2$ as 7 is a prime.) $\qquad\square$

Another object derived from the distances in a graph is the distance degree matrix. It is constructed as follows: For each node v and for all $i \in \mathbb{N}$, count the nodes w with $d(v, w) = i$. We now give the exact definition of this matrix.

Definition 3.30 Let $\mathcal{G} = (\mathcal{V}, \mathcal{R})$ be a connected and simple undirected graph; let $\mathcal{V} = \{v^{(1)}, \ldots, v^{(n)}\}$ be the set of all nodes in $\mathcal{G}$.

For all nodes $v^{(\nu)}$ and all $i = 1, \ldots, d = d(\mathcal{G})$, let $\deg_i \left(v^{(\nu)} \right)$ denote the number of nodes $v^{(\nu')}$ that have the distance i from $v^{(\nu)}$; more formally,

$$\deg_i \left(v^{(\nu)} \right) \;:=\; \left| \left\{ v^{(\nu')} \in V \,\middle|\, d\left(v^{(\nu)}, v^{(\nu')} \right) = i \right\} \right| \quad (i = 1, 2, \ldots, d, \; \nu = 1, \ldots, n) \,.$$

The quantity $\deg_i \left(v^{(\nu)} \right)$ is called the i-th distance degree of $v^{(\nu)}$. For example, the first distance degree $\deg_1 \left(v^{(\nu)} \right)$ of any node $v^{(\nu)}$ is equal to the usual degree $\deg \left(v^{(\nu)} \right)$.

The distance degree matrix $\mathbf{Deg}(\mathcal{G})$ contains all distance degrees of all nodes. More precisely,

$$\mathbf{Deg}(\mathcal{G}) \;:=\; \begin{pmatrix} \deg_1 \left(v^{(1)} \right) & \cdots & \deg_d \left(v^{(1)} \right) \\ \vdots & & \vdots \\ \deg_1 \left(v^{(n)} \right) & \cdots & \deg_d \left(v^{(n)} \right) \end{pmatrix} \qquad (\text{with } d = d(\mathcal{G})) \,.$$

In [BlKQ81], the matrix $\mathbf{Deg}(\mathcal{G})$ is called the "distance degree sequence" of $\mathcal{G}$. ☐

We now show how the sequence of distances can be computed if the distance degree matrix is given. This result means that $\mathbf{Deg}(\mathcal{G})$ contains at least as much information about the distances in $\mathcal{G}$ as the distance distribution $D(\mathcal{G})$ does.

Theorem 3.31 *Given a connected and simple undirected graph $\mathcal{G}$ with n nodes. Let $D(\mathcal{G}) = (D_1, \ldots, D_d)$ be the sequence of distances, and let $\mathbf{Deg}(\mathcal{G}) = \left(\deg_i(v^{(\nu)}) \right)$ be the distance degree matrix.*

Then $D_i = \sum\limits_{\nu=1}^{n} \deg_i \left(v^{(\nu)} \right)$ *for all* $i = 1, \ldots, d(\mathcal{G})$.

P r o o f : Let $i \in \{1, \ldots, d(\mathcal{G})\}$. For each ν, we define the following set:

$$M^{(\nu)} \;:=\; \left\{ \left(v^{(\nu)}, v^{(\nu')} \right) \,\middle|\, v^{(\nu')} \in V, \; d\left(v^{(\nu)}, v^{(\nu')} \right) = i \right\} \,.$$

When listing the elements of the sets $M^{(1)}, \ldots, M^{(n)}$ then each ordered pair $\left(v^{(\nu)}, v^{(\nu')} \right)$ with $d(v^{(\nu)}, v^{(\nu')}) = i$ appears exactly once (as an element of $M^{(\nu)}$); this is used in $(*)$ of the equation at the end of this proof; relationship $(**)$ in this equation follows from the fact that the sets $M^{(\nu)}$ are pairwise disjoint, and relationship $(***)$ follows from $\left| M^{(\nu)} \right| = \deg_i(v^{(\nu)})$, $\nu = 1, \ldots, n$. So, the following assertion is true for all $i = 1, \ldots, d$:

$$D_i \overset{(*)}{=} \left| M^{(1)} \cup \ldots \cup M^{(n)} \right| \overset{(**)}{=} \left| M^{(1)} \right| + \ldots + \left| M^{(n)} \right| = \deg_i \left(v^{(1)} \right) + \ldots + \deg_i \left(v^{(n)} \right) \,. \quad ☐$$

The papers [BlKQ81] and [HiNo84] are about undirected graphs with special distance degrees, namely DDR graphs and DDI graphs.

A so-called *DDR graph* (= *distance degree regular graph*) is defined as a graph with the property that all nodes have the same distance degrees; more precisely, for each i, there exists a unique distance degree ψ_i with $\psi_i = \deg_i(v) = \deg_i(w)$ for all nodes v, w. In this case, 3.31 implies that $\psi_i = D_i/n$ for all i.

Any DDR graph is regular in the usual sense because $\deg(v) = \deg_1(v)$ ($v \in V$).

Obviously, the DDR property is equivalent to the fact that all rows of $\mathbf{Deg}(\mathcal{G})$ are equal. On the other hand, the DDR property has nothing to do with the uniform distance distribution, which has been defined in 3.28; the difference between these properties is the following: Distance degree regularity says something about the *rows* of $\mathbf{Deg}(\mathcal{G})$. In contrast to this, uniform distance distribution means that all *columns* of $\mathbf{Deg}(\mathcal{G})$ have the same sum.

A *DDI graph (distance degree injective graphs).* is defined as an undirected graph with the property that all rows of $\mathbf{Deg}(\mathcal{G})$ are pairwise distinct. Hence, distance degree injectivity is the opposite of distance degree regularity.

We now quote several results about DDR and DDI graphs.

Theorem 3.32 *Given a DDR graph $\mathcal{G}$ with diameter ≥ 2. Let $\psi_i := \deg_i(v)$ for each i and for all nodes v.*
Then $3\,\psi_i \;\geq\; 2(\psi_1 + 1)$ for all $i = 1, 2, \ldots, d(\mathcal{G}) - 1$. $\qquad\square$

This result is proven in [HiNo84]. Moreover, [HiNo84] describes the structure of the graph $\mathcal{G}$ if there exists an i such that $3\,\psi_i = 2\,(\psi_i + 1)$ in Theorem 3.32.

The following results about DDR graph and DDI graphs are proven in [BlKQ81].

Theorem 3.33
a) *Let $\mathcal{G}$ be simple, node transitive graph. (This property is defined in 3.93.) Then $\mathcal{G}$ is a DDR graph. The proof of this assertion is straight-forward.*
b) *The implication in part a) cannot be reversed. It is even possible to find DDR graphs automorphism group consists of the identity.*
c) *If a graph $\mathcal{G}$ has the DDI-property then the identity is the only automorphism on $\mathcal{G}$.*
d) *The reverse implication is not always true; i.e., there exist graphs $\mathcal{G}$ without the DDR property, and the only automorphism of $\mathcal{G}$ is the identity.* $\qquad\square$

Of course, the most detailed information about all distances in a graph is given by the *distance matrix*, which is defined as follows:
$$\mathbf{D}(\mathcal{G}) \;:=\; \begin{pmatrix} d\left(v^{(1)}, v^{(1)}\right) & \cdots & d\left(v^{(1)}, v^{(n)}\right) \\ \vdots & & \vdots \\ d\left(v^{(n)}, v^{(1)}\right) & \cdots & d\left(v^{(n)}, v^{(n)}\right) \end{pmatrix}.$$
The characteristic polynomial $p_{\mathcal{G}}$ of the distance matrix is called the *distance polynomial* of $\mathcal{G}$. More formally, let $\mathbf{I}_n$ denote the unit matrix of dimension n, and let det denote the determinant function; then we define $p_{\mathcal{G}}(\xi) \;:=\; det(\mathbf{I}_n \cdot \xi - \mathbf{d}(\mathcal{G}))$. The properties of this polynomial are investigated in [KrTr83].

The author of [Merr90] investigates the eigenvalues of $\mathbf{D}(\mathcal{G})$ where $\mathcal{G}$ is a tree.

The usual point of view is that a graph is given and then its distance matrix is constructed. The inverse situation, however, is that a (distance) matrix is given and its graph shall be reconstructed. More precisely, given an $N \times N$-matrix $\mathbf{A} = (a_{ij})$; decide whether there exists a graph $\mathcal{G}$ with $n \geq N$ nodes $v^{(1)}, \ldots, v^{(n)}$ such that $d\left(v^{(i)}, v^{(j)}\right) = a_{ij}$ for all $1 \leq i, j \leq N$; if yes then construct $\mathcal{G}$.

If such a graph exists then **A** is called *realizable*; if G may even be chosen as a tree then **A** is called *tree realizable*.

The paper [Simo90] is about optimal graphs G for given realizable matrices **A**. Tree realizable matrices **A** are investigated in [ImSc83], [Simo87], and [CuRu89]. In several cases, the entries $d(u, v)$ in the distance matrix are replaced with the values $d_H(u, v)$ where H is an additive path function. (The notation $d_H((u, v)$ was defined in Definition/Remark 2.135.) By the way, in [ImSc83], there are investigations about distances in trees and in groups.

A further problem related to the system of all shortest paths in an undirected graph is to investigate the properties of the sets $\mathcal{P}_{Min}(u, v)$, $u, v \in V$. In particular, the following questions are interesting:

1) How many shortest connections between u and v exist? That means, what is the cardinality of $|\mathcal{P}_{Min}(u, v)|$?

2) For which nodes $u_1, v_1, \ldots, u_j, v_j$ exist pairwise edge disjoint or node disjoint paths $P_i \in \mathcal{P}_{Min}(u_i, v_i)$, $i = 1, \ldots, j$?

3) How do the sets $\mathcal{P}_{Min}(u, v)$ (and consequently the distances $d(u, v)$) behave if edges or nodes are deleted from the given graph?

We shall see that the answer to one of the questions 1) – 3) frequently imply an answer to another of these three questions. Moreover, many results about the sets $\mathcal{P}_{Min}(u, v)$ are modified Menger theorems. To formulate the next results precisely we now define geodetic properties of graphs.

Definition/Remark 3.34 Let G be a connected and simple undirected graph with n nodes. G is called *j-geodetic* if for each pair (u, v) of nodes, there are at most j shortest u-v-connections. Every 1-geodetic graph is called *geodetic*. G is called *j-geodetically node connected* if removing an arbitrary set S of nodes with $|S| = (j - 1)$ does not increase any distance $d(u, v)$ with $u, v \in V \backslash S$.

Consequently, every j-geodetic node connected graph is j-node connected; the reason is that removing $(j - 1)$ nodes does not destroy all shortest u-v-connections so that u and v remain connected.

G is called *j-geodetically edge connected* if removing $(j - 1)$ arbitrary edges does not increase any of the distances $d(u, v) \geq 2$.

Not every j-geodetically edge connected graph is also j-edge connected. The reason is that removing $(j-1)$ edges may destroy all u-v-connections if $d(u, v) = 1$. For example, consider an undirected graph G with exactly two nodes a, b, which are connected by an edge $\{a, b\}$. This graph is 2-geodetically edge connected because G has no nodes of distance 2. Nevertheless, removing the edge $\{a, b\}$ destroys the connectivity of G; therefore, G is not 2-edge connected.

Given $2j$ pairwise distinct nodes $x_1, \ldots, x_j, y_1, \ldots, y_j$ and a j-tuple $\tau = (P_1, \ldots, P_j)$ such that P_i is a shortest connection between x_i and y_i, $i = 1, \ldots, j$. Then τ is called a *node disjoint [edge disjoint] shortest connection* for $(x_1, y_1, \ldots, x_j, y_j)$ if the paths P_i are pairwise node disjoint [edge disjoint]. $\qquad\square$

Remark 3.35 Let $\mathcal{G}$ be a simple and geodetic graph. Then the geodetic node and edge connectivity of $\mathcal{G}$ are almost always bad. To see this we suppose that $\mathcal{G}$ has two nodes u, v with $d(u, v) \geq 2$. The assumption that $\mathcal{G}$ is geodetic implies the existence of exactly one u-v-path P with $\ell(P) = d(u, v)$. Let $x \in V \backslash \{u, v\}$ be a node of P; such a node x exists because $\ell(P) = d(u, v) \geq 2$; let $e \in \mathcal{R}$ be an edge of P. Then removing x or e increases the distance between u and v.

That means: If a geodetic graph $\mathcal{G}$ has any nodes of distance ≥ 2 then $\mathcal{G}$ is not 2-geodetically node connected and not 2-geodetically edge connected. It remains to the reader to investigate the geodetic node and edge connectivity of the complete undirected graph $\mathcal{K}_\mathcal{V}$; this graph is geodetic, too, but it does not have any nodes u, v with $d(u, v) \geq 2$. $\qquad\square$

We now quote several results about geodetic connectivity.

Theorem 3.36 (*see* [Schw89], *Theorem B,* [EnJS77])
Let G be a finite and simple connected graph $\mathcal{G}$; let $j \geq 2$.
Then $\mathcal{G}$ is j-geodetically node connected if $\mathcal{G}$ is j-geodetically edge connected. $\qquad\square$

Theorem 3.37 *Let $\mathcal{G}$ be simple and $(3j - 2)$-geodetically node connected undirected graph with n nodes.*
Then there exists a node disjoint shortest connection $\tau = (P_1, \ldots, P_j)$ for each $2j$-tuple $(x_1, y_1, \ldots, x_j, y_j)$ of pairwise distinct nodes.
This result follows from [EnSa84]. $\qquad\square$

Remark 3.38 Theorem 3.37 is a modified Menger theorem as described in 3.16 because an implication (a)$\Rightarrow$(b) is claimed about the following assertions (a) and (b):

(a) *For $k := (3j - 3)$, for all sets $S = V \backslash \{u, v\}$ and for $\mathcal{P} := \mathcal{P}_{Min}(u, v)$, the following is true: If $k = (3j - 3)$ arbitrary nodes are removed from S then the induced subgraph has at least one path in $\mathcal{P}$.*

(b) *For $l := j$ and for all sets $Q_i = \mathcal{P}_{Min}(u_i, v_i)$ with pairwise distinct nodes $x_1, y_1, \ldots, x_l, y_l$, the following is true: There exist pairwise node disjoint x_i-y_i-paths P_i $(i = 1, \ldots, l)$ where each path P_i has minimal length; that means that there exist $l = j$ pairwise node disjoint paths $P_i \in Q_i$, $i = 1, \ldots, l$.* $\qquad\square$

Theorem 3.39 *The following assertions are equivalent:*
(a) *$\mathcal{G}$ is a simple and j-geodetically edge connected graph.*
(b) *For each $2j$-tuple $(x_1, y_1, \ldots, x_j, y_j)$ of pairwise distinct nodes, there exists an edge disjoint shortest connection.*
This fact follows from Theorem E in [Schw89]. $\qquad\square$

Remark 3.40 Theorem 3.39 is a modified Menger theorem as described in 3.16 because an the equivalence of the following two assertions (a),(b) is claimed:

(a) *If $k := (j - 1)$ arbitrary edges are removed from $S := \mathcal{R}$ then the induced sub-*

graph has still a path belonging to $\mathcal{P}$:= $\mathcal{P}_{Min}(u, v)$ *where* u, v *are two arbitrary nodes in* $\mathcal{G}$.

(b) *For* l := j *and for all sets* $\mathcal{Q}_1$:= $\mathcal{P}_{Min}(u_1, v_1)$, ..., $\mathcal{Q}_l$:= $\mathcal{P}_{Min}(u_l, v_l)$ *with pairwise distinct nodes* $x_1, y_1, \ldots, x_l, y_l$, *the following is true: There exist* l *pairwise edge disjoint paths* $P_1 \in \mathcal{Q}_1, \ldots, P_l \in \mathcal{Q}_l$. □

Several articles, for example [Ples84] and [SrOA87], are about the construction of j-geodetic graphs (for $j = 1$ or $j \geq 1$). A simple characterization of geodetic graphs is given in [Sri87].

The following problem is considered in [PPSp90]: Construct an undirected graph $\mathcal{G} = (\mathcal{V}, \mathcal{R})$ with $\mathcal{V} = \{0, \pm 1, \pm 2, \ldots\}^2$ such that the distance $d(u, v)$ in $\mathcal{G}$ is a good approximation to the Euclidean distance $\|u - v\|_2$.

An axiomatic characterization of the system of all shortest paths is given in [Ne94].

The paper [KiSa80] is about intersections of shortest paths, that is, about their common nodes and edges.

In general, node or edge disjoint shortest paths are of great relevance for parallel computing. For example, let $\mathcal{G}$ be a network of processors. We assume that j messages $M_1, \ldots, M_j$ are simultaneously transmitted from the processors $x_1, \ldots, x_j$ to the processors $y_1, \ldots, y_j$, respectively. Then the following circumstances are very helpful to make the transmission fast and safe:

- For all $i = 1, \ldots, j$, the path P_i used by the message M_i is as short as possible; more formally, $P_i \in \mathcal{P}_{Min}(x_i, y_i)$ for all i.
- The paths P_i should be pairwise node or at least edge disjoint.
  (This condition helps to avoid collisions of different messages M_i and $M_{i'}$.)

These two condition mean that $\tau = (P_1, \ldots, P_j)$ is a shortest node disjoint or edge disjoint connection of $(x_1, y_1, \ldots, x_j, y_j)$.

The convexity of a graph is a further term in relation to a system of shortest paths. Recall the geometric definition of convexity: A subset A in the Euclidian plane $\mathbb{R}^2$ is *convex* if and only if the following is true for all $a, b \in A$: All points x of the shortest a-b-connection (i.e. the line segment $\overline{a, b}$) belong to A. The definition of a convex graph is analogous.

Definition/Remark 3.41 Let $\mathcal{G}$ be a simple undirected graph. A subset $A \subseteq \mathcal{V}$ (or the subgraph induced by A) is called *convex in* $\mathcal{G}$ if for all nodes $a, b \in A$, the following is true: All nodes x on any shortest a-b-connection in $\mathcal{G}$ belong to A.

Convex subgraphs are investigated in [Ples84]. □

The domination number and the distance irredundance are studied in [HaHe94]; both quantities are defined with the help of distances.

The paper [AH92] is about equivalence classes of edges, and these equivalence classes are defined with the help of distances.

A new measure of distance has been introduced in [JLee89]. Let S be a set of nodes then the S-distance $d_S(u, v)$ is defined as the minimal length of all u-v-paths that visit all nodes of S.

Other generalizations and modifications of the standard measure of distance are the Steiner distance in [ChOeTZ], the massive propagator distance in [Filk92], the k-distance in [HsLu], and the multiply sure distance in [McCaWi]. Further graph theoretical measures of distance are studied in [ChJa93].

3.4 Definitions and results in connexion with paths of arbitrary lengths

The current chapter 3 of this book is almost exclusively about very short or very long paths. This section, however, is addressed to paths that may have arbitrary lengths. For example, the path length distribution described on page 143 is based on the all paths in a graph and not only on very long or very short ones. Many terms in this section are similar to terms in Sections 3.2 and 3.3. For example the definition of the path length distribution resembles the one of the distance distribution on page 3.28.

First, we consider **digraphs**.

The paper [AmMa90] gives sufficient conditions for the existence of cycles and acyclic paths with various lengths.

The paper [GaCh91] investigates the *sequence of cycle lengths* in digraphs; this sequence is defined as follows: Let n be the number of nodes in a digraph $\mathcal{G}$. Then the sequence of cycle lengths is the n-tuple $\mathbf{C}(\mathcal{G}) = (c_1, \ldots, c_n)$ with the property that $\mathcal{G}$ contains exactly c_i node injective cycles of length i $(i = 1, \ldots, n)$. The sequence of cycle lengths is also called the *cycle length distribution*.

Results on special digraphs are given in [ChPS86] and in [HaBe81]. One of the results in [ChPS86] is the following: Given a digraph $\mathcal{G}$ with a c-clique, i.e. a set of c nodes which are pairwise connected by arcs. Then there exist two nodes u, v such that for all $k \in \mathbb{N}$ the following is true: There are at least c^k paths of length k from u to v. The paper [HaBe81] describes structural (for example, category theoretical) properties of paths in directed graphs.

Next, we describe several typical results about paths in **undirected graphs**.

Let $C(\mathcal{G})$ be the set of all cycle lengths existing in a graph $\mathcal{G}$. The papers [GyKS84] and [GPSV85] estimate the quantity $\sum\limits_{i \in C(\mathcal{G})} 1/i$.

More detailed information about all cycles in an undirected graph is given by the *sequence of cycle lengths* (= *cycle length distribution*); the definition of this sequence in undirected graphs is analogic to the definition in the directed case. The papers [Gern86], [LiuShi], and [WuShi] are about the sequence of cycle lengths.

A similar construction is considered in [FaSc78]: The *path length distribution* is an

n-tuple $(X_1, X_2, \ldots, X_n)$ such that each X_λ is the number of all node injective paths P with $\ell(P) = \lambda - 1$ and $\alpha(P) \neq \omega(P)$. The paper gives answers to the question whether a given n-tuple $\xi = (x_1, \ldots, x_n)$ is the path length distribution of a graph.

The *path length degree matrix* $\chi(\mathcal{G})$ is defined as follows: Let $\mathcal{V} = \{v^{(1)}, \ldots, v^{(n)}\}$ be the set of all nodes in $\mathcal{G}$. For all $i, \nu = 1, \ldots, n$, let $\chi_i\left(v^{(\nu)}\right)$ denote the number of all cycles or other paths that start at $v^{(\nu)}$ and have the length i. Then $\chi(\mathcal{G})$ is the matrix of all numbers $\chi_i\left(v^{(\nu)}\right)$ where $i, \nu = 1, \ldots, n$.
The path length degree matrix is similar to the distance degree matrix described in 3.30. Both types or matrices are investigated in [BlKQ81].

The paper [ChiH92] is about the following question: Given a bipartite undirected graph $\mathcal{G}$, can $\mathcal{G}$ be divided into a a system of paths $P_1, \ldots, P_j$ with $\ell(P_i) = i$, $i = 1, \ldots, j$?

Next, we quote several modified Menger theorems.

Theorem 3.42 (*see* [FOSJ87])
Given a connected and simple undirected graph G with n nodes and m edges. Let $j, K \in \{1, \ldots, n\}$. Moreover, let $\delta_1(\mathcal{G}) \geq \lfloor (n - m + 2) / \lfloor (K + 4)/3 \rfloor \rfloor + m - 2$.
Then (a)$\Rightarrow$(b) where (a) and (b) are the following assertions:
  (a) *G is j-node connected.*
  (b) *For each pair (x, y) of nodes, there are j pairwise internally node disjoint x-y-connections of a length $\leq K$.*   □

This result is a modified Menger theorem as described in 3.16. This can be seen by formulating (a) and (b) in Theorem 3.42 as follows:
  (a) *For $k = j - 1$, for all sets $S = \mathcal{V}\setminus\{x, y\}$ and for all sets $\mathcal{P} = P(x, y)$ with $x, y \in \mathcal{V}$, the following is true: If $k = j - 1$ arbitrary nodes are removed from S then the induced subgraph will have at least one path in $\mathcal{P} = P(x, y)$.*
  (b) *For $l = j$ and for all sets $\mathcal{Q}_1 = \ldots = \mathcal{Q}_l = \{P \in P(x, y) \mid \ell(P) \leq K\}$ with $x, y \in \mathcal{V}$, the following is true: There exist l pairwise internally node disjoint paths $P_1 \in \mathcal{Q}_1, \ldots, P_l \in \mathcal{Q}_l$.*

The next result is a Menger-type theorem as described in 3.13.

Theorem 3.43 *Let $\mathcal{G}$ be a connected and simple undirected graph with n nodes. Let $K, K' \in \mathbb{N}$, let $x, y \in \mathcal{V}$ with $d(x, y) \leq K$, and let $a(x, y)$ and $b(x, y)$ be the following numbers:*

$a(x, y)$ *is defined as the minimal cardinality of a set $\mathcal{V}' \subseteq \mathcal{V}\setminus\{x, y\}$ with the property that $d(x, y) > K$ in the subgraph induced by $\mathcal{V}\setminus\mathcal{V}'$. That means that removing $(a(x, y) - 1)$ arbitrary nodes from $\mathcal{V}\setminus\{x, y\}$ does not destroy all x-y-paths of a length $\leq K$; it is, however, possible to find $a(x, y)$ nodes such that all these short x-y-connections are interrupted.*

$b(x, y)$ *is the maximum number of pairwise node disjoint x-y-paths in $\mathcal{G}$ of a length $\leq K'$.*

Then $K' = \begin{pmatrix} a(x,y)+K-2 \\ K-2 \end{pmatrix} + \begin{pmatrix} a(x,y)+K-3 \\ K-2 \end{pmatrix}$ implies that $b(x, y) \geq a(x, y)$.

P r o o f : The assertion is shown in [PyTu93].   □

We next quote the following Menger-type theorem:

Theorem 3.44 *Let $\mathcal{G}$ be a connected and simple undirected graph with n nodes. Let $x, y \in V$, and let $K, K' \in \mathbb{N}$ with $K' = K$. We define $a(x, y)$ and $b(x, y)$ as in 3.43. Then there exists no constant C with the following properties:*
- *C is independent of $\mathcal{G}$, x, y, and K.*
- *$a(x, y) \le C \cdot \sqrt{K} \cdot b(x, y)$ for all $\mathcal{G}$, x, y, and K.*

P r o o f : A counterexample is given in [BoEx82]. Boyles and Exoo describe a family of graphs $\mathcal{G}_K$ with nodes x_K and y_K such that $a(x_K, y_K) = \frac{1}{4}(K - 3) \cdot b(x_K, y_K)$ for all K. That means that for any given C, the value $a(x_K, y_K)$ is greater than the value $C \cdot \sqrt{K} \cdot b(x, y)$ if K is sufficiently large.   □

This result may indeed be considered as a modified Menger theorem as described in 3.13; the assertion about $a(x, y)$ and $b(x, y)$ is that a particular relationship between these two quantities is not always given.

Theorem 3.45 *The following conjecture is false for all $K \ge 2$:*

> *Let $\mathcal{G}$ be a connected and simple undirected graph with n nodes. Let $K \ge 2$.*
> *Then (a)$\Rightarrow$(b) where (a) and (b) are the following assertions:*
> *(a) If $x, y \in V$ with $d(x, y) \le 3K - 2$ and if K arbitrary edges are removed then $d(x, y) \le 3K - 2$ in the resulting subgraph.*
> *(b) For all nodes x, y with $d(x, y) \le 3K - 2$, there exist $j = 2$ edge-disjoint x-y paths of a length $\le 3K - 2$.*   (3.5)

P r o o f : The conjecture is disproven in [Exoo83]. For each K, the author of that paper constructs a graph $\mathcal{H}_K$ with two nodes x_K, y_K such that the following are true: $d(x_K, y_K) \le 3K - 2$ after removing K arbitrary edges, and there do not exist two edge disjoint x_K-y_K-paths P_1, P_2 with $\ell(P_1), \ell(P_2) \le 3K - 2$.   □

Conjecture (3.5) has the structure described in 3.16. This can be seen by reformulating (a) and (b) as follows:
- (a) *For $k := K$, for $S := \mathcal{R}$, and for all sets $\mathcal{P} = \{P \in \mathcal{P}(x, y) \mid \ell(P) \le 3K - 2\}$ with $x, y \in V$ and $d(x, y) \le 3K - 2$, the following is true: If k arbitrary edges are removed from S then the induced subgraph has at least one path in $\mathcal{P}$.*
- (b) *For the number $l = j = 2$ and for all sets $\mathcal{Q}_1 = \mathcal{Q}_2 = \{P \in \mathcal{P}(x, y) \mid \ell(P) \le 3K - 2\}$ with $x, y \in V$ and $d(x, y) \le 3K - 2$, the following is true: There exist l pairwise edge disjoint paths $P_1 \in \mathcal{Q}_1$, $P_2 \in \mathcal{Q}_2$.*

Next, we describe several distance measures arising from the system of all paths in a graph.

Definition 3.46 Let $\mathcal{G}$ be a finite undirected graph.
The *chain distance* $d_C(u, v)$ between two nodes u, v is defined as the sum of the lengths of all node injective u-v-paths.
The *Boolean distance* $d_B(u, v)$ of two nodes u, v is defined as the set of all vertices of all

u-v-paths. A further measure of distance is defined as follows: $\rho_B(u, v) := |d_B(u, v)|$ for all nodes u, v. The properties of $d_B(u, v)$ or $\rho_B(u, v)$ are investigated in [MeTo81] and in [HMPT82].

It seems that the word "path" in the definition of Boolean distances must be understood as "node injective path" or "edge injective path". Otherwise, the definition of $d_B(u, v)$ has the following absurd consequence: Let $\mathcal{G}$ be connected, and let x be an arbitrary node in $\mathcal{G}$; then there exists an u-v-path visiting x; hence, $x \in d_B(u, v)$ for all nodes x so that $d_B(u, v) = \mathcal{V}$. ∎

3.5   Long cycles and long acyclic paths in digraphs

We start with long cycles in digraphs. A survey about this subject is given in [BeTh81].

The sources [PBSh91], [Man92], [Hägg93] are about Hamiltonian cycles in digraphs.

The next result gives lower bounds for the longest node injective cycle. We do not intend to find the best possible bounds; instead of this, we demonstrate that structural properties of a digraph $\mathcal{G}$ have an influence on the maximal cycle length in $\mathcal{G}$.

Theorem 3.47 *Let $\mathcal{G}$ be a finite, simple, and strongly connected digraph. Let P be a node injective cycle of maximal length.*

   a) *P has a length $\geq \delta_1^{\max}(\mathcal{G}) + 1$.*
   b) *If $\mathcal{G}$ is oriented then P has a length $\geq \delta_1^{\max}(\mathcal{G}) + 2$.*
   c) *If $\mathcal{G}$ is bipartite then P has a length $\geq 2\,\delta_1^{\max}(\mathcal{G})$.*
   d) *If $\mathcal{G}$ is oriented and bipartite then P has a length $\geq 2\,\delta_1^{\max}(\mathcal{G}) + 2$.*

P r o o f : We only consider the case that $\delta_1^{\max} = \delta_1^+$. Otherwise, construct all paths appearing in the following proof backwards and not forwards; that means that in all recursive constructions, v_j is chosen as a predecessor and not as a successor of the current node v_{j-1}.

The following proof is valid for all assertions a) – d).

We recursively construct an infinite path $Q := [v_0, v_1, v_2 \ldots)$ and start with an arbitrary arc (v_0, v_1).

Let $j \geq 2$, and let the nodes $v_0, \ldots, v_{j-1}$ be given. The idea is to construct the new node v_j by avoiding the last K nodes $v_{j-1}, v_{j-2}, \ldots, v_{j-K}$ where K is as great as possible; thus, we avoid short cycles. More precisely, we define the sets U_j and V_j as follows:

$U_j := \{v_k | j - \delta_1^+ \leq k \leq j - 2 \text{ and } k \geq 0\}$, $V_j := \{v_{j-1}\}$ in the proof of a),
$U_j := \{v_k | j - \delta_1^+ - 1 \leq k \leq j - 3 \text{ and } k \geq 0\}$, $V_j := \{v_{j-1}, v_{j-2}\}$ in the proof of b),
$U_j := \{v_k | k = j - 2, j - 4, \ldots, j - 2(\delta_1^+ - 1) \text{ and } k \geq 0\}$ and
$\quad V_j := \{v_k | k = j - 1, j - 3, j - 5, \ldots, j - 2\delta_1^+ + 1 \text{ and } k \geq 0\}$ in the proof of c),
$U_j := \{v_k | k = j - 4, j - 6, \ldots, j - 2\,\delta_1^+ \text{ and } k \geq 0\}$ and
$\quad V_j := \{v_{j-2}\} \cup \{v_k | k = j - 1, j - 3, j - 5, \ldots, j - 2\delta_1^+ + 1 \text{ and } k \geq 0\}$ in the proof of d).

Then we choose v_j such that $(v_{j-1}, v_j) \in \mathcal{R}$ and that $v_j \notin U_j$; this is possible because v_{j-1} has at least δ_1^+ successors and $|U_j|$ has at most $(\delta_1^+ - 1)$ elements. The

specific assumptions in assertion a), b), c), and d) imply that v_j is automatically not in V_j; in particular, $v_j \neq v_{j-1}$ because $\mathcal{G}$ is simple, $v_j \neq v_{j-2}$ if $\mathcal{G}$ is oriented, and $v_j \neq v_{j-1}, v_{j-3}, v_{j-5}, \ldots$ if $\mathcal{G}$ is bipartite.
We have seen that $v_j \notin U_j \cup V_j$ for all $j \geq 2$. This implies that the following nodes are pairwise distinct:

in the proof of a): the nodes $v_{j-\delta_1^+}, v_{j-\delta_1^+ +1}, v_{j-\delta_1^+ +2}, \ldots, v_j$ if $j \geq \delta_1^+$,

in the proof of b): the nodes $v_{j-\delta_1^+ -1}, v_{j-\delta_1^+}, v_{j-\delta_1^+ +1}, \ldots, v_j$ if $j \geq \delta_1^+ + 1$,

in the proof of c): the nodes $v_{j-2\delta_1^+ +1}, v_{j-2\delta_1^+}, v_{j-2\delta_1^+ +1}, \ldots, v_j$ if $j \geq 2\delta_1^+ - 1$,

$$(3.6)$$

in the proof of d): the nodes $v_{j-2\delta_1^+ -1}, v_{j-2\delta_1^+}, v_{j-2\delta_1^+ +1}, \ldots, v_j$ if $j \geq 2\delta_1^+ + 1$,

We choose P' as a path of maximal length among all node injective cycles $P'' \subseteq Q$. (P' exists because $\mathcal{G}$ is finite.) Fact (3.6) implies that $\ell(P') \geq \delta_1^+ + 1$ in the proof of a), $\ell(P') \geq \delta_1^+ + 2$ in the proof of b), $\ell(P') \geq 2\delta_1^+$ in the proof of c), and $\ell(P') \geq \delta_1^+ + 2$ in the proof of d). Hence, the assertions a) – d) about the maximal cycle P are true. $\square$

The lower bound in 3.47 b) is tight for the digraph on the right side; this graph has the properties that $\delta_1^{\max} = 1$ and that the longest node injective cycles are of length 3.

This example has been given in [Ja81]; the same paper describes further digraphs $\mathcal{G}$ for which $\delta_1^{\max} + 2$ is the maximal length of any node injective cycle; the values $\delta_1^{\max}(\mathcal{G})$ are greater than 1.

Next, we quote several further results about long cycles in digraphs:

Theorem 3.48 *Let $\mathcal{G}$ be an oriented and strongly connected digraph with n nodes. Let $\delta_2 + 2 \leq n$.*

Then $\mathcal{G}$ has a node injective cycle P with $\ell(P) \geq \left\lfloor \dfrac{2(n-1)}{2n - 2\delta_2 - 1} \right\rfloor + 1$.

P r o o f : We employ the following result of Heydemann, which has been quoted in [Ja81], page 155:

> Let $1 < h < n - 1$, and let $\deg(u) + \deg(v) \geq 2n - 2h - 1$ for all pairs of nodes with $(u,v) \notin \mathcal{R}, (v,u) \notin \mathcal{R}$.
> Then $\mathcal{G}$ contains a cycle of a length $\geq \lfloor (n-1)/h \rfloor + 1$ and an acyclic path of a length $\geq \lfloor (n-1)/h \rfloor + \lfloor (n-2)/h \rfloor$.

$$(3.7)$$

Let $h := (2n - 2\delta_2 - 1)/2$. The assumption $\delta_2 + 2 \leq n$ implies that $1 < h < n - 1$. Moreover, the following is true for all nodes u, v with $(u,v) \notin \mathcal{R}$ and $(v,u) \notin \mathcal{R}$:

$$\begin{aligned} \deg(u) + \deg(v) &= (\deg^+(u) + \deg^-(v)) + (\deg^+(v) + \deg^-(u)) \\ &\geq 2\delta_2 = 2n - 2h - 1 \end{aligned}$$

$$(3.8)$$

Consequently, the assertion follows by applying (3.7) to this special h. $\square$

Theorem 3.49 *Let $\mathcal{G}$ be a bipartite digraph with n nodes.*
   a) *Then $\mathcal{G}$ has a node injective cycle P with $\ell(P) \geq 2\,\delta_1^{\max}$.*
      This follows from [Ayel83], Theorem 1 and from [Ayel81].
   b) *If $\mathcal{G}$ is even oriented and if $\delta_1^{\min} \geq \frac{1}{4}n - 1$ then $\mathcal{G}$ contains a node injective cycle P with $\ell(P) \geq 4\,\delta_1^{\min}$.*
      This result from [So86] is quoted in [Zha87], page 348.   □

The following papers are give special results about long cycles.

Longest cycles in tournaments are described in [BoHä90], in [Shen88], and in other sources. (A *tournament* is a digraph constructed by orientating each edge of the complete undirected graph $\mathcal{K}_n$; more precisely, each edge $\{u, v\}$ is replaced with one of the arcs (u, v), (v, u).)

It is proven in [Benh84] that particular digraphs have Hamiltonian cycles with the property that exactly one arc is traversed in the wrong direction. The paper [Thom87] describes the behaviour of the maximal cycle length if the direction of an arc in a digraph is inverted.

Longest cycles in regular digraphs are investigated in [Darb82]. (The case of regular undirected graphs is described in 3.71 b).)

The next result gives lower bounds to the length of simple cycles and of acyclic paths.

Theorem 3.50 *Let $\mathcal{G}$ be an oriented digraph with n nodes.*

   a) *Suppose that $\delta_1^{\min} \geq 3$. Then there exists a node injective cycle P in $\mathcal{G}$ with $\ell(P) \geq \delta_1^+ + \delta_1^- + 1$, or there exists an acyclic path P with $\ell(P) \geq \delta_1^+ + \delta_1^- + 3$.*
This result has been proven in [Song86].
(A similar result is given in [Zha81]; the author omits the assumption that $\delta_1^{\min} \geq 3$ and guarantees a cycle of a length $\geq 2\,\delta_1^{\min} + 1$ or a path of length $2\,\delta_1^{\min} + 2$. That means that Zhang draws a weaker conclusion from a weaker assumption than Song does.)

   b) *If $\mathcal{G}$ is oriented and bipartite then $\mathcal{G}$ has a cycle P with $\ell(P) \geq 2(\delta_1^+ + \delta_1^-)$ or another path P with $\ell(P) \geq 2(\delta_1^+ + \delta_1^-) + 3$.*
This is Theorem 2 in [Zha87], and it is an improvement of Theorem 2 in [Ayel83]. □

Next, we consider long acyclic paths in digraphs. They are of great relevance when planning projects. The length of a particular longest path can often be used to calculate the accomplish a project.

Many practical projects can be represented as CPM- and PERT-networks, which are described in [Neu75] and other sources. We now give a simple example for the practical relevance of longest paths.

Example 3.51 Given a large project with elementary tasks $s, t_1, t_2, \ldots, t_n$. We assume that each task can be accomplished in exactly one day. Moreover, we assume that not all tasks can be done simultaneously; in particular, the task s must have been accomplished before any other task may begin.
We represent this situation by an acyclic, simple digraph $\mathcal{G} = (\mathcal{V}, \mathcal{R})$ where

$V := \{s, t_1, \ldots, t_n\}$. We define $\mathcal{R}$ as the set of all arcs (v, w) with the following property: w must be done later than v. In particular, $(s, v) \in \mathcal{R}$ for all $v \neq s$. Let $D : V \to \mathbb{N}$; then we interpret $D(v)$ as the day when $v \in V$ is executed. We say that D is *admissible* if D satisfies the following conditions:

- $D(s) = 1$; that means that the task s is finished on the first day.
- $D(v) < D(w)$ for all $(v, w) \in \mathcal{R}$; that means that D takes respect of the requirement that w must be begun after finishing v.

Then the last day of complete project has the number
$$MAX(D) := \max\{D(v) \mid v \in V\}.$$
Consequently, minimizing the duration of a project means to find an admissible function D such that $MAX(D)$ is minimal. □

The next result describes the close relationship between the duration of a project and longest paths in digraphs.

Theorem 3.52 *Given the situation in 3.51. For all nodes $v \in V$, let $L(v)$ denote the maximal length of any s-v-path. Then the following are true:*

a) $D(v) \geq L(v) + 1$ *for each admissible function D and for all nodes v.*
b) *The function $\widetilde{D}(v) := L(v) + 1$, $v \in V$, is admissible.*
c) *Let $L^{\max}$ be the length of the longest path in $\mathcal{G}$ that starts with s. Then the whole project can be accomplished in $(L^{\max} + 1)$ days, and this is the minimum duration of the project.*

P r o o f   o f   a) :   Let $P = [s = v_0, v_1, \ldots, v_l = v]$ be an s-v-path of maximal length. Then $D(s) = 1$, and $D(v_i + 1) \geq D(v_i) + 1$ for all i because D is admissible and $D(v) \in \mathbb{N}$ for all $v \in V$. Consequently, $D(v_i) \geq i + 1$ for all i. This implies that $D(v) = D(v_l) \geq l + 1 = L(v) + 1$.

P r o o f   o f   b) :   It is clear that $\widetilde{D}(s) = L(s) + 1 = 1$. Let $(v, w) \in \mathcal{R}$, and let P be an s-v-path with $\ell(P) = L(v)$. Then we obtain the following inequality, in which $(*)$ follows from the definition of $L(w)$ as the maximal length of all s-w-paths.

$$\widetilde{D}(v) \;=\; L(v) + 1 \;=\; \ell(P) + 1 \;=\; \ell(P \oplus (v, w)) \overset{(*)}{\leq} L(w) \;<\; \widetilde{D}(w).$$

P r o o f   o f   c) :   Part b) says that it does not cause conflicts if each task v is finished on the day with number $\widetilde{D}(v) = L(v) + 1$. Hence, $(L^{\max} + 1)$ days are sufficient for the complete project. On the other hand, there exists a task $v^{\max}$ such that $L(v^{\max}) = L^{\max}$. It follows from a) that this task must be accomplished on the day with number $(L(v^{\max}) + 1) = (L^{\max} + 1)$ or later. □

Now we show how longest paths depend on structural properties of the underlying digraph.

Theorem 3.53 *Given a simple, finite, strongly connected digraph $\mathcal{G}$. Let x be an arbitrary node in $\mathcal{G}$, and let P be an acyclic path of maximal length among all candidates with start node x.*

a) *Then $\ell(P) \geq \delta_1^+$.*
b) *If $\mathcal{G}$ is oriented then $\ell(P) \geq \delta_1^+ + 1$.*
c) *If $\mathcal{G}$ is bipartite then $\ell(P) \geq 2\,\delta_1^+ - 1$.*
d) *If $\mathcal{G}$ is oriented and bipartite then $\ell(P) \geq 2\,\delta_1^+ + 1$.*

P r o o f : The proof is almost the same as the one of Theorem 3.47. We show assertions a) – d) simultaneously.

For this purpose, we recursively construct an infinite path $Q := [v_0, v_1, v_2 \ldots)$; we start with $v_0 := x$ and an arbitrary arc $(v_0, v_1) = (x, v_1)$.
If $v_0 = x, v_1, \ldots, v_{j-1}$ are given, we define the sets U_j, V_j and the node v_j in the same way as in the proof of Theorem 3.47.
We apply (3.6) to $j := \delta_1^+$ in the proof of a), to $j := \delta_1^+ + 1$ in the proof of b), to $j := 2\,\delta_1^+ - 1$ in the proof of c), and to $j := 2\,\delta_1^+ + 1$ in the proof of d). Thus, we see that the following nodes are pairwise distinct:

in the proof of a): the nodes $v_0 = x, v_1, \ldots, v_{\delta_1^+}$,

in the proof of b): the nodes $v_0 = x, v_1, \ldots, v_{\delta_1^+ + 1}$,

in the proof of c): the nodes $v_0 = x, v_1, \ldots, v_{2\,\delta_1^+ - 1}$,

in the proof of d): the nodes $v_0 = x, v_1, \ldots, v_{2\,\delta_1^+ + 1}$.

Consequently, Q has an acyclic prefix P' with $\ell(P') = \delta_1^+$, $\ell(P') = \delta_1^+ + 1$, $\ell(P') = 2\,\delta_1^+ - 1$, and $\ell(P') = 2\,\delta_1^+ + 1$, respectively. This and the relationship $\ell(P) \geq \ell(P')$ imply the assertions a) – d).   □

Exchanging the rôles of start and end nodes, we obtain the following modification of 3.53:

Theorem 3.54 *Given a simple, finite, strongly connected digraph $\mathcal{G}$. Let x be an arbitrary node in $\mathcal{G}$, and let P be a path of maximal length among all candidates with end node x.*

a) *Then $\ell(P) \geq \delta_1^-$.*
b) *If $\mathcal{G}$ is oriented then $\ell(P) \geq \delta_1^- + 1$.*
c) *If $\mathcal{G}$ is bipartite then $\ell(P) \geq 2\,\delta_1^- - 1$.*
d) *If $\mathcal{G}$ is oriented and bipartite then $\ell(P) \geq 2\,\delta_1^- + 1$.*

P r o o f : We define the digraph $\mathcal{G}^- = (V, \mathcal{R}^-)$ by reversing the directions of all arcs in $\mathcal{G}$, i.e., $\mathcal{R}^- := \{(w, v) \,|\, (v, w) \in \mathcal{R}\}$. If $\mathcal{G}$ is oriented then $\mathcal{G}^-$ is oriented, and if $\mathcal{G}$ is bipartite then $\mathcal{G}^-$ is bipartite. Therefore, we may apply Theorem 3.53 a), b), c), d), respectively, when proving assertion a), b), c), and d); we obtain a path $P^- = [v_0 = x, v_1, \ldots, v_l] \in \mathcal{P}(\mathcal{G}^-)$ with

a) $l = \delta_1^+(\mathcal{G}^-) = \delta_1^-(\mathcal{G})$,   b) $l = \delta_1^+(\mathcal{G}^-) + 1 = \delta^-(\mathcal{G}) + 1$,
c) $l = 2\,\delta_1^+(\mathcal{G}^-) - 1 = 2\,\delta_1^-(\mathcal{G}) - 1$,   d) $l = 2\,\delta_1^+(\mathcal{G}^-) + 1 = 2\,\delta_1^-(\mathcal{G}^-) + 1$,

respectively. The assertions a) – d) follow from the fact that P is at least as long as the path $P' := [v_l, v_{l-1}, \ldots, v_0 = x] \in \mathcal{P}(\mathcal{G})$.   □

Corollary 3.55 *Given a simple, finite, strongly connected digraph* $\mathcal{G}$. *Let* P *be an acyclic path in* $\mathcal{G}$ *with maximal length.*

a) *Then* $\ell(P) \geq \delta_1^{\max}$ $(= \max\{\delta_1^+, \delta_1^-\})$.
b) *If* $\mathcal{G}$ *is oriented then* $\ell(P) \geq \delta_1^{\max} + 1$.
c) *If* $\mathcal{G}$ *is bipartite then* $\ell(P) \geq 2\,\delta_1^{\max} - 1$.
d) *If* $\mathcal{G}$ *is oriented and bipartite then* $\ell(P) \geq 2\,\delta_1^{\max} + 1$.

P r o o f : The result follows from 3.53 if $\delta_1^{\max} = \delta_1^+$ and from 3.54 if $\delta_1^{\max} = \delta_1^-$. □

We now quote several results from literature.

Theorem 3.56 (*see* [Ayel83], *Theorem A and* [BGHS81])
Let $\mathcal{G}$ *be a strongly connected, simple digraph with* n *nodes. Then* $\mathcal{G}$ *contains a path* P *of a length* $\ell(P) \geq \min(n-1, \delta_1^+ + \delta_1^-)$. □

Theorem 3.57 *Let* $\mathcal{G}$ *be a strongly connected, oriented digraph with* $\delta_1^{\min} \geq 3$. *Then* $\mathcal{G}$ *contains a path of a length* $\ell(P) \geq \delta_1^+ + \delta_1^-$.

P r o o f : Recall Theorem 3.50. If $\mathcal{G}$ contains an acyclic path Q of a length $\geq \delta_1^+ + \delta_1^- + 3$, let $P := Q$. If $\mathcal{G}$ contains a node injective cycle Q of a length $\geq \delta_1^+ + \delta_1^- + 1$, generate P by removing the last arc of Q. □

Remark 3.58 In most cases, Theorem 3.57 is at least as good as 3.53 b), 3.54 b), and 3.55 b); there are, however, two exceptions:

- $\delta_1^{\min} \leq 2$ (I.e., the condition $\delta_1^{\min} \geq 3$ of 3.57 not satisfied.)
- The start or the end node of P is prescribed (as in 3.53 b) and in 3.54 b). □

Theorem 3.59 *Let* $\mathcal{G}$ *be strongly connected and oriented. We assume that* $\delta_1^{\min} \geq 2$.
Let $\delta^* := \max\left(2\,\delta_1^{\min} + 2, \, \delta_1^+ + \delta_1^-\right)$.
Then there exists a path P *with* $\ell(P) \geq \min(n-1, \delta^*)$.

P r o o f : Let P be an acyclic path of maximal length. The assertion is correct if $\ell(P) = n - 1$.
We consider the case that $\ell(P) < n - 1$. Then $\ell(P) \geq 2\,\delta_1^{\min} + 2$ by Theorem 5 of [Ja81]; moreover, $\ell(P) \geq \delta_1^+ + \delta_1^-$ by Theorem 3.56. Consequently, $\ell(P) \geq \delta^*$. □

Remark 3.60 Given the situation in 3.59

a) In most cases, $\delta^* = \delta_1^+ + \delta_1^-$; only if $|\delta_1^+ - \delta_1^-| \leq 1$ then $2\delta_1^{\min} > \delta_1^+ + \delta_1^-$.
b) If $\delta_1^{\min} \geq 6$ then the lower bound $2\delta_1^{\min} + 2$ may be replaced with $2\delta_1^{\min} + 3$.
     This follows from [Song86]. □

The next theorem, which has been shown in [Song89], gives a lower bound of cycle lengths even if $\mathcal{G}$ is not oriented.

Theorem 3.61 *Given a strongly connected digraph* $G = (V, R)$ *with* $n \geq 7$ *nodes and at least* $\frac{3}{4} n^2 - \frac{3}{2} n + \frac{3}{4}$. *Let* $\delta_1^{\min} \geq (n-3)/2$. *Then* G *contains a node injective cycle of a length* $\geq (n-1)$.

$\square$

The following lower bound of maximal path lengths depends on δ_2; the result is analogous to 3.48.

Theorem 3.62 *Let* G *be a strongly connected, oriented digraph with* n *nodes. Let*
$$\delta_2 + 2 \leq n$$
Then G *contains an acyclic path* P *with*

$$\ell(P) \geq \left\lfloor \frac{2 \cdot (n-1)}{2n - 2\delta_2 - 1} \right\rfloor + \left\lfloor \frac{2 \cdot (n-2)}{2n - 2\delta_2 - 1} \right\rfloor.$$

P r o o f : Define $h := (2n - 2\delta_2 - 1)/2$. The assumption $\delta_2 + 2 \leq n$ implies that
$$1 < h < n - 1. \quad \text{Then } \deg(u) + \deg(v) \overset{(3.8)}{\geq} 2n - 2h - 1 \text{ for all nodes } u, v \text{ with } (u,v) \notin R$$
and $(v,u) \notin R$. This and (3.7) imply the assertion.

$\square$

The next assertions are about bipartite digraphs.

Remark 3.63 Let G be a simple, bipartite digraph with n nodes.

a) Theorem 1 in [Ayel83] yields an acyclic path P of a length $\ell(P) \geq 2\delta_1^{\max} - 1$. This path is constructed as follows: Theorem 1 in [Ayel83] says that there exists a node injective cycle P' of a length $\ell(P') \geq 2\delta_1^{\max}$; remove the last arc of P then resulting path P has the desired properties.

b) The second part of Theorem 1 in [Ayel83] gives a sufficient condition for the existence of an acyclic path P of length $\ell(P) \geq \min\left(n - 1, 2(\delta_1^+ + \delta_1^-) - 2\right)$.

c) If, in addition, G is oriented then 3.50 b) yields an acyclic path P of length $\ell(P) \geq 2(\delta_1^+ + \delta_1^-) - 1$.

Moreover, Theorem 1 in [Zha87] yields an acyclic path P of a length $\ell(P) \geq 2\delta_2 + 1$. More precisely, this theorem yields an acyclic path of length $2\delta - 1$ if $\delta_2 \geq \delta - 1$. Choosing $\delta := \delta_2 + 1$, we obtain the desired path length.

$\square$

The article [Darb82] describes longest paths in regular digraphs.

The authors of [WoWo88] study Hamiltonian paths in bipartite digraphs.

The paper [KlW89] and [GuPf92] investigate the lengths of s-v-paths in rooted trees where s is the root and v is a leaf; these paths are maximal in the sense that they cannot be extended. The maximal length of any path in a rooted tree G is equal to the *height* $\mathbf{h}(G)$ of G. If G is a computation tree or a decision tree then $\mathbf{h}(G)$ describes the worstcase time of computation. Therefore, the height of a tree is studied in many sources about data structures and about complexity theory. More special results can be found in the sources [FGOR] and [Kemp90]. The difference between the longest and the shortest path from the root to any leaf is investigated in [DesP94] and in [CamW94].

3.6 Long cycles and other long paths in undirected graphs

This section is mainly about lower bounds for the maximal lengths of cycles or other paths in undirected graphs. In 3.2, the length of the longest node injective cycle was called the circumference of a graph.

The papers [BVMS90] and [BSV88] give an overview over the research on long cycles in undirected graphs.

A special type of long paths are Hamiltonian cycles or acyclic Hamiltonian paths. The author of [Liu91] describes how to compute the number of Hamiltonian cycles and paths of a given graph. The paper [Blum86] describes techniques to prove that a given graph has a Hamiltonian cycle. Further sources about Hamiltonian cycles or acyclic Hamiltonian paths are given in [Gou91] and in the current section of this book (for example, in Theorems 3.76, 3.78 and immediately after 3.78).

3.6.1 Relationships between long paths and degrees of nodes

This subsection is mainly about sufficient conditions for the existence of node injective cycles or of acyclic paths with a length $\geq \lambda$; the sufficient conditions or the lower bound λ are expressed with the help of the quantities δ_k, $\bar{\delta}_k$, and $\bar{\bar{\delta}}_k$ defined in 3.8. It is clear that λ is also a lower bound of the maximal length of all node injective cycles and of all acyclic paths, respectively. In particular, we shall often quote the following type of result:

> Given an undirected graph $\mathcal{G}$ (possibly with additional properties).
> Then the following is true for a particular constant C and a particular $k \in \{1, 2, 3\}$:
> There exists a node injective cycle P [an acyclic path P, resp.] with $\ell(P) \geq C \cdot \delta_k$.

It is clear that $C \cdot \delta_k$ is also a lower bound for the maximal length of cycles or other paths in $\mathcal{G}$.

Next, we show Lemma 3.64; it is about a relationship between $\delta_k(\mathcal{G})$ and $\delta_l(\mathcal{G})$ where $k \leq l$. We shall use this relationship in Lemma 3.65; this result says that a lower bound of path lengths can be expressed with the help of δ_k if it can be expressed with the help of δ_l.

Lemma 3.64 *Let $\mathcal{G}$ be a simple undirected graph with n nodes, and let $1 \leq k \leq l \leq n$. Then $\delta_l \geq \frac{l}{k} \delta_k$.*

P r o o f : First, we consider the case $l = k + 1$. Given a set of independent nodes $\{v_1, \ldots, v_l\} \subseteq V$ with the property that $\sum_{i=1}^{l} \deg(v_i) = \delta_l$. We define $S_j := \sum_{i=1}^{l} \deg(v_i) - \deg(v_j)$ for all j with $1 \leq j \leq l = k + 1$.
Then $\sum_{j=1}^{l} S_j = l \cdot \sum_{i=1}^{l} \deg(v_i) - \sum_{j=1}^{l} \deg(v_j) = k \cdot \sum_{i=1}^{l} \deg(v_i)$ because $k = l - 1$; that means that $k \cdot \sum_{i=1}^{l} \deg(v_i) = \sum_{j=1}^{l} S_j$. Moreover, the definitions of δ_k and of S_j imply that $S_j \geq \delta_{l-1} = \delta_k$ for all $j = 1, \ldots, l$; consequently, $\sum_{j=1}^{l} S_j \geq l \cdot \delta_k$.

It follows that $k \cdot \delta_l = k \cdot \sum_{i=1}^{l} \deg(v_i) = \sum_{j=1}^{l} S_j \geq l \cdot \delta_k$. Hence, $\delta_l \geq \frac{l}{k} \delta_k$.

Let $l - k > 1$. Successively applying the result in the above case, we obtain

$$\delta_l \geq \frac{l}{l-1} \cdot \delta_{l-1} \geq \frac{l}{l-1} \cdot \frac{l-1}{l-2} \cdot \delta_{l-2} \geq \dots \geq \frac{l}{l-1} \cdot \frac{l-1}{l-2} \cdot \dots \cdot \frac{k+2}{k+1} \cdot \frac{k+1}{k} \cdot \delta_k = \frac{l}{k} \cdot \delta_k \cdot \square$$

Lemma 3.65 *Let $k, l, n \in \mathbb{N}$ with $1 \leq k \leq l \leq n$. Moreover, let $F : \mathbb{N} \to \mathbb{N} \cup \{0\}$ and $G : \mathbb{N} \times \mathbb{N} \times \mathbb{N} \to \mathbb{N}$; we assume that G is monotonically increasing in the second argument. Recall that $\alpha(\mathcal{G})$ denotes the independence number and $\tau(\mathcal{G})$ denotes the toughness of a given graph $\mathcal{G}$. Let $K, t \in \mathbb{N} \cup \{0\}$.*

Suppose that the following assertion is true:

 (a) Each simple undirected graph $\mathcal{G}$ with n nodes and with $\delta_l(\mathcal{G}) \geq F(n)$, $\kappa(\mathcal{G}) \geq K$ and $\tau(\mathcal{G}) \geq t$ has a node injective cycle P [an acyclic path P] of the length $\ell(P) \geq G(n, \delta_l(\mathcal{G}), \alpha(\mathcal{G}))$.

Then the following assertion is true as well:

 (b) Each simple undirected graph $\mathcal{G}$ with n nodes and with $\delta_k(\mathcal{G}) \geq \frac{k}{l} \cdot F(n)$, $\kappa(\mathcal{G}) \geq K$, and $\tau(\mathcal{G}) \geq t$ has a node injective cycle P [an acyclic path P] of the length $\ell(P) \geq G\left(n, \frac{l}{k} \cdot \delta_k, \alpha(\mathcal{G})\right)$.

(It may occur that no condition is formulated about some of the parameters δ_l, $\kappa(\mathcal{G})$, and $\tau(\mathcal{G})$; this can be expressed by choosing $F(n) = 0$, $K = 0$, and $t = 0$, respectively.)

P r o o f : Let $\mathcal{G}$ be a simple undirected graph with n nodes; let $\delta_k \geq \frac{k}{l} \cdot F(n)$, $\kappa(\mathcal{G}) \geq K$, and $\tau(\mathcal{G}) \geq t$. Then Lemma 3.64 implies that $\delta_l \geq \frac{l}{k} \cdot \delta_k \geq F(n)$. Moreover, assumption (a) yields a node injective cycle P [an acyclic path P] with $\ell(P) \geq G(n, \delta_l, \alpha(\mathcal{G}))$. Moreover, $G(n, \delta_l, \alpha(\mathcal{G})) \geq G\left(n, \frac{l}{k} \cdot \delta_k, \alpha(\mathcal{G})\right)$ because $\delta_l \overset{\text{Lemma 3.64}}{\geq} \frac{l}{k} \cdot \delta_k$ and G is monotone in the second variable. Consequently, $\ell(P) \geq G\left(n, \frac{l}{k} \cdot \delta_k, \alpha(\mathcal{G})\right)$, and assertion (b) is verified. $\square$

Next, we give a list of results about maximal cycle and path lengths. We start with an assertion that is similar to 3.47 and that is equivalent to Theorem 2 in [Dir52].

Theorem 3.66 *Let $\mathcal{G}$ be a simple and connected undirected graph with n nodes. Then $\mathcal{G}$ has a node injective cycle P of a length $\ell(P) \geq \delta_1 + 1$.*

P r o o f : The proof is almost the same as the one of Theorem 3.47 a). We recursively construct an infinite path $Q := [v_0, v_1, v_2 \dots)$ and start with an arbitrary edge $\{v_0, v_1\}$. If $j \geq 2$ and if the nodes $v_0, \dots, v_{j-1}$ are already given we define $U_j := \{v_k | j - \delta_1 \leq k \leq j - 2 \text{ and } k \geq 0\}$ and $V_j := \{v_{j-1}\}$ Then we choose v_j such that $\{v_{j-1}, v_j\} \in R$ and that $v_j \notin U_j$; this is possible because v_{j-1} has at least δ_1 neighbours and $|U_j|$ has at most $(\delta_1 - 1)$ elements. Moreover, $v_j \neq v_{j-1}$ because $\mathcal{G}$ is simple; consequently, $v_j \notin V_j$. We have seen that $v_j \notin U_j \cup V_j$ for all $j \geq 2$. This implies that the nodes $v_{j-\delta_1}, v_{j-\delta_1+1}, v_{j-\delta_1+2}, \dots, v_j$ are pairwise distinct if $j \geq \delta_1$.

We choose P' as a path of maximal length among all node injective cycles $P'' \subseteq Q$. (P' exists because $\mathcal{G}$ is finite.) It follows that $\ell(P) \geq \ell(P') \geq \delta_1 + 1$. □

Remark 3.67 Let $\delta \geq 2$. The complete undirected graph $\mathcal{G} := K_{\delta+1}$ has minimal degree $\delta_1(\mathcal{G}) = \delta$, and it has no node injective cycle with more than $(\delta + 1) = (\delta_1 + 1)$ edges. Consequently, the lower bound in 3.66 is tight as long as $\mathcal{G}$ does not belong to a special class of undirected graphs. □

Remark 3.68 The bound δ_1 is also tight if $\delta_1(\mathcal{G}) = \delta$ is fixed and the number n of nodes may be arbitrarily great. The following example is given in [Dir52]. Let $\delta \geq 2$. Then we choose an arbitrary number $N \in \mathbb{N}$ and N pairwise disjoint sets $\mathcal{V}_1, \ldots, \mathcal{V}_N$ with $|\mathcal{V}_\nu| = \delta + 1$ for all ν. Moreover, we define the complete graphs $\mathcal{G}_\nu := K_{\mathcal{V}_\nu}$, $\nu = 1, \ldots, N$. In each graph $\mathcal{G}_\nu$, we choose a node $v_{0,\nu}$. Then we define $\mathcal{G}$ as the union of all graphs $\mathcal{G}_\nu$ where all nodes $v_{0,\nu}$ are replaced with a single node v_0. More precisely, let $\mathcal{V}'_\nu := \mathcal{V}_\nu \cup \{v_0\}\backslash\{v_{0,\nu}\}$, and let $\mathcal{G} := (\mathcal{V}, \mathcal{R})$ with $\mathcal{V} := \bigcup_{\nu=1}^n \mathcal{V}'_\nu$ and $\mathcal{R} := \bigcup_{\nu=1}^n \{\{v, w\} \mid v \neq w, \, v, w \in \mathcal{V}'_\nu\}$.
Then $\delta_1(\mathcal{G}) = \delta$. Moreover, $\mathcal{G}$ contains no node injective cycle P of a length greater than $\delta + 1$; the reason is that such a long cycle P must visit two different sets $\mathcal{V}'_\mu \neq \mathcal{V}'_\nu$, and this is only possible if P visits the node v_0 more than once; hence, P is not node injective if $\ell(P) > \delta + 1$. This and the equation $\delta_1(\mathcal{G}) = \delta$ imply that $\delta_1(\mathcal{G}) + 1$ is a lower bound on the length of all node injective cycles in $\mathcal{G}$. □

Next, we give many lower bounds λ on the circumference $\mathbf{c}(\mathcal{G})$. The tables in 3.69, 3.72 and 3.74 describe lower bounds depending on δ_1, δ_2 and δ_3, respectively.

Each of these tables is organized as follows: When reading it from top to bottom then the (approximated) constant C in the bound $\lambda = C \cdot \delta_k$ becomes greater and greater; this makes it easy to compare different results with each other if they are listed in the same table. On the other hand, lower bounds in *different* tables cannot be compared with the help of their constants. For example, consider the lower bounds $\lambda_1 := 1.4\,\delta_1$ and $\lambda_2 := 0.8\,\delta_2$. Which is better? Comparing the constants implies that λ_1 is better because $1.4 > 0.8$. In reality, however, λ_2 is better because $\lambda_2 = 0.8\,\delta_2 \overset{\text{Lemma 3.64}}{\geq} 0.8 \cdot 2\,\delta_1 \geq 1.4\,\delta_1 = \lambda_1$.

The conditions in square brackets describe the situation in which the given constant C is valid.

The last column of each table in 3.69, 3.72, and 3.74 describes the origin of the current result. We use the following abbreviations:

"A, B" means: Part B of source A; an example is [Alon86], Theorem 1.1.

The symbol $\langle A \rangle$ means: The current result follows from the corresponding theorem of this book, possibly by applying Lemma 3.65. For example the origin of Theorem 3.69 d) is given as "$\langle 3.72\,c)\rangle$". This theorem says that $\mathbf{c}(\mathcal{G}) \geq \min(n, \delta_2 + 1)$. Applying Lemma 3.65 with $k = 1$ and $l = 2$ yields $\mathbf{c}(\mathcal{G}) \geq \min(n, 2\delta_1 + 2)$, and this is the assertion of Theorem 3.69 d).

"A; B" means: Each of the sources A and B may be used to derive the current theorem.

"A ↘ B" means: The current result is mentioned in source A, and A quotes this result from source B.

The word "Theorem" is abbreviated as "Th.", and the word "page" is abbreviated as "p.".

Theorem 3.69 *Let G be a connected and simple undirected graph with $n \geq 3$ nodes. Then the assertions a) – j) in the following table are true.* □

	Assumptions about δ_1	Further assumptions	Lower bound λ for $c(G)$	$C \approx$	Origin of the result
a)	—	—	$\delta_1 + 1$	1	[Dir52], Th. 2; ⟨3.66⟩
b)	$\delta_1 \geq n/k$ $(k \in \mathbb{N})$	—	$\dfrac{n}{k-1}$	$\dfrac{k}{k-1} \in (1,2]$	[Alon86], Th. 1.1; ⟨3.72 a)⟩
c)	—	$\kappa \geq 2$	$\min(n, 2\delta_1)$	2	[Dir52], Th. 4; ⟨3.72 b)⟩
d)	—	$\tau \geq 1$	$\min(n, 2\delta_1 + 2)$	2	⟨3.72 c)⟩
e)	$\delta_1 \geq \dfrac{1}{2}n - 2$	$\tau \geq 1$ $n \geq 11$	n	2	⟨3.72 d)⟩
f)	$\delta_1 \geq \dfrac{1}{3}n$	$\tau \geq 1$	$\min\left(n, \dfrac{1}{2}n + \delta_1\right)$	$\dfrac{5}{2}$ $\left[\delta_1 = \dfrac{n}{3}, \ c(G) = \dfrac{5n}{6}\right]$	⟨3.72 e)⟩
g)	$\delta_1 \geq \dfrac{1}{3}n$	$\tau \geq 1$	$\min\left(n, \dfrac{\tau}{\tau+1}n + \delta_1\right)$	$3\dfrac{\tau}{\tau+1} + 1 \geq$ $\dfrac{3}{2} + 1 = \dfrac{5}{2}$ $\left[\delta_1 = \dfrac{1}{3}n\right]$	⟨3.72 f)⟩
h)	$\delta_1 \geq \dfrac{1}{3}n + \dfrac{2}{3}$	$\kappa \geq 2$	$\min(n, n + \delta_1 - \alpha)$	3 $\left[\delta_1 = \dfrac{n}{3}, \ c(G) = n\right]$	⟨3.72 g)⟩
i)	$\delta_1 \geq \dfrac{1}{3}n$	$\tau \geq 1$	$\min(n, n + \delta_1 - \alpha)$	3 $\left[\delta_1 = \dfrac{n}{3}, \ c(G) = n\right]$	⟨3.72 h)⟩
j)	$\delta_1 \geq \dfrac{1}{3}n$	$\tau \geq 1$	$\min(n, n + \delta_1 - \alpha + 1)$	3 $\left[\delta_1 = \dfrac{n}{3}, \ c(G) = n\right]$	[BVMS90], Th. 14

Remark 3.70   a) Assertion 3.69 b) is a generalization of Theorem 3 in [Dir52], which is only about the case $k = 2$. Result 3.69 b) is obtained by applying Lemma 3.65 to result 3.72 a).

b) The lower bound in 3.69 c) cannot be improved in any case. This follows from [ErGa59], page 344. Erdös and Galai used the bipartite undirected graph in 3.83 as an example. This graph is 2-node connected; its circumference is $2 \, \delta_1$ because each node injective cycle in $\mathcal{G}$ can visit a node $v \in \mathcal{V}'$ at most $\delta_1 = |\mathcal{V}'|$ times.

c) On the other hand, Theorem 3.69 c) can be improved by replacing the lower bound $\min(n, 2 \, \delta_1)$ by another bound of the form $\min(n, 2 \, \widetilde{\delta}_1)$ where always $\widetilde{\delta}_1(\mathcal{G}) \geq \delta_1(\mathcal{G})$ (and $\widetilde{\delta}_1(\mathcal{G}) = \delta_1(\mathcal{G})$ for some graphs $\mathcal{G}$.) For example, it has been proven in [Feng88] that every 2-node connected graph $\mathcal{G}$ contains a cycle P of length $\ell(P) \geq \min(n, 2 \, \widetilde{\delta}_1)$ if the quantity $\widetilde{\delta}_1$ is defined as
$$\widetilde{\delta}_1 := \underbrace{\min\{\max(\deg(u), \deg(v)) \mid d(u, v) = 2\}}_{\geq \, \delta_1}.$$

d) Another improved version of 3.69 c) follows from [Alon86], Lemma 2.3: Given a 2-node connected and simple undirected graph $\mathcal{G}$ with n nodes. Let v_0 be a node of $\mathcal{G}$. Suppose that not all nodes x, y with $x \neq y$ can be connected with an acyclic x-y-path of length $\delta_1 - 1$. Then we can find a cycle P of a length $\ell(P) \geq \min(n, 2 \, \delta_1)$ such that P even visits the given node v_0.
A further improvement of this result follows from [Eno84] (see also Remark 3.73 b)): If $\kappa(\mathcal{G}) = 2$ then in any case, a cycle P exists such that $\ell(P) \geq \min(n, 2\delta_1)$ and P visits a given node v_0; this follows from Proposition 2 of [Eno84] with $m := 2 \, \delta_1$; moreover, if $\kappa(\mathcal{G}) \geq 3$ and P' is an acyclic path in $\mathcal{G}$ with $\ell(P') = \kappa - 2$ then there exists a cycle P with $\ell(P) \geq \min(n, 2\delta_1 - \ell(P'))$ such that the given path P' is an infix of P; this result follows from Theorem 3 in [Eno84] with $m := 2 \, \delta_1$.
A similar result is shown in [EgGL91]: Let $\kappa(\mathcal{G}) \geq 2$ and $\delta_1 \leq n/2$; given a set $\mathcal{V}'$ of $\kappa(\mathcal{G})$ nodes; then there exists a cycle P with $\ell(P) \geq 2 \, \delta_1$ such that P visits all nodes of $\mathcal{V}'$. A further source about this type of results is [Ota95].
It is obvious that the results in [Eno84], [EgGL91], and [Ota95] are similar to the ones of [Frai86] (see Theorem 3.19).

e) An immediate consequence of 3.69 f) is Corollary 4 in [BVMS90]. It says: If $\mathcal{G}$ is 1-tough and $\delta_1(\mathcal{G}) \geq \frac{1}{3} n$ then there exists a cycle P of a length $\ell(P) \geq \frac{5}{6} n$.

f) The condition in 3.69 c) is weaker than the one in 3.69 d), and the condition in 3.69 d) is weaker than the ones in 3.69 e); this can be seen by recalling the implication $\tau \geq 1 \Rightarrow \kappa \geq 2$, which was proven in 3.7. The sharper assumptions yield better lower bounds on $c(\mathcal{G})$; to see this we write the lower bounds as follows:
$$\min(n , 2 \, \delta_1) \text{ in Theorem 3.69 c)},$$
$$\min(n , 2 \, \delta_1 + 2) \text{ in Theorem 3.69 d), and}$$
$$n = \min(n , 2\delta_1 + 4) \text{ in 3.69 e)}$$

g) It follows from [BLC91] that every graph contains a Hamiltonian cycle if $n \geq 30$, $\tau \geq 2$, and $\delta_1 \geq \frac{1}{2} n - 3$. We leave it to the reader to compare this result to Theorem 3.69 f).

h) We compare Theorems 3.69 d) and 3.69 f). The lower bound λ jumps from $\approx 2\delta_1$ to $\approx 2.5\delta_1$ as soon as $\delta_1 \geq \frac{1}{3} n$. This behaviour of λ has been described in [BVMS90], page 61.

i) Assertion 3.69 f) is an immediate consequence of Theorem 3.69 g). Moreover, 3.69 f) also follows from 3.69 j) because $\alpha(\mathcal{G}) \leq \frac{1}{2} n$ for all 1-tough graphs; this inequality is mentioned in [BVMS90] between Theorem 9 and 10.
Nevertheless, we have listed the result 3.69 f) because it is a simple example for a lower bound of cycle lengths.

j) In reality, assertion 3.69 g) is only interesting if $\tau = 2$; in this case, 3.69 g) yields a Hamiltonian cycle, and a better result cannot be obtained for $\tau > 2$.

k) Suppose that $\mathcal{G}$ is 2-node connected and that $\delta_1 \geq \max\left(\frac{1}{3}(n+2), \alpha(\mathcal{G})\right)$. Then $\delta_1 \geq \alpha$ so that $n + \delta_1 - \alpha \geq n$. This and assertion 3.69 h) imply that $\mathcal{G}$ has a Hamiltonian cycle. This result Nash-Williams; it is quoted in [BVMS90], Theorem 12.

l) Let $\mathcal{G}$ be 1-tough and $\delta_1 \geq \max(\frac{1}{3} n, \alpha - 1)$. Then 3.69 i) yields a node injective cycle of a length $\geq \min(n, n + \delta_1 - (\alpha - 1)) = n$, i.e. a Hamiltonian cycle. This result is given in [BVMS90], Theorem 13.

m) The term *dominating cycle* is defined in [BVMS90].
If $\mathcal{G}$ is 1-tough and $\delta_1 \geq \frac{1}{3} n$ then each node injective cycle of maximal length is dominating; this is a result of Bigalke and Jung, and it is quoted in [BVMS90], Theorem 6.
Given the assumptions in 3.69 h); then each node injective cycle of maximal length is dominating as well. This follows from [BVMS90], Theorem 7.

n) Theorem 3.69 j) yields a somewhat better lower bound of $c(\mathcal{G})$ than 3.69 i); that means that using Theorem 14 of [Bau74] is better than applying Lemma 3.65 to Theorem 3.72 h). $\qquad\square$

Remark 3.71 (Long cycles in special graphs)

a) The papers [Jack81], [Jack83] use $\delta_1(\mathcal{G})$ to express the lengths of longest cycles in bipartite undirected graphs.

b) Longest cycles in regular graphs are investigated in [Fan85] and in [Yu94]; if $\mathcal{G}$ is regular then all nodes have the same degree δ_1.
Corollary 1.2 in [Fan85] says that any regular graph $\mathcal{G}$ has a cycle P with $\ell(P) \geq \min(n, 3\delta_1)$ if $\mathcal{G}$ is 2-node connected and $\delta_1 \geq \frac{1}{3} n - \frac{4}{3}$. A consequence of this result is that a Hamiltonian cycle P exists if $\delta_1 = \frac{1}{3} n$ and $\tau \geq 1$. In contrast

to this, if G is not regular then $\delta_1 = \frac{1}{3}n$ and $\tau \geq 1$ only imply the existence of a cycle with a length $\geq \frac{5}{6}n$; this follows from 3.69 f).
The main result in [Yu94] is that any regular graph G has a cycle P with $\ell(P) \geq \min\{n, 3\,\delta_1 + 5\}$ if $\delta_1 \geq 71$ and G is 3-node connected. □

Theorem 3.72 *Let G be a simple and connected undirected graph with $n \geq 3$ nodes. Then the assertions* a) – h) *in the following table are true.* □

	Assumptions about δ_2	Further assumptions	Lower bound λ for $\mathbf{c}(G)$	$C \approx$	Origin of the result
a)	$\delta_2 \geq \dfrac{2n}{k}$ $(k \in \mathbb{N})$	—	$\dfrac{n}{k-1}$	$\dfrac{1}{2}\dfrac{k}{k-1}$	[EgMi89]
b)	—	$\kappa \geq 2$	$\min(n, \delta_2)$	1	[Eno84], p. 287; ↘[Dir78]
c)	—	$\tau \geq 1$	$\min(n, \delta_2 + 2)$	1	[BVMS90], Th. 1;
d)	$\delta_2 \geq n - 4$	$\tau \geq 1,$ $n \geq 11$	n	1	[BVMS90], Th. 2; ↘[Jung73]
e)	$\delta_2 \geq \dfrac{2}{3}n$	$\tau \geq 1$	$\min\left(n, \frac{1}{2}n + \frac{1}{2}\delta_2\right)$	$\dfrac{5}{4}$ $\left[\begin{array}{c}\delta_2 = \frac{2}{3}n, \\ \mathbf{c}(G) = \frac{1}{2}n + \frac{1}{2}\delta_2\end{array}\right]$	[BVMS90], Th. 3
f)	$\delta_2 \geq \dfrac{2}{3}n$	$\tau \geq 1$	$\min\left(n, \frac{\tau}{\tau+1}n + \frac{1}{2}\delta_2\right)$	$\dfrac{3}{2}\dfrac{\tau}{\tau+1} + \dfrac{1}{2} \geq$ $\dfrac{3}{2}\cdot\dfrac{1}{2} + \dfrac{1}{2} = \dfrac{5}{4}$ $[\delta_2 = \frac{2}{3}n]$	⟨3.74 b)⟩
g)	$\delta_2 \geq \frac{2}{3}n + \frac{4}{3}$	$\kappa \geq 2$	$\min(n, n + \frac{1}{2}\delta_2 - \alpha)$	$\dfrac{3}{2}$ $\left[\begin{array}{c}\delta_2 = \frac{2}{3}n, \\ \frac{1}{2}\delta_2 - \alpha \geq 0 \\ \mathbf{c}(G) = n\end{array}\right]$	⟨3.74 c)⟩
h)	$\delta_2 \geq \frac{2}{3}n$	$\tau \geq 1$	$\min(n, n + \frac{1}{2}\delta_2 - \alpha)$	$\dfrac{3}{2}$ $\left[\begin{array}{c}\delta_2 = \frac{2}{3}n, \\ \frac{1}{2}\delta_2 - \alpha \geq 0, \\ \mathbf{c}(G) = n\end{array}\right]$	⟨3.74 d)⟩

Remark 3.73   a) *(see also* 3.70 b))
The lower bound in 3.72 c) cannot be improved in every case. The graph in 3.83 can again be used as an example.

b) (*see also* 3.70 d))
The result in 3.72 b) is improved in [Eno84], Proposition 2. It says that we can force the cycle of a length $\geq \min(n, \delta_2)$ to visit a given node v_0. Moreover, Theorem 3 of that article implies that the following is true: If $\kappa \geq 3$ and P' is an acyclic path with $\ell(P') \leq \kappa - 2$ then there exists a cycle P with $\ell(P) \geq \min(n, \delta_2 - \ell(P'))$ such that P' is an infix of P.
All 2-connected graphs with $\mathbf{c}(\mathcal{G}) \leq \delta_2 + 1$ are classified in [BJS89].

c) (*see also* 3.70 f))
The condition in Theorem 3.72 b) is weaker than the one in Theorem 3.72 c), and the condition in Theorem 3.72 c) is weaker than the ones in Theorem 3.72 d). The sharper assumptions yield better lower bounds of $\mathbf{c}(\mathcal{G})$; this can be seen by writing the lower bounds as follows:
$$\min(n, \delta_2) \text{ in } 3.69 \text{ c)},$$
$$\min(n, \delta_2 + 2) \text{ in } 3.69 \text{ d)}, \text{ and}$$
$$n = \min(n, \delta_2 + 4) \text{ in } 3.69 \text{ e)}$$

d) (*see also* 3.70 g))
It has been shown in [BLC91] that every undirected graph is Hamiltonian if $n \geq 30$, $\tau \geq 2$, and $\delta_2 \geq n - 7$. We leave it to the reader to compare this result to 3.72 d).

e) (*see also* 3.70 h))
When comparing 3.72 c) and 3.72 e), the following can be observed: The lower bound λ jumps from $\approx 1 \cdot \delta_2$ to $\approx 1.25 \cdot \delta_2$ as soon as $\delta_2 \geq \frac{2}{3} n$.

f) (*see also* 3.70 i))
Assertion 3.72 e) is an immediate consequence of 3.72 f). Moreover, 3.72 e) also follows from 3.72 h) because $\alpha(\mathcal{G}) \leq \frac{1}{2} n$ for all 1-tough graphs; this inequality is mentioned in [BVMS90] between Theorem 9 and 10.
Nevertheless, we have listed the result 3.72 e) because it is a simple example for a lower bound of cycle lengths.

g) (*see also* 3.70 j))
The assertion 3.72 f) yields a Hamiltonian cycle if $\tau = 2$; this result cannot be improved, even if $\tau \geq 3$.

h) (*see also* 3.70 m))
If $\mathcal{G}$ is 1-tough and $\delta_2 \geq \frac{2}{3} n$ then each node injective cycle of maximal length is dominating; this follows from Theorem 5 in [BVMS90].
Given the conditions in 3.72 g); then each node injective cycle of maximal length is dominating as well. This follows from [BVMS90], Theorem 7.   □

The main result in [Tian93] says that $\mathbf{c}(\mathcal{G}) \geq \min(n, 2\delta_2 - 2\delta_1 + 2)$ if $\mathcal{G}$ is 1-tough.

Next, we describe the circumference of an undirected graph $\mathcal{G}$ in terms of δ_3.

Theorem 3.74 *Let $\mathcal{G}$ be a simple and connected undirected graph with $n \geq 3$ nodes. Then the assertions a) - d) in the following table are true.* □

	Assumptions about δ_3	Further assumptions	Lower bound λ for $c(\mathcal{G})$	$C \approx$	Origin of the result
a)	$\delta_3 \geq n$	$\tau \geq 1$	$\min(n, \frac{1}{2}n + \frac{1}{3}\delta_3)$	$\frac{5}{6}$ $\left[\begin{array}{c}\delta_3 = n, \\ c(\mathcal{G}) = \frac{n}{2} + \frac{\delta_3}{3}\end{array}\right]$	[BVMS90], Th. 3
b)	$\delta_3 \geq n$	$\tau \geq 1$	$\min\left(n, \frac{\tau}{\tau+1}n + \frac{1}{3}\delta_3\right)$	$\frac{\tau}{\tau+1} + \frac{1}{3} \geq$ $\frac{1}{2} + \frac{1}{3} = \frac{5}{6}$ $[\delta_3 = n]$	[BVMS90], Kor. 15
c)	$\delta_3 \geq n + 2$	$\kappa \geq 2$	$\min(n, n + \frac{1}{3}\delta_3 - \alpha)$	1 $\left[\begin{array}{c}\delta_3 = n, \\ \frac{1}{3}\delta_3 - \alpha \geq 0, \\ c(\mathcal{G}) = n\end{array}\right]$	[BVMS90], Th. 10
d)	$\delta_3 \geq n$	$\tau \geq 1$	$\min(n, n + \frac{1}{3}\delta_3 - \alpha)$	1 $\left[\begin{array}{c}\delta_3 = n, \\ \frac{1}{3}\delta_3 - \alpha \geq 0, \\ c(\mathcal{G}) = n\end{array}\right]$	[BVMS90], Th. 9

Remark 3.75   a) *(see also* 3.70 i))
Assertion 3.74 a) is an immediate consequence of 3.74 b). Moreover, 3.72 a) also follows from 3.72 d) because $\alpha(\mathcal{G}) \leq \frac{1}{2}n$ for all 1-tough undirected graphs; this inequality is mentioned in [BVMS90] between Theorems 9 and 10.
Nevertheless, we have listed the result 3.72 a) because it is a simple example for a lower bound of cycle lengths.

b) *(see also* 3.70 j))
Assertion 3.74 b) yields a Hamiltonian cycle if $\tau = 2$; this result cannot be improved if $\tau \geq 3$.

c) Theorem 3.74 c) implies the following result, which has been proven in [Bon80] and quoted in in [BVMS90] as Theorem 11.
Let $\mathcal{G}$ be 2-node connected with $\delta_3 \geq n + 2$. Then $\alpha(\mathcal{G}) \leq n - \frac{1}{3}\delta_3$, and consequently, there exists a node injective cycle P of a length $\ell(P) \geq \min\left(n, n + \frac{1}{3}\delta_3 - \alpha\right)$
$\geq \min\left(n, n + \frac{1}{3}\delta_3 - (n - \frac{1}{3}\delta_3)\right) = \min\left(n, \frac{2}{3}\delta_3\right)$.

d) Conjecture 1 in [BVMS90] says the following:
Given the assumptions in 3.74 d); then there exists a cycle of length n.
This conjecture is false. It has been disproven in [BVMS90], page 60.

e) (*see* 3.70 m))

    If $\mathcal{G}$ is 1-tough and $\delta_3 \geq n$ then each longest cycle is dominating. This is a result
of Bigalke and Jung, and it is quoted in [BVMS90], Theorem 6.

    Given the assumptions in 3.74 c); then every longest cycle is dominating as well.
This follows from [BVMS90], Theorem 7. $\square$

The following result is similar to 3.74 c).

Theorem 3.76 *Given a simple undirected graph $\mathcal{G}$ with n nodes; we assume that
$\kappa(\mathcal{G}) \geq 2$ and $\delta_3(\mathcal{G}) \geq n + \kappa(\mathcal{G})$. Then $\mathcal{G}$ contains a Hamiltonian cycle.*

P r o o f : The theorem is shown in [BBVL89]. $\square$

The article [Sku87] gives answers to the question whether or not a Hamiltonian cycle
exists in an 1-tough graph with $\delta_3 \geq n$, and the author of [Wei93a] shows that every
1-tough graph with $n \geq 3$ nodes and with $\delta_3 \geq n$ has a Hamiltonian cycle.

We have given lower bounds for the circumference in terms of δ_1, δ_2, and δ_3. The
lower bound in the next result, however, may also depend on δ_j with $j \geq 4$; this result
has been proven in [FoFr85].

Theorem 3.77 *Given a simple and connected undirected graph $\mathcal{G}$ with n nodes;
let $k \geq 2$. If $\kappa(\mathcal{G}) \geq k$ then $\mathcal{G}$ contains a node injective cycle P with*
$\ell(P) \geq \min\left(n, \frac{2}{k+1}\delta_{k+1}\right)$. $\square$

For example, if $\kappa \geq k = 2$ then a node injective cycle of length $\ell(P) \geq \min\left(n, \frac{2}{3}\delta_3\right)$
$\overset{\text{Lemma 3.64}}{\geq} \min(n, \delta_2) \overset{\text{Lemma 3.64}}{\geq} \min(n, 2\delta_1)$ is guaranteed. The relations $\ell(P) \geq$
$\min(n, 2\delta_1)$ and $\ell(P) \geq \min(n, \delta_2)$ are equivalent to Theorems 3.69 c) and 3.72 b),
respectively.

The next result says that $\mathcal{G}$ is Hamiltonian if the quantities $\overline{\delta}_k$ or $\overline{\overline{\delta}}_k$ satisfy particular
conditions.

Theorem 3.78 *Let $\mathcal{G}$ be a simple undirected graph with n nodes and with $\kappa(\mathcal{G}) \geq 2$.*

    a) *If $\overline{\delta}_2(\mathcal{G}) \geq n - \delta_1(\mathcal{G})$ then $\mathcal{G}$ contains a Hamiltonian cycle.*

    b) *Let $k \geq 2$. If $\overline{\overline{\delta}}_k \geq n/2 + 8 k^3$ then $\mathcal{G}$ contains a Hamiltonian cycle.*

P r o o f :   Assertion a) is shown in [FGJL91], and assertion b) is shown in
[FGJL92] $\square$

Further sufficient conditions for the existence of Hamiltonian cycles are given in
[Sch89], [Ry90], [Chen91] [Yin91], [Song92], [BlG93], and [Chen93]; those conditions
are described with the help of neighbourhoods or degrees of nodes. The author of
[Mu92] has constructed a graph with a particular degree sequence and without a
Hamiltonian cycle.

One of the results in [Li89] is the following: If $n \geq 8 \lfloor(\delta_1 - 1)/2\rfloor^2$ and $\delta_2 \geq n$ then

$\mathcal{G}$ contains $\lfloor(\delta_1 - 1)/2\rfloor$ edge disjoint Hamiltonian cycles. A further result about the existence of several edge disjoint Hamiltonian cycles is given in [Ega93].

Also, there exist sources about expressing the circumference of a graph in terms of the maximum degree Δ_1 and the connectivity κ; one of these sources is [JaWo93].

In [Dang93], the circumference of graphs is described in terms of two parameters δ_0 and δ^*; these two values are defined with the degrees of the nodes $v \in V$. The same or similar parameters are used in the articles [Fan84], [Zhao92], [BCS93], [LTSh93], and [ChS94] to describe the circumference of a graph.

Next, we give several results about longest acyclic paths in undirected graphs. There are many results of literature about longest cycles, but it seems that there are not so many results about longest acyclic paths. The reason may be that each lower bound of the maximal cycle length yields a lower bound of the maximal path length; this follows from the next remark.

Remark 3.79   a) Given a simple undirected graph $\mathcal{G}$ and a number $c \geq 2$. We assume that there exists a node injective cycle P in $\mathcal{G}$ with $\ell(P) \geq c$. Then $\mathcal{G}$ contains an acyclic path of a length $\geq c - 1$. Such a path can be generated by removing the last arc of P.

b) The reversed implication is not always true. For example, Theorem 3.82 implies that every graph $\mathcal{G}$ with a fixed δ_1 and arbitrary $n > 2\delta_1$ contains a path of length $2\delta_1$. On the other hand, $\mathcal{G}$ need not have a cycle of length $2\delta_1 + 1$; for example, the cycle length in the graphs in Remarks 3.67 and 3.68 is bounded by $\delta_1 + 1$. □

The next result is analogous to Theorems 3.47, 3.53, and 3.66.

Theorem 3.80 *Let $\mathcal{G}$ be an simple and connected undirected graph with n nodes. Let x be an arbitrary node in $\mathcal{G}$, and let P be an acyclic path with maximal length among all candidates with start node x. Then $\ell(P) \geq \delta_1$.*

P r o o f : We recursively construct an infinite path $Q := [v_0, v_1, v_2 \ldots)$ with $\alpha(Q) := x$. For this purpose, let $v_0 := x$, and let $\{v_0, v_1\}$ be an arbitrary edge with endpoint v_0. If $j \geq 2$ and if the nodes $v_0, \ldots, v_{j-1}$ are already given we define the same sets U_j, V_j as in the proof of 3.66. Then we choose v_j such that $\{v_{j-1}, v_j\} \in \mathcal{R}$ and that $v_j \notin U_j$; this is possible because v_{j-1} has at least δ_1 neighbours and $|U_j|$ has at most $(\delta_1 - 1)$ elements. Moreover, $v_j \neq v_{j-1}$ because $\mathcal{G}$ is simple; consequently, $v_j \notin V_j$. We have seen that $v_j \notin U_j \cup V_j$ for all $j \geq 2$. Consequently, the nodes of the path $P' := [v_0 = x, v_1, \ldots, v_{\delta_1}]$ are pairwise distinct so that $\ell(P) \geq \ell(P') \geq \delta_1$. □

Remark 3.81 The complete graph $\mathcal{G} := \mathcal{K}_{\delta_1 + 1}$ has the minimal degree δ_1, and it has no acyclic path longer than δ_1. Hence, the lower bound in 3.80 is tight as long as the graph $\mathcal{G}$ does not satisfy additional conditions. □

The graph $\mathcal{G}$ in 3.81 has only $(\delta_1 + 1)$ vertices. The question arises whether there exist graphs with maximal path length δ_1 and arbitrarily many nodes. The answer is "no" as shown in the next result; it can be interpreted as follows: If $\mathcal{G}$ is not isomorphic to $\mathcal{K}_{\delta_1+1}$ (and consequently, $\delta_1 < n - 1$) then $\mathcal{G}$ contains an acyclic path with more than δ_1 edges.

Theorem 3.82 *Let $\mathcal{G}$ be a connected and simple undirected graph with n nodes. Then $\mathcal{G}$ contains an acyclic path P with $\ell(P) \geq \min(n - 1, 2\delta_1)$.*

P r o o f : If $\delta_1 \geq \frac{1}{2}n$ then 3.69 b) yields a Hamiltonian cycle P'. We generate P by removing the last arc of P'. Then $\ell(P) = n - 1$, and it is $n - 1 = \min(n - 1, 2\delta_1)$ because $\delta_1 \geq \frac{1}{2}n$. If, however, $\delta_1 < \frac{1}{2}n$ then $2\delta_1 < n$ so that $2\delta_1 \leq n - 1$. Then Theorem (1.14) in [ErGa59] yields a path P of a length $\geq 2\delta_1$, and it is $2\delta_1 = \min(n - 1, 2\delta_1)$ because $\delta_1 < \frac{1}{2}n$. $\qquad\square$

Theorem 3.83 *The bound in 3.82 is tight for all n and for all δ_1. More precisely, for any given $n \geq 2$ and $\delta \geq 2$, there exists a graph $\mathcal{G}$ with n nodes and $\delta_1(\mathcal{G}) = \delta$ such that the longest acyclic path has at most $\min(n - 1, 2\delta_1)$ edges.*

P r o o f : If $\delta \geq \frac{1}{2}n$ then all undirected graphs $\mathcal{G}$ with $\delta_1(\mathcal{G}) = \delta$ have the the property that $\min(n - 1, 2\delta_1(\mathcal{G})) = \min(n - 1, 2\delta) = n - 1$; this is the upper bound for the length of each acyclic path in $\mathcal{G}$.

If $\delta < \frac{1}{2}n$ then we construct an example according to [ErGa59], Theorem (1.15): We choose $\mathcal{G}$ as the complete bipartite graph $\mathcal{K}_{\mathcal{V}',\mathcal{V}''}$ with $|\mathcal{V}'| = \delta$ and $|\mathcal{V}''| = n - \delta$; this graph contains an edge $\{v', v''\}$ for each $v' \in \mathcal{V}'$ and $v'' \in \mathcal{V}''$. It follows from $\delta < \frac{1}{2}n$ that $\delta < n - \delta$ so that $|\mathcal{V}'| < |\mathcal{V}''|$. Moreover, $\delta_1(\mathcal{G}) = \delta$ because $\deg(v) = \delta$ for all $v \in \mathcal{V}''$. Each path P in $\mathcal{G}$ visits $\mathcal{V}'$ and $\mathcal{V}''$ alternately. Moreover, P can enter $\mathcal{V}'$ at most $|\mathcal{V}'| = \delta$ times if P is acyclic. Hence, each acyclic path P of maximal length has the following structure: P starts with a node in $\mathcal{V}''$; the second, the fourth, ..., the 2δ th node lies in $\mathcal{V}'$, and P ends with a node in $\mathcal{V}''$. Consequently, P visits $(2\delta + 1)$ nodes so that P has the length $\ell(P) = 2\delta = 2\delta_1$. $\qquad\square$

Remark 3.84
The graphs with maximal path length $\leq 2\delta_1$ have been characterized in [Dir59]. $\quad\square$

Next, we quote several results about long paths with prescribed start and end nodes. All assertions follows from [Jung86] and the sources listed in that paper. In [Jung86], the length of a path P is defined as the number of nodes visited by P (and not as the number of edges of P). Therefore, the path lengths given in [Jung86] itself are by 1 greater than the ones given in our quotations of [Jung86].

Theorem 3.85 *Let $\mathcal{G}$ be a connected and simple undirected graph with n nodes. Let $x \neq y$ be two nodes of $\mathcal{G}$, let P be an acyclic x-y-path of maximal length, and let $V(P)$ be the set of all nodes of P. Let $k \in \{2, 3, 4, 5\}$ with the following properties:*

(i) $\mathcal{G}$ *is k-connected.*

(ii) *The subgraph of $\mathcal{G}$ induced by $V \backslash V(P)$ has a connected component with at least $(k-2)$ nodes.*
     (This condition is trivial if $k \in \{2,3\}$, and it is substantial if $k \in \{4,5\}$.)

Then P has a length $\ell(P) \geq (k-1)(\delta_1 - k + 2)$.

P r o o f : This follows from [Jung86], Theorem 3.   □

Remark 3.86   a) Theorems 3.82 and 3.85 say that an acyclic path P with $\ell(P) \approx 2\,\delta_1$ exists if one of the following situations is given:

   – The nodes $\alpha(P)$ and $\omega(P)$ are not prescribed, and $\delta_1(\mathcal{G}) \leq \frac{1}{2}n$.
      (Then 3.82 yields a path P of a length $\approx 2\,\delta_1(\mathcal{G})$.)
   – The nodes $\alpha(P) = x$ and $\omega(P) = y$ are prescribed, and $\kappa(\mathcal{G}) = 3$.
      (Then Theorem 3.85 with $k := 3$ yields an x-y-path P of a length $\approx 2\,\delta_1$.)

 b) If $k \leq 5$ then Theorem 3.85 will often yield a path of a length $\approx (k-1) \cdot \delta_1$. If, however, $k \geq 6$ then a high connectivity $\kappa(\mathcal{G}) = k$ does not always guarantee an x-y-path that is substantially longer than $4 \cdot \delta_1$. Such a graph $\mathcal{G}$ is described in [Jung86], page 128.

 c) The following result has been formulated in Theorem 1 of [Jung86], and it has been proven in [Eno84], page 297.

$$\text{If } \kappa(\mathcal{G}) \geq 3 \text{ then } \mathcal{G} \text{ has an } x\text{-}y\text{-path } P$$
$$\text{with } \ell(P) \geq \min(n-1, \delta_2(\mathcal{G}) - 2). \tag{3.9}$$

We now derive fact (3.9) from Remark 3.73 b). For this purpose, we define $\widetilde{\mathcal{G}} := (V, \widetilde{\mathcal{R}})$ where $\widetilde{\mathcal{R}} := \mathcal{R} \cup \{\{x,y\}\}$. Obviously, $\kappa(\widetilde{\mathcal{G}}) \geq \kappa(\mathcal{G}) \geq 3$ and $\delta_2(\widetilde{\mathcal{G}}) \geq \delta_2(\mathcal{G})$. Let $P' := [x,y]$; then $\ell(P') \leq \kappa - 2$. So, we may apply Enemoto's result in Remark 3.73 b) to $\widetilde{\mathcal{G}}$. We obtain a cycle $\widetilde{P}$ in $\widetilde{\mathcal{G}}$ that uses the edge $\{x,y\}$ and has a length $\ell(\widetilde{P}) \geq \min(n-1, \delta_2 - 1)$. Without loss of generality, $\{x,y\}$ is the last edge of the cycle $\widetilde{P}$. Then we obtain the desired path P by removing $\{x,y\}$ from $\widetilde{P}$.

By the way, Corollary 2.2 in [Jung86] is a slightly improved version of fact (3.9); H.A.Jung compares (3.9) and his result 2.2 on page 127 of his paper.   □

The following result is proven in [EgGL91]: Let $\kappa(\mathcal{G}) \geq 3$ and $\delta_1 \leq (n+1)/2$; given a set V' of (δ_1) nodes and two vertices x, y with $v \neq w$; then there exists an acyclic x-y-path P with $\ell(P) \geq 2\,\delta_2$ such that P visits all nodes of V'.

The paper [Hu87] describes a relationship between $\delta_2(\mathcal{G})$ and longest paths in $\mathcal{G}$.

One of the results in [Fan91] is the following: Given a simple and 2-node connected undirected graph $\mathcal{G}$ with n nodes; moreover, given two distinct nodes x,y. Then $\mathcal{G}$ contains a x-y-path whose length is at least the average of the degrees of the other vertices in $\mathcal{G}$.

The papers [Wei93b] and [GYu94] give sufficient conditions for the existence of a

Hamiltonian path in a graph; these conditions are formulated with the help of neighbourhoods of nodes.

3.6.2   Describing the maximal length of cycles and other paths without using the degrees of nodes

In the last subsection, we gave lower bounds of the maximal length of cycles or other paths; almost all of these bounds depended on the quantities $\delta_1(\mathcal{G})$, $\delta_2(\mathcal{G})$, $\delta_3(\mathcal{G})$ etc., and these numbers are computed with the help of the degrees of nodes. The lower bounds in this subsection, however, are expressed with other quantities than the δ_k's, or they arise from the special structure of the underlying graph.

We start with several results about long cycles.

Theorem 3.87 *Given a 2-node connected and simple undirected graph $\mathcal{G}$ with n nodes. Then $\mathcal{G}$ has a node injective cycle P of a length $\ell(P) \geq \frac{2\,(n+\alpha(\mathcal{G})-2)}{\alpha(\mathcal{G})}$.*
($\alpha(\mathcal{G})$ means the independence number of $\mathcal{G}$.)
This result is given in [Four82].   □

Remark 3.88 The lower bound in 3.87 is tight. To see this we recall the graph $\mathcal{G}$ given in 3.68. We constructed $\mathcal{G}$ by defining N complete graphs $\mathcal{G}_\nu = \mathcal{K}_{\mathcal{V}_\nu}$ and merging them at one cutpoint v_0. In order to generate a 2-node connected graph we insert a new node v_1 and connect it to all vertices of $\mathcal{G}$. The resulting graph $\mathcal{G}'$ is indeed 2-node connected because each path via v_0 can be replaced with a path via v_1.
If $\mu \neq \nu$ then all connections from $\mathcal{V}_\mu$ to $\mathcal{V}_\nu$ must use v_0 or v_1. This implies that every node injective cycle can visit at most two different subsets $\mathcal{V}_\mu$, $\mathcal{V}_\nu$ and that every node injective cycle of maximal length must visit v_0 and v_1.
Given a node injective cycle $P = [w_0, \ldots, w_{\lambda-1}, w_\lambda = v_0, w_{\lambda+1}, \ldots, w_l]$ of maximal length. In order to calculate $\ell(P)$ we generate a new cycle Q with the additional property that $\alpha(Q) = v_0$; more formally, $Q := [v_0 = w_\lambda, w_{\lambda+1}, \ldots, w_{l-1}, w_l = w_0, w_1, \ldots, w_\lambda]$. Then there exist $\mu \neq \nu$ with the following property: First, Q visits v_0; then Q visits the δ nodes of $\mathcal{V}_\mu\backslash\{v_0\}$; then Q visits v_1; then Q visits the δ nodes of $\mathcal{V}_\nu\backslash\{v_0\}$; at last, Q returns to v_0. This implies that

$$\mathbf{c}(\mathcal{G}') \;=\; \ell(P) \;=\; \ell(Q) \;=\; 2\delta + 2. \tag{3.10}$$

Moreover, the independence number of $\mathcal{G}'$ is N because each system of nodes $u_\nu \in \mathcal{V}_\nu\backslash\{v_0\}$, $\nu = 1, \ldots, N$, is independent. Moreover, $\mathcal{G}'$ has exactly $n = N \cdot \delta + 2$ nodes. Consequently, Fournier's lower bound described in 3.87 is equal to $\frac{2\,(n+\alpha(\mathcal{G}')-2)}{\alpha(\mathcal{G}')} = \frac{2\,([N\cdot\delta+2]+N-2)}{N} = 2\delta + 2$. This and (3.10) imply that the actual circumference $\mathbf{c}(\mathcal{G}')$ is equal to Fournier's lower bound $\frac{2\,(n+\alpha(\mathcal{G}')-2)}{\alpha(\mathcal{G}')}$.   □

Remark 3.89 If the graph $\mathcal{G}$ in 3.87 need not be 2-node connected then it may occur that the circumference of $\mathcal{G}$ is not much greater than $n/\alpha(\mathcal{G})$. (Note that $n/\alpha(\mathcal{G})$ is approximately the half of the lower bound given in 3.87.) An example of a graph $\mathcal{G}$

with $\mathbf{c}(\mathcal{G}) \approx n/\alpha$ is the graph in 3.68. It has $n = N \cdot \delta + 1$ nodes and the independence number $\alpha(\mathcal{G}) = N$. Its circumference is $\mathbf{c}(\mathcal{G}) = \delta + 1 \approx \frac{N \cdot \delta + 1}{N} = \frac{n}{\alpha(\mathcal{G})}$. $\quad\square$

A relationship between circumference and toughness is given in [Broe93]. The paper [Di90] describes the asymptotic behaviour of the circumference of 1-tough planar graphs.

Upper bounds of the maximal cycle length are studied in [Marc83]. Marcu uses matroid theoretic methods to prove a necessary and sufficient condition for the assertion that all cycle lengths (and consequently the maximal cycle length) in a graph is smaller than a given number k.

The next results describe the cycle lengths in special undirected graphs.

Definition 3.90 Let $N \geq k \geq 1$; the *Kneser graph* $\mathcal{G} = K(N,k)$ is defined as the graph with the following sets of nodes and edges:
$$\mathcal{V} := \{X \subseteq \{1,\ldots,N\} \mid |X| = k\} ; \quad \mathcal{R} := \{\{X,Y\} \mid X \neq Y \wedge X \cap Y \neq \emptyset\} . \quad \square$$

Theorem 3.91 *Let $k \geq 4$ and $N = 2k + 4$.*
Then the Kneser graph $K(N,k)$ contains a cycle P of a length $\ell(P) \geq 15 \cdot 2^{k-1}$.
Moreover, the cycle P is diagonal free.[2]
This result has been given in [AlPo89] $\quad\square$

The next result describes longest diagonal free acyclic paths in Kneser graphs.

Theorem 3.92 *Let $p(N,k)$ denote the maximal length of any diagonal free path P in the Kneser graph $K(N,k)$.*

Then for each $k \geq 3$, there exists an N_k such that $3 \cdot \left(\frac{5}{2}\right)^{k-1} \leq p(N,k) \leq \binom{2k}{k}$
for all $N \geq N_k$. This is proven in the paper [AlPo89], which was also quoted in 3.91.
Note that both the lower and the upper bound on $p(N,k)$ do not depend on N. $\quad\square$

Definition 3.93 An undirected graph $\mathcal{G}$ is called *node transitive* if for all nodes v, w, there exists a graph automorphism Ψ with $\Psi(v) = w$. $\quad\square$

Theorem 3.94 *Let $\mathcal{G}$ be a node transitive graph with n nodes. Then $\mathbf{c}(\mathcal{G}) \geq \sqrt{3n}$.*
This is shown in [Bab79]. $\quad\square$

Further results about node transitive graphs can be found in [Alsp80]. The article [BoLo85] gives a detailed description of Halin graphs and their circumferences. Hamiltonian properties of grid graphs are proven in [Zamf92], and Hamiltonian properties of other special graphs are investigated in [AlHM87], [CIJS90], [Li90a], and in [Yuc89].

---

[2] A path P is *diagonal free* if all nodes v, w of P with $\{v, w\} \in \mathcal{R}$ have the property that $\{v, w\}$ is an edge of P; in other words, if the edge $\{v, w\}$ does not appear in P then v and w are not connected in $\mathcal{G}$.

The circumferences of other special graphs are described in [Sav93].

Next we quote several results about long acyclic paths. As described in 3.79, we can transform any node injective cycle P into an acyclic path P' by removing the last edge of P. Therefore, we can use most of the results 3.87 – 3.94 to get a lower bound on the maximal length of acyclic paths. We next quote several further results.

First, we define divisor graphs and quote several results about longest paths in these graphs.

Definition 3.95 Let $n \geq 2$. The *divisor graph* D_n is defined as $D_n := (\mathcal{V}, \mathcal{R})$ such that $\mathcal{V} := \{1, \ldots, n\}$ and two nodes $i \neq j$ are connected if and only if i is a divisor of j or j is a divisor of i. ☐

Remark 3.96 For all n, let $F(n)$ be the maximal length of any acyclic path in the divisor graph D_n.
  a) It follows from [Poll83] that $F(n) \geq n \cdot e^{-(2+\varepsilon)\sqrt{\log n \, \log\log n}}$.
  b) It has been shown in [Pomer83] that $\lim_{n\to\infty} F(n)/n = 0$. That means that for all constants c, there exists an n_c such that $F(n) \leq c \cdot n$ for all $n \geq n_c$.
  c) A somewhat stronger result has been given in [Pomer83]. Let $D'_n := (\mathcal{V}, \mathcal{R}')$ be the graph with $\mathcal{V} := \{1, \ldots, n\}$ and the set $\mathcal{R}'$ of all edges $\{i, j\}$ for which the smallest common multiple $scm(i, j)$ is smaller than or equal to n. Let $F'(n)$ be the maximal length of any acyclic path in D'_n.
  Then D_n is a subgraph of D'_n so that $F(n) \leq F'(n)$. One of the results of [Pomer83] is that $\lim_{n\to\infty} F'(n)/n = 0$. This implies indeed that $\lim_{n\to\infty} F(n)/n = 0$. ☐

Next, we define hypercubes, which form a further special class of graphs; then we quote a result about longest paths in hypercubes.

Definition 3.97 Let $N \leq 1$. The *N-dimensional hypercube* Q_N is defined as follows: The set of all nodes of Q_N is $\{0, 1\}^N$; two N-tuples $(t_1, \ldots, t_N)$ and $(t'_1, \ldots, t'_N)$ are connected if and only if $t_\nu \neq t'_\nu$ for exactly one ν. ☐

Theorem 3.98 *Given the hypercube Q_N. Let $s(N)$ denote the maximal length of an acyclic, diagonal free path P in Q_N.* (Diagonal free paths have been defined in 3.91.)
Then [Deim85] says that $s(N) \leq 2^{N-1} - \dfrac{2^{N-1}}{N^2 - 5N + 7}$ *for all $N \geq 6$.* ☐

Next, we define locally Hamiltonian graphs, and we quote an article about longest paths in these graphs.

Definition 3.99 Let $\mathcal{G} = (\mathcal{V}, \mathcal{R})$ be an undirected graph. For each node v, let $\overline{\mathcal{N}}(v) := \{w \in \mathcal{V} \mid w = v$ or w is a neighbour of $v\}$. Then $\mathcal{G}$ is *locally Hamiltonian* if each set $\overline{\mathcal{N}}(v)$, $v \in \mathcal{V}$, induces a subgraph with a Hamiltonian cycle. ☐

Theorem 3.100 *Every locally Hamiltonian graph with n nodes contains a path of a length* $\geq \frac{\ln n}{6 \ln(1,5)}$. *On the other hand, there exist examples where the maximal path length does not exceed the value* $24\sqrt{\frac{n}{3}} + 4$.
These results are proven in [EnMK82].   □

We now describe several further articles.

The paper [GyRS84] gives upper bounds on the maximal path length in bipartite undirected graphs. More precisely, [GyRS84] describes a function $r : \mathbb{N}^3 \to \mathbb{N}$ such that always the following is true: Let $\mathcal{V}', \mathcal{V}'' \neq \emptyset$ be two disjoint sets of nodes, and let $L \in \mathbb{N}$; then the complete bipartite undirected graph $\mathcal{K}_{\mathcal{V}',\mathcal{V}''}$ has the following properties:

- If a subgraph $\mathcal{G}$ of $\mathcal{K}_{\mathcal{V}',\mathcal{V}''}$ has at most $r(|\mathcal{V}'|, |\mathcal{V}''|, L)$ edges then the maximal length of all acyclic paths in $\mathcal{G}$ is at most L.
- There exists a subgraph $\mathcal{G}$ of $\mathcal{K}_{\mathcal{V}',\mathcal{V}''}$ with more than $r(|\mathcal{V}'|, |\mathcal{V}''|, L)$ edges and an acyclic path of a length $L + 1$.

The paper [BoHa81] is about longest *edge injective* paths.

The authors of [AjKS81] investigate longest paths in random graphs.

The main result in [Maru] is that each graph with a particular algebraic property contains a Hamiltonian path. The existence of Hamiltonian paths in so-called domination-critical graphs is proven in [Wojc90]. It is shown in [FeMy92] that the shuffle exchange network has a Hamiltonian path.

3.6.3   Further results about long cycles and other long paths

The previous results were mainly about lower or upper bounds of the maximal lengths of cycles or other paths. Now, we consider several other combinatorial results about long paths. Theorem 3.102 is about paths with a maximal value $H(P)$ where H is an additive path function. After this, we shall describe several further papers; they are about longest paths but they do not give bounds on their length.

Next, we define several terms appearing in Theorem 3.102.

Definition 3.101 Given an undirected graph $\mathcal{G} = (\mathcal{V}, \mathcal{R})$ and an edge function $h : \mathcal{R} \to \mathbb{R}$.
For each node $v \in \mathcal{V}$, the *weighted degree* of v is defined as $\deg^{(h)}(v) := \sum_{\{v,v'\} \in \mathcal{R}} h(v,v')$. Moreover, we define the *weighted minimum degree* in analogy to the usual minimum degree: $\delta_1^{(h)} := \delta_1^{(h)}(\mathcal{G}) := \min\left\{ \deg^{(h)}(v) \,\middle|\, v \in \mathcal{V} \right\}$.
Let $H^{(h)} := SUM_h$; we interpret $H^{(h)}(P)$ as the *weight* of P.   □

The following results from [BoFa85] are about lower bounds of the maximal value $H^{(h)}(P)$, these bounds depend on $\delta^{(h)}(\mathcal{G})$; two of these results are generalizations of Theorems 3.69 c) and 3.80, respectively.

Theorem 3.102 *Given a connected and simple undirected graph $\mathcal{G} = (\mathcal{V}, \mathcal{R})$ with n nodes. Given an edge function $h : \mathcal{R} \to \mathbb{R}$ with $h(r) > 0$ for all edges r. We use the terminology of Definition 3.101. Then the following assertions are true:*

a) *If $\delta_1^{(h)}(\mathcal{G}) \geq 2$ then $\mathcal{G}$ contains a cycle P with $H^{(h)}(P) \geq \delta_1^{(h)}$.*
   This follows from Theorem 1.(ii) in [BoFa85].

b) *If $\mathcal{G}$ is even 2-node connected then there exists a cycle P with $H^{(h)}(P) \geq 2\delta_1^{(h)}$, or the $H^{(h)}$-maximal cycle is Hamiltonian.*
   This is Theorem 4 in [BoFa85]. This result is a generalization of Theorem 3.69 c) where $h(r) = 1$ for all $r \in \mathcal{R}$.

c) *For any given node $v_0 \in \mathcal{G}$, there exists an acyclic path P with $\alpha(P) = v_0$ and with $H^{(h)}(P) \geq \delta_1^{(h)}$.*
   This follows from Theorem 1.(i) in [BoFa85], and it is a generalization of 3.80 where all weights $h(r)$ were equal to 1.

d) *If $\mathcal{G}$ is even 2-node connected then we may prescribe the start node v_0 and the end node $v_1 \neq v_0$ of the path P with $H^{(h)}(P) \geq \delta_1^{(h)}$.*
   This follows from Theorem 3 in the paper [BoFa85]. $\square$

Another result in this context is the following: Every 2-edge-connected and simple undirected graph $\mathcal{G} = (\mathcal{V}, \mathcal{R})$ with n nodes contains an acyclic path P with $H^{(h)}(P) \geq \frac{2}{n} \sum_{r \in \mathcal{R}} h(r)$; this assertion is proven in [Frie92].

Algebraic properties of longest cycles in graphs are investigated in [Grö84]. The paper [KlPe90] is about the intersections of longest paths.

Hamiltonian cycles in random graphs are studied in [BoFF90] and in [McDi]

The article [Zeli82], too, is related to longest paths in graphs. Zelinka investigates the *elongation* of two nodes v, w, which is defined as the maximal length of any acyclic v-w-path. The main results of [Zeli82] are upper or lower bounds for the maximum or the minimum elongation appearing in a given graph.

The so-called detour distance between two nodes is investigated in [ChJT93]; the definition of the detour distance is similar to the one of the elongation.

The article [CaVi91] is about long cycles in subgraphs induced by all vertices of degree $\geq d$ where d is a given number.

Chapter 4

Algorithmic search for optimal paths in graphs and for other optimal discrete objects

This chapter is mainly about algorithms to find optimal or almost optimal paths in graphs. This is the difference between the current chapter and the previous ones; in Chapter 1 – 3, we only described paths in graphs but we did not search for good candidates.

This chapter is very long. That is the reason why it is divided into many sections. In most cases, each section is addressed to a particular class of optimization problems. The only exceptions are the introductory section 4.1 and the section 4.6, which is about a special class of algorithms and not about a special class of problems.

Section 4.1 gives an introduction to optimal path problems. We start with a simple optimal path problem and its solution with the help of Dijkstra's algorithm. Moreover, we describe three well-known problems related to the search for optimal paths. Also, we discuss general aspects of optimal path algorithms.

In the subsequent sections 4.2 – 4.4, we mainly present and study search strategies that are generalized versions of the well-known algorithms of Dijkstra, of Ford and Bellman, and of Floyd. In most cases, we try to find C-minimal paths where the path function C has particular structural properties (in particular, additivity, order preservation, Bellman properties); these properties make the search for C-optimal paths easier.

The main difference between Section 4.2 and Sections 4.3, 4.4 is that we are searching for a single optimal path in 4.2 whereas we are searching for a system of optimal paths in 4.3 and in 4.4. A further difference is the following: Section 4.2 often describes the search for C-minimal paths where $C = (C_0)_{\max}$ and C_0 is order preserving or even additive. Sections 4.3 and 4.4, however, mainly describe the search for C-optimal paths where C itself is order preserving or even additive.

Most of the problems in Sections 4.5 – 4.8 are easier than the ones in 4.2 – 4.4.

In Section 4.5, the search for optimal paths is made easier by additional properties of the underlying graph, for example its planarity.

Another method to make the search problem easier is to use a powerful tool; this is done in Section 4.6 about finding optimal paths with parallel computers.

The problems in Section 4.7 about optimal paths in random graphs may also be considered as easier versions of the problems in Sections 4.2 – 4.4. The reason is that in 4.2 – 4.4, the worst-case complexity of algorithms is studied; i.e., we consider the most complicated problem instances; in Section 4.7, however, the average complexity of search algorithms is studied; that means that we also consider the easier instances of a search problem.

Several search problems in Section 4.8 are easy because we do not require to obtain an optimal path; it suffices that a search strategy outputs an almost optimal result.

Many problems in Section 4.8 and most problems in the sections 4.9 – 4.12 are more difficult than the ones in Sections 4.2 – 4.4. One of the difficult tasks in Section 4.8 is to find all paths P whose costs $C(P)$ are sufficiently close to the optimal value $C(P^*)$. A further complicated problem in Section 4.8 is to find not only the best path but also the second best candidate, the third best, and so on.

The problems in Section 4.9 are similar to the ones in Section 4.2; a single C-optimal path must be found. The difference between the sections 4.2 and 4.9 is the structure of the given path function. In 4.2, the function C is often of the form $C = C_0$ or $C = (C_0)_{\max}$ where C_0 is order preserving. In 4.9, however, C is has the more complicated form $C(P) = C_1(P)/C_2(P)$. Even if C_1 and C_2 are additive the search strategies in Section 4.2 cannot be directly applied to find C-minimal paths; that was one of the reasons why Section 4.8 is not included in 4.2. The other reason is that the optimal paths in 4.8 have particular relationships to optimal solutions of problems in Approximation Theory.

The search for paths in Section 4.10 is difficult because the optimal candidate must meet side constraints. A typical case is the Traveling Salesman Problem where the desired path P^* must visit all nodes of the graph exactly once. Another task is to find a C_1-optimal path among all candidates with a fixed C_2-value; solutions to this search problem are often used to optimize the value $C_1(P)/C_2(P)$ in Section 4.9.

The search problems in Section 4.11 are so difficult that they can (probably) not be solved in polynomial time by a deterministc algorithm. There are many relationships between the sections 4.11 and 4.10 because the great complexity of a path problem often arises from side constraints. An example is the Traveling Salesman Problem, which is $\mathcal{NP}$-complete.

Section 4.12 is mainly about the search for (optimal) graphs. That makes the difference of Section 4.12 to Sections 4.2 – 4.11 about paths. Nevertheless, the search for optimal paths is often closely related to optimal path problems.

In the last part of Section 4.12, we shall describe combinatorial problems with a remote relationship to optimal paths in graphs. For example, one of these problems is to find all s-γ-paths in a graph; this is a modification of the problem to find an optimal s-γ-path in a graph.

A survey over optimal paths in graphs is given in [DeoP84]. The article contains several diagrams illustrating the relationships between different versions of the optimal path problem. Moreover, Deo and Pang give a large list of literature. A more recent overview over shortest path algorithms is given in [EvMi92].

4.1 Introduction to optimal path problems

In this section, we shall describe a simple and well-known optimal path problem and its solution with Dijkstra's algorithm. Moreover, we shall discuss three well-known graph theoretic problems that are related to the search for optimal paths in graphs. Also, we shall describe measures of complexity for optimal path algorithms, and we shall outline the idea to represent the instructions of such algorithms as elements of a formal language.

4.1.1 A simple optimal path problem and Dijkstra's algorithm

One of the simplest and most well-known optimal path problem is the following:

Problem 1 *Given a finite digraph* $\mathcal{G} = (\mathcal{V}, \mathcal{R})$ *and two nodes* $s, \gamma \in \mathcal{V}$. *Given an additive cost function* $C = SUM_h$ *where* $h(r) \geq 0$ *for all arcs* r.

    a) *Find a* C*-minimal* s-γ-path.

    b) *Compute the minimum of all costs* $C(P)$ *where* P *an* s-γ-path.
       (This can be done without finding an optimal s-γ-path.)

Probably, the most well-known method to solve this problem is Dijkstra's algorithm. It is described in [Chri75], [Nolt76], [PaSt82], [EvMi92], and in many other sources. We give a somwhat extended version of this search strategy; the instructions labelled with (1) or (2) may be removed; the resulting versions of Dijkstra's algorithm will be described in Corollary 4.2.

The following objects appear in Dijkstra's algorithm:

    $f(v)$: the currently minimal C-value of any s-v-path;

    $P(v)$: an s-v-path constructed by Dijkstra's algorithm; it is always
           $f(v) = C(P(v))$.

  $pred(v)$: a particular predecessor of v defined by Dijkstra's algorithm.

    $Q(v)$: the s-v-path reconstructed by visiting the nodes v, $pred(v)$,
           $pred(pred(v))$, and so on, until s is reached.

Moreover, Dijkstra's algorithm uses the sets OPEN, CLOSED, and SUCC of nodes.

Algorithm 1   (Dijkstra's Algorithm)

1. (* *Initialization* *)

   OPEN := $\{s\}$; CLOSED := $\emptyset$; $f(s) := 0$;

   (1) $P(s) := [s]$;

2. **If** OPEN is empty **then** stop unsuccessfully, that is, without finding a solution path **else** go to step 3.

3. Select a node $\bar{v} \in$ OPEN such that $f(\bar{v}) = \min \{f(v) \mid v \in \text{OPEN}\}$.

   If $\bar{v} \neq \gamma$, $\gamma \in$ OPEN, and $f(\gamma) = f(\bar{v})$ **then** let $\bar{v} := \gamma$.

   OPEN := OPEN $\setminus \{\bar{v}\}$;

   CLOSED := CLOSED $\cup \{\bar{v}\}$;

4. **If** $\bar{v} = \gamma$ **then**

   (1) Output $P(\bar{v}) = P(\gamma)$;

   (2) Output $Q(\bar{v}) = Q(\gamma)$;

   Output $f(\bar{v}) = f(\gamma)$ and terminate successfully.

   Else goto step 5.

5. Expand $\bar{v}$ by generating the set of all successors of $\bar{v}$, that is,

   SUCC := $\mathcal{N}^{+}(\bar{v}) = \{v' \in \mathcal{V} \mid (\bar{v}, v') \in \mathcal{R}\}$;

6. **For each successor** $v' \in$ SUCC, **do** the following:

   a) **begin**

   $\widehat{f}(v') := f(\bar{v}) + h(\bar{v}, v')$;

   (1) $\widehat{P}(v') := P(\bar{v}) \oplus (\bar{v}, v')$;

   b) (* *These steps are executed if* v' *has never been in OPEN before; in this case,* v' *has newly been detected by Dijkstra's algorithm.* *)

   I f $v' \notin$ OPEN $\cup$ CLOSED   **t h e n**

   begin

   $f(v') := \widehat{f}(v')$;

   (1) $P(v') := \widehat{P}(v')$ ($= P(\bar{v}) \oplus (\bar{v}, v')$);

   (2) $pred(v') := \bar{v}$;

   OPEN := OPEN $\cup \{v'\}$;

   end

   c) (* *These instructions are executed if* v' *has been found earlier by Dijkstra's algorithm; in this case, there already exist an old* s-v'-*path* $P(v')$ *and a value* $f(v')$. *If* $\widehat{f}(v') < f(v')$ *then the algorithm replaces* $f(v')$ *by* $\widehat{f}(v')$; *moreover,* $P(v')$ *is replaced by* $\widehat{P}(v')$, *and the new predecessor of* v' *is* $\bar{v}$. *)

   e l s e

   begin

   If $\widehat{f}(v') < f(v')$ **then**

   begin

   $f(v') := \widehat{f}(v')$;

   (1) $P(v') := \widehat{P}(v')$ ($= P(\bar{v}) \oplus (\bar{v}, v')$);

   (2) $pred(v') := \bar{v}$;

   end;

   end;

   end;

7. Goto step 2.

Dijkstra's algorithm has the following properties:

Theorem 4.1
  a) $f(v) = C(P(v))$ *for all nodes* v *for which* $f(v)$ *and* $P(v)$ *exist.*
  b) $P(v) = Q(v)$ *for all nodes* v *for which* $P(v)$ *and* $Q(v)$ *exist.*
  c) *The paths* $P(\gamma)$ *and* $Q(\gamma)$ *in step 4 are* C-*minimal* s-γ-*paths; the value* $f(\gamma)$
     is the minimal cost of all s-γ-*paths.*

P r o o f : Assertion a) is proven by a simple induction on the number of the iterations
of Dijkstra's algorithm. Assertion b) is a consequence of Lemma 4.54; the algorithm
BF* mentioned in that lemma is a generalized version of Dijkstra's algorithm. Result
c) follows from Theorem 4.57 where Γ is chosen as $\{\gamma\}$.   □

Corollary 4.2
  a) *Dijkstra's algorithm solves Problem* 1 a) *if either all instructions* (1) *or all in-*
     structions (2) *are removed.*
  b) *Dijkstra's algorithm solves Problem* 1 b) *if all instructions* (1) *and all instructions*
     (2) *are removed.*   □

4.1.2   Three well-known problems and their relationships to op-
          timal path problems

We describe three well-known optimization problems in Graph Theory: the Traveling
Salesman Problem, the search for cost minimal flows, and the search for minimal
spanning trees. We shall show the relationships between these problems and the
optimization of paths in graphs.

The *Traveling Salesman Problem* (*TSP*) is an $\mathcal{NP}$-hard optimal path problem. Its
hardness is caused by a side constraint saying how often each node must be visited.

Problem 2 (Traveling Salesman Problem, TSP)
Given the complete undirected graph $\mathcal{G} = (\mathcal{V}, \mathcal{R}) = \mathcal{K}_\mathcal{V}$ *with* n *nodes. Given the edge*
function $h : \mathcal{R} \to \mathbb{N}$, *and let* $C = SUM_h$.
Find a C-*minimal path* P^* *among all node injective cycles that visit each node of* $\mathcal{G}$.
(The nodes of $\mathcal{G}$ are interpreted as cities. The travelling salesman must visit each city exactly once
before returning to his starting point, and he tries to minimize the cost of his tour.)

More details about the Traveling Salesman Problem can be found in Subsection 4.11.2
of this book.

One of the most important graph theoretic problem is the search for *cost minimal*
flows. We focus on a special case that is very closely related to optimal paths in
graphs. The following definition introduces several notations in the context of flows
in graphs.

Definition 4.3 Given a digraph $\mathcal{G}$. Let $s \neq \gamma$ be two nodes such that $\mathcal{N}^-(s) = \emptyset$ and
$\mathcal{N}^+(\gamma) = \emptyset$. Given the arc functions $\beta : \mathcal{R} \to \mathbb{R}$ and $h : \mathcal{R} \to \mathbb{R}$.

We define $\beta^+(v) := \sum\limits_{\alpha(r)=v} \beta(r)$, $\beta^-(v) := \sum\limits_{\omega(r)=v} \beta(r)$, and $\beta(v) := \beta^+(v) - \beta^-(v)$

for all nodes v.

We shall call β a *flow (with source s and sink γ)* if $\beta^+(s) = \beta^-(\gamma)$ and $\beta(v) = 0$ for all $v \notin \{s, \gamma\}$. We define $\mathcal{B}$ as the set of all flows with $\beta(r) \in [0,1]$ $(r \in \mathcal{R})$ and $\beta^+(s) = \beta^-(\gamma) = 1$. The *cost of a flow* β is defined as $c_h(\beta) := \sum\limits_{r \in \mathcal{R}} \beta(r) \cdot h(r)$. $\square$

We formulate the problem of finding cost minimal flows.

Problem 3 *Given the situation as described in Definition 4.3.*
Find a flow $\beta^* \in \mathcal{B}$ with minimal cost $c_h(\beta^*)$ among all candidates $\beta \in \mathcal{B}$.

This problem is solved in [Nolt76], in [PaSt82], and in many other sources.

We now describe the relationship between cost minimal flows and optimal paths in digraphs. Let P be a node injective s-γ-path in $\mathcal{G}$. We define the flow β_P as follows: If an arc r is used by P then $\beta_P(r) := 1$, otherwise $\beta_P(r) := 0$. Let h and c_h be the same functions as in Definition 4.3. We define the additive path function $C := SUM_h$. Let P^* be an C-optimal s-γ-path. It is plausible to conjecture that $c_h(\beta)$ is minimal if $\beta = \beta_{P^*}$. A heuristic argument for this is the following: Interpret each arc as a road which can be used by 1000 cars per hour. We imagine that $\beta(r)$ means that the $1000 \cdot \beta(r)$ cars actually use this road. Let $h(r)$ be the cost caused by 1000 cars driving along the road r. Then $c_h(\beta)$ is the cost of moving 1000 cars per hour from s to γ such that each road r is used by $1000 \cdot \beta(r)$ cars. It is obvious that $c_h(\beta)$ is minimal if all cars use the cheap path P^*.

In the next theorem we describe a situation in which our conjecture is true.

Theorem 4.4 *Let $\mathcal{G} = (\mathcal{V}, \mathcal{R})$ be a finite, simple and acyclic digraph, and let s, γ be two distinct nodes of $\mathcal{G}$; we assume that no arc ends with s and no arc starts with γ. Let h and c_h be the same functions as in Definition 4.3 and in Problem 3. We define the path function $C := SUM_h$. Let P^* be a C-minimal s-γ-path.*
Then β_{P^*} is a c_h-minimal flow.

P r o o f : We extend $\mathcal{G}$ to a digraph $\mathcal{G}'$ by adding the arc $r_0 := (\gamma, s)$; more formally, $\mathcal{R}' := \mathcal{R} \cup \{r_0\}$ and $\mathcal{G}' := (\mathcal{V}, \mathcal{R}')$.

We say that $\beta' : \mathcal{R}' \to \mathbb{R}$ is a *circulation* in $\mathcal{G}'$ if $\sum\limits_{\alpha(r)=v} \beta'(v) = \sum\limits_{\omega(r)=v} \beta'(v)$ for all nodes $v \in \mathcal{V}$. For each arc $r \in \mathcal{R}'$, we now define the lower bound $\underline{\sigma}(r)$ and the upper bound $\overline{\sigma}(r)$ of the capacity of r; if $r \in \mathcal{R}$ then $\underline{\sigma}(r) := 0$ and $\overline{\sigma}(r) := 1$; if $r = r_0$ then $\underline{\sigma}(r) := \overline{\sigma}(r) := 1$. A circulation β' is called *feasible* if $\underline{\sigma}(r) \leq \beta'(r) \leq \overline{\sigma}(r)$ for all $r \in \mathcal{R}'$.

We extend h from $\mathcal{R}$ to $\mathcal{R}'$ by defining $h(r_0) := 0$. If β' is a circulation in $\mathcal{G}'$ we define the cost of β' as $c'_h(\beta') := \sum\limits_{r \in \mathcal{R}'} \beta(r) \cdot h(r)$.

Each flow $\beta \in \mathcal{B}$ in $\mathcal{G}$ can be extended to a feasible circulation β' in $\mathcal{G}'$ by defining $\beta'(r) := \beta(r)$, $r \neq r_0$ and $\beta'(r_0) := 1$. This extension is called β'_P if it arises from a flow $\beta = \beta_P$ with $P \in \mathcal{P}(s, \gamma)$.

Conversely, if β' is a feasible circulation in $\mathcal{G}'$ then the restriction $\beta'|_{\mathcal{R}}$ is an element of $\mathcal{B}$. Moreover, $c'_h(\beta') = c_h(\beta'|_{\mathcal{R}})$ because $h(r_0) = 0$. For these reasons, it is sufficient to show that β'_{P^*} is a c'_h-minimal circulation.

We prove this with an idea of E. Triesch [Trie]. Consider all integer circulations, i.e. all circulations $\beta' : \mathcal{R}' \to \{0, \pm 1, \pm 2, \pm 3, \ldots\}$. Obviously, β' is integer and admissible if and only if $\beta'(r) \in \{0, 1\}$ ($r \in \mathcal{R}$) and $\beta(r_0) = 1$; this is true if and only if $\beta' = \beta'_P$ for some s-γ-path P. Hence, there exist only finitely many integer and feasible circulations β' because there are only finitely many s-γ-paths in the acyclic digraph $\mathcal{G}$.

Suppose that Klein's algorithm is used to find a cost minimal circulation. (This algorithm is described in [Nolt76], [PaSt82], and in many other sources.) We start with a circulation $\beta'_0 := \beta'_{P_0}$ where P_0 is an arbitrary s-γ-path. Klein's algorithm finds exclusively integer feasible circulations, i.e. circulations of type β'_P where $P \in \mathcal{P}(s, \gamma)$. Each new circulation has lower costs than the previous one. Hence, Klein's algorithm does not consider any circulation twice; moreover, only finitely many circulations β_P exist; so, Klein's algorithm will terminate and output a c'_h-optimal circulation $\beta'_{P^{**}}$ with $P^{**} \in \mathcal{P}(s, \gamma)$. Consequently, the following inequality is valid, in which $(*)$ and $(**)$ follow from the optimality of $\beta_{P^{**}}$ and of P^*, respectively:

$$C(P^{**}) = c'_h\left(\beta'_{P^{**}}\right) \overset{(*)}{\leq} c'_h\left(\beta'_{P^*}\right) = C(P^*) \overset{(**)}{\leq} C(P^{**}).$$

This implies that $c'_h\left(\beta'_{P^{**}}\right) = c'_h\left(\beta'_{P^*}\right)$ so that β'_{P^*} is optimal, too. $\qquad\square$

Remark 4.5 Theorem 4.4 is not always true if $\mathcal{G}$ has a cycle. E.g., let $\mathcal{G} = (\mathcal{V}, \mathcal{R})$ with $\mathcal{V} = \{s, \gamma, x_0, x_1, x_2\}$ and $\mathcal{R} := \{(s, \gamma), (s, x_0), (x_0, x_1), (x_1, x_2), (x_2, x_0)\}$. Let $h(s, \gamma) := 1$, $h(x_0, x_1) := -1$, and $h(r) := 0$ for all other arcs r. Moreover, let $C := SUM_h$. We define $\beta_0(r) := 0$ if $r = (s, x_0)$ and $\beta_0(r) := 1$ if $r \neq (s, x_0)$.

Then the path $P^* := [s, \gamma]$ is the only C-minimal s-γ-path in $\mathcal{G}$. Nevertheless, β_{P^*} is not cost minimal because $c_h(\beta_0) = 0 < 1 = c_h(\beta_{P^*})$.

The cycle $[x_0, x_1, x_2, x_0]$ is C-negative. Perhaps, Theorem 4.4 is true if $\mathcal{G}$ is not acyclic but $C(P) \geq 0$ for all cycles P. $\qquad\square$

Remark 4.6 Let $\mathcal{G}$ be acyclic. Suppose that several C-minimal s-γ-paths $P_1^*, \ldots, P_k^*$ exist in $\mathcal{G}$. We conjecture that the set of all c_h-minimal s-γ-flows β consists of all convex combinations of the form $\sum_{\kappa=0}^k \lambda_\kappa \beta_{P_\kappa^*}$ with $\lambda_1, \ldots, \lambda_k \geq 0$ and $\lambda_1 + \ldots + \lambda_k = 1$. It is left to the reader to prove or disprove this conjecture; probably, it is shown or refuted in literature. $\qquad\square$

In [Pol92], an optimal flow algorithm is given that uses a modification of Floyd's optimal path algorithm.

A further relationship between optimal paths and optimal flows is described in [FoFu56]. A more detailed description of this source will be given on page 329 of this book.

Next, we consider the search for *minimal spanning trees*.

Problem 4 *Given an undirected graph $\mathcal{G}$ and an edge function $h : \mathcal{R} \to \mathbb{R}$. We define the cost $c_h(\mathcal{G}') := \sum_{r \in \mathcal{G}'} h(r)$ for each subgraph $\mathcal{G}' \subseteq \mathcal{G}$.[1]*
Find a spanning tree $\mathcal{T} \subseteq \mathcal{G}$ for which $c_h(\mathcal{T})$ is minimal.

This problem can be solved with Kruskal's greedy algorithm. It is described in [PaSt82], [Koz91], [EvMi92], and many other books about Graph Theory.

A variant is the search for a minimal spanning rooted tree in a strongly connected digraph. This problem is solved in [Lov85].

The paper [HaTa95] is about the following problem: Given an undirected graph, find a spanning tree of minimal diameter.

We next describe several relationships between extremal spanning trees and optimal paths in graphs.

Remark 4.7   a) The results in [MaPl87] and [Rohn91] imply that the maximal spanning tree of a graph contains all paths of maximal capacity; a more detailed description of these paths can be found in Subsection 4.5.1 of this book.

b) The following search strategies may be considered as special versions of a general dynamic programming algorithm:
- the greedy algorithm for minimal spanning trees,
- the Ford-Bellman method to find extremal paths in acyclic digraphs.

This is a consequence of Theorem 3.1 in [Hck92] and of Remarks 3.1 and 3.2 in [Hck93].

c) We can define nonadditive cost measures for subgraphs in almost the same way as nonadditive path functions. More precisely, it is easy to define cost measures c such that there is no $h : \mathcal{R} \to \mathbb{R}$ with $c(\mathcal{G}') = \sum_{r \in \mathcal{G}'} h(r)$ for each subgraph $\mathcal{G}' \subseteq \mathcal{G}$.
It is an interesting problem to define structural properties for cost measures of subgraphs; examples of these properties are order preservation, Bellman conditions or matroid principles. A further task is to to develop algorithms finding optimal spanning trees by using a given structural property of c.   □

We have considered minimal flow and the minimal spanning tree problem. There exist many other combinatorial problems with relationships to optimal paths in graphs. We shall discuss several of these problems in Sections 4.5 and 4.12.

4.1.3   Measures for the complexity of optimal path algorithms

The most usual measure of complexity for optimal path algorithm is the worstcase time of computation. This measure of complexity is often described as a function $T(n)$ where n is the number of nodes of the given graph.

Sometimes, however, the time of computation is expressed with other quantities, for example, the number of arcs of a graph, its diameter or the maximal degree of all nodes. The article [Mahr80] gives an overview over different measures of complexity

---

[1]This cost measure is similar to additive path functions.

for optimal path algorithms.

A further measure of complexity is the number of nodes or arcs processed by the search algorithm. More details about this measure of complexity can be found in in Chapter 4 of [DePe85] as well as in Subsection 4.2.1.4 and in Theorem 4.64 of this book.

4.1.4   Formal descriptions of optimal path algorithms

All programs **P** for concrete computers or abstract automata can be generated as elements of a formal language **L** ; examples are PASCAL programs and programs for Turing machines. All these programs **P** consist of a fixed set of elementary operations.

Probably, one can give a similar formal description of optimal path algorithms or of particular classes of such algorithms. The elementary operations of such an abstract machine should be appending arcs to paths, constructing infixes of paths, or other operations for paths in graphs. It is surely interesting to investigate the power of restricted automata that are forbidden to perform a particular type of graph theoretic operation.

4.2   The Single-Source-Single-Target problem

This section is about algorithms that find good or even optimal paths from a start node s to a goal node or to a set of goal nodes. In particular, we shall study solutions of Problem 1, of Problem 5, and of Problem 7 (see page 173 – 184); we are mainly interested in Problem 5:

Problem 5 *Given a digraph* $\mathcal{G}$*, a node* $s \in V$*, and a set* Γ *of nodes. Given a cost function* $C : \mathcal{P}(s) \to (\mathbf{R}, \preceq)$ *where* $\preceq$ *is total. (E.g.,* $\mathbf{R}$ *may be equal to* $\mathbb{R}$*, and* ″$\preceq$″ *may be equal to* ″$\leq$″*.) Given an order preserving path function* $C_0 : \mathcal{P}(s) \to (\mathbf{R}, \preceq)$ *such that* $C = C_0$ *or* $C = (C_0)_{\max}$ *.*
   a) *Find a* C*-minimal* s*-*Γ*-path.*
   b) *Compute the minimum of all costs* $C(P)$ *where* P *is a* s*-*Γ*-path.*
      (This can be done without constructing a C-minimal s-γ-path.)

This problem is a general version of Problem 1, in which $C = C_0$ is additive and $\Gamma = \{\gamma\}$. Moreover, the problems 1 and 29 resemble each other; the structure of C, however, is different in the two problems.

The node s will often be called *start node*, and the elements of Γ will often be called *goal nodes*. Every path from s to Γ is called a *goal path* or a *solution path*.

Instances of Problem 5 appear in many practical situations. The simplest case is that the shortest path must be found in a network of streets or of rails. Another situation is that an optimal sequence of financial transactions must be found; an example has been given in [HuRu90]. A further instance of Problem 5 is the $n \times n$-puzzle in [Nils82]; an initial configuration must be transformed into a final configuration with the minimal number of moves. In general, the following type of problem can often be reduced to an instance of Problem 5:

Problem 6 *Given an initial state s and a final state γ. Let $\mathcal{V}'$ be a set of possible intermediate states.*
Find an optimal sequence $(v_0, v_1, \ldots, v_{k-1}, v_k)$ with the following properties: $v_0 = s$, $v_k = \gamma$, $v_1, \ldots, v_{k-1} \in \mathcal{V}'$, and for each κ, the state $v_{\kappa-1}$ can be directly transformed into the state v_κ.

The reduction works as follows: A digraph $\mathcal{G} = (\mathcal{V}, \mathcal{R})$ is defined with $\mathcal{V} := \{s\} \cup \mathcal{V}' \cup \{\gamma\}$. Two states $u, v \in \mathcal{V}$ are connected by an arc $(u, v) \in \mathcal{R}$ if and only if the state u can be directly transformed into the state v. Consequently, there exists a one-to-one correspondence between the sequences $(s, v_1, \ldots, v_{k-1}, \gamma = v_k)$ of states and the s-γ-paths $[s = v_0, v_1, \ldots, v_{k-1}, v_k = \gamma]$ in the digraph $\mathcal{G}$.

Now we study general algorithmic concepts to solve Problem 1, Problem 5, and similar problems.

4.2.1 Abstract search strategies for paths in graphs

Here we describe the two abstract algorithmic concepts "Branch and Bound" and "Best-First search". Both concepts are closely related to Problems 1 and 5.

4.2.1.1 Branch and Bound

Branch-and-Bound algorithms are described in [LaWo66], [Iba77], [NKK84], and in many other sources. These algorithms consist mainly of branching and of bounding steps. The branching steps decompose a given (sub-)problem **P** into easier subproblems $\mathbf{P}_1, \ldots, \mathbf{P}_k$. The bounding steps assign each problem an (estimated) cost value, and they decide not to solve a subproblem $\mathbf{P}_\kappa$ if its (estimated) cost exceeds a given bound. In addition, several Branch-and-Bound algorithms use a *dominance relation* in order to decide which subproblems shall preferably be solved; if $\mathbf{P}_i$ dominates $\mathbf{P}_j$ the search strategy concludes that the solution of $\mathbf{P}_i$ seems to be more urgent than the one of $\mathbf{P}_j$; for this reason, the algorithm first solves $\mathbf{P}_i$; problem $\mathbf{P}_j$ is not solved, or it is solved later than $\mathbf{P}_i$.

Many Best-First search strategies for optimal s-γ-paths in graphs can be considered as Branch-and-Bound algorithms with dominance relations; an example is Dijkstra's algorithm (see Algorithm 1). The given optimal path problem and all of its subproblems are formulated as follows:

P: Find a C-minimal s-γ-path P with a given prefix Q.

The path Q equals $[s]$ in the original problem, and Q is a substantial path in all subproblems. Given a problem **P** and a prefix Q; let $r_1, \ldots, r_k$ be the arcs with start node $\omega(Q)$; then each subproblem $\mathbf{P}_\kappa$ ($\kappa = 1, \ldots, k$) is defined as the problem to find a C-minimal s-γ-path with the prefix $Q \oplus r_\kappa$. This is the branching step. Bounding steps are of minor importance. Whether or not a problem shall be solved is decided by a dominance relation. For this purpose, we choose a path function F (possibly $F = C$). Let $\mathbf{P}'$ and $\mathbf{P}''$ be the problem to find an optimal s-γ-path with the prefixes Q' and Q'', respectively. Then we say that $\mathbf{P}'$ *dominates* $\mathbf{P}''$ if $F(Q') \leq F(Q'')$. Roughly speaking, if $F(Q)$ is very small then it is assumed that the C-optimal s-γ-path will have Q as its prefix; for this reason, the algorithm focuses on the search of s-γ-paths with prefix Q.

4.2.1.2 Introduction to Best-First search and the abstract Best-First search strategy BF***

The problems 1, 5, and similar problems can often be solved with Best-First search methods. An example is Dijkstra's algorithm (see Algorithm 1). The main steps of all Best-First search strategies are the following:
- Select a node $\bar{v}$ with the currently best s-$\bar{v}$-connection; let P be this s-$\bar{v}$-path. (See step 3 of Dijkstra's algorithm.)
- Let v' be a successor of $\bar{v}$. Replace the previous s-v'-path P' by $P \oplus (\bar{v}, v')$ if this path is better than P'. (See step 6c) of Algorithm 1.)

In particular, we shall discuss the Best-First search strategies BF***, BF**, BF*, DIJKSTRA-BF*, A*, and Dijkstra's algorithm in this section 4.2. The next remark says which of these search strategies are more general and which are more special.

Remark 4.8 The search strategy BF*** is more general than BF**, and this algorithm is more general than BF*; the algorithm BF* is more general than DIJKSTRA-BF* and A*, and both algorithms are more general than Dijkstra's algorithm. Furthermore, BF*** is a generalization of GRAPHSEARCH in [Nils82]. □

Next, we introduce the abstract search strategy BF***. We assume that a locally finite[2] digraph $\mathcal{G} = (\mathcal{V}, \mathcal{R})$ with a start node $s \in \mathcal{V}$ and a set $\Gamma \subseteq \mathcal{V}$ of goal nodes are given. In addition, the following objects appear in BF***:

OPEN, CLOSED, SUCC: sets of nodes.

$P(v)$: the previous s-v-path recorded by BF***.

$\widehat{P}(v)$: the candidate for the new path $P(v)$.

$pred(v)$: the current predecessor of v if $v \neq s$.

$Q(v)$: the path $[w_k = s, w_{k-1}, \ldots, w_0 = v]$ with the property $w_{i+1} = pred(w_i)$, $i = 0, \ldots, k-1$; we define $Q(s) := s$; it is possible that $Q(v)$ is not defined or that $Q(v) \neq P(v)$; these cases will be discussed in 4.9 and in 4.14.

Several steps of BF*** are not exactly defined. The reason is that BF*** is meant to be an abstract concept with many concrete search algorithms as special cases.

The listing of BF*** can be found on page 182.

Remark 4.9
a) The objects $pred(v)$ and $Q(v)$ do not appear in all versions of BF***.
b) Let $pred(v)$ be defined in a particular version of BF***; then it is possible that the recursion formula $w_0 := v$, $w_{i+1} := pred(w_i)$, $i = 0, 1, 2, 3, \ldots$, produces a substantial cycle, i.e. $w_i = w_j$ for some $i \neq j$. In this case, the path $Q(v)$ is not defined; in particular, $Q(s)$ is not defined if $pred(s)$ is exists.

---

[2]This property was defined on page 15.

c) The paths $Q(v)$ form a rooted tree with root s.

d) The paths $P(v)$, however, do not always form a rooted tree; this will be shown in Remark 4.14.

For all these reasons, the paths $P(v)$ are not always the same as the paths $Q(v)$. □

Algorithm 2 (Search Strategy BF***)

1. (* *Initialization* *)
   OPEN := $\{s\}$; CLOSED := $\emptyset$; initialize further variables if necessary.

2. **If** OPEN is empty then stop unsuccessfully, i.e., without finding a solution path;
   else go to step 3.

3. Select a node $\bar{v} \in$ OPEN.
   (* *In most cases, $\bar{v}$ is found by comparing the paths $P(v)$ or $Q(v)$ ($v \in$ OPEN) with each other and choosing a node $\bar{v}$ for which $P(\bar{v})$ or $Q(\bar{v})$ is optimal.* *)
   OPEN := OPEN $\setminus \{\bar{v}\}$;
   CLOSED := CLOSED $\cup \{\bar{v}\}$;

4. **If** $\bar{v} \in \Gamma$ **then** write $P(\bar{v})$ or $Q(\bar{v})$ and terminate successfully; **else** go to step 5.

5. Expand the node[3] $\bar{v}$ by generating the set of all successors of $\bar{v}$:
   $$\text{SUCC} \quad := \quad \mathcal{N}^+(v) \quad = \quad \{v' \in \mathcal{V} \mid (\bar{v}, v') \in \mathcal{R}\} \,.$$

6. **For** each of these successors $v' \in$ SUCC, **do** the following:

   a) **begin**
      $$\widehat{P}(v') \ := \ P(\bar{v}) \oplus (\bar{v}, v');$$

   b) (* *These steps are executed if v' has been in OPEN never before; in this case, v' has newly been detected by BF***.* *)
      I f $v' \notin$ OPEN $\cup$ CLOSED **t h e n**
      > **begin**
      > $P(v') := \widehat{P}(v') \ (= P(\bar{v}) \oplus (\bar{v}, v'))$;
      > $pred(v') := \bar{v}$;
      > OPEN := OPEN $\cup \{v'\}$;
      > **end**

   c) (* *These steps are executed if BF*** has already detected v'; in this case, an old path $P(v')$ has existed. BF*** compares the qualities of $P(v')$ and $\widehat{P}(v')$ (e.g. with the help of a path function). The old path $P(v')$ is replaced by the new path $\widehat{P}(v')$ if $\widehat{P}(v')$ is better than $P(v')$. Moreover, v' is moved from CLOSED to OPEN if v' has already been in CLOSED.* *)
      e l s e
      > **begin**
      > **If** $\widehat{P}(v')$ is "better" than $P(v')$ **then**
      > > **begin**
      > > $P(v') := \widehat{P}(v') \ (= P(\bar{v}) \oplus (\bar{v}, v'))$;
      > > $pred(v') := \bar{v}$;
      > > CLOSED := CLOSED $\setminus \{v'\}$;
      > > OPEN := OPEN $\cup \{v'\}$.
      > > **end;**
      > **end;**
   end;

7. Goto step 2.

---

[3] We also say that the path $P(\bar{v})$ is expanded.

4.2.1.3 General definitions in the context of Best-First search strategies

First, we define the digraph that consists of all objects that have already been detected by a search strategy.

Definition 4.10 Given a directed or an undirected graph $\mathcal{G} = (\mathcal{V}, \mathcal{R})$. We assume that a search strategy for paths is applied to $\mathcal{G}$ (e.g. Dijkstra's algorithm, BF***, or a Ford-Bellman algorithm). Then the *explicit (di-)graph* $\mathcal{E}(\mathcal{G}) = (\mathcal{E}(\mathcal{V}), \mathcal{E}(\mathcal{R}))$ is the graph that consists of all nodes and of all arcs or edges that are known to the search strategy.

In particular, if a special version of BF*** is applied to a digraph $\mathcal{G}$ then the subgraph $\mathcal{E}(\mathcal{G})$ of $\mathcal{G}$ is induced by the set $\mathcal{E}(\mathcal{R}) \subseteq \mathcal{R}$ of arcs, which is constructed as follows:

Initialize $\mathcal{E}(\mathcal{R}) := \emptyset$.
In each iteration of step 2 – step 7 of BF***, do the following at the beginning of step 5:
$$\mathcal{E}(\mathcal{R}) := \mathcal{E}(\mathcal{R}) \cup \{r \in \mathcal{R} \,|\, \alpha(r) = \overline{v}\}.$$
A consequence of this definition is that $\mathcal{E}(\mathcal{V}) := \mathcal{E}(\mathcal{V}) \cup \text{SUCC}$ in each iteration.

Our definition of explicit graphs is similar to the one in [BaMa85]. □

Next we introduce a term that is often used in the context of Best-First search algorithms and other search strategies.

Definition 4.11 Given two objects x_1 and x_2, for example two elements $x_1, x_2 \in$ OPEN. We assume that an algorithm must choose between them; e.g., this situation is given when when deciding whether $\overline{v} := x_1$ or $\overline{v} := x_2$ in step 3 of BF***.

We say that there is a *tie* between x_1 and x_2 if x_1 and x_2 are equally good; in this case, the given algorithm cannot decide whether to choose x_1 or x_2. *Breaking ties* means to decide for one of the candidates x_1, x_2 by applying an additional rule, which is called *tie breaking rule*.

For example, Dijkstra's algorithm and many other versions of BF*** use the following tie breaking rule when performing step 3:

If two nodes x_1 and x_2 are of equal value then prefer a candidate in Γ to a node outside Γ; that means, choose $v' := x_1$ if $x_1 \in \Gamma$ and $x_2 \notin \Gamma$. □

4.2.1.4 Measuring the complexity of Best-First search strategies by considering the expanded nodes.

It has been mentioned in Subsection 4.1.3 that the complexity of an optimal path algorithm can be described with the help of the nodes processed by that algorithm; this measure of complexity is sometimes used when comparing Best-First search strategies. For example, Theorem 4.64 gives a sufficient condition for the fact that one optimal path algorithm expands fewer nodes than another one.

Similar measures of complexity are formulated in page 521 of [DePe85] in order to compare classes of Best-First search algorithms; the structure of all these criteria of complexity has been analyzed in [Huck92] immediately after Definition 4.5.

4.2.2   The algorithm BF**

Here, we study the Best-First search algorithm BF**; it is more special than BF***
but much more general than Dijkstra's algorithm. BF** was introduced in [HuRu90].
All path functions in that paper were of the form $H : \mathcal{P}(\mathcal{G}) \to (\mathbb{R}, \leq)$; we now consider
the case that all path functions are of the form $H : \mathcal{P}(s) \to (\mathbf{R}, \preceq)$ where $\preceq$ is *total*
and *identitive*. (A special case is that $(\mathbf{R}, \preceq) = (\mathbb{R}, \leq)$.) It is easy to translate all
results of [HuRu90] into the current setting.

The main result about BF** is Theorem 4.24. It says that BF** solves the following
problem:

Problem 7 *Given a locally finite digraph* $\mathcal{G} = (\mathcal{V}, \mathcal{R})$. *Let* $s \in \mathcal{V}$ *and* $\Gamma \subseteq \mathcal{V}$.
Given two path functions $C^{(1)}, C^{(2)} : \mathcal{P}(s) \to (\mathbf{R}, \preceq)$.
Find an s-γ-*path* P^{**} *with* $C^{(1)}(Q) \preceq C^{(2)}(Q)$ *for all prefixes* $Q \leq P^{**}$.

Special versions of BF** are able to solve Problem 5. This will be shown in Theorems
4.34, 4.39, 4.89, and in many further results.

Now we describe Algorithm BF** explicitly. Recall the objects appearing in BF***;
the algorithm BF** uses the following additional functions:

$F^{(1)} : \mathcal{P}(s) \to \mathbf{R}$:   expansion function; it helps to decide which node is expanded in step 3 of BF**.

$F^{(2)} : \mathcal{P}(s) \to \mathbf{R}$:   redirection function; this function is used in step 6c) of
BF**; $F^{(2)}$ helps to decide whether the current s-v-path
$P(v)$ is redirected; that means, the function $F^{(2)}$ is used to
decide whether $P(v)$ is replaced by $\widehat{P}(v)$.

$f^{(1)}, f^{(2)}$:   the $F^{(1)}$- and $F^{(2)}$-values of the current s-v-connection
$P(v)$; more formally, $f^{(1)}(v) := F^{(1)}(P(v))$ and $f^{(2)}(v) :=$
$F^{(2)}(P(v))$.

$\widehat{f}^{(1)}, \widehat{f}^{(2)}$:   the $F^{(1)}$- and $F^{(2)}$-values of the s-v-connection $\widehat{P}(v)$; more
formally, $\widehat{f}^{(1)}(v) := F^{(1)}(\widehat{P}(v))$ and $\widehat{f}^{(2)}(v) := F^{(2)}(\widehat{P}(v))$.

The instructions of Algorithm 3 are given on page 185.

We shall often write the objects appearing in BF** with a subscript, for example,
OPEN_t, $P_t(v)$, etc. In most cases, this index describes the state of a variable immediately before the tth iteration[4] of BF**. Exceptions are $\bar{v}_t$ and $\widehat{P}_t(v')$, respectively,
which denote the node $\bar{v}$ and the path $\widehat{P}(v')$ generated *during* the tth iteration. For
example, OPEN_1, CLOSED_1 and $P_1(s)$, respectively, are equal to OPEN, CLOSED
and $P(s)$ immediately after the initialization and immediately before the first iteration; consequently, $\text{OPEN}_1 = \{s\}$, $\text{CLOSED}_1 = \emptyset$ and $P_1(s) = [s]$.
The notation $\mathcal{E}_t(\mathcal{G})$ denotes the explicit digraph $\mathcal{E}(\mathcal{G})$ immediately before the tth iteration. In accordance to Definition 4.10, we assume that $\mathcal{E}_{t+1}(\mathcal{G})$ is available immediately
after step 5 of the tth iteration.

Also, we shall often write $\text{BF}^{**}(F', F'')$ if BF** uses F' as expansion function and F''

---

[4]The expression "iteration" means "iteration of step 2 – step 7".

as redirection function; e.g., Algorithm 3 is formulated as $BF^{**}(F^{(1)}, F^{(2)})$.

Algorithm 3 (Algorithm BF**)

1. (* *Initialization* *)
   OPEN := $\{s\}$; CLOSED := $\emptyset$; $P(s) := [s]$;

2. **If** OPEN is empty then terminate with failure; that means, no path is output;
   else go to step 3.

3. Select $\bar{v}$ such that $f^{(1)}(\bar{v}) = \min \{f^{(1)}(v) \mid v \in \text{OPEN}\}$.

   If even a goal node $\tilde{v} \in$ OPEN exists with $f^{(1)}(\tilde{v}) = f^{(1)}(\bar{v})$ **then** $\bar{v} := \tilde{v}$.
   OPEN := OPEN $\setminus \{\bar{v}\}$;
   CLOSED := CLOSED $\cup \{\bar{v}\}$.

4. **If** $\bar{v} \in \Gamma$ **then** output the path $P(\bar{v})$ and stop; **else** go to step 5.

5. Expand the node $\bar{v}$ by generating the set of all successors of $\bar{v}$.
   $$\text{SUCC} := \mathcal{N}^+(\bar{v}) = \{v' \in \mathcal{V} \mid (\bar{v}, v') \in \mathcal{R}\} .$$

6. **For** each successor $v' \in$ SUCC, **do** the following:

   a) **begin**
      $\widehat{P}(v') := P(\bar{v}) \oplus (\bar{v}, v')$;
      $\widehat{f}^{(1)}(v') := F^{(1)}(\widehat{P}(v'))$;
      $\widehat{f}^{(2)}(v') := F^{(2)}(\widehat{P}(v'))$;

   b) (* *These instructions are executed only if v' has never been in OPEN before; in this case, v' has been newly detected by BF**.* *)
      I f $v' \notin$ OPEN $\cup$ CLOSED **t h e n**
         begin
            $P(v') := \widehat{P}(v') \, (= P(\bar{v}) \oplus (\bar{v}, v'))$;
            $pred(v') := \bar{v}$;
            OPEN := OPEN $\cup \{v'\}$;
         end

   c) (* *These instructions are executed if v' has been detected earlier by BF**; in this case, an old s-v'-path $P(v')$ existes. If the $F^{(2)}$-value of the new path $\widehat{P}(v')$ is better than the $F^{(2)}$-value of $P(v')$ then BF** will replace $P(v')$ by $\widehat{P}(v')$ and define $\bar{v}$ as the new node $pred(v')$. If v' has been in CLOSED then v' is removed from CLOSED and moved to OPEN; these steps are marked with "(1)" and "(2)", respectively.* *)
      e l s e
         begin
            If $\widehat{f}^{(2)}(v') \prec f^{(2)}(v')$ **then**
               begin
                  $P(v') := \widehat{P}(v') \, (= P(\bar{v}) \oplus (\bar{v}, v'))$;
                  $pred(v') := \bar{v}$;
                  (1) CLOSED := CLOSED$\setminus\{v'\}$;
                  (2) OPEN := OPEN $\cup \{v'\}$;
               end;
         end;
      end;

7. Goto step 2.

Remark 4.12 The description of BF** is not quite clear if the following situation is given:

> The node $\bar{v} = \bar{v}_t$ is expanded in the tth iteration, and the digraph $\mathcal{G}$ has a loop $(\bar{v}, \bar{v})$. The path $P(\bar{v})$ is redirected, i.e., $P_t(\bar{v}_t)$ is replaced by $P_{t+1}(\bar{v}_t) = \widehat{P}_t(\bar{v}_t) = P_{t+1}(\bar{v}_t) \oplus (\bar{v}_t, \bar{v}_t)$.

Let $v' \neq \bar{v}_t$. Then the definition of the paths $\widehat{P}(v') = P(\bar{v}) \oplus (\bar{v}, v')$ in step 6a) is ambiguous because it is not clear whether $P(\bar{v})$ means $P_t(\bar{v})$ or $P_{t+1}(\bar{v})$. (The first interpretation is correct if $P(\bar{v})$ is redirected after generating $\widehat{P}(v')$, and the second is correct if $P(\bar{v})$ is redirected before generating $\widehat{P}(v')$.) To avoid this ambiguity we decide for the first alternative, i.e., $\widehat{P}(v') := P_t(\bar{v}) \oplus (\bar{v}, v')$ for all successors v' of $\bar{v}$.

By the way, the above situation can be avoided by applying BF** only to simple digraphs. □

Next, we study the behaviour of BF**, and we start with the following example, which will give us information about the paths $P(v)$ and $Q(v)$ constructed by BF**.

Example 4.13 Let $\mathcal{G} = (\mathcal{V}, \mathcal{R})$ be the digraph in Figure 5; in particular, $\mathcal{V} = \{s, x, y, z\}$, and $\mathcal{R} = \{(s, x), (s, y), (x, y), (y, z)\}$; let $\Gamma := \{z\}$. We define $h(s, y) := h(y, z) := 4$, $h(s, x) := 6$, and $h(x, y) := -3$. Let $F := SUM_h$ and $F^{(1)} := F^{(2)} := F$.

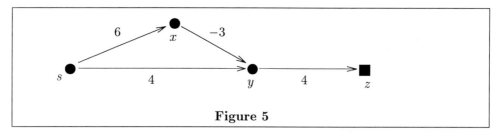

Figure 5

Most objects generated by $\mathrm{BF}^{**}\big(F^{(1)}, F^{(2)}\big) = \mathrm{BF}^{**}(F, F)$ are shown in the following table; moreover, $\widehat{P}_t(x) = [s, x]$ for all $t = 1, \ldots, 5$. The mark $'\circ'$ at a path $P_t(v)$ means that $v \in \mathrm{OPEN}$ at the beginning of the tth iteration. Recall that the node $\bar{v}_t$ is chosen such that $F(P_t(\bar{v}_t))$ is minimal among all candidates $F(P_t(v))$, $v \in \mathrm{OPEN}_t$. Moreover, recall that all paths $\widehat{P}_t(v)$ are constructed as $P_t(\bar{v}_t) \oplus (\bar{v}_t, v)$ and that $pred_t(v)$ is always the penultimate node of $P_t(v)$.

$t =$	$P_t(x)$	$P_t(y)$	$\widehat{P}_t(y)$	$P_t(z)$	$\widehat{P}_t(z)$	$pred_t(x)$	$pred_t(y)$	$pred_t(z)$	$\bar{v}_t$
1	—	—	$[s, y]$	—	—	—	—	—	s
2	$[s, x]^\circ$	$[s, y]^\circ$	—	—	$[s, y, z]$	s	s	—	y
3	$[s, x]^\circ$	$[s, y]$	$[s, x, y]$	$[s, y, z]^\circ$	—	s	s	y	x
4	$[s, x]$	$[s, x, y]^\circ$	—	$[s, y, z]^\circ$	$[s, x, y, z]$	s	x	y	y
5	$[s, x]$	$[s, x, y]$	—	$[s, x, y, z]^\circ$	—	s	x	y	z

The following remark describes several properties of the paths $P(v)$ and $Q(v)$; we use Example 4.13 to verify these properties.

Remark 4.14 a) (*see also* [HuRu90], *Remark* 2.1) The system of the paths $P_t(v)$ does not always form a rooted tree. For example, there exists no rooted tree containing the paths $P_4(y) = [s, x, y]$ and $P_4(z) = [s, y, z]$ in Example 4.13.

b) As mentioned in 4.9, the equation $P_t(v) = Q_t(v)$ is not always valid. To see this choose $t := 4$ and $v := z$ in Example 4.13. It is $pred_4(z) = y$, $pred_4(y) = x$, and $pred_4(x) = s$; consequently, $Q_4(z) = [s, x, y, z] \neq [s, x, z] = P_4(z)$. More details about the paths $Q(v)$ can be found in Lemma 4.45.

c) It is possible that $P_{t+1}(x) \neq P_t(x)$ but $pred_{t+1}(x) = pred_t(x)$. An example is given in 4.13 where $P_5(z) \neq P_4(z)$ but $pred_5(z) = pred_4(z)$.
We shall always call a replacement of $P_t(v)$ by another path $P_{t+1}(v)$ a redirection, independent of whether $pred_{t+1}(v) \neq pred_t(v)$ or $pred_{t+1}(v) = pred_t(v)$. □

Remark 4.15 The difference between the paths $P(v)$ and $Q(v)$ has the following consequence: If we use the values $F^{(1)}(Q(v))$, $F^{(2)}(Q(v))$ instead of $F^{(1)}(P(v))$, $F^{(2)}(P(v))$ in steps 3 and 6 of BF** then the resulting algorithm will do other things than the original algorithm BF**. More precisely, a node $\bar{v} = v$ with minimal value $F^{(1)}(Q(v))$ is selected by step 3. Moreover, step 6 generates the path $\widehat{Q}(v') := Q(\bar{v}) \oplus (\bar{v}, v')$; the value $F^{(2)}(\widehat{Q}(v'))$ is compared with $F^{(2)}(Q(v'))$; if $F^{(2)}(\widehat{Q}(\bar{v})) \prec F^{(2)}(Q(v'))$ then $Q(v')$ is redirected by defining $pred(v') := \bar{v}$; this operation has the consequence that $Q(v') = \widehat{Q}(v')$. Moreover, if $w \in V$ and $Q(w) = Q(v') \oplus U$ before the redirection then $Q(w) = \widehat{Q}(v') \oplus U$ after the redirection. That means that the redirection of v' is automatically propagated[5] to w. (In contrast to this, the redirection of $P(v')$ does not influence other paths $P(w)$.) □

To describe the output path of BF** more precisely, we introduce the following terms:

Definition 4.16 Given a path $P \in \mathcal{P}(s)$ and two path functions $H', H'' : \mathcal{P}(\mathcal{G}) \to \mathbf{R}$. Moreover, let $\overline{P}$ be an infinite path with start node s.
We say that P is H''-bounded (*with regard to* H') if $H'(Q) \preceq H''(Q)$ for all prefixes $Q \leq P$. We say that P is weakly H''-bounded (*with regard to* H') if $H'(Q) \preceq H''(Q)$ for all proper prefixes $Q \leq P$, $Q \neq P$.
The weak H''-boundedness of $\overline{P}$ is defined in the same way.
If $H''(P)$ is equal to a constant $M \in \mathbf{R}$ then any [weakly] H''-bounded path P is said to be [weakly] M-bounded. (This definition of M-boundedness is almost the same as the one in [DePe85], page 514.)
The weak M-boundedness of $\overline{P}$ is defined in the same way. □

We now describe two obvious properties of H''-bounded paths.

---

[5]Principles of propagation are described on page 202 – 208 of [Huck92] and in paragraph 4.2.3.5 of this book.

Remark 4.17 Given the same situation as in Definition 4.16.
  a) P is H''-bounded if and only if P is weakly H''-bounded and $H'(P) \preceq H''(P)$.
  b) Let H'' be equal to a constant M. Then P is M-bounded if and only if $H'_{\max}(P) \preceq M$.
  $\square$

The results, remarks, and examples from Lemma 4.18 to Example 4.29 are similar to the ones in [HuRu90]; they describe the behaviour of BF**. In particular, BF** generates a $C^{(2)}$-bounded path with respect to $C^{(1)}$ if the given path functions $C^{(1)}$, $C^{(2)}$ and the decision functions $F^{(1)}$, $F^{(2)}$ of BF** satisfy particular conditions; this will be shown in Theorem 4.24.

The following lemma describes the nodes in OPEN and in CLOSED.

Lemma 4.18 *Given a simple, locally finite digraph $\mathcal{G} = (\mathcal{V}, \mathcal{R})$, a start node s, and a set $\Gamma \subseteq \mathcal{V}$ of goal nodes. We assume that $BF^{**}(F^{(1)}, F^{(2)})$ is applied to $\mathcal{G}$. Given a path $P = [v_0, v_1, \ldots, v_k]$ in $\mathcal{G}$ with $v_0 = s$. Then the following are true:*
  a) *If not all vertices $v_\kappa \in CLOSED_t$ then the smallest $\overline{\kappa}$ with $v_{\overline{\kappa}} \notin CLOSED_t$ has the property that $v_{\overline{\kappa}} \in OPEN_t$.*
  b) *If $t \geq 2$ and if $P = P_t(v_k)$ then all vertices v_κ, $\kappa < k$, have already been selected by step 3.*

P r o o f o f a) : The node s is detected in step 1 of BF**. Hence, $s \in OPEN_t \cup CLOSED_t$ for all t. If $s = v_0 \in OPEN_t$ then assertion a) is true (with $\overline{\kappa} = 0$). The other case ist that $s = v_0$ is in $CLOSED_t$. Then the minimality of $\overline{\kappa}$ implies that $v_{\overline{\kappa}-1} \in CLOSED_t$. This is only possible if $v_{\overline{\kappa}-1}$ was already expanded; since this event, $v_{\overline{\kappa}} \in CLOSED \cup OPEN$ so that $v_{\overline{\kappa}}$ can only be in $OPEN_t$.

P r o o f o f b) : We show the result by an induction on t. The only relevant paths for $t = 2$ are of the form $P = P_1(v_1) = [v_0, v_1] = [s, v_1]$; then assertion b) is true because the node $v_0 = s$ has indeed been expanded in the first iteration.
We assume that our result is true for t. We must show assertion b) if $P = P_{t+1}(v_k)$. This assertion follows immediately from the induction hypothesis if $P_{t+1}(v_k) = P_t(v_k)$. Let now $P = P_{t+1}(v_k) \neq P_t(v_k)$. Then $P_{t+1}(v_k)$ has been defined as $\widehat{P}_t(v_k) = P_t(v_{k-1}) \oplus (v_{k-1}, v_k)$ in the tth iteration. Hence, $P_t(v_{k-1}) = [s, v_1, \ldots, v_{k-1}]$. The induction hypothesis implies that all nodes v_κ, $\kappa < k-1$ have been expanded before the tth iteration, and v_{k-1} has been expanded in the iteration itself; so, all nodes $v_0 = s, v_1, \ldots, v_{k-1}$ have been expanded before the $(t+1)$st iteration.
  $\square$

Next, we give a sufficient condition for the situation that BF** outputs a path P from s to a goal node after finitely many iterations; we do not say anything about the quality of P. In Lemma 4.19 – Remark 4.21, we compare $f_t^{(2)}(v) = F^{(2)}(P_t(v))$ with the $F^{(2)}$-values of other paths Q between s and v.

Lemma 4.19 *Given a simple, locally finite digraph $\mathcal{G} = (\mathcal{V}, \mathcal{R})$ and a start node $s \in \mathcal{V}$. Let $\Gamma \subseteq \mathcal{V}$. We assume that the redirection function $F^{(2)} : \mathcal{P}(s) \to (\mathbf{R}, \preceq)$ of BF^{**} is order preserving and that $\preceq$ is total and identitive. Given a path $Y = [v_0, v_1, \ldots, v_k]$ in $\mathcal{G}$ with $v_0 = s$; let $Y_\kappa := [v_0, \ldots, v_\kappa]$ for any κ. Let $t \in \mathbb{N}$*

be a number of an iteration; we assume that the following are true:

(i) $f_t^{(2)}(v_{k-1}) = F^{(2)}(P_t(v_{k-1})) \preceq F^{(2)}(Y_{k-1})$,

(ii) v_{k-1} *is in* $CLOSED_t$. (Then $v_k \in OPEN_t \cup CLOSED_t$ by Lemma 4.18 a)).

Then $f_t^{(2)}(v_k) = F^{(2)}(P_t(v_k)) \preceq F^{(2)}(Y_k)$.

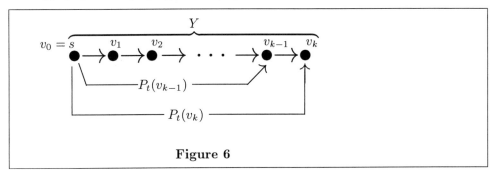

Figure 6

P r o o f : The situation of this lemma is illustrated in Figure 6. The result is true if $t = 1$ because $CLOSED_1 = \emptyset$ and premise (ii) is false. Let $t \geq 2$. We show the result by an induction on t and start with the situation after the first iteration (i.e. $t = 2$); then the premise that $v_{k-1} \in CLOSED$ is only true if Y is of the form $Y = [s, v_1]$, i.e. $k = 1$. But then $f_t^{(2)}(v_1) = F^{(2)}(Y)$.

Next, we assume that our result is true for $t \geq 2$ and for all paths Y. Now the situation after the tth iteration must be considered. Given a path $Y = [v_0, v_1, \ldots, v_k]$ with $v_0 = s$. Then we distinguish between two cases about the vertex $\bar{v}_t$ expanded in the tth iteration:

Case 1: $\bar{v}_t \neq v_{k-1}$.

Our assumptions say that after the tth iteration, $v_{k-1} \in CLOSED_{t+1}$ and $f_{t+1}^{(2)}(v_{k-1}) = F^{(2)}(P_{t+1}(v_{k-1})) \preceq F^{(2)}(Y_{k-1})$. Note now that $f_{t+1}^{(2)}(v_{k-1}) \prec f_t^{(2)}(v_{k-1})$ would only be possible if v_{k-1} had been moved to OPEN (step 6c). Hence, the assumption that $v_{k-1} \in CLOSED_{t+1}$ implies that $f_t^{(2)}(v_{k-1}) = f_{t+1}^{(2)}(v_{k-1}) \overset{\text{see above}}{\preceq} F^{(2)}(Y_{k-1})$. Moreover, v_{k-1} has already been in $CLOSED_t$ because the only vertex moving from OPEN to CLOSED is $\bar{v}_t \neq v_{k-1}$.

We have seen that $f_t^{(2)}(v_{k-1}) \preceq F^{(2)}(Y_{k-1})$ and that v_{k-1} has been in $CLOSED_t$. This and the induction hypothesis imply that $f_t^{(2)}(v_k) \preceq F^{(2)}(Y)$. Consequently, $f_{t+1}^{(2)}(v_k) \preceq f_t^{(2)}(v_k) \preceq F^{(2)}(Y)$.

Case 2: $\bar{v}_t = v_{k-1}$.

Then $v_{k-1} = \bar{v}_t$ must have been moved from OPEN to CLOSED by step 3 of the tth iteration.

If $f_{t+1}^{(2)}(v_{k-1}) \succ F^{(2)}(Y_{k-1})$ then our assertion is true because our premise is false. We now assume that

$$f_{t+1}^{(2)}(v_{k-1}) \preceq F^{(2)}(Y_{k-1}). \tag{4.1}$$

Then our assumptions about v_{k-1} and Y_{k-1} are true, and we must really show that $f_{t+1}^{(2)}(v_k) \preceq F^{(2)}(Y)$.

The $f_t^{(2)}$-value of the expanded node $\overline{v}_t = v_{k-1}$ has not changed. (Note that $\mathcal{G}$ is simple so that $(\overline{v}_t, v_{k-1}) \notin \mathcal{R}$.) Consequently, $F^{(2)}(P_t(v_{k-1})) = f_t^{(2)}(v_{k-1}) = f_{t+1}^{(2)}(v_{k-1}) \overset{(4.1)}{\preceq} F^{(2)}(Y_{k-1})$ so that $F^{(2)}(P_t(v_{k-1})) \preceq F^{(2)}(Y_{k-1})$. This and the order preservation of $F^{(2)}$ imply relation $(*)$ in the following inequality:

$$F^{(2)}(\widehat{P}_t(v_k)) = F^{(2)}\big(P_t(v_{k-1}) \oplus (v_{k-1}, v_k)\big) \overset{(*)}{\preceq} F^{(2)}(Y_{k-1} \oplus (v_{k-1}, v_k)) = F^{(2)}(Y). \quad (4.2)$$

I f v_k has been detected for the first time t h e n $P_{t+1}(v_k)$ is defined as $\widehat{P}_t(v_k)$. Hence, $f_{t+1}^{(2)}(v_k) = F^{(2)}(P_{t+1}(v_k)) = F^{(2)}(\widehat{P}_t(v_k)) \overset{(4.2)}{\preceq} F^{(2)}(Y)$.

I f , however, v_k was detected earlier t h e n $f_t^{(2)}(v_k) = F^{(2)}(P_t(v_k))$ is compared with $F^{(2)}(\widehat{P}_t(v_k))$; this is done in step 6c. Consequently, $f_{t+1}^{(2)}(v_k) \preceq F^{(2)}(\widehat{P}_t(v_k)) \overset{(4.2)}{\preceq} F^{(2)}(Y)$.

That means that in any case, $f_{t+1}^{(2)}(v_k) \preceq F^{(2)}(Y)$ if our premises about v_{k-1} and Y_{k-1} are true.   □

Corollary 4.20 *Given a simple, locally finite digraph $\mathcal{G} = (V, \mathcal{R})$, and a start node $s \in V$ and a set Γ of goal nodes. We assume that the redirection function $F^{(2)} : \mathcal{P}(s) \to (\mathbf{R}, \preceq)$ of BF** is order preserving and that $\preceq$ is total and identitive. Let $t \in \mathbb{N}$, and let $Y = [v_0, v_1, \ldots, v_k]$ be a path with $v_0 = s$ and with $v_0, \ldots, v_{k-1} \in \mathrm{CLOSED}_t$.*

Then $f_t^{(2)}(v_k) = F^{(2)}(P_t(v_k)) \preceq F^{(2)}(Y)$.

P r o o f : Let $Y_\kappa := [v_0, \ldots, v_\kappa]$, $\kappa = 0, \ldots, k$. We show that $f_t^{(2)}(v_\kappa) \preceq F^{(2)}(Y_\kappa)$ for all $\kappa = 0, \ldots, k$.

The induction on κ starts with $\kappa = 0$. We obtain an inequality in which $(*)$ follows from the fact that $F^{(2)}(P_{t'}(v)) \preceq F^{(2)}(P_{t''}(v))$ for all $t' \geq t''$ and all nodes v. It is

$$f_t^{(2)}(v_0) = F^{(2)}(P_t(s)) \overset{(*)}{\preceq} F^{(2)}(P_1(s)) = F^{(2)}([s]) = F^{(2)}(Y_0).$$

Let now $1 \leq \kappa \leq k$; we assume that already $f_t^{(2)}(v_{\kappa-1}) \preceq F^{(2)}(Y_{\kappa-1})$. Then all assumptions of Lemma 4.19 are true for $v_{\kappa-1}$ and $Y_{\kappa-1}$. Consequently, $f_t^{(2)}(v_\kappa) \preceq F^{(2)}(Y_\kappa)$.

We have shown that $f_t^{(2)}(v_\kappa) \preceq F^{(2)}(Y_\kappa)$ for all $\kappa = 0, \ldots, k$; the result follows by choosing $\kappa = k$.   □

Remark 4.21 Corollary 4.20 of this book is similar to Lemma 1 in [DePe85]; Dechter's and Pearl's lemma says that $F^{(2)}(P_t(v_k)) \preceq F^{(2)}(Y)$ for all paths $Y \in \mathcal{P}(s, v_k)$ belonging to the explicit digraph at time t; Dechter and Pearl have assumed that $F^{(1)}, F^{(2)} : \mathcal{P}(s) \to \mathbb{R}$ and $F^{(1)} = F^{(2)}$. If, however, $F^{(1)} \neq F^{(2)}$ then Lemma 1 in [DePe85] is false. A counterexample is the digraph in Figure 7. The arc (x_3, x_6) has been drawn twice because it belongs to $X_1 := [s, x_1, x_2, x_3, x_6]$ and to $X_2 := [s, x_1, x_4, x_3, x_6]$.

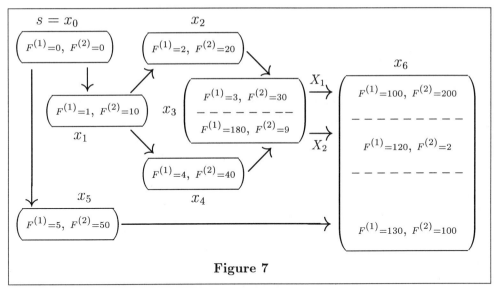

Figure 7

Each node v is marked with the $F^{(1)}$- and the $F^{(2)}$-values of all s-v-paths; the upper and the lower line of x_3, respectively, show the values $F^{(1)}(Q)$, $F^{(2)}(Q)$ of the "upper" path $Q = [s, x_1, x_2, x_3]$ and the "lower" path $Q = [s, x_1, x_4, x_3]$; the lines in x_6 have been arranged analogically. It is obvious that $F^{(2)}$ is order preserving.
It is easy to see that x_{t-1} is selected for expansion if $t = 1, \ldots, 6$. The path $X_1 = [s, x_1, x_2, x_3, x_6]$ is completed after the fourth iteration, i.e. $P_5(x_6) = X_1$. In the fifth iteration, x_4 is expanded. That means that all arcs of $Y := X_2 = [s, x_1, x_4, x_3, x_6]$ have now been generated. In the sixth iteration, the algorithm notices that $F^{(2)}([s, x_5, x_6]) = 100 < 200 = F^{(2)}(Y)$. Consequently $P_7(x_6) := [s, x_5, x_6]$.
It is $F^{(2)}(Y) = 2$ and $F^{(2)}(P_7(x_6)) = 100$. That means that not $F^{(2)}(P_7(x_6)) \leq F^{(2)}(Y)$ although all arcs of Y have been detected before generating $P_7(x_6)$.
It is obvious that this counterexample is based on the fact that the order of expansions caused by $F^{(1)}$ is independent of $F^{(2)}$. Also, the fact that $F^{(2)}(Y) < F^{(2)}(P_7(x_6))$ is not a counterexample to Corollary 4.20; the reason is that $P(x_3)$ has been redirected when expanding x_4; hence, $x_3 \in OPEN_7$; hence, the last two nodes x_3, x_6 of Y are in OPEN so that Y does not satisfy the condition of 4.20. □

Corollary 4.20 says that $F^{(2)}(P(\omega(Y))) \preceq F^{(2)}(Y)$ if almost all nodes of Y are in CLOSED; The following lemma, however, says that $F^{(2)}(P(\omega(Y))) \preceq F^{(2)}(Y)$ if Y is a prefix of a path $P(x)$.

Lemma 4.22 *Given a simple, locally finite digraph $\mathcal{G} = (\mathcal{V}, \mathcal{R})$, a node $s \in \mathcal{V}$, and a set $\Gamma \subseteq \mathcal{V}$. We assume that BF** uses an arbitrary expansion function $F^{(1)} : \mathcal{P}(s) \to (\mathbf{R}, \preceq)$ and an arbitrary redirection function $F^{(2)} : \mathcal{P}(s) \to (\mathbf{R}, \preceq)$ where $\preceq$ is total and identitive. ($F^{(2)}$ need not be order preserving.) Let x be a node for which $P(x)$ already exists. Let Y be a prefix of $P(x)$; let $y = \omega(Y)$.*

Then $F^{(2)}(P(y)) \preceq F^{(2)}(Y).$[6]

P r o o f : We prove the assertion by an induction on the number t of iterations.

If $t = 1$ then $P_1(s) = [s]$ is the only existing path $P(x)$. Therefore, the assertion is trivially true.

We assume that the assertion is true for t, and we investigate the effects of the tth iteration of BF**. We distinguish between two cases:

Case 1: $P_{t+1}(x) = P_t(x)$.
Then each prefix $Y \leq P_{t+1}(x)$ is also a prefix of $P_t(x)$; hence, we obtain the following inequality where $(*)$ is a consequence of the induction hypothesis:
$$F^{(2)}(P_{t+1}(y)) \preceq F^{(2)}(P_t(y)) \overset{(*)}{\preceq} F^{(2)}(Y).$$

Case 2: $P_{t+1}(x)$ is not equal to $P_t(x)$.
This situation occurs if x is newly detected or if $P_t(x)$ is redirected. In any case, there exists a node $\overline{x}$ such that $P_{t+1}(x) = P_t(\overline{x}) \oplus (\overline{x}, x)$. ($\overline{x}$ is the node expanded in the current iteration.)

Let $Y \leq P_{t+1}(x)$ and $y = \omega(Y)$. The first case is that $Y \leq P_t(\overline{x})$. Then $F^{(2)}(P_t(y)) \preceq F^{(2)}(Y)$ by applying the induction hypothesis to $\overline{x}$. Hence, $F^{(2)}(P_{t+1}(y)) \preceq F^{(2)}(P_t(y)) \preceq F^{(2)}(Y)$. The other case is that $Y = P_{t+1}(x)$. Then $y = x$. These two equalities imply that $F^{(2)}(P_{t+1}(y)) = F^{(2)}(P_{t+1}(x)) = F^{(2)}(Y)$. $\qquad\square$

The next lemma says that all paths $P(v)$ are acyclic if the redirection function does not cause negative cycles. We prove this result with 4.22. Lemma 4.23 will be used in the proofs of 4.39 and 4.91.

Lemma 4.23 *Given a simple, locally finite digraph $\mathcal{G} = (\mathcal{V}, \mathcal{R})$, a node $s \in \mathcal{V}$, and a set $\Gamma \subseteq \mathcal{V}$. Let $\preceq$ be total and identitive. Let BF** use an expansion function $F^{(1)} : \mathcal{P}(s) \to (\mathbf{R}, \preceq)$ and a redirection function $F^{(2)} : \mathcal{P}(s) \to (\mathbf{R}, \preceq)$ where $\preceq$ is total and identitive. We assume that $\mathcal{G}$ contains no $F^{(2)}$-negative cycles (see Definition/Remark 2.128).*
*Then all paths $P(v)$ generated by BF**$\big(F^{(1)}, F^{(2)}\big)$ are acyclic.*

P r o o f : We show the result by an induction of the number t of iterations. Our assertion is trivial for $t = 1$ because $P_1(s) = [s]$ is the only path generated by BF**$(F^{(1)}, F^{(2)})$ before the first iteration.

We assume that the assertion is true immediately before the tth iteration and that $\overline{v}_t$ is currently selected for expansion. Let $v \in \mathcal{V}$. Then there are three cases:

Case 1: $P_{t+1}(v) = P_t(v)$.
Then $P_{t+1}(v) = P_t(v)$ is acyclic by the induction hypothesis.

Case 2: The node v is newly detected.
Then $v \notin P_t(\overline{v}_t)$ because all nodes of this path have already been found by BF**. This and the induction hypothesis imply that $P_{t+1}(v) = P_t(\overline{v}_t) \oplus (\overline{v}_t, v)$ is acyclic.

---

[6]Note that $P(y)$ exists, too; this can easily be shown with Lemma 4.18.

Case 3: $P_t(v)$ is redirected.

Then $P_t(v)$ is replaced by $\widehat{P}_t(v) = P_t(\overline{v}_t) \oplus (\overline{v}_t, v)$. This can only occur if

$$F^{(2)}(\widehat{P}_t(v)) = F^{(2)}(P_t(\overline{v}_t) \oplus (\overline{v}_t, v)) \prec F^{(2)}(P_t(v)). \tag{4.3}$$

Suppose that $P_{t+1}(v) = \widehat{P}_t(v)$ has a substantial cycle X. The path $P_t(\overline{v}_t)$ is acyclic by the induction hypothesis. Therefore, the cycle X must be caused by the arc $(\overline{v}_t, v)$. Hence, v is visited by $P_t(\overline{v}_t)$.

Let Y be a prefix of $P_t(\overline{v}_t)$ with $v = \omega(Y)$. Let X' be the suffix of $\widehat{P}_t(v)$ with $\widehat{P}_t(v) = Y \oplus X'$. Then $\alpha(X') = \omega(Y) = v = \omega(\widehat{P}_t(v)) = \omega(X')$ so that X' is a cycle; it is possible that $X' = X$.

Then $F^{(2)}(P_t(v)) \preceq F^{(2)}(Y)$ by 4.22. This and (4.3) imply that $F^{(2)}(Y \oplus X') = F^{(2)}(\widehat{P}_t(v)) \overset{(4.3)}{\prec} F^{(2)}(P_t(v)) \preceq F^{(2)}(Y)$. Hence, $F^{(2)}(Y \oplus X' \oplus [v]) \prec F^{(2)}(Y \oplus [v])$ so that X' is a negative cycle. That is a contradiction to our assumption. Consequently, $P_{t+1}(v) = \widehat{P}_t(v)$ is acyclic. □

We now prove our main result about the quality of the solution paths found by BF**. It says that BF** will output a $C^{(2)}$-bounded path with respect to $C^{(1)}$ if the cost functions $C^{(1)}$, $C^{(2)}$, the expansion function $F^{(1)}$, and the redirection function $F^{(2)}$ satisfy particular conditions.

Theorem 4.24 *Given a simple and locally finite digraph $\mathcal{G} = (\mathcal{V}, \mathcal{R})$, a start node $s \in \mathcal{V}$, and a set $\Gamma \subseteq \mathcal{V}$ of goal nodes.*

Given two path functions $C^{(1)}, C^{(2)} : \mathcal{P}(s) \to (\mathbf{R}, \preceq)$ where $\preceq$ is total and identitive. (For example, $\mathbf{R}$ may be equal to $\mathbb{R}$, and $\preceq$ may be equal to $\leq$.)

We assume that BF** uses the path functions $F^{(1)}$, $F^{(2)} : \mathcal{P}(s) \to (\mathbf{R}, \preceq)$ for expansion and redirection, respectively. Moreover, we make the following assumptions about $C^{(1)}$, $C^{(2)}$, $F^{(1)}$, $F^{(2)}$:

A.1) $(\forall\, P, Q \in \mathcal{P}(s)\,,\ \omega(P) \neq \omega(Q)\,)$
$$\left[F^{(1)}(Q) \preceq F^{(1)}(P) \ \wedge \ P \text{ is } C^{(2)}\text{-bounded} \right] \ \Rightarrow \ C^{(1)}(Q) \preceq C^{(2)}(Q).$$

A.2) $(\forall\, P, Q \in \mathcal{P}(s)\,,\ \omega(P) = \omega(Q)\,)$
$$\left[F^{(2)}(Q) \preceq F^{(2)}(P) \ \wedge \ C^{(1)}(P) \preceq C^{(2)}(P) \right] \ \Rightarrow \ C^{(1)}(Q) \preceq C^{(2)}(Q).$$

B) *There exists a $C^{(2)}$-bounded path $P^* = [w_0, w_1, \ldots, w_l]$ with $w_0 = s$ and $\gamma^* := w_l \in \Gamma$.*

C) *No infinite path $P = [s, v_1, v_2, v_3, \ldots)$ is weakly $C^{(2)}$-bounded.*[7]

D) *$F^{(2)}$ is order preserving.*

Then BF** with expansion function $F^{(1)}$ and redirection function $F^{(2)}$ will output a $C^{(2)}$-bounded solution path P^{**} after finitely many steps. (It is possible that $P^{**} \neq P^*$.)

---

[7] Note that $\preceq$ is total; therefore, this condition can be formulated as follows: For each infinite path P, there exists an i such that $C^{(1)}([s, v_1, \ldots, v_i]) \succ C^{(2)}([s, v_1, \ldots, v_i])$.

P r o o f : Let $P_i^* := [w_0, w_1, \ldots, w_i]$ for all $i = 0, \ldots, l$. We show several auxiliary assertions. The first is similar to Lemma 2 in [DePe85].

> Let $t \in \mathbb{N}$, and let BF** execute the tth iteration. Then there exists a node w of P^* such that $w \in \text{OPEN}_t$ and $C^{(1)}(P_t(w)) \preceq C^{(2)}(P_t(w))$. (4.4)

Proof of (4.4): First, we see that $\gamma^* \notin \text{CLOSED}_t$. Otherwise, there exists an iteration $t' < t$ such that step 3 of the t'th iteration has transported γ^* from OPEN to CLOSED; this and $\gamma^* \in \Gamma$ imply that BF* must have stopped in step 4 of the t'th iteration, and that is a contradiction to the assumption that the tth iteration is executed.

So, not all nodes of P^* are elements of CLOSED_t. We choose the minimal number λ such that $w_\lambda \notin \text{CLOSED}_t$. Lemma 4.18 a) implies that $w_\lambda \in \text{OPEN}_t$. Then Corollary 4.20 says that $F^{(2)}(P_t(w_\lambda)) = f_t^{(2)}(w_\lambda) \preceq F^{(2)}(P_\lambda^*)$. Moreover, $C^{(1)}(P_\lambda^*) \overset{B)}{\preceq} C^{(2)}(P_\lambda^*)$.

We apply assumption A.2) to $P := P_\lambda^*$ and $Q := P_t(w_\lambda)$. For this end, we observe that $\omega(P) = w_\lambda = \omega(Q)$; this and the inequalities $F^{(2)}(Q) \preceq F^{(2)}(P)$ and $C^{(1)}(P) \preceq C^{(2)}(P)$ imply that $C^{(1)}(Q) \preceq C^{(2)}(Q)$. That means that $C^{(1)}(P_t(w_\lambda)) \preceq C^{(2)}(P_t(w_\lambda))$. Hence, fact (4.4) is true for $w := w_\lambda$.

The next assertion says that all paths $P_t(v)$ are weakly $C^{(2)}$-bounded. More precisely, we show that the following are true for all $t \in \mathbb{N}$ and for all vertices v:

> a) If $P_t(v)$ exists then $P_t(v)$ is weakly $C^{(2)}$-bounded.
> b) If v is equal to the node $\bar{v}_t$ selected in step 3 of the tth iteration then the whole path $P_t(v)$ is $C^{(2)}$-bounded. (4.5)
> (That means that $C^{(1)}(P_t(\bar{v}_t)) \preceq C^{(2)}(P_t(\bar{v}_t))$ in addition to a).)

Proof of (4.5): We prove both assertions a) and b) simultaneously by an induction on t. For $t = 1$, the only relevant path is $P_1(s) = [s]$, which is indeed $C^{(2)}$-bounded because it is a prefix of the $C^{(2)}$-bounded path P^*.

We now assume that the assertions (4.5 a) and (4.5 b) are correct for all paths $P_t(v)$. We now prove that fact (4.5 a) remains true after the tth iteration. Obviously, the only critical situation arises if $P_{t+1}(v)$ is not equal to $P_t(v)$. Then the following must have occurred: Either v and $P_{t+1}(v)$ have been detected in step 6b) or $P_t(v)$ has been replaced by $P_{t+1}(v)$ in step 6c). In both cases, $P_{t+1}(v) = P_t(\bar{v}_t) \oplus (\bar{v}_t, v)$ where $\bar{v}_t$ is the vertex expanded in the tth iteration. Part (4.5b) of the induction hypothesis says that $P_t(\bar{v}_t)$ is $C^{(2)}$-bounded. This implies that $P_{t+1}(v)$ is weakly $C^{(2)}$-bounded.

We have proven that fact (4.5 a) is valid for all paths $P_{t+1}(v)$. We must show that assertion (4.5 b) is also true for $t + 1$; that means, $P_{t+1}(\bar{v}_{t+1})$ is $C^{(2)}$-bounded where $\bar{v}_{t+1}$ is the node selected in step 3 of the $(t + 1)$st iteration. Fact (4.5 a) implies that $P_{t+1}(\bar{v}_{t+1})$ is weakly $C^{(2)}$-bounded; therefore, we must show that $C^{(1)}(P_{t+1}(\bar{v}_{t+1})) \preceq C^{(2)}(P_{t+1}(\bar{v}_{t+1}))$.

Our proof of this fact resembles the one of Lemma 3 (b) in [DePe85], page 510. Let $w \in \text{OPEN}_{t+1}$ with $C^{(1)}(P_{t+1}(w)) \preceq C^{(2)}(P_{t+1}(w))$. (The existence of w follows from fact (4.4) about $t + 1$.) Then

$$F^{(1)}(P_{t+1}(\overline{v}_{t+1})) \;=\; f^{(1)}_{t+1}(\overline{v}_{t+1}) \;\preceq\; f^{(1)}_{t+1}(w) \;=\; F^{(1)}(P_{t+1}(w)) \qquad (4.6)$$

because $f^{(1)}_{t+1}(w)$ and $f^{(1)}_{t+1}(\overline{v}_{t+1})$ are compared in step 3 of the $(t+1)$st iteration.
Moreover, $P_{t+1}(w)$ is weakly $C^{(2)}$-bounded by fact (4.5 a), and $C^{(1)}(P_{t+1}(w)) \overset{(4.4)}{\preceq} C^{(2)}(P_{t+1}(w))$. Hence,

$$P_{t+1}(w) \text{ is } C^{(2)}-\text{bounded.} \qquad (4.7)$$

We now consider two cases: The first is that $\overline{v}_{t+1} = w$. Then (4.7) says that $P_{t+1}(\overline{v}_{t+1})$ is $C^{(2)}$-bounded, and (4.5 b) is proven.
The other case is that $\overline{v}_{t+1} \neq w$; then the paths $P := P_{t+1}(w)$ and $Q := P_{t+1}(\overline{v}_{t+1})$ satisfy the assumptions of A.1): It is $\omega(P) = w \neq \overline{v}_{t+1} = \omega(Q)$; moreover, (4.6) implies that $F^{(1)}(Q) = F^{(1)}(P_{t+1}(\overline{v}_{t+1})) \preceq F^{(1)}(P_{t+1}(w)) = F^{(1)}(P)$, and (4.7) says that $P = P_{t+1}(w)$ is $C^{(2)}$-bounded. Applying A.1), we see that $C^{(1)}(P_{t+1}(\overline{v}_{t+1})) = C^{(1)}(Q) \preceq C^{(2)}(Q) = C^{(2)}(P_{t+1}(\overline{v}_{t+1}))$. Consequently, $C^{(1)}(P_{t+1}(\overline{v}_{t+1})) \preceq C^{(2)}(P_{t+1}(\overline{v}_{t+1}))$; moreover, $P_{t+1}(\overline{v}_{t+1}))$ is weakly $C^{(2)}$-bounded as seen in (4.5 a); that means that $P_{t+1}(\overline{v}_{t+1})$ is $C^{(2)}$-bounded, and (4.5 b) is proven also in this case.

Fact (4.5 b) has the following consequence: Suppose that BF** outputs a solution path P^{**} after selecting some node $\gamma^{**} \in \Gamma$ in the Tth iteration where $T \in \mathbb{N}$; then $P^{**} = P_T(\gamma^{**})$ is $C^{(2)}$-bounded.

Now we show *that* step 3 of BF** will actually select a goal node γ^{**} after finitely many iterations. For this purpose we prove the following claim:

There exist only finitely many $C^{(2)}$-bounded paths $Q \in \mathcal{P}(s)$. $\qquad (4.8)$

Proof of (4.8): The proof is similar to the one of the Infinity Lemma (see [Knu82], Vol. 1). Suppose that infinitely many $C^{(2)}$-bounded paths $Q \in \mathcal{P}(s)$ exist. Then we construct an infinite path $X := [v_0, v_1, v_2, \ldots)$ with $v_0 = s$ and with the following property:

For each k, the path $X_k := [s, v_1, \ldots, v_k]$ is a prefix of infinitely many $C^{(2)}$-bounded paths Q. $\qquad (4.9)$

The nodes $v_0, v_1, v_2, \ldots$ are constructed recursively. We define $v_0 := s$. Then fact (4.9) about X_0 follows from the assumption that (4.8) is false. Let $v_0, \ldots, v_j$ be given, and let (4.9) be true for all $k = 1, \ldots, j$. Then X_j is a prefix of infinitely many $C^{(2)}$-bounded paths. On the other hand, v_j has only finitely many successors in $\mathcal{G}$ because $\mathcal{G}$ is locally finite. Hence, there exists an arc $(v_j, x) \in \mathcal{G}$ such that infinitely many $C^{(2)}$-bounded paths begin with the nodes $s, v_1, \ldots, v_j, x$. Let then $v_{j+1} := x$.
It follows from (4.9) that each path X_k is the prefix of at least one $C^{(2)}$-bounded path Q; hence, $C^{(1)}(X_k) \preceq C^{(2)}(X_k)$ for all k. That means that X is weakly $C^{(2)}$-bounded, and that is a contradiction to assumption C).

Next, we show the following fact:

If a path $P(v)$ is expanded by BF** then $P(v)$ will never be expanded again. $\qquad (4.10)$

Proof of (4.10): Let $t' < t''$. Suppose that v is expanded in the t'th and in the t''th iteration (i.e. $\overline{v}_{t'} = \overline{v}_{t''} = v$); we must show that $P_{t'}(v) \neq P_{t''}(v)$. For this purpose, we observe that v has moved from OPEN to CLOSED in the t'th iteration and that $v = \bar{v}_{t''} \in$ OPEN immediately before the t''th iteration. So, there exists an iteration $t \in \{t'+1, t'+2, \dots, t''\}$ that has moved v from CLOSED back to OPEN. This is only possible if $P(v)$ has been redirected, and this implies that $F^{(2)}(P_{t'}(v)) \preceq F^{(2)}(P_t(v)) \prec F^{(2)}(P_{t+1}(v)) \preceq F^{(2)}(P_{t''+1}(v))$. Hence, $P_{t'}(v) \neq P_{t''}(v)$.

We now accomplish the proof of our theorem. Fact (4.5 b) says that BF** only expands $C^{(2)}$-bounded paths $P(v)$. Fact (4.8) says that only finitely many $C^{(2)}$-bounded paths exist, and fact (4.10) implies that each of them can be expanded at most once. Hence, BF** executes only a finite number T of iterations. Moreover, fact (4.4) has the consequence that $\text{OPEN}_T \neq \emptyset$. Hence, BF** can only terminate because of selecting a goal node γ^{**} in step 3 of the Tth iteration. That means that BF** will output an s-Γ-path P^{**} after finitely many iterations. We have already seen that P^{**} is $C^{(2)}$-bounded.   □

Next, we describe special cases of Theorem 4.24.

Remark 4.25 Given a simple, locally finite digraph $\mathcal{G} = (\mathcal{V}, \mathcal{R})$, a start node s, and a set Γ of goal nodes. We assume that the path functions $C^{(1)}, C^{(2)}, F^{(1)}, F^{(2)} : \mathcal{P}(s) \to (\mathbb{R}, \leq)$ satisfy all conditions A.1), A.2), B), C), D) of Theorem 4.24.

a) The following special case of 4.24 is given in the proof of 4.34:

$F^{(1)} = F^{(2)} = C^{(1)}$ and $C^{(2)}(P) = M$ for all paths P where

$$M := \min \left\{ C^{(1)}_{\max}(Q) \,\middle|\, Q \text{ is a solution path} \right\}.$$

In this situation, $\text{BF}^{**}(F^{(1)}, F^{(2)}) = \text{BF}^{**}(C^{(1)}, C^{(1)})$ will output a $C^{(1)}_{\max}$-minimal s-Γ-path P^{**}; this path is also $F^{(1)}_{\max}$-minimal and $F^{(2)}_{\max}$-minimal. The same special case of Theorem 4.24 arises when defining $(\forall x \in \mathbb{R})\ \psi(x) := x$ in Theorems 1 and 2 of [DePe85].

b) The following special case of 4.24 is given in the proof of 4.89:

$F^{(2)} = C^{(1)}$, $F^{(1)} = C^{(1)}_{\max} = F^{(2)}_{\max}$, and $C^{(2)}(P) = M$ for all paths P where

$$M := \min \left\{ C^{(1)}_{\max}(Q) \,\middle|\, Q \text{ is a solution path} \right\}.$$

In this situation, $\text{BF}^{**}(F^{(1)}, F^{(2)}) = \text{BF}^{**}(C^{(1)}_{\max}, C^{(1)})$ will output a $C^{(1)}_{\max}$-minimal s-Γ-path P^{**}; this path is also $F^{(1)}$-minimal and $F^{(2)}_{\max}$-minimal. A special version of $\text{BF}^{**}(F^{(1)}, F^{(2)})$ with $F^{(1)} = F^{(2)}_{\max}$ is Algorithm A** in [DePe85].   □

Next, we replace the complicated conditions A.1), A.2) of Theorem 4.24 by simpler ones. First, we formulate the following two conditions:

A.1') $(\forall\, P, Q \in \mathcal{P}(s))$

$\qquad \left[F^{(1)}(Q) \preceq F^{(1)}(P) \ \wedge \ C^{(1)}(P) \preceq C^{(2)}(P) \right] \ \Rightarrow \ C^{(1)}(Q) \preceq C^{(2)}(Q).$

A.2') $(\forall\, P, Q \in \mathcal{P}(s))$

$\qquad \left[F^{(2)}(Q) \preceq F^{(2)}(P) \ \wedge \ C^{(1)}(P) \preceq C^{(2)}(P) \right] \ \Rightarrow \ C^{(1)}(Q) \preceq C^{(2)}(Q).$

Then A.1') implies A.1), and A.2') implies A.2). To see this note that the conditions $\omega(P) \neq \omega(Q)$ and $\omega(P) = \omega(Q)$ do not appear in A.1') and A.2'), respectively; moreover, the premise of A.1) implies the one of A.1').

Let now $(\mathbf{R}, \preceq) = (\mathbb{R}, \leq)$. We transform A.1') and A.2') into two conditions about real functions. For this purpose, we assume that $F^{(2)}$ is injective; that means that $F^{(2)}(P) \neq F^{(2)}(Q)$ for different paths $P, Q \in \mathcal{P}(s)$. Then there exist three real functions $\mathbf{f}, \mathbf{c}^{(1)}, \mathbf{c}^{(2)} : \mathbb{R} \to \mathbb{R}$ such that

$$F^{(1)} = \mathbf{f} \circ F^{(2)}, \quad C^{(1)} = \mathbf{c}^{(1)} \circ F^{(2)}, \quad \text{and} \quad C^{(2)} = \mathbf{c}^{(2)} \circ F^{(2)}. \quad (4.11)$$

We now formulate two conditions A.1") and A.2") about the real functions $\mathbf{f}$, $\mathbf{c}^{(1)}$, $\mathbf{c}^{(2)}$; condition A.1") implies A.1') (and consequently A.1)), and condition A.2") implies A.2') (and consequently A.2)); this can be seen by defining $x := F^{(2)}(P)$ and $y := F^{(2)}(Q)$.

A.1") $(\forall x, y \in \mathbb{R}) \left[\mathbf{f}(y) \leq \mathbf{f}(x) \wedge \mathbf{c}^{(1)}(x) \leq \mathbf{c}^{(2)}(x) \right] \Rightarrow \mathbf{c}^{(1)}(y) \leq \mathbf{c}^{(2)}(y)$.

A.2") $(\forall x, y \in \mathbb{R}) \left[y \leq x \wedge \mathbf{c}^{(1)}(x) \leq \mathbf{c}^{(2)}(x) \right] \Rightarrow \mathbf{c}^{(1)}(y) \leq \mathbf{c}^{(2)}(y)$.

These conditions are analysed in the next result.

Theorem 4.26 *Given the functions* $\mathbf{c}^{(1)}, \mathbf{c}^{(2)}, \mathbf{f} : \mathbb{R} \to \mathbb{R}$. *Let* $U := \{u \in \mathbb{R} \,|\, \mathbf{c}^{(1)}(u) \leq \mathbf{c}^{(2)}(u)\}$. *We assume that* $\mathbf{c}^{(1)}, \mathbf{c}^{(2)}, \mathbf{f}$ *are continuous, that* $U \neq \emptyset$, *and that* $X := \sup(U) < \infty$. *Then the following assertions* (a) *and* (b) *are equivalent:*

(a) *The assertions* A.1") *and* A.2") *are true.*

(b) *For all* $z \in \mathbb{R}$, *the following assertions are equivalent:*

(b1) $z \leq X$ (b2) $\mathbf{c}^{(1)}(z) \leq \mathbf{c}^{(2)}(z)$ (i.e. $z \in U$).   (b3) $\mathbf{f}(z) \leq \mathbf{f}(X)$.

(The equivalence (b1)⇔(b3) means that the graph of $\mathbf{f}$ lies in the left lower and in the right upper quadrant with respect to the point $(X, \mathbf{f}(X))$; this is illustrated in Figure 9 in Example 4.28.)

P r o o f o f (a)⇒(b) : Let A.1") and A.2") be true.
We show (b1)⇒(b2). Let $z \leq X$. The definition of U and the continuity of $\mathbf{c}^{(1)}, \mathbf{c}^{(2)}$ imply that $\mathbf{c}^{(1)}(X) = \mathbf{c}^{(2)}(X)$. Hence, the premise of A.2") is true for $y := z$ and $x := X$. Then A.2") implies (b2).
We show (b2)⇒(b3). Let $\mathbf{c}^{(1)}(z) \leq \mathbf{c}^{(2)}(z)$. Suppose that $\mathbf{f}(z) > \mathbf{f}(X)$. The continuity of $\mathbf{f}$ implies that a $\delta > 0$ exists with $\mathbf{f}(z) > \mathbf{f}(X + \delta)$. That means that the premise of A.1") is true for $y := X + \delta$ and $x := z$. Then A.1") implies that $\mathbf{c}^{(1)}(X + \delta) \leq \mathbf{c}^{(2)}(X + \delta)$, i.e. $X + \delta \in U$; that is a contradiction to $X = \sup(U)$. Hence, $\mathbf{f}(z) \leq \mathbf{f}(X)$.
We show (b3)⇒(b1). Let $\mathbf{f}(z) \leq \mathbf{f}(X)$. We have seen that $\mathbf{c}^{(1)}(X) = \mathbf{c}^{(2)}(X)$. Hence, the premise of A.1") is true for $y := z$ and $x := X$. Then A.1") implies that $z \in U$. Hence, $z \leq \sup(U) = X$.

P r o o f o f (b)⇒(a) : Let (b1),(b2),(b3) be equivalent for all $z \in \mathbb{R}$.
We show A.1"). Let $\mathbf{f}(y) \leq \mathbf{f}(x)$ and $\mathbf{c}^{(1)}(x) \leq \mathbf{c}^{(2)}(x)$. Then (b2) is true for $z := x$. Hence, (b3) is true for $z := x$ so that $\mathbf{f}(y) \leq \mathbf{f}(x) \leq \mathbf{f}(X)$. Hence, (b3) is also true for $z := y$ so that $\mathbf{c}^{(1)}(y) \leq \mathbf{c}^{(2)}(y)$ by (b3)⇒(b2).
We show A.2"). Let $y \leq x$ and $\mathbf{c}^{(1)}(x) \leq \mathbf{c}^{(2)}(x)$. Then (b2) is true for $z := x$. Hence, (b1) is true for $z := x$ so that $y \leq x \leq X$. Hence, (b1) is also true for $z := y$ so that $\mathbf{c}^{(1)}(y) \leq \mathbf{c}^{(2)}(y)$ by (b1)⇒(b2). $\qquad \square$

Remark 4.27 Given the situation of Theorem 4.26. We assume that A.1") and A.2") are true. Let $Q \in \mathcal{P}(s)$, and let $z := F^{(2)}(Q)$. Then equation (4.11) implies that $F^{(1)}(Q) = \mathbf{f}(z)$, $C^{(1)}(Q) = \mathbf{c}^{(1)}(z)$, and $C^{(2)}(Q) = \mathbf{c}^{(2)}(z)$. These facts and Theorem 4.26 have the following consequence:

$$F^{(2)}(Q) \leq X \iff C^{(1)}(Q) \leq C^{(2)}(Q) \iff F^{(1)}(Q) \leq \mathbf{f}(X). \tag{4.12}$$

A consequence of fact (4.12) is that the following assertions are equivalent for all paths $P \in \mathcal{P}(s)$:

$F_{\max}^{(2)}(P) \leq X$.
P is $C^{(2)}$-bounded with respect to $C^{(1)}$.
$F_{\max}^{(1)}(P) \leq \mathbf{f}(X)$.

That means: If all conditions A.1"), A.2"), B), C), D) are satisfied then Theorem 4.24 says that BF** will output a path P^{**} with the following properties:

P^{**} is $C^{(2)}$-bounded with respect to $C^{(1)}$,
$F_{\max}^{(2)}(P^{**}) \leq X$,
$F_{\max}^{(1)}(P^{**}) \leq \mathbf{f}(X)$.

This path P^{**} is found with the help of an expansion function $F^{(1)}$ that is not necessarily order preserving. □

The next two examples will show the behaviour of BF**.

Example 4.28 In this example, we consider a digraph $\mathcal{G}$ with the properties A.1"), A.2"), B), C), D).

Let $\mathcal{G}$ be the digraph in Figure 8. The goal nodes are drawn as rectangles and the other nodes are drawn as ovals. We assume that $C^{(1)}(P) \leq C^{(2)}(P)$ if and only if $P = [s]$ or P ends with a bold arc; hence, the only paths with $C^{(1)}(P) > C^{(2)}(P)$ are $\widetilde{P}_i := [s, x_i]$, $i = 5, 6, 7$.

Each node v has been decorated with the $F^{(1)}$- and the $F^{(2)}$-values of the path(s) between s and v; the left [middle, right] values at $v = x_i$ ($i = 3, 4$) describe the $F^{(1)}$- and $F^{(2)}$-value of the s-v-path starting with (s, x_1), [with (s, x_3), with (s, x_2), respectively]. The decision functions $\acute{F}^{(1)}$ and $\acute{F}^{(2)}$ will be considered in Example 4.29.

Next, we introduce the real functions $\mathbf{c}^{(1)}$, $\mathbf{c}^{(2)}$, and $\mathbf{f}$. For this purpose, we observe that $F^{(2)}(P) \leq 10 \Leftrightarrow C^{(1)}(P) \leq C^{(2)}(P)$ for all $P \in \mathcal{P}(s)$. Hence, there exist two continuous functions $\mathbf{c}^{(1)}, \mathbf{c}^{(2)} : \mathbb{R} \to \mathbb{R}$ such that $\mathbf{c}^{(i)}\left(F^{(2)}(P)\right) = C^{(i)}(P)$ for all $i = 1, 2$, $P \in \mathcal{P}(s)$ and that $U = \{z \in \mathbb{R} \mid \mathbf{c}^{(1)}(z) \leq \mathbf{c}^{(2)}(z)\} = (-\infty, 11]$. This implies that $X = \sup(U) = 11$.

We define the function $\mathbf{f}$ as shown in Figure 9. The thick black circles symbolize the points $\left(F^{(2)}(P), F^{(1)}(P)\right) = \left(F^{(2)}(P), \mathbf{f}\left(F^{(2)}(P)\right)\right)$ where $P \in \mathcal{P}(s)$. For example, the circle at the position $(4, 3)$ comes from $P = [s, x_2, x_3, x_4]$. Moreover, $\mathbf{f}(X) = \mathbf{f}(11) := 10$.

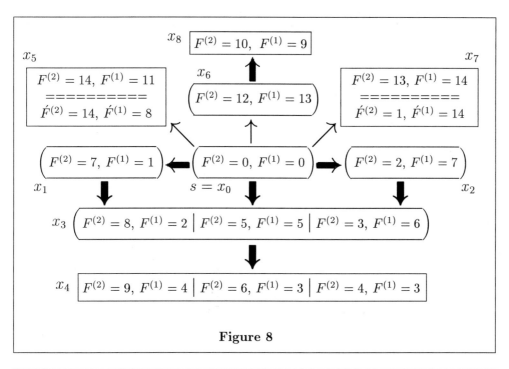

Figure 8

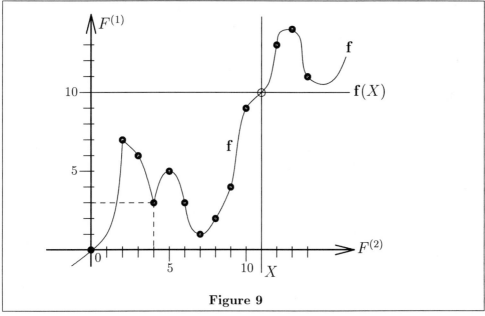

Figure 9

Now we show that $\mathcal{G}$ and $\mathbf{c}^{(1)}, \mathbf{c}^{(2)}, \mathbf{f}$ satisfy all conditions A.1"), A.2"), B), C), D). For this purpose, we observe that $z \leq X = 11 \overset{U=(-\infty, X]}{\Longleftrightarrow} \mathbf{c}^{(1)}(z) \leq \mathbf{c}^{(2)}(z)$ and that $z \leq X \overset{\text{Def. of } \mathbf{f}}{\Longleftrightarrow} \mathbf{f}(z) \leq \mathbf{f}(X) = 10$. Hence, conditions (b1), (b2), (b3) of Theorem 4.26 are equivalent, and this implies that A.1") and A.2") are true. — Condition B) follows from the fact that $[s, x_3, x_4]$ is a $C^{(2)}$-bounded goal path. Condition C) is satisfied because no infinite path exists in the acyclic graph $\mathcal{G}$. Condition D) follows from the facts that $F^{(2)}([s, x_2, x_3]) \leq F^{(2)}([s, x_3]) \leq F^{(2)}([s, x_1, x_3])$ and that $F^{(2)}([s, x_2, x_3, x_4]) \leq F^{(2)}([s, x_3, x_4]) \leq F^{(2)}([s, x_1, x_3, x_4])$.

It follows from Theorem 4.24 that BF** will output a $C^{(2)}$-bounded solution path P^{**}.

We check this result by considering the behaviour of BF** in $\mathcal{G}$. The algorithm expands s and finds the paths $P(x_i) = [s, x_i]$ for $i = 1, 2, 3, 5, 6, 7$. Then BF** expands x_1 and compares the path $\widehat{P}(x_3) = [s, x_1, x_3]$ with $P(x_3) = [s, x_3]$; BF** does not redirect $P(x_3)$ because $F^{(2)}\left(\widehat{P}(x_3)\right) = 8 > 5 = F^{(2)}(P(x_3))$.

Then BF** expands x_3 because $F^{(1)}(P(x_3)) = F^{(1)}([s, x_3]) = 5$ is better than $F^{(1)}([s, x_2]) = 7$, $F^{(1)}([s, x_5]) = 11$, $F^{(1)}([s, x_6]) = 13$, and $F^{(1)}([s, x_7]) = 14$. After this, BF** selects x_4 for expansion and outputs the path $P^{**} := [s, x_3, x_4]$. This path consists of bold arcs, and that means that P^{**} is indeed $C^{(2)}$-bounded.

Moreover, $F^{(1)}_{\max}(P^{**}) = 5 \leq 10 = \mathbf{f}(X)$, and $F^{(2)}_{\max}(P^{**}) = 6 \leq 11 = X$; that means that P^{**} has the properties described in Remark 4.27.

The values $F^{(1)}_{\max}(P^{**})$ and $F^{(2)}_{\max}(P^{**})$ are bounded by $\mathbf{f}(X) = 10$ and by $X = 11$, respectively; but $F^{(1)}_{\max}(P^{**})$ and $F^{(2)}_{\max}(P^{**})$ are not minimal. This follows from $F^{(1)}_{\max}([s, x_1, x_3, x_4]) = 4 < 5 = F^{(1)}_{\max}(P^{**})$ and $F^{(2)}_{\max}([s, x_2, x_3, x_4]) = 4 < 6 = F^{(2)}_{\max}(P^{**})$. $\qquad\square$

Example 4.29 In this example, all conditions A.1), A.2), B), C), D) are satisfied, but the assertions A.1'), A.2') (and consequently A.1"), A.2")) are not true.

Let $\mathcal{G}$ be the same digraph as in in Example 4.28. We define $\acute{F}^{(1)}(P) := F^{(1)}(P)$ and $\acute{F}^{(2)}(P) := F^{(2)}(P)$ for all paths P up to $\widetilde{P}_i = [s, x_i]$ $(i = 5, 7)$; for these paths, we define $\acute{F}^{(1)}(\widetilde{P}_5) := 8$, $\acute{F}^{(2)}(\widetilde{P}_5) := 14$, $\acute{F}^{(1)}(\widetilde{P}_7) := 14$, $\acute{F}^{(2)}(\widetilde{P}_7) := 1$.

Then A.1') and A.2') are not true:
If $Q := [s, x_5]$ and $P := [s, x_6, x_8]$ then $\acute{F}^{(1)}(Q) \leq \acute{F}^{(1)}(P)$, $C^{(1)}(P) \leq C^{(2)}(P)$ but not $C^{(1)}(Q) \leq C^{(2)}(Q)$ in conflict with A.1').
If $Q = [s, x_7]$ and $P = [s, x_6, x_8]$ then $\acute{F}^{(2)}(Q) \leq \acute{F}^{(2)}(P)$, $C^{(1)}(P) \leq C^{(2)}(P)$ but not $C^{(1)}(Q) \leq C^{(2)}(Q)$ in conflict with A.2').

It follows that no functions $\mathbf{c}^{(1)}, \mathbf{c}^{(2)}, \mathbf{f}$ exist such that A.1") and A.2") are true.

On the other hand, condition A.1) is satisfied. To see this we assume that $P, Q \in \mathcal{P}(s)$ with $\omega(P) \neq \omega(Q)$. The implication in A.1) is true if P is not $C^{(2)}$-bounded or if $C^{(1)}(Q) \leq C^{(2)}(Q)$. The remaining case is that P consists of bold arcs and Q ends with a thin arc; then $\omega(P) \in \{s, x_1, x_2, x_3, x_4\}$ and $Q \in \{[s, x_5], [s, x_6], [s, x_7]\}$; it follows that $\acute{F}^{(1)}(Q) \geq 8 > \acute{F}^{(1)}(P)$ so that the premise of A.1) is false and the implication itself is true.

Moreover, condition A.2) is satisfied. To see this we assume that $P, Q \in \mathcal{P}(s)$ with $w(P) = w(Q)$. If $w(P) = w(Q) \notin \{x_3, x_4\}$ then A.2) is true because $P = Q$. If, however, $w(P) = w(Q) \in \{x_3, x_4\}$ the A.2) is true because $C^{(1)}(Q) \preceq C^{(2)}(Q)$. — The counterexample to A.2') does not conflict with A.2) since $w([s, x_7]) \neq w([s, x_6, x_8])$. Hence, Theorem 4.24 can be applied. It says that BF** also outputs a $C^{(2)}$-bounded path if the decision functions $F^{(1)}$ and $F^{(2)}$ are replaced by $\acute{F}^{(1)}$ and $\acute{F}^{(2)}$. Consequently, BF** does not output the paths $\tilde{P}_5$ and $\tilde{P}_7$.

Let $X^{(1)} := 11$ and $X^{(2)} := 10$. In Example 4.28, we saw that $BF^{**}\big(F^{(1)}, F^{(2)}\big)$ outputs an $X^{(i)}$-bounded s-Γ-path with regard to $F^{(i)}$ ($i = 1, 2$). Here, however, we have seen that $BF^{**}\big(\acute{F}^{(1)}, \acute{F}^{(2)}\big)$ does not output the s-Γ-path $\tilde{P}_5$ with $\acute{F}^{(1)}_{\max}(\tilde{P}_5) \leq X^{(1)}$ or the ($\acute{F}^{(2)}_{\max}$-minimal) s-Γ-path $\tilde{P}_7$ with $\acute{F}^{(2)}_{\max}(\tilde{P}_7) \leq X^{(2)}$.   □

Remark 4.30 (Illustrating properties of the functions $C^{(1)}$, $C^{(2)}$, $F^{(1)}$, $F^{(2)}$ in Theorem 4.24)

Remark 4.9 of [Huck92] is about so-called path diagrams, which are similar to Figure 9. Path diagrams contain the points $a_P = \big(F^{(1)}(P), F^{(2)}(P)\big)$ and several further objects. Many properties of $F^{(1)}$ and $F^{(2)}$ like order preservation can be seen from path diagrams. Perhaps, path diagrams will also be described in a second edition of this book.   □

The next two theorems say that in several cases, the expansion and the redirection functions may be replaced by other functions that satisfy restricted conditions. Both results can also be found in [HuRu90].

In the following theorem, we replace a given redirection function by a new one.

Theorem 4.31 *Given a simple, locally finite digraph $\mathcal{G} = (V, \mathcal{R})$, a start node $s \in V$, and a set $\Gamma \subseteq V \backslash \{s\}$ of goal nodes. We assume that there are only finitely many s-v-paths for each node v.[8] Given two decision functions $F^{(1)}, F^{(2)} : \mathcal{P}(s) \to \mathbb{R}$. We assume that $BF^{**}\big(F^{(1)}, F^{(2)}\big)$ outputs a solution path Q^{**} ending with a node $\gamma \in \Gamma$.*

Then there exists a redirection function $\overline{F}^{(2)}$ with the following properties:

a) *$BF^{**}(F^{(1)}, \overline{F}^{(2)})$ expands the same vertices and redirects the same paths as the given algorithm $BF^{**}(F^{(1)}, F^{(2)})$.*

b) *$\overline{F}^{(2)}$ is prefix monotone in the following sense:*
   If $P = [v_0, v_1, \ldots, v_k]$ is a path with $v_0 = s$ and with

$$s, \ldots, v_{k-1} \notin \Gamma. \tag{4.13}$$

   then $\overline{F}^{(2)}(Q) < \overline{F}^{(2)}(P)$ for all proper prefixes Q of P.[9]

---

[8]This is not true for all locally finite digraphs: If $\mathcal{G}$ consists of the vertices $x_0 = s, x_1, \ldots, y_0, y_1, \ldots$ and the arcs (x_i, x_{i+1}), (x_i, y_i), (y_{i+1}, y_i), $i \in \mathbb{N} \cup \{0\}$, then each path of the form $[x_0 = s, x_1, \ldots, x_k, y_k, y_{k-1}, \ldots, y_0]$, $k \in \mathbb{N}$, is a path between $s = x_0$ and y_0.

[9]This implies that $\overline{F}^{(2)}(P) = \overline{F}^{(2)}_{\max}(P)$.

c) $\overline{F}_{\max}^{(2)}(Q^{**}) \leq \overline{F}_{\max}^{(2)}(P)$ *for all s-Γ-paths P that satisfy condition* (4.13).[10]

P r o o f : The proof is given in [HuRu90], Theorem 6.2.   □

In the next theorem, we replace the expansion function $F^{(1)}$ and the redirection function $F^{(2)}$ by a single expansion and redirection function F. Thus, we transform $BF^{**}(F^{(1)}, F^{(2)})$ into the algorithm BF^*, which will be studied in Subsection 4.2.2.1.

Theorem 4.32 *Given a simple, locally finite digraph $\mathcal{G} = (\mathcal{V}, \mathcal{R})$, a start node $s \in \mathcal{V}$, and a set $\Gamma \subseteq \mathcal{V}$ of goal nodes. Given the path functions $F^{(1)}, F^{(2)} : \mathcal{P}(s) \to \mathbb{R}$. We assume that $F^{(1)}$ have the following property: For each $v \in \mathcal{V}$, there exists an open, bounded interval I_v such that $F^{(1)}(P) \in I_v$ for all paths P ending with v; moreover, we assume that $I_v \cap I_{v'} = \emptyset$ if $v \neq v'$.*
Then $F^{(1)}$ and $F^{(2)}$ can be replaced by a single expansion and redirection function F. More precisely, there exists a path function F such that $BF^{**}(F, F)$ always takes the same decisions as the given algorithm $BF^{**}(F^{(1)}, F^{(2)})$.

P r o o f : The proof is given in [HuRu90], Theorem 6.1.   □

Remark 4.33 Theorem 4.32 is probably true if general path functions $F^{(1)}, F^{(2)} : \mathcal{P}(s) \to (\mathbf{R}, \preceq)$ are given where $\preceq$ is total and identitive. In this case, we redefine the term "interval", which appears in 4.32. Let $a, b \in \mathbf{R}$; then we define the interval (a, b) as the set $\{\varrho \in \mathbf{R} \,|\, a \preceq \varrho \preceq b\}$.   □

4.2.2.1   The algorithm BF*

BF^* is a special version of BF^{**}. More precisely, if BF^{**} uses the same expansion and redirection function then we shall call the resulting algorithm BF^*.

Our algorithm BF^* is almost the same as the one in in [DePe85]. The main differences are the following: Our algorithm BF^* outputs $P(\overline{v})$ in step 4 whereas Dechter's and Pearl's version of BF^* outputs the path $Q(\overline{v})$. Moreover, all path functions appearing in [DePe85] are real-valued whereas our expansion and redirection function may have an almost arbitrary domain of values.

The notation $BF^*(F)$ means that BF^* uses F as expansion and redirection function; we call F the *decision function* of $BF^*(F)$. Moreover, we define $f(v) := F(P(v))$ and $\widehat{f}(v) := F(\widehat{P}(v))$ for all nodes v.

4.2.2.1.1   BF* with an order preserving decision function

This paragraph is about the usual situation that BF^* uses an order preserving decision function. We shall describe how BF^* can be used to find good paths. Moreover, we

---

[10]The path Q^{**} itself satisfies condition (4.13). This can be seen as follows: Let t be the iteration in which Q^{**} is output. Then $Q^{**} = P_t(\gamma)$ for some $\gamma \in \Gamma$. Lemma 4.18 b) says that all nodes $v \in Q^{**}$ with $v \neq \gamma$ have been selected earlier by step 3 of $BF^{**}(F^{(1)}, F^{(2)})$; hence, they are no goal nodes because otherwise $BF^{**}(F^{(1)}, F^{(2)})$ would have stopped before selecting γ.

shall give several results about the structure of the paths generated by BF*.

The next result is a generalization of Theorem 1 in [DePe85] and a consequence of Theorem 4.24 on page 193.

Theorem 4.34 *Given a simple, locally finite digraph* $\mathcal{G} = (\mathcal{V}, \mathcal{R})$*, a start node* $s \in \mathcal{V}$*, and a set* Γ *of goal nodes. Given the cost functions* $C, C_0 : \mathcal{P}(s) \to (\mathbf{R}, \preceq)$ *where* $\preceq$ *is total and identitive; let* $C = (C_0)_{\max}$*. (E.g.,* $\mathbf{R}$ *may be equal to* $\mathbb{R}$*, and* "$\preceq$" *may be equal to* "$\leq$".*) Let* $\mathcal{M} \subseteq \mathbf{R}$ *be the following set:* $\mathcal{M} := \{C(Q) \mid Q \text{ is solution path}\}$*. We make the following assumptions:*

   (i) *There exists a* C*-minimal goal path; that means that* $\mathcal{M}$ *has a minimum* M*.*

   (ii) *There is no upper bound on the sequence* $\big(C_0([s, v_1, \ldots, v_i])\big)_{i \in \mathbb{N}}$
       if $[s, v_1, v_2, v_3, \ldots)$ *is an infinite path in* $\mathcal{G}$*.* [11]

   (iii) C_0 *is order preserving.*

Then $BF^*(F)$ *with the decision function* $F := C_0$ *will output an* s*-*Γ*-path* P^{**} *with minimal* C*-value. In other words,* $BF^*(F)$ *outputs an* $F_{\max}$*-minimal path* P^{**} *from* s *to* Γ*.*

P r o o f : We define $C^{(1)}(P) := F(P) = C_0(P)$, $C^{(2)}(P) := M$, $F^{(1)}(P) := F(P) = C_0(P)$, and $F^{(2)}(P) := F(P) = C_0(P)$ for all paths $P \in \mathcal{P}(s)$. We verify that the path functions $C^{(1)}$, $C^{(2)}$, $F^{(1)}$, $F^{(2)}$ satisfy all conditions A.1), A.2), B), C), D) of Theorem 4.24.

Condition A.1): Let $F^{(1)}(Q) \preceq F^{(1)}(P)$ and let P be $C^{(2)}$-bounded with respect to $C^{(1)}$. Then $C^{(1)} = F = F^{(1)}$ implies that $C^{(1)}(Q) \preceq C^{(1)}(P)$. Moreover, $C^{(1)}(P) \preceq C^{(2)}(P) = M$ because P is $C^{(2)}$-bounded. Hence, $C^{(1)}(Q) \preceq C^{(1)}(P) \preceq C^{(2)}(P) = M = C^{(2)}(Q)$.

Condition A.2): Let $F^{(2)}(Q) \preceq F^{(2)}(P)$ and $C^{(1)}(P) \preceq C^{(2)}(P)$. Then $C^{(1)}(Q) = F(Q) = F^{(2)}(Q) \preceq F^{(2)}(P) = F(P) = C^{(1)}(P) \preceq C^{(2)}(P) = M = C^{(2)}(Q)$ so that $C^{(1)}(Q) \preceq C^{(2)}(Q)$.

Condition B): This condition follows from assumption (i).

Condition C): Let $P = [s, v_1, v_2, v_3, \ldots)$ be an infinite path. Then (ii) says that there exists a prefix $Q = [s, v_1, \ldots, v_i]$ of P such that $C^{(1)}(Q) = C_0(Q) > M = C^{(2)}(Q)$. Hence, P is not weakly $C^{(2)}$-bounded.

Condition D): This condition follows from (iii) and from $F^{(2)} = C_0$.

Applying Theorem 4.24, we see that $BF^{**}\big(F^{(1)}, F^{(2)}\big) = BF^*(F)$ will output a $C^{(2)}$-bounded solution path P^{**}. Consequently, $C^{(1)}_{\max}(P^{**}) \preceq C^{(2)}(P^{**}) = M$ because P^{**} is $C^{(2)}$-bounded and $C^{(2)}$ is a constant function.
On the other hand, M is the minimum of $\mathcal{M}$ so that $C^{(1)}_{\max}(P^{**}) = (C_0)_{\max}(P^{**}) = C(P^{**}) \succeq M$.
Hence, $C^{(1)}_{\max}(P^{**}) = M$; this and the relationships $C^{(1)} = F$, $C = F_{\max}$ imply that

$$C(P^{**}) = F_{\max}(P^{**}) = C^{(1)}_{\max}(P^{**}) = M \,.$$

---

[11] Note that $\preceq$ is total; therefore, condition (ii) can be formulated as follows: For each $\varrho \in \mathbf{R}$, there exists an i such that $C_0([s, v_1, \ldots, v_i]) \succ \varrho$.

That means that BF*(F) will output a path P^{**} with minimal C-value. $\square$

Next, we prove several modified versions of Theorem 4.34; most of them are formally stronger than 4.34, but they can easily be concluded from this result. Roughly speaking, we shall replace the condition $C = (C_0)_{\max}$ by other assumptions.

Theorem 4.35 *Given a simple, locally finite digraph* $\mathcal{G} = (\mathcal{V}, \mathcal{R})$, *a start node* $s \in \mathcal{V}$, *and a set* Γ *of goal nodes. Given two cost functions* $C, C_0 : \mathcal{P}(s) \to (\mathbf{R}, \preceq)$ *where* $\preceq$ *is total and indentitive. (E.g.,* $\mathbf{R}$ *may be equal to* $\mathbb{R}$, *and* "$\preceq$" *may be equal to* "$\leq$".) *We make the following assumptions:*
  (i)   *There exists a* $(C_0)_{\max}$-*minimal goal path.*
  (ii) *There is no upper bound on the sequence* $(C_0([s, v_1, \ldots, v_i]))_{i \in \mathbb{N}}$
         if $[s, v_1, v_2, v_3, \ldots)$ *is an infinite path in* $\mathcal{G}$.
  (iii) C_0 *is order preserving.*
  (iv) *Each* $(C_0)_{\max}$-*minimal solution path is* C-*minimal.*
Then BF*(F) with $F := C_0$ *will output a* C-*minimal solution path* P^{**}.[12]

P r o o f : Let $\widetilde{C} := (C_0)_{\max}$ and $\widetilde{C}_0 := C_0$. Then $\widetilde{C}$ and $\widetilde{C}_0$ satisfy all conditions (i) – (iii) of Theorem 4.34; in particular, conditions (i) of 4.34 follows from the assumptions (i) and (iv) of the current result.

Theorem 4.34 says that BF*(F) will output an $\widetilde{C}$-optimal solution path P^{**}; this path is $(C_0)_{\max}$-minimal by the definition $\widetilde{C} := (C_0)_{\max}$. Then assumption (iv) implies that P^{**} is C-minimal. $\square$

The next result describes a situation where BF* with decision function C_0 outputs a (C_0)-minimal path. The result is similar to Corollary 1 in [DePe85].

Theorem 4.36 *Given a simple, locally finite digraph* $\mathcal{G} = (\mathcal{V}, \mathcal{R})$, *a node* $s \in \mathcal{V}$, *and a set* $\Gamma \subseteq \mathcal{V}$. *Given the cost functions* $C, C_0 : \mathcal{P}(s) \to (\mathbf{R}, \preceq)$ *where* $\preceq$ *is total and identitive. (E.g.,* $\mathbf{R}$ *may be equal to* $\mathbb{R}$, *and* "$\preceq$" *may be equal to* "$\leq$".) *We make the following assumptions:*
  (i)   *There exists a* C_0-*minimal* s-Γ-*path* P^* *with* $C_0(P^*) = (C_0)_{\max}(P^*)$.
  (ii) *There is no upper bound on the sequence* $(C_0([s, v_1, \ldots, v_i]))_{i \in \mathbb{N}}$
         if $[s, v_1, v_2, v_3, \ldots)$ *is an infinite path in* $\mathcal{G}$.
  (iii) C_0 *is order preserving.*
  (iv) *Each* C_0-*minimal goal path is also* C-*minimal.*
Then BF*(F) with $F := C_0$ *will output an* s-Γ-*path* P^{**} *that is minimal with respect to* C *and* $C_0 = F$.

P r o o f : Let $\widetilde{C} := (C_0)_{\max}$. We show that $\widetilde{C}$ satisfies condition (i) in 4.34. Let Q be a solution path. Then the following inequality is valid where $(*)$ follows from the C_0-minimality of P^*: $(C_0)_{\max}(Q) \overset{(*)}{\succeq} C_0(Q) \overset{(*)}{\succeq} C_0(P^*) \overset{(i)}{=} (C_0)_{\max}(P^*)$.

---

[12]In particular, assumptions (i) – (iv) of this theorem are valid if the conditions (i) – (iii) of Theorem 4.34 are satisfied and if $C(P) = (C_0)_{\max}(P)$ for all s-Γ-paths P; therefore, Theorem 4.35 is somewhat stronger than 4.34.

Hence, $\widetilde{C}(P^*) = (C_0)_{\max}(P^*)$ is the minimum of all values $\widetilde{C}(Q) = (C_0)_{\max}(Q)$.

The path function $\widetilde{C}$ and C_0 satisfy all conditions of Theorem 4.34. Consequently, $\mathrm{BF}^*(F)$ will output an s-Γ-path P^{**} with minimal value with respect to $\widetilde{C} = (C_0)_{\max}$. Then the following inequality is valid:

$$C_0(P^{**}) \preceq (C_0)_{\max}(P^{**}) \preceq (C_0)_{\max}(P^*) \overset{(i)}{=} C_0(P^*).$$

This and the C_0-minimality of P^* imply that $C_0(P^{**})$ is minimal as well; so, P^{**} is a C_0-minimal s-Γ-path. This and (iv) imply that P^{**} is also C-minimal. □

Remark 4.37 Let C_0 be prefix monotone. Then $(C_0)_{\max} = C_0$. Therefore, we may omit the condition $C_0(P^*) = (C_0)_{\max}(P^*)$ in assumption (i) of Theorem 4.36. □

The next result describes a case in which C itself may be used as decision function.

Theorem 4.38 *Given a simple, locally finite digraph $\mathcal{G} = (\mathcal{V}, \mathcal{R})$, a node $s \in \mathcal{V}$, and a set $\Gamma \subseteq \mathcal{V}$. Given the cost function $C : \mathcal{P}(s) \to (\mathbf{R}, \preceq)$ where $\preceq$ is total and identitive. (E.g., $\mathbf{R}$ may be equal to $\mathbb{R}$, and "$\preceq$" may be equal to "$\leq$".) We assume that C have the following properties:*
   (i) *There exists a C-minimal s-Γ-path P^* with $C(P^*) = C_{\max}(P^*)$.*
   (ii) *There is no upper bound on the sequence $\left(C([s, v_1, \ldots, v_i]) \right)_{i \in \mathbb{N}}$ if $[s, v_1, v_2, v_3, \ldots)$ is an infinite path in $\mathcal{G}$.*
   (iii) *C is order preserving.*
Then $\mathrm{BF}^*(F) = \mathrm{BF}^*(C)$ will output a C-minimal goal path.

P r o o f : Let $C_0 := C$. Then C and C_0 satisfy all conditions of in 4.36 are satisfied; therefore, this result yields the desired assertion. □

The next theorem describes the behaviour of $\mathrm{BF}^*(F)$ if F is order preserving and $F(P) = F_{\max}(P)$ for all solution paths. In this case, we may relax the conditions of Theorem 4.34. In Remark 4.40, we shall compare the assumptions of Theorem 4.39 with the ones of Theorem 4.34.

Theorem 4.39 *Given a simple, locally finite digraph $\mathcal{G} = (\mathcal{V}, \mathcal{R})$, a node $s \in \mathcal{V}$, and a set $\Gamma \subseteq \mathcal{V}$. Given the cost functions $C, C_0 : \mathcal{P}(s) \to (\mathbf{R}, \preceq)$ where $\preceq$ is total and identitive. (E.g., $\mathbf{R}$ may be equal to $\mathbb{R}$, and "$\preceq$" may be equal to "$\leq$".) We assume that the following assertions are true:*

   (i) *There exists an s-Γ-path P^+.*

   (ii) *There is no upper bound on the sequence $\left(C_0([s, v_1, \ldots, v_i]) \right)_{i \in \mathbb{N}}$ if $[s, v_1, v_2, v_3, \ldots)$ is an infinite acyclic path in $\mathcal{G}$.*

   (iii) *C_0 is order preserving.*

   (iv) *The following equations are valid for all s-Γ-paths P:*
$$(\alpha) \;\; C_0(P) = (C_0)_{\max}(P), \qquad (\beta) \;\; C_0(P) = C(P).$$
   (By the way, condition (α) is equivalent to the condition that $C_0(Q) \preceq C_0(P)$ for all s-Γ-paths P and all prefixes $Q \leq P$.)

   (v) *$\mathcal{G}$ has no C_0-negative cycles (see Definition 2.128).*
Then $\mathrm{BF}^*(F)$ with $F := C_0$ will output a C-minimal goal path.

P r o o f : Our assumptions are not strong enough to apply 4.34. We therefore show the current result explicitly. The proof is partially the same as the one of 4.24. We start with the following assertion:

All paths $P(v)$ generated by BF*(F) are acyclic.   (4.14)

Proof of (4.14): This fact follows from (v) and from Lemma 4.23.

Let $\widehat{M} := F(P^+)$. We next prove the following fact:

There are only finitely many acyclic paths $P \in \mathcal{P}(s)$ with $F_{\max}(P) \preceq \widehat{M}$.   (4.15)

Proof of (4.15): Let $H := F_{\max} = (C_0)_{\max}$. Then H is prefix monotone. Moreover, let P be an acyclic infinite path with start node s, and let Q be a finite prefix of P. Then $H(Q) = (C_0)_{\max}(Q) \succeq C_0(Q)$. This and (ii) imply that there is no upper bound on the values $H(Q)$, $Q \leq P$. Fact (4.15) follows from 2.125.

Now we show the following fact:

There exists a $(C_0)_{\max}$-minimal goal path P^*, and this path is
also C_0-minimal and acyclic.   (4.16)

Proof of (4.16): The path function $H = (C_0)_{\max}$ satisfies all conditions of Lemma 2.133; in particular, $\mathcal{G}$ has no H-negative cycles; this follows from (v) and from Lemma 2.131 (with $\widetilde{H} := C_0$ and $\widetilde{\widetilde{H}} := H$). Hence, there exists a $(C_0)_{\max}$-minimal, acyclic goal path P^*. It follows from (iv)(α) that P^* is also C_0-minimal.

Let now $M := C_0(P^*) = F(P^*)$. Then

$$M = C_0(P^*) \overset{(4.16)}{\preceq} C_0(P^+) = F(P^+) = \widehat{M}.$$ (4.17)

Let $P^* = [w_0, w_1, w_2, \ldots, w_l]$ with $w_0 = s$ and $\gamma^* := w_l \in \Gamma$. We define $P_j^* := [w_0, w_1, \ldots, w_j]$ for all $j = 0, \ldots, l$. We now show the following assertion:

Let $t \in \mathbb{N}$, and let BF** execute the tth iteration. Then there exists a
node w of P^* such that $w \in \mathrm{OPEN}_t$ and $F(P_t(w)) \preceq M$.   (4.18)

Proof of (4.18): It is $\gamma^* \notin \mathrm{CLOSED}_t$; the reasons are the same as in the proof of (4.4). So, not all nodes of P^* are elements of CLOSED_t. We choose the minimal number λ such that $w_\lambda \notin \mathrm{CLOSED}_t$. Then $w_\lambda \in \mathrm{OPEN}_t$ by Lemma 4.18 a). Note that BF* uses the expansion function $F^{(1)} := F$ and the redirection function $F^{(2)} := F$. This and Corollary 4.20 imply that $F(P_t(w_\lambda)) = F^{(2)}(P_t(w_\lambda)) \preceq F^{(2)}(P_\lambda^*) = F(P_\lambda^*)$. Moreover, $F(P_\lambda^*) \preceq F_{\max}(P^*) \overset{(iv)(\alpha),\, F = C_0}{=\!=\!=\!=} F(P^*) = M$ so that $F(P_t(w_\lambda)) \preceq F(P_\lambda^*) \preceq M$. Hence, $F(P_t(w)) \preceq M$ for $w := w_\lambda$.

The next assertions say that all paths $P_t(v)$ are weakly M-bounded with respect to F. More precisely, the following are true for all $t \in \mathbb{N}$ and for all vertices v:

If $P_t(v)$ exists then $P_t(v)$ is weakly M-bounded with respect to $F = C_0$.   (4.19)

If v is equal to the node $\bar{v}_t$ selected in step 3 of the tth iteration then
the whole path $P_t(v)$ is M-bounded. (That means that $F(P_t(\bar{v}_t)) \preceq M$ in   (4.20)
addition to (4.19) and that $F_{\max}(P_t(v)) = M$.)

Proof of (4.19) *and* (4.20): We prove both assertions (4.19) and (4.20) simultaneously by an induction on t. For $t = 1$, the only relevant path is $P_1(s) = [s]$, which is indeed M-bounded because $F_{\max}([s]) = F([s] \overset{[s] \leq P^*}{\preceq} F_{\max}(P^*) = (C_0)_{\max}(P^*) \overset{(iv)(\alpha)}{=}$ $C_0(P^*) = M$.

We now assume that the assertions (4.19) and (4.20) are true for all paths $P_t(v)$, and we next prove that fact (4.19) remains true after the tth iteration. Obviously, the only critical situation arises if $P_{t+1}(v)$ is not equal to $P_t(v)$. Then the following must have occurred: Either v and $P_{t+1}(v)$ have been detected in step 6b) or $P_t(v)$ has been replaced by $P_{t+1}(v)$ in step 6c). In both cases, $P_{t+1}(v) = P_t(\overline{v}_t) \oplus (\overline{v}_t, v)$ where $\overline{v}_t$ is the vertex expanded in the tth iteration. But assertion (4.20) of the induction hypothesis says that $P_t(\overline{v}_t)$ is M-bounded. This implies that $P_{t+1}(v)$ is weakly M-bounded with respect to F.

We have proven that fact (4.19) is valid for all paths $P_{t+1}(v)$. We must show that assertion (4.20) for $t + 1$ is true as well; that means, $P_{t+1}(\overline{v}_{t+1})$ is M-bounded where $\overline{v}_{t+1}$ is the node selected in step 3 of the $(t + 1)$st iteration.

The path $P_{t+1}(\overline{v}_{t+1})$ is weakly M-bounded by fact (4.19). It remains to be shown that $F(P_{t+1}(\overline{v}_{t+1})) \preceq M$. For this purpose, we recall (4.18); it says that there exists a node $w \in$ OPEN with $F(P_{t+1}(w)) \preceq M$. The path $P_{t+1}(\overline{v}_{t+1})$ is compared with $P_{t+1}(w)$ in step 3 of the $(t + 1)$st iteration. This implies that $F(P_{t+1}(\overline{v}_t)) \preceq F(P_{t+1}(w)) \preceq M$.

Next, we show the following assertion with the help of fact (4.20):

If BF* outputs an s-Γ-path P^{**} then P^{**} is M-bounded with respect to $F = C_0$.   (4.21)

Proof of (4.21): We assume that P^{**} is output in the Tth iteration where $T \in \mathbb{N}$; let $\gamma^{**} := \omega(P^{**})$. Then $P^{**} = P_T(\gamma^{**})$ is M-bounded by fact (4.20).

Next we show *that* step 3 of BF* will actually select a goal node γ^{**} after finitely many iterations. It follows from (4.15) and (4.17) that there exist only finitely many acyclic paths $P \in \mathcal{P}(s)$ with $F_{\max}(P) \preceq M$; facts (4.14) and (4.20) imply that BF* cannot expand other paths than these finitely many ones. Moreover, assertion (4.10) is also true in the current situation. (Replace $F^{(2)}$ by F in the proof of (4.10).) Hence, BF* will only execute a finite number T of iterations. Moreover, fact (4.18) implies that OPEN$_T \neq \emptyset$ so that BF* can only terminate by selecting a goal node γ^{**}.

We now accomplish the proof of our theorem. Fact (4.21) implies that BF* will output a path P^{**} that is M-bounded with regard to C_0. Consequently, $C_0(P^{**}) \preceq (C_0)_{\max}(P^{**}) \preceq M = C_0(P^*)$; this implies that $C_0(P^{**}) \preceq C_0(P^*)$; moreover, P^* is C_0-minimal by fact (4.16); so, P^{**} is C_0-minimal as well as P^*. Hence, P^{**} is C-minimal by assumption (iv)(β).   □

Next, we discuss the assumptions of Theorem 4.39.

Remark 4.40 The assumptions in 4.39 are not so strong as the ones in 4.34.

In condition (i) of 4.39, we only require the existence of an s-Γ-path; in condition (i) of 4.34, however, we have required the existence of a C-minimal goal path.

Moreover, if the conditions (ii) and (iii) in 4.34 are satisfied then the conditions (ii), (iii), and (v) in Theorem 4.39 are satisfied as well. (Condition (v) can be verified with the help of Lemma 2.130.) The reverse implication, however, is false. To see this we define $C_0(X) := 0$ for all paths X. Let U be a cycle with $\ell(U) > 0$ and $\alpha(U) = \omega(U) = s$. We define the infinite path $P := U \oplus U \oplus U \oplus \cdots$; then $C_0(Q) = 0$ for all finite prefixes $Q \leq P$. This situation is not in conflict with (ii), (iii), and (v) in Theorem 4.39 because P is not acyclic and C_0 is an order preserving path function causing no negative cycles. But this situation is not in accordance to condition (ii) of 4.34 because the values $C_0(Q)$ with $Q \leq P$, $Q \neq P$, are bounded. $\qquad\square$

Remark 4.41 Condition (ii) of 4.39 may be replaced by the assumption that $\mathcal{G}$ is finite. In this case, $\mathcal{G}$ has no infinite node injective paths.

If, however, $\mathcal{G}$ is infinite then condition (ii) must not be omitted. For example, let $\mathcal{G} := (\mathcal{V}, \mathcal{R})$ with
$$\mathcal{V} := \{s, \gamma, x_1, x_2, x_3, \ldots\} \text{ and } \mathcal{R} := \{(s, \gamma), (s, x_1), (x_1, x_2), (x_2, x_3), (x_3, x_4), \ldots\}.$$
Let $\Gamma := \{\gamma\}$. For all paths Q, we define $C(Q) := C_0(Q) := 1$ if $Q = [s, \gamma]$ and $C(Q) := C_0(Q) := 0$ if $Q \neq [s, \gamma]$.
Then all conditions of 4.39 with exception of (ii) are satisfied; in particular, (v) is true because the digraph $\mathcal{G}$ is acyclic.
The algorithm BF*(F) with $F := C_0$ tries to expand all nodes $x_1, x_2, x_3, \ldots$ and never expands γ; the reason for this behaviour is that
$$F(P(x_i)) = F([s, x_1, x_2, \ldots, x_i]) = 0 < 1 = F([s, \gamma]) = F(P(\gamma))$$
for all $i = 1, 2, 3, \ldots$. $\qquad\square$

Remark 4.42 Condition (v) in 4.39 must not be omitted, too. For example, let $\mathcal{G} := (\mathcal{V}, \mathcal{R})$ with $\mathcal{V} := \{s, \gamma, x_1, x_2, x_3\}$ and $\mathcal{R} := \{(s, \gamma), (s, x_1), (x_1, x_2), (x_2, x_3), (x_3, x_1)\}$. Let $\Gamma := \{\gamma\}$. Let $C := C_0 := SUM_h$ where $h(s, \gamma) := 1$, $h(x_3, x_1) := -1$, $h(r) := 0$ for all other arcs. We define the paths $X_1 := [x_1, x_2, x_3, x_1]$, $X_2 := [x_2, x_3, x_1, x_2]$, $X_3 := [x_3, x_1, x_2, x_3]$.

Then all paths X_i are negative cycles with $C_0(X_i) = C(X_i) = -1$, $i = 1, 2, 3, 4$. All conditions of 4.39 up to (v) are satisfied; in particular, (ii) is true because $\mathcal{G}$ contains no infinite node injective paths.
Then Algorithm BF*(F) with $F := C_0$ will never expand the node Γ. Instead of this, the algorithm first expands x_1, x_2 and x_3. In all further iterations, one of the paths $P(x_i)$ is redirected because $\widehat{P}(x_i)$ is of the form $P(x_i) \oplus X_i$ with $F(\widehat{P}(x_i)) = F(P(x_i)) + F(X_i) = F(P(x_i)) - 1 < F(P(x_i))$. Therefore, BF* will expand and redirect the paths $P(x_1)$, $P(x_2)$, $P(x_3)$ again and again, and γ will never selected by step 3. $\qquad\square$

Now, we describe a situation in which BF* outputs a path P^{**} with a bounded C-value; that means that $C(P^{**})$ is not necessarily minimal but $C(P^{**}) \preceq \overline{M}$ for an upper bound $\overline{M}$.

Theorem 4.43 *Given a simple, locally finite digraph $\mathcal{G} = (\mathcal{V}, \mathcal{R})$, a node $s \in \mathcal{V}$, and a set $\Gamma \subseteq \mathcal{V}$. Given the cost functions C, $C_0 : \mathcal{P}(s) \to (\mathbf{R}, \preceq)$ where $\preceq$ is total and* *identitive; let* $C = (C_0)_{\max}$. *(E.g.,* $\mathbf{R}$ *may be equal to* $\mathbb{R}$, *and* "$\preceq$" *may be equal to* "$\leq$".) *Let* $\overline{M} \in \mathbf{R}$. *Let* $F(P) := \max(C_0(P), \overline{M})$ *for all* $P \in \mathcal{P}(s)$. *Moreover, we make the following assumptions:*

(i') *There exists a* C*-minimal solution path* P^*.

(ii') *There is no upper bound on the sequence* $\big(C_0([s, v_1, \ldots, v_i])\big)_{i \in \mathbb{N}}$ *if* $[s, v_1, v_2, v_3, \ldots)$ *is an infinite path.*

(iii') *The decision function* F *is order preserving.*

Then exactly one of the following assertions is true:

(a) $BF^*(F)$ *outputs a goal path* P^{**} *with* $C(P^{**}) \preceq \overline{M}$.

(b) *There exists no goal path* P *with* $C(P) = (C_0)_{\max}(P) \preceq \overline{M}$, *and* $BF^*(F)$ *outputs an* s-Γ-path P^{**} *with* $C(P^{**}) \succ \overline{M}$.

P r o o f : Let $\widehat{C} := F_{\max}$ and $\widehat{C}_0 := F$. We show that $\widehat{C}$ and $\widehat{C}_0$ satisfy the conditions (i) and (ii) in 4.34.

In order to show condition (i) we prove that that the C-minimal goal path P^* in (i') is also $\widehat{C}$-minimal. Let $Q_{\max}$ be a prefix of P^* with $C_0(Q_{\max}) = (C_0)_{\max}(P^*) = C(P^*)$. We distinguish between two cases:

Case 1: $C_0(Q_{\max}) \succ \overline{M}$. Then the condition of this case implies that

$$C(P^*) = (C_0)_{\max}(P^*) = C_0(Q_{\max}) = F(Q_{\max}) \tag{4.22}$$

Moreover, $C_0(Q_{\max}) \succeq C_0(Q)$ for all $Q \leq P^*$; this and $C_0(Q_{\max}) \succeq \overline{M}$ imply that

$$(\forall Q \leq P^*) \ C_0(Q_{\max}) \succeq \max(C_0(Q), \overline{M}) = F(Q). \tag{4.23}$$

Consequently, $F_{\max}(P^*) \succeq F(Q_{\max}) \overset{(4.22)}{=} C_0(Q_{\max}) \overset{(4.23)}{\succeq} F_{\max}(P^*)$ so that $F_{\max}(P^*) = C_0(Q_{\max}) \overset{(4.22)}{=} (C_0)_{\max}(P^*)$. It follows that $\widehat{C}(P^*) = F_{\max}(P^*) = (C_0)_{\max}(P^*) = C(P^*)$ so that $\widehat{C}(P^*) = C(P^*)$

Let now P be an s-Γ-path. Then the following inequality is valid where (∗) follows from $\widehat{C}(P^*) = C(P^*)$:

$$\widehat{C}(P) = F_{\max}(P) \succeq (C_0)_{\max}(P) = C(P) \overset{(i')}{\succeq} C(P^*) \overset{(*)}{=} \widehat{C}(P^*).$$

Hence, $\widehat{C}(P^*)$ is minimal among all values $\widehat{C}(P)$ if $C_0(Q_{\max}) \succ \overline{M}$.

Case 2: $C_0(Q_{\max}) \preceq \overline{M}$. Then $C_0(Q) \preceq C_0(Q_{\max}) \preceq \overline{M}$ for all $Q \leq P^*$ so that $F(Q) = \overline{M}$ for all prefixes Q. Hence, $\widehat{C}(P^*) = F_{\max}(P^*) = \overline{M}$. If P is an arbitrary goal path then $\widehat{C}(P) = F_{\max}(P) \succeq F(P) \succeq \overline{M} = F_{\max}(P^*) = \widehat{C}(P)$ so that $F_{\max}(P^*)$ is again minimal.

Now we have shown that $\widehat{C}$ satisfies condition (i) of 4.34.

The path function $\widehat{C}_0 = F$ satisfies condition (ii) of 4.34. This follows from (ii') and from the fact that $\widehat{C}_0(P) = F(P) \succeq C_0(P)$ for all paths P.

Moreover, $\widehat{C}_0 = F$ is order preserving by (iii').

Consequently, $\widehat{C}$ and $\widehat{C}_0$ satisfy all conditions of Theorem 4.34. This result implies that $BF^*(F)$ outputs a goal path P^{**} such that

$$\widehat{C}(P^{**}) = F_{\max}(P^{**}) \preceq F_{\max}(P) = \widehat{C}(P) \text{ for all } s\text{-}\Gamma\text{-paths } P. \tag{4.24}$$

Next, we show that exactly one of the assertions (a), (b) is true. For this purpose, we prove the following claims:

$$\text{If } \widehat{C}(P^{**}) = F_{\max}(P^{**}) = \overline{M} \text{ then } C(P^{**}) = (C_0)_{\max}(P^{**}) \preceq \overline{M}. \qquad (4.25)$$

$$\text{If } \widehat{C}(P^{**}) = F_{\max}(P^{**}) \succ \overline{M} \text{ then there is no } s\text{-}\Gamma\text{-path } P \text{ with } C(P) = \qquad (4.26)$$
$$(C_0)_{\max}(P) \preceq \overline{M}.$$

Proof of (4.25): Note that $C_0(Q) \preceq F(Q)$ for all $Q \in \mathcal{P}(s)$. Hence, $C(P^{**}) = (C_0)_{\max}(P^{**}) \preceq (F)_{\max}(P^{**}) = \overline{M}$.

Proof of (4.26): We assume that

$$\widehat{C}(P^{**}) = F_{\max}(P^{**}) \succ \overline{M}. \qquad (4.27)$$

If (4.26) were false then there existed a solution path P with $C(P) = (C_0)_{\max}(P) \preceq \overline{M}$. This fact would imply $(*)$ of the following equation:

$$F_{\max}(P) = \max\{F(Q) \mid Q \leq P\} = \max\{\max(C_0(Q), \overline{M}) \mid Q \leq P\} =$$
$$\max\left(\{C_0(Q) \mid Q \leq P\} \cup \{\overline{M}\}\right) = \max\left((C_0)_{\max}(P), \overline{M}\right) \overset{(*)}{=} \overline{M}.$$

Consequently, $F_{\max}(P) = \overline{M} \overset{(4.27)}{\prec} F_{\max}(P^{**})$; that is a contradiction to the minimality of $F_{\max}(P^{**})$.

(4.25) and (4.26) imply that at least one of the assertions (a), (b) is correct. Moreover, (a) and (b) are not true at the same time. Hence, exactly one of the assertions (a), (b) is correct.   ☐

We now reformulate the order preservation of the redirection function in 4.43.

Remark 4.44 Given the same path functions C, C_0 and F as in 4.43. We have required F to be order preserving. That means that

$$(\forall P_1, P_2, Q) \; \left(F(P_1) \preceq F(P_2) \;\Rightarrow\; F(P_1 \oplus Q) \preceq F(P_2 \oplus Q)\right). \qquad (4.28)$$

Now we give equivalent formulations of this condition.

The relation $F = \max(C_0, \overline{M})$ implies that the following assertion is equivalent to (4.28):

$$(\forall P_1, P_2, Q) \left(\begin{array}{l} \max(C_0(P_1), \overline{M}) \preceq \max(C_0(P_2), \overline{M}) \;\Rightarrow \\ \max(C_0(P_1 \oplus Q), \overline{M}) \preceq \max(C_0(P_2 \oplus Q), \overline{M}). \end{array} \right) \qquad (4.29)$$

To reformulate this condition we observe the following equivalence, which is valid for all elements a_1, a_2, $x \in \mathbf{R}$:

$$\max(a_1, x) \preceq \max(a_2, x) \iff \max(\max(a_1, x), \max(a_2, x)) = \max(a_2, x) \iff$$
$$\max(a_1, a_2, x) = \max(a_2, x) \iff (a_1 \preceq a_2) \lor (a_1 \preceq x).$$

Consequently, condition (4.29) is equivalent to the following assertion:

$$(\forall P_1, P_2, Q) \left(\begin{array}{l} C_0(P_1) \preceq C_0(P_2) \;\lor\; C_0(P_1) \preceq \overline{M} \;\Longrightarrow \\ C_0(P_1 \oplus Q) \preceq C_0(P_2 \oplus Q) \;\lor\; C_0(P_1 \oplus Q) \preceq \overline{M} \end{array} \right). \qquad (4.30)$$

It remains to the reader to analyse this condition carefully.   ☐

The following question arises from 4.43: Replace $F(P) = \max(C_0(P), \overline{M})$ by $F(P) = \max(C_0(P), \overline{C}_0(P))$ where $\overline{C}_0(P))$ is path function; what are the properties of the path P^{**} output by $BF^*(F)$?

4.2.2.1.2 Properties of the paths $Q(v)$ generated by BF*

Here, we mainly investigate the properties of the paths $Q(v)$ generated by BF*. Many of our results will be applied later.

The following theorem gives several relationships between the paths $P(v)$ and $Q(v)$. Recall that $P(v)$ exists if and only if v has already been detected by BF* and that $Q(v)$ exists if and only if s can be reached by successively applying $pred$ to v.

Lemma 4.45 *Given a simple, locally finite digraph* $\mathcal{G} = (\mathcal{V}, \mathcal{R})$, *a start node* $s \in \mathcal{V}$, *and a set* $\Gamma \subseteq \mathcal{V}$. *We assume that BF* uses the order preserving decision function* $F : \mathcal{P}(s) \to (\mathbf{R}, \preceq)$ *where* $\preceq$ *is total and identitive. (E.g.,* $\mathbf{R}$ *may be equal to* $\mathbb{R}$, *and* "$\preceq$" *may be equal to* "$\leq$".)
   a) *For each node* y *and its current predecessor* $x = pred(y)$, *the following is true:*
      $F(P(x) \oplus (x, y)) \preceq F(P(y))$.
   b) *Given a path* $Q := [u_0, \ldots, u_k = v]$ *with the following property:* $u_{\kappa-1} = pred(u_\kappa)$ *for all* $\kappa = k, k - 1, \ldots, 1$. *(It may be that* $u_i = u_j$ *for some* $i \neq j$.)
      Then $F(P(u_0) \oplus Q)) \preceq F(P(v))$.
   c) *As long as* $P(s) = [s]$, *the following assertion is correct:*
      $F(Q(v)) \preceq F(P(v))$ *for all nodes for which the paths* $P(v)$ *and* $Q(v)$ *exist.*
   d) *If, in addition,* F *is* $\prec$-*preserving and does not cause negative cycles then the path* $Q(v)$ *exists for each node* v *that has already been detected by BF*.*

P r o o f   o f   a) : We shall show the claim by an induction on the number t of iterations. The assertion is trivial for $t = 1$ because BF* only knows the node s and the path $P_1(s) = [s]$. We now assume that the assertion is true for the paths $P_t(x)$ and $P_t(y)$, and we consider the tth iteration.

If the node y is newly detected then $pred(y) = x$ where x is the node expanded in the tth iteration; then the following are true:

$$(\alpha) \quad P_{t+1}(y) = P_t(x) \oplus (x, y), \quad (\beta) \quad F(P_{t+1}(x)) \preceq F(P_t(x)). \tag{4.31}$$

Then the order preservation of F implies that

$$F(P_{t+1}(x) \oplus (x,y)) \overset{(4.31)(\beta)}{\preceq} F(P_t(x) \oplus (x,y)) \overset{(4.31)(\alpha)}{=} F(P_{t+1}(y)).$$

If y has been detected earlier and $P_t(y)$ is redirected then we can again use the statements $(4.31)(\alpha), (\beta)$ to prove that $F(P_{t+1}(x) \oplus (x,y)) \preceq F(P_{t+1}(y))$.

The last case is that y has been detected earlier and $P_{t+1}(y) = P_t(y)$. Then fact $(4.31)(\beta)$ is true. This and the order preservation of F yield relation $(*)$ of the following assertion; relation $(**)$ follows from the induction hypothesis.

$$F(P_{t+1}(x) \oplus (x,y)) \overset{(*)}{\preceq} F(P_t(x) \oplus (x,y)) \overset{(**)}{\preceq} F(P_t(y)) = F(P_{t+1}(y)).$$

P r o o f   o f   b) : Let $Q_\kappa := [u_0, \ldots, u_\kappa]$ for all $\kappa = 0, \ldots, k$. We show the following fact by an induction on κ:

$$(\forall\,\kappa = 0,1,\ldots,k)\;\; F(P(u_0)\oplus Q_\kappa)) \preceq F(P(u_\kappa)). \tag{4.32}$$

This assertion is trivial for $\kappa = 0$. Let (4.32) be true for $\kappa < k$. Then the order preservation of F implies that $F(P(u_0)\oplus Q_{\kappa+1}) \preceq F(P(u_\kappa)\oplus(u_\kappa,u_{\kappa+1}))$. Applying part a) to $x := u_\kappa$ and $y := u_{\kappa+1}$, we obtain $F(P(u_\kappa)\oplus(u_\kappa,u_{\kappa+1})) \preceq F(P(u_{\kappa+1}))$. Consequently, $F(P(u_0)\oplus Q_{\kappa+1}) \preceq F(P(u_\kappa)\oplus(u_\kappa,u_{\kappa+1})) \preceq F(P(u_{\kappa+1}))$. Assertion b) is equivalent to fact (4.32) for $\kappa := k$.

P r o o f   o f   c) : This assertion follows from part b). Let $u_0 := s$. Then $P(u_0) = [s]$, and the path Q in part b) is equal to $Q(v)$; in particular, $\alpha(Q) = \alpha(Q(v)) = s = u_0$. Hence, $F(Q(v)) = F(P(u_0)\oplus Q) \overset{\text{b)}}{\preceq} F(P(v))$.

P r o o f   o f   d) : Let F be $\prec$-preserving, and let F cause no negative cycles. We show that the following:

> If $Q = [u_0,\ldots,u_k]$ is a path as described in part b) then Q is node injective. $\qquad$ (4.33)

Otherwise, we consider the earliest iteration t in which BF* generates a path Q with a substantial cycle; we show that then $\mathcal{G}$ contains an F-negative cycle. Let $\bar{v}_t = u_j$ be the node expanded in the tth iteration[13]. Then the cycle in Q is generated by defining $pred(u_i) := \bar{v}_t = u_j$ where $i < j$. That means that $P_t(u_i)$ has been replaced by $\hat{P}_t(u_i) = P_t(\bar{v}_t)\oplus(\bar{v}_t,u_i) = P_t(u_j)\oplus(u_j,u_i)$. This is only possible if

$$F(\,P_t(u_j)\oplus(u_j,u_i)\,) = F(\hat{P}_t(u_i)) \prec F(P_t(u_i)). \tag{4.34}$$

Let $Q' := [u_i,u_{i+1},\ldots,u_j]$. Then relationship $(*)$ of inequality (4.35) follows from (4.34) and the $\prec$-preservation of F. Relationship $(**)$ follows by applying part b) to $P_t(u_i)$ and Q'.

$$F(\,P_t(u_j)\oplus(u_j,u_i)\oplus Q'\,) \overset{(*)}{\prec} F(P_t(u_i)\oplus Q') \overset{(**)}{\preceq} F(P_t(u_j)). \tag{4.35}$$

That means that $(u_j,u_i)\oplus Q'$ is an F-negative cycle, which is a contradiction to our assumption.

So, fact (4.33) is true. Consequently, all nodes v, $pred(v)$, $pred(pred(v))$, etc. are pairwise distinct. Moreover, the sequence of these nodes cannot be infinite because BF* has defined $pred(x)$ only for finitely many nodes x. Consequently, the repeated iteration of $pred$ must end by reaching a node x without $pred(x)$. This is only possible if $x = s$. Hence, the sequence of nodes v, $pred(v)$, $pred(pred(v))$,$\ldots$ will reach the node s after finitely many steps. That means that $Q(v)$ exists. $\qquad\square$

Corollary 4.46 *Let $\mathcal{G} = (V,\mathcal{R})$ be a simple, locally finite digraph. Let $s,\gamma \in V$ and $\Gamma = \{\gamma\}$. Let $t^* \in \mathbb{N}$. We assume that BF* uses the order preserving decision function $F : \mathcal{P}(s) \to (\mathbf{R},\preceq)$ where $\preceq$ is total and identitive. (E.g., $\mathbf{R}$ may be equal to $\mathbb{R}$, and "$\preceq$" may be equal to "$\leq$".) Moreover, we make the following assumptions:*

> *(i) There exists an $F_{\max}$-minimal s-γ-path P^*.*
> *(ii) The sequence $\big(F([s,v_1,v_2,\ldots v_i])\big)_{i\in\mathbb{N}}$ has no upper bound if $[s,v_1,v_2,v_3,\ldots)$ is an infinite path $\mathcal{G}$ in $\mathcal{G}$.*
> *(iii) The node γ is selected in step 3 of the t^*th iteration*

---

[13]We may assume that $\bar{v}_t$ is a node of Q because a cycle of Q can only be generated by defining $x := pred(y)$ where $x,y \in Q$.

(iv) $P_{t^*}(s) = [s]$.

(v) *The path $Q_{t^*}(\gamma)$ exists. (This can often be shown with 4.45 d).)*

Then $Q_{t^*}(\gamma)$ is an $F_{\max}$-minimal s-γ-path.

P r o o f : Define $M := F_{\max}(P^*)$. We show the following property of the path $Q_{t^*}(\gamma)$, which exists by assumption (v):

$$F(Q_{t^*}(u)) \preceq M \text{ for all nodes } u \text{ of } Q_{t^*}(\gamma). \tag{4.36}$$

For this purpose, we define the path functions C, C_0, $C^{(1)}$, $C^{(2)}$, $F^{(1)}$ and $F^{(2)}$ as follows: Let $C(P) := F_{\max}(P)$, $C_0(P) := F(P)$, $C^{(1)}(P) := F(P)$, $C^{(2)}(P) := M$, $F^{(1)}(P) := F(P)$, $F^{(2)}(P) := F(P)$ for all paths $P \in \mathcal{P}(s)$. Then (i), (ii) and the order preservation of F imply that C and C_0 satisfy all conditions of Theorem 4.34. Following the proof of 4.34, we see that $C^{(1)}, C^{(2)}, F^{(1)}$, and $F^{(2)}$ satisfy all conditions A.1), A.2), B), C), D) of Theorem 4.24. In particular, fact (4.4) is true. Consequently, if $t \leq t^*$ then there exists a node $w \in \mathrm{OPEN}_t$ with $F(P_t(w)) = C^{(1)}(P_t(w)) \preceq C^{(2)}(P_t(w)) = M$. This implies that each iteration $t \leq t^*$ expands a node $\overline{v}_t$ with

$$F(P_t(\overline{v}_t)) \preceq M. \tag{4.37}$$

If $u \in Q_{t^*}(\gamma)$ then u must have been expanded at a time $t' \leq t^*$ so that $u = \overline{v}_{t'}$. Consequently,

$$F(Q_{t^*}(u)) \overset{\text{(iv), 4.45 c)}}{\preceq} F(P_{t^*}(u)) \preceq F(P_{t'}(u)) = F(P_{t'}(\overline{v}_{t'})) \overset{(4.37)}{\preceq} M.$$

This proves fact (4.36).

The set of all prefixes of $Q_{t^*}(\gamma)$ is the same as the set $\{Q_{t^*}(u) \mid u \in Q_{t^*}(\gamma)\}$. This implies that $F_{\max}(Q_{t^*}(\gamma)) = \max \{F(Q_{t^*}(u)) \mid u \in Q_{t^*}(\gamma)\} \overset{(4.36)}{\preceq} M$. Consequently, $F_{\max}(Q_{t^*}(\gamma)) = M$ because M is the minimal $F_{\max}$-value of all s-γ-paths. □

4.2.2.1.3   Replacing a given path function by an order preserving one

We consider the following problem: Given two cost functions C, C_0 with $C = (C_0)_{\max}$; find a C-optimal goal path if C_0 is not order preserving.

In the next result, we solve this problem with the help of Theorems 2.37 and 4.34.

Theorem 4.47 *Given a simple, finite digraph $\mathcal{G} = (\mathcal{V}, \mathcal{R})$, a node $s \in \mathcal{V}$ and a set $\Gamma \subseteq \mathcal{V}$ of goal nodes; we assume that each node v of $\mathcal{G}$ can be reached by an s-v-path. Given two cost functions $C, C_0 : \mathcal{P}(s) \to (\mathbb{R}, \leq)$ with $C = (C_0)_{\max}$; let $H := C_0$. We assume that H satisfies all conditions of Theorem 2.37. We construct the graph $\mathcal{G}' = (\mathcal{V}', \mathcal{R}')$, the node s' and the graph homomorphism*

$$\Psi : (\mathcal{V}' \cup \mathcal{R}' \cup \mathcal{P}(s')) \to (\mathcal{V} \cup \mathcal{R} \cup \mathcal{P}(s))$$

in the same way as described in 2.37. Let $\Gamma' := \{x \in \mathcal{V}' \mid \Psi(x) \in \Gamma\}$. We define the path functions $C_0' := C_0 \circ \Psi$ and $H' := C_0'$. (Then $H' = H \circ \Psi$ because $H = C_0$.)

We make the following assumptions:

(i) *There exists a $(C_0')_{\max}$-minimal goal path in $\mathcal{G}'$.*

(ii) *There is no upper bound on the sequence $(C_0'([s', v_1', \ldots, v_i']))_{i \in \mathbb{N}}$ if $[s', v_1', v_2', v_3' \ldots)$ is an infinite path in $\mathcal{G}'$.*

(iii) $H' = C'_0$ is order preserving.[14]

Then the following algorithm will output a C-minimal goal path P^{**} in $\mathcal{G}$:

1) Apply BF* to $\mathcal{G}'$ using the decision function $F' := C'_0 = H'$.
2) Construct $P^{**} := \Psi(P')$ where P' is the path found by
   $$\text{BF*}(F') = \text{BF*}(C'_0) = \text{BF*}(H').$$

P r o o f : Let P be a path in $\mathcal{G}'$ with start node s'; then define $C' := F'_{\max} = (C'_0)_{\max} = H'_{\max}$. Then C' and C'_0 satisfy all conditions of Theorem 4.34. Hence, BF*(F') will find an s'-Γ'-path P' for which $(C'_0)_{\max}(P) = H'_{\max}(P')$ is minimal. It follows from the last part of Remark 2.38 that $P^{**} := \Psi(P')$ is a $H_{\max}$-minimal s-Γ-path in $\mathcal{G}$; this and the relationship $C = (C_0)_{\max} = H_{\max}$ imply that P^{**} is C-minimal.   □

Next, we consider a similar problem: Let $C = (C_0)_{\max}$ where $H := C_0$ satisfies the conditions in 2.39. Find a goal path P^{**} with minimal value $C(P^{**}) = (C_0)_{\max}(P^{**}) = H_{\max}(P^{**})$.

This problem is much more difficult than the one in 4.47. It cannot be solved with algorithm $\mathcal{A}$ in 2.41 even if $H = C_0$ is a multiplicative path function.

In this special case, however, one can use another graph to find a $(C_0)_{\max}$-minimal s-Γ-path. This is shown in the following example.

Example 4.48 Let $\mathcal{G} = (\mathcal{V}, \mathcal{R})$ be a simple and finite digraph, let $s \in \mathcal{V}$, and let $\Gamma \subseteq \mathcal{V}$. Let $C, C_0 : \mathcal{P}(s) \to \mathbb{R}$; we assume that $C = (C_0)_{\max}$ and that C_0 is multiplicative; moreover, we assume that the underlying arc function h has the property that $h(r) \neq 0$ for all arcs r.

Let $H := C_0$ then $C = H_{\max}$. We construct the graph $\widetilde{\mathcal{G}}'$ and order preserving the path function $\widetilde{H}'$ in the same way as in 2.43. Moreover, we define $\widetilde{\Gamma}' := \{(\mu, v, i) \mid \mu = 1, 2, v \in \Gamma, i \in \{\pm 1\}\}$.

It is easy to see that the following is true for all paths $P \in \mathcal{P}(s)$:
$$\text{If } X \leq P \text{ and } H(X) = H_{\max}(P) \text{ then } H(X) > 0. \tag{4.38}$$

This follows from the fact that $H(X) \geq H([s]) = 1$.

The analogic assertion is true for all paths $\widetilde{P}' \in \mathcal{P}(\widetilde{s}')$:
$$\text{If } \widetilde{Y}' \leq \widetilde{P}' \text{ and } \widetilde{H}'(\widetilde{Y}') = \widetilde{H}'_{\max}(\widetilde{P}') \text{ then } \widetilde{H}'(\widetilde{Y}') > 0. \tag{4.39}$$

This follows from the fact that $\widetilde{H}'(\widetilde{Y}') \geq \widetilde{H}'([\widetilde{s}']) = 1$.

Let $\widetilde{P}'$ be a path in $\widetilde{\mathcal{G}}'$ with start node $\widetilde{s}'$. Then we show the following equation, which is similar to Theorem 2.18 b):
$$\widetilde{H}'_{\max}(\widetilde{P}') = H_{\max}(\Psi(\Lambda(\widetilde{P}'))). \tag{4.40}$$

We cannot show (4.40) with the help of Theorem 2.18 b) beause not $\widetilde{H}' = H \circ \Psi \circ \Lambda$. Therefore, we must prove (4.40) explicitly. For this end, we define $P := \Psi(\Lambda(\widetilde{P}'))$.

---

[14] A sufficient condition for this is given in 2.37 c).

Let X be a prefix of P with $H_{\max}(P) = H(X)$. Moreover, let $\widetilde{X'}$ be the prefix $\widetilde{P'}$ with $\Psi(\Lambda(\widetilde{X'})) = X$. Then $H(X) = H_{\max}(P) \overset{(4.38)}{>} 0$; consequently, $\widetilde{H'}(\widetilde{X'}) \overset{(2.26)}{=} H(X)$,

and $\widetilde{H}'_{\max}(\widetilde{P'}) \overset{(\widetilde{X'} \preceq \widetilde{P'})}{\geq} \widetilde{H'}(\widetilde{X'}) = H(X) = H_{\max}(P)$. So, $\widetilde{H}'_{\max}(\widetilde{P'}) \geq H_{\max}(P)$.

Let $\widetilde{Y'}$ be a prefix of $\widetilde{P'}$ with $\widetilde{H}'_{\max}(\widetilde{P'}) = \widetilde{H'}(\widetilde{Y'})$. Moreover, let $Y := \Psi(\Lambda(\widetilde{Y'}))$; then Y is a prefix of P. It is $\widetilde{H'}(\widetilde{Y'}) = \widetilde{H}'_{\max}(\widetilde{P'}) \overset{(4.39)}{>} 0$; consequently, $\widetilde{H'}(\widetilde{Y'}) \overset{(2.26)}{=} H(Y)$,

and $\widetilde{H}'_{\max}(\widetilde{P'}) = \widetilde{H'}(\widetilde{Y'}) = H(Y) \overset{(Y \preceq P)}{\leq} H_{\max}(P)$. So, $\widetilde{H}'_{\max}(\widetilde{P'}) \leq H_{\max}(P)$.

We have seen that $\widetilde{H}'_{\max}(\widetilde{P'}) \geq H_{\max}(P)$ and that $\widetilde{H}'_{\max}(\widetilde{P'}) \leq H_{\max}(P)$; these inequalities imply fact (4.40).

Assertion (4.40) implies that the following algorithm will find a $H_{\max}$-minimal s-Γ-path P^{**} in $\mathcal{G}$:

1) Apply BF* to the digraph $\widetilde{\mathcal{G}}'$ with the set $\widetilde{\Gamma}'$ of goal nodes; use the order preserving decision function $\widetilde{F'} := \widetilde{H'}$.

2) Let $\widetilde{P'}$ be the path found by BF*. Then let $P^{**} := \Psi(\Lambda(\widetilde{P'}))$. $\qquad\qquad$ □

Remark 4.49 The search strategy in Example 4.48 has been generalized in Theorem 4.18 – Remark 4.22 of [Huck92]; the algorithm in Remark 4.19 of [Huck92] uses the graph $\widetilde{\mathcal{G}}'$ to find $(C_0)_{\max}$-minimal paths for a large class of path functions C_0 in $\mathcal{G}$; this class contains all multiplicative path functions described in Example 4.48 of this book. $\qquad\qquad$ □

4.2.2.1.4 Comparing BF*(F_1) with BF*(F_2) if F_1, F_2 are two decision functions

Our next result is Theorem 4.50; it says that BF*(F_2) does not expand more nodes than BF*(F_1) if the decision functions F_1 and F_2 satisfy particular conditions. Theorem 4.50 is a generalization of Result 6 in [Nils82], which will be formulated as Theorem 4.64 in this book. A discussion about Theorem 4.50 can be found in Remark 4.67.

In the following result, we formulate several conditions about two decision functions F_1, F_2. The following example makes these conditions more comprehensive: Let C be a real-valued path function, and let p_1, p_2 be two real-valued node functions. We assume that $p_1(v) < p_2(v)$ if $v \notin \Gamma$ and $p_1(v) = p_2(v)$ if $v \in \Gamma$. Then the path functions $F_i(P) := C(P) + p_i(\omega(P))$ satisfy all conditions in 4.50.

Theorem 4.50 *Given a simple, locally finite digraph $\mathcal{G} = (\mathcal{V}, \mathcal{R})$, a node $s \in \mathcal{V}$, and a set $\Gamma \subseteq \mathcal{V} \backslash \{s\}$. Let $C : \mathcal{P}(s) \to (\mathbf{R}, \preceq)$ be a cost function where $\preceq$ is total and injective. (E.g., $\mathbf{R}$ may be equal to $\mathbb{R}$, and "$\preceq$" may be equal to "$\leq$".) Let $F_1, F_2 : \mathcal{P}(s) \to (\mathbf{R}, \preceq)$ be two further path functions, which are used as decision functions of BF*. We assume that the following assertions are true:*

(i) F_1 *and* F_2 *are order preserving.*

(ii) *If $X \in \mathcal{P}(s)$ is not a goal path then $F_1(X) \prec F_2(X)$.*

(iii) *If $X \in \mathcal{P}(s)$ is a goal path then $F_1(X) = F_2(X) = C(X)$ and $F_2(Y) \preceq F_2(X)$ for all prefixes $Y \leq X$.*

(iv) *If $X, Y \in \mathcal{P}(s)$ with $\omega(X) = \omega(Y)$ then $F_1(X) \preceq F_1(Y) \Rightarrow F_2(X) \preceq F_2(Y)$.*

Moreover, we assume that each algorithm $BF^*(F_i)$, $i = 1, 2$, is successful and outputs a goal path P_i^* with minimal $(F_i)_{\max}$-value.

Then the following assertions are true:

a) *P_1^* and P_2^* are C-minimal goal paths.*

b) *If a path $P \in \mathcal{P}(s)$ with $\omega(P) \notin \Gamma$ is expanded by $BF^*(F_2)$ then $BF^*(F_1)$ will expand a path Q with the properties $\omega(Q) = \omega(P)$ and $F_2(Q) \preceq F_2(P)$.*
   (This implies that $BF^*(F_2)$ expands a subset of the nodes expanded by $BF^*(F_1)$.)

P r o o f   o f   a) : The conditions (iii) and (ii) imply that $F_1(Y) \preceq F_2(Y)$ for all paths $Y \in \mathcal{P}(s)$, independent of whether or not Y is a goal path. This fact and condition (iii) yield the following inequalities for all goal paths X and all prefixes $Y \leq X$:

$$F_1(Y) \preceq F_2(Y) \preceq C(X) = F_1(X), \quad F_2(Y) \preceq C(X) = F_2(X).$$

Consequently, $(F_1)_{\max}(X) = C(X) = (F_2)_{\max}(X)$ for all goal paths X. This and the $(F_i)_{\max}$-minimality of P_i^* imply that P_i^* is indeed C-minimal for all $i = 1, 2$.

We define $M := C(P_1^*) = C(P_2^*)$.

P r o o f   o f   b) : Our proof resembles the one of [Nils82], Result 6.

The assertion is true for $P = [s]$ because $BF^*(F_1)$ expands $Q := [s]$.

We assume that $P \neq [s]$ is the shortest counterexample to our assertions. More precisely, let P be the shortest path with the following properties:

$$\omega(P) \notin \Gamma, \ P \text{ is expanded by } BF^*(F_2), \text{ but no path } Q \text{ with} \atop \omega(Q) = \omega(P) \text{ and } F_2(Q) \preceq F_2(P) \text{ is expanded by } BF^*(F_1). \tag{4.41}$$

Then P has at least one arc because $P \neq [s]$. Let v be the last node of P, let (v', v) be the last arc of P, and let P' be the prefix of P with $P = P' \oplus (v', v)$.

We have assumed that P is expanded by $BF^*(F_2)$. This is only possible if $BF^*(F_2)$ has generated this path by expanding P' earlier than P. Hence, $v' = \omega(P') \notin \Gamma$ because otherwise $BF^*(F_2)$ would have terminated before expanding P.

Since $\ell(P') < \ell(P)$ and $\omega(P') \notin \Gamma$, assertion b) is true for P'. That means that $BF^*(F_1)$ expands a path $Q' \in \mathcal{P}(s, v')$ with

$$F_2(Q') \preceq F_2(P'). \tag{4.42}$$

Let t be the iteration in which $BF^*(F_1)$ expands Q'. $BF^*(F_1)$ has detected v in this iteration or in an earlier one. Hence, the path $P_{t+1}(v)$ exists. Let $t' \leq t$ be the iteration in which $BF^*(F_1)$ has generated the path $P_{t+1}(v)$ ($= P_{t'+1}(v)$). Then v has been moved to OPEN in the t'th iteration. We show that (4.41) implies the following assertion:

v remains in OPEN for all the time after the t'th iteration of $BF^*(F_1)$.   (4.43)

To see this we assume that $t'' \geq t'$ (for example, $t'' = t'$). Let $Q := P_{t''+1}(v)$. (Note that Q' is not necessarily a prefix of Q.) Applying (ii) to $v \notin \Gamma$, we see that Q has the following property:

$$F_1(Q) = F_1(P_{t''+1}(v)) \overset{t'' \geq t'}{\underset{\sim}{\succeq}} F_1(P_{t'+1}(v)) = F_1(P_{t+1}(v)) \overset{Q' = P_{t+1}(v')}{\underset{\sim}{\preceq}} \qquad (4.44)$$

$$F_1(Q' \oplus (v',v)) \overset{(ii)}{\prec} F_2(Q' \oplus (v',v)) \overset{(i),(4.42)}{\underset{\sim}{\preceq}} F_2(P).$$

Fact (4.44) implies that $F_1(Q) \preceq F_1(Q' \oplus (v,v'))$; this and (iv) imply that $F_2(Q) \preceq F_2(Q' \oplus (v,v'))$; consequently, $F_2(Q) \preceq F_2(Q' \oplus (v',v)) \overset{(4.44)}{\underset{\sim}{\preceq}} F_2(P)$.
Then (4.41) implies that $Q = P_{t''+1}(v)$ is never expanded by $\mathrm{BF}^*(F_1)$. Hence, v is not moved from OPEN to CLOSED in the t''th iteration, and this is true for all $t'' \geq t'$. This proves fact (4.43).

It follows from (4.43) that v is still in OPEN when $\mathrm{BF}^*(F_1)$ outputs the solution path P_1^*. Let t^* be the number of this iteration and let $Q^* := P_{t^*}(v)$. Then the following statement is true:

$$F_1(Q^*) = F_1(P_{t^*}(v)) \succeq F_1(P_1^*) \overset{(iii)}{=} C(P_1^*) = M. \qquad (4.45)$$

Moreover, $t^* > t$; this follows from the assumptions that $\mathrm{BF}^*(F_1)$ expands Q' with $\omega(Q') = v' \notin \Gamma$ in the tth iteration and that $\mathrm{BF}^*(F_1)$ outputs the goal path P_1^* in the t^*th iteration. Hence, $t' \leq t < t^*$ so that $t'' := t^* - 1 \geq t'$. So, we may apply fact (4.44) to this special t'' and obtain

$$F_1(Q^*) = F_1(P_{t^*}(v)) = F_1(P_{t''+1}(v)) \overset{(4.44)}{\prec} F_2(P). \qquad (4.46)$$

We consider $\mathrm{BF}(F_2)$. When expanding P there exists a node $w \in$ OPEN with $F_2(P(w)) \preceq F_2(P_2^*) = M$. (This follows by applying fact (4.4) in the proof of 4.24 to the following path functions: $C^{(1)}(X) := F_2(X)$, $C^{(2)}(X) := F_2(P_2^*) = M$, $F^{(1)}(X) := F^{(2)}(X) := F_2(X)$ $(X \in \mathcal{P}(s))$.) This and the fact that $\mathrm{BF}^*(F_2)$ expands P imply that $F_2(P) \preceq F_2(P(w)) \preceq F_2(P_2^*) = M \overset{(4.45)}{\preceq} F_1(Q^*)$ so that $F_2(P) \preceq F_1(Q^*)$; that is a contradiction to fact (4.46), which says that $F_1(Q^*) \prec F_2(P)$. So, no counterexample P to our assertion exists, and our result is correct. $\qquad \square$

4.2.2.1.5 BF* with order preserving and prefix monotone decision functions

We study the case that the cost function C is order preserving and prefix monotone; the simplest example is an additive function $C = SUM_h$ with a non-negative arc function $h : \mathcal{R} \to [0, \infty)$.

If C is order preserving and prefix monotone we may use $F := C$ itself as a decision function of BF^*; then the algorithm will output an optimal solution path with regard to $F_{\max} = F = C$; this is the assertion of Theorem 4.57. We shall show this result with Theorem 4.39, and we shall give another and more elementary proof.

Moreover, we shall see that step 6c) of BF^* can be made simpler. The resulting search method DIJKSTRA-BF^* is a generalization of Dijkstra's algorithm (see Algorithm 1).

The next result says that BF^* with an order preserving and prefix monotone decision function F always expands F-minimal paths.

Lemma 4.51 *Given a simple, locally finite digraph $\mathcal{G} = (\mathcal{V}, \mathcal{R})$, a node $s \in \mathcal{V}$, and a set $\Gamma \subseteq \mathcal{V}$. We assume that $BF^*(F)$ is applied to $\mathcal{G}$ where F is an order preserving and prefix monotone path function.*

Then the following is true for all iterations t: The path $P_t(\overline{v}_t)$ is an F-minimal s-$\overline{v}_t$-connection.[15]

P r o o f : If $t = 1$ then s is expanded, and $P_1(s) = [s]$ is an F-minimal connection from s to $\overline{v}_1 = s$ because F is prefix monotone.

Suppose that $t \geq 2$ is the first iteration for which $P(\overline{v}_t)$ is not F-minimal. Then there exists a path $P \in \mathcal{P}(s, \overline{v}_t)$ with

$$F(P) \prec F(P_t(\overline{v}_t)). \tag{4.47}$$

Let $P = [s = u_0, u_1, \ldots, u_l = \overline{v}_t]$. Then $s \in \mathrm{CLOSED}_t$[16] and $u_l \in \mathrm{OPEN}_t$. Hence, there exists a $k < l$ such that $u_0, \ldots, u_k \in \mathrm{CLOSED}_t$ and $u_{k+1} \notin \mathrm{CLOSED}_t$; then $u_{k+1} \in \mathrm{OPEN}_t$ because u_{k+1} has been detected at the latest when u_k was expanded. Let $P' := [s = u_0, u_1, \ldots, u_k]$ and $Q' := [u_{k+1}, u_{k+2}, \ldots, u_l]$; then $P = P' \oplus (u_k, u_{k+1}) \oplus Q'$; it is possible that $u_{k+1} = \overline{v}_t$ and $Q' = [\overline{v}_t]$.

The node $u_k \in \mathrm{CLOSED}_t$ was expanded by an iteration $t' < t$. By the minimality of t, the path $P_{t'}(u_k)$ was F-minimal. In particular, $F(P_{t'}(u_k)) \preceq F(P')$. This and the order preservation of F imply

$$F(\widehat{P}_{t'}(u_{k+1})) \;=\; F(P_{t'}(u_k) \oplus (u_k, u_{k+1})) \;\preceq\; F(P' \oplus (u_k, u_{k+1})). \tag{4.48}$$

Note that $F(\widehat{P}_{t'}(u_{k+1}))$ was compared with $F(P_{t'}(u_{k+1}))$ when generating $P_{t'+1}(u_{k+1})$; moreover, the value $F(P(u_{k+1}))$ can only have become smaller but never greater after the t'th iteration; consequently, $F(P_t(u_{k+1})) \preceq F(P_{t'+1}(u_{k+1})) \preceq F(\widehat{P}_{t'}(u_{k+1}))$ so that $F(P_t(u_{k+1})) \preceq F(\widehat{P}_{t'}(u_{k+1}))$. This fact is used in $(*)$ of the following inequality:

$$F(P_t(u_{k+1})) \overset{(*)}{\preceq} F(\widehat{P}_{t'}(u_{k+1})) \overset{(4.48)}{\preceq} F(P' \oplus (u_k, u_{k+1})). \tag{4.49}$$

The prefix monotonicity of F implies that

$$F(P' \oplus (u_k, u_{k+1})) \preceq F(P' \oplus (u_k, u_{k+1}) \oplus Q') = F(P) \overset{(4.47)}{\prec} F(P_t(\overline{v}_t))$$

so that $F(P' \oplus (u_k, u_{k+1})) \prec F(P_t(\overline{v}_t))$. This fact is used in $(*)$ of the following inequality:

$$F(P_t(u_{k+1})) \overset{(4.49)}{\preceq} F(P' \oplus (u_k, u_{k+1})) \prec F(P_t(\overline{v}_t)). \tag{4.50}$$

On the other hand, $\overline{v}_t$ is expanded in step 3 of the tth iteration. This is only possible if $F(P_t(\overline{v}_t)) \preceq F(P_t(u_{k+1}))$, which is a contradiction to (4.50).

Hence, no iteration t exists in which $P(\overline{v}_t)$ is not F-minimal, and our assertion is proven. $\qquad\square$

---

[15] Recall that $\overline{v}_t$ is defined as the node expanded in the tth iteration.

[16] Note s has never been transported back from CLOSED to OPEN because $P(s)$ has never been redirected; this follows from the fact that F is prefix monotone so that always $F(P(s)) = F([s]) \preceq F(\widehat{P}(s))$.

Corollary 4.52 *Let G, s, Γ, F be the same objects as in Lemma 4.51. Let $v \in V$.*
   a) If v is selected for expansion then $P(v)$ will never be redirected after this event.
   b) The node v is never moved from CLOSED to OPEN.
   c) If $P(v)$ is redirected then $v \in$ OPEN at that moment.
   d) The node v is expanded at most once.

P r o o f   o f   a) :   Let X be the current path $P(v)$ when v is selected for expansion. Then X is F-minimal by Lemma 4.51. Hence, BF*(F) will never construct a path $\widehat{P}(v)$ with $F(\widehat{P}(v)) \preceq F(X) = F(P(v))$. Consequently, no redirection occurs.

P r o o f   o f   b) :   If $v \in$ CLOSED then v must have already been selected for expansion. This and assertion a) imply that v cannot return to OPEN.

P r o o f   o f   c) :   It follows from b) that v has never been selected for expansion when $P(v)$ is redirected. Hence, $v \in$ OPEN.

P r o o f   o f   d):   Assertion b) says that $v \in$ CLOSED at any time after the first expansion of v. Hence, $v \notin$ OPEN so that v cannot be selected again.   □

Example 4.53 The order preservation of the decision function F in Lemma 4.51 must not be omitted, even if F is prefix monotone. To see this we consider the digraph G in Figure 10. Let $\Gamma := \{x_3\}$. Let $F : \mathcal{P}(s) \to \mathbb{R}$ be defined as follows: $F([s]) := 0$, and
$F([s, x_1]) := 1,$ $F([s, x_1, x_2]) := 4,$ $F([s, x_1, x_2, y]) := 5,$ $F([s, x_1, x_2, y, x_3]) := 6,$
$F([s, x_2]) := 3,$ $F([s, x_2, y]) := 20,$ $F([s, x_2, y, x_3]) := 23,$
$F([s, x_3]) := 18$.

Then F is prefix monotone; but F is not order preserving because $F([s, x_1, x_2]) = 4 > 3 = F([s, x_2])$ and $F([s, x_1, x_2, y]) = 5 < 20 = F([s, x_2, y])$. The first 4 iterations of Algorithm BF*(F) expand the paths $[s], [s, x_1], [s, x_2],$ and $[s, x_3]$ and do not execute any redirection. In particular, $[s, x_2]$ is not replaced by $[s, x_1, x_2]$ so that BF* does not detect the good s-y-path $[s, x_1, x_2, y]$. Therefore, BF* will not find the path $[s, x_1, x_2, y, x_3]$ but the worse path $[s, x_3]$ with $F([s, x_3]) = 18$.   □

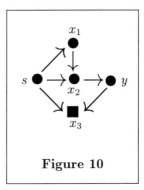

Figure 10

Next, we show that the paths $P(v)$ and $Q(v)$ are the same.

Lemma 4.54 *Let G be a simple, locally finite digraph; let $s \in V$ and $\Gamma \subseteq V$. If BF* uses an order preserving and prefix monotone path function F then $P(v) = Q(v)$ for all nodes for which $P(v)$ exists.*

(Recall that $Q(v)$ is the path obtained by recursively applying the function *pred*.)

P r o o f :   Let $v \in V$, and let t be the number of an iteration. Then the following is true:

     If v is in OPEN$_t$ then there is no node $x \neq v$ such that v lies on $Q_t(x)$.   (4.51)

To see this we assume that $v \in \text{OPEN}_t$. Then v has never been expanded before because otherwise, $v \in \text{CLOSED}_t$ by Corollary 4.52. So, BF* has not yet detected any arc with start node v; even deleting all these arcs would not influence the iterations $1, 2, \ldots, t-1$. Consequently, $Q_t(x)$ cannot have any arc with start node v. If $v \in Q_t(x)$ then v must be the end node of this path, and that means that $v = x$.

To show the current result we count the following critical events:

>   Type 1: A node v' is newly detected.
>   Type 2: The path $P(v')$ from s to a node v' is redirected.

It is clear that these are the only events concerning any path $P(v)$ or $Q(v)$.

We show that the following is true for all k:

>   When k events of type 1 or type 2 have occurred then $P(v) = Q(v)$ for all nodes.   (4.52)

The proof is an induction on k. Let $k = 0$. Then $P(s) = Q(s) = [s]$ is the only path known to BF*.

We assume that (4.52) is true for k. Let $\bar{v}$ be the node that is expanded when the $(k+1)$st event of type 1 or type 2 occurs. We distinguish between two cases.

Case 1: The $(k+1)$st critical event is of type 1.
Then v' must be a successor of $\bar{v}$. Then assertion $P(v) = Q(v)$ for $v = v'$ is shown with the following equation, in which relationship $(*)$ follows from the induction hypothesis:

$$P(v') \;=\; P(\bar{v}) \oplus (\bar{v}, v') \;\overset{(*)}{=}\; Q(\bar{v}) \oplus (\bar{v}, v') \;=\; Q(v'). \qquad (4.53)$$

The assertion $P(v) = Q(v)$ for $v \neq v'$ can be seen as follows: As v' has been newly detected, the construction of $P(v')$ and $Q(v')$ has no influence on any path $P(v)$ or $Q(v)$ with $v \neq v'$; hence, $P(v) = Q(v)$ for all nodes for which these paths exist.

Case 2: The $(k+1)$st critical event is of type 2.
Then v' must be a successor of $\bar{v}$.
The assertion $P(v') = Q(v')$ follows again from (4.53).
We show that $P(v) = Q(v)$ for all nodes $v \neq v'$. First, we note that no path $P(v)$, $v \neq v'$, is influenced by the redirection of $P(v')$. The same is true for $Q(v)$. To see this we apply Corollary 4.52 c); it says that the node v' is still lying in OPEN when $P(v')$ is redirected. This and (4.51) imply that no path $Q(v)$ with $v \neq v'$ visits v', and this is true before the redirection of $P(v')$ and after this event. Hence, $Q(v)$ is not changed.
We have seen that the redirection of $P(v')$ does not influence any path $P(v)$ or $Q(v)$, $v \neq v'$; consequently, the equality $P(v) = Q(v)$ remains valid.   □

Remark 4.55 In Remark 4.14 b), we described a situation in which $P(v)$ and $Q(v)$ are different. The question arises why such a situation cannot occur if BF** uses a single order preserving and prefix monotone function F as expansion and redirection function.

We give the answer with the help of Example 4.13. The transformation of $P_3(y)$ into $P_4(y)$ in that example was executed later than the first expansion of y, which was carried out in the second iteration. That means that $P(y)$ was redirected after the first expansion of y. Such a sequence of events, however, is not possible if $BF^{**} = BF^*(F)$ where F is order preserving and prefix monotone. This follows from Corollary 4.52 a).

$\square$

Remark 4.56 If BF^* uses an order preserving and prefix monotone decision function, the following are true:
  a) The system of all paths $P(v)$ forms a rooted tree with root s.
     This follows from 4.54 and from the fact that the paths $Q(v)$ form a tree.
  b) Many versions of Dijkstra's algorithm use the paths $Q(v)$ instead of $P(v)$; in particular, they output the path $Q(v^{**})$ instead of $P(v^{**})$ where v^{**} is the first expanded goal node. Examples of such a Dijkstra algorithm are given in [EvMi92], in [Nolt76], and in this book (see Algorithm 1, remove all instructions marked with (1)). — It follows from 4.54 that it makes no difference whether Dijkstra's algorithm processes the paths $P(v)$ or the paths $Q(v)$.

$\square$

Next, we give sufficient conditions for the correctness of BF^* with a prefix monotone and order preserving decision function.

Theorem 4.57 *Given a simple, locally finite digraph* $\mathcal{G} = (\mathcal{V}, \mathcal{R})$, *a node* $s \in \mathcal{V}$, *and a set* $\Gamma \subseteq \mathcal{V}$ *of goal nodes. Let* $C : \mathcal{P}(s) \to (\mathbf{R}, \preceq)$ *be a cost function where* $\preceq$ *is total and identitive. (E.g.,* $\mathbf{R}$ *may be equal to* $\mathbb{R}$, *and* "$\preceq$" *may be equal to* "$\leq$".)

We make the following assumptions:
  (i) *There exists an* s-Γ-path P^+.
  (ii) *There is no upper bound on the sequence* $\left(C([s, v_1, \ldots, v_i]) \right)_{i \in \mathbb{N}}$
       if $[s, v_1, v_2, v_3, \ldots)$ *is an infinite and acyclic path in* $\mathcal{G}$.
  (iii) C *is order preserving and prefix monotone.*

Let $F := C$ *be the decision function of* BF^*.

Then $BF^*(F) = BF^*(C)$ *will output a* C-*minimal goal path* P^{**}.

P r o o f : We give two versions of the proof. The first is based on Theorem 4.39, and the second is independent of 4.39. We define $C_0 := C$.

V e r s i o n   1 : We apply Theorem 4.39 and check conditions (i) – (v) of that result.

Conditions (i), (ii), (iii): They are equivalent to conditions (i), (ii), (iii) in the current theorem, respectively.

Condition (iv): It is $C_0 = (C_0)_{\max}$ because $C_0 = C$ is prefix monotone; moreover, $C_0 = C$ by the definition of C_0.

Condition (v): We have assumed that $C_0 = C$ is order preserving and prefix monotone. Lemma 2.129 implies that $\mathcal{G}$ has no C_0-negative cycles.

Then Theorem 4.39 says that BF^* with decision function $F = C = C_0$ will output a C-minimal goal path.

V e r s i o n   2 : First, we show that BF^* will output a goal node at all. Suppose that this is not true. Let $P^+ = [s = u_0, u_1, \ldots, u_k]$. ($P^+$ has been introduced in (i).)

We prove the following fact by an induction on κ:

$$\text{BF* selects each node } u_\kappa \text{ for expansion.} \tag{4.54}$$

This assertion is true for $\kappa = 0$ because $u_0 = s$ is selected for expansion in any case.

We assume that fact (4.54) is true for $\kappa - 1$. Let t be the iteration in which $u_{\kappa-1}$ is expanded. If u_κ has been selected before the tth iteration then (4.54) is true for κ. The other case is that u_κ has not yet been selected for expansion. Then u_κ has been newly detected in the tth iteration or earlier. Hence, $P_{t+1}(u_\kappa)$ exists, and $u_\kappa \in \text{OPEN}_t$ because u_κ has not yet been selected for expansion. Let $M := C(P_{t+1}(u_\kappa)) = F(P_{t+1}(u_\kappa))$. Suppose that $t' \geq t + 1$ and that BF* does not expand u_κ in the iterations $t + 1, \ldots, t'$. Then $u_\kappa \in \text{OPEN}_{t'}$ so that $F(P_{t'}(\overline{v}_{t'})) \preceq F(P_{t'}(u_\kappa)) \preceq F(P_{t+1}(u_\kappa)) = M$. Moreover, $P_{t'}(\overline{v}_{t'})$ is node injective by 4.56 a). It follows from Lemma 2.125 that there are only finitely many node injective paths $P \in \mathcal{P}(s)$ with $F(P) \preceq M$, and it follows from fact (4.10) (with $F^{(2)} := F$) that each path can only be expanded once. Consequently, the situation that $t' \geq t + 1$ and $\overline{v}_{t'} \neq u_\kappa$ can only occur finitely many times. So, the node u_κ is selected for expansion at the latest when all node injective paths P with $F(P) \preceq M$ have been expanded. Note that BF* does not intermediately stop because we have assumed that BF* does not find a goal path and because $u_\kappa \in \text{OPEN} \neq \emptyset$.

Now we have proven fact (4.54). This assertion implies that BF* will select the node u_k for expansion. But $u_k \in \Gamma$ because P^+ is a goal path. Hence, BF* does find the goal path $P(u_k)$, which is a contradiction to our assumption.

Next, we show that the s-Γ-path P^{**} output by BF* is C-minimal. If Γ has exactly one element then the assertion follows immediately from Lemma 4.51. We consider the case that Γ has more than one element. Suppose that P^{**} is not optimal. Then there exists a goal path P with $F(P) = C(P) \prec C(P^{**}) = F(P^{**})$.

Let t^{**} be the number of the iteration in which BF* outputs P^{**}. The path P has not yet been expanded because otherwise the algorithm would have stopped earlier. Hence, there exists a prefix $P' := [s = u_0 = s, u_1, \ldots, u_k, u_{k+1}]$ of P with $u_0, \ldots, u_k \in \text{CLOSED}_{t^{**}}$ and $u_{k+1} \in \text{OPEN}_{t^{**}}$. Let $t' < t^{**}$ be the iteration in which u_k was moved to CLOSED. It follows from 4.51 that BF* had found an optimal s-u_k-path $P_{t'}(u_k)$ at that moment. Consequently, $F(P_{t'}(u_k)) \preceq F([u_0, \ldots, u_k])$; this and the order preservation of F imply that

$$F(P_{t'}(u_k) \oplus (u_k, u_{k+1})) \preceq F([u_0, \ldots, u_k] \oplus (u_k, u_{k+1})) = F(P'). \tag{4.55}$$

When expanding u_k, a path $P_{t'+1}(u_{k+1})$ has been generated with

$$F(P_{t'+1}(u_{k+1})) \preceq F(\widehat{P}_{t'}(u_{k+1})) = F(P_{t'}(u_k) \oplus (u_k, u_{k+1})). \tag{4.56}$$

Consequently, $F(P_{t'+1}(u_{k+1})) \overset{(4.56)}{\preceq} F(P_{t'}(u_k) \oplus (u_k, u_{k+1})) \overset{(4.55)}{\preceq} F(P')$ so that $F(P_{t'+1}(u_{k+1})) \preceq F(P')$. This is used in relation $(\diamond)$ of the following inequality; relationship $(*)$ follows from the fact that $F(P_t(u_{k+1}))$ can only become smaller when t becomes greater; relationship $(**)$ follows from the prefix monotonicity of F.

$$F(P_{t^{**}}(u_{k+1})) \overset{(*)}{\preceq} F(P_{t'+1}(u_{k+1})) \overset{(\diamond)}{\preceq} F(P') \overset{(**)}{\preceq} F(P) \prec F(P^{**}).$$

Hence, $F(P_{t^{**}}(u_{k+1})) \prec F(P')$. Consequently, BF* does not expand P' in the

t^{**}th iteration because $F(P_{t^{**}}(u_{k+1}))$ is a better candidate. That is a contradiction to the assumption that P^* is expanded in the t^{**}th iteration. □

Remark 4.58 Corollary 4.52 says that that of BF* never moves a node from CLOSED back to OPEN; moreover, Lemma 4.54 says that $P(v) = Q(v)$ for all v. Therefore, the instructions (1) or (2) of BF**$(F, F) = $ BF*(F) may be omitted. We shall call the simplified algorithm *DIJKSTRA-BF**. If F is additive then DIJKSTRA-BF* is essentially the same as Dijkstra's algorithm (see Algorithm 1. □

Remark 4.58 implies a complexity theoretical result:

Theorem 4.59 *Given the same situation as in Theorem 4.57. In addition, we assume that the digraph $\mathcal{G}$ is finite and has n nodes.*
Then the worstcase time of BF* and DIJKSTRA-BF* is $O(n^2)$.

P r o o f : Both search strategies do not execute more than n iterations because each of the n nodes of $\mathcal{G}$ can only be expanded once. Each iteration consists mainly of searching for the node to be expanded and of the expansion process; both actions do not take more than $O(n)$ time. That means that BF* stops after $O(n^2)$ steps. □

The next remark says that in our current situation, paths with cycles are irrelevant; we can therefore replace the given digraph by an acyclic one.

Remark 4.60 Given a digraph $\mathcal{G} = (\mathcal{V}, \mathcal{R})$ with n nodes. Let $s \in \mathcal{V}$ and $\Gamma \subseteq \mathcal{V}$. Let $C : \mathcal{P}(s) \rightarrow (\mathbf{R}, \preceq)$ be an order preserving and prefix monotone cost function where $\preceq$ is total and identitive. We use BF* with decision function $F := C$ to find a C-minimal goal path P^{**}.

Then 2.132 implies that we must only search for an optimal path among all node injective candidates. This is the key observation to replace $\mathcal{G}$ by an acyclic digraph. All node injective paths in $\mathcal{G}$ have n or fewer nodes. This and a) imply that a C-minimal goal path can be found among all candidates of a length $< n$.
For this reason, we may replace $\mathcal{G}$ by the digraph $\mathcal{G}^{[n]}$, which was described on page 18. We choose $(s, 0)$ as start node and $\Gamma^{[n]} := \Gamma \times \{0, \dots, n-1\}$ as the set of all goal nodes. Let Ψ be the canonical graph homomorphism from $\mathcal{G}^{[n]}$ to $\mathcal{G}$, which was described on page 18 as well. We define the path function $C^\bullet := C \circ \Psi$. Then $C^\bullet$ is order preserving by 2.36 a), and $C^\bullet$ is obviously prefix monotone.

Therefore, it is easy to see that the following search strategy will output a C-minimal solution path P^{**} in $\mathcal{G}$:

Apply BF* with the decision function $F^\bullet := C^\bullet$ to $\mathcal{G}^{[n]}$.
Let $P^\bullet$ be the path found by this procedure. Then let $P^{**} := \Psi(P^\bullet)$. □

4.2.2.1.6 Algorithms for path problems with additive cost measures: The algorithms A and A*, Dijkstra's algorithm, and Breadth-First search

This paragraph is about the search of C-minimal goal paths where C is additive. This problem is very often solved by using the algorithms A, A*, Dijkstra's algorithm or

Breadth-First search. Each of these search strategies may be considered as BF*(F) where F belongs to a particular class of decision functions. Algorithm A may use the most general class of decision functions; this class becomes smaller and smaller for Algorithm A*, for Dijkstra's algorithm and for the Breadth-First search. Hence, Algorithm A is more general than Algorithm A*, this is more general than Dijkstra's algorithm, and this is more general than Breadth-First search.

We give a short description of these four search algorithms:

- **Algorithm A:** A search strategy BF*(F) is called *Algorithm A* if F is of the form: $F = C + p$ where $p : V \to \mathbb{R}$ is an arbitrary node function.
This node function is called a *heuristic function* (see also page 103). In many cases, $p(v)$ approximates the C-value of the best v-Γ-path; then $F(P) = C(P) + p(\omega(P))$ is the estimated cost of the best s-Γ-path with prefix P.
Algorithm A has been described in [Nils82] and in other sources of literature; Nilsson's version of Algorithm A outputs the path $Q(v^*)$ and not $P(v^*)$ where $v^* \in \Gamma$ is the last node selected from OPEN.

- **Algorithm A*:** If Algorithm A uses an *admissible* heuristic function then we call it *Algorithm A**. (Admissibility of heuristic functions has been defined in 2.139 a) on page 108.)
Algorithm A* may output $P(v^*)$ or $Q(v^*)$ where v^* is the last node selected from OPEN. Both versions work correctly. This will be shown in Theorem 4.61 b).
It is motivated at the end of Remark 2.140 why it is good to choose an admissible and consistent heuristic function.

- **Dijkstra's algorithm:** Dijkstra's algorithm was introduced in [Dijk59], and it was formulated as Algorithm 1 on page 174 in this book.
Dijkstra's algorithm may also be considered as a special version of Algorithm A*. Let C be additive and prefix monotone, and let $p(v) := 0$ for all nodes v. Then p is admissible because C is prefix monotone. Consequently, BF*(F) with $F := C$ is a special version of Algorithm A*. Comparing BF*(F) to Algorithm 1 shows that both search methods do essentially the same things.
Moreover, Dijkstra's algorithm may be considered as a special version of Algorithm DIJKSTRA-BF*. The reason is that BF*(F) with the prefix monotone decision function $F = C$ never carries out the steps (1) and (2) of BF**.
It follows from 4.54 that always $P(v) = Q(v)$ for BF*(F) with $F = C$. Hence, $P(v) = Q(v)$ is also true for all paths $P(v)$, $Q(v)$ generated by DIJKSTRA-BF* with decision function $F = C$. That means that the following six algorithms do essentially the same; therefore, we call each of them a *Dijkstra algorithm*.

- Algorithm 1 with instructions (1) and without instructions (2),
- Algorithm 1 with instructions (2) and without instructions (1),
- the algorithm BF* with $F = C$,
- the modified version of BF* that uses $F = C$ as decision function and generates, processes, and outputs $Q(v)$ instead of $P(v)$,
- the algorithm DIJKSTRA-BF* with $F = C$,
- the modified version of DIJKSTRA-BF* that uses $F = C$ as decision function and generates, processes, and outputs $Q(v)$ instead of $P(v)$.

• **Breadth-First search (BFS):** A Dijkstra algorithm is called *Breadth-First search algorithm* if it uses the special decision function $C = F = \ell$. (Recall that $\ell(P)$ denotes the length of P.) In Definition 4.74, we shall define the Breadth-First search algorithm DIJKSTRA-BFS as Algorithm DIJKSTRA-BF* with decision function $F := \ell$.

In literature, there exist versions of the algorithms A and A* using $F^{(1)} := C + p$ as expansion function and $F^{(2)}(P) := C(P)$. These versions of A and A^* are no BF*-algorithms because they their expansion function is not equal to their redirection function. But it is easy to transform these algorithms into equivalent BF*-algorithms by using the single decision function $F := F^{(1)}$; that means that we replace the redirection function $F^{(2)}$ by $F^{(1)}$. The behaviour of the algorithms A and A* is independent of the choice of $F^{(1)}$ or $F^{(2)}$ as redirection function. The reason is that the paths $P(v')$ and $\widehat{P}(v')$ compared in step 6c) have the same end nodes; therefore, it does not make any difference whether comparing the values $F^{(1)}(P(v')) = C(P(v')) + p(v')$ with $F^{(1)}(\widehat{P}(v')) = C(\widehat{P}(v')) + p(v')$ or the values $F^{(2)}(P(v')) = C(P(v'))$ with $F^{(2)}(\widehat{P}(v')) = C(\widehat{P}(v'))$. We therefore do not cause any loss of generality when assuming that the algorithms A and A* use the same function $F = F^{(1)}$ for expansion and redirection.

Now we give the well-known result about the correctness of A*.

Theorem 4.61 *Given a simple, locally finite digraph* $\mathcal{G} = (\mathcal{V}, \mathcal{R})$ *with* $s \in \mathcal{V}$ *and* $\Gamma \subseteq \mathcal{V}$. *Given an additive cost function* $C : \mathcal{P}(s) \to \mathbb{R}$ *and a node function* $p : \mathcal{V} \to \mathbb{R}$ *with the following properties:*
  (i) *There exists an* s-Γ-*path* P^+.
  (ii) *There is no upper bound on the sequence* $\big(C([s, v_1, \ldots, v_i]) + p(v_i)\big)_{i \in \mathbb{N}}$
       if $[s, v_1, v_2, v_3, \ldots)$ *is is an infinite and node injective path in* $\mathcal{G}$.
  (iii) p *is admissible.*
  (iv) $p(v) = 0$ *for all* $v \in \Gamma$.
  (v) $\mathcal{G}$ *has no* C-*negative cycle.*

Then the following are true:
  a) *Algorithm* A^* *with decision function* $F := C + p$ *will output a* C-*minimal goal path* P^{**}.
  b) *Let* $v^* \in \Gamma$ *be the first element of* Γ *selected in step 3 of* A^*.
     Then $Q(v^*)$ *is well-defined, and* $Q(v^*)$ *is likewise a* C-*minimal* s-Γ-*path.*

P r o o f   o f   a) :   Let $C_0 := F = C + p$. We show that C, C_0 and F satisfy all conditions (i) – (v) of Theorem 4.39.

Conditions (i) *and* (ii): They follow from conditions (i) and (ii) of the current theorem.

Condition (iii): The order preservation of $C_0 = F = C + p$ can be shown by a simple computation; note that C is additive.

Condition (iv): Let X be a goal path, and let $Y \leq X$. Let Q be the suffix of X with the property $X = Y \oplus Q$, and let $v := \alpha(Q) = \omega(Y)$. Then

$$C_0(Y) = C(Y) + p(v) \overset{(iii)}{\leq} C(Y) + C(Q) = C(X) \overset{(iv)}{=} C(X) + p(\omega(X)) = C_0(X).$$

Consequently, $C_0(Y) \leq C(X) = C_0(X)$ for all prefixes $Y \leq X$, and the equations (α) and (β) in condition (iv) follow immediately.

Condition (v): We must show the following fact:

$$\mathcal{G} \text{ has no } C_0\text{-negative cycles.} \qquad (4.57)$$

The idea of the proof is that the difference between C_0 and C is a node function, which is irrelevant for the negativity if cycles. More formally, let U, V be two paths and let X be a cycle with $\omega(U) = \alpha(X) = \omega(X) = \alpha(V)$. Then the following inequality is valid, in which $(*)$ follows from assumption (v) in the current result:

$$C_0(U \oplus V) \; = \; C(U \oplus V) + p(\omega(V)) \overset{(*)}{\leq} C(U \oplus X \oplus V) + p(\omega(V)) = C_0(U \oplus X \oplus V).$$

That means that no cycle X is C_0-negative.

Then Theorem 4.39 implies that Algorithm $A^* = BF^*(F)$ will output a C-minimal s-Γ-path P^{**}.

P r o o f o f b) : We shall employ Lemma 4.45. For this purpose, we show that always $P(s) = [s]$. Otherwise there were an iteration t such that $P_t(s) = [s]$ and $P_{t+1}(s) \neq [s]$. This can only occur if $F(P_{t+1}(s)) < F(P_t(s))$. Consequently, $C(P_{t+1}(s)) = F(P_{t+1}(s)) - p(s) < F(P_t(s)) - p(s) = C(P_t(s)) = 0$ so that $C(P_{t+1}(P(s)) < 0$. That means that $\mathcal{G}$ has the C-negative cycle $P_{t+1}(s)$, and that is in conflict with condition (v).

We now apply Lemma 4.45 c) and d). There exist no F-negative cycles because of (4.57) and $F = C_0$. Moreover, $F = C_0$ is $<$-preserving. Consequently, $Q(v^*)$ exists by Lemma 4.45 d). Assertion c) of 4.45 implies that $F(Q(v^*)) \leq F(P(v^*))$. Moreover, $p(v^*) \overset{(iv)}{=} 0$ so that $C(Q(v^*)) \leq C(P(v^*))$. It is clear that the path P^{**} output by $BF^*(F)$ is exactly the path $P(v^*)$, and part a) of the current theorem says that $P(v^*) = P^{**}$ is C-minimal. Consequently, $Q(v^*)$ is C-minimal, too. □

We now discuss the assumptions of 4.61.

Remark 4.62   a) (*see also Remark* 4.41) Condition (ii) in Theorem 4.61 may be omitted if $\mathcal{G}$ is finite.

b) The conditions (iii) and (iv) in 4.61 imply that $F(Q) \geq 0$ for all paths Q with $\alpha(Q), \omega(Q) \in \Gamma$. To see this we define $v := \alpha(Q)$. Then $0 = p(v)$ by condition (iv), and $p(v) \leq C(Q)$ by condition (iii) so that $0 \leq C(Q)$.

c) Condition (iv) of 4.61 may be relaxed to

$$\text{(iv')} \quad p(v) \geq 0 \text{ for all } v \in \Gamma.$$

The original condition (iv) follows from (iii) and (iv'). To see this we assume that $v \in \Gamma$ and $Q := [v]$. Then $0 \overset{(iv')}{\leq} p(v) \overset{(iii)}{\leq} C(Q) = 0$, and that means that $p(v) = 0$ if $v \in \Gamma$.

d) Condition (iv) in Theorem 4.61 must not be omitted. To see this we mislead Algorithm A^* by defining a bad function p. Let $\mathcal{G} = (\mathcal{V}, \mathcal{R})$ with $\mathcal{V} = \{s, x, \gamma\}$ and $\mathcal{R} = \{(s, x), (x, \gamma), (s, \gamma)\}$. Let $\Gamma := \{\gamma\}$. We define $h(s, x) := h(x, \gamma) := 1$ and $h(s, \gamma) := 3$. Moreover, we define $p(s) := p(x) := 0$ and $p(\gamma) := -10$. Let $C := SUM_h$, and let $F := C + p$ be the decision function of A^*. Then the first iteration of A* expands s and constructs the paths $P(x) = [s, x]$ and $P(\gamma) = [s, \gamma]$. The second iteration of A* selects γ because $F(P(\gamma)) = -7 < 1 = F(P(x))$. Hence, A* will output the path $[s, \gamma]$ with $C(s, \gamma) = 3$ and not the optimal path $[s, x, \gamma]$ with $C([s, x, \gamma]) = 2$.

e) The algorithm C in [BaMa83] outputs optimal paths even if using particular heuristic functions that are not admissible; a more detailed description of Algorithm C is given in Theorem 4.95. □

Definition 4.63 Suppose that two admissible heuristic functions have the properties that $p_1(v) < p_2(v)$, for all $v \notin \Gamma$ and $p_1(v) = p_2(v) = 0$ for all $v \in \Gamma$. Then we say that p_2 *is better informed than* p_1. □

This terminology is easy to understand. Let Q be a C-minimal path from $v \notin \Gamma$ to Γ. Then $p_1(v) < p_2(v) \leq C(Q)$ if the situation in 4.63 is given. That means that $p_2(v)$ is a better approximation of $C(Q)$ than $p_1(v)$. This justifies the term "better informed".

Next, we describe the positive effect of using better informed heuristic functions. If p_2 is better informed than p_1 then A* with p_2 will expand at most as many nodes as A* with p_1. This assertion is the same as Result 6 in [Nils82].

A better informed heuristic function, however, does not always make A* work faster. The following situation may occur: A* with a badly informed heuristic function expands many nodes, and each of them is processed once; A* with a better informed heuristic function expands few nodes, but each of them is processed very often. An example is given in Theorem 4.66; a comment on this example can be found in Remark 4.67.

Theorem 4.64 *Given a simple, locally finite digraph* $\mathcal{G} = (\mathcal{V}, \mathcal{R})$, *a start node* $s \in \mathcal{V}$, *and a set* $\mathcal{V} \subseteq \Gamma$ *of goal nodes. Given the additive cost function* $C : \mathcal{P}(\mathcal{G}) \to \mathbb{R}$ *and the heuristic functions* p_1, p_2. *Let* $F_i(P) := C(P) + p_i(\omega(P))$ *for all paths* P *and all* $i = 1, 2$. *We define* A_i^* *as Algorithm A* * with decision function* F_i $(i = 1, 2)$. *Moreover, let* $X_i \subseteq \mathcal{V} \backslash \Gamma$ *be the set of all nodes that are expanded by* A_i^* *and do not belong to* Γ. *We assume that* p_1 *and* p_2 *satisfy all conditions of Theorem 4.61. (Consequently,* A_1^* *and* A_2^* *will output a* C-*optimal goal path.)*
Moreover, we assume that $p_2(v)$ *is better informed than* $p_2(v)$.
Then the following is true: $X_2 \subseteq X_1$.

P r o o f : We apply Theorem 4.50 and show that all conditions (i) – (iv) in that result are satisfied.

Condition (i): The functions F_i, $i = 1, 2$, are semi-additive and therefore order preserving.

Condition (ii): Let $X \in \mathcal{P}(s)$ be a path with $\omega(X) \notin \Gamma$. The $p_1(\omega(X)) < p_2(\omega(X))$ because p_2 is better informed than p_1. Consequently, $F_1(X) < F_2(X)$.

Condition (iii): Let $X \in \mathcal{P}(s)$ with $\omega(X) \in \Gamma$. Then $p_1(\omega(X)) = p_2(\omega(X)) = 0$ because p_2 is better informed than p_1. Consequently, $F_1(X) = F_2(X) = C(X)$. Moreover, let Y be a prefix of X, and let Q be the suffix of X with $X = Y \oplus Q$.

Then $p_2(\omega(Y)) \leq C(Q)$ because p_2 is admissible. Hence,

$$F_2(Y) \;=\; C(Y) + p_2(\omega(Y)) \;\leq C(Y) + C(Q) \;=\; C(X) \;=\; F_2(X)\,.$$

Condition (iv): Given two paths $X, Y \in P(s)$ with the same end node. Let $w :=$ $\omega(X) = \omega(Y)$. If $F_1(P) \leq F_1(Q)$ then $C(P) + p_1(w) \leq C(Q) + p_1(w)$; adding $p_2(w) - p_1(w)$ yields $C(P) + p_2(w) \leq C(Q) + p_2(w)$, and that means that $F_2(P) \leq F_2(Q)$.

The conditions of Theorem 4.50 have been proven, and we may apply this result. Let $v \in X_2$. Then Algorithm A_2^* expands a path P with end node v. By 4.50, A_1^* expands a path Q with $\omega(Q) = \omega(P) = v$. Consequently, $v \in X_1$. □

Next, we describe the complexity of A*. For this purpose, we show the following auxiliary result:

Lemma 4.65 *Given a locally finite digraph* $\mathcal{G} = (\mathcal{V}, \mathcal{R})$, *and let* $s \in \mathcal{V}$ *and* $\Gamma \subseteq \mathcal{V}$. *We assume that no substantial cycle of* $\mathcal{G}$ *visits* s.
Given additive cost functions $C_i = SUM_{h_i}$ *with* $h_i : \mathcal{R} \to \mathbb{R}$, $(i = 1, 2)$. *We assume that there exists a constant* c *such that the following is true for all arcs* r:

$$h_2(r) \;=\; \begin{cases} h_1(r) + c, & \text{if } \alpha(r) = s, \\ h_1(r)\,, & \text{if } \alpha(r) \neq s. \end{cases}$$

Given a node functions $p : \mathcal{V} \to \mathbb{R}$. *Let* $F_i := C_i + p$ *for all* $i = 1, 2$. *We define Algorithm* A_i *as Algorithm A with decision function* F_i, $i = 1, 2$.

Then Algorithm A_1 *will expand the same nodes and generate the same paths* $P(v)$ *and* $\widehat{P}(v)$ *as Algorithm* A_2.

P r o o f : The assertion follows from the following observation:

For all paths $P \neq [s]$, the difference between $F_1(P) = C_1(P) + p(\omega(P))$ and $F_2(P) = C_2(P) + p(\omega(P))$ is equal to the constant c.   (4.58)

Therefore, each comparison between $F_1(P)$ and $F_1(Q)$ has the same outcome as the one between $F_2(P)$ and $F_2(Q)$.

Assertion (4.58) is false if $c \neq 0$ and $\mathcal{G}$ does contain a substantial cycle X visiting the start node s. For example, let X be a node injective cycle with $\alpha(X) = \omega(X) = s$; let r be the first arc of X. Then r appears twice in the path $P := X \oplus r$. Consequently, $F_2(P) - F_1(P) \;=\; C_2(P) - C_1(P) \;=\; 2\,c$.   □

Theorem 4.66 *The worstcase time of Algorithm* A^* *is exponential. More precisely, for all* $n \geq 4$, *there exist*

- *a digraph* $\mathcal{G} = (\mathcal{V}, \mathcal{R})$ *with* n *nodes,*
- *an additive cost function* C,
- *an admissible heuristic function* p

such that A^* *with decision function* $C + p$ *executes* $\Omega(2^n)$ *iterations.*

P r o o f : We show the assertion by an induction on n. Our proof is a detailed version of the argumentation in [BaMa83], page 20.

We start our induction with a digraph $\mathcal{G} = (V, \mathcal{R})$ with $n = 4$ nodes s, m_1, m_2, γ and the arcs $(s, m_1), (m_1, m_2), (m_2, \gamma)$. Let $\Gamma := \{\gamma\}$.

We define $h(r) := 3$ for all $r \in \mathcal{R}$ and $C := SUM_h$. (Then h has property (4.59).) Let $p(v) := 0$ for all nodes $v \in V$. Then A* with the heuristic function p must select each node for expansion before outputting the only goal path $[s, m_1, m_2, \gamma]$ in $\mathcal{G}$. Consequently, A* will execute $4 \geq \frac{1}{8} \cdot 2^4 + 1$ iterations.

Let $\mathcal{G}_n$ be a digraph with n nodes; we assume that A* executes at least $\frac{1}{8} \cdot 2^n + 1$ iterations until selecting a goal node for expansion. Our idea is to construct a digraph $\mathcal{G}_{n+1}$ that contains two copies of $\mathcal{G}_n$; each copy of $\mathcal{G}_n$ will cause approximately $\frac{1}{8} \cdot 2^n$ of A*; hence, processing $\mathcal{G}_{n+1}$ takes approximately double as much time as processing $\mathcal{G}_n$.

- *The given digraph $\mathcal{G}_n$.*

We assume that the given digraph is of the form $\mathcal{G}_n = (V, \mathcal{R})$ such that

$$V = \{s, m_1, \ldots, m_{n-2}, \gamma\}, \mathcal{R} = \mathcal{R}_0 \cup \mathcal{R}_1 \cup \mathcal{R}_2 \text{ with}$$
$$\mathcal{R}_0 = \{(s, m_i) \mid i = 1, \ldots, n-2\},$$
$$\mathcal{R}_1 = \{(m_i, m_j) \mid 1 \leq j < i \leq n-2\},$$
$$\mathcal{R}_2 = \{(m_1, \gamma)\}.$$

We assume that $C = SUM_h$ is given where $h : \mathcal{R} \to \mathbb{N}$. (In particular, $h(r) \geq 1$ for all r.) Let $c_j := h(s, m_j)$ $(j = 1, \ldots, n-2)$; moreover, we define
$$c_{\min}(h) := \min\left(h(s, m_1), \ldots, h(s, m_{n-2})\right) = \min(c_1, \ldots, c_{n-2}),$$
$$c_{\max}(h) := \max\left(h(s, m_1), \ldots, h(s, m_{n-2})\right) = \max(c_1, \ldots, c_{n-2}).$$

We assume that h has the following property:
$$c_{\max}(h) \leq 2 \cdot c_{\min}(h) - 3. \tag{4.59}$$
Moreover, we assume that $p : V \to \mathbb{N}$ is an admissible heuristic function forcing A* to execute at least $(\frac{1}{8} 2^n + 1)$ iterations.

- *The definition of $\mathcal{G}_{n+1}$*

Now we define the digraph $\mathcal{G}_{n+1} = (\widetilde{V}, \widetilde{\mathcal{R}})$; it is illustrated in Figure 11. (The arcs $r \in \mathcal{R}_1$ do not appear in this figure.)

We define $\widetilde{V}$ and $\widetilde{\mathcal{R}}$ by replacing n by $n+1$ in the definition of V and $\mathcal{R}$, respectively.

$$\widetilde{V} := \{s, m_1, \ldots, m_{n-1}, \gamma\}, \widetilde{\mathcal{R}} := \widetilde{\mathcal{R}}_0 \cup \widetilde{\mathcal{R}}_1 \cup \widetilde{\mathcal{R}}_2 \text{ where}$$
$$\widetilde{\mathcal{R}}_0 := \{(s, m_i) \mid i = 1, \ldots, n-1\},$$
$$\widetilde{\mathcal{R}}_1 := \{(m_i, m_j) \mid 1 \leq j < i \leq n-1\},$$
$$\widetilde{\mathcal{R}}_2 := \{(m_1, \gamma)\}.$$

The set of goal nodes in $\mathcal{G}_{n+1}$ is defined as $\widetilde{\Gamma} := \Gamma = \{\gamma\}$. Moreover, let $s' := m_{n-1}$.

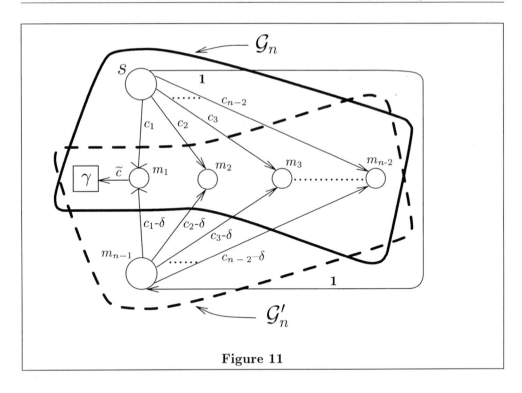

Figure 11

- *Relationships between* $\mathcal{G}_{n+1}$ *and* $\mathcal{G}_n$

It is clear that $\mathcal{V} \subseteq \tilde{\mathcal{V}}$ and $\mathcal{R} \subseteq \tilde{\mathcal{R}}$. Moreover, $\mathcal{G}_{n+1}$ contains another subgraph $\mathcal{G}'_n = (\mathcal{V}', \mathcal{R}')$ that is isomorphic to $\mathcal{G}_n$. This subgraph is defined as follows:

$$\mathcal{V}' := \{\, s', m_{n-1}, m_1, \ldots, m_{n-2}, \gamma \,\}, \quad \mathcal{R}' := \mathcal{R}'_0 \cup \mathcal{R}'_1 \cup \mathcal{R}'_2 \text{ with}$$

$$\begin{aligned}
\mathcal{R}'_0 &:= \{\, (s', m_i) \,|\, i = 1, \ldots, n-2 \,\}, \\
\mathcal{R}'_1 &:= \{\, (m_i, m_j) \,|\, 1 \le j < i \le n-2 \,\}, \\
\mathcal{R}'_2 &:= \{(m_1, \gamma)\}.
\end{aligned}$$

Then $\mathcal{G}'_n$ is indeed isomorphic to $\mathcal{G}_n$ because the following function $\Psi : (\mathcal{V} \cup \mathcal{R}) \to (\mathcal{V}' \cup \mathcal{R}')$ is an isomorphism: For all $v \in \mathcal{V}$ and $r \in \mathcal{R}$, let

$$\Psi(v) := \begin{cases} s' = m_{n-1} & \text{if} \quad v = s, \\ v & \text{if} \quad v \ne s, \end{cases} \qquad \Psi(r) := \big(\, \Psi(\alpha(r)),\, \Psi(\omega(r)) \,\big).$$

- *The path function* $\tilde{C}$ *and the heuristic function* $\tilde{p}$ *for* $\mathcal{G}_{n+1}$

We define an arc function $\tilde{h} : \mathcal{R} \to \mathbb{N}$, the path function $\tilde{C} := SUM_{\tilde{h}}$ and a heuristic function $\tilde{p}$ such that A* with decision function $\tilde{F} := \tilde{C} + \tilde{p}$ will execute at least $(\frac{1}{8} \cdot 2^{n+1} + 1)$ iterations.

Let $\delta := c_{\max}(h) - c_{\min}(h) + 2$. It seems to be most reasonable to define the values of $\tilde{h}$ and $\tilde{p}$ in the following order:

$\widetilde{h}(r) := h(r)$ for all $r \in \mathcal{R}$, $r \neq (m_1, \gamma)$,

$\widetilde{p}(v) := p(v)$ for all nodes $v \in \mathcal{V}$,

$\widetilde{h}(s, m_{n-1}) := 1$,

$\widetilde{h}(m_{n-1}, m_j) := h(s, m_j) - \delta = c_j - \delta$ for all $j = 1, \ldots, n - 2$.

$\widetilde{p}(m_{n-1}) := 1 + \max\{C(P) + p(\omega(P)) \mid P \in \mathcal{P}(s) \text{ is a path in } \mathcal{G}\}.$[17]

$\widetilde{h}(m_1, \gamma) := 2 + \max\{\widetilde{p}(v) \mid v \in \widetilde{\mathcal{V}}\}$. (This definition will make $\widetilde{p}$ admissible.)

Moreover, let $\widetilde{c} := \widetilde{h}(m_1, \gamma)$. We define $\widetilde{C} := SUM_{\widetilde{h}}$. Also, we define

$c_{\min}(\widetilde{h}) := \min(\widetilde{h}(s, m_1), \ldots, \widetilde{h}(s, m_{n-1}))$, $c_{\max}(\widetilde{h}) := \max(\widetilde{h}(s, m_1), \ldots, \widetilde{h}(s, m_{n-1}))$.

Then $\widetilde{p}(v) \geq 0$ for all $v \in \widetilde{\mathcal{V}}$. Moreover, $\widetilde{h}(r) \geq 1$ for all $r \in \widetilde{\mathcal{R}}$; this is easy to see if $r \in \mathcal{R} \cup \{(s, m_1)\}$; if, however, $r = (m_{n-1}, m_j)$ for some $j = 1, \ldots, n - 2$ then

$$\widetilde{h}(r) = c_j - \delta \geq c_{\min}(h) - \delta = 2\,c_{\min}(h) - c_{\max}(h) - 2 \overset{(4.59)}{\geq} 1.$$ Consequently, $\widetilde{h}(r) \in \mathbb{N}$ for all $r \in \widetilde{\mathcal{R}}$.

Moreover, $\widetilde{p}$ is admissible. To see this we assume that $v \in \widetilde{\mathcal{V}} \backslash \widetilde{\Gamma}$ and $Q \in \mathcal{P}(v, \gamma)$; then Q must use the arc (m_1, γ). This and the definition of $\widetilde{h}(m_1, \gamma)$ imply that $\widetilde{p}(v) \leq \widetilde{h}(m_1, \gamma) \leq \widetilde{C}(Q)$.

It may be that $c_{\min}(\widetilde{h})$ and $c_{\max}(\widetilde{h})$ do not yet have property (4.59); at the end of the proof, we shall show how to avoid this disadvantage of $\widetilde{h}$.

We define $\widetilde{F} := \widetilde{C} + \widetilde{p}$ for all paths P in $\mathcal{G}_{n+1}$.

- *The behaviour of A^* when applied to $\mathcal{G}_n$, to $\mathcal{G}'_n$, and to $\mathcal{G}_{n+1}$*

Let $A^*[\mathcal{G}_n]$ and $A^*[\mathcal{G}'_n]$ denote the application of A^* to $\mathcal{G}_n$ and $\mathcal{G}'_n$, respectively. We compare $A^*[\mathcal{G}_n]$ to $A^*[\mathcal{G}'_n]$. The following is true:

> $A^*[\mathcal{G}_n]$ and $A^*[\mathcal{G}'_n]$ do the same things up to isomorphism. More precisely, whenever $A^*[\mathcal{G}_n]$ processes a node v, an arc r and a path P, respectively, then $A^*[\mathcal{G}'_n]$ will process the node $\Psi(v)$, the arc $\Psi(r)$ and the path $\Psi(P)$ in the same way. $\qquad$ (4.60)

To see this we observe that the acyclic digraphs $\mathcal{G}_n$ and $\mathcal{G}'_n$ have no substantial cycles that visit s and s', respectively. Moreover, let $h_1 := \widetilde{h}\big|_{\mathcal{R}}$ and $h_2 := \widetilde{h}\big|_{\mathcal{R}'}$. Then the following is true:

$$(\forall\, r \in \mathcal{R}) \quad h_2(r) = \begin{cases} h_1(\Psi^{-1}(r)) - \delta, & \text{if } \alpha(r) = s', \\ h_1(\Psi^{-1}(r)), & \text{if } \alpha(r) \neq s'. \end{cases}$$

In particular, if $\alpha(r) = s'$ then $r = (s', m_j)$ with $j \in \{1, \ldots, n - 2\}$. Hence, $\Psi^{-1}(r) = (s, m_j)$, and $h_2(r) = \widetilde{h}(s', m_j) = \widetilde{h}(s, m_j) - \delta = h_1(\Psi^{-1}(r)) - \delta$. Assertion (4.60) follows from Lemma 4.65.

Next, we apply A^* with decision function $\widetilde{F}$ to $\mathcal{G}_{n+1}$. We call this procedure $A^*[\mathcal{G}_{n+1}]$.

---

[17]This maximum exists because $\mathcal{G}$ is acyclic and contains finitely many paths with start node s. We have defined a great value $\widetilde{p}(m_{n-1})$ in order to force Algorithm A^* to process the entire digraph $\mathcal{G}_n$ before expanding the path $[s, m_{n-1}]$.

∗ *Phase 1:* First, procedure A*$[\mathcal{G}_{n+1}]$ executes the same expansions and redirections as A*$[\mathcal{G}_n]$; the simulation of A*$[\mathcal{G}_n]$ goes on until only m_{n-1} and γ are lying in OPEN.

The reason for this behaviour of A*$[\mathcal{G}_{n+1}]$ is that the great heuristic value $\tilde{p}(m_{n-1})$ and the high cost $\tilde{h}(m_{n-1}, \gamma)$ prevent an earlier expansion of the nodes m_{n-1} and γ.

Moreover, the following is true for each $j \in \{1, \ldots, n-2\}$: The path $P(m_j)$ generated by procedure A*$[\mathcal{G}_n]$ starts with one of the arcs $(s, m_{j'})$, $j' = 1, \ldots, n-2$. (Both cases $j' = j$ and $j' \neq j$ may occur.) Consequently,

$$(\forall\, j = 1, \ldots, n-2)\quad c_{\min}(h) \leq h(s, m_{j'}) = \tilde{h}(s, m_{j'}) \leq \tilde{C}(P(m_j)). \qquad (4.61)$$

immediately after the simulation of A*$[\mathcal{G}_n]$.

∗ *Phase 2:* When procedure A*$[\mathcal{G}_{n+1}]$ has simulated A*$[\mathcal{G}_n]$, the nodes m_{n-1} and γ are the only elements of OPEN. Procedure A*$[\mathcal{G}_{n+1}]$ has generated the path $P(m_{n-1}) = [s, m_{n-1}]$ and a path $P(\gamma)$. We compare the values $\tilde{F}(P(m_{n-1}))$ and $\tilde{F}(P(\gamma))$ to decide which is the next node to be selected for expansion. First, we compute $\tilde{F}(P(m_{n-1}))$.

$$\tilde{F}(P(m_{n-1})) = \tilde{F}([s, m_{n-1}]) = \tilde{C}([s, m_{n-1}]) + \tilde{p}(m_{n-1}) = 1 + \tilde{p}(m_{n-1}).$$

Next, we estimate $\tilde{F}(P(\gamma))$. The path $P(\gamma)$ uses the arc (m_1, γ); therefore, $\tilde{F}(P(\gamma)) \geq \tilde{h}(m_1, \gamma)$. Moreover, $\tilde{h}(m_1, \gamma) \geq 2 + \tilde{p}(m_{n-1})$ by the definition of $\tilde{h}(m_1, \gamma)$. Consequently, $\tilde{F}(P(\gamma)) \geq 2 + \tilde{p}(m_{n-1}) = 1 + \tilde{F}(P(m_{n-1}))$.

This implies that A*$[\mathcal{G}_{n+1}]$ selects m_{n-1} and not γ for the next expansion. That means that all nodes $m_1, \ldots, m_{n-2}$ are processed again.

In particular, procedure A*$[\mathcal{G}_{n+1}]$ generates the paths $\hat{P}(m_j) = [s, m_{n-1}] \oplus (m_{n-1}, m_j)$ for all $j = 1, \ldots, n-2$; these paths $\hat{P}(m_j)$ have the following property:

$$\begin{aligned} \tilde{F}(\hat{P}(m_j)) &= \tilde{h}(s, m_{n-1}) + \tilde{h}(m_{n-1}, m_j) + \tilde{p}(m_j) = 1 + (c_j - \delta) + \tilde{p}(m_j) \\ &\leq (1 + c_{\max}(h) - \delta) + \tilde{p}(m_j) = (c_{\min}(h) - 1) + \tilde{p}(m_j) \overset{(4.61)}{<} \tilde{C}(P(m_j)) + \tilde{p}(m_j) \\ &= \tilde{F}(P(m_j)). \end{aligned}$$

That means that $\tilde{F}(\hat{P}(m_j)) < \tilde{F}(P(m_j))$ for all $j = 1, \ldots, n-2$ so that all paths $P(m_j)$ are redirected. The new paths $P(m_j)$, $(j = 1, \ldots, n-2)$ are equal to $[s, m_{n-1}, m_j] = [s, s', m_j]$.

∗ *Phase 3:* After this, procedure A*$[\mathcal{G}_{n+1}]$ does almost the same as A*$[\mathcal{G}'_n]$. The reason is that A*$[\mathcal{G}_{n+1}]$ has generated almost the same situation as A*$[\mathcal{G}'_n]$ after expanding $s' = m_{n+1}$. There are only two slight differences between A*$[\mathcal{G}'_n]$ and the current phase of A*$[\mathcal{G}_{n+1}]$:

– If A*$[\mathcal{G}'_n]$ generates a path $Q' := P(v)$ and $Q' := \hat{P}(v)$, respectively, then A*$[\mathcal{G}_{n+1}]$ will generate the path $Q := (s, m_{n-1}) \oplus P(v)$ and $Q := (s, m_{n-1}) \oplus \hat{P}(v)$. The difference between $\tilde{h}(Q')$ and $\tilde{h}(Q)$ is equal to the constant $\tilde{h}(s, m_{n-1})$. Therefore, the difference between the paths generated by A*$[\mathcal{G}'_n]$ and A*$[\mathcal{G}_{n+1}]$ does not influence the fact that both procedures take analogous decisions about expanding and redirecting.

– Procedure A*$[\mathcal{G}_{n+1}]$ has detected the node γ earlier than the procedure A*$[\mathcal{G}'_n]$. More precisely, A*$[\mathcal{G}_{n+1}]$ has detected γ when processing $\mathcal{G}_n$. Hence, A*$[\mathcal{G}_{n+1}]$ knows the node γ when beginning the simulation of A*$[\mathcal{G}'_n]$ whereas A*$[\mathcal{G}'_n]$ itself has not yet detected this node at this moment.

Therefore, the following actions of A*$[\mathcal{G}_{n+1}]$ and A*$[\mathcal{G}'_n]$ are not completely analogous:

A*$[\mathcal{G}'_n]$ executes its first expansion of m_1. The procedure detects γ and generates the first path $P(\gamma)$, which is equal to $[s', m_1, \gamma]$.

A*$[\mathcal{G}_{n+1}]$ executes its first expansion of m_1 after expanding $s' = m_{n-1}$. The procedure replaces the given path $P(\gamma) = [s, m_1, \gamma]$ by the better path $P(\gamma) := [s, m_{n-1}, m_1, \gamma]$.[18]

In particular, A*$[\mathcal{G}_{n+1}]$ does not expand γ earlier than A*$[\mathcal{G}'_n]$ because the cost $\tilde{c} = \tilde{h}(m_1, \gamma)$ is very high.

Next, we calculate the time of computation consumed by A*$[\mathcal{G}_{n+1}]$. Phase 1 is a simulation of A*$[\mathcal{G}_n]$ without the expansion of γ. Hence, A*$[\mathcal{G}_{n+1}]$ executes at least $\left(\frac{1}{8} \cdot 2^n\right)$ iterations. In phase 2, the node $m_{m-1} = s'$ is expanded; this is 1 iteration. At last, A*$[\mathcal{G}_{n+1}]$ simulates A*$[\mathcal{G}'_n]$ in the situation after the expansion of s'. By (4.60), this requires at least $\left(\frac{1}{8} \cdot 2^n\right)$ iterations. Consequently, the number of iterations of A*$[\mathcal{G}_{n+1}]$ is at least $\left(\frac{1}{8} \cdot 2^n\right) + 1 + \left(\frac{1}{8} \cdot 2^n\right) = \left(\frac{1}{8} \cdot 2^{n+1}\right) + 1$.

As mentioned above, the function $\tilde{h}$ does not necessarily have property (4.59). We avoid this problem as follows: For each $\Delta \in \mathbb{N}$ and all $r \in \mathcal{R}$, we define $\tilde{h}_\Delta(r) := \tilde{h}(r) + \Delta$ if $\alpha(r) = s$ and $\tilde{h}_\Delta(r) := \tilde{h}(r)$ if $\alpha(r) \neq s$. Moreover, we define $c_{\min}(\tilde{h}_\Delta)$ and $c_{\max}(\tilde{h}_\Delta)$ in analogy to $c_{\min}(\tilde{h})$ and $c_{\max}(\tilde{h})$, respectively. Then $c_{\min}(\tilde{h}_\Delta) = c_{\min}(\tilde{h}) + \Delta$ and $c_{\max}(\tilde{h}_\Delta) = c_{\max}(\tilde{h}) + \Delta$ for all $\Delta \in \mathbb{N}$. Hence, $c_{\max}(\tilde{h}_\Delta) \leq 2 \cdot c_{\min}(\tilde{h}_\Delta) - 3$ if Δ is sufficiently great. Moreover, A*$[\mathcal{G}_{n+1}]$ with arc function $\tilde{h}_\Delta$ does the same things as A*$[\mathcal{G}_{n+1}]$ with arc function $\tilde{h}$; this follows from Lemma 4.65. So, if $\tilde{h}$ does not have property (4.59) then replace $\tilde{h}$ by $\tilde{h}_\Delta$ where Δ is sufficiently great. $\square$

Remark 4.67 The graphs $\mathcal{G}_n$ in Theorem 4.66 are examples for the fact that better informed heuristic functions may increase the complexity of A*.

To see this we define $p_1(v) := 0$ for all nodes v of $\mathcal{G}_n$. Let p_2 be equal to the heuristic function p in 4.66. Moreover, let $F_i(P) := C(P) + p_i(\omega(P))$ for all paths P and for all $i = 1, 2$. Then the following assertions are true:

- Algorithm A* executes at most n iterations if using the heuristic function p_1. The reason is that $F_1 = C$ is prefix monotone.
- Algorithm A* executes at least $\frac{1}{8} 2^n$ iterations when using the heuristic function p_2. This follows from Theorem 4.66.
- p_2 is better informed than p_1.

This situation seems to be a contradiction to Theorem 4.64, which says that A*(F_2)

---

[18]It is $\widetilde{F}([s, m_{n-1}, m_1]) = \widetilde{F}([s, m_1]) + (1 - \delta) < \widetilde{F}([s, m_1])$; therefore, $[s, m_{n-1}, m_1]$ is better than $[s, m_1]$ so that $[s, m_{n-1}, m_1, \gamma]$ is indeed better than $[s, m_1, \gamma]$.

does not select more nodes than $A^*(F_1)$.

In reality, however, there is no conflict with Theorem 4.64. The reason is that this theorem does not say how often a node is selected for expansion. The high complexity of $A^*(F_2)$ is caused by the fact that it expands several nodes many times. Hence, Algorithm $A^*(F_2)$ executes much more iterations than Algorithm $A^*(F_1)$, which never expands a node more than once. □

Remark 4.68 Section 4.3 of [DePe85] is about the question whether A^* has optimal complexity with respect to the set of expanded nodes; these criteria of complexity have been defined on page 521 of [DePe85], and they have been mentioned in Subsection 4.2.1.4 of this book.

Corollary 2 of [DePe85] says that A^* is optimal with respect to a particular complexity criterion and that A^* is not optimal with respect to particular other complexity criterion; similar results are given in the table on page 531 of [DePe85]. □

Next, we focus on **variants of Algorithm A**. (This subject will also be discussed in Section 4.2.3 of this book.) Many variants of Algorithm A are given in literature, for example, the algorithms B, C, PropA, MarkA, and A**. These algorithms are discussed in the following parts of this book:

Algorithm B	:	Theorem 4.94
Algorithm C	:	Theorem 4.95
Algorithm PropA	:	Remark 4.96
Algorithm MarkA	:	Subsection 4.2.5
Algorithm A**	:	Theorem 4.92, Remark 4.93

Remark 4.69 The following table shows in which variant of Algorithm A should be usually applied in a given situation; in special cases, however, it may be better to use another search strategy than the strategy suggested in the table.

We have recommended almost the same algorithms as Bagchi and Mahanti have done in [BaMa83], page 26; in the second row of the table, however, we have proposed another search strategy than Bagchi and Mahanti. We shall give reasons for this in Remark 4.96 c).

Is $\mathcal{G}$ acyclic?	Is p consistent?	Is p admissible ?	
yes	yes	yes	MarkA
no	yes	yes	A*
○ [1) see below]	no	yes	PropA
○	○	no	C

1) The symbol "○" means: arbitrary answer "yes" or "no"

When deciding which algorithm is good and which is not we have used the following criteria: the quality of the output path and the time of computation. We have not regarded the sets of nodes processed by the search strategies; this, however, has been done in the table in [DePe85], page 531, which is based on the complexity criteria mentioned in Subsection 4.2.1.4 of this book. □

Another modification of A* is presented in [DePe85], page 513. The algorithm uses a decision function $\varphi\big(C(P), p(\omega(P))\big)$ where C is the given cost function, p is a heuristic function, and φ is of the form $\varphi : \mathbb{R} \times \mathbb{R} \dashrightarrow \mathbb{R}$. (In most cases, φ is choosen as $\varphi(x,y) := x + y$.)

We now describe this type of decision functions more precisely.

Remark 4.70 Let $\mathcal{G} = (\mathcal{V}, \mathcal{R})$ be a simple, locally finite digraph. Let $s \in \mathcal{V}$ and $\Gamma \subseteq \mathcal{V}$. Let $C : \mathcal{P}(\mathcal{G}) \to \mathbb{R}$ be an additive path function with $C(P) \geq 0$ for all paths. (This is equivalent to the fact that C is based on an arc function with exclusively non-negative values.) Moreover, let p be an admissible heuristic function.

Now we define a decision function F with the help of C and p. For this purpose, we choose an arbitrary strictly increasing, bijective, and concave[19] function $\phi : [0, \infty) \to [0, \infty)$ with $\phi(0) = 0$. Then we define $\varphi(u,v) := \phi\big(\phi^{-1}(u) + \phi^{-1}(v)\big)$ for all $u, v \geq 0$ and $F(P) := \varphi\big(C(P), p(\omega(P))\big)$ for all paths $P \in \mathcal{P}(\mathcal{G})$. Dechter and Pearl consider the following examples:

- $\phi(x) := x$ for all x.
  Then ϕ is concave. Moreover, $\varphi(u,v) = u + v$, and $F(P) = C(P) + p(\omega(P))$, which is the decision function of A*.

- $\phi(x) := \sqrt{x}$ for all x.
  Then $\varphi(u,v) = \sqrt{u^2 + v^2}$, and $F(P) = \sqrt{C(P)^2 + p(\omega(P))^2}$.   □

We will show that BF*(F) will output C-optimal paths if F is defined as in Remark 4.70. For this purpose, we show several auxiliary results about the function ϕ.

Lemma 4.71 Let $\phi : [0, \infty) \to [0, \infty)$ be strictly increasing, bijective and concave. Let $\varphi(x,y) := \phi\big(\phi^{-1}(x) + \phi^{-1}(y)\big)$ for all $x, y \geq 0$. Then the following are true:

  a) $\phi(0) = 0$, $\phi^{-1}(0) = 0$.
  b) $\varphi(x,0) = x$ for all $x \geq 0$.
  c) $\varphi(y, x - y) \leq \varphi(x, 0)$ for all $x \geq y \geq 0$.
  d) $\varphi(u,v) \leq \varphi(u + v, 0) = u + v$.
  e) φ is strictly increasing in each argument.

P r o o f o f a) : Obviously, $\phi^{-1}(x) \geq 0$ for all $x \geq 0$. This and the monotonicity of ϕ imply that $0 = \phi\big(\phi^{-1}(0)\big) \geq \phi(0)$. Hence, $0 \geq \phi(0)$ so that $\phi(0) = 0$. This equation implies $\phi^{-1}(0) = 0$.

P r o o f o f b) : Let $x \geq 0$. The definition of φ implies that $\varphi(x,0) = \phi\big(\phi^{-1}(x) + \phi^{-1}(0)\big)$. By part a), $\phi\big(\phi^{-1}(x) + \phi^{-1}(0)\big) = \phi\big(\phi^{-1}(x)\big) = x$. Hence, $\varphi(x,0) = x$ for all $x \geq 0$.

P r o o f o f c): First, we show the following assertion:
$$\phi^{-1}(\lambda \cdot x) \leq \lambda \cdot \phi^{-1}(x) \text{ for all } x \geq 0 \text{ and all } \lambda \in [0, 1]. \qquad (4.62)$$

To see this we assume that $x \geq 0$ and $\lambda \in [0, 1]$. We obtain the following inequality,

---

[19] That means that $\phi(\lambda x + (1 - \lambda) y) \geq \lambda \phi(x) + (1 - \lambda) \phi(y)$ for all x, y and all $\lambda \in [0, 1]$. An example is $\phi(x) = \sqrt{x}$.

in which $(*)$ follows from the assumption that that ϕ is concave.

$$\lambda \cdot x \overset{\text{a)}}{=} \lambda \cdot \phi\left(\phi^{-1}(x)\right) + (1-\lambda) \cdot \phi\left(\phi^{-1}(0)\right) \overset{(*)}{\leq}$$
$$\phi\left(\lambda \cdot \phi^{-1}(x) + (1-\lambda) \cdot \phi^{-1}(0)\right) \overset{\text{a)}}{=} \phi\left(\lambda \cdot \phi^{-1}(x)\right) .$$

This and the monotonicity of ϕ^{-1} imply that $\phi^{-1}(\lambda \cdot x) \leq \lambda \cdot \phi^{-1}(x)$, and fact (4.62) is proven.

Let $x \geq y \geq 0$. Define $\lambda := \frac{y}{x}$ if $x \neq 0$ and $\lambda := 1$ if $x = 0$. In any case, $y = \lambda \cdot x$ and $x - y = (1-\lambda) \cdot x$. Then we obtain the following inequality:

$$\varphi(y, x-y) = \varphi(\lambda x, (1-\lambda) x) = \phi\left(\phi^{-1}(\lambda \cdot x) + \phi^{-1}((1-\lambda) \cdot x)\right)$$
$$\overset{(4.62)}{\leq} \phi\left(\lambda \cdot \phi^{-1}(x) + (1-\lambda) \cdot \phi^{-1}(x)\right) = \phi\left(\phi^{-1}(x)\right) = x \overset{\text{b)}}{=} \varphi(x, 0) .$$

P r o o f   o f   d) : Let $x := u + v$, $y := u$. Then $\varphi(u, v) \leq \varphi(u+v, 0)$ by part c), and $\varphi(u+v, 0) = u + v$ by part b).

P r o o f   o f   e) : This follows immediately from the fact that ϕ and ϕ^{-1} are strictly increasing. $\square$

Now we show that BF* will output a C-minimal path if it uses a decision function according to Remark 4.70. Our assertion is similar to Theorem 4.61. The advantage of 4.61 is that C may attain negative values, which is not allowed in Theorem 4.72. The advantage of 4.72, however, is that BF* may use a large class of decision functions $F(P) = \varphi\left(C(P), p(\omega(P))\right)$ whereas in 4.61, we only consider the case that $\varphi(u, v) = u + v$.

Theorem 4.72 *Given a simple, locally finite digraph $\mathcal{G} = (\mathcal{V}, \mathcal{R})$, a node $s \in \mathcal{V}$ and a set $\Gamma \subseteq \mathcal{V}$. Let $C : \mathcal{P}(\mathcal{G}) \to (\mathbb{R}, \leq)$ be an additive cost function. We assume that $C(P) \geq 0$ for all paths P. (This is equivalent to the fact that C is based on an arc functions with exclusively non-negative values.)*
Let ϕ, φ, and p be the same functions as in 4.70. Let $C_0(P) := \varphi\left(C(P), p(\omega(P))\right)$ for all $P \in \mathcal{P}(\mathcal{G})$. Moreover, we make the following assumptions:
  (i)   There exists an s-Γ-path P^+.
  (ii) There is no upper bound on the sequence $\left(C_0([s, v_1, \dots, v_i])\right)_{i \in \mathbb{N}}$
       if $[s, v_1, v_2, v_3, \dots)$ is an infinite node injective path.
  (iii) p is admissible.
  (iv) $p(v) = 0$ for all nodes $v \in \Gamma$.
  (v) $\mathcal{G}$ has no C-negative cycles.
Then BF(F) with decision function $F := C_0$ will output a C-minimal goal path.*

P r o o f: We check the conditions of 4.39.

Conditions (i), (ii): They are equivalent to conditions (i) and (ii) in the current theorem, respectively.

Condition (iii): We must show that C_0 is order preserving.
Given three paths P_1, P_2, Q with $\alpha(P_1) = \alpha(P_2) = s$ and $w := \omega(P_1) = \omega(P_2) = \alpha(Q)$. Let $C_0(P_1) \leq C_0(P_2)$. Then $\varphi\left(C(P_1), p(w)\right) \leq \varphi\left(C(P_2 \oplus Q), p(w)\right)$ by the definition

of C_0. This and the strict monotonicity of φ imply that $C(P_1) \leq C(P_2)$; consequently, $C(P_1 \oplus Q) \leq C(P_2 \oplus Q)$ since C is additive. We apply the strict monotonicity of φ again and obtain the following inequality:

$$C_0(P_1 \oplus Q) = \varphi\big(C(P_1 \oplus Q), p(w(Q))\big) \leq \varphi\big(C(P_2 \oplus Q), p(w(Q))\big) = C_0(P_2 \oplus Q).$$

Condition (iv): We start with (iv)(α) and show that $C_0(P) = (C_0)_{\max}(P)$ for all s-Γ-paths P. For this end, we prove that $C_0(Q) \leq C_0(P)$ for all prefixes $Q \leq P$. Let Q' be the suffix of P with $P = Q \oplus Q'$. Let $w := w(Q) = \alpha(Q')$. Then $p(w) \leq C(Q')$ because p is admissible, and Lemma 4.71 d) implies that $\varphi(C(Q), C(Q')) \leq \varphi(C(Q) + C(Q'), 0)$. The fact $p(w) \leq C(Q')$ is used in relationship $(*)$ of the following inequality, and the fact $\varphi(C(Q), C(Q')) \leq \varphi(C(Q) + C(Q'), 0)$ is used in $(**)$.

$$C_0(Q) = \varphi\big(C(Q), p(w)\big) \overset{(*)}{\leq} \varphi\big(C(Q), C(Q')\big) \overset{(**)}{\leq}$$

$$\varphi(C(Q) + C(Q'), 0) = \varphi(C(P), 0) \overset{(iv)}{=} \varphi\big(C(P), p(w(P))\big) = C_0(P).$$

Consequently, $C_0(Q) \leq C_0(P)$ for all prefixes $Q \leq P$ so that $(C_0)_{\max}(Q) = C_0(P)$.

We next show equation (iv)(β) of 4.39 by proving that $C(P) = C_0(P)$ for all goal paths P. This relationship results from the following equation:

$$C_0(P) = \varphi\big(C(P), p(w(P))\big) \overset{(iv)}{=} \varphi(C(P), 0) \overset{\text{Lemma 4.71b)}}{=} C(P).$$

Condition (v): Let X be a cycle in $\mathcal{G}$; we show that X is not C_0-negative. Let U, V be two paths with $w(U) = \alpha(X) = w(X) = \alpha(V)$. Let $w := w(V)$. Then $C(U \oplus V) \leq C(U \oplus X \oplus V)$ because C is an additive path function with a non-negative arc function. We use the monotonicity of φ and obtain

$$C_0(U \oplus V) = \varphi\big(C(U \oplus V), p(w)\big) \leq \varphi\big(C(U \oplus X \oplus V), p(w)\big) = C_0(U \oplus X \oplus V).$$

Hence, X is not a C_0-negative cycle.

All conditions of Theorem 4.39 are satisfied. This result implies that $\mathrm{BF}^*(F)$ with $F = C_0$ will output a C-minimal s-Γ-path. $\qquad\square$

Next, we concentrate on **the shortest path problem and Breadth-First search**. We describe the behaviour of algorithms that output C-minimal solution paths where the cost measure C is equal to the length function ℓ. Moreover, we compare these algorithms to the Breadth-First search algorithm that marks nodes in trees.

Theorem 4.73 *Let $\mathcal{G} = (V, \mathcal{R})$ be a simple, locally finite digraph. Let s be a node of $\mathcal{G}$, and let Γ be a set of goal nodes. Given the cost function $C := \ell$.*
If an s-Γ-path exists at all then DIJKSTRA-BF* with decision function $F := C = \ell$ will output an s-Γ-path of minimal length.

P r o o f: The claim is an immediate consequence of 4.57 and of 4.58. $\qquad\square$

Next, we describe relationships between the algorithm DIJKSTRA-BF* with decision function ℓ and Breadth-First search.

Definition/Remark 4.74   We define *DIJKSTRA-BFS* as the algorithm DIJKSTRA-BF* with cost and decision function $C := F := \ell$; indeed, we may consider this algorithm as a Breadth-First search strategy.

In contrast to this, we define *BFS* as the Breadth-First search procedure to visit and mark all nodes of a given digraph (e.g., see Remark 2.168).   □

If $\mathcal{G}$ equals the rooted tree $\mathcal{T}(N, k)$ in Remark 2.168 we can formulate the similarities between DIJKSTRA-BFS and BFS more precisely.

Theorem 4.75 *Given the terminology of Remark 2.168. We assume that the algorithm DIJKSTRA-BFS uses the following tie-breaking rule: If $\overline{v}^{(1)}, \overline{v}^{(2)}, \ldots, \overline{v}^{(j)} \in$OPEN and $\ell\left(P(\overline{v}^{(1)})\right) = \ldots = \ell\left(P(\overline{v}^{(j)})\right)$ is minimal then DIJKSTRA-BFS expands the path with the minimal $\mathbf{N}_h$-value.[20]*
Also, we assume that $\Gamma = \emptyset$ so that DIJKSTRA-BFS will expand all nodes of $\mathcal{G}$.
Let BFS mark the nodes of $\mathcal{G}$ as described in Remark 2.168.

Then the following is true for all nodes $v \in V$ and for all numbers $t \in \mathbb{N}$:

DIJKSTRA-BFS expands v in the tth iteration if and only if BFS marks v in the tth iteration.

P r o o f (Sketch) : The proof is an induction on t. The main idea is to assume that DIJKSTRA-BFS organizes the set OPEN as a linear list with the following property: If the list contains the nodes $x_1, \ldots, x_j$ in this order then $\mathbf{N}_h(x_1) < \mathbf{N}_h(x_2) < \ldots < \mathbf{N}_h(x_j)$. Then DIJKSTRA-BFS automatically processes the list OPEN as a FIFO-list (FIFO = First In First Out); nodes are inserted on the right side of OPEN and removed on the left side. Consequently, DIJKSTRA-BFS expands each node v exactly when v is marked by BFS.
The details of this proof are left to the reader.   □

Next, we compare DIJKSTRA-BFS to A*.

Theorem 4.76 *Given the notations Remark 2.168.*
We assume that DIJKSTRA-BFS uses the same tie-breaking rule as in 4.75, and we assume that $\Gamma = \emptyset$.
Let $c < 1$ and $p : V \to [0, c)$ be a heuristic function. We assume that $p(x) < p(y) \iff \mathbf{N}_h(x) < \mathbf{N}_h(y)$ for all nodes x, y of the same height. Let $F := \ell + p$.

Then DIJKSTRA-BFS and A with decision function F will expand the same nodes during the same iterations.*

P r o o f : Let $x, y \in V$. DIJKSTRA-BFS selects x and not y for expansion if $\ell(P(x)) < \ell(P(y))$ or if $\ell(P(x)) = \ell(P(y))$ and $\mathbf{N}_h(P(x)) < \mathbf{N}_h(P(y))$. If $\ell(P(x)) < \ell(P(y))$ then $F(P(x)) < \ell(P(y)) \leq F(P(y))$ so that A* prefers x to y. If $\ell(P(x)) = \ell(P(y))$ and $\mathbf{N}_h(P(x)) < \mathbf{N}_h(P(y))$ then $p(x) < p(y)$ so that $F(P(x)) < F(P(y))$ and A* prefers again x to y.

---

[20]This is also the candidate with minimal $\mathbf{n}_h$-value because all candidates $P(\overline{v}^{(i)})$ have the same length.

That means that DIJKSTRA-BFS and A* use equivalent rules to select a node for expansion. This implies the assertion.   □

Next, we compare DIJKSTRA-BFS to Algorithm BF* with decision function $\mathbf{N}_h$.

Remark 4.77 Given a simple, finite, digraph $\mathcal{G} = (\mathcal{V}, \mathcal{R})$, a node $s \in \mathcal{V}$ and a set $\Gamma \subseteq \mathcal{V}$.
Let $N > 1$, $k \geq 1$, and let $h : \mathcal{R} \to \{1, \ldots, N\}$. Then the following are true:
If there exists an s-Γ-path P^+ at all then $\mathrm{BF}^*(\mathbf{N}_h)$ will output an $\mathbf{N}_h$-minimal goal path. To see this we define $C := C_0 := F := \mathbf{N}_h$; then C, C_0, and F satisfy all conditions in Theorem 4.39. This follows from Theorem 2.167.
If even h is injective and $\mathcal{G} = \mathcal{T}(N, k)$ then $\mathrm{BF}^*(\mathbf{N}_h)$ makes the same expansions as DIJKSTRA-BFS with the tie-breaking rule described in 4.75.   □

4.2.2.1.7   Relationships between Depth-First search and BF*

This paragraph is about finding paths with $\mathrm{BF}^*(-\ell)$ or similar search strategies. Moreover, we shall compare $\mathrm{BF}^*(-\ell)$ to Depth-First search (DFS).

Now we consider the paths output by $\mathrm{BF}^*(-\ell)$. At first glance, the following analogy seems to be valid: $\mathrm{BF}^*(\ell)$ always outputs ℓ-minimal s-Γ-paths, and $\mathrm{BF}^*(-\ell)$ always outputs $(-\ell)$-minimal s-Γ-paths, i.e. longest candidates. The next example, however, shows that this is not the case.

Example 4.78 It may occur that $\mathrm{BF}^*(-\ell)$ outputs a goal path of maximual, of medium and of minimal length, respectively. The following figure shows three rooted trees; for each tree, the set Γ is defined as the set of its leaves.

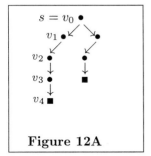

Figure 12A

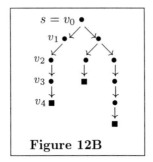

Figure 12B

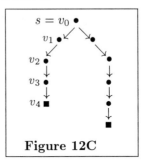

Figure 12C

We assume that $\mathrm{BF}^*(-\ell)$ resolves ties by expanding the leftmost possible candidate. That means that in each of the three examples, $\mathrm{BF}^*(-\ell)$ will expand the nodes $v_0 = s$, $v_1, \ldots, v_4$ in this order. When selecting v_4, the algorithm $\mathrm{BF}^*(-\ell)$ outputs the goal path $P^{**} := [v_0, v_1, \ldots, v_5]$.
This path is the longest s-Γ-path in Figure 12A, it is an s-Γ-path of medium length in Figure 12B, and it is an s-Γ-path of minimal length in Figure 12C.   □

Next, we explain why $\mathrm{BF}^*(\ell)$ always outputs shortest paths but $\mathrm{BF}^*(-\ell)$ does not always output longest paths.

Remark 4.79 The problem to find shortest s-Γ-paths means a minimization of $\ell = \ell_{\max}$, and that is a minimax problem.

The problem to find longest s-Γ-paths, however, means a maximization of $\ell = \ell_{\max}$, and that is a maximax problem.

Even when replacing ℓ by $(-\ell)$ we obtain a minimin problem and not a minimax problem, which can be solved with BF*-strategies.

Note that Example 4.34 is not in conflict with Theorem 4.34; that result says that BF*(F) outputs an $F_{\max}$-minimal s-Γ-path. But each s-Γ-path P is $F_{\max}$-minimal if $F = -\ell$ because $F_{\max}(P) = (-\ell)_{\max}(P) = -\ell([s]) = 0$.   □

Next, we show how to modify the given digraph $\mathcal{G}$ to make BF* output C-maximal paths if $C = \ell$ or C is another additive path function. Our construction works even if there exists a path that visits two nodes of Γ; this is the case in Figure 13A.

Remark 4.80 Given a finite acyclic digraph $\mathcal{G} = (\mathcal{V}, \mathcal{R})$ with n nodes. Let $s \in \mathcal{V}$ and $\Gamma \subseteq \mathcal{V}$. Moreover, let $C := SUM_h$ be an additive path function. We are searching for a C-maximal s-Γ-path.

To find such a path we define a new digraph $\widetilde{\mathcal{G}} := (\widetilde{\mathcal{V}}, \widetilde{\mathcal{R}})$ as follows: Let $\Gamma = \{x_1, \ldots, x_k\}$; then we generate a new set $\widetilde{\Gamma} = \{\widetilde{x}_1, \ldots, \widetilde{x}_k\}$ of goal nodes. Let $\widetilde{\mathcal{V}} := \mathcal{V} \cup \widetilde{\Gamma}$, and let $\widetilde{\mathcal{R}} := \mathcal{R} \cup \{(x_1, \widetilde{x}_1), \ldots, (x_k, \widetilde{x}_k)\}$. We define $\psi(P) := P \oplus (x_i, \widetilde{x}_i)$ for all $i = 1, \ldots, k$ and all $P \in \mathcal{P}(s, x_i)$; then ψ is a bijection from set of all s-Γ-paths in $\mathcal{G}$ onto the set of all s-$\widetilde{\Gamma}$-paths in $\widetilde{\mathcal{G}}$; the path $\psi^{-1}(\widetilde{P})$ is generated by cutting off the last arc of $\widetilde{P}$.

Next, we define $\widetilde{h} : \widetilde{\mathcal{R}} \to \mathbb{R}$. Let $\Delta := 1 + 2 \cdot \sum_{r \in \mathcal{R}} |h(r)|$. We define $\widetilde{h}(r) := -h(r)$ for all $r \in \mathcal{R}$, and $\widetilde{h}(x_i, \widetilde{x}_i) := \Delta$ for all $i = 1, \ldots, k$.

The construction of $\widetilde{\mathcal{G}}$ is illustrated in Figure 13A and 13B. Each arc r of $\mathcal{G}$ and of $\widetilde{\mathcal{G}}$ are marked with $h(r)$ and with $\widetilde{h}(r)$, respectively; in particular, $\Delta = 1 + 2 \cdot (|1| + |2| + |5| + |-3|) = 23$. Any node v is drawn as a 'O', as a '□', and as a '■' if $v \in \mathcal{V} \backslash \Gamma$, $v \in \Gamma$, and $v \in \widetilde{\Gamma}$, respectively.

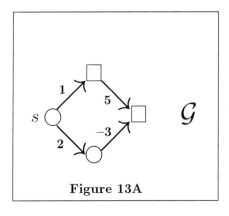

Figure 13A

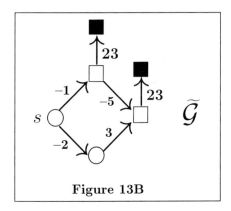

Figure 13B

Let $F := SUM_{\widetilde{h}}$. Then $F(\widetilde{P}) = \Delta - C(\psi^{-1}(\widetilde{P}))$ for all s-$\widetilde{\Gamma}$-paths $\widetilde{P}$; to see this let r be the last arc of $\widetilde{P}$; then r is the only arc of $\widetilde{P}$ that reaches $\widetilde{\Gamma}$, and $\psi^{-1}(\widetilde{P})$ is a path in $\mathcal{G}$. Consequently, $F(\widetilde{P}) = F(\psi^{-1}(\widetilde{P})) + \widetilde{h}(r) = -C(\psi^{-1}(\widetilde{P})) + \Delta$.

Moreover, $F_{\max}(\widetilde{P}) = F(\widetilde{P})$ for all s-$\widetilde{\Gamma}$-paths $\widetilde{P}$. To see this we assume that Q be a prefix of $\psi^{-1}(\widetilde{P})$. Then $-C(Q) \leq \sum_{r \in \mathcal{R}} |h(r)|$, and $C(\psi^{-1}(\widetilde{P})) \leq \sum_{r \in \mathcal{R}} |h(r)|$ because $\mathcal{G}$ is acyclic. Consequently, $-C(Q) + C(\psi^{-1}(\widetilde{P})) < \Delta$. This implies that $F(Q) = -C(Q) < \Delta - C(\psi^{-1}(\widetilde{P})) = F(\widetilde{P})$ for all prefixes $Q \leq \psi^{-1}(\widetilde{P})$. Consequently, $F_{\max}(\widetilde{P}) = F(\widetilde{P})$.

We have seen that $F_{\max}(\widetilde{P}) = F(\widetilde{P}) = \Delta - C(\psi^{-1}(\widetilde{P}))$ for all s-$\widetilde{\Gamma}$-paths $\widetilde{P}$; moreover, ψ is bijective. Consequently, we can find a C-maximal s-Γ-path P^* as follows: We use $BF(F)$ to find an $F_{\max}$-minimal s-$\widetilde{\Gamma}$-path $\widetilde{P}^*$, and we construct $P^* := \psi^{-1}(\widetilde{P}^*)$. ☐

Remark 4.81 Given the following problem: Let $\mathcal{G}$ be an acyclic digraph, let C_0 be order preserving, and let $C = (C_0)_{\max}$; find a C-maximal s-Γ-path.
If C_0 is prefix monotone and additive then $C = C_0$, and the problem can be solved as described in Remark 4.80
We leave it to the reader to develop a search strategy for C-maximal goal paths if C_0 is not prefix monotone or not additive. ☐

Next, we describe relationships between the algorithm BF* with decision function $(-\ell)$ and Depth-First search.

Definition/Remark 4.82 We define *BF*-DFS* as the algorithm BF* with cost and decision function $C := F := -\ell$; indeed, we may consider this algorithm as a Depth-First search strategy.
In contrast to this, we define *DFS* as the Deapth-First search procedure that visits and marks all nodes of a given digraph (e.g., see Remark 2.168). ☐

If $\mathcal{G}$ equals the rooted tree $\mathcal{T}(N, k)$ in Remark 2.168 we can formulate the similarities between BF*-DFS and DFS more precisely.

Theorem 4.83 *Given the notations of Remark 2.168. We assume that BF*-DFS uses the following tie-breaking rule: If $\overline{v}^{(1)}, \overline{v}^{(2)}, \ldots, \overline{v}^{(j)} \in OPEN$ and $\ell\left(P(\overline{v}^{(1)})\right) = \ldots = \ell\left(P(\overline{v}^{(j)})\right)$ is minimal then BF*-DFS expands the path with the minimal $\mathbf{n}_h$-value.[21] Also, we assume that $\Gamma = \emptyset$ so that BF*-DFS will expand all nodes of $\mathcal{G}$.*
Let DFS mark the nodes of $\mathcal{G}$ as described in Remark 2.168.
Then the following is true for all nodes $v \in V$ and for all numbers $t \in \mathbb{N}$:
BF-DFS expands v in the tth iteration if and only if DFS marks v in the tth iteration.*

P r o o f (Sketch) : The proof is an induction on t. The main idea is to assume that BF*-DFS organizes the set $OPEN_t$ as a linear list with the following property:

---

[21]This is also the candidate with minimal $\mathbf{N}_h$-value because all candidates $P(\overline{v}^{(i)})$ have the same length.

If the list contains the nodes $x_1, \ldots, x_j$ in this order then $\mathbf{n}_h(x_1) < \mathbf{n}_h(x_2) < \ldots < \mathbf{n}_h(x_j)$. It turns out that BF* autimatically processes the list OPEN as a LIFO-list (LIFO = Last In First Out); any insertion or removal of a node is done on the left side of OPEN. Consequently, BF*-DFS expands each node v exactly when v is marked by DFS. — The details of this proof are left to the reader.   □

Next, we compare BF*-DFS to Algorithm A.

Theorem 4.84 *Given the notations Remark 2.168.*
We assume that BF-DFS uses the same tie-breaking rule as in 4.83, and we assume that $\Gamma = \emptyset$.*
Let $c < 1$, and let $p : V \to [0, c)$ be a heuristic function. We assume that $p(x) < p(y) \iff \mathbf{n}_h(x) < \mathbf{n}_h(y)$ for all nodes x, y of the same height. Let $F := -\ell + p$.
Then BF-DFS and Algorithm A [22] with decision function F will expand the same nodes during the same iterations.*

P r o o f : Let $x, y \in V$. The algorithm BF*-DFS selects x and not y for expansion if $-\ell(P(x)) < -\ell(P(y))$ or if $-\ell(P(x)) = -\ell(P(y))$ and $\mathbf{n}_h(P(x)) < \mathbf{n}_h(P(y))$. If $-\ell(P(x)) < -\ell(P(y))$ then $F(P(x)) < -\ell(P(y)) \le F(P(y))$ so that Algorithm A prefers x to y. If $\ell(P(x)) = \ell(P(y))$ and $\mathbf{n}_h(P(x)) < \mathbf{n}_h(P(y))$ then $p(x) < p(y)$ so that $F(P(x)) < F(P(y))$ and Algorithm A prefers again x to y.
That means that BF*-DFS and Algorithm A use equivalent rules to select a node for expansion. This implies the assertion.   □

Next, we compare BF*-DFS to Algorithm BF* with decision function $\mathbf{n}_h$.

Remark 4.85 Given a finite, acyclic digraph $\mathcal{G} = (\mathcal{V}, \mathcal{R})$, a node $s \in \mathcal{V}$ and a set $\Gamma \subseteq \mathcal{V}$. Let $N > 1$, $k \ge 1$, and let $h : \mathcal{R} \to \{1, \ldots, N\}$ be injective.

Then the following are true:

If there exists an s-Γ-path at all then BF*$(\mathbf{n}_h)$ will output an $\mathbf{n}_h$-minimal goal path. To see this we define $C := F := \mathbf{n}_h$; then F and C satisfy all conditions in Theorem 4.39; in particule, $\mathbf{n}_h$ is order preserving by Theorem 2.165.
If even $\mathcal{G} = \mathcal{T}(N, k)$ then BF*$(\mathbf{n}_h)$ executes the same expansions as BF*-DFS with the tie-breaking rule described in 4.75.   □

Next, we describe a situation in which Depth-Fist search is preferable to Breadth-First search.

Remark 4.86 We have seen in 4.78 that BF*-DFS cannot be used to find goal paths of extremal length. The advantage of BF*-DFS, however, is the following: When searching for an arbitrary goal path then BF*-DFS often requires less space than DIJKSTRA-BFS.

For example, let $\mathcal{G} = (\mathcal{V}, \mathcal{R})$ be a finite rooted N-ary tree with n nodes and a height $\mathbf{h}(\mathcal{G}) \in O(n)$. We define s as the root of $\mathcal{G}$ and Γ as the set of all leaves of $\mathcal{G}$. Let P^*

---

[22] Note that p is not admissible.

be an s-Γ-path of minimal length, and let W be the set of all nodes in $\mathcal{G}$ whose height is equal to $\ell(P^*) - 1$.

The set OPEN of DIJKSTRA-BFS may require $\Omega(n)$ space. To see this we consider the situation immediately before DIJKSTRA-BFS selects $\omega(P^*)$ for expansion. If $w \in W$ then OPEN contains either w itself or all children of w. That means that OPEN contains $|W|$ nodes, and it is possible that $|W| \approx n/N$.

The set OPEN of BF*-DFS, however, only requires logarithmic space. We outline the reason for this space complexity. It can be shown that the following is true in any step of BF*-DFS: If two nodes $v', v'' \in$ OPEN have the same height > 0 then v' and v'' have the same father. That means that at most N nodes of height h can be in OPEN for each $h = 0, 1, \ldots, \mathbf{h}(\mathcal{G})$. Hence, OPEN contains $\approx N \cdot \mathbf{h}(\mathcal{G})$ nodes. That means that OPEN only requires logarithmic space. $\qquad\square$

4.2.2.2 The algorithm A** and related BF**-strategies

This paragraph is about BF**-strategies whose expansion function $F^{(1)}$ and redirection function $F^{(2)}$ have the property that $F^{(1)} = F^{(2)}_{\max}$; if, in addition, $F^{(2)} = C + p$ where C is additive and p is an admissible heuristic function then we shall call this BF**-strategy *Algorithm A **. Our algorithm A** is almost the same as the one in [DePe85]; the main difference is that our version of A** outputs $P(\overline{v})$ in step 4 whereas Dechter's and Pearl's algorithm A** outputs $Q(\overline{v})$.

The decision function $F^{(1)} = F^{(2)}_{\max}$ of A** has the advantage that it describes the quality of a path P at least as exactly as $F^{(2)} = C + p$. To see this we assume that P be a path with start node s and that P^* be C-minimal among all s-Γ-paths $\widetilde{P}$ with $P \leq \widetilde{P}$. The admissibility of p implies that $C(Q) + p(\omega(Q)) \leq C(P^*)$ for all prefixes $Q \leq P \leq P^*$; consequently,

$$C(P) + p(\omega(P)) \;\leq\; \max_{Q \leq P} \big(C(Q) + p(\omega(Q))\big) \;\leq\; C(P^*).$$

That means that $F^{(2)}_{\max}(P) = \max_{Q \leq P}\big(C(Q) + p(\omega(Q))\big)$ approximates $C(P^*)$ at least as well as $F^{(2)}(P) = C(P) + p(\omega(P))$.

On the other hand, $(C + p)_{\max}$ is not always order preserving as shown in Theorem 4.87; it is therefore a much worse redirection function than $(C + p)$

That means that each decision function $F^{(1)} = (C + p)_{\max}$ and $F^{(2)} = C + p$ of A** has the following advantages:

- The value $F^{(1)}(P)$ of the expansion function $F^{(1)}$ often approximates the cost of P^* more exactly than $F^{(2)}$.
- The redirection function $F^{(2)}$ is order preserving.

Next, we study several results about $\mathrm{BF}^{**}\big(F^{(1)}, F^{(2)}\big)$ with $F^{(1)} = F^{(2)}_{\max}$. The following theorem says that the expansion function $F^{(1)} = H_{\max}$ is not always order preserving even if the redirection function $F^{(2)} = H$ has this property.

Theorem 4.87 *The path function $H_{\max}$ is not always order preserving even if H is additive.*

P r o o f : An example is given in Figure 14. Each arc r is marked with a weight $h(r)$. We define $H := SUM_h$.

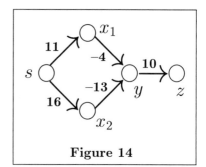

Figure 14

Then H attains the following values:

$$H([s]) = 0,$$

$$H([s, x_1]) = 11, \qquad H([s, x_2]) = 16,$$
$$H([s, x_1, y]) = 7, \qquad H([s, x_2, y]) = 3,$$
$$H([s, x_1, y, z]) = 17, \quad H([s, x_2, y, z]) = 13.$$

Consequently, $H_{\max}([s, x_1, y]) = 11 < 16 = H_{\max}([s, x_2, y])$,

whereas $H_{\max}([s, x_1, y, z]) = 17 > 16 = H_{\max}([s, x_2, y, z])$.

That means that $H_{\max}([s, x_1, y]) < H_{\max}([s, x_2, y])$ but $H_{\max}([s, x_1, y] \oplus [y, z]) > H_{\max}([s, x_2, y] \oplus [y, z])$. Hence, $H_{\max}$ is not order preserving. $\qquad\square$

Remark 4.88 The graph in Remark 4.87 contains arcs with $h(r) < 0$; it is not possible to construct a similar example with $h(r) \geq 0$ for all arcs. The reason is that $H_{\max} = H$ for all prefix monotone path functions. Consequently, if H is prefix monotone and order preserving then $H_{\max} = H$ is order preserving, too. $\qquad\square$

Next, we describe the paths output by $BF^{**}\big(F^{(1)}, F^{(2)}\big)$ if $F^{(1)} = F^{(2)}_{\max}$. The result is similar to 4.34.

Theorem 4.89 *Given a simple, locally finite digraph $\mathcal{G} = (\mathcal{V}, \mathcal{R})$, a start node $s \in \mathcal{V}$ and a set $\Gamma \subseteq \mathcal{V}$ of goal nodes. Given the cost functions $C, C_0 : \mathcal{P}(s) \to (\mathbf{R}, \preceq)$ where $\preceq$ is total and identitive; let $C = (C_0)_{\max}$. (E.g., $\mathbf{R}$ may be equal to $\mathbb{R}$, and "$\preceq$" may be equal to "$\leq$".) Let $\mathcal{M} \subseteq \mathbf{R}$ be the following set: $\mathcal{M} := \{C(Q) \,|\, Q$ is an s-Γ-path$\}$. We make the following assumptions:*

     (i) *There exists a C-minimal goal path; that means that $\mathcal{M}$ has a minimum M.*

     (ii) *There is no upper bound on the sequence $\big(C_0([s, v_1, \ldots, v_i])\big)_{i \in \mathrm{I\!N}}$*

           if $[s, v_1, v_2, v_3, \ldots)$ is an infinite path in $\mathcal{G}$. [23]

     (iii) *C_0 is order preserving.*

Then $BF^{**}\big(F^{(1)}, F^{(2)}\big)$ with $F^{(1)} := C$ and $F^{(2)} := C_0$ will output a goal path P^{**} with minimal C-value; in other words, $BF^{**}\big(F^{(1)}, F^{(2)}\big)$ finds an $F^{(2)}_{\max}$-minimal path P^{**} from s to Γ.

(The relationship $F^{(1)} = F^{(2)}_{\max}$ follows from $F^{(1)} = C = (C_0)_{\max} = F^{(2)}_{\max}$.)

P r o o f : We show that all conditions A.1), A.2), B), C), D) of Theorem 4.24 are satisfied. Let $C^{(1)}(P) := F^{(2)}(P)$ and $C^{(2)}(P) := M$ for all paths $P \in \mathcal{P}(s)$.

---

[23] Note that $\preceq$ is assumed to be total; therefore, condition (ii) can be formulated as follows: For each $\varrho \in \mathbf{R}$, there exists an i such that $C_0([s, v_1, \ldots, v_i]) \succ \varrho$.

Condition A.1): Let $P, Q \in \mathcal{P}(s)$. Let $F^{(1)}(Q) \preceq F^{(1)}(P)$ and let P be $C^{(2)}$-bounded with regard to $C^{(1)}$. The assumption $F^{(1)}(Q) \preceq F^{(1)}(P)$ and the equation $C^{(1)} = F^{(2)}$ imply that $C^{(1)}(Q) = F^{(2)}(Q) \preceq F^{(2)}_{\max}(Q) = F^{(1)}(Q) \preceq F^{(1)}(P)$ so that $C^{(1)}(Q) \preceq F^{(1)}(P)$.

Since P is $C^{(2)}$-bounded, the following is true for all prefixes X of P: $F^{(2)}(X) = C^{(1)}(X) \preceq C^{(2)}(X) = M$. Consequently, $F^{(1)}(P) = F^{(2)}_{\max}(P) \preceq M = C^{(2)}(Q)$ so that $F^{(1)}(P) \preceq C^{(2)}(Q)$. This and the above relationship $C^{(1)}(Q) \preceq F^{(1)}(P)$ imply that $C^{(1)}(Q) \preceq F^{(1)}(P) \preceq C^{(2)}(Q)$. Hence, condition A.1) is satisfied.

Condition A.2): If $F^{(2)}(Q) \preceq F^{(2)}(P)$ and $C^{(1)}(P) \preceq C^{(2)}(P)$ then

$$C^{(1)}(Q) = F^{(2)}(Q) \preceq F^{(2)}(P) = C^{(1)}(P) \preceq C^{(2)}(P) = M = C^{(2)}(Q).$$

Conditions B), C), D): They follow from assumption (i), (ii), and (iii) respectively; recall that $C^{(1)} = F^{(2)} = C_0$ and $C^{(2)} \equiv M$.

Theorem 4.24 implies that Algorithm BF**$\left(F^{(1)}, F^{(2)}\right)$ will output a $C^{(2)}$-bounded s-Γ-path P^{**}. That means that $F^{(2)}(X) = C^{(1)}(X) \preceq C^{(2)}(X) = M$ for all prefixes $X \leq P^{**}$. This fact is used in $(*)$ of the inequality $C(P^{**}) = F^{(2)}_{\max}(P^{**}) \overset{(*)}{\preceq} M$. On the other hand, $C(P^{**}) \succeq M$ because $M = \min(\mathcal{M})$. Consequently, $C(P^{**}) = M$. That means that the path P^{**} ouput by BF**$\left(F^{(1)}, F^{(2)}\right)$ is a C-minimal s-Γ-path. $\qquad\square$

The next result follows from 4.89 and is similar to 4.35.

Theorem 4.90 *Given a simple, locally finite digraph $\mathcal{G} = (\mathcal{V}, \mathcal{R})$, a start node $s \in \mathcal{V}$ and a set $\Gamma \subseteq \mathcal{V}$ of goal nodes. Given the cost function $C : \mathcal{P}(s) \to (\mathbf{R}, \preceq)$ where $\preceq$ is total and identitive. (E.g., $\mathbf{R}$ may be equal to $\mathbb{R}$, and "$\preceq$" may be equal to "$\leq$".) We assume that BF** uses an expansion function $F^{(1)}$ and a redirection function $F^{(2)}$ with $F^{(1)} = \left(F^{(2)}\right)_{\max}$. Moreover, we make the following assumptions:*

(i) *There exists an $F^{(1)}$-minimal s-Γ-path.[24]*
(ii) *There is no upper bound on the sequence $\left(F^{(2)}([s, v_1, \ldots, v_i])\right)_{i \in \mathbb{N}}$ if $[s, v_1, v_2, v_3, \ldots)$ is an infinite path in $\mathcal{G}$.*
(iii) *$F^{(2)}$ is order preserving.*
(iv) *Each $F^{(1)}$-minimal solution path is C-minimal.*

*Then BF**$\left(F^{(1)}, F^{(2)}\right)$ will output a C-minimal solution path P^{**}.[25]*

P r o o f :   Theorem 4.89 implies that BF**$\left(F^{(1)}, F^{(2)}\right)$ outputs an $F^{(1)}$-optimal solution path P^{**}; this path is also C-minimal by our assumption. $\qquad\square$

The next result is analogous to 4.39.

---

[24]This path is automatically $\left(F^{(2)}\right)_{\max}$-minimal because $\left(F^{(2)}\right)_{\max} = F^{(1)}$.
[25]In particular, conditions (i) – (iv) of this theorem are satisfied if all assumptions of Theorem 4.89 are given; therefore, Theorem 4.90 is somewhat stronger than 4.89.

Theorem 4.91 *Given a simple and locally finite digraph* $G = (V, \mathcal{R})$, *a start node* $s \in V$, *and a set* $\Gamma \subseteq V$ *of goal nodes. Let* $C, C_0 : \mathcal{P}(s) \to (\mathbf{R}, \preceq)$ *where* $\preceq$ *is total and identitive. (E.g.,* $\mathbf{R}$ *may be equal to* $\mathbb{R}$, *and* "$\preceq$" *may be equal to* "$\leq$".) *We make the following assumptions:*

  (i)   *There exists an* s-Γ-path P^+.

  (ii) *There is no upper bound on the sequence* $\left(C_0([s, v_1, \ldots, v_i])\right)_{i \in \mathbb{N}}$
        if $[s, v_1, v_2, v_3, \ldots]$ *is an infinite acyclic path in* G.

  (iii) C_0 *is order preserving.*

  (iv) *The following are true for all* s-Γ-*paths* P:
        $$(\alpha) \;\; C_0(P) = (C_0)_{\max}(P), \quad (\beta) \;\; C_0(P) = C(P).$$

        (By the way, condition (α) is equivalent to the condition that
        $C_0(Q) \preceq C_0(P)$ for all s-Γ-paths P and all prefixes $Q \leq P$.)

  (v) G *has no* C_0-*negative cycles (see Definition 2.128).*

Let $F^{(2)} := C_0$ *and* $F^{(1)} := F^{(2)}_{\max}$.

Then $BF^{**}\!\left(F^{(1)}, F^{(2)}\right)$ *will output a* C-*minimal goal path.*

P r o o f : The proof has a similar structure as the one of 4.39. We start with the following assertion:

> All paths $P(v)$ generated by $BF^{**}\!\left(F^{(1)}, F^{(2)}\right)$ are acyclic.   (4.63)

Proof of (4.63): This fact follows from (v) and from Lemma 4.23.

Let $\widehat{M} := F^{(1)}(P^+) = F^{(2)}_{\max}(P^+)$. We now prove the following:

> There are only finitely many acyclic paths $P \in \mathcal{P}(s)$ with
> $F^{(1)}(P) \preceq \widehat{M}$.   (4.64)

Proof of (4.64): The function $F^{(1)} = F^{(2)}_{\max}$ is prefix monotone. Moreover, let P be a node injective infinite path with start node s, and let Q be a finite prefix of P; then $F^{(1)}(Q) = F^{(2)}_{\max}(Q) \succeq F^{(2)}(Q) = C_0(Q)$. This and (ii) imply that there is no upper bound on the values $F^{(1)}(Q)$, $Q \leq P$. Fact (4.64) follows from Lemma 2.125 about $H := F^{(1)}$.

Next, we show the following:

> There exists a $(C_0)_{\max}$-minimal goal path P^*, and this path is
> also C_0-minimal and acyclic.   (4.65)

Proof of (4.65): The path function $H := F^{(1)} = (C_0)_{\max}$ satisfies all conditions of Lemma 2.133; in particular, G has no H-negative cycles; this follows from (v) and from Lemma 2.131 (with $\widetilde{H} := C_0$ and $\widetilde{\widetilde{H}} := H$.) Hence, there exists an H-minimal, acyclic goal path P^*. Then P^* is $(C_0)_{\max}$-minimal because $(C_0)_{\max} = H$. It follows from (iv)(α) that P^* is also C_0-minimal.

Let now $M := C_0(P^*)$. Then $M = C_0(P^*) \overset{(iv)(\alpha)}{=} (C_0)_{\max}(P^*) = F^{(1)}(P^*)$. Moreover,

$$M \overset{(4.65)}{=} (C_0)_{\max}(P^*) \preceq (C_0)_{\max}(P^+) = F^{(1)}(P^+) = \widehat{M}. \qquad (4.66)$$

Let $P^* = [w_0, w_1, w_2, \ldots, w_l]$ with $w_0 = s$ and $\gamma^* := w_l \in \Gamma$. We define $P_j^* := [w_0, w_1, \ldots, w_j]$ for each $j = 0, \ldots, l$. We now show the following assertion:

Let $t \in \mathbb{N}$, and let BF** execute the tth iteration. Then there exists a node w of P^* such that $w \in \text{OPEN}_t$ and $F^{(2)}(P_t(w)) \preceq M$. $\qquad (4.67)$

Proof of (4.67): It is $\gamma^* \notin \text{CLOSED}_t$; the argumentation is the same as in the proof of (4.4). So, not all nodes of P^* are elements of CLOSED_t. We choose the minimal number λ such that $w_\lambda \notin \text{CLOSED}_t$. It follows from Lemma 4.18 a) that $w_\lambda \in \text{OPEN}_t$. Then Corollary 4.20 implies that $F^{(2)}(P_t(w_\lambda)) \preceq F^{(2)}(P_\lambda^*)$. Moreover, $F^{(2)}(P_\lambda^*) \preceq F_{\max}^{(2)}(P^*) = (C_0)_{\max}(P^*) \overset{(iv)(\alpha)}{=} C_0(P^*) = F^{(2)}(P^*) = M$ so that $F^{(2)}(P_t(w_\lambda)) \preceq F^{(2)}(P_\lambda^*) \preceq M$.

The next assertions say that all paths $P_t(v)$ are weakly M-bounded with respect to $F^{(2)}$. More precisely, the following are true for all $t \in \mathbb{N}$ and for all vertices v:

If $P_t(v)$ exists then $P_t(v)$ is weakly M-bounded with respect to $F^{(2)} = C_0$. $\qquad (4.68)$

If v is equal to the node $\bar{v}_t$ selected in step 3 of the tth iteration then the whole path $P_t(v)$ is M-bounded with respect to $F^{(2)}$. $\qquad (4.69)$
(That means that $F^{(1)}(P_t(v)) = F_{\max}^{(2)}(P_t(v)) = M$.)

Proof of (4.68) *and* (4.69): We show (4.68) and (4.69) simultaneously by an induction on t. For $t = 1$, the only relevant path is $P_1(s) = [s]$. This path is M-bounded because

$$F^{(1)}([s]) = F_{\max}^{(2)}([s]) \overset{[s] \leq P^*}{\preceq} F_{\max}^{(2)}(P^*) = (C_0)_{\max}(P^*) \overset{(4.66)}{=} M.$$

We now assume that the assertions (4.68) and (4.69) are true for all paths $P_t(v)$, and we first prove that fact (4.68) remains true after the tth iteration. Obviously, the only critical situation arises if $P_{t+1}(v)$ is not equal to $P_t(v)$. Then the following must have occurred: Either v and $P_{t+1}(v)$ have been detected in step 6b) or $P_t(v)$ has been replaced by $P_{t+1}(v)$ in step 6c). In both cases, $P_{t+1}(v) = P_t(\bar{v}_t) \oplus (\bar{v}_t, v)$ where $\bar{v}_t$ is the vertex expanded in the tth iteration. Assertion (4.69) of the induction hypothesis says that $P_t(\bar{v}_t)$ is M-bounded. This implies that $P_{t+1}(v)$ is weakly M-bounded.

We have proven that fact (4.68) is valid for all paths $P_{t+1}(v)$. We must show that assertion (4.69) about $t + 1$ is true as well; that means, $P_{t+1}(\bar{v}_{t+1})$ is M-bounded where $\bar{v}_{t+1}$ is the node selected in step 3 of the $(t + 1)$st iteration.

For this purpose, we recall (4.67); it says that there exists a node $w \in \text{OPEN}$ with $F^{(2)}(P_{t+1}(w)) \preceq M$. Moreover, $P_{t+1}(w)$ is weakly M-bounded by fact (4.68) about $t + 1$. Hence,

$P_{t+1}(w)$ is M-bounded with respect to $F^{(2)}$. $\qquad (4.70)$

Then $F^{(1)}(P_{t+1}(\bar{v}_{t+1})) \preceq F^{(1)}(P_{t+1}(w))$ because the paths $P_{t+1}(\bar{v}_{t+1})$ and $P_{t+1}(w)$ are compared with each other in step 3 of the $(t + 1)$st iteration. Moreover, it is $F^{(1)}(P_{t+1}(w)) \preceq M$ by (4.70), and it is $F^{(1)} = F_{\max}^{(2)}$. Consequently, $F^{(1)}(P_{t+1}(\bar{v}_{t+1})) \preceq F^{(1)}(P_{t+1}(w)) \preceq M$ so that $F_{\max}^{(2)}(P_{t+1}(\bar{v}_{t+1})) = F^{(1)}(P_{t+1}(\bar{v}_{t+1})) \preceq M$. Therefore, the path $P_{t+1}(\overline{v}_{t+1})$ is M-bounded with respect to $F^{(2)}$.

Assertion (4.69) has the following consequence:

> If $BF^{**}\big(F^{(1)}, F^{(2)}\big)$ outputs a solution path P^{**} then P^{**} is M-bounded with respect to $F^{(2)}$. $\qquad(4.71)$

Proof of (4.71): We assume that P^{**} is output immediately after selecting some node $\gamma^{**} \in \Gamma$ in step 3 of the the Tth iteration; then $P^{**} = P_T(\gamma^{**})$ is M-bounded by fact (4.69).

Next, we show *that* BF^{**} will actually expand a goal node γ^{**} after finitely many iterations. It follows from (4.64) and (4.66) that there exist only finitely many acyclic paths $P \in \mathcal{P}(s)$ with $F^{(1)}(P) \preceq M$. Facts (4.63) and (4.69) imply that BF^{**} cannot expand other paths than these finitely many ones. Moreover, assertion (4.10) is also true in the current situation. Hence, BF^{**} will only execute a finite number T of iterations. Moreover, fact (4.67) implies that $OPEN_T \neq \emptyset$ so that BF^{**} can only terminate by selecting a goal node γ^{**}.

We now accomplish the proof of our theorem. Fact (4.71) implies that BF^{**} will output an s-Γ-path P^{**} which is M-bounded with regard to $F^{(2)} = C_0$. Consequently, $C_0(P^{**}) \preceq (C_0)_{\max}(P^{**}) \preceq M = C_0(P^*)$; this implies that $C_0(P^{**}) \preceq C_0(P^*)$; moreover, P^* is C_0-minimal by fact (4.65); so, P^{**} is C_0-minimal as well as P^*. Hence, P^{**} is C-minimal by condition (iv)(β). $\qquad\square$

Theorem 4.91 is used in the proof Theorem 4.92; this result says that Algorithm A^{**} works correctly if A^{**} outputs $P(\overline{v})$ and not $Q(\overline{v})$; Theorem 4.92 is analogous to 4.61.

Theorem 4.92 *Given a simple, locally finite digraph $\mathcal{G} = (V, \mathcal{R})$ with $s \in V$ and $\Gamma \subseteq V$. Given the additive cost function $C : \mathcal{P}(s) \to \mathbb{R}$ and the node function $p : V \to \mathbb{R}$. We make the following assumptions:*
  (i) *There exists an s-Γ-path P^+.*
  (ii) *There is no upper bound on the sequence $\big(C([s, v_1, \dots, v_i]) + p(v_i)\big)_{i \in \mathbb{N}}$ if $[s, v_1, v_2, v_3, \dots)$ is an acyclic infinite path in $\mathcal{G}$.*
  (iii) *p is admissible.*
  (iv) *$p(v) = 0$ for all $v \in \Gamma$.*
  (v) *$\mathcal{G}$ has no C-negative cycle.*

Let $C_0 := C + p$, $F^{(2)} := C_0 = C + p$, and $F^{(1)} := F_{\max}^{(2)} = (C_0)_{\max}$. Then Algorithm $BF^{**}\big(F^{(1)}, F^{(2)}\big)$ will output a C-minimal goal path P^{**}.

P r o o f : We show that C and C_0 satisfy all conditions (i) – (v) of Theorem 4.91.

Conditions (i) *and* (ii): They follow from conditions (i) and (ii) of the current theorem.

Condition (iii): The order preservation of $C_0 = C + p$ can be proven by an easy computation.

Condition (iv): Let X be a goal path and let $Y \leq X$. Let Q be the suffix of X with $X = Y \oplus Q$, and let $v := \alpha(Q) = \omega(Y)$. Then $C_0(Y) = C(Y) + p(v) \overset{(iii)}{\leq} C(Y) + C(Q) = C(X) \overset{(iv)}{=} C(X) + p(\omega(X)) = C_0(X)$. Consequently, $C_0(Y) \leq C(X) = C_0(X)$, and this implies the equations (iv)(α) and (β) of Theorem 4.91.

Condition (v): We must show that $\mathcal{G}$ has no C_0-negative cycles; this can be done in the same way as in 4.61.

It follows from Theorem 4.91 that Algorithm A** = BF**$\left(F^{(1)}, F^{(2)}\right)$ with $F^{(2)} = C_0$ and $F^{(1)} = F^{(2)}_{\max}$ will output a C-minimal s-Γ-path. $\qquad\square$

Remark 4.93 Corollary 7 in [DePe85] says that A** is optimal with respect several criteria of complexity; these criteria have been defined on page 521 of [DePe85], and they have been mentioned in Subsection 4.2.1.4 of this book. $\qquad\square$

4.2.2.3 The algorithm BF**$\left(F^{(1)}, F^{(2)}\right)$ with an expansion function of the form $F^{(1)}(P) = f\left(\omega(P), F^{(2)}(P)\right)$

In literature, there exist several versions of BF** with the following property: The value $F^{(1)}(P)$ of the expansion function depends on the end point of P and the value $F^{(2)}(P)$ of the redirection function.

Examples are given in [BaMa83], page 18. The authors of the paper use an additive redirection function $F^{(2)}$, a path dependent heuristic $\bar{p}$ (see page 103), and an expansion function $F^{(1)}(P) := F^{(2)}(P) + \bar{p}\left(\omega(P), F^{(2)}(P)\right)$. Then $F^{(1)}$ is of the form $f\left(\omega(P), F^{(2)}(P)\right)$; this follows from defining $f(v, x) := x + \bar{p}(v, x)$ for all $v \in V$ and all $x \in \mathbb{R}$.

Another search algorithm with a redirection function $F^{(2)}$ and an expansion function $F^{(1)}(P) = f\left(\omega(P), F^{(2)}(P)\right)$ is Algorithm GENGOAL in Section 9 of [LT91] and [LTh91]; we shall discuss GENGOAL in Section 4.2.4.

4.2.3 Modifications of BF** and BF***

Here, we describe modified versions of BF** or even of BF***. The new algorithms are not always special cases of BF**; that is the reason why we describe them in a new subsection and not in 4.2.2. A detailed description of modified BF**- and BF***- algorithms is given in [Huck92], page 197 – 209, and perhaps in a second edition of this book.

4.2.3.1 Modified conditions for expansion and redirection

The original algorithm BF** searches for the minimum of the values $F^{(1)}(P(v))$, $v \in$ OPEN, to find the node to be expanded in the current iteration; BF** redirects the path $P(v)$ if $F^{(2)}(P(v)) \prec F^{(2)}(\widehat{P}(v))$. Other criteria for expansion are used by the search algorithms B and C in [Mart77] and [BaMa83]. Generalized rules of expansion and redirection are formulated in [Huck92], page 197 – 198.

Now, we describe several properties the algorithms B and C. The assumptions (i), (ii) of Theorem 4.94 and of Theorem 4.95 can be found on page 3 of [BaMa83].

Theorem 4.94 (*see also* 4.61)
Given a simple, locally finite digraph $\mathcal{G} = (\mathcal{V}, \mathcal{R})$. *Let* $s \in \mathcal{V}$ *and* $\Gamma \subseteq \mathcal{V}$.
Let $C = SUM_h$ *be a cost function basing on an arc function* $h : \mathcal{R} \to \mathbb{R}$. *Given a node function* $p : \mathcal{V} \to \mathbb{R}$. *We make the following assumptions:*

   (i)   *There exists an* s-Γ-*path* P^+.
   (ii) *There is a* $\delta > 0$ *such that* $h(r) > \delta$ *for all* $r \in \mathcal{R}$.
         Moreover, $p(v) \geq 0$ *for all* $v \in \mathcal{V}$, *and* $p(v) = 0$ *for all* $v \in \Gamma$.
   (iii) p *is admissible.*

Then Algorithm B will output a C-*minimal* s-γ-*path. Moreover, if* $\mathcal{G}$ *is finite and* $n = |\mathcal{V}|$ *then the algorithm will not execute more than* $O(n^2)$ *expansions so that its time of computation is bounded by* $O(n^3)$.

P r o o f :   See [Mart77], Theorems 3.5, 3.2 in [BaMa83], and Remark (iii) in [BaMa83].   □

Theorem 4.95 *Given a simple, locally finite digraph* $\mathcal{G} = (\mathcal{V}, \mathcal{R})$. *Let* $s \in \mathcal{V}$ *and* $\Gamma \subseteq \mathcal{V}$. *Given the additive cost function* $C : \mathcal{P}(s) \to [0, \infty)$ *and the node function* $p : \mathcal{V} \to \mathbb{R}$. *We make the following assumptions:*

   (i)   *There exists an* s-Γ-*path* P^+.

   (ii) *There is a* $\delta > 0$ *such that* $h(r) > \delta$ *for all* $r \in \mathcal{R}$.
         Moreover, $p(v) \geq 0$ *for all* $v \in \mathcal{V}$, *and* $p(v) = 0$ *for all* $v \in \Gamma$.

Then the following assertions are true:

   a) *If* $|\mathcal{V}| < \infty$ *and* $n = |\mathcal{V}|$ *then Algorithm[26]* C *will not execute more than* $O(n^2)$ *expansions so that its time of computation is bounded by* $O(n^3)$.
      This follows from [BaMa83], Theorem 3.3.

   b) *If* p *is admissible then Algorithm* C *will output a* C-*minimal solution path.*
      This follows from Theorems 3.5, 3.7 in [BaMa83], and from Remark (iii) on page 11 of the same source.

   c) *If, however,* p *is not admissible then Algorithm* C *will output at least as good paths as Algorithms A and B. More precisely, we can formulate tie breaking rules for Algorithm* C *such that the following is true: If* P_1^{**}, P_2^{**} *and* P_3^{**} *are the output paths of the algorithms A, B and C, respectively, then* $C(P_3^{**}) \leq C(P_1^{**})$ *and* $C(P_3^{**}) \leq C(P_2^{**})$.
      This result follows from Theorems 3.5 and 3.7 of [BaMa83].

   d) *If* p *is not admissible but proper then the search strategy* C *will even output a* C-*minimal goal path; moreover,* C *does not execute more than* $O(n)$ *expansions so that its time of computation is bounded by* $O(n^2)$.
      This result follows from Theorems 4.4 and 4.5 of [BaMa83].

   e) *Bagchi and Mahanti have introduced path dependent heuristic functions (see page 103). If Algorithm* C *uses particular path dependent heuristic functions then it will work in* $O(n^2)$ *time.*
      This resule follows from Theorem 5.1 in [BaMa83].   □

---
[26]The algorithm C should not be confused with the cost function C.

4.2.3.2 Modifications of the lists OPEN and CLOSED

The original algorithm BF** contains several rules for entering or leaving the lists OPEN and CLOSED; these rules are formulated in the steps 1, 3, 6b) and 6c). Several other algorithms, however, use other rules of entering or leaving OPEN and CLOSED, or they process further lists of nodes.

An example is the algorithm DIJKSTRA-BF*, which does not allow a node to return from CLOSED to OPEN. The algorithms IDA* in [Korf85] and IDA*-CR(b) in [SCGS], [SaChGS] use likewise a list of nodes that will not be expanded again.

An additional list with name QUEUE is used by the propagating algorithm PropA in [BaMa85]. (A more detailed description of this search strategy is given in 4.96.)

A detailed description of modified lists OPEN and CLOSED can be found in [Huck92], page 199.)

4.2.3.3 Modified conditions for termination or for the output

The steps 2) and 4) of BF*** have the effect that BF*** stops when the expanded node $\overline{v}$ lies in Γ and that the current path $P(\overline{v})$ or $Q(\overline{v})$ must be output. But it may be reasonable to modify these steps of BF***. For example, the algorithms IDA* in [Korf85] and IDA*-CR(b) in [SCGS], [SaChGS] can expand several nodes in Γ.

4.2.3.4 Multistage and nested Best-First search

A *multistage* Best-First search algorithm consists of several Best-First search procedures $\mathcal{A}_1, \mathcal{A}_2, \ldots, \mathcal{A}_k$, which are executed in succession. For example, the search algorithm in [ChGh89] successively executes two modifications of Algorithm A. Further examples of multistage Best-First search are the algorithms IDA*, IDA*-CR(b), and IDA*-CRM(b, M) in [Korf85], [SCGS], [SaChGS].

A *nested* Best-First search strategy consists of several Best-First search procedures $\mathcal{A}_1, \ldots, \mathcal{A}_k$; each procedure $\mathcal{A}_\kappa$, $\kappa = 1, \ldots, k-1$ can call procedure $\mathcal{A}_{\kappa+1}$. For example the propagating algorithm PropA in [BaMa85] is a nested search strategy.

It is possible to develop combinations of multistage and nested search strategies; for example, this idea is realized by applying a sequence $\mathcal{A}_1, \ldots, \mathcal{A}_k$ of Best-First search algorithms where each $\mathcal{A}_\kappa$ calls a sequence $\mathcal{A}_{\kappa, 1}, \ldots \mathcal{A}_{\kappa, k_\kappa}$ of Best-First search strategies.

More details about multistage and nested search strategies can be found in Subsections 4.2.2.5 and 4.2.2.7 of [Huck92].

4.2.3.5 Search methods with propagation

This paragraph is mainly about modifications of BF** with the following property: If a path $P_t(v)$ is replaced by a "better" path $P_{t+1}(v) = \hat{P}_t(v)$ then this event is "propagated" to other nodes w. More precisely, for particular v-w-paths Q, the path $P_t(w)$ is compared with the path $P_{t+1}(v) \oplus Q$; if $P_{t+1}(v) \oplus Q$ is "better" than $P_t(w)$ then the previous path $P_t(w)$ is replaced by $P_{t+1}(w) := P_{t+1}(v) \oplus Q$.

The following example shows that this idea is reasonable. Let BF* use a $\prec$-preserving redirection function $F^{(2)}$. Moreover, let $P_t(v)$ be a prefix of $P_t(w)$ and let Q be the v-w-path with $P_t(w) = P_t(v) \oplus Q$. If the path $P_t(v)$ is redirected then $F^{(2)}(P_{t+1}(v)) \prec F^{(2)}(P_t(v))$. This implies that $F^{(2)}(P_{t+1}(v) \oplus Q) \prec F^{(2)}(P_t(v) \oplus Q) = F^{(2)}(P_t(w))$. Therefore, it is reasonable to replace $P_t(w)$ by the path $P_{t+1}(w) := P_{t+1}(v) \oplus Q$, which has a better $F^{(2)}$-value.

An example of a propagating algorithm is Algorithm PropA in [BaMa85]. At first sight, this algorithm is more complicated than necessary. This complicated structure has been carefully[27] motivated in Subsection 4.2.2.6 of [Huck92]; several simple propagating algorithms are suggested in this source; all these algorithms have heavy disadvantages; these disadvantages are avoided by the more complicated algorithm PropA. Moreover, a generalization of PropA is given in Subsection 4.2.2.6 of [Huck92]; this algorithm is called PropBF**. The decision function of PropBF** need not be additive; the correctness of PropBF**, however, has not been proven in [Huck92].

We now describe several properties of Algorithm PropA in [BaMa85].

Remark 4.96 Given a simple digraph $\mathcal{G} = (\mathcal{V}, \mathcal{R})$ with n nodes. Let $C = SUM_h$ where $h(r) > 0$ for all arcs. Let p be a heuristic function.

a) PropA is a nested Best-First search strategy; it uses three lists of nodes: OPEN, CLOSED, and QUEUE. Moreover, PropA uses the function $(C + p)$ to assess the quality of a path.

b) The worstcase complexity of PropA is $O(n^3)$. This is an immediate consequence of Theorem 2.2 b) in [BaMa85]; this result says that a selection of a node from OPEN or from QUEUE can only occur $O(n^2)$ times.[28]

c) The algorithm PropA should be used if the following three conditions are satisfied simultaneously:

$\mathcal{G}$ contains cycles, p is not consistent (see Definition 2.135), p is admissible.

This proposal is in accordance to recommendation (i) in [BaMa85], page 26; the reason for this recommendation is the short running time of PropA.

Bagchi and Mahanti propose to use PropA also in the case that p is consistent. In this situation, however, the original algorithm A seems to be better. The advantage of Algorithm A is that it is simpler than PropA. Moreover, the decision function $C + p$ is prefix monotone by Theorem 2.137. Consequently, Algorithm A does not execute more than $O(n^3)$ steps by Lemma 4.51 and Theorem 4.59. Therefore, Algorithm A is as fast as PropA. This and the simple structure of Algorithm A makes this algorithm be better than PropA. $\square$

---

[27] The assertion in Remark 4.59 c) of [Huck92] seems to be false.

[28] Bagchi and Mahanti claim that no node is selected more than once from QUEUE (see [BaMa85], page 6, line 4); this assertion, however, is not proven. Perhaps, an exact proof of this statement is similar to the one of Lemma 4.51; we must use the fact that nodes from QUEUE are selected with the help of a prefix monotone and order preserving decision function.

Next, we describe other search strategies with propagating concepts.

The following type of propagation is performed in Algorithm BF** $\left(F^{(1)}, F^{(2)}\right)$ with $F^{(1)} = F_{\max}^{(2)}$: When computing $F_{\max}^{(2)}(P)$ then the maximal value $F^{(2)}(Q)$, $Q \leq P$ is "propagated" from $\omega(Q)$ to $\omega(P)$.

Also, propagating processes appear in many Ford-Bellman algorithms (see Section 4.3): If a node v has the predecessors $u_1, \ldots, u_k$ then information about the currently best paths $P(u_1), \ldots, P(u_k)$ is "propagated" to v in order to construct a good path $P(v)$.

4.2.3.6   Bidirectional search

When searching for an optimal s-γ-path, most Best-First search strategies start from s and construct a path in direction to γ. A *bidirectional* search strategy, however, constructs particular paths $P^+(v)$ forwards from s to v, and it constructs particular paths $P^-(v)$ backwards from γ to v; candidates for an optimal s-γ-paths are constructed by appending $P^-(v)$ at $P^+(v)$.

Examples of bidirectional search strategies are given in [Kwa89], in [Jeya83] and in [LuRa89]. All these papers describe the search for a C-optimal path where $C = SUM_h$ and all values of $h(r) \geq 0$ for all arcs r. The authors of [LuRa89] study the average running time of their algorithm for random graphs.

There exist further sources of literature about bidirectional search; it seems, however, that there exist no literature about bidirectional search in the situation that the cost measure C is not additive.

4.2.4   The algorithm GENGOAL: a search strategy whose redirection function is not order preserving

4.2.4.1   Introduction of GENGOAL

The algorithm GENGOAL in in [LT91] and [LTh91] can find optimal paths even if the redirection function is not order preserving. Among all algorithms in [LT91], GENGOAL is the search strategy that is most similar to BF**. That is the reason why we describe GENGOAL in more detail than other optimal path algorithms from literature.

Given a cost structure $(\mathbf{R}, \odot, \varrho_0, \preceq)$ (see Definition 2.11). We assume that $\preceq$ is total. Let h be an arc function and let $C_0 : \mathcal{P}(s) \rightarrow (\mathbf{R}, \preceq)$ be a cost measure. (By the way, the cost measure C_0 is called λ in [LT91] and [LTh91].) We assume that C_0 can be computed as described in Definition 2.21 b). (We shall mean this property when saying the C_0 is *embedded into the cost structure* $(\mathbf{R}, \odot, \varrho_0, \preceq)$.)

Now we give the listing of GENGOAL; the numbers before the instructions are the same as in [LT91]. The statements "begin$_i$" and "end$_i$" belong together.

Algorithm 4 (GENGOAL)

(1) d : array [vertex] subset of $\mathbf{R}$;
(2) g : array [vertex] subset of $\mathcal{V} \times \mathbf{R} \times \mathbf{R}$;
(3) b : boolean;
(4) v : vertex;
(5) ϱ, ϱ' : element of $\mathbf{R}$;
(6) D: priority queue of $\mathcal{V} \times \mathbf{R}$.
(7) $d[s] := \{\varrho_0\}$; $g[s] := \{(s, \varrho_0, \varrho_0)\}$;
(8) **For** all $v \neq s$ **do**
    $begin_0$
     (9) $d[v] := \{\infty\}$; $g[v] := \emptyset$;
    end_0;
(10) **While** $D \neq \emptyset$ **do**
    $begin_1$
    (11) $(v, \varrho) :=$ extract smallest from D; (* *This instructed is described on page 255.* *)
    (12) **If** $v = \gamma$ **then** stop;
    (13) **For** all $(v, w) \in \mathcal{R}$ **do**
      $begin_2$
      (14) $b :=$ false;
      (15) **For** all $\varrho' \in d[w]$
        $begin_3$
        (16) **If** $\varrho' \preceq_m \varrho \odot h(v, w)$ **then**
          (17) $b :=$ true;
        (18) **else if** $\varrho \odot h(v, w) \prec_m \varrho'$ **then**
          $begin_4$
          (19) $d[w] := (d[w]\backslash\{\varrho'\}) \cup \{\varrho \odot h(v, w)\}$;
          (20) $f[w] := (f[w]\backslash\{(*, \varrho', *)\}) \cup \{(v, \varrho \odot h(v, w), \varrho)\}$;
          (21) **If** $(w, \varrho \odot h(v, w)) \notin D$ **then**
            (22) Insert $(w, \varrho \odot h(v, w))$ into D
          (23) **else**
            (24) Replace (w, ϱ') by $(w, \varrho \odot h(v, w))$
          end_4
        end_3 ;
      (25) $b =$ false **then**
        $begin_5$
        (26) $d[w] := d[w] \cup \{\varrho \odot h(v, w)\}$;
        (27) $f[w] := f[w] \cup \{(v, \varrho \odot h(v, w), \varrho\}$;
        (28) Insert $(w, \varrho \odot h(v, w))$ into D;
        end_5
      end_2 ;
    end_1 ;

The set D consists of pairs (v, ϱ) where $\varrho = C_0(P(v, \varrho))$ and $P(v, \varrho)$ is an s-v-path already found by GENGOAL.

The *redirection function* of GENGOAL is $F^{(2)} := C_0$. This follows from the conditions of step (18) and from the equations $\varrho \circ h(v, w) = C_0(P(v, \varrho) \oplus (v, w))$ and $\varrho' = C_0(P(w, \varrho'))$. These equations say that $F^{(2)} = C_0$ is used to decide whether $P(w, \varrho')$ must be replaced by $P(v, \varrho) \oplus (v, w)$.

The *expansion function* is $F^{(1)}(P) := f(\omega(P), F^{(2)}(P))$. This can be seen as follows: Lengauer and Theune assume that the pairs $(v, \varrho) \in D$ are sorted with respect to the lexicographic order of the pairs $(f(v, \varrho), \varrho)$. That means that a candidate with minimal value $f(v, \varrho)$ is extracted from D; if two minimal candidates $f(v_1, \varrho_1)$ and $f(v_2, \varrho_2)$ exist with $\varrho_1 \prec \varrho_2$ then (v_1, ϱ_1) will be preferred. It is $\varrho = C_0(P(v, \varrho)) = F^{(2)}(P(v, \varrho))$ for all $(v, \varrho) \in D$ so that $f(v, \varrho) = f\left(\omega(P(v, \varrho)), F^{(2)}(P(v, \varrho))\right) = F^{(1)}(P(v, \varrho))$. Consequently, step (11) selects a pair (v, ϱ) for which $F^{(1)}(P(v, \varrho)) = f(v, \varrho) = f\left(\omega(P(v, \varrho)), F^{(2)}(P(v, \varrho))\right)$ is minimal with respect to $\preceq$; ties are broken in favour of the path $P(v, \varrho)$ with the smaller value $\varrho = F^{(2)}(P(v, \varrho))$. So, we may consider $F^{(1)}$ as expansion function of GENGOAL.

It is not assumed that $\varrho_1 \preceq \varrho_2 \Rightarrow \varrho_1 \odot \varrho \preceq \varrho_2 \odot \varrho$ for all $\varrho_1, \varrho_2, \varrho \in \mathbf{R}$. So, it may occur that there exist two paths $P_1, P_2 \in \mathcal{P}(s)$ and an arc r such that

$$C_0(P_1) \preceq C_0(P_2) \text{ but } C_0(P_1 \oplus r) = C_0(P_1) \odot h(r) \succ C_0(P_2) \odot h(r) = C(P_2 \oplus r).$$

That means that the cost function C_0 and the redirection function $F^{(2)} = C_0$ are not order preserving. In this case, Lengauer and Theune replace $\preceq$ by a monotone reduction $\preceq_m$ (see Remark 2.45). Then $F^{(2)} = C_0$ is order preserving with respect to $\preceq_m$.

The relation $\preceq_m$, however, is not necessarily total. That means: If $\mathcal{P}$ is a set of paths then the minimum of all values $C_0(P)$, $P \in \mathcal{P}(\mathcal{G})$, with regard to $\preceq_m$ is not always unique. Therefore, it is not enough to record a single path $P(v)$ as the currently optimal candidate. Algorithm GENGOAL often records simultaneously several current paths $P(v, \varrho_1), \ldots, P(v, \varrho_k)$ in its memory; all these paths are C_0-minimal among all previous s-v-paths, and the values $C_0(P(v, \varrho_\kappa))$ are pairwise incomparable with respect to $\preceq_m$.

A refined version of this idea is to divide $\mathbf{R}$ into ordered chains $\mathcal{C}^{(1)}, \ldots, \mathcal{C}^{(k)}$ with respect to $\preceq_m$; then $P(v, \varrho_\kappa)$ is defined as the C_0-minimal path among all previous s-v-paths P with $C_0(P) \in \mathcal{C}^{(\kappa)}$. This idea is described in [LT91], page 21 and in [LTh91], page 321.

4.2.4.2   Evaluation functions with special properties

In order to describe more details of GENGOAL we study special properties of evaluation functions. Several of these properties are formulated in Definition 12 of [LT91] and of [LTh91].

Definition 4.97 Given a simple and finite digraph $\mathcal{G} = (V, \mathcal{R})$ Let $s, \gamma \in \mathcal{G}$ and $\Gamma = \{\gamma\}$. Given a cost structure $(\mathbf{R}, \odot, \varrho_0, \preceq)$.
Let $\preceq_m$ be a monotone reduction of $\preceq$ as defined in 2.45. For all $a, b \in \mathbf{R}$, we write

$a \prec_m b$ if $a \preceq_m b$ and not $b \preceq_m a$; we write $a \equiv_m b$ if $a \preceq_m b$ and $b \preceq_m a$.

Moreover, given a cost function $C_0 : \mathcal{P}(s) \to (\mathbf{R}, \preceq)$ and a function $f : \mathcal{V} \times \mathbf{R} \to \mathbf{R}_{\equiv}$. (Then f is an evaluation function; these functions were defined in 2.13.) Let $F^{(1)}(P) := f(\omega(P), C_0(P))$ and $F^{(2)}(P) := C_0(P)$. (These functions $F^{(1)}$ and $F^{(2)}$ are the decision functions of GENGOAL.) We define $C_0 \langle s, \gamma \rangle := \{C_0(P) \mid P \in \mathcal{P}(s, \gamma)\}$.

f is called *monotone* if $\varrho_1, \varrho_2 \in \mathbf{R}$ and for all $v \in \mathcal{V}$,
$$\varrho_1 \preceq_m \varrho_2 \implies f(v, \varrho_1) \preceq f(v, \varrho_2).$$
f is called *strictly monotone* if for all $\varrho_1, \varrho_2 \in \mathbf{R}$ and for all $v \in \mathcal{V}$,
$$\varrho_1 \prec_m \varrho_2 \implies f(v, \varrho_1) \prec f(v, \varrho_2).$$
Note that strict monotonicity does not always imply monotonicity. It is possible that f is strictly monotone and that there exist $v \in \mathcal{V}$ and $\varrho_1 \equiv_m \varrho_2$ such that $f(v, \varrho_1) \succ f(v, \varrho_2)$; in this case, f is not monotone.

We say that f *bounds C_0 from below* if the following is true for all nodes $v \in \mathcal{V}$, for all $\varrho \in \mathbf{R}$ and for all v-γ-paths Q: $f(v, \varrho) \preceq \varrho \odot C_0(Q)$.

We say that f is *consistent* if f has the properties described in Definition 2.136.

We say that f is *regular*[29] if $f(v, \varrho) \succeq \varrho$ for all v, ϱ.

Let $\phi : C_0 \langle s, \gamma \rangle \to \mathbf{R}$ be strictly increasing. Then f is called a *ϕ-measure* if for the goal node γ and for all s-γ-paths P, the following is true:
$$F^{(1)}(P) = f(\gamma, C_0(P)) = \phi(C_0(P)).$$
Note that f is a ϕ-measure for at least one function ϕ if f has the follwing property:
$$(\forall\, \varrho_1, \varrho_2 \in \mathbf{R}) \quad \varrho_1 \prec \varrho_2 \implies f(\gamma, \varrho_1) \prec f(\gamma, \varrho_2).$$
In this case, let $\phi(\varrho) := f(\gamma, \varrho)$ for all $\varrho \in C_0 \langle s, \gamma \rangle$. Then ϕ is strictly increasing.

f is called *admissible (with respect to C_0)* if f has the properties described in Definition 2.139 b) (with $H := C_0$).

A relationship between the admissibility of f and the admissibility of heuristic functions has been given in Theorem 2.141.   □

The next result describes relationships between properties of evaluation functions.

Theorem 4.98 *Given a simple and finite digraph $\mathcal{G} = (\mathcal{V}, \mathcal{R})$; let $s, \gamma \in \mathcal{V}$, $\gamma \neq s$, and $\Gamma = \{\gamma\}$. Given a cost structure $(\mathbf{R}, \odot, \varrho_0, \preceq)$ where $\preceq$ is total. Let $f : \mathcal{V} \times \mathbf{R} \to \mathbf{R}_{\equiv}$. We make the following assumptions:*
  (i) *The operation $\circ$ is associative, and ϱ_0 is a neutral element.*
  (ii) *f bounds C_0 from below.*

Then the following assertions are true:
  a) *The function f is admissible.*
  b) *If, in addition, f is regular then $f(\omega(P''), C_0(P'')) \equiv C_0(P'')$ for all s-Γ-paths P''.*

P r o o f   o f   a) : Let P'' be a goal path and $P' \leq P''$. Let Q' be the suffix of P'' with $P'' = P' \oplus Q'$. Then $f(\omega(P'), C_0(P')) \preceq C_0(P') \odot C_0(Q')$ by assumption (ii). Moreover, $C_0(P'') = C_0(P') \odot C_0(Q')$ by (i) and Theorem 2.23. (Here, we use that

---

[29]This property of f is described in [LT91] and in [LTh91] immediately before Definition 12; but this property has not been called "regular" in those sources.

ϱ_0 is a neutral element.) So, $f(\omega(P'), C_0(P')) \preceq C_0(P') \odot C_0(Q') = C_0(P'')$, and this implies the admissibility of f.

P r o o f   o f   b) : Let P'' be an s-Γ-path; then $\gamma = \omega(P'')$. Assertion a) implies that $f(\omega(P''), C_0(P'')) \preceq C_0(P'')$, and the regularity of f implies that $C_0(P'') \preceq f(\omega(P''), C_0(P''))$. Hence, $f(\omega(P''), C_0(P'')) \equiv C_0(P'')$. $\qquad\square$

4.2.4.3   Lengauer's and Theune's main results about GENGOAL

We now quote Lengauer's and Theune's main results about GENGOAL.

Theorem 4.99 *Given a simple and finite digraph $\mathcal{G} = (\mathcal{V}, \mathcal{R})$ with n nodes and m arcs. Let $s, \gamma \in \mathcal{V}$ and let $\Gamma := \{\gamma\}$ be the set of the single goal node γ. We assume that there exists at least one s-γ-path.*
Let $(\mathbf{R}, \odot, \varrho_0, \preceq)$ be a cost structure where $\odot$ is associative, ϱ_0 is a neutral element, and $\preceq$ is total.
Let h be an arc function, and let $C, C_0 : \mathcal{P}(\mathcal{G}) \to (\mathbf{R}, \preceq)$ be two cost functions; we assume that $C(P) \in MAX\{C_0(Q) \mid Q \leq P\}$ for all paths P. Also, we assume that C_0 can be computed as described in Definition 2.21 b) and that that $\mathcal{G}$ does not have C_0-negative cycles.[30]
Let $\preceq_m$ be a monotone reduction of $\preceq$; we assume that $\preceq_m$ has a finite width q. (The word 'width' was defined on page 20).
Moreover, let $f : \mathcal{V} \times \mathbf{R} \to \mathbf{R}_\equiv$ be an evaluation function. We assume that GENGOAL uses the expansion function $F^{(1)}(P) := f(\omega(P), C_0(P))$ and the redirection function $F^{(2)} := C_0$.

Then the following are true:

a) *If f is monotone then GENGOAL will output a C-minimal s-γ-path P^*.*

b) *Suppose that f is monotone and a ϕ-measure for some ϕ; moreover, suppose that there exists a C_0-minimal path P^+ with $C(P^+) = C_0(P^+)$.*
Then GENGOAL will output a C_0-minimal s-γ-path P^*.

c) *Suppose that f bounds C_0 from below, that f is regular, and that f is a ϕ-measure*

for some ϕ. Then GENGOAL will output a C_0-minimal s-γ-path P^*.

d) *If f is monotone, strictly monotone, and consistent then the running time of GENGOAL is in $O(mq^2 + n \log n)$; there exists a fast version of GENGOAL with a running time $O(mq + n \log n)$.*

P r o o f   o f   a) : This result follows from part 3 of Lemma 7 and from Theorem 10 in [LT91]; see also part 1 of Lemma 9 in [LTh91].

P r o o f   o f   b) : This result follows from Theorem 10 and 11 in [LT91].

P r o o f   o f   c) : This result follows from Theorem 10 and 12 in [LT91].

P r o o f   o f   d) : This result is given in [LT91] at the end of Subsection 9.2. $\qquad\square$

---

[30]It follows from Lemma 2.131 that $\mathcal{G}$ has no C-negative cycles. This implies the existence of C_0-minimal and of C-minimal s-Γ-paths because we need only consider acyclic candidate paths.

We now compare the results about GENGOAL with the ones about BF** and BF*.

Remark 4.100 Lengauer's and Theune's results have three advantages.

The first advantage is that these results are valid even if $\preceq$ is not identitive.

The second advantage is that these results are valid if the redirection function is not order preserving.

The third advantage is that these results give an upper bound of the complexity of GENGOAL.   □

For these reasons, Lengauer's and Theune's results can be considered as improvements of Theorems 4.24, 4.34, and 4.39 in this book.

On the other hand, there are several disadvantages of the results about GENGOAL when comparing them with our assertions about BF** and BF*.

Remark 4.101

The first disadvantage of Lengauer's and Theune's results is that the monotone reduction $\preceq_m$ may become pathological if $F^{(2)}$ is not order preserving. (This situation was described in 2.45.) In these cases, GENGOAL requires much time and space because it must generate and record many currently optimal paths whose $F^{(2)}$-values are pairwise $\preceq_m$-incomparable.

The second disadvantage of Lengauer's and Theune's results is that they are only valid if the values $F^{(1)}(P)$ depend on $F^{(2)}(P)$ and $\omega(P)$.
In contrast to this, BF** may use decision functions $F^{(1)}$ and $F^{(2)}$ with the property that $F^{(1)}(P)$ is independent of $F^{(2)}(P)$ and $\omega(P)$; for example, it is possible that $F^{(1)}(P_1) \neq F^{(1)}(P_2)$ although $F^{(2)}(P_1) = F^{(2)}(P_2)$ and $\omega(P_1) = \omega(P_2)$. In this case, it seems to be difficult or even impossible to conclude Theorem 4.24 about $\mathrm{BF}^{**}\big(F^{(1)}, F^{(2)}\big)$ from the assertions in [LT91] and [LTh91] about GENGOAL.
If two decision functions $F^{(1)}$, $F^{(2)}$ are given, it is often difficult or even impossible to find two decision functions $\widetilde{F}^{(1)}$, $\widetilde{F}^{(2)}$ with the following properties:

- $\widetilde{F}^{(1)}$ and $\widetilde{F}^{(2)}$ effect (almost) the same decisions as $F^{(1)}$ and $F^{(2)}$,
- $\widetilde{F}^{(1)}$ can be written as $\widetilde{F}^{(1)}(P) = f(\omega(P), \widetilde{F}^{(2)}(P))$,
- f has a special structural property (see Definition 4.97).

The hardness of the problem to find such functions $\widetilde{F}^{(1)}$ and $\widetilde{F}^{(2)}$ has been demonstrated in Remark 4.63 and Theorem 4.64 of [Huck92].

The third disadvantage of Lengauer's and Theune's results is that they only describe the case that GENGOAL outputs optimal paths. Theorem 4.24, however, is also valid for particular cases in which the output path of BF** is not extremal. An example is given in 4.28.   □

4.2.5   A further algorithm for extremal goal paths: MarkA

Here, we consider the algorithm MarkA, which has another structure than BF***. Mark A was introduced in [BaMa85]; also, MarkA was described in Subsection 4.2.4 of [Huck92] but that description seems to be not correct; that is the one of the reasons why we now study MarkA in detail.

MarkA is applied to acyclic graphs $\mathcal{G}$ [31] with a start node s and a set Γ of goal nodes. The algorithm outputs C-minimal goal paths if $C = SUM_h$ and $h(r) > 0$ for all arcs r; like Algorithm A*, the search strategy MarkA uses an admissible heuristic function p.

MarkA assigns several arcs a mark. For each node v, there exists at most one currently marked arc r with $\alpha(r) = v$. The *potential solution path* (*psp*) of v is defined as the path $P(v) = [v_0, v_1, \ldots, v_k]$ with $v_0 = v$ and with the property that all arcs $(v_\kappa, v_{\kappa+1})$ are marked and that no currently marked arc with start node v_k exists. Also, MarkA assigns each node v of the explicit graph a value $b(v)$, which can be considered as an improved version of the heuristic value $p(v)$. A more exact description of $b(v)$ is given in 4.108.

The following listing of MarkA is quoted from [BaMa85] almost word-for-word.

Algorithm 5 (MarkA)

1. (* *Initialization; at the beginning, the explicit graph $\mathcal{G}'$ consists solely of the start node s.* *)
   Set $b(s) := p(s)$. If s is a goal node, label s SOLVED.

2. Repeat the following steps until s is labelled SOLVED. Then exit with $b(s)$ as the solution cost.

   2.1 Find the last node u of the psp $P(s)$. Expand u by generating all its immediate successors, if any, and update the explicit graph. For each successor u_i of u not earlier present in the explicit graph and now newly introduced, set $b(u_i) := p(u_i)$. Label SOLVED all successors of u that are goal nodes.

   2.2 Generate a set Z of nodes initially only containing node u.

   2.3 Repeat the following steps until Z is empty.

      2.3.1 Remove from Z a node v such that no descendant of v in the explicit graph $\mathcal{G}'$ appears in Z.

      2.3.2 **If** v has no successors **then** $e(v) := \infty$.

         If, however, v has the immediate successors $v_1, \ldots, v_k$ in $\mathcal{G}'$ **then** set
         $$e(v) := \min_{1 \leq i \leq k} [h(v, v_i) + b(v_i)].$$

         Let the minimum occur for $i = i_0$. (In all cases, resolve ties arbitrarily, but always in favour of a SOLVED node.) Mark the arc (v, v_{i_0}), and remove any mark of an arc (v, v_i) with $i \neq i_0$. Label v SOLVED if v_{i_0} is SOLVED.

      2.3.3 **If** $b(v) < e(v)$ **then** let $b(v) := e(v)$.

      2.3.4 Add to Z all immediate predecessors of v along marked arcs. (Ignore predecessors of v that are not connected with v by a marked arc.)

---

[31] The authors of [BaMa85] assume that $\mathcal{G}$ has no "loops". It is not quite clear whether a "loop" is meant to be an arc (v, v) or a path P with $\alpha(P) = \omega(P)$. It seems, however, that the second interpretation is correct. The reason is that Bagchi and Mahanti use a topologic ordering when proving Lemma 4.3; such an ordering exists only in acyclic graphs.

We shall use the subscripts t and $t+1$, respectively, to describe the objects generated by MarkA immediately before and after the tth iteration. In particular, we shall use the notations $b_t(v)$, $P_t(v)$, $\mathcal{G}'_t$ and $b_{t+1}(v)$, $P_{t+1}(v)$, $\mathcal{G}'_{t+1}$.

The following example shows how MarkA works.

Example 4.102   a) Let $\mathcal{G} = (\mathcal{V}, \mathcal{R})$ be a digraph with $\mathcal{V} := \{s, x, z_0, z_1\}$ and $\mathcal{R} := \{(s, z_0), (s, x), (x, z_1)\}$. Let $\Gamma := \{z_0\}$. We define $h(s, z_0) := h(x, z_1) := 1$ and $h(s, x) := 2$. Let $C := SUM_h$, and let $p(v) := 0$ for all v.
The initialization of MarkA has the effect that $b_1(s) := p(s) = 0$. The first iteration expands s. The values $b_2(z_0) := p(z_0) = 0$ and $b_2(x) := p(x) = 0$ are computed, z_0 is marked with SOLVED, and s is moved to Z. The values $e(s) := 1$ and $b_2(s) := e(s) = 1$ are caused by (s, z_0); therefore, (s, z_0) is marked and $P(s)$ is defined as $[s, z_0]$. Moreover, the label SOLVED is copied from z_0 to s. Hence, MarkA terminates after outputting the optimal path $[s, z_0]$.

b) We use the same example as in part a) up to the following exception: We define $\Gamma := \{z_1\}$. Then again $b_1(s) := p(s) = 0$ by the initialization. The first iteration expands s. The values $b_2(z_0) := p(z_0) = 0$ and $b_2(x) := p(x) := 1$ are computed, and s is moved to Z. The value $e(s) := 1$ and $b_2(s) := e(s) = 1$ are caused by (s, z_0) so that (s, z_0) is marked and $P(s)$ is defined as $[s, z_0]$. Here, however, neither z_0 nor s are marked with SOLVED. Hence, the second iteration occurs. The node z_0 is expanded and moved to Z. The values $e(z_0) := \infty$ and $b_3(z_0) := e(z_0) := \infty$ are defined and s is moved to Z. The new psp $P_3(s) = [s, x]$ is found, and it is $b_3(s) = 2$. The third iteration of MarkA expands x and finds the goal path $P_4(s) = [s, x, z_1]$.

c) It may occur that an arcs gets and looses its mark again and again. For example, let $k \in \mathbb{N}\backslash\{1\}$. Moreover, let $\mathcal{G} = (\mathcal{V}, \mathcal{R})$ with
$\mathcal{V} := \{s, x_1, \ldots, x_k, y_1, \ldots, y_k, \gamma\}$,
$\mathcal{R} := \{(s, x_1), (s, y_1)\} \cup \{(x_\kappa, x_{\kappa+1}), (y_\kappa, y_{\kappa+1}) \mid \kappa = 1, \ldots, k-1\} \cup \{(x_k, \gamma), (y_k, \gamma)\}$.
Let $\Gamma := \{\gamma\}$. Let $C := SUM_h$ where h is defined as follows:
$$h(s, x_1) := 1, \quad h(s, y_1) := 2, \qquad h(x_k, \gamma) := 1, \quad h(y_k, \gamma) := 2$$
$$h(x_\kappa, x_{\kappa+1}) := 3 \text{ if } \kappa \text{ is even}, \qquad h(x_\kappa, x_{\kappa+1}) := 1 \text{ if } \kappa \text{ is odd},$$
$$h(y_\kappa, y_{\kappa+1}) := 1 \text{ if } \kappa \text{ is even}, \qquad h(y_\kappa, y_{\kappa+1}) := 3 \text{ if } \kappa \text{ is odd}.$$
Then it occurs approximately k times that a mark is moved from (s, x_1) to (s, x_2) or from (s, x_2) to (s, x_1).   □

Example 4.102 b) shows that $b(v)$ may be infinite. We leave it to the reader to characterize the situations in which $b(v) = \infty$ for a given node v.

Next, we give an exact description of $b(v)$. Let $F(P) := C(P) + p(\omega(P))$ for all paths P. We will show that the following is true for all nodes v of the explicit graph $\mathcal{G}'$: If $b(v) < \infty$ then $b(v) = F_{\max}(P(v))$ for all nodes of the explicit graph $\mathcal{G}'$. For this purpose, we show the auxiliary results 4.103 – 4.107.

Lemma 4.103 Let $t \in \mathbb{N}$. Let $v \in \mathcal{G}'_{t+1}$, and let r be a marked arc in the graph $\mathcal{G}'_{t+1}$. Then $v \in \mathcal{G}'_t$.

P r o o f : Let t' be the iteration in which r has been marked. Then $v \in \mathcal{G}'_{t'}$ because MarkA cannot do the following things during the same iteration: finding a node w and marking an arc starting from w. Moreover, we have assumed that r is a marked arc immediately before the $(t+1)$st iteration; hence, $t' < t+1$. This and $v \in \mathcal{G}'_{t'}$ have the consequence that $v \in \mathcal{G}'_t$. $\square$

Lemma 4.104 *Let (v, w) be marked immediately before the tth iteration of step 2. Let w be loaded into Z during the tth iteration. Then v is likewise loaded into Z during the tth iteration.*

P r o o f : Suppose that v does not enter Z during the tth iteration. Then the mark (v, w) is not removed in the tth iteration. In particular, (v, w) is still marked when step 2.3.4 is executed for w. Consequently, v enters Z at this moment. That is a contradiction to the assumption. $\square$

Lemma 4.105 *Let v be a node of $\mathcal{G}'_t$, and let $b_{t+1}(v) \neq b_t(v)$ or $P_{t+1}(v) \neq P_t(v)$. Then v is loaded into Z during the tth iteration.*

P r o o f : If $b_{t+1}(v) \neq b_t(v)$ then $b(v)$ is changed in the tth iteration; this is only possible if v has entered Z.

Let $P_{t+1}(v) \neq P_t(v)$, and let $P_t(v) = [v = x_0, x_1, \ldots, x_l]$. Then all arcs $(x_\lambda, x_{\lambda+1})$ have been marked before the tth iteration. Moreover, $P_t(v) \neq P_{t+1}(v)$ implies that some node x_{λ_0} must enter Z during the tth iteration. If $\lambda_0 = 0$, we are ready because $x_{\lambda_0} = x_0 = v$. If $\lambda_0 > 0$, a repeated application of 4.104 implies that all nodes $x_{\lambda_0 - 1}$, $x_{\lambda_0 - 2}, \ldots, x_0 = v$ must enter Z as well. $\square$

Lemma 4.106 *Let v be a node of $\mathcal{G}'_t$. We make the following assumptions:*
  (i) *$\mathcal{G}'_{t+1}$ has a marked arc r with $\alpha(r) = v$.*
  (ii) *v is not expanded during the tth iteration.*
Then $\mathcal{G}'_t$ has had a marked arc r' with $\alpha(r') = v$. (It is possible that $r' = r$.)

P r o o f : Suppose that such an arc r' does not exist. Then $P_{t+1}(v) \neq P_t(v)$ because $P_{t+1}(v)$ contains r whereas $\ell(P_t(v)) = 0$. Lemma 4.105 implies that v has entered the set Z during the tth iteration. This is only possible if one of the following cases occurs:
  (1) *v is expanded in the tth iteration.*
  (2) *A marked arc r' with $\alpha(r') = v$ has existed before the tth iteration.*
      (This is a consequence of instruction 2.3.4.)
Case (1) is in conflict with assumption (ii), and case (2) is in conflict with the assumption that r' does not exist. Hence, this assumption is false, and r' does exist. $\square$

Lemma 4.107 *Let v be a node in $\mathcal{G}'_t$. We make the following assumption:*
  (+) *There exists a marked arc r in $\mathcal{G}'_t$ with $\alpha(r) = v$.*
Then all successors of v belong to $\mathcal{G}'_t$.

P r o o f : Let $t_0 \leq t$ be the first iteration in which a marked arc with start node v has been generated. Then v lies in $\mathcal{G}'_{t_0}$, and the following are true:

$$\mathcal{G}'_{t_0+1} \text{ has a marked arc } r_0 \text{ with } \alpha(r_0) = v. \tag{4.72}$$
$$\mathcal{G}'_{t_0} \text{ has no marked arc } r'_0 \text{ with } \alpha(r'_0) = v. \tag{4.73}$$

Now we show that the following is true:

$$v \text{ has been expanded in the } t_0 \text{th iteration.} \tag{4.74}$$

Suppose that (4.74) were false. Then (4.72) and $\neg$(4.74) imply that the graph $\mathcal{G}'_{t_0}$ satisfies all conditions of Lemma 4.106. Consequently, a marked arc r'_0 with $\alpha(r'_0) = v$ has existed in $\mathcal{G}'_{t_0}$. That is a contradiction to (4.73).

Note that $t \neq t_0$ by (+) and (4.73). Consequently, $t > t_0$. This and (4.74) imply that v was expanded before the tth iteration has started. Therefore, all successors of v are elements of $\mathcal{G}'_t$. ☐

Next, we give an exact description of $b(v)$.

Theorem 4.108 *Let* $t \in \mathbb{N}$, *and let* v *be a node of the explicit graph* $\mathcal{G}'_t$. *Then* $b_t(v) = F_{\max}(P_t(v))$ *or* $b_t(v) = \infty$.

P r o o f : The proof is an induction on t. If $t = 1$ then s is the only node in $\mathcal{G}'_t$. Moreover, $b_t(s) = p(s) = F_{\max}(P_t(s))$ because $P_t(s) = [s]$.

Let $t \geq 1$. We assume that our result is true for all $t' \leq t$, i.e.,

$$(\forall\, t' \leq t)(\forall\, v \in \mathcal{G}'_{t'}) \quad b_{t'}(v) < \infty \;\Rightarrow\; b_{t'}(v) = F_{\max}(P_{t'}(v)). \tag{4.75}$$

We must prove assertion (4.75) for $t + 1$, i.e.,

$$(\forall\, v \in \mathcal{G}'_{t+1}) \quad b_{t+1}(v) < \infty \;\Rightarrow\; b_{t+1}(v) = F_{\max}(P_{t+1}(v)). \tag{4.76}$$

Let $v_1, \ldots, v_k$ be the nodes of $\mathcal{G}'_{t+1}$ sorted in an reverse topological order; that means that $(v_i, v_j) \Rightarrow i > j$ for all arcs $(v_i, v_j) \in \mathcal{G}'_{t+1}$. Then (4.76) is equivalent to the following statement:

$$(\forall\, i = 1, \ldots, k) \quad b_{t+1}(v_i) < \infty \;\Rightarrow\; b_{t+1}(v_i) = F_{\max}(P_{t+1}(v_i)). \tag{4.77}$$

We show (4.77) by an induction on i. Let $i = 1$. Then no arc $(v_i, v_j) \in \mathcal{G}'_{t+1}$ exists because otherwise $j < i = 1$, and this is not possible. Hence, $v_i = v_1$ is a tip node, and $P_{t+1}(v_i) = [v_i]$. Consequently, one of the following cases occurs: $b_{t+1}(v_i) = \infty$ or $b_{t+1}(v_i) = p(v_i) = F_{\max}([v_i]) = F_{\max}(P_{t+1}(v_i))$.

We assume that (4.77) is true for all $i' \leq i$, i.e.,

$$(\forall\, i' = 1, \ldots, i) \quad b_{t+1}(v_{i'}) < \infty \;\Rightarrow\; b_{t+1}(v_{i'}) = F_{\max}(P_{t+1}(v_{i'})). \tag{4.78}$$

We must show that $b_{t+1}(v_{i+1}) < \infty \;\Rightarrow\; b_{t+1}(v_{i+1}) = F_{\max}(P_{t+1}(v_{i+1}))$. For this purpose, we assume that

$$b_{t+1}(v_{i+1}) \;<\; \infty. \tag{4.79}$$

We use (4.75), (4.78), and (4.79) to show the following:

$$b_{t+1}(v_{i+1}) \;=\; F_{\max}(P_{t+1}(v_{i+1})). \tag{4.80}$$

We distinguish between five cases.

Case 1: There exists no marked arc r in $\mathcal{G}'_{t+1}$ with $\alpha(r) = v_{i+1}$.

Then $b_{t+1}(v_{i+1}) \stackrel{(4.79)}{=} p(v_{i+1}) = F_{\max}(P_{t+1}(v_{i+1}))$ because $P_{t+1}(v_{i+1}) = [v_{i+1}]$.

Case 2: v_{i+1} is expanded in the tth iteration.
Let (v_{i+1}, v_j) be the arc marked by the tth iteration. Then $j < i+1$ by the reverse topological order. When expanding v_{i+1}, this node is assigned the value $b_{t+1}(v_{i+1}) = \max(p(v_{i+1}),\, h(v_{i+1}, v_j) + b_{t+1}(v_j))$. This and (4.79) imply that $b_{t+1}(v_j) < \infty$; consequently, $b_{t+1}(v_j) \stackrel{(4.78)}{=} F_{\max}(P_{t+1}(v_j))$. Then $(*)$ in the following equation results from the fact that $P_{t+1}(v_{i+1}) = (v_{i+1}, v_j) \oplus P_{t+1}(v_j)$; it is $b_{t+1}(v_{i+1}) =$

$$\max(p(v_{i+1}),\, h(v_{i+1}, v_j) + b_{t+1}(v_j)) = \max(p(v_{i+1}),\, h(v_{i+1}, v_j) + F_{\max}(P_{t+1}(v_j))) \stackrel{(*)}{=}$$
$$F_{\max}(P_{t+1}(v_{i+1}))\,.$$

Next, we assume that the conditions in the cases 1 and 2 are not satisfied. Then there exists a marked arc r in $\mathcal{G}'_{t+1}$ with $\alpha(r) = v_{i+1}$, and v_{i+1} has not been expanded in the tth iteration. Then Lemma 4.103 implies that $v_{i+1} \in \mathcal{G}'_t$. Consequently, we may apply Lemma 4.106 to see that a marked arc r' with $\alpha(r') = v_{i+1}$ has existed in $\mathcal{G}'_t$. Let $r' = (v_{i+1}, v_{j(t)})$ and $r = (v_{i+1}, v_{j(t+1)})$; it is possible that $j(t+1) = j(t)$. The topological ordering of $\mathcal{G}'_{t+1}$ implies that

$$j(t) \le i \quad \text{and} \quad j(t+1) \le i\,. \tag{4.81}$$

We continue the case distinction.

Case 3: The conditions of the cases 1 and 2 are not satisfied. Moreover, $b_{t+1}(v_{i+1}) = b_t(v_{i+1})$, and $P_{t+1}(v_{i+1}) = P_t(v_{i+1})$.
Then $b_{t+1}(v_{i+1}) = b_t(v_{i+1}) \stackrel{(4.79),(4.75)}{=} F_{\max}(P_t(v_{i+1})) = F_{\max}(P_{t+1}(v_{i+1}))\,.$

Case 4: The conditions of the cases 1 and 2 are not satisfied. Moreover, $b_{t+1}(v_{i+1}) \ne b_t(v_{i+1})$ or $P_{t+1}(v_{i+1}) \ne P_t(v_{i+1})$. Moreover, $b_t(v_{i+1}) = p(v_{i+1})$.
Instruction 2.3.1 of MarkA is executed such that $b_{t+1}(v_{j(t+1)})$ and $P_{t+1}(v_{j(t+1)})$ have been determined before $b_{t+1}(v_{i+1})$ and $P_{t+1}(v_{i+1})$. Consequently, the following equation is valid:

$$b_{t+1}(v_{i+1}) = \max\big(b_t(v_{i+1}),\, h(v_{i+1}, v_{j(t+1)}) + b_{t+1}(v_{j(t+1)})\big)\,. \tag{4.82}$$

This fact and (4.79) imply that $b_{t+1}(v_{j(t+1)}) < \infty$. Applying (4.78) to $j(t+1) \stackrel{(4.81)}{\le} i$, we see that $b_{t+1}(v_{j(t+1)}) = F_{\max}(P_{t+1}(v_{j(t+1)}))$. This fact and current assumption $b_t(v_{i+1}) = p(v_{i+1})$ are used in assertion $(*)$ of the next equation; relationship $(**)$ follows from $P_{t+1}(v_{i+1}) = (v_{i+1}, v_{j(t+1)}) \oplus P_{t+1}(v_{j(t+1)})$.

$$b_{t+1}(v_{i+1}) \stackrel{(4.82)}{=} \max\big(b_t(v_{i+1}),\, h(v_{i+1}, v_{j(t+1)}) + b_{t+1}(v_{j(t+1)})\big) \stackrel{(*)}{=}$$
$$\max\big(p(v_{i+1}),\, h(v_{i+1}, v_{j(t+1)}) + F_{\max}(P_{t+1}(v_{j(t+1)}))\big) \stackrel{(**)}{=} F_{\max}(P_{t+1}(v_{i+1}))\,.$$

Case 5: The conditions of the cases 1 and 2 are not satisfied. Moreover, $b_{t+1}(v_{i+1}) \ne b_t(v_{i+1})$ or $P_{t+1}(v_{i+1}) \ne P_t(v_{i+1})$. Moreover, $b_t(v_{i+1}) \ne p(v_{i+1})$.
In this case, there was an iteration $\tilde{t} < t$ that assigned v_{i+1} the current value $b_t(v_{i+1})$ and that marked the arc $(v_{i+1}, v_{j(t)})$. The formulation of step 2.3.1 implies that $b_{\tilde{t}+1}(v_{j(t)})$ was computed earlier than $b_{\tilde{t}+1}(v_{i+1})$. This and the current assumption

$b_t(v_{i+1}) \neq p(v_{i+1})$ have the following consequence:

$$b_t(v_{i+1}) = b_{\widetilde{t}+1}(v_{i+1}) = h\left(v_{i+1}, v_{j(t)}\right) + b_{\widetilde{t}+1}\left(v_{j(t)}\right) . \tag{4.83}$$

The steps 2.1 and 2.3.3 have the effect that always $b(v) \geq p(v)$; consequently,

$$p(v_{i+1}) \leq b_t(v_{i+1}) . \tag{4.84}$$

We now show the following fact:

$$v_{j(t+1)} \in \mathcal{G}'_{\widetilde{t}+1} . \tag{4.85}$$

This assertion is true if v_{i+1} has been expanded by the $\widetilde{t}$th iteration. If this is not the case then the conditions of Lemma 4.106 are satisfied because $\mathcal{G}'_{\widetilde{t}+1}$ contains the marked arc $r := (v_{i+1}, v_{j(t)})$ and $v_{i+1} \in \mathcal{G}'_{\widetilde{t}}$ by Lemma 4.103. It follows from Lemma 4.106 that a marked arc $r' \in \mathcal{G}'_{\widetilde{t}}$ exists with $\alpha(r') = v_{i+1}$. Then Lemma 4.107 implies that $v_{j(t+1)}$ even belongs to $\mathcal{G}'_{\widetilde{t}}$.

It follows from (4.85) that $v_{j(t+1)}$ has been detected in the $\widetilde{t}$th iteration or earlier; this fact and $i+1 > j(t+1)$ imply that step 2.3.2 of MarkA has marked the arc $\left(v_{i+1}, v_{j(t)}\right)$ after comparing it with $\left(v_{i+1}, v_{j(t+1)}\right)$. Moreover, the condition in step 2.3.1 implies that $b_{\widetilde{t}+1}\left(v_{j(t+1)}\right)$ was computed earlier than $b_{\widetilde{t}+1}(v_{i+1})$. These arguments are used in assertion $(*)$ of the following inequality. Relation $(**)$ follows from the fact that the b-value is never made smaller by MarkA.

$$b_t(v_{i+1}) \overset{(4.83)}{=} h\left(v_{i+1}, v_{j(t)}\right) + b_{\widetilde{t}+1}\left(v_{j(t)}\right) \overset{(*)}{\leq}$$
$$h\left(v_{i+1}, v_{j(t+1)}\right) + b_{\widetilde{t}+1}\left(v_{j(t+1)}\right) \overset{(**)}{\leq} h\left(v_{i+1}, v_{j(t+1)}\right) + b_{t+1}\left(v_{j(t+1)}\right) . \tag{4.86}$$

Consequently,

$$p(v_{i+1}) \overset{(4.84)}{\leq} b_t(v_{i+1}) \overset{(4.86)}{\leq} h\left(v_{i+1}, v_{j(t+1)}\right) + b_{t+1}\left(v_{j(t+1)}\right) . \tag{4.87}$$

The formulation of instruction 2.3.1 has the consequence that $b_{t+1}\left(v_{j(t+1)}\right)$ and $P_{t+1}\left(v_{j(t+1)}\right)$ must have been generated earlier than $b_{t+1}(v_{i+1})$ and $P_{t+1}(v_{i+1})$. Consequently, the following assertion is true:

$$b_{t+1}(v_{i+1}) = \max\left(b_t(v_{i+1}), \, h(v_{i+1}, v_{j(t+1)}) + b_{t+1}(v_{j(t+1)})\right) \tag{4.88}$$

This equation and assumption (4.79) imply that $b_{t+1}\left(v_{j(t+1)}\right) < \infty$. Applying (4.78) to $j(t+1) \overset{(4.81)}{\leq} i$, we see that $b_{t+1}(v_{j(t+1)}) = F_{\max}(P_{t+1}(v_{j(t+1)}))$. This fact is used in assertion $(*)$ of the following equation; relationship $(**)$ follows from $P_{t+1}(v_{i+1}) = [v_{i+1}, v_{j(t+1)}] \oplus P_{t+1}(v_{j(t+1)})$.

$$b_{t+1}(v_{i+1}) \overset{(4.88)}{=} \max\left(b_t(v_{i+1}), \, h(v_{i+1}, v_{j(t+1)}) + b_{t+1}(v_{j(t+1)})\right) \overset{(4.86)}{=}$$
$$h(v_{i+1}, v_{j(t+1)}) + b_{t+1}(v_{j(t+1)}) \overset{(4.87)}{=} \max\left(p(v_{i+1}), \, h\left(v_{i+1}, v_{j(t+1)}\right) + b_{t+1}\left(v_{j(t+1)}\right)\right)$$
$$\overset{(*)}{=} \max\left(p(v_{i+1}), \, h\left(v_{i+1}, v_{j(t+1)}\right) + F_{\max}\left(P_{t+1}\left(v_{j(t+1)}\right)\right)\right) \overset{(**)}{=} F_{\max}(P_{t+1}(v_{i+1})) .$$

We have shown that (4.80) is true in all cases $1 - 5$, and our proof is complete. □

Next, we give a generalized version Mark_2A of MarkA; the algorithm Mark_2A uses an arbitrary path function F. Mark_2A has the following properties:

- If $F = SUM_h + p$ for some functions h and p then Mark_2A does essentially the same things as MarkA itself.
- If Mark_2A generates a value $b(v) < \infty$ then $b(v) = F_{\max}(P(v))$.

The term $F_{\max'}(P)$ in the listing of Mark_2A is defined as the maximum of all values $F(Q)$ where Q is a prefix of P and $Q \neq [\alpha(P)]$.

Algorithm 6 (Mark_2A)

1. (* *Initialization; at the beginning, the explicit graph $\mathcal{G}'$ consists solely of the start node s.* *)
   Let $b(s) := F([s])$. **If** s is a goal node **then** label s SOLVED.

2. Repeat the following steps until s is labelled SOLVED. Then exit with $b(s)$ as the solution cost.

   2.1 Find the last node u of the psp $P(s)$. Expand u by generating all its immediate successors, if any, and update the explicit graph. For each successor u_i of u not earlier present in the explicit graph and now newly introduced, set $b(u_i) := P([u_i])$. Label SOLVED any successors of u that are goal nodes.

   2.2 Generate a set Z of nodes initially only containing node u.

   2.3 Repeat the following steps until Z is empty.

   2.3.1 Remove from Z a node v such that no descendant of v in the explicit graph $\mathcal{G}'$ appears in Z.

   2.3.2 **If** v has no successors **then** $e := \infty$.
   If v has the immediate successors $v_1, \ldots, v_k$ in $\mathcal{G}'$ **and if** $b(v_1) = \ldots = b(v_k) = \infty$ **then** let $e(v) := \infty$.
   If a SOLVED node v_{i_0} exists **then** mark the arc (v, v_{i_0}); **else** mark an arbitrary arc (v, v_{i_0}). Remove the mark of any arc (v, v_i) with $i \neq i_0$.

   If v has the immediate successors $v_1, \ldots, v_k$ in $\mathcal{G}'$ **and if** there exists a κ with $b(v_\kappa) < \infty$ **then** let
   $$e(v) := \min\{F_{\max'}((v, v_i) \oplus P(v_i)) \mid b(v_i) < \infty\}.$$
   Let the minimum occur for $i = i_0$. (Resolve ties arbitrarily, but in favour of a SOLVED node.) Mark the arc (v, v_{i_0}), and remove any mark (v, v_i) with $i \neq i_0$. Label v SOLVED is v_{i_0} is SOLVED.

   2.3.3 Let $b(v) := \max(F([v]), e(v))$.

   2.3.4 Add to Z all immediate predecessors of v along marked arcs. (Ignore predecessors of v that are not connected with v by a marked arc.)

Next, we comment on the literature about MarkA.

Remark 4.109   a) The authors of [BaMa85] define a quantity $\widehat{Q}(v)$, and Lemma 4.3 of [BaMa85] says that $b(v) \leq \widehat{Q}(v)$. This assertion seems to be false; it seems to be possible that $h(v) = p(v) > \widehat{Q}(v)$.
b) As mentioned above, the description of MarkA in [Huck92] is partially false. □

Next, we compare MarkA with other algorithms.

Remark 4.110   a) As mentioned above, MarkA has another structure than the
best-first search method BF*** on page 182. There are the following differences:

    1.   MarkA automatically expands the tip node.   The Best-First search
strategies, however, select the next expanded node as the best candidate
from a list OPEN.

    2.   The paths $P(v)$ of Algorithm MarkA start with v; the paths $P(v)$, $\widehat{P}(v)$,
and $Q(v)$ generated by BF***, however, end with v.

b) The authors of [BaMa85] write that MarkA is similar to the search strategy AO*
for AND/OR graphs. (Two papers about AO* are quoted on page 394.)

c) When searching for a new psp $P(v)$, MarkA executes a minimization over the
candidates $(u, v_\kappa) \oplus Y(v_\kappa)$.   This process is similar to the minimization
over the paths $P(u_\kappa) \oplus (u_\kappa, v)$ in the Ford-Bellman strategies (see Section 4.3).

d) MarkA may be considered as a propagating search strategy because the
information about a new psp $P(v)$ is propagated backwards to s.   □

4.2.6   Further problems

The *quickest path problem* is the following:

Problem 8 *Given a digraph* $\mathcal{G} = (\mathcal{V}, \mathcal{R})$, *and two nodes* s, γ. *Given two arc functions*
$c_1, c_2 : \mathcal{R} \to (0, \infty)$ *and a number* $\sigma > 0$; *for all paths* $P = [v_0, \ldots, v_l]$, *let* $C(P) := 0$
if $l = 0$ *and* $C(P) := \displaystyle\sum_{i=0}^{l-1} c_1(v_i, v_{i+1}) + \max_{i=0,\ldots,l-1} \left(\dfrac{\sigma}{c_2(v_i, v_{i+1})} \right)$ *if* $l > 0$.

Find a C-minimal s-γ-*path.*

The path function C gives a realistic description of the cost of transmitting data;
therefore, finding quickest paths is of great practical relevance. It is difficult to develop quickest path algorithms because C does not have the strong prefix oriented
Bellman property of type 0.

In [ChCh90], an algorithm is presented that finds quickest s-γ-paths if two nodes
$s, \gamma \in V$ are given; the algorithm works in $O(m^2 + nm \log m)$ time. Further sources
about the quickest path problem are quoted in Remark 4.172.

An inverse shortest path problem is solved in [BuT94]. Roughly speaking, the problem
is the following: Given a set $\mathcal{P}$ of s-γ-paths, reconstruct C such that C is additive and
$\mathcal{P}$ is the set of all C-minimal s-γ-paths.

An extended version of optimal path problems is solved in [OrR90], [OrR91], and
[Phil94]. The authors of these papers consider path problems with additive cost func-
tions $C_\tau : \mathcal{P}(s, \gamma) \to \mathbb{R}$ that depend on a node function $\tau : \mathcal{V} \to [0, \infty)$; the value $\tau(v)$
is interpreted as the delay when visiting v. The problem is to find P and τ such that
$C_\tau(P)$ is minimal among all candidates $P \in \mathcal{P}(s, \gamma)$ and $\tau : \mathcal{V} \to [0, \infty)$.

The papers [Na94] and [Sti92] describe the search for paths in graphs in the situation
that the cost function may change during the search process.

4.3   The Single-Source-All-Targets problem

This section is about the following problem:

Problem 9 *Given a directed or undirected graph* $\mathcal{G} = (\mathcal{V}, \mathcal{R})$, *whose nodes are* $v^{(1)}, \ldots, v^{(n)}$. *We assume that all nodes can be reached from the start node* $s = v^{(1)}$. *Moreover, let* $C : \mathcal{P}(\mathcal{G}) \to \mathbb{R}$ *be a real-valued path function.*

   a) *Find an* s-$v^{(\nu)}$*-path* $P^*\left(v^{(\nu)}\right)$ *for each node* $v^{(\nu)}$ *such that* $P^*\left(v^{(\nu)}\right)$ *has the minimal* C*-value among all paths from* s *to* $v^{(\nu)}$.
   b) *Compute the minimal* C*-value of all* s-$v^{(\nu)}$*-paths for each node* $v^{(\nu)}$.
      (*This can be done without constructing optimal* s-$v^{(\nu)}$*-paths.*)

The papers [GaPa83] and [GKPS85] give overviews over this problem and its solutions; the papers [ShWi81], [Pall84], and [GaPa88] present several search strategies and compare them with each other. A result in [SpiP73] says that $\Omega(n^2)$ is a lower bound for the complexity of Problem 9 a).

In this section, we study several search strategies in detail. One of the most well-known algorithms was created by Ford and Bellman. It is described in [Ford46], [Moo57], [Bell58], [Nolt76], [EvMi92], and in many other sources.

The original search method of Ford and Bellman works correctly if the path function C is additive. We, however, shall consider more general algorithms that solve Problem 9 a) and modifications of it even if C is not additive. In particular we present three search strategies: FORD-BELLMAN 1, FORD-BELLMAN 2, and FORD-BELLMAN 3; these algorithms solve Problem 9 a) in various situations. Moreover, we introduce the two search algorithms FORD-BELLMAN 1' and FORD-BELLMAN 2' to solve Problem 9 b).

All Ford-Bellman algorithms are based on Dynamic Programming; in particular, FORD-BELLMAN 2 is a graph theoretic version of Dynamic Programming. More about this technique of optimization can be found in [KaHe67], [Helm86], [Hck92], [MaCh93], [Hck93], [deMo94], and in many other sources.

We make the following technical conventions:
Let $\mathcal{A}$ be one of the algorithms 7 – 11. Then $P_t(v)$, $\chi_t(v)$, $\mathcal{Q}_t(v)$, respectively, are the objects $P(v)$, $\chi(v)$, $\mathcal{Q}(v)$ generated immediately before step 2 of $\mathcal{A}$; moreover, $\widehat{P}_t(v)$, $\widehat{\chi}_t(v)$, $\widehat{\mathcal{Q}}_t(v)$, respectively, are the objects $\widehat{P}(v)$, $\widehat{\chi}(v)$, $\widehat{\mathcal{Q}}(v)$ generated during step 2 of $\mathcal{A}$; this terminology is analogous to the one introduced on page 184. The *explicit* graph is defined as the graph consisting of all nodes and of all arcs or edges that are currently known by $\mathcal{A}$. Moreover, we shall consider the initialization step as the zeroth iteration of $\mathcal{A}$.

4.3.1   The algorithm FORD-BELLMAN 1

The algorithm FORD-BELLMAN 1 has almost the same structure as the usual Ford-Bellman algorithm. Two things in FORD-BELLMAN 1 are not yet specified: the condition COND in the while-loop and the output procedure WRITE; we shall consider several versions of COND and WRITE; in many special cases, it will be possible to replace the while-loop by a for-to-loop.

We now give the listing of FORD-BELLMAN 1.

Algorithm 7 (FORD-BELLMAN 1)

1. (∗ *Initialization* ∗)

    Let $P(s) := [s]$; the path $P(v)$ is still undefined if $v \neq s$.
    (∗ $P(v)$ *is the currently best path from s to v.* ∗)

    Let $W := \{s\}$.
    (∗ W *is the set of all nodes for which a path $P(v)$ has been defined.* ∗)

    Let $S := \{s\}$.
    (∗ S *is the set of all nodes for which $P(v)$ has been changed immediately before the current iteration.* ∗)

    Let $\tau := 0$. (∗ τ *is used as a counter of the iterations.* ∗)

2. **While** condition COND is satisfied **do**

    2.1. **For** $i := 1$ **to** n **do**
    If $\mathcal{N}^-(v^{(i)}) \cap S \neq \emptyset$, **then** choose a node $\widehat{u} \in \mathcal{N}^-(v^{(i)}) \cap S$
    such that $\widehat{P}(v^{(i)}) := P(\widehat{u}) \oplus (\widehat{u}, v^{(i)})$ has the following property:

    $$C\left(\widehat{P}(v^{(i)})\right) = \min\left\{C\left(P(u) \oplus (u, v^{(i)})\right) \mid u \in \mathcal{N}^-(v^{(i)}) \cap S\right\}.$$

    2.2. Let $S := \emptyset$.

    2.3. **For** $i := 1$ **to** n **do**
    If the path $\widehat{P}(v^{(i)})$ exists but no path $P(v^{(i)})$ has still been constructed **then**

    $$P(v^{(i)}) := \widehat{P}(v^{(i)}); \quad W := W \cup \{v^{(i)}\}; \quad S := S \cup \{v^{(i)}\};$$

    If both $\widehat{P}(v^{(i)})$ and $P(v^{(i)})$ exist **and** $C(\widehat{P}(v^{(i)})) < C(P(v^{(i)}))$ **then**

    $$P(v^{(i)}) := \widehat{P}(v^{(i)}); \quad S := S \cup \{v^{(i)}\}.$$

    2.4 $\tau := \tau + 1$.

3. Perform the output procedure WRITE.

Now we study the behaviour of FORD-BELLMAN 1 and begin with several elementary facts.

Theorem 4.111 *Given a digraph $\mathcal{G} = (\mathcal{V}, \mathcal{R})$ with $\mathcal{V} = \{v^{(1)} = s, v^{(2)}, \ldots, v^{(n)}\}$. Let $C : \mathcal{P}(s) \to \mathbb{R}$.*

a) *The following is true for all $t = 1, \ldots, n$, and all nodes v: If the path $P_t(v)$ exists then the length of $P_t(v)$ is at most $t - 1$.*

b) *Let $v \in \mathcal{V}$, and let $d(v)$ be the minimal length of all s-v-paths. Then the algorithm FORD-BELLMAN 1 will generate the first path $P(v)$ in the $d(v)$-th iteration (and not earlier or later).*

c) *Let $t \in \mathbb{N}$, and let $S = \emptyset$ immediately after the tth iteration. Then $t > d(v)$ for all $v \in \mathcal{V}$.*

d) *FORD-BELLMAN 1 can only delete a given path $P(v)$ if replacing it by another one. Consequently, when $P(v)$ has been generated then a path $P(v)$ will exist in all later iterations. This implies that always $S \subseteq W$.*

e) *FORD-BELLMAN 1 can only replace a given path $P(v)$ by a better one. This implies that $C(P_{t+1}(v)) \leq C(P_t(v))$ for all t and all v.*

f) *When FORD-BELLMAN 1 has found a C-minimal path $P(v)$ then $P(v)$ is never changed any more.*

P r o o f : The proof to a) is a simple induction on t.

We now show assertion b) with an induction on $d(v)$.

Let $d(v) = 1$. Then v is a successor of s. This node is in S immediately before the first iteration. Consequently, v is detected, and the path $P(v)$ is constructed.

Let $t := d(v) > 1$. We assume that the assertion is true for all v' with $d(v') = t - 1$. Let P be an s-v-path with $\ell(P) = d(v) = t$ and let u be the last node on P before v. Then $d(u) = t - 1$. This and the induction hypothesis imply that the first path $P(u)$ has been generated in the $(t - 1)$st iteration. Consequently, u is in S immediately before the tth iteration, and a path $P(v)$ is generated.

That means that the first path $P(v)$ is constructed in the $d(v)$-th iteration at the latest. Fact a), however, says that $P(v)$ is not constructed earlier. Hence, the first path $P(v)$ is generated exactly in the $d(v)$-th iteration.

We show assertion c). Let $J := \max\{d(v) | v \in \mathcal{V}\}$. Then there exists a node w with $d(w) = J$ and an s-w-path $[x_0 = s, x_1, \ldots, x_J = w]$ of minimal length. Let $j \leq J$. Then $[s, x_1, \ldots, x_j]$, too, is an s-x_j-path of minimal length; hence, $d(x_j) = j$. Part b) of the current result implies that the first path $P(x_j)$ is found during the jth iteration; consequently, $x_j \in S$ so that $S \neq \emptyset$ immediately after the jth iteration. This is true for all $j = 0, \ldots, J$. So, if $S = \emptyset$ after the tth iteration then $t > J$.

The assertions d) – f) are trivial.   □

Next, we describe a situation in which FORD-BELLMAN 1 solves Problem 9 a).

Theorem 4.112 *Given a digraph $\mathcal{G} = (\mathcal{V}, \mathcal{R})$ with $\mathcal{V} = \{v^{(1)} = s, v^{(2)}, \ldots, v^{(n)}\}$. Let $C : \mathcal{P}(s) \to \mathbb{R}$. We assume that C have the weak prefix oriented Bellman property of type 0 and that C be $=$-preserving.*

Let $v \in \mathcal{V}$ with the property that there exists a C-minimal s-v-connection. Let $P = [s = x_0, x_1, \ldots, x_t = v]$ be a Bellman path from s to v. (Such a path exists because of the weak Bellman principle.) We assume that P have the following property:

$$P \text{ has minimal length among all Bellman paths from } s \text{ to } v. \tag{4.89}$$

Then the path $P_{t+1}(v)$ immediately is a C-minimal s-v-path.

(We must not claim that $P_{t+1}(v)$ is even a Bellman path. The following situation is possible: There exists a C-minimal s-v-path Q which is not a Bellman path; if Q is shorter than all Bellman paths from s to v then FORD-BELLMAN 1 may generate $P(v) = Q$ before constructing any Bellman path.)

P r o o f : We show the result with an induction on t.

Let $t = 0$. The only existing path before the first iteration is $P(s) = [s]$. This path is C-minimal for the following reason: By the weak Bellman principle, there exists a Bellman path Q from s to $v = s$; then $[s]$ is a prefix of Q so that $[s]$ is indeed C-minimal.

We assume that the result is true for $t - 1$.

We show that the prefix $P' := [s = x_0, \ldots, x_{t-1}]$ of P is also a Bellman path of minimal length. First, the path P' is a Bellman path because P has this property. If P' did not have minimal length then there would exist a shorter Bellman path P'' from s to x_{t-1}; in particular, $C(P'') = C(P')$. Then $C(P'' \oplus (x_{t-1}, x_t)) = C(P' \oplus (x_{t-1}, x_t)) = C(P)$ by the $=$-preservation of C; moreover, $\ell(P'' \oplus (x_{t-1}, x_t)) < C(P' \oplus (x_{t-1}, x_t)) = \ell(P)$. Consequently, $P'' \oplus (x_{t-1}, x_t)$ is a shorter Bellman path from s to $x_t = v$ than P. That is a contradiction to (4.89). Therefore, the path P' must be a Bellman path of minimal length from s to x_{t-1}.

The induction hypothesis says that immediately before the tth iteration, there exists a C-minimal s-x_{t-1}-path $P_t(x_{t-1})$. That means that $C(P_t(x_{t-1})) = C(P')$. This and the $=$-preservation have the following consequence:

$$C(P_t(x_{t-1}) \oplus (x_{t-1}, x_t)) = C(P' \oplus (x_{t-1}, x_t)) = C(P). \qquad (4.90)$$

Let $\tau \le t - 1$ be the iteration in which $P_t(x_{t-1})$ was generated. Then $x_{t-1} \in S$ immediately after the τth iteration. In the $(\tau+1)$st iteration, a path $\widehat{P}_{\tau+1}(v)$ has been generated with $C(\widehat{P}_{\tau+1}(v)) \le C(P_{\tau+1}(x_{t-1}) \oplus (x_{t-1}, x_t)) = C(P_t(x_{t-1}) \oplus (x_{t-1}, v))$. Then FORD-BELLMAN 1 has defined the path $P_{\tau+2}(v)$ such that

$$C(P_{\tau+2}(v)) \le C(\widehat{P}_{\tau+2}(v)) \le C(P_t(x_{t-1}) \oplus (x_{t-1}, x_t)) \overset{(4.90)}{=} C(P).$$

Then $C(P_{t+1}(v)) \le C(P_{\tau+2}(v)) \le C(P)$ because $t + 1 \ge \tau + 2$. Hence, $P_{t+1}(v)$ is indeed an optimal s-v-connection. $\square$

We discuss Theorem 4.112 in Remarks 4.113 – 4.115.

Remark 4.113 In Theorem 4.112, we must not omit the assumption that C is $=$-preserving, even if C has the strong prefix oriented Bellman property of type 0.

A counterexample is the digraph $\mathcal{G}$ in Figure 15. We define the path function C as follows:

$C([s]) := 0,$

$C([s, v^{(2)}, v^{(3)}, v^{(5)}]) := 4,$

$C([s, v^{(2)}, v^{(3)}, v^{(5)}, v^{(6)}]) := 4,$

$C([s, v^{(4)}, v^{(5)}]) := 4,$

$C([s, v^{(4)}, v^{(5)}, v^{(6)}]) := 6,$

$C(P) := 0$ for all other paths $P \in \mathcal{P}(s).$

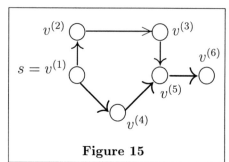

Figure 15

C has even the strong prefix oriented Bellman property of type 0. In particular, all prefixes of the optimal path $C\left(\left[s, v^{(2)}, v^{(3)}, v^{(5)}, v^{(6)}\right]\right)$ are C-minimal, and it is easy to see that all other C-minimal paths have C-minimal prefixes as well.

Moreover, C is not =-preserving because $C\left(\left[s, v^{(2)}, v^{(3)}, v^{(5)}\right]\right) = 4 = C\left(\left[s, v^{(4)}, v^{(5)}\right]\right)$ but $C\left(\left[s, v^{(2)}, v^{(3)}, v^{(5)}, v^{(6)}\right]\right) = 4 \neq 6 = C\left(\left[s, v^{(4)}, v^{(5)}, v^{(6)}\right]\right)$.

In the second iteration, FORD-BELLMAN 1 detects the path $P\left(v^{(5)}\right) := \left[s, v^{(4)}, v^{(5)}\right]$; this path will never be changed because it is already an optimal s-$v^{(5)}$-connection. Consequently, the third iteration of FORD-BELLMAN 1 generates the path $P\left(v^{(6)}\right) := \left[s, v^{(4)}, v^{(5)}, v^{(6)}\right]$ with $C\left(P\left(v^{(6)}\right)\right) = 6$ and not the optimal s-$v^{(6)}$-connection $P^* := \left[s, v^{(2)}, v^{(3)}, v^{(5)}, v^{(6)}\right]$ with $C(P^*) = 4$.

By the way, P^* will not be found in any later iteration of FORD-BELLMAN 1. The reason is that the path $\left[s, v^{(4)}, v^{(5)}\right]$ will never be replaced by $\left[s, v^{(2)}, v^{(3)}, v^{(5)}\right]$. □

Remark 4.114 In literature, Theorem 4.112 is sometimes replaced by the following assertion:

If $t \geq 1$, the tth iteration of the given Ford-Bellman algorithm generates a path $P(v)$ that is optimal among all s-v-connections of length $\leq t$.

(The path $P(v)$ need not be C-minimal among all s-v-paths; this is the difference between $P(v)$ and the path P in Theorem 4.112). $\qquad$ (4.91)

This assertion is not always true for FORD-BELLMAN 1 even if C is =-preserving and has the strong prefix oriented Bellman property of type 0. The digraph $\mathcal{G}$ in Figure 16 is a counterexample.

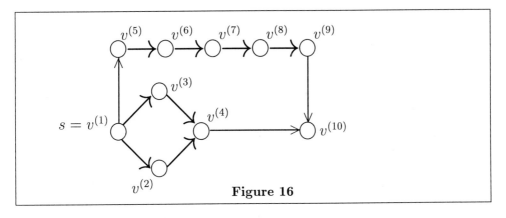

Figure 16

We define $C : \mathcal{P}(s) \to \mathbb{R}$ as follows:
$$C\left(\left[s, v^{(5)}, v^{(6)}, v^{(7)}, v^{(8)}, v^{(9)}, v^{(10)}\right]\right) := 2,$$
$C\left(\left[s, v^{(3)}, v^{(4)}\right]\right) := 4, \quad C\left(\left[s, v^{(3)}, v^{(4)}, v^{(10)}\right]\right) := 5,$
$C\left(\left[s, v^{(2)}, v^{(4)}\right]\right) := 1, \quad C\left(\left[s, v^{(2)}, v^{(4)}, v^{(10)}\right]\right) := 8,$
$C(P) := 0$ for all other paths $P \in \mathcal{P}(s)$.

Then C has even the strong Bellman property of type 0 because the following is true for all C-minimal paths $P \in \mathcal{P}(s)$ and for all proper prefixes $Q \leq P$ of P: Q is the only path in $\mathcal{G}$ from s to $\omega(Q)$. Moreover, C is $=$-preserving because C is injective on all sets $\mathcal{P}(s, v)$, $v \in \mathcal{V}$.

We consider the second and the third iteration of FORD-BELLMAN 1. In the second iteration, the path $P\left(v^{(4)}\right) := \left[s, v^{(2)}, v^{(6)}\right]$ is generated because it has a better C-value than $\left[s, v^{(3)}, v^{(6)}\right]$. In the third iteration, the only candidate for $P\left(v^{(10)}\right)$ is $P\left(v^{(4)}\right) \oplus \left(v^{(4)}, v^{(10)}\right) = \left[s, v^{(2)}, v^{(4)}, v^{(10)}\right]$. (The optimal s-$v^{(10)}$-path $\left[s, v^{(5)}, v^{(6)}, v^{(7)}, v^{(8)}, v^{(9)}, v^{(10)}\right]$ will be found much later.) Hence, the path $P\left(v^{(10)}\right) := \left[s, v^{(2)}, v^{(4)}, v^{(10)}\right]$ is generated during the third iteration. This path, however, is not C-minimal among all s-$v^{(10)}$-paths of a length ≤ 3; the path $\left[s, v^{(3)}, v^{(4)}, v^{(10)}\right]$ with $C\left(\left[s, v^{(3)}, v^{(4)}, v^{(10)}\right]\right) = 5$ is better than $P\left(v^{(10)}\right)$ with $C\left(P\left(v^{(10)}\right)\right) = 8$.

Consequently, not all paths $P(v)$ constructed in the third iteration are C-minimal among all s-v-paths of a length ≤ 3. $\Box$

Remark 4.115 Assertion (4.91) is true if C is $<$-preserving and order preserving. This follows from Theorem 4.121 a); in particular, (4.91) is true if C is additive, which is assumed in almost all descriptions about Ford-Bellman algorithms in literature.

We conjecture that (4.91) is also true if $C : \mathcal{P}(\mathcal{G}) \to \mathbb{R}$ satisfies the following condition:

If P is C-minimal among all paths P' from s to $\omega(P)$ with $\ell(P') = \ell(P)$ then each prefix $Q \leq P$ is C-minimal among all paths Q' from s to $\omega(Q)$ with $\ell(Q') = \ell(Q)$.

This is the strong Bellman $B_{\leq}^{(0)}$-condition of type B1 with respect to ℓ. This can be seen by defining $H := C$ and $H_* := \ell$ in Definition 2.111. (The set $B^{(0)}$ was defined in 2.74 on page 66.) $\Box$

Theorem 4.116 Let $\mathcal{G} = (\mathcal{V}, \mathcal{R})$ be a digraph with $\mathcal{V} = \left\{v^{(1)} = s, v^{(2)}, \dots, v^{(n)}\right\}$, and let $C : \mathcal{P}(s) \to \mathbb{R}$ be a path function. Let $n' \geq n$. We assume that C has the weak prefix oriented Bellman property of type 0 and that C is $=$-preserving. Moreover, we make the following assumption:

For each node v, there exists a C-minimal s-v-path $P^*(v)$.   (4.92)

We apply FORD-BELLMAN 1 with the following specifications:

- COND := $(\tau < n - 1)$;[32]   • WRITE := (Write all paths $P(v)$, $v \in \mathcal{V}$).

Then FORD-BELLMAN 1 will construct a C-minimal path $P(v)$ for each node v. Moreover, FORD-BELLMAN 1 executes $O(n^3)$ steps if $n' = n$.

P r o o f : By assumption (4.92) and Theorem 2.116, there even exists an acyclic Bellman path $P^*(v)$ from s to v; then $\ell(P^*(v)) \leq n - 1$.

It follows that all conditions of Theorem 4.112 are satisfied. Consequently, all paths $P(v)$, $v \in \mathcal{V}$, are C-minimal immediately after the $(n-1)$st iteration, and they are not changed by any further iteration.

---

[32]That means that step 2 is executed exactly $(n' - 1)$ times.

We estimate the time of computation. The time of each execution of step 2 is $O(n^2)$ because $\leq n$ predecessors are considered for each of the $\leq n$ elements of S. Consequently, the computation does not require more than $O(n^3)$ time units because step 2 is executed $(n-1)$ times.

□

Remark 4.117 Let $\mathcal{G} = (\mathcal{V}, \mathcal{R})$ be a digraph with $\mathcal{V} = \left\{ v^{(1)} = s, v^{(2)}, \ldots, v^{(n)} \right\}$, and let $C : \mathcal{P}(s) \to \mathbb{R}$ be a path function with the weak prefix oriented Bellman property of type 0. We assume that each node s has a C-minimal s-v-path $P^*(v)$.
Then we may replace the application of FORD-BELLMAN 1 to $\mathcal{G}$ by the application of this algorithm (or of FORD-BELLMAN 2) to the *acyclic* digraph $\mathcal{G}^{[n]}$; this digraph was defined on page 18.

To see this we apply Theorem 2.116. It says that for each node v, there exists an acyclic Bellman path $P^*(v)$ from s to v; this path has a length $\leq n-1$.
Consequently, we can find a C-minimal Bellman path $P^*(v)$ for all v by exclusively considering candidates of a length $\leq n-1$. All these paths appear in the acyclic digraph $\mathcal{G}^{[n]}$. was defined on page 18. We define the cost function $C^{\bullet}$ for $\mathcal{G}^{[n]}$ as follows: For all paths $P^{\bullet} \in \mathcal{P}(\mathcal{G}^{[n]})$, let $C^{\bullet}(P^{\bullet}) = C(\Psi(P^{\bullet}))$ where Ψ is the canonical homomorphism from $\mathcal{G}^{[n]}$ to $\mathcal{G}$. Then $C^{\bullet}$ has the weak prefix oriented Bellman property of type 0 if C has this property.
Consequently, we can solve Problem 9 in $\mathcal{G}$ by solving the same problem in the acyclic digraph $\mathcal{G}^{[n]}$.

□

Corollary 4.118 *Let $\mathcal{G} = (\mathcal{V}, \mathcal{R})$ be a digraph with $\mathcal{V} = \left\{ v^{(1)} = s, v^{(2)}, \ldots, v^{(n)} \right\}$, and let $C : \mathcal{P}(\mathcal{G}) \to \mathbb{R}$ be a path function with the following properties:*
  (i) *C is injective.*
  (ii) *C has the x-Y-minimality property [33] for all nodes x and for all paths Y.*
  (iii) *C is $\mathbb{L}_0$-minimum preserving [34].*

Let $n' \geq n$. We apply FORD-BELLMAN 1 with the following specifications:

- COND $:= (\tau < n' - 1)$; [35]
- WRITE $:=$ (Write all paths $P(v)$, $v \in \mathcal{V}$).

Then FORD-BELLMAN 1 will output a C-minimal s-v-path $P(v)$ for each $v \in \mathcal{V}$. Moreover, FORD-BELLMAN 1 executes $O(n^3)$ steps if $n' = n$.

P r o o f : The function C is $=$-preserving because it is injective. Moreover, for each node v, there exists a C-minimal s-v-path P; this follows from assumption (ii) about $x := s$ and $Y := [v]$.

We next show that C has even the strong prefix oriented Bellman property of type 0. Applying (a)$\Rightarrow$(b) of Corollary 2.103, we conclude that C has the strong prefix oriented Bellman property of type 2; by Theorem 2.91 a), this Bellman principle implies the strong prefix oriented Bellman property of type 0.

---

[33] This property was introduced in Definition/Remark 2.5.
[34] This property was described in Example 2.54, and it was only defined in the case that $\mathrm{def}(C) = \mathcal{P}(\mathcal{G})$.
[35] That means that step 2 is executed exactly $(n' - 1)$ times.

Then $C|_{\mathcal{P}(s)}$ satisfies all conditions of Theorem 4.116. Consequently, all paths $P(v)$ generated by FORD-BELLMAN 1 after the $(n-1)$st iteration are C-minimal, and they are not changed in any later iteration. Moreover, FORD-BELLMAN 1 works in $O(n^3)$ time units. □

Remark 4.119   a) Condition (iii) in 4.118 may be replaced by the assumption that the function C is order preserving. The reason is that by Theorem 2.52, order preservation is at least as strong as $\mathbb{L}_0$-minimum preservation.

b) Corollary 4.118 shows that it was reasonable to define modifications of the standard order preserving principle. In particular, the $\mathbb{L}_0$-minimum preservation plays an important rôle in 4.118, and replacing this property of C by the original order preservation would make condition (iii) in 4.118 unnecessarily sharp. □

Now we give answers to the following question: What does FORD-BELLMAN 1 do if condition (4.92) is not satisfied? In this case, FORD-BELLMAN 1 should give an error message. Unfortunately, FORD-BELLMAN 1 does not always notice that condition (4.92) is not satisfied; this is even true if C has all properties described in 4.116, i.e., C is $=$-preserving and has the weak prefix oriented Bellman property of type 0. If, however, C is $=$-preserving and $<$-preserving then FORD-BELLMAN 1 can decide whether or not each node v has a C-minimal s-v-connection.

We now give an example that FORD-BELLMAN 1 does not notice that the digraph $\mathcal{G}$ does not satisfy condition (4.92).

Example 4.120 Let $\mathcal{G}$ be the digraph shown in Figure 17.

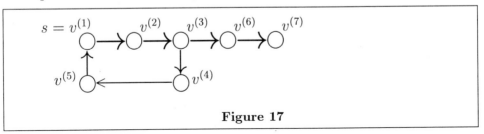

Figure 17

We define $C_1(P) := \begin{cases} \ell(P), & \text{if} \quad \ell(P) \leq 10, \\ -\ell(P), & \text{if} \quad \ell(P) > 10, \end{cases}$ for all paths $P \in \mathcal{P}(s)$, and we define $C_2 := \ell$. (The function ℓ was defined on page 15.)

For each node v, there exists no C_1-minimal s-v-path because we can construct an s-v-path with arbitrarily small C_1-value; we must only use the cycle $[s, v^{(2)}, v^{(3)}, v^{(4)}, v^{(5)}, v^{(1)}]$ very often.

Then C_1 satisfies the even strong prefix oriented Bellman condition of type 0 for the following reason: The Bellman principle describes a property of C_1-minimal paths, and these paths do not exist.

Moreover, C_1 is injective on all sets $\mathcal{P}(s, v)$. Therefore, C_1 is $=$-preserving. That means that all conditions of 4.116 are satisfied up to condition (4.92).

The following relationship between C_1 and $C_2 = \ell$ is obvious.

Let $P \in \mathcal{P}(s)$. If P has a length smaller than 10 then $C_1(P) = C_2(P)$. Consequently, the assertion $C_1(P) \neq C_2(P)$ is only true if $\ell(P) \geq 10$. $\qquad$ (4.93)

We next show that FORD-BELLMAN 1 does not notice that there exist no C_1-minimal s-v-paths. The rough idea is that FORD-BELLMAN 1 cannot decide which of the cost measures C_1, C_2 is used because the algorithm will never construct sufficiently long paths.

Let FORD-BELLMAN 1 use C_1 or C_2. In any case, the first seven iterations of FORD-BELLMAN 1 generate only paths of a length ≤ 7. We now see that the following is true:

The path $P_8(v)$ is an ℓ-minimal s-v-path.

(Recall that $P_8(v)$ has been defined as the path $P(v)$ immediately after the seventh and $\qquad$ (4.94) immediately before the eighth iteration)

To see this we apply (4.93); it implies that for each $v \in V$, the path $P_8(v)$ generated by FORD-BELLMAN 1 is independent of the cost function C_1 or C_2 used by that algorithm. It follows from Theorem 4.116 says that each path $P_8(v)$ is C-minimal for $C := C_2 = \ell$.

We next consider the eighth iteration. If FORD-BELLMAN 1 generates any paths $\widehat{P}(v)$ then all these paths are of a length ≤ 8. This implies that

$$C_1(\widehat{P}(v)) \overset{4.93}{=} C_2(\widehat{P}(v)) = \ell(\widehat{P}(v)) \overset{(4.94)}{\geq} \ell(P_8(v)) = C_2(P_8(v)) \overset{4.93}{=} C_1(P_8(v)).$$ Hence, FORD-BELLMAN 1 does not change the path $P_8(v)$ in the eighth iteration, and this is independent of the cost function C_1 or C_2 used by this algorithm.

For the same reason, the paths $P_8(v)$, $v \in V$, are not changed in the ninth, tenth, eleventh iteration, and so on. Consequently, FORD-BELLMAN 1 will never find a path of a length ≥ 10. Therefore, the algorithm cannot decide whether
- the underlying cost measure is C_1, which does not satisfy condition (4.92) of 4.116,
- the underlying cost measure is $C_2 = \ell$, which does satisfy condition (4.92) of 4.116.

In particular, FORD-BELLMAN 1 will not give an error message if the cost measure C_1 is used. $\qquad\square$

The next result says that FORD-BELLMAN 1 can decide whether all current paths $P(v)$ are C-optimal if C is order preserving.

Theorem 4.121 *Given a digraph $\mathcal{G} = (\mathcal{V}, \mathcal{R})$ with $\mathcal{V} = \{v^{(1)} = s, v^{(2)}, \ldots, v^{(n)}\}$. Let $C : \mathcal{P}(s) \to \mathbb{R}$ be order preserving. For each node v, let $d(v)$ be the minimal length of all s-v-paths.*

Then the following assertions are true for all $t = 0, 1, 2, 3, \ldots$:

a) If $v \in V$ and $t \geq d(v)$ then the tth iteration of FORD-BELLMAN 1 will generate a path $P_{t+1}(v)$ with the following property: $P_{t+1}(v)$ has minimal C-value among all s-v-paths of a length $\leq t$.[36]

---

[36]It may be that $t > n$; in 4.124, an example is give where the C-minimal s-v-connection has a length greater than n.

b) *Let $S = \emptyset$ after the tth iteration.*
    Then all current paths $P_{t+1}(v)$ are C-optimal.

P r o o f o f a) : We shall show the assertion by an induction on t.

If $t = 0$ then the only existing path $P_{t+1}(v)$ is $P_1(s) = [s]$. This path is the only s-s-path of length 0; consequently, $P_0(s)$ is the C-minimal s-s-path of length 0.

We assume that the assertion is true for $t - 1$; that means, each existing path $P_t(v)$ is C-minimal among all s-v-paths of a length $\leq t - 1$.

Suppose that the assertion is not true for t. Then there exists a node x such that $t \geq d(x)$ and $P_{t+1}(x)$ is not C-minimal among all candidates of a length $\leq t$. (The path $P_{t+1}(x)$ exists by Theorem 4.111 b).)

Since $P_{t+1}(v)$ is not C-minimal, there is an s-x-path X with $\ell(X) \leq t$ such that
$$C(X) < C(P_{t+1}(x)).\qquad\qquad(4.95)$$

Then $X \neq [s]$ because otherwise, $x = s$ and $C(P_1(s)) = C([s]) = C(X) \overset{(4.95)}{<} C(P_{t+1}(s))$. That is in conflict with Theorem 4.111 e).

The fact that $X \neq [s]$ implies that the path X has at least one arc. Let (w, x) be the last arc of X, and let $X = Y \oplus (w, x)$. Then $t - 1 \geq \ell(X) - 1 = \ell(Y) \geq d(w)$ so that $t - 1 \geq d(w)$.

The induction hypothesis implies that $P_t(w)$ is C-minimal among all s-w-paths of a length $\leq t - 1$. Hence, $C(P_t(w)) \leq C(Y)$. This and the order preservation of C have the following consequence:
$$C(P_t(w) \oplus (w, x)) \leq C(Y \oplus (w, x)) = C(X).\qquad(4.96)$$

Let $\tau \leq t - 1$ be the iteration in which the path $P_t(w)$ was constructed. Then x was in S immediately after the τth iteration. That means that in the $(\tau + 1)$st iteration, the path $P_{\tau+1}(w) \oplus (w, x) = P_t(w) \oplus (w, x)$ was one of the candidates for $\hat{P}_{\tau+1}(x)$. Consequently,
$$C(P_{\tau+2}(x)) \leq C(P_t(w) \oplus (w, x)).\qquad(4.97)$$

Moreover, $C(P_{t+1}(x)) \leq C(P_{\tau+2}(x))$ because $\tau + 2 \leq t + 1$. Hence,
$$C(P_{t+1}(x)) \leq C(P_{\tau+2}(x)) \overset{(4.97)}{\leq} C(P_t(w) \oplus (w, x)) \overset{(4.96)}{\leq} C(X).$$
Consequently, $C(P_{t+1}(x)) \leq C(X)$; that is a contradiction to (4.95).

P r o o f o f b) : Let $S = \emptyset$ after the tth iteration. Then Theorem 4.111 c) implies that $t - 1 \geq d(v)$ for all nodes v; it follows from Theorem 4.111 b) that the paths $P_t(v)$ and $P_{t+1}(v)$ exist.

We show that a contradiction arises if each of the following assertions is true:

There exists a node x such that $P_{t+1}(x)$ is not C-minimal.   (4.98)

$S = \emptyset$ immediately after the tth iteration.   (4.99)

By assumption (4.99), the following is true:
$$P_{t+1}(v) = P_t(v) \text{ for all nodes } v.\qquad(4.100)$$

Assumption (4.98) implies that there exists a path $X \in \mathcal{P}(s,x)$ such that

$$C(X) \;<\; C(P_{t+1}(x)).\tag{4.101}$$

We assume that we have chosen the node x and the path X such that the following is true:

If $x' \in V'$ and if X' is an s-x'-path with $C(X') < C(P_{t+1}(x'))$
then $\ell(X) \leq \ell(X')$. $\qquad(4.102)$

That means that X has minimal length among all paths $X' \in \mathcal{P}(x')$ with $C(X') < C(P_{t+1}(\omega(X')))$.

Part a) of the current result has the consequence that the path $P_{t+1}(x)$ is C-minimal among all s-x-paths of a length $\leq t$. This and (4.101) imply that $\ell(X) > t$ so that $\ell(X) > 0$. So, there are a path $Y \in \mathcal{P}(s)$ and a node $w \in V$ such that $X = Y \oplus (w,x)$.

Then Y is shorter than X. This and fact (4.102) imply that $C(Y) \geq C(P_{t+1}(w)) \overset{(4.100)}{=} C(P_t(w))$. Using the order preservation of C, we obtain:

$$C(X) \;=\; C(Y \oplus (w,x)) \;\geq\; C(P_t(w) \oplus (w,x)).\tag{4.103}$$

Let $\tau \leq t-1$ be the iteration in which $P_t(w)$ was generated. Then $w \in S$ immediately before the $(\tau + 1)$st iteration. In that iteration, $P_{\tau+1}(w) \oplus (w,x) = P_t(w) \oplus (w,x)$ was one of the candidates for $\widehat{P}_{\tau+1}(x)$ so that $C(P_t(w) \oplus (w,x)) \geq C(\widehat{P}_{\tau+1}(x)) \geq C(P_{\tau+2}(x))$. This and $\tau + 2 \leq t+1$ imply that

$$C(P_t(w) \oplus (w,x)) \;\geq\; C(P_{\tau+2}(x)) \;\geq\; C(P_{t+1}(x)).\tag{4.104}$$

Consequently, $C(X) = C(Y \oplus (w,x)) \overset{(4.103)}{\geq} C(P_t(w) \oplus (w,x)) \overset{(4.104)}{\geq} C(P_{t+1}(x))$. The resulting assertion $C(X) \geq C(P_{t+1}(x))$ is a contradiction to (4.101). $\qquad\square$

Next, we show the following: If C is $=$-preserving and $<$-preserving then FORD-BELLMAN 1 can decide whether all nodes v have an optimal s-v-connection.

Theorem 4.122 *Given a digraph $\mathcal{G} = (V, \mathcal{R})$ with $V = \{v^{(1)} = s, v^{(2)}, \ldots, v^{(n)}\}$. Let $C : \mathcal{P}(s) \to \mathbb{R}$ be $<$-preserving and $=$-preserving. (E.g., C may be additive.)*

Then the following are true immediately after n iterations of FORD-BELLMAN 1:
a) If $S = \emptyset$ then all paths $P(v)$ are C-optimal.
b) If $S \neq \emptyset$ then there are nodes v for which no C-minimal s-v-connection exists.
c) Suppose that COND *and* WRITE *are specified as follows:*

  - COND $:= (\tau < n);$ [37]

  - WRITE $:=$ $\left(\begin{array}{l} \textbf{If } S = \emptyset \textbf{ then } \text{output all paths } P(v),\ v \in V, \\ \textbf{else } \text{give an error message.} \end{array} \right)$.

  Then FORD-BELLMAN 1 decides correctly whether or not a C-minimal s-v-path exists for all nodes v. If YES then FORD-BELLMAN 1 will output these paths, if NO then FORD-BELLMAN 1 will write an error message.

P r o o f   o f   a) : The path function C is order preserving because it is $=$-preserving and $<$-preserving.

---

[37]That means that step 2 is executed exactly n times.

Let $S = \emptyset$ after the n-th iteration. Then $P_n(v) = P_{n-1}(v)$ for all nodes v. Then the optimality of all paths $P_n(v)$ follows from Theorem 4.121 b) for $t := n$.

P r o o f   o f   b) : By Theorem 2.95, the $<$-preservation of C implies the strong Bellman property of type 0. Then Theorem 2.116 implies that for all nodes v the following is true: If there exists a C-minimal s-v-path $P^*(v)$ then there exists a C-minimal and acyclic s-v-path $P_*(v)$; in this case, $\ell(P_*(v)) \leq n - 1$.

The path function C is order preserving as mentioned above. Therefore, we may apply Theorem 4.121 a). It says that each path $P_{n-1}(v)$, $v \in V$, is C-optimal among all s-v-paths of a length $\leq n$. One of these paths is $P_*(v)$ as far as a C-minimal s-v-connection exists at all; consequently, $C(P_{n-1}(v)) \leq C(P_*(v))$. That means: If a C-minimal s-v-path exists then $P_{n-1}(v)$ is C-minimal. In this case, $P(v)$ is not changed in the nth iteration of FORD-BELLMAN 1 so that $v \notin S$. Conversely, if $w \in S$ after the nth iteration then there does not exist any C-minimal s-w-path $P_*(w)$.

Consequently, if $S \neq \emptyset$ after the nth iteration then there are nodes w that do not have a C-optimal s-w-connections.

P r o o f   o f   c) : This assertion follows immediately from a) and b).   □

Remark 4.123 We next show the following: It is very unlikely that any simple modification of FORD-BELLMAN 1 can decide the existence of a C-minimal s-v-path individually for each node v. The main idea of our argumentation is that FORD-BELLMAN 1 sometimes notices very late that no optimal s-v-path exists.

To see this we define the following digraph $\mathcal{G} = (\mathcal{V}, \mathcal{R})$ with $n = 6$ nodes:

$\mathcal{V} := \{s, x, y_0, y_1, y_2, z\}$, $\mathcal{R} := \{(s, x), (x, z), (s, y_0), (y_0, y_1), (y_1, y_2), (y_2, y_0), (y_0, z)\}$.

Let $k \in \mathbb{N}$. We define $h_k : \mathcal{R} \to \mathbb{R}$ as follows:

$$h_k(s, x) := h_k(x, z) := 1/2, \quad h_k(s, y_0) := h_k(y_0, z) := 1,$$

$$h_k(y_0, y_1) := h_k(y_1, y_2) := h_k(y_2, y_0) := \frac{-1}{3k}.$$

Let $C := SUM_{h_k}$ and $Y := [y_0, y_1, y_2, y_0]$. Then $C([s, x, z]) = 1$, $C([s, y_0, z]) = 2$, $C(Y) = -1/k$, and $C([s, y_0] \oplus Y^\kappa \oplus [y_0, z]) = 2 - (\kappa + 1)/k$ for all $\kappa \in \mathbb{N} \cup \{0\}$.

If κ is very great then $C([s, y_0] \oplus Y^\kappa \oplus [y_0, z])$ is very small. Therefore, there exists no C-minimal s-z-path.

The algorithm FORD-BELLMAN 1 generates the path $P(z) = P' := [s, x, z]$ in the second iteration because $C([s, x, z]) = 1 < 2 = C([s, y_0, z])$. The shortest among all better s-z-paths is $P'' := [s, y_0] \oplus Y^{k+1} \oplus [y_0, z]$ with $C(P'') = 2 - (k+1)/k < 1 = C(P')$ and $\ell(P'') = 2 + 3 \cdot (k+1) = 5 + 3 \cdot k$. That means that it takes more than $3k$ iterations until $P(z) = P'$ will be replaced by the better s-z-path P''.

Recall that k is independent of n. We therefore cannot express the iterations changing $P(z)$ in terms of n; more precisely, there exists no function $f : \mathbb{N} \to \mathbb{N}$ such that the following statement is always true:

If $P(v)$ is not changed by a sequence of $f(n)$ iterations of FORD-BELLMAN 1 then $P(v)$ will be never changed any more, and $P(v)$ is a C-minimal s-v-path.

That means that FORD-BELLMAN 1 cannot decide the existence of a C-minimal s-v-path only by observing when v enters and leaves the set S. The only method is to wait until $S = \emptyset$ or to make a detailed analysis of C. □

Next, we study the behaviour of FORD-BELLMAN 1 if the cost function C is order preserving. That means that we only assume $\leq$-preservation but no $<$-preservation and no Bellman condition. The following remark shows that the C-minimal paths may be very long.

Remark 4.124 Given a digraph $\mathcal{G} = (\mathcal{V}, \mathcal{R})$ with $\mathcal{V} = \{v^{(1)} = s, v^{(2)}, \ldots, v^{(n)}\}$. Let $C : \mathcal{P}(s) \to \mathbb{R}$ be order preserving and $=$-preserving.

Then the following situation is possible: Each node v has a C-minimal s-v-connection $P^*(v)$, but some of these paths have a length greater than n.

An example is the digraph $\mathcal{G}$ in Figure 18 . It has $n = 5$ nodes. Each arc r is marked with a function $\sigma_r(\xi)$. Let $p(v) := 0$ for all nodes v. We define $C := \sigma \circ\circ p$.

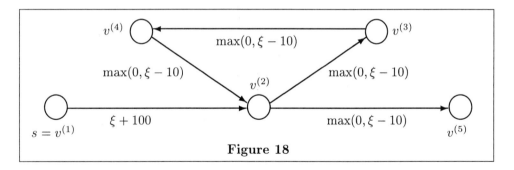

Figure 18

All functions $\xi \mapsto \sigma_r(\xi)$ are monotonically increasing. Therefore, the path function C is order preserving.

It is $C([s, v^{(2)}]) = 100$. If the path uses one of the further arcs then the C-value is reduced by ten until the C-value 0 is reached. Consequently, there exists a C-minimal s-v-connection $P^*(v)$ for each node $v \in \{v^{(2)}, v^{(3)}, v^{(4)}, v^{(5)}\}$. The only way to generate $P^*(v)$ is to move along the cycle $[v^{(2)}, v^{(3)}, v^{(4)}, v^{(2)}]$ at least three times. Consequently, $C(P^*(v)) > n = 5$.

Our example shows that order preservation alone is not sufficient to solve Problem 9 a) with n iterations.

By the way, the digraph $\mathcal{G}$ is also a counterexample to the following assertion: If C is order preserving then the system of all C-minimal s-v-connections is a rooted tree. This situation was mentioned in 2.114 c). □

Theorem 4.125 *Given a digraph $\mathcal{G} = (\mathcal{V}, \mathcal{R})$ with $\mathcal{V} = \{v^{(1)} = s, v^{(2)}, \ldots, v^{(n)}\}$. Let $C : \mathcal{P}(s) \to \mathbb{R}$ be order preserving. Suppose that FORD-BELLMAN 1 is applied to $\mathcal{G}$ where* COND *and* WRITE *are specified as follows:*

- COND := $(S \neq \emptyset)$;
- WRITE := (Write all paths $P(v)$, $v \in \mathcal{V}$).

Then the following assertions are true:

a) *If each node v has a C-minimal s-v-connection $P^*(v)$ then the algorithm FORD-BELLMAN 1 terminates after finitely many steps and outputs exclusively C-minimal paths $P(v)$.*

b) *If not all nodes can be reached by a C-minimal path then FORD-BELLMAN 1 will never stop.*

P r o o f   o f   a) : Let L be the maximal length of all paths $P^*(v)$ with $v \in \mathcal{V}$. We distinguish between two cases.
The first case is that FORD-BELLMAN 1 stops before the Lth iteration. The set S is empty when the algorithm stops. Therefore, all current paths $P(v)$ are optimal by Theorem 4.121 b).
The second case is that FORD-BELLMAN 1 executes the Lth iteration. By 4.121 a), each path $P_{L+1}(v)$ is C-minimal among all s-v-paths of length $\leq L$; in particular, $C(P_{L+1}(v)) \leq C(P^*(v))$. Consequently, $P_{L+1}(v)$ is optimal as well as $P^*(v)$. Therefore, FORD-BELLMAN 1 does not change any path $P(v)$ in its $(L+1)$st iteration so that $S = \emptyset$. That means that the algorithm stops and outputs C-minimal paths $P(v) = P_{L+1}(v)$, $v \in \mathcal{V}$.

P r o o f   o f   b) : Let x be a node without a C-minimal s-x-connection. We show that FORD-BELLMAN 1 does not terminate. The only case in which the algorithm stops is that $S = \emptyset$. In this case, all current paths $P(v)$ are C-optimal by 4.121 b). But there is no C-optimal s-v-path for $v = x$. Therefore, S will never be empty, and FORD-BELLMAN will never stop.   □

Remark 4.126 We show the practical relevance of the example in Remark 4.124 and of similar examples. For this purpose, we interpret the digraph $\mathcal{G}$ in 4.124 as a model of a cleaning process, for example in a car wash.
We assume that the arc $(s, v^{(2)})$ represents a car entering a car wash with $C(r_1) = 100$ grams of dust. Let all further arcs of $\mathcal{G}$ describe elementary cleaning processes. If $\omega(r') = \alpha(r'')$ for two arcs r', r'' then the process r'' can be started immediately after finishing the process r'. Each function $\sigma_r(\xi) := \xi - a_r$ means that the cleaning process r removes a_r grams of dust if $\xi \geq a_r$; if $\xi < a_r$ then the cleaning process removes all dirt. The arc (x_1, z) symbolizes the final cleaning process.
A C-minimal s-γ-path corresponds to a cleaning process that removes as much dirt as possible. In our case, we can reach the value $C = 0$, i.e. perfect cleanness.
Many properties of cleaning processes can be expressed with functions σ_r. For example, let r is associated with an elementary cleaning process that can usually remove a_r grams of dirt up to the last A_r grams; then we use $\sigma_r(\xi) := \min(A_r, \xi - a_r)$ to describe this. Another case is that r represents a cleaning process that does not

remove a fixed quantity but a fixed percentage of dirt; this case can be symbolized with a function $\sigma_r(\xi) := \min(A_r, a_r \cdot \xi)$ with $A_r \geq 0$ and $0 < a_r < 1$; for example, $\sigma_r(\xi) := \min(11, 0.7 \cdot \xi)$ means that the cleaning process r usually removes 30% of the dust, but the last 11 grams are not removed. □

The next remark says that applying FORD-BELLMAN 1 may be useful even if C is not order preserving and has no Bellman properties.

Remark 4.127 Given a digraph $\mathcal{G} = (\mathcal{V}, \mathcal{R})$ with $\mathcal{V} = \{v^{(1)} = s, v^{(2)}, \ldots, v^{(n)}\}$. Let $C : \mathcal{P}(s) \to \mathbb{R}$. We assume that each node $v \in \mathcal{V}$ can be reached by a C-minimal s-v-path. We do not assume that C is order preserving or has a Bellman property.

Then we can find C-minimal s-v-connections for all nodes v if the following assertions are true:
- The path function $H := C$ satisfies the conditions of Theorem 2.37.
- All restrictions $\phi_r|_{I_\mu}$ described in 2.37 are monotonically increasing.

To find the desired optimal s-v-paths we first construct the digraph $\mathcal{G}'$ and the path function H' as described in 2.37; by Theorem 2.37 c), the path function H' is order preserving. Suppose that an H'-minimal s'-v'-connection in $\mathcal{G}'$ exists for each node v'. Then $\mathcal{G}'$ and H' satisfy all conditions in Theorem 4.125. That means that FORD-BELLMAN 1 will find an H'-minimal s'-v'-path $P(v')$ for all nodes v' in $\mathcal{G}'$. Remark 2.38 says how to construct the C-minimal s-v-paths in $\mathcal{G}$ from the H'-minimal s'-v'-paths in $\mathcal{G}'$.

Also, we can find C-minimal s-v-connections for all nodes v if $H := C$ satisfies all conditions of Theorem 2.39. This is described in Remark 2.41.
One of the conditions in 2.41 is that for each node $\tilde{v}$ in $\tilde{\mathcal{G}}$, there exists a $\tilde{H}$-minimal and a $\tilde{H}$-maximal path from $\tilde{s}$ to $\tilde{v}$. It remains to the reader to check whether or not this condition automatically follows from the assumption that each node v in $\mathcal{G}$ has a C-minimal s-v-connection.
If all conditions of Remark 2.41 are satisfied then we apply FORD-BELLMAN 1 to $\tilde{\mathcal{G}}$. Thus, we can find an $\tilde{H}$-minimal $\tilde{s}$-$\tilde{v}$-path $P(\tilde{v})$ and a $\tilde{H}$-maximal $\tilde{s}$-$\tilde{v}$-path $\overline{P}(\tilde{v})$ for all nodes $\tilde{v}$ in $\tilde{\mathcal{G}}$. (The paths $\overline{P}(\tilde{v})$ can be found by searching for $(-\tilde{H})$-minimal $\tilde{s}$-$\tilde{v}$-paths with the help of FORD-BELLMAN 1.) It is described in 2.41 how to find C-minimal s-v-paths in $\mathcal{G}$ when the paths $P(\tilde{v})$ and $\overline{P}(\tilde{v})$ have been found. □

Next, we discuss a modification of FORD-BELLMAN 1 that makes this algorithm simpler.

Remark 4.128 Perhaps, FORD-BELLMAN 1 works precisely if processing the paths $Q(v)$ (see page 181) instead of the paths $P(v)$. More precisely, we consider a modification of Algorithm 7 generating a current predecessor $pred(v)$ for each $v \neq s$; the path $Q(v)$ can be constructed backwards by following the nodes v, $pred(v)$, $pred(pred(v))$, etc. until s is reached. We assume that the node $pred(v)$, the cost $C(Q(v))$ but not the path $Q(v)$ are recorded by FORD-BELLMAN 1.

We outline how $pred(v)$ is constructed. First, FORD-BELLMAN 1 chooses a $\widehat{u} \in$ $\mathcal{N}^-(v) \cap S$ such that

$$C(Q(\widehat{u}) \oplus (\widehat{u}, v)) = \min\{C(Q(u) \oplus (u, v))|u \in \mathcal{N}^-(v) \cap S\}.$$

(Here, we assume that $C(Q(u) \oplus (u, v))$ can be computed without knowing $Q(u)$ itself.)

Then FORD-BELLMAN 1 defines $pred(v) := \widehat{u}$ if $C(Q(\widehat{u}) \oplus (\widehat{u}, v)) < C(Q(v))$.

We leave it to the reader to check whether this simpler version of FORD-BELLMAN 1 works correctly in the situations of Theorems 4.116 and 4.122. Perhaps, one can prove a similar result as 4.54 about the equality of $P(v)$ and $Q(v)$.

If, however, C is only order preserving then the simplified algorithm does not always work. The reason is that all paths $Q(v)$ must be acyclic. Remark 4.124, however, shows an example where all C-minimal s-v-paths have cycles. $\square$

4.3.2   The algorithm FORD-BELLMAN 2

We shall now describe the search strategy FORD-BELLMAN 2. It is a simplified version of FORD-BELLMAN 1. Moreover, Algorithm FORD-BELLMAN 2 is a special version of the dynamic programming paradigm.

FORD-BELLMAN 2 may only be applied to digraphs that have a Bellman-Ford-enumeration, which is now defined.

Definition/Remark 4.129   A *Bellman-Ford-enumeration* is a node function $p : \mathcal{V} \to \{1, \ldots, n\}$ such that $p(x_1) < p(x_2) < \cdots < p(x_t)$ for all acyclic paths $[x_1, x_2, \ldots, x_t]$ with the fixed start node $x_1 = s$.

If $\mathcal{G}$ is acyclic then any topological sorting function p is a Bellman-Ford-enumeration. There exist, however, Bellman-Ford-enumerable digraphs with cycles. An example is the digraph $\mathcal{G}$ in 2.117. Let $p\left(v^{(i)}\right) := i$ for all $i = 1, \ldots, 5$. Then p is a Ford-Bellman enumeration, which can be seen by checking all acyclic paths in $\mathcal{P}(s)$, namely

$$[s], \left[s, v^{(2)}\right], \left[s, v^{(2)}, v^{(3)}\right], \left[s, v^{(2)}, v^{(3)}, v^{(4)}\right], \left[s, v^{(2)}, v^{(5)}\right].$$

In [HaS88], an algorithm $\mathcal{A}$ with the following properties is given:

- The algorithm $\mathcal{A}$ decides whether a given digraph has a Bellman-Ford-enumeration.

- If YES then the algorithm $\mathcal{A}$ computes such an enumeration.

- The running time of $\mathcal{A}$ is $O(n^2)$. $\square$

Next, we describe Algorithm FORD-BELLMAN 2 in detail. The main difference between FORD-BELLMAN 1 and FORD-BELLMAN 2 is the following: FORD-BELLMAN 1 can define and redefine a path $P(v)$ several times; FORD-BELLMAN 2, however, defines the path $P(v)$ only once for each node v. We assume that the node function $p\left(v^{(i)}\right) := i$ $(i = 1, \ldots, n)$ is a Bellman-Ford enumeration.

Algorithm 8 (FORD-BELLMAN 2)

1. (* *Initialization* *)

   Let $P(s) := [s]$; the path $P(v)$ is not yet defined for $v \neq s$.
   (* $P(v)$ *is the currently best path from* s *to* v. *)

2. **For** $i := 2$ **to** n **do**

   2.1 Choose a node $\widehat{u} \in \mathcal{N}^- \left(v^{(i)} \right)$

   such that $\widehat{P}\left(v^{(i)} \right) := P(\widehat{u}) \oplus \left(\widehat{u}, v^{(i)} \right)$ has the following property:

   $$C\left(P\left(v^{(i)} \right) \right) = \min \left\{ C\left(P(u) \oplus (u, v^{(i)}) \right) \mid u \in \mathcal{N}^- \left(v^{(i)} \right) \right\}.$$

3. **For** $i := 1$ **to** n **do**
   Write $P\left(v^{(i)} \right)$.

Now we show that FORD-BELLMAN 2 works correctly.

Theorem 4.130 *Given a digraph* $\mathcal{G} = (\mathcal{V}, \mathcal{R})$ *with* $\mathcal{V} = \left\{ v^{(1)} = s, v^{(2)}, \ldots, v^{(n)} \right\}$.
We assume that $p\left(v^{(i)} \right) := i$ $(i = 1, \ldots, n)$ *be a Bellman-Ford-enumeration of* $\mathcal{V}$.
Let $C : \mathcal{P}(s) \to \mathbb{R}$ *be a path function. We assume that* C *is* $=$-*preserving and has the weak prefix oriented Bellman property of type 0. Moreover, we assume that* $\mathcal{G}$ *and* C *satisfy condition* (4.92).
Then FORD-BELLMAN 2 will output a C-*minimal* s-$v^{(i)}$-*path* $P\left(v^{(i)} \right)$ *for all nodes* $v^{(i)} \in \mathcal{V}$. *Moreover, the running time is in* $O(n^2)$.

P r o o f : Assumption (4.92) and the weak Bellman property imply that for each node v, a Bellman path from s to v exists, and Theorem 2.116 says that there exists even an acyclic Bellman path $P^*(v)$.

We prove the minimality of all paths $P\left(v^{(i)} \right)$ with an induction on i.

Let $i = 1$. Then $v^{(i)} = v^{(1)} = s$ and $P\left(v^{(i)} \right) = P(s) = [s]$. This path is C-minimal for the following reason: By the weak Bellman principle, there exists a Bellman path Q from s to $v = s$; then $[s]$ is a prefix of Q so that $[s]$ is indeed C-minimal.

Let now $i \geq 2$. We assume that $P\left(v^{(j)} \right)$ is C-minimal for all $j < i$, and we must show that $P\left(v^{(i)} \right)$ is C-minimal.

The node $v^{(i)}$ is different from s because $i \geq 2$. Therefore, the acyclic Bellman path $P^*\left(v^{(i)} \right)$ has at least one arc. Let $\left(v^{(i')}, v^{(i)} \right)$ be the last arc of $P^*\left(v^{(i)} \right)$, and let P' be the prefix of $P^*\left(v^{(i)} \right)$ with $P^*\left(v^{(i)} \right) = P' \oplus \left(v^{(i')}, v^{(i)} \right)$.

Then $i' < i$ because p is a Bellman-Ford enumeration and $P^*\left(v^{(i)} \right)$ is acyclic. By the induction hypothesis, the path $P\left(v^{(i')} \right)$ is C-minimal, and the same is true for P' because P' is a prefix of the Bellman path $P^*\left(v^{(i)} \right)$. Consequently, $C\left(P\left(v^{(i')} \right) \right) = C(P')$. This and the $=$-preservation of C imply that

$$C\left(P\left(v^{(i')} \right) \oplus \left(v^{(i')}, v^{(i)} \right) \right) = C\left(P' \oplus \left(v^{(i')}, v^{(i)} \right) \right) = C\left(P^*\left(v^{(i)} \right) \right). \quad (4.105)$$

A further consequence of the relationship $i' < i$ is that the path $P\left(v^{(i')} \right)$ has been generated before constructing $P\left(v^{(i)} \right)$. Therefore, the path $P\left(v^{(i')} \right) \oplus \left(v^{(i')}, v^{(i)} \right)$ is one of the candidates for $P(v^{(i)})$. Hence, FORD-BELLMAN 2 will generate a path

$$P(v^{(i)}) \text{ with } C(P(v^{(i)})) \leq C\big(P(v^{(i')}) \oplus (v^{(i')}, v^{(i)})\big) \overset{(4.105)}{\leq} C(P^*(v^{(i)})). \text{ This and}$$

the optimality of $P^*(v^{(i)})$ imply that $P(v^{(i)})$ is C-optimal as well.

We now estimate the time required by FORD-BELLMAN 2. For each $i = 2, \ldots, n$, the algorithm processes at most n predecessors of the current node $v^{(i)}$. Therefore, the running time of FORD-BELLMAN 2 is $O(n^2)$. $\qquad\square$

Remark 4.131 If $\mathcal{G} = (\mathcal{V}, \mathcal{R})$ is an arbitrary finite digraph then the following search strategy is sensible:

    1. Test whether $\mathcal{G}$ has a Bellman-Ford enumeration p. Compute p if possible.

    2. If $\mathcal{G}$ is not Bellman-Ford-enumerable then apply FORD-BELLMAN 1.

    3. If $\mathcal{G}$ is Bellman-Ford-enumerable then apply FORD-BELLMAN 2.

If $\mathcal{G}$ has a Bellman-Ford enumeration then the complete search strategy works in $O(n^2)$ time units caused by the algorithm $\mathcal{A}$ (see 4.129) and by FORD-BELLMAN 2. If $\mathcal{G}$ does not have a Bellman-Ford enumeration then the complete search strategy works in $O(n^3)$ time units, which are required by FORD-BELLMAN 1.

Hence, the above search strategy has the following advantage: It is substantially faster than FORD-BELLMAN 1 if $\mathcal{G}$ has a Bellman-Ford enumeration; if this is not the case than the above search strategy is almost as fast as FORD-BELLMAN 1. $\qquad\square$

Corollary 4.132 *Given an acyclic digraph* $\mathcal{G} = (\mathcal{V}, \mathcal{R})$ *with* $\mathcal{V} = \{v^{(1)}, \ldots, v^{(n)}\}$ *and* $s = v^{(1)}$. *We assume that* $p\,(v^{(i)}) := i \quad (i = 1, \ldots, n)$ *be a Bellman-Ford-enumeration of* $\mathcal{V}$. *Let* $C : \mathcal{P}(\mathcal{G}) \to \mathbb{R}$ *be order preserving.*

Then all paths $P(v)$ *generated by FORD-BELLMAN 2 are* C-*minimal.*

P r o o f : Since $\mathcal{G}$ is finite and acyclic there exists a C-minimal s-v-path for each node v. Moreover, Theorem 2.99 implies that C has the weak prefix oriented Bellman property of type 0. Also, the path function C is $=$-preserving because C is order preserving.
Consequently, all conditions in 4.130 are given so that we may apply this theorem. $\square$

Remark 4.133 If C is not $=$-preserving then it may be that FORD-BELLMAN 2 does not solve Problem 9 a).

This can be seen by applying FORD-BELLMAN 2 to the digraph $\mathcal{G}$ in 2.117. The algorithm can only generate paths of a length $\leq n-1$ so that it cannot find the optimal s-$v^{(5)}$-path $(s, v^{(2)}) \oplus X^2 \oplus (v^{(2)}, v^{(5)})$ $\qquad\square$

4.3.3   The algorithm FORD-BELLMAN 3

Here, we describe the algorithm FORD-BELLMAN 3. This search strategy may also be used if C is not $=$-preserving; we assume that the digraph $\mathcal{G}$ is acyclic and that the node function $p\left(v^{(i)}\right) := i,\ i = 1, \ldots, n$, is a topological ordering of $\mathcal{G}$.

FORD-BELLMAN 3 has almost the same structure as FORD-BELLMAN 2. The main difference between both algorithms is the following: FORD-BELLMAN 2 constructs a single optimal path $P(v)$ for each $v \in \mathcal{V}$, whereas FORD-BELLMAN 3 constructs a set $\mathcal{Q}(v)$ of optimal paths.

This procedure is necessary if C is not $=$-preserving. This can be seen as follows: Let $u \in \mathcal{V}$ and P_1 and P_2 two optimal s-u-paths. Then it may occur that $C(P_1 \oplus (u,v)) \neq C(P_2 \oplus (u,v))$ although $C(P_1) = C(P_2)$. It is therefore possible that exactly one of the paths $P_1 \oplus (u,v)$ and $P_2 \oplus (u,v)$ is C-minimal. For this reason, the search algorithm FORD-BELLMAN 3 must construct both candidates $P_i \oplus (u,v),\ i = 1, 2$, and this is only possible by recording both P_1 and P_2 in a set $\mathcal{Q}(u)$ of optimal s-u-paths.

Algorithm 9 (FORD-BELLMAN 3)

1. (* *Initialization* *)

   Let $\mathcal{Q}(s) := \{\,[s]\,\}$; the set $\mathcal{Q}(v)$ is not yet defined for $v \neq s$.
   (* $\mathcal{Q}(v)$ *is a set of the currently optimal paths from s to v.* *)

2. **For** $i := 2$ **to** n **do**

   2.1. Let $\widehat{\mathcal{Q}}\left(v^{(i)}\right) := \left\{ P' \oplus \left(v^{(\nu)}, v^{(i)}\right) \,\middle|\, \nu < i,\ P' \in \mathcal{Q}\left(v^{(\nu)}\right) \right\}$.

   2.2. Construct the set $\mathcal{Q}\left(v^{(i)}\right)$ as the set of all paths in $\widehat{\mathcal{Q}}\left(v^{(i)}\right)$ with minimal C-value. More formally,
   $$\mathcal{Q}\left(v^{(i)}\right) := \left\{ P \in \widehat{\mathcal{Q}}\left(v^{(i)}\right) \,\middle|\, C(P) = \min\left\{ C(\widehat{P}) \,\middle|\, \widehat{P} \in \widehat{\mathcal{Q}}\left(v^{(i)}\right) \right\} \right\}.$$

3. **For** $i := 1$ **to** n **do**
   Write a path $P\left(v^{(i)}\right) \in \mathcal{Q}\left(v^{(i)}\right)$.

Now, we show that FORD-BELLMAN 3 works correctly.

Theorem 4.134 *Let* $\mathcal{G} = (\mathcal{V}, \mathcal{R})$ *be an acyclic digraph with* $\mathcal{V} = \left\{ v^{(1)}, v^{(2)}, \ldots, v^{(n)} \right\}$; *let* $s = v^{(1)}$. *We assume that* $p\left(v^{(i)}\right) := i$ $(i = 1, \ldots, n)$ *is a topological ordering of* $\mathcal{V}$. *Let* $C : \mathcal{P}(s) \to \mathbb{R}$ *be a path function with the weak prefix oriented Bellman property of type 0. We assume that condition (4.92) is satisfied.*

Then the following assertions are true:

a) *Let* $i \in \{1, \ldots, n\}$; *if* P *is a Bellman path from* s *to* $v^{(i)}$ *then* $P \in \mathcal{Q}\left(v^{(i)}\right)$.

b) *For each node* v, *the path* $P(v)$ *output by FORD-BELLMAN 3 is* C*-minimal.*

c) *FORD-BELLMAN 3 requires at least* 2^{n-3} *time units in the worst case.*

P r o o f o f a) : We show the assertion by an induction on i.

If $i = 1$ then $v^{(i)} = s$. Since $\mathcal{G}$ is acyclic the only s-$v^{(i)}$-path is $[s]$. This path is an element of $\mathcal{Q}\left(v^{(i)}\right) = \mathcal{Q}(s)$.

Let $i \geq 2$. We assume that the assertion is true for all $j < i$. Let P be a Bellman path from s to $v^{(i)}$. Then P must have at least one arc because $\omega(P) = v^{(i)}$ and $i \geq 2$. Let $\left(v^{(i')}, v^{(i)}\right)$ be the last arc of P, and let P' be the prefix of P with $P = P' \oplus \left(v^{(i')}, v^{(i)}\right)$. Then $i' < i$ because p is a topological ordering. Moreover, P' is a Bellman path because it is a prefix of the Bellman path P. By the induction hypothesis, $P' \in \mathcal{Q}(v^{(i')})$. Consequently, $P = P' \oplus \left(v^{(i')}, v^{(i)}\right) \in \widehat{\mathcal{Q}}(v^{(i)})$. Since P is a Bellman path, P has the minimal C-value among all s-$v^{(i)}$-paths; consequently, P has the minimal C-value among all candidates in $\widehat{\mathcal{Q}}(v^{(i)})$. Therefore, P is added to $\mathcal{Q}(v^{(i)})$.

P r o o f o f b) : By the Bellman property of C, there exists a Bellman path $P^*(v)$ from s to v for each node v. Part a) implies that $P^*(v) \in \mathcal{Q}(v)$. Moreover, $\mathcal{Q}(v)$ is constructed such that all elements of this set have the same C-values. Consequently, FORD-BELLMAN 3 outputs a path $P(v)$ with $C(P(v)) = C(P^*(v))$; that means that $P(v)$ is C-minimal.

P r o o f o f c) : Let $n \geq 3$, and let $\mathcal{G} = (\mathcal{V}, \mathcal{R})$ be the digraph with $\mathcal{V} := \left\{s = v^{(1)}, \ldots, v^{(n)}\right\}$ and $\mathcal{R} := \left\{\left(v^{(i)}, v^{(j)}\right) \mid 1 \leq i < j \leq n\right\}$.

Then $\mathcal{G}$ is acyclic. Let $C(P) := 0$ for all $P \in \mathcal{P}(s)$. Then C has the strong prefix oriented Bellman property of type 0 and is even =-preserving.

We now show that the time of computation is very long. Let P be an arbitrary path from s to a node v. Then P is a Bellman path. This and part a) imply that P is an element of $\mathcal{Q}(v)$. Consequently,

$$\mathcal{Q}(v) = \mathcal{P}(s, v) \text{ for all nodes } v. \tag{4.106}$$

When constructing $\widehat{\mathcal{Q}}\left(v^{(n)}\right)$, FORD-BELLMAN 3 generates all paths $P' \oplus \left(v^{(i')}, v^{(n)}\right)$ where $P' \in \mathcal{Q}\left(v^{(1)}\right) \cup \ldots \cup \mathcal{Q}\left(v^{(n-1)}\right)$. In particular, all paths $P' \in \mathcal{Q}\left(v^{(n-1)}\right) \overset{(4.106)}{=} \mathcal{P}\left(s, v^{(n-1)}\right)$ are processed. Consequently,

$$\text{FORD-BELLMAN 3 requires at least } \left|\mathcal{P}\left(s, v^{(n-1)}\right)\right| \text{ time units.} \tag{4.107}$$

The set $\mathcal{P}\left(s, v^{(n-1)}\right)$ has the exponential cardinality $\left|\mathcal{P}\left(s, v^{(n-1)}\right)\right| = 2^{n-2}$. To see this we define $\mathcal{L}$ as the set of all subsets of $\mathcal{V} \setminus \left\{s, v^{(n-1)}, v^{(n)}\right\}$. Moreover, we define the function $\psi : \mathcal{P}\left(s, v^{(n-1)}\right) \to \mathcal{L}$ as follows: For all paths $P \in \mathcal{P}\left(s, v^{(n-1)}\right)$, let

$$\psi(P) := \left\{v \in \mathcal{V} \setminus \{s, v^{(n-1)}, v^{(n)}\} \mid v \text{ is visited by } P\right\}.$$

It is easy to see that ψ is bijective. Therefore, $\left|\mathcal{P}\left(s, v^{(n-1)}\right)\right| = |\mathcal{L}| = 2^{n-3}$.

This fact and (4.107) imply our complexity theoretical result. $\square$

Remark 4.135 If C is not $=$-preserving then it may be that FORD-BELLMAN 3 does not solve Problem 9 a).

This can be seen by applying the algorithm to the graph $\mathcal{G}$ in 2.117. FORD-BELLMAN 3 can only generate paths of a length $\leq n - 1$ so that it cannot find the optimal s-$v^{(5)}$-path $\left(s, v^{(2)}\right) \oplus X^2 \oplus \left(v^{(2)}, v^{(5)}\right)$ $\qquad\qquad\square$

4.3.4   Algorithms computing minimal costs of paths

In the previous subsections, we searched for an optimal s-v-path for each node v. Here, however, we are only interested in the costs of this optimal s-v-path; that means that we are interested in a solution of Problem 9 b). We describe the two algorithms FORD-BELLMAN 1' and FORD-BELLMAN 2', which are simpler versions of FORD-BELLMAN 1 and FORD-BELLMAN 2, respectively; the new search algorithms only compute the values $C(P(v))$ and $C(\widehat{P}(v))$ instead of constructing the paths $P(v)$ or $\widehat{P}(v)$ themselves. The advantage of FORD-BELLMAN 1' and FORD-BELLMAN 2' is that they require less space than the our previous Ford-Bellman strategies.

We do not transform Algorithm FORD-BELLMAN 3 into a simpler version FORD-BELLMAN 3'. The reason is the following: Our simplifications are only possible if C is $=$-preserving. In this case, however, FORD-BELLMAN 3 is completely useless because FORD-BELLMAN 2 is simpler. Therefore, transforming FORD-BELLMAN 3 into a simpler algorithm FORD-BELLMAN 3' is not reasonable.

At the end of this subsection we show that computing minimal costs of s-v-paths can be replaced by solving a system of equations.

Now we describe the simplified Ford-Bellman algorithms in detail. Let $C : \mathcal{P}(s) \to \mathbb{R}$ be $=$-preserving. If follows from Theorem 2.19 that for each arc r, there exists a function $\phi_r : \mathbb{R} \dashrightarrow \mathbb{R}$ such that the following is true:

$$(\forall\, P \in \mathcal{P}(s)\,,\ r \in \mathcal{R})\quad C(P \oplus r) \;=\; \phi_r(C(P))\,.$$

For example, if C is additive and $C = SUM_h$, let $\phi_r(\xi) := \xi + h(r)$ for all arcs r and all $\xi \in \mathbb{R}$; then $C(P \oplus r) = C(P) + h(r)$ for all P and r.

We define $\chi(v) := C(P(v))$ and $\widehat{\chi}(v) := C(\widehat{P}(v))$ for all nodes v. Then the following is true for all arcs $r = (v, w)$:

$$C(P(v) \oplus (v, w)) \;=\; C(P(v) \oplus r) \;=\; \phi_r(C(P(v))) \;=\; \phi_r(\chi(v))\,.$$

This equation is helpful to understand step 2.1 of the algorithms FORD-BELLMAN 1' and FORD-BELLMAN 2'.

Next, we concentrate on FORD-BELLMAN 1'. We now give the list of instructions of this algorithm.

Algorithm 10 (FORD-BELLMAN 1')

1. (* *Initialization* *)

   Let $\chi(s) := C([s])$; the value $\chi(v)$ is not yet defined if $v \neq s$.
   (* $\chi(v)$ *is the currently minimal cost of any path from* s *to* v. *)

   Let $W := \{s\}$.
   (* W *is the set of all nodes for which a path* $\chi(v)$ *has been defined.* *)

   Let $S := \{s\}$.
   (* S *is the set of all nodes for which* $\chi(v)$ *has been changed immediately before the current iteration.* *)

   Let $\tau := 0$. (* τ *is used as a counter of the iterations.* *)

2. **While** condition COND is satisfied **do**

   2.1. **For** $i := 1$ **to** n **do**
   If $\mathcal{N}^-(v^{(i)}) \cap S \neq \emptyset$ **then** let

   $$\widehat{\chi}\left(v^{(i)}\right) := \min\left\{\phi_r(\chi(u)) \,\middle|\, u \in \mathcal{N}^-\left(v^{(i)}\right) \cap S, \; r = \left(u, v^{(i)}\right)\right\}.$$

   2.2. Let $S := \emptyset$.

   2.3. **For** $i := 1$ **to** n **do**
   If the value $\widehat{\chi}\left(v^{(i)}\right)$ exists but no value $\chi\left(v^{(i)}\right)$ has still been computed **then**

   $$\chi\left(v^{(i)}\right) := \widehat{\chi}\left(v^{(i)}\right); \quad W := W \cup \left\{v^{(i)}\right\}; \quad S := S \cup \left\{v^{(i)}\right\};$$

   If both $\widehat{\chi}\left(v^{(i)}\right)$ and $\chi(v^{(i)})$ exist **and** $\widehat{\chi}\left(v^{(i)}\right) < \chi\left(v^{(i)}\right)$ **then**

   $$\chi\left(v^{(i)}\right) := \widehat{\chi}\left(v^{(i)}\right); \quad S := S \cup \left\{v^{(i)}\right\}.$$

   2.4 $\tau := \tau + 1$.

3. Perform the output procedure WRITE.

Remark 4.136 Obviously, the following is true: Whenever FORD-BELLMAN 1 constructs a path $P(v)$ and a path $\widehat{P}(v)$ then FORD-BELLMAN 1' computes the values $\chi(v) = C(P(v))$ and $\widehat{\chi}(v) = C(\widehat{P}(v))$, respectively. $\qquad\square$

The next theorem says that FORD-BELLMAN 1' works correctly.

Theorem 4.137 *Given a digraph* $\mathcal{G} = (\mathcal{V}, \mathcal{R})$ *with* $\mathcal{V} = \left\{v^{(1)} = s, v^{(2)}, \ldots, v^{(n)}\right\}$. *Let* $C : \mathcal{P}(s) \to \mathbb{R}$ *be a path function. The following assertions are true:*

a) (*see Theorem 4.116*) *Let* C *have the weak prefix oriented Bellman property of type 0, and let* C *be* =-*preserving. Moreover, let each node* v *have a* C-*minimal* s-v-*connection.*
   Suppose that FORD-BELLMAN 1 with the following specifications is applied:
   - COND := $(\tau < n - 1)$;[38]   • WRITE := (Write all values $\chi(v)$, $v \in \mathcal{V}$.)
   Then FORD-BELLMAN 1' will compute the minimal C-*value of all* s-v-*paths for each* $v \in \mathcal{V}$. *Moreover, FORD-BELLMAN 1 executes* $O(n^3)$ *steps.*

---

[38]That means that step 2 is executed exactly $(n - 1)$ times.

b) (*see Theorem 4.122*) *Let* C *be* $<$*-preserving and* $=$*-preserving. Suppose that FORD-BELLMAN 1' with the following specifications is applied:*

- COND := $(\tau < n);$ [39]

- WRITE := $\left(\begin{array}{l} \textbf{If } S = \emptyset \textbf{ then } \text{write all values } \chi(v), \ v \in \mathcal{V}, \\ \textbf{else } \text{give an error message.} \end{array} \right).$

Then FORD-BELLMAN 1' can decide whether or not each node v *has a* C*-minimal* s-v *connection. If YES then the algorithm writes the minimal* C*-values* $\chi(v)$ *of all* s-v*-paths, if NO then the algorithm gives an error message.*

c) (*see Theorem 4.125 a*)) *Let* C *be order preserving. Moreover, let each node* v *have a* C*-minimal* s-v*-connection. We apply FORD-BELLMAN 1' to* $\mathcal{G}$ *and specify* COND *and* WRITE *as follows:*

- COND := $(S \neq \emptyset);$ • WRITE := (Write all values $\chi(v), \ v \in \mathcal{V}$).

Then FORD-BELLMAN 1' terminates after finitely many steps and computes the minimal C*-value of all* s-v*-connections for all nodes* v.

d) (*see Theorem 4.125 b*)) *Given the same situation as in part* c) *up to the following exception: Not all nodes* v *can be reached by a* C*-minimal* s-v*-path. Then FORD-BELLMAN 1' will never stop.*

P r o o f : The assertions a), b), c), and d) are immediate consequences of Remark 4.136 and of Theorems 4.116, 4.122, 4.125 a), and 4.125 b), respectively.   □

Next, we concentrate on FORD-BELLMAN 2'. We give the listing of this algorithm.

Algorithm 11 (FORD-BELLMAN 2')

1. (* *Initialization* *)

   Let $\chi(s) := C([s])$; the value $\chi(v)$ is not yet defined for $v \neq s$.
   (* $\chi(v)$ *is the currently minimal* C*-value of all* s-v*-paths.* *)

2. **For** $i := 2$ **to** n **do**

   2.1. Compute the following value $\chi\left(v^{(i)}\right)$

   $$\chi\left(v^{(i)}\right) \; := \; \min\left\{ \phi_r(\chi(u)) \,\middle|\, u \in \mathcal{N}^-\left(v^{(i)}\right), \; r = \left(u, v^{(i)}\right) \right\} .$$

3. **For** $i := 1$ **to** n **do**
   Write $\chi\left(v^{(i)}\right)$.

Remark 4.138

It is obvious that FORD-BELLMAN 2' has the following property: Whenever FORD-BELLMAN 2 constructs a path $P(v)$ then FORD-BELLMAN 2' will compute the value $\chi(v) = C(P(v))$.   □

---

[39] That means that step 2 is executed exactly n times.

Now we show that FORD-BELLMAN 2' works correctly.

Corollary 4.139 (*see 4.130*)
Given a digraph $\mathcal{G} = (\mathcal{V}, \mathcal{R})$ with $\mathcal{V} = \{v^{(1)} = s, v^{(2)}, \ldots, v^{(n)}\}$.
We assume that $p\left(v^{(i)}\right) := i$ $(i = 1, \ldots, n)$ is a Bellman-Ford-enumeration of $\mathcal{V}$.
Let $C : \mathcal{P}(s) \to \mathbb{R}$ be a path function. We assume that C is $=$-preserving and has the weak prefix oriented Bellman property of type 0. Moreover, we assume that $\mathcal{G}$ and C satisfy condition (4.92).
Then FORD-BELLMAN 2' will output the minimal C-value of all s-$v^{(i)}$-paths for each node $v^{(i)} \in \mathcal{V}$. Moreover, the running time is in $O(n^2)$.

P r o o f : The assertion is an immediate consequence of 4.125 and 4.138.   □

Next, we show that the costs of optimal s-v-paths can be computed by solving a system of equations.

Remark 4.140 Let $\mathcal{G} = (\mathcal{V}, \mathcal{R})$ be the digraph with $\mathcal{V} = \{v^{(1)} = s, v^{(2)}, \ldots, v^{(n)}\}$ and $\mathcal{R} = \{(v^{(i)}, v^{(j)}) \mid i, j \in \{1, \ldots, n\}\}$. We define the arc $r_{ij} := (v^{(i)}, v^{(j)})$ for all i and j.

Let $C : \mathcal{P}(s) \to \mathbb{R}$ be a path function with the following properties: C is $=$-preserving, C has the weak prefix oriented Bellman property of type 0, and there exists a C-minimal s-v-path for each node v. We assume that $C = \phi \circ o \circ p$ where $p : \mathcal{V} \to \mathbb{R}$ and $\phi_r : \mathbb{R} \to \mathbb{R}$ for each arc $r \in \mathcal{R}$; in particular, we assume that

$$\phi_r = \text{identity for all arcs } r = (v, v) \text{ with } v \in \mathcal{V}. \tag{4.108}$$

For all i, j, we define the function $\phi_{i,j}$ as the function ϕ_r for $r = (v^{(i)}, v^{(j)})$. Then (4.108) implies that $\phi_{i,i} = id_{\mathbb{R}}$ for all $i = 1, \ldots, n$.

For each node v, let $\chi(v)$ be the minimal C-value of all s-v-paths. Applying (2.62) to the current situation, we obtain the following assertion:

$$\begin{aligned}(\forall i = 1, \ldots, n) \quad \chi\left(v^{(i)}\right) &= \min\left\{\phi_{j,i}\left(\chi\left(v^{(j)}\right)\right) \mid j = 1, \ldots, n\right\} \\ &= \min\left\{\chi\left(v^{(i)}\right), \phi_{j,i}\left(\chi\left(v^{(j)}\right)\right) \mid j \neq i\right\}.\end{aligned} \tag{4.109}$$

Defining $x_i := \chi\left(v^{(i)}\right)$ for all i, we rewrite (4.109) as follows:

$$(\forall i = 1, \ldots, n) \quad x_i = \min\{\phi_{j,i}(x_j) \mid j = 1, \ldots, n\} \tag{4.110}$$

$$(\forall i = 1, \ldots, n) \quad x_i = \min\{x_i, \phi_{j,i}(x_j) \mid j \neq i\}. \tag{4.111}$$

Consequently, finding the minimal path costs $\chi\left(v^{(i)}\right)$ for all i means to find the unknown values $x_1, \ldots, x_n$ in (4.110) or in (4.111).

There are two methods of solving (4.110) and (4.111).

To apply the f i r s t method we reformulate conditions (4.110) and (4.111) as a fixed point problem. Let $F : \mathbb{R}^n \to \mathbb{R}^n$ be the following function:

$$(\forall z_1, \ldots, z_n) \quad F(z_1, \ldots, z_n) := (w_1, \ldots, w_n) \text{ with}$$
$$w_i := \min\{\phi_{j,i}(z_j) \mid 1 \leq j \leq n\} = \min\{z_i, \phi_{j,i}(z_j) \mid j \neq i\}, \ i = 1, \ldots, n.$$

Then (4.110) and (4.111) are equivalent to searching for a fixed point of F. This problem can be solved with a method of Approximation Theory: We choose a start vector $z = (z_1, \ldots, z_n) \in \mathbb{R}^n$ and obtain the desired solution $x = (x_1, \ldots, x_n)$ as the limit of the sequence z, $F(z)$, $F(F(z))$, $F(F(F(z)))$, ...

The iteration of F can be simulated by a slight modification of FORD-BELLMAN 1'. We assume that the modified version of FORD-BELLMAN 1' initializes $\chi(v^{(i)})$ as $C([s, v^{(i)}])$ for $i = 1, \ldots, n$. Moreover, we replace step 2.1 of FORD-BELLMAN 1' by the following step $\widetilde{2.1}$, which is independent of the set S:

$\widetilde{2.1}$. **For** $i := 1$ **to** n **do**

$\qquad$ Let $\widehat{\chi}\left(v^{(i)}\right) := \min \left\{ \phi_r(\chi(u)) \mid u \in \mathcal{N}^-\left(v^{(i)}\right), \; r = \left(u, v^{(i)}\right) \right\}$.

Then the initialization of the modified algorithm FORD-BELLMAN 1' will generate the start vector $(z_1, \ldots, z_n)$ with $z_i := \chi\left(v^{(i)}\right) = C\left([s, v^{(i)}]\right)$, $i = 1, \ldots, n$. Each further iteration of this algorithm effects an application of F to $\left(\chi\left(v^{(1)}\right), \ldots, \chi\left(v^{(n)}\right)\right)$;

to see this we observe that $\chi\left(v^{(i)}\right) \overset{(4.108)}{=} \phi_{i,i}\left(\chi\left(v^{(i)}\right)\right)$ is one of the candidate when computing $\widehat{\chi}\left(v^{(i)}\right)$ in step $\widetilde{2.1}$; this implies that always $\widehat{\chi}_t\left(v^{(i)}\right) \leq \chi_t\left(v^{(i)}\right)$; consequently, step 2.3 will generate the cost vector

$\left(\chi_{t+1}\left(v^{(1)}\right), \ldots, \chi_{t+1}\left(v^{(n)}\right)\right) = \left(\widehat{\chi}_t\left(v^{(1)}\right), \ldots, \widehat{\chi}_t\left(v^{(n)}\right)\right) \overset{(*)}{=} F\left(\chi_t\left(v^{(1)}\right), \ldots, \chi_t\left(v^{(n)}\right)\right)$; equa-

tion $(*)$ follows from the definition of F and the formulation of step $\widetilde{2.1}$.

The s e c o n d method of solving (4.110) and (4.111) can only be applied if C is additive. Let $C = SUM_h$ where $h : \mathcal{R} \to \mathbb{R}$. Then we can choose the functions ϕ_r such that

$$\phi_r(\xi) = \xi + h(r) \text{ for all arcs } r \in \mathcal{R} \text{ and all arguments } \xi \in \mathbb{R}. \tag{4.112}$$

Moreover, let $h_{i,j} := h\left(v^{(i)}, v^{(j)}\right)$ for all $i, j \in \{1, \ldots, n\}$. Then $h_{1,1} = h_{2,2} = \ldots = h_{n,n} = 0$ by (4.112) and (4.108).

Then we formulate (4.110) and (4.111) as follows:

$$(\forall \, i = 1, \ldots, n) \quad x_i = \min\{x_j + h_{j,i} \mid j = 1, \ldots, n\} \tag{4.113}$$
$$(\forall \, i = 1, \ldots, n) \quad x_i = \min\{x_i, \, x_j + h_{j,i} \mid j \neq i\}. \tag{4.114}$$

We concentrate on (4.113), which is equivalent to (4.114). As described in Section 3.10 of [Ju94], we can write the conditions in (4.113) as a system of pseudolinear equations. For this purpose, we define the "addition" $\boxplus$ and the "multiplication" $\boxtimes$ as follows:

$$(\forall \, x, y \in \mathbb{R}) \quad x \boxplus y := \min(x, y), \quad x \boxtimes y := x + y.$$

(Then the quadruple $(\mathbb{R}, \mathbb{R}, \boxplus, \boxtimes)$ is even a proper dioid.)

We now formulate (4.113) as follows:

$$(\forall \, i = 1, \ldots, n) \quad x_i = \overset{n}{\underset{j=1}{\boxplus}} \; h_{ji} \boxtimes x_i. \tag{4.115}$$

The system (4.115) of equations can be solved with the help of matrix algorithms; they even yield the minimal costs of all $v^{(i')}$-$v^{(i)}$ paths for each pair (i', i) (and not only for the case $i' = 1$). We shall describe these matrix algorithms in Section 4.4. $\square$

4.3.5 Finding C-minimal s-v-paths if $C : \mathcal{P}(s) \to (\mathbf{R}, \preceq)$ and if $\preceq$ is neither total nor identitive

In Section 3 – 5 of [LT91], [LTh91], two generalized Ford-Bellman algorithms are presented. These search methods work correctly even if the cost measure C is of the form $C = \lambda : \mathcal{P}(s) \to (\mathbf{R}, \preceq)$ where $\preceq$ is not total and identitive.[40]

More precisely, let $v \in \mathcal{V}$. It is possible that the set of all values $C(P)$, $P \in \mathcal{P}(s, v)$, does not have a unique minimum. For this reason, we define the set $dist(v)$ containing all these minima:

$$dist(v) := \{ C(P) \mid P \in \mathcal{P}(s, v), \ C(P) \in MIN\{C(Q) \mid Q \in \mathcal{P}(s, v)\} \}.$$

Then part 3 of Theorem 2 in [LT91], [LTh91] says that Lengauer's and Theune's generalized Ford-Bellman strategy has the following properties:

- It computes the set $dist(v)$ for all nodes v.
- It constructs a path $Q(d, v)$ with $C(Q(d, v)) = d$ for all $v \in \mathcal{V}$ and all $d \in dist(v)$.

Next, we compare the Ford-Bellman strategies in [LT91], [LTh91] with the ones in this book. The common feature of these algorithms is the method to improve a currently optimal s-v-path P:

> Let u be a predecessor of v, and let P' be a currently optimal s-u-path. Replace P by $P' \oplus (u, v)$ if $C(P' \oplus (u, v)) \prec C(P)$. $\qquad$ (4.116)

On the other hand, there are several differences between the Ford-Bellman algorithms in this book and the ones in [LT91], [LTh91]:

- The underlying cost measures of our algorithms have real values; the binary relation $\preceq$ is the same as $\leq$, which is total and identitive. That means that the currently optimal costs $\chi(v)$ are unique for each node v. In FORD-BELLMAN 1 and FORD-BELLMAN 2, $\chi(v)$ can be defined as $C(P(v))$; in FORD-BELLMAN 3, $\chi(v)$ can be defined as $C(P)$ for any $P \in Q(v)$ because all paths in $Q(v)$ have the same C-value; in FORD-BELLMAN 1' and FORD-BELLMAN 2', the values $\chi(v)$ are explicitly given.

In contrast to this, the algorithms of Lengauer and Theune must often record several minimal path costs $\chi_1(v), \ldots, \chi_k(v)$; the reason is that $\preceq$ is not total; therefore, a set of C-values may have several minima, which cannot be compared with each other.

- The algorithms FORD-BELLMAN 1, 2, 3 give an explicit construction of s-v-paths. It is possible that these s-v-paths contain cycles; for example, this occurs when applying FORD-BELLMAN 1 to the digraph in 4.124.

The search strategies in [LT91] and [LTh91], however, use a backtracking method similar to the one in Remark 4.128.

- All Ford-Bellman strategies in this book contain fixed rules saying which predecessors $u \in \mathcal{N}^-(v)$ must be used to improve the current s-v-connection $P(v)$ (e.g., all nodes $u \in \mathcal{N}^-(v) \cap S$ in step 2.1 of Algorithm 7.).

Such a rule does not exist in Lengauer's and Theune's search strategies.

---

[40]Lengauer and Theune require C to satisfy the so-called condition F1; this condition is essentially the local order preserving principle described on page 61 of this book.

A consequence of this difference is that each step of our Ford-Bellman algorithms generates a fixed explicit graph whereas the explicit graph generated by Lengauer's and Theune's algorithms depends on the choice of $u \in \mathcal{N}^-(v)$. Moreover, our results about optimal paths are independent from explicit graphs whereas Lengauer's and Theune's results are not. For example, Theorems 4.112 and 4.121 a) in our book say the following:

> At a particular moment, the given Ford-Bellman strategy has generated an s-v-connection P that is optimal among all candidates $P' \in \mathcal{P}(s, v)$ (perhaps of a bounded length).

In contrast to this, part 2 of Lemma 1 in [LT91], [LTh91] says the following:

> At a particular moment, the given Ford-Bellman strategy has found an s-v-connection P that is optimal among all candidates $P' \in \mathcal{P}(s, v)$ inside the explicit graph $\mathcal{G}' \subseteq \mathcal{G}$.

4.3.6 Using Best-First Search instead of Ford-Bellman strategies

When searching for C-minimal s-v-paths for each node v, we may also use the algorithm BF* with $F := C$ as decision function. We only must have the algorithm run until OPEN is empty.

We now describe several details of this method.

Remark 4.141 If $C = F$ is order preserving then all paths $P(v)$ of BF*(F) are indeed C-minimal at the moment when BF* terminates. To see this we assume that P be a C-minimal s-v-path. When OPEN is empty then all nodes of P are in CLOSED. Corollary 4.20 about $F^{(2)} := C = F$ implies that $C(P(v)) = F(P(v)) = C(P^{(2)}(v)) \leq F^{(2)}(P) = F(P) = C(P)$ so that $P(v)$ is C-minimal as well.

Now, we study the complexity of this and other Best-First search strategies.

a) If C is only order preserving and has no additional properties then the Best-First-Search is sometimes very slow; it has been shown in Theorem 4.66 that Algorithm BF* may execute $\Omega(2^n)$ iterations until OPEN is empty. In this case and in similar situations, it is better to use FORD-BELLMAN 1 or FORD-BELLMAN 2 instead of BF*.

The reason why BF* is often much slower than the Ford-Bellman algorithms is the following: The Ford-Bellman algorithms choose its paths $P(v)$ more carefully than BF*. The consequence is that BF* must often change the current paths $P(v)$ until they are (almost) optimal. In contrast, the path $P(v)$ of FORD-BELLMAN 1 or FORD-BELLMAN 2 are (almost) C-optimal in an early stage of the algorithm (see for example Theorem 4.121). The difference is very great if the acyclic digraph $\mathcal{G}$ in Remark 4.66 is given. Algorithm BF* executes many iterations and changes the paths $P(v)$ very often; in contrast, the first path $P(v)$ generated by FORD-BELLMAN 2 is already an an optimal s-v-connection for each node v.

b) If $F = C$ is order preserving and prefix monotone then we can use DIJKSTRA-BF*. The running time is bounded by $O(n^2)$ because each node is expanded exactly once and each expansion cases $O(n)$ time of computation. Therefore, it may be reasonable to solve Problem 9 with a Best-First search strategy if C is order preserving and prefix monotone.

There exist several articles about solving Problem 9 with Best-First search strategies; most of these sources make the assumption that C is even additive. A typical example is [HuDi88]; Hung and Divoky describe five variants of Dijkstra's algorithm and compare them with each other by experiments. □

4.3.7 Further results from literature

We quote further results from literature about finding a minimal s-v-path to each node v.

Remark 4.142 Given a digraph $\mathcal{G} = (\mathcal{V}, \mathcal{R})$ with $\mathcal{V} = \{v^{(1)} = s, v^{(2)}, \ldots, v^{(n)}\}$.

a) Let $C : \mathcal{P}(\mathcal{G}) \to \mathbb{R}$ be additive. Then the simplex method of Linear Programming finds C-optimal s-v-paths in a short time; this is shown in [GoHK90].

b) Let $C : \mathcal{P}(\mathcal{G}) \to \mathbb{R}$ be additive. Then the dynamic breadth-first search algorithm in [GoHK91] finds C-optimal s-v-paths for all nodes v, or the algorithm finds a C-negative cycle. The search strategy in [GoHK91] works in $O(n\,m)$ time.

c) The article [Cri91] is about finding C-optimal s-v-paths if $C : \mathcal{P}(s) \to \mathbb{R}$ is additive and no C-negative cycles exist. The main advantage of Crippa's algorithm is that a new complete search procedure is avoided if the cost function C is changed.

d) If C is additive but not prefix monotone then the Best-First search strategy in [Kun80] solves Problem 9 in $O(n^2)$ time units; Kundu makes the additional assumption that no arc r of a weight $h(r) < 0$ appears in a cycle; this is equivalent to the condition that

$$C(P \oplus Y) \; \leq \; C(P \oplus X) \tag{4.117}$$

for all paths $P \in \mathcal{P}(s)$, for all cycles X and for all prefixes $Y \leq X$. Moreover, this condition implies that $\mathcal{G}$ has no C-negative cycles.

It remains to be checked whether a generalized version of Kundu's algorithm can be applied in the situation that C is not additive and has property (4.117).

The following case is considered in In [Gold93]: C is additive, not prefix monotone, and the weights of all arcs are elements of $\{0, \pm 1, \pm 2, \ldots\}$. Goldberg presents an $O(\sqrt{n}\,m \log N)$ algorithm where N is essentially the absolute value of the most negative arc length.

e) The randomized and partially parallel algorithms in [Coh94] solve Problem 9 very fast if $\mathcal{G}$ is undirected and C is additive and prefix monotone. □

Remark 4.143 An special case of Problem 9 a) is that $C = SUM_h$ where $h(r) \in \{0, 1, 2, \ldots\}$ for all arcs r. Let D be the maximum of all weights $h(r)$, $r \in \mathcal{R}$, and let $m = |\mathcal{R}|$. Then fast algorithms can be created by organizing OPEN as a data structure of D elements.

a) In [Hans80], there is an algorithm that finds a C-minimal s-v-connection for all nodes in a time of $O(m \cdot \log D)$. An improved algorithm in [KaPo83] solves the same problem in $O(m \cdot \log \log D)$ time units.

b) The authors of [AMOT90] have found a search algorithm that only requires a time of $O\left(m + \frac{n \log D}{\log \log D}\right)$.

If $m = \Theta(n^2)$ and $D = \Theta(n)$ then the bound in [AMOT90] is better than the one in [KaPo83] because $O(m \cdot \log D) = O(n^2 \cdot \log \log n)$ whereas $O\left(m + \frac{n \log D}{\log \log D}\right) = O(n^2)$. If however, $m = \Theta(n)$ and $D = \Theta(n)$ then the result in [KaPo83] is better than [AMOT90] because $O(m \cdot \log D) = O(n \cdot \log \log n)$ whereas $O\left(m + \frac{n \log D}{\log \log D}\right) = O\left(n \cdot \frac{\log n}{\log \log n}\right)$. ☐

The following problem is solved in [YTO91]: Given two additive path functions C_1, C_2 : $P(\mathcal{G})$; for each $\lambda \in \mathbb{R}$, find a system of $(C_1 - \lambda C_2)$-minimal s-v-paths $(v \in \mathcal{V})$.

In [ChCh90], an algorithm was presented that finds quickest s-v-paths for all node $v \in \mathcal{V}$; the algorithm works in $O(m^2 + nm \log m)$ time. (The quickest path problem for a single pair (s, γ) was formulated in Subsection 4.2.6.) Further sources about the quickest path problem are quoted in Remark 4.172.

Next, we quote a paper about finding minimax paths in undirected networks.

Remark 4.144 Let $\mathcal{G} = (\mathcal{V}, \mathcal{R})$ be an undirected graph, and let $h : \mathcal{R} \to \mathbb{R}$ be an edge function. For all paths $P = r_1 \oplus \cdots \oplus r_k$, we define $H(P) := h(r_k)$ and $C(P) := H_{\max}(P) = \max\{h(r_1), \ldots, h(r_k)\}$.
Then we can use an algorithm from [Bard91] to find a C-minimal s-v-path for each node $v \in \mathcal{V}$. ☐

4.3.8 Finding C-maximal s-v-paths for all nodes v

Here we describe algorithms that solve the following problem: For each node v, find a C-maximal s-v-connection (and not a C-minimal one).

Let C be order preserving; then the same is true for $-C$. So, Theorem 4.125 implies that FORD-BELLMAN 1 with path function $(-C)$ will output $(-C)$-minimal s-v-connections, which are C-maximal. An example is described in the last part of Remark 4.127. — Moreover, we may use BF* as described in Subsection 4.3.6; we must choose $F := G := -C$ as decision function.

Let C have the weak prefix oriented Bellman property of type 0. Then we must not apply Theorem 4.112 to show that FORD-BELLMAN 1 with path function $-C$ outputs $(-C)$-minimal (i.e. C-maximal) paths. The reason is that a Bellman property of C does not always imply the same property of $-C$; this follows from Remark 2.82.

Let C be additive and prefix monotone. Then C is order preserving so that we may apply FORD-BELLMAN 1 with path function $-C$.
Another algorithm, however, is given in [Nolt75]. Noltemeier's search strategy starts with a Depth-First search procedure that generates a rooted tree $\mathcal{T}$ with root s. After this, a postprocessing is applied to all transversals of $\mathcal{T}$; transversals are arcs (u, v) in $\mathcal{G}$ for which no u-v-path in $\mathcal{T}$ exists.

The relationship between Depth-First search trees and longest paths is also investigated in [FrIm89]. The main result in the paper is that a system of longest paths can be used to decide whether a Depth-First search tree without transversals exists. That means that Fraysseix and Imai use maximal paths to investigate Depth-First search trees whereas Noltemeier uses Depth-Fist search trees to find maximal paths.

Let $C = \ell$. Then C is additive so that we can use FORD-BELLMAN 1, BF* and Noltemeier's algorithm to find paths of maximal length. In this case, BF* uses the path function $-\ell$ and is consequently equivalent to the version of BF*-DFS that works until OPEN is empty. (The algorithm BF*-DFS was described in paragraph 4.2.2.1.7.) It is, however, probable that BF*-DFS will often require much more time of computation than necessary; therefore, FORD-BELLMAN 1 and Noltemeier's search strategy seem to be better if $C = \ell$.

4.4 The All-Pairs-Optimal-Path problem

This section is about finding optimal (v, w)-paths for all pairs (v, w) of nodes; an easier version of this problem is to compute the cost of each optimal (v, w)-connection (and not the (v, w)-paths themselves). We formulate the problem explicitly:

Problem 10

Given a directed or an undirected graph $G = (V, R)$ and a cost function $C : P(G) \to \mathbb{R}$.
   a) *Find a C-minimal v-w-path for each pair (v, w) of nodes.*
   b) *Compute the minimal C-value of all v-w-paths for each pair (v, w) of nodes.*

We shall describe several algorithms that solve this problem. Most of these techniques use matrices of paths or of costs. For example, the algorithms in Section 4.4.1 simulate the steps of Ford-Bellman algorithms by matrix operations.

Standard algorithms for problem 10 are given in [Chri75], [Nolt76], [PaSt82], [EvMi92], and in many other sources. More recent algorithms are studied in [Taka92]. Almost all results in these sources make the assumption that the cost function C is additive. We, however, shall develop matrix algorithms in situations where the cost function C need not be additive.

To make the formulation of matrix algorithms easier we shall assume that $G = (V, R)$ is the complete digraph[41] with $V = \{1, \ldots, n\}$ and $R = V \times V$.

Moreover, we assume that
$$C(X \oplus (i, i) \oplus Y) = (X \oplus Y) \tag{4.118}$$
for all loops (i, i) and all paths X, Y.

(Remark 4.146 will show how to generate a complete digraph G with monotone or even additive cost function C if the given digraph G' is not complete.)

---

[41]In Remark 4.146, we shall show how to generate a complete digraph G with monotone or even additive cost function C if the given digraph G' is not complete.

We assume that for each arc (i,j), a function $\phi_{(i,j)} : \mathbb{R} \dashrightarrow \mathbb{R}$ is given such that the following is true:
$$(\forall\, P \in \mathcal{P}(\mathcal{G})\,,\ (i,j) \in \mathcal{R})\quad C(P \oplus (i,j)) \;=\; \phi_{(i,j)}(C(P))\,.$$
Then $C(P_1) = C(P_2) \Rightarrow C(P_1 \oplus Q) = C(P_2 \oplus Q)$ for all paths P_1, P_2, Q with $\omega(P_1) = \omega(P_2) = \alpha(Q)$. This property is stronger than the $=$-preservation of C because the start nodes of P_1 and P_2 may be different.

Let $\xi \in \mathbb{R}$ and $(i,j) \in \mathcal{R}$; we define $\xi^{\phi_{(i,j)}} := \phi_{(i,j)}(\xi)$.

An immediate consequence of (4.118) is that we may choose $\phi_{(1,1)}, \dots, \phi_{(n,n)}$ as the identity function on $\mathbb{R}$.

A further consequence of (4.118) is the following:

Lemma 4.145 *Let $i \in V$. Given the i-j-path $Q(j)$ for each $j \in V$. Let $\widehat{Q}(j)$ be a path with the following property:*
$$C(\widehat{Q}(j)) \;=\; \min_{k=1,\dots,n}\, \{C(Q(k) \oplus (k,j))\}\,.$$
Then $C(\widehat{Q}(j)) \leq C(Q(j))$, and in consequence, $C(\widehat{Q}(j)) = \min\{C(Q(j)), C(\widehat{Q}(j))\}$.

P r o o f : The definition of $\widehat{Q}(j)$ implies that $C(\widehat{Q}(j)) \leq C(Q(j) \oplus (j,j))$, and Assumption (4.118) says that $C(Q(j) \oplus (j,j)) = C(Q(j))$. $\qquad\square$

The next remark shows how to transform a given digraph $\mathcal{G}'$ into a complete digraph $\mathcal{G}$ such that $\mathcal{G}$ has the same optimal paths as $\mathcal{G}'$.

Remark 4.146 Let $\mathcal{G}' = (V, \mathcal{R}')$ be a digraph without loops; let $V = \{1, \dots, n\}$. Let $\phi'_{(i,j)} : \mathbb{R} \to \mathbb{R}$ be a monotonically increasing mapping for each arc $(i,j) \in \mathcal{R}'$, and let $p(i) := 0$ for all nodes i. We assume that $C' = \phi' \circ p$. Moreover, let $\mathcal{G} := (V, \mathcal{R})$ with $\mathcal{R} = V \times V$.

Then we define a function $\phi_{(i,j)}$ for each $(i,j) \in \mathcal{R}$ and a bound M such that the complete digraph $\mathcal{G}$ with the cost function $C := \phi \circ\circ p$ has two properties: C satisfies condition (4.118), and the following is true for all $i, j \in V$ and all i-j-paths $P \in \mathcal{P}(\mathcal{G})$ without loops:

$$\begin{aligned} &P \text{ is a } C'\text{-minimal } i\text{-}j\text{-path in } \mathcal{G}' \text{ if and only if} \\ &P \text{ is a } C\text{-minimal } i\text{-}j\text{-path in } \mathcal{G} \text{ and } C(P) < M. \end{aligned} \qquad (4.119)$$

For this purpose, we define $\mathcal{R}_0 := \{(i,i)\,|\,i \in V\}$ and $\mathcal{R}_1 := \mathcal{R}\backslash(\mathcal{R}' \cup \mathcal{R}_0)$. Part a) of this remark is about the general case that the functions ϕ'_r are increasing, and the parts b), c) are about the special case that C' is additive. Our idea is the following: If a path P uses an arc $r \in \mathcal{R}_1$ then $C(P)$ will be very great.

a) In the general case, we define $\phi_{(i,j)}(\xi)$ as follows: For all $(i,j) \in \mathcal{R}$ and $\xi \in \mathbb{R}$,
$$\phi_{(i,j)}(\xi) \;:=\; \begin{cases} \arctan\left(\phi'_{(i,j)}(\tan(\xi))\right) & \text{if}\quad (i,j) \in \mathcal{R}' \text{ and } |\xi| < \pi/2, \\ e^{\xi} + 2 & \text{if}\quad (i,j) \in \mathcal{R}' \text{ and } |\xi| \geq \pi/2, \\ \xi & \text{if}\quad (i,j) \in \mathcal{R}_0, \\ e^{\xi} + 2 & \text{if}\quad (i,j) \in \mathcal{R}_1. \end{cases}$$

Moreover, let $M := \pi/2$.

Then $C := \phi \circ \circ p$ has the following properties: $C(P) = \arctan(C'(P)) < M$ if $P \in \mathcal{P}(\mathcal{G}')$, and $C(P) > M$ if $P \in \mathcal{P}(\mathcal{G})$ uses an arc $r \in \mathcal{R}_1$. These two properties imply fact (4.119) about $M := \pi/2$. Moreover, each $\phi_{(i,j)}$ is monotonically increasing so that C is order preserving.

b) Let $C' = SUM_{h'}$ be additive, and let $\mathcal{G}'$ be strongly connected. Then we define an arc function $h : \mathcal{R} \to \mathbb{R}$ and the additive path function $C := SUM_h$.
For this purpose, let $M := 1 + n^2 \cdot (\max_{r \in \mathcal{R}'} |h'(r)|)$. We define $h(r) := h'(r)$ for all $r \in \mathcal{R}'$, $h(r) := 0$ for all $r \in \mathcal{R}_0$, and $h(r) := M + 1$ for all $r \in \mathcal{R}_1$.

It is obvious that $C(P) = C'(P)$ for all paths P in $\mathcal{G}'$. Moreover, the following is true for all nodes i, j and for all i-j-paths $P \in \mathcal{P}(\mathcal{G})$:

> There exists an i-j-path $Q \in \mathcal{P}(\mathcal{G}')$ with $C(Q) \leq C(P)$; moreover, if P uses an arc $r \in \mathcal{R}_1$ then Q can be chosen such that $C(Q) < C(P)$.   (4.120)

The path Q can be constructed by removing all loops of P and by replacing each arc $(k_1, k_2) \in \mathcal{R}_1$ by an arc injective k_1-k_2-path $Q(k_1, k_2) \in \mathcal{G}'$; the existence of $Q(k_1, k_2)$ follows from the assumption that $\mathcal{G}'$ is strongly connected, and the relationship $C(Q) < C(P)$ follows from $C(Q(k_1, k_2)) = C'(Q(k_1, k_2)) = SUM_{h'}(Q(k_1, k_2)) < h(k_1, k_2)$ for all $(k_1, k_2) \in \mathcal{R}_1$.

Then fact (4.119) can be seen as follows:

Let P be a C'-minimal i-j-path in $\mathcal{G}'$. Then P is C-minimal in $\mathcal{G}$. Otherwise there existed a i-j-path $P_0 \in \mathcal{P}(\mathcal{G}')\backslash\mathcal{P}(\mathcal{G})$ with $C(P_0) < C(P)$, and by (4.120), there existed an i-j-path $Q \in \mathcal{P}(\mathcal{G}')$ with $C'(Q) = C(Q) \leq C(P_0) < C(P) = C'(P)$. That is a contradiction to the optimality of P. Moreover, $C'(P) < M$ for the following reasons: There exists an arc injective i-j-path $Q \in \mathcal{P}(\mathcal{G}')$ because $\mathcal{G}'$ is strongly connected; then $C'(P) \leq C'(Q)$ because P is C'-optimal, and $C'(Q) < M$ because Q is arc injective; hence, $C'(P) \leq C'(Q) < M$.
Let P be a C-minimal i-j-path in $\mathcal{G}$ with $C(P) < M$; we must show that P is C'-minimal in $\mathcal{G}'$. We have assumed that P contains no loops. Moreover, P does not contain any arcs $r \in \mathcal{R}_1$ because otherwise there existed a path Q with $C(Q) \overset{(4.120)}{<} C(P)$, which is a contradiction to the C-minimality of P. Consequently, P is a path in $\mathcal{G}'$. This and the C-minimality of P imply that P is a C'-minimal i-j-connection.

c) If $C' = SUM_{h'}$ is additive and $\mathcal{G}'$ is not strongly connected, then an additive path function C for $\mathcal{G}$ can be defined as follows: Let $h(r) := h'(r)$ if $r \in \mathcal{R}'$, let $h(r) := 0$ if $r \in \mathcal{R}_0$, and let $h(r) := \infty$ if $r \in \mathcal{R}_1$. Moreover, let $C := SUM_h$, and let $M := \infty$.
Then C has the following properties: $C(P) < M = \infty$ if $P \in \mathcal{P}(\mathcal{G}')$, and $C(P) = M = \infty$ if $P \in \mathcal{P}(\mathcal{G})$ uses an arc $r \in \mathcal{R}_1$. These two properties imply fact (4.119).

The slight disadvantage of C is that its domain of values is $\mathbb{R} \cup \{\infty\}$ and not $\mathbb{R}$. But this disadvantage cannot always be avoided. For example, let $\mathcal{G}' = (\mathcal{V}, \mathcal{R}')$ with $\mathcal{V} := \{v^{(1)}, v^{(2)}, v^{(3)}, v^{(4)}\}$ and
$$\mathcal{R}' := \{(v^{(1)}, v^{(2)}), (v^{(1)}, v^{(3)}), (v^{(3)}, v^{(4)}), (v^{(4)}, v^{(3)})\}.$$

We define $X := \left[v^{(3)}, v^{(4)}, v^{(3)}\right]$ and $P := \left[v^{(1)}, v^{(2)}\right]$. Let $C' := SUM_{h'}$ with $h'\left(v^{(1)}, v^{(2)}\right) := h'\left(v^{(1)}, v^{(3)}\right) := 1$ and $h'\left(v^{(3)}, v^{(4)}\right) := h'\left(v^{(4)}, v^{(3)}\right) := -1$.

Moreover, let $\mathcal{G} = (V, V \times V)$, and let $C := SUM_h$ where h is an extension of h' to $V \times V$; we assume that $h(r) < \infty$ for all $r \in V \times V$.

Then P is the only (and in consequence, the C'-optimal) $v^{(1)}$-$v^{(2)}$-path in $\mathcal{G}'$. But P is not a C-minimal $v^{(1)}$-$v^{(2)}$-path in $\mathcal{G}$ because such a path does not exist at all; the reason is that the values $C\left((v^{(1)}, v^{(3)}) \oplus X^\nu \oplus (v^{(4)}, v^{(2)})\right)$, $\nu \in \mathbb{N}$, can be arbitrarily small.

It follows that in this case, fact (4.119) is false if $h(r) < \infty$ for all $r \in V \times V$. $\quad\square$

Next, we define a relation for matrices of paths.

Definition 4.147 Let $n \in \mathbb{N}$, $V := \{1, \dots, n\}$ and $\mathcal{G} := (V, V \times V)$. Let $C : \mathcal{P}(\mathcal{G}) \to \mathbb{R}$ be a path function, and let $\mathbf{A} = (P_{i,j})$ and $\mathbf{B} = (Q_{i,j})$ be two $n \times n$-matrices of paths. We say that $\mathbf{A}$ is C-equivalent to $\mathbf{B}$ ($\mathbf{A} \cong_C \mathbf{B}$) if $C(P_{i,j}) = C(Q_{i,j})$ for all i, j. $\quad\square$

The next remark describes how the solution of Problem 10 b) can be used to solve two other problems.

Remark 4.148 Computing the costs of shortest paths is a good method to solve the following problems:

- Constructing transitive reflexive closures.
- Testing whether two nodes are strongly connected.

More precisely, let $\tilde{\mathcal{G}} = (\tilde{V}, \tilde{\mathcal{R}})$ be a digraph with $\tilde{V} = \{v^{(1)}, \dots, v^{(n)}\}$ and $\tilde{\mathcal{R}} \subseteq \tilde{V} \times \tilde{V}$. (We do not require that $\tilde{\mathcal{R}} = \tilde{V} \times \tilde{V}$.) The *transitive reflexive closure* of $\tilde{\mathcal{G}}$ is defined as the digraph $\tilde{\mathcal{G}}^* := (\tilde{V}, \tilde{\mathcal{R}}^*)$ with

$$\tilde{\mathcal{R}}^* := \left\{ \left(v^{(i)}, v^{(j)}\right) \mid \text{There exists a } v^{(i)} - v^{(j)} - \text{path in } \tilde{\mathcal{G}} \right\}.$$

The transtive reflexive closure can be constructed as follows: Define $\mathcal{G} := (V, \mathcal{R})$ with $V := \{1, \dots, n\}$ and $\mathcal{R} := V \times V$. Moreover, we define $h : \mathcal{R} \to \mathbb{R}$ as follows: For all (i, j), let $h(i, j) := 0$ if $i = j$ or $\left(v^{(i)}, v^{(j)}\right) \in \tilde{\mathcal{R}}$; let $h(i, j) := 1$ otherwise. Let $C := SUM_h$, and let $\chi^*(i, j)$ be the minimal cost of all i-j-paths in $\mathcal{G}$. Then $\left(v^{(i)}, v^{(j)}\right) \in \tilde{\mathcal{R}}^*$ if and only if $\chi^*(i, j) = 0$.

The values $\chi^*(i, j)$ can also be used to test whether to nodes $v^{(i)}$ and $v^{(j)}$ are strongly connected because this is true if and only if $\chi^*(i, j) = \chi^*(j, i) = 0$. $\quad\square$

4.4.1 Matrix algorithms simulating Ford-Bellman algorithms

Here, we describe several matrix algorithms solving Problem 10 a) or b). These algorithms are generated by transforming the operations of FORD-BELLMAN 1 into matrix operations. The advantage of the matrix algorithms in this subsection is that they can be used for general classes of cost functions C; the slight disadvantage of these algorithms is that they are not very fast.

4.4.1.1   Transforming FORD-BELLMAN 1 into a matrix algorithm

Here, we show how FORD-BELLMAN 1 can be transformed into a matrix algorithm. The next remark describes a modification of FORD-BELLMAN 1; this modification can be more easily transformed into a matrix algorithm than the original version of algorithm 7.

Introductory Remark 4.149 Given the digraph $G = (V, V \times V)$ with $V = \{1, \ldots, n\}$. Given a cost measure $C : \mathcal{P}(G) \to \mathbb{R}$; we assume that $C = \phi \circ \circ p$ where $p : V \to \mathbb{R}$ and $\phi_{(j,j')} : \mathbb{R} \dashrightarrow \mathbb{R}$ for all arcs (j, j'). Let $s = i$ be a fixed start node of G.

First, we modify Algorithm FORD-BELLMAN 1 as follows:

a) In the *original* version of FORD-BELLMAN 1, the path $P(s)$ is initialized as $[s]$, the paths $P(j)$, $j \neq s$, were not defined.
The *modified* version defines $P(j) := [s, j]$ for all nodes j (including $j = s$). Moreover, W is initalized as $W := V$.

b) In the *original* version of FORD-BELLMAN 1, the path $\widehat{P}(j)$ is defined as a C-minimal path among all candidates $P(k) \oplus (k, j)$ with $k \in S$. The only reason for the restriction to $k \in S$ was to reduce the time of computation.
The *modified* version defines $\widehat{P}(j)$ as the C-minimal path among all candidates $P(k) \oplus (k, j)$ with $k \in V$;[42] the following tie breaking rule is used:

> If $k' < k''$ and if $P(k') \oplus (k', j)$ and $P(k'') \oplus (k'', j)$ are two equivalent candidates for $\widehat{P}_t(j)$ then the one with the smaller k' is preferred.

For example, this can be done by the following procedure:

$\widehat{P}(j) := P(1) \oplus (1, j)$;
For $k := 2$ **to** n **do**
   If $C(P(k) \oplus (k, j)) < C(\widehat{P}(j)$ **then** $\widehat{P}(j) := P(k) \oplus (k, j)$.

c) In the *original* version of FORD-BELLMAN 1, the path $P_t(j)$ is replaced by $\widehat{P}_t(j)$ if and only if $C(\widehat{P}_t(j)) < C(P(j))$; the new path P_{t+1} has the property that $C(P_{t+1}(j)) = \min\{C(P_t(j)), C(\widehat{P}_t(j))\}$.
Moreover, j is added to S if and only if $P_{t+1}(j) \neq P_t(j)$.

The *modified* version replaces $P_t(j)$ in any case by $\widehat{P}_t(j)$. Then $C(P_{t+1}(j)) = \min\{C(P_t(j)), C(\widehat{P}_t(j))\}$ by Lemma 4.145.
Moreover, j is added to S if and only if $C(P_{t+1}(j)) < C(P_t(j))$.

These modifications do not influence the correctness of the algorithm. A consequence of the modified initialization, however, is that all results in Section 4.3 should be revised if they are about path lengths.

For example, the *new* version of Theorem 4.121 a) says the following: The path $P_{t+1}(v)$ generated by the modified algorithm FORD-BELLMAN 1 is C-minimal all s-v-paths of length $\leq t + 1$ (and not only of length $\leq t$). To see this we start with $t = 0$.

---

[42]That means that the definition of $\widehat{P}(j)$ is independent of S; nevertheless, it may be reasonable to update S because the test '$S = \emptyset$?' can be used to terminate FORD-BELLMAN 1 (see Theorem 4.125). — In the current section the test '$S = \emptyset$?' will be replaced by an explicit test whether $P_{t+1}(j) = P_t(j)$ ($j \in V$) or by a similar test.

If $v = s$ then $[s]$ and $[s,s]$ are the only s-s-paths of a length ≤ 1, and it is $C(P_1(s)) = C([s,s]) \overset{(4.118)}{=} C([s])$. If $v \neq s$ then $P_1(v) = [s,v]$ is the only candidate of length ≤ 1; hence, $P_1(v)$ is C-minimal among all candidates of length ≤ 1. The induction step $t \to t+1$ is mainly the same as in Theorem 4.121.

By the way, $d(s) = 0$ and $d(v) = 1$ $(v \neq s)$ in the graph $\mathcal{G}$. $\qquad\square$

Next, we give simple recursion formulas concerning the paths that are constructed by the modified version of FORD-BELLMAN 1.

Introductory Remark 4.150 Given the same situation as in Remark 4.149. We now give several recursion formulas; they describe the paths $P_t(j)$ constructed by the modified version of FORD-BELLMAN 1, and they describe the cost values $C(P_t(j))$. The following formula is valid for t and for all j:

$$P_1(j) = [i,j], \qquad P_{t+1}(j) = P_t(k'') \oplus [k'',j]$$
$$\text{where } k'' = \min\{k' \mid C(P_t(k') \oplus [k',j]) = \min_{k=1,\dots,n} C(P_t(k) \oplus [k,j])\}. \tag{4.121}$$

Moreover, the following is true for all t and j:

$$C(P_1(j)) = C([i,j]), \qquad C(P_{t+1}(j)) = \min_{k=1,\dots,n} C(P_t(k) \oplus [k,j]).$$

Defining $\chi(j) := C(P(j))$ and $\chi_t(j) := C(P_t(j))$ for all j and t, we obtain the following recursion formula:

$$(\forall\, t,\, j) \quad \chi_1(j) = C([i])^{\phi(i,j)}, \qquad \chi_{t+1}(j) = \min_{k \in \mathcal{V}} \chi_t(k)^{\phi(k,j)}. \tag{4.122}$$

This formula produces the same results as the iteration of the function F in 4.140.

Up to now we have fixed the start node $s = i$. Our current problem, however, is to find optimal i-j-paths for each start node i. For this reason, we now imagine that i is variable and not fixed. We replace the names $P(j)$ and $P_t(j)$ by the names $P(i,j)$ and $P_t(i,j)$, respectively; moreover, we define we define $\chi(i,j) := C(P(i,j))$ and $\chi_t(i,j) := C(P_t(i,j))$. Then (4.121) and (4.122) imply the following recursion formulas, which are valid for all start nodes i, all end nodes j, and for all iteration numbers t:

$$P_1(i,j) = [i,j], \qquad P_{t+1}(i,j) = P_t(i,k'') \oplus [k'',j]$$
$$\text{where } k'' = \min\{k' \mid C(P_t(i,k') \oplus [k',j]) = \min_{k=1,\dots,n} C(P_t(i,k) \oplus [k,j])\}. \tag{4.123}$$

$$\chi_1(i,j) = C([i])^{\phi(i,j)} \text{ and } \chi_{t+1}(i,j) = \min_{k \in \mathcal{V}} \chi_t(i,k)^{\phi(k,j)}. \tag{4.124} \quad\square$$

Next, we transform the formulas (4.123) and (4.124) into matrix operations. These operations are applied to the following matrices:

$$\begin{pmatrix} P_t(1,1) & \cdots & P_t(1,n) \\ \vdots & & \vdots \\ P_t(n,1) & \cdots & P_t(n,n) \end{pmatrix}, \quad \begin{pmatrix} \chi_t(1,1) & \cdots & \chi_t(1,n) \\ \vdots & & \vdots \\ \chi_t(n,1) & \cdots & \chi_t(n,n) \end{pmatrix}.$$

Introductory Remark 4.151 Here, we introduce matrix operations, which look like matrix multiplications.

First, we define the following binary operations $\boxplus$ and $\boxtimes$ for *paths*:

$$\boxplus : \mathcal{P}(\mathcal{G}) \times \mathcal{P}(\mathcal{G}) \to \mathcal{P}(\mathcal{G}), \quad P' \boxplus P'' := \begin{cases} P' & \text{if } C(P') \leq C(P''), \\ P'' & \text{if } C(P') > C(P''). \end{cases} \tag{4.125}$$

$$\boxtimes : \mathcal{P}(\mathcal{G}) \times \mathcal{P}(\mathcal{G}) \dashrightarrow \mathcal{P}(\mathcal{G}), \quad P \boxtimes Q := P \oplus Q.$$

(These operations were defined in 2.170, and it was shown that $\boxplus$ is associative; the result of $P_1 \boxplus \ldots \boxplus P_k$ is a path with minimal C-value; ties are resolved in favour to the path P_κ with the smallest index κ. Of course, the operation $\boxtimes$ is associative, too.)

Second, we define the binary operations $\boxplus$ and $\boxtimes$ for *real numbers*; let $Part(\mathbb{R}, \mathbb{R})$ be the set of all partial functions from $\mathbb{R}$ to $\mathbb{R}$; then

$$\boxplus : \mathbb{R} \times \mathbb{R} \to \mathbb{R}, \quad x \boxplus y := \min(x, y),$$
$$\boxtimes : \mathbb{R} \times Part(\mathbb{R}, \mathbb{R}) \dashrightarrow \mathbb{R}, \quad x \boxtimes f := x^f = f(x). \tag{4.126}$$

By the way, the quadruples $(\mathcal{P}(\mathcal{G}), \mathcal{P}(\mathcal{G}), \boxplus, \boxtimes)$ and $(\mathbb{R}, Part(\mathbb{R}, \mathbb{R}), \boxplus, \boxtimes)$ are weak dioids as defined in 2.169.

Using the above operations $\boxplus$ and $\boxtimes$, we can formulate (4.123) and (4.124) such that they look like the definition of matrix multiplication; the following are true for all i, j and for all t:

$$P_1(i, j) = [i, j], \quad P_{t+1}(i, j) = \overset{n}{\underset{k=1}{\boxplus}} P_t(i, k) \boxtimes [k, j], \tag{4.127}$$

$$\chi_1(i, j) = C(i)^{\phi_{(i,j)}}, \quad \chi_{t+1}(i, j) = \overset{n}{\underset{k=1}{\boxplus}} \chi_t(i, k) \boxtimes \phi_{(k,j)}. \tag{4.128}$$

Then we can write the application of (4.127) and (4.128) to each pair (i, j) as matrix multiplications:

$$\begin{pmatrix} P_{t+1}(1,1) & \cdots & P_{t+1}(1,n) \\ \vdots & & \vdots \\ P_{t+1}(n,1) & \cdots & P_{t+1}(n,n) \end{pmatrix} = \begin{pmatrix} P_t(1,1) & \cdots & P_t(1,n) \\ \vdots & & \vdots \\ P_t(n,1) & \cdots & P_t(n,n) \end{pmatrix} \boxtimes \begin{pmatrix} [1,1] & \cdots & [1,n] \\ \vdots & & \vdots \\ [n,1] & \cdots & [n,n] \end{pmatrix}, \tag{4.129}$$

$$\begin{pmatrix} \chi_{t+1}(1,1) & \cdots & \chi_{t+1}(1,n) \\ \vdots & & \vdots \\ \chi_{t+1}(n,1) & \cdots & \chi_{t+1}(n,n) \end{pmatrix} = \begin{pmatrix} \chi_t(1,1) & \cdots & \chi_t(1,n) \\ \vdots & & \vdots \\ \chi_t(n,1) & \cdots & \chi_t(n,n) \end{pmatrix} \boxtimes \begin{pmatrix} \phi_{(1,1)} & \cdots & \phi_{(1,n)} \\ \vdots & & \vdots \\ \phi_{(n,1)} & \cdots & \phi_{(n,n)} \end{pmatrix}. \tag{4.130}$$

We write these equations in a more compact form and obtain the following formulas for all t:

$$\big(P_{t+1}(i,j)\big) = \big(P_t(i,j)\big) \boxtimes \big([i,j]\big), \tag{4.131}$$

$$\big(\chi_{t+1}(i,j)\big) = \big(\chi_t(i,j)\big) \boxtimes \big(\phi_{(i,j)}\big). \tag{4.132}$$

The formulas (4.129), (4.130), (4.131) and (4.132) are simpler if C is easily computable of type B. Then there exist a binary operation $\odot : \mathbb{R} \times \mathbb{R} \to \mathbb{R}$ and an arc function

$h : \mathcal{R} \to \mathbb{R}$ such that $\phi_{(i,j)}(C(P)) = C(P \oplus (i,j)) = C(P) \odot h(i,j)$ for all paths P and for all arcs (i,j). For example, '$\odot$' equals '$+$' if C is additive.

We define the following operations $\boxplus$ and $\boxtimes$:

$$\boxplus : \mathbb{R} \times \mathbb{R} \to \mathbb{R}, \quad x \boxplus y := \min(x,y),$$
$$\boxtimes : \mathbb{R} \times \mathbb{R} \to \mathbb{R}, \quad x \boxtimes y := x \odot y. \tag{4.133}$$

Then the following recursion formulas (4.134), (4.135), (4.136) are equivalent to (4.128), (4.130), (4.132), respectively; the matrix multiplication in (4.135) and (4.136) means that formula (4.134) is applied to each pair (i,j).

$$\chi_{t+1}(i,j) = \overset{n}{\underset{k=1}{\boxplus}} \chi_t(i,k) \boxtimes h(k,j), \tag{4.134}$$

$$\begin{pmatrix} \chi_{t+1}(1,1) & \cdots & \chi_{t+1}(1,n) \\ \vdots & & \vdots \\ \chi_{t+1}(n,1) & \cdots & \chi_{t+1}(n,n) \end{pmatrix} = \begin{pmatrix} \chi_t(1,1) & \cdots & \chi_t(1,n) \\ \vdots & & \vdots \\ \chi_t(n,1) & \cdots & \chi_t(n,n) \end{pmatrix} \boxtimes \begin{pmatrix} h(1,1) & \cdots & h(1,n) \\ \vdots & & \vdots \\ h(n,1) & \cdots & h(n,n) \end{pmatrix}, \tag{4.135}$$

$$\big(\chi_{t+1}(i,j)\big) = \big(\chi_t(i,j)\big) \boxtimes \big(h(i,j)\big). \tag{4.136}$$

Formula (4.134) has a simpler structure than formula (4.128) because in (4.134), all operands of $\boxtimes$ are real numbers. Moreover, the formulas (4.135) and (4.136) are simpler than the formulas (4.130) and (4.132), respectively, because in (4.135) and (4.136), two matrices of real numbers are multiplied. □

Now, we formulate an algorithm to solve Problem 10. The algorithm will apply formula (4.129) to construct matrices of paths, and it will apply formula (4.130) or (4.135) to construct matrices of costs. The purpose of the matrix $\mathbf{Y}$ is to compare the previous cost matrix with the new one.

Algorithm 12

1. (* Initialization *)

   $\mathbf{A} := \mathbf{A}' := \big([i,j]\big); \quad \mathbf{X} := \big(C(i,j)\big); \quad \Phi := \big(\phi_{(i,j)}\big);$

   If C is easily computable of type B then $\mathbf{H} := \big(h(i,j)\big);$

   $\tau := 0;$

   (* The variable τ is used to count the iterations. *)

   $b := false;$

   (* The variable b says whether the current iteration has changed the cost matrix $\mathbf{X}$. *)

2. **While** COND **do**

   2.1. Let $\mathbf{A} := \mathbf{A} \boxtimes \mathbf{A}'$ (where "$\boxtimes$" means the matrix multiplication in (4.129)).

   2.2. **If** C is not easily computable of type B **then**
   let $\mathbf{Y} := \mathbf{X} \boxtimes \Phi$ (where "$\boxtimes$" means the matrix multiplication in (4.130)).

   2.3. **If** C is easily computable of type B **then**
   let $\mathbf{Y} := \mathbf{X} \boxtimes \mathbf{H}$ (where "$\boxtimes$" means the matrix multiplication in (4.135)).

   2.4. **If** $\mathbf{Y} = \mathbf{X}$ **then** $b := true$ **else** $b := false;$

   2.5 $\mathbf{X} := \mathbf{Y};$

   2.6. $\tau := \tau + 1;$

3. Perform the output procedure WRITE.

4.4.1.2 The correctness of Algorithm 12

Now we describe situations in which Algorithm 12 constructs optimal paths and computes minimal costs of paths. As seen in Remarks 4.149 – 4.151, Algorithm 12 simulates a slightly modified version of FORD-BELLMAN 1 for each start node $i \in \mathcal{V}$. Therefore, most of the results in Subsection 4.3.1 remain true if we translate them into the setting of matrix algorithms.

The next result is analogous to 4.116.

Theorem 4.152 *Given the digraph* $\mathcal{G}$, *the cost measure* C, *and the functions* p *and* $\phi_{(i,j)}$ *as described in* 4.149. *(Then each restriction* $C|_{\mathcal{P}(i)}$ *is automatically =-preserving.) We assume that* C *has the weak prefix oriented Bellman property of type 0. Moreover, we make the following assumption:*

   (i) *For all nodes* i, j, *there exists a* C-*minimal* i-j-*connection.*

We define the following $(n \times n)$-*matrices* $\mathbf{A}_1$, $\mathbf{A}'$, $\mathbf{X}_1$, Φ, *and* $\mathbf{H}$: [43]

$$\mathbf{A}_1 := \mathbf{A}' := \big([i,j]\big), \quad \mathbf{X}_1 := \big(C([i,j])\big), \quad \Phi := \big(\phi_{(i,j)}\big), \quad \mathbf{H} := \big(h(i,j)\big).$$

Moreover, the matrices $\mathbf{A}_2$, $\mathbf{A}_3$, $\mathbf{A}_4,\ldots$ *and* $\mathbf{X}_2$, $\mathbf{X}_3$, $\mathbf{X}_4,\ldots$ *are recursively defined in the following formula; the matrix multiplications in these definitions are the same as the ones in* (4.129), (4.130), (4.135), *respectively; for all* $t = 1, 2, 3, \ldots$, *we define*

$$\mathbf{A}_{t+1} := \mathbf{A}_t \boxtimes \mathbf{A}', \quad \mathbf{X}_{t+1} := \mathbf{X}_t \boxtimes \Phi, \quad \mathbf{X}_{t+1} := \mathbf{X}_t \boxtimes \mathbf{H} \text{ (if } C \text{ is easily computable).}$$

Let $n' \geq n$.

Then all paths $P_{n'}(i,j)$ *in* $\mathbf{A}_{n'}$ *are* C-*minimal, and the costs* $\chi_{n'}(i,j)$ *in* $\mathbf{X}_{n'}$ *are minimal.*

This implies that Algorithm 12 will solve Problem 10 if we specify COND *and* WRITE *as follows:*

   • COND $:= (\tau < n' - 1)$ • WRITE $:=$ (Write the matrices $\mathbf{A}$ and $\mathbf{X}$.)

P r o o f (S k e t c h) : The proof is analogous to the one of 4.116. ☐

The next result is similar to 4.121 and 4.122.

Theorem 4.153 *Given the digraph* $\mathcal{G}$, *the cost measure* C, *and the functions* p, $\phi_{(i,j)}$ *as described in* 4.149. *We assume that all functions* $\phi_{(i,j)}$ *are strictly increasing. (Then* C *is automatically* <-*preserving and* =-*preserving.) We define the same matrices as in* 4.152. *Then the following assertions are true:*

   a) *Let* $t = 0, 1, 2, 3, \ldots$; *then the path* $P_{t+1}(i,j)$ *in the matrix* $\mathbf{A}_{t+1}$ *has the minimal* C-*value among all* i-j-*paths of length* $\leq t+1$, *and the value* $C(P_{t+1}(i,j))$ *is equal to the entry* $\chi_{t+1}(i,j)$ *in the matrix* $\mathbf{X}_{t+1}$.

---

[43]The matrix $\mathbf{H}$ is only defined if $\mathbf{C}$ is easily computable of type B with the help of $\odot$ and the function $h : \mathcal{R} \to \mathbb{R}$; in this case, we may choose the functions $\phi_{(i,j)}$ such that $\phi_{(i,j)}(\xi) = \xi \odot h(i,j)$ for all $\xi \in \mathbb{R}$ and all $(i,j) \in \mathcal{R}$.

b) If $\mathbf{X}_{n'} = \mathbf{X}_{n'-1}$ for some $n' \geq n$ then all paths $P_{n'}(i,j)$ in $\mathbf{A}_{n'}$ are C-optimal, and the costs $\chi_{n'}(i,j)$ in $\mathbf{X}_{n'}$ are minimal.

c) If $\mathbf{X}_{n'} \neq \mathbf{X}_{n'-1}$ for some $n' \geq n$ then it is not possible to connect each pair (i,j) by a C-minimal i-j-path.

d) It follows that Algorithm 12 can decide whether or not each pair (i,j) can be connected by a C-optimal i-j-path; if YES then Algorithm 12 will output an optimal i-j-path and write its cost for each pair (i,j).
For this purpose, we specify COND and WRITE as follows:
- COND $:= (\tau < n)$.
- WRITE $:= \left(\begin{array}{ll} \textbf{If } b = \text{true} & \textbf{then} \text{ write } \mathbf{A} \text{ and } \mathbf{X} \\ \textbf{else} \text{ write :} & \text{"Not all pairs } (i,j) \text{ have} \\ & a \text{ } C\text{-minimal } i\text{-}j\text{-path."} \end{array} \right)$.

e) Algorithm 12 requires $\Theta(n^4)$ time.

P r o o f (S k e t c h) : The proof is analogous to the ones of 4.121 and 4.122. Result a) about $t = 0$ can be shown with the arguments given at the end of Remark 4.149. Assertion e) follows from the fact that $\Theta(n)$ formal matrix multiplications are executed, and each of them takes $O(n^3)$ time units. $\qquad\square$

The next result is analogous to 4.125.

Theorem 4.154 *Given the digraph $\mathcal{G}$, the cost measure C, and the functions p, $\phi_{i,j}$ as described in 4.149. We assume that C is order preserving.*
We define the same matrices as in 4.152.

Then the following assertions are true:

a) Let $t = 0, 1, 2, \dots$. Then each path $P_{t+1}(i,j)$ in the matrix $\mathbf{A}_{t+1}$ has the minimum C-value among all i-j-paths with length $\leq t+1$. The value $C(P_{t+1}(i,j))$ is equal to the entry $\chi_{t+1}(i,j)$ in the matrix $\mathbf{X}_{t+1}$.

b) If $\mathbf{X}_{t+1} = \mathbf{X}_t$ for some t then all paths $P_t(i,j)$ in $\mathbf{A}_t$ are C-optimal, and the costs $\chi_t(i,j)$ in $\mathbf{X}_t$ are minimal.

c) Hence, Algorithm 12 can find an optimal i-j-path and compute its cost for each pair (i,j) if a C-minimal i-j-path exists for all pairs (i,j); for this purpose, we define COND and WRITE as follows:
- COND $:= (b = \text{false})$.   • WRITE $:= (\text{Write } \mathbf{A} \text{ and } \mathbf{X})$.
If not each pair (i,j) can be connected by a C-minimal path then Algorithm 12 will never terminate.

P r o o f : The proof is similar to the one of 4.125. When showing assertion a) about $t = 0$, it is sensible to follow the argumentation at the end of Remark 4.121. When showing assertion e) the following observation is helpful: Condition COND in Theorem 4.125 means that FORD-BELLMAN 1 has found a better path in the current iteration, and condition COND in the current result has the same meaning for Algorithm 12. $\qquad\square$

4.4.1.3   Comparing our recursions formulas with similar ones in literature

We compare the formulas (4.134) – (4.136) with particular formulas in [PanR89] and [Rote90].

Remark 4.155 Pan and Reif [PanR89], and Rote [Rote90] have given recursion formulas that are similar but not equivalent to the formulas (4.134) – (4.136). In order to understand the difference between our formulas and the ones in [PanR89] and [Rote90] we shall translate the equations (2) of [PanR89] and (7') of [Rote90] into our terminology.

For this purpose, we define the function η as follows: For all nodes i, j, let $\eta(i,j) := 0$ if $i = j$ and $\eta(i,j) := \infty$ if $i \neq 0$. Moreover, we define the matrix $\mathbf{I} := \big(\eta(i,j)\big)$. We define the matrix operation $\boxplus$ as the componentwise application of the operation $\boxplus$ defined in (4.133); all other operations are the same as in (4.133) – (4.136).

We now can write the equations (2) in [PanR89] and (7') in [Rote90] in our terminology. We give three equivalent formulations (4.137), (4.138) and (4.139) of Pan's, Reif's and Rote's equations; the formulas (4.137), (4.138), and (4.139) are written in the same style as (4.134), (4.135), and (4.136), respectively.

$$\chi_{t+1}(i,j) \;=\; \eta(i,j) \;\boxplus\; \left(\overset{n}{\underset{k=1}{\boxplus}} \; \chi_t(i,k) \boxtimes h(k,j) \right), \qquad (4.137)$$

$$\begin{pmatrix} \chi_{t+1}(1,1) & \cdots & \chi_{t+1}(1,n) \\ \vdots & & \vdots \\ \chi_{t+1}(n,1) & \cdots & \chi_{t+1}(n,n) \end{pmatrix} =$$

$$\begin{pmatrix} \eta(1,1) & \cdots & \eta(1,n) \\ \vdots & & \vdots \\ \eta(n,1) & \cdots & \eta(n,n) \end{pmatrix} \boxplus \begin{pmatrix} \chi_t(1,1) & \cdots & \chi_t(1,n) \\ \vdots & & \vdots \\ \chi_t(n,1) & \cdots & \chi_t(n,n) \end{pmatrix} \boxtimes \begin{pmatrix} h(1,1) & \cdots & h(1,n) \\ \vdots & & \vdots \\ h(n,1) & \cdots & h(n,n) \end{pmatrix}, \qquad (4.138)$$

$$\big(\chi_{t+1}(i,j)\big) = \big(\eta(i,j)\big) \boxplus \big(\chi_t(i,j)\big) \boxtimes \big(h(i,j)\big) = \mathbf{I} \boxplus \big(\chi_t(i,j)\big) \boxtimes \big(h(i,j)\big). \qquad (4.139)$$

Now we answer the question why Pan, Reif and Rote give other recursion formulas than we do. The reason is that these authors do not automatically assume that the underlying digraph $\mathcal{G}$ has all loops; moreover, they do not assume that (4.118) is true for all loops (v,v). For this reason, they must take care of the case that for some node i, the path $[i]$ of length 0 is the only C-minimal i-i-path.

In contrast to this, no cost of a path of length 0 appears in the formulas (4.134) – (4.136). But this is not necessary because we assume that (4.118) is true. Consequently, all t-fold concatenations $[i,i]^t$, $(t > 0)$ are C-minimal if $[i]$ is C-minimal, and these paths are indeed considered in the (4.134) – (4.136).   □

4.4.1.4   The associative law for modified matrix multiplications

We return to the situation that (4.118) is true. The next remark says that the modified matrix multiplications are not always associative.

Remark 4.156 Unlike the original matrix multiplication, the matrix operation $\boxtimes$ in equation (4.129) is not always associative. Not even the following fact is always valid for the special matrix $\mathbf{A} := \big([i,j]\big)$:

$$(\mathbf{A} \boxtimes \mathbf{A}) \boxtimes \mathbf{A} \ \cong_{C} \ \mathbf{A} \boxtimes (\mathbf{A} \boxtimes \mathbf{A}). \tag{4.140}$$

At first sight, it is surprising that $\boxtimes$ is not associative although the associative law is valid for the operations $\boxplus$ and $\boxtimes$ in 4.151. Nevertheless, it is easy to understand that $\boxtimes$ is not always associative. Let C have the weak prefix oriented Bellman property of type 0 but not the analogous suffix oriented property. We multiply the matrix $\mathbf{A} = \big([i,j]\big)$ successively on the left and on the right side, respectively:

$$(\ldots((\mathbf{A} \boxtimes \mathbf{A}) \boxtimes \mathbf{A})\ldots) \boxtimes \mathbf{A} \quad \text{and} \quad \mathbf{A} \boxtimes (\ldots \boxtimes (\mathbf{A} \boxtimes (\mathbf{A} \boxtimes \mathbf{A}))\ldots).$$

Roughly speaking, the left formula constructs paths by generating longer and longer prefixes whereas the right formula produces paths by generating longer and longer suffixes; consequently, the rôles of prefixes and suffixes must be exchanged when comparing the left construction with the right one. By the Bellman property of C, the left formula will finally produce a matrix of C-optimal i-j-connections; this, however, is not always true for the right formula because C does not have the weak suffix oriented Bellman property of type 0.

We now give an example in which (4.140) is not true although C has even the strong prefix and suffix oriented Bellman property of type 0.

Let $n := 8$, and let $\mathcal{G} = (\mathcal{V}, \mathcal{R})$ with $\mathcal{V} := \{1, \ldots, n\}$ and $\mathcal{R} := \mathcal{V} \times \mathcal{V}$. We define the following sets of arcs:

$\mathcal{R}_0 := \{(i,i) \mid i \in \mathcal{V}\};$ $\qquad\qquad$ $\mathcal{R}_1 := \{(1,2),(2,3),\ldots,(7,8),(8,1)\};$
$\mathcal{R}_2 := \{(1,3),(3,5),(5,8),(3,6),(6,8)\};$ $\quad$ $\mathcal{R}_3 := \mathcal{R} \backslash (\mathcal{R}_0 \cup \mathcal{R}_1 \cup \mathcal{R}_2).$

Next, we define a function ϕ_r for each arc r.
If $r \in \mathcal{R}_0$ then $\phi_r(x) := x$ for all x. If $r \in \mathcal{R}_1$ then $\phi_r(x) := 100\,x$ for all x. If $r \in \mathcal{R}_2$ then we define $\phi_r(x)$, $x \in \mathbb{R}$, as follows:

$$\phi_{(1,3)}(x) := \begin{cases} 1 & \text{if} \quad x = 0, \\ 100 & \text{else.} \end{cases}$$

$$\phi_{(3,5)}(x) := \begin{cases} 2 & \text{if} \quad x = 1, \\ 100 & \text{else.} \end{cases} \qquad \phi_{(5,8)}(x) := \begin{cases} 3 & \text{if} \quad x = 2, \\ 100 & \text{else.} \end{cases}$$

$$\phi_{(3,6)}(x) := \begin{cases} 4 & \text{if} \quad x = 0, \\ 100 & \text{else.} \end{cases} \qquad \phi_{(6,8)}(x) := \begin{cases} 5 & \text{if} \quad x = 4, \\ 100 & \text{else.} \end{cases}$$

If $r \in \mathcal{R}_3$ then $\phi_r(x) := 100$ for all $x \in \mathbb{R}$.

Let $p(v) := 0$ for all v, and let $C := \phi \circ\circ p$. Then C has the following properties:

$C(P) = 0$ for all paths P that only use arcs in $\mathcal{R}_0 \cup \mathcal{R}_1$.
$C(P) > 0$ for all other paths P. $\qquad\qquad\qquad\qquad\qquad\qquad$ (4.141)
In particular, $C(P) \geq 100$ if P uses an arc of $\mathcal{R}_3$.

$$C([1,3]) = 1, \quad C([1,3,5]) = 2, \quad C([1,3,5,8]) = 3 \quad C([1,3,6,8]) = 100$$
$$C([3,6]) = 4, \quad C([3,6,8]) = 5, \quad C([3,5,8]) = 100. \tag{4.142}$$

We define the paths $P_1(i,j)$ and $P_2(i,j)$ $(i,j \in V)$ such that $\mathbf{A} = \big(P_1(i,j)\big)$ and $\mathbf{A}^2 = \big(P_2(i,j)\big)$; in particular, $P_1(i,j) = [i,j]$ for all i,j. Moreover, we define the paths $Q'(i,j)$ and $Q''(i,j)$ such that $(\mathbf{A} \boxtimes \mathbf{A}) \boxtimes \mathbf{A} = \big(Q'(i,j)\big)$ and $\mathbf{A} \boxtimes (\mathbf{A} \boxtimes \mathbf{A}) = \big(Q''(i,j)\big)$.

The path function C satisfies the strong infix oriented Bellman condition of type 0. To see this we assume that $i,j \in V$, that P is an C-minimal i-j-connection, and that $Q \subseteq P$. Then (4.141) implies that all arcs of P are elements of $\mathcal{R}_0 \cup \mathcal{R}_1$. The same is true for Q so that $C(Q) = 0$ is minimal, too.

Next, we disprove assertion (4.140). Let P be a 1-5-path of length 2. If $P \neq [1,3,5]$ then P uses an arc $r \in \mathcal{R}_3$; consequently, $C(P) \overset{(4.141)}{\geq} 100 > 2 \overset{(4.142)}{=} C([1,3,5])$ so that

$$P_2(1,5) = [1,3,5]. \tag{4.143}$$

Let P be a path from 1 to 8 path with $\ell(P) = 3$. Then $C(P) \overset{(4.141)}{\geq} 100$ if P contains an arc $r \in \mathcal{R}_3$. Moreover, $P = [1,3,5,8]$ and $P = [1,3,6,8]$ are the only 1-8-paths without any arc $r \in \mathcal{R}_3$. Consequently, the following is true:

$$\begin{array}{l} [1,3,6,8] \text{ is the only 1-8-path with minimal } C\text{-value} \\ \text{among all candidates of length 3.} \end{array} \tag{4.144}$$

The relationships (4.143) and (4.144) imply that

$$Q'(1,8) = [1,3,5,8]. \tag{4.145}$$

In order to describe $Q''(1,8)$ we construct $P_2(3,8)$. Let P be a 3-8-path of length 2. If $P \neq [3,5,8]$ and $P \neq [3,6,8]$ then P must use an arc $r \in \mathcal{R}_3$; consequently, $C(P) \overset{(4.141)}{\geq} 100$. So, $[3,6,8]$ is is the only 3-8-path P with $\ell(P) = 2$ and $C(P) < 100$ so that

$$P_2(3,8) = [3,6,8]. \tag{4.146}$$

The path $Q''(1,8)$ must have a suffix of the form $P_2(i,8)$ where $i \in V$. If $Q''(1,8)$ were equal to $Q'(1,8) \overset{(4.145)}{=} [1,3,5,8]$ then $P_2(3,8)$ were equal to $[3,5,8]$, and that is a contradiction to (4.146). So, we may conclude that $Q'(1,8) \neq Q''(1,8)$. This and (4.144) imply that $C(Q''(1,8)) > C(Q'(1,8))$ so that (4.140) is not true. $\qquad\square$

We now define two versions of the associative law.

Definition 4.157 Given the same matrix operation $\boxtimes$ as in equation (4.129).
   a) We call the operation $\boxtimes$ *associative* if the following is true for all matrices $\mathbf{A}, \mathbf{B}$, and $\mathbf{C}$ of paths: $(\mathbf{A} \boxtimes \mathbf{B}) \boxtimes \mathbf{C} = \mathbf{A} \boxtimes (\mathbf{B} \boxtimes \mathbf{C})$.
   b) We call the operation $\boxtimes$ *weakly associative* if the following is true for all matrices $\mathbf{A}, \mathbf{B}$, and $\mathbf{C}$ of paths: $(\mathbf{A} \boxtimes \mathbf{B}) \boxtimes \mathbf{C} \cong_C \mathbf{A} \boxtimes (\mathbf{B} \boxtimes \mathbf{C})$; we call the this formula the *weak associative law*. $\qquad\square$

We now give a sufficient condition for the weak associativity of $\boxtimes$.

Theorem 4.158 *Let* $n \in \mathbb{N}$, $\mathcal{V} := \{1, \ldots, n\}$, *and* $\mathcal{G} := (\mathcal{V}, \mathcal{V} \times \mathcal{V})$. *Moreover, let* $C : \mathcal{P}(\mathcal{G}) \to \mathbb{R}$ *be a path function of the form* $C = \phi \circ \circ p$. *(Then* C *is automatically* =-*preserving.) We assume that* C *has property (4.118).*
Given the matrices $\mathbf{U} = (U_{i,j})$, $\mathbf{V} = (V_{i,j})$ *and* $\mathbf{W} = (W_{i,j})$ *of paths . We define the path matrices* $\mathbf{X}' = (X'_{i,j})$, $\mathbf{X}'' = (X''_{i,j})$ *and the path matrices* $\mathbf{Y}' = (Y'_{i,j})$, $\mathbf{Y}'' = (Y''_{i,j})$ *as follows:*

$$\mathbf{X}' := \mathbf{U} \boxtimes \mathbf{V}, \quad \mathbf{X}'' := (\mathbf{U} \boxtimes \mathbf{V}) \boxtimes \mathbf{W} = \mathbf{X}' \boxtimes \mathbf{W},$$
$$\mathbf{Y}' := \mathbf{V} \boxtimes \mathbf{W}, \quad \mathbf{Y}'' := \mathbf{U} \times (\mathbf{V} \times \mathbf{W}) = \mathbf{U} \boxtimes \mathbf{Y}'.$$

Moreover, we define $\chi_{i,l} := \min_{j,k \in \mathcal{V}} C(U_{i,j} \oplus V_{j,k} \oplus W_{k,l})$ *for all* $i, l \in \mathcal{V}$.

a) *Let* C *be* <-*preserving in the usual sense; that means that*

(i) $C(P_1) < C(P_2) \implies C(P_1 \oplus Q'') < C(P_2 \oplus Q'')$
for all paths P_1, P_2, Q'' *with* $\alpha(P_1) = \alpha(P_2)$ *and* $\omega(P_1) = \omega(P_2) = \alpha(Q'')$.

Then $C(X''_{i,l}) = \chi_{i,l}$ *for all* $i, l \in \mathcal{V}$.

b) *Let* C *be* <-*preserving in the following sense:*

(ii) $C(P_1) < C(P_2) \implies C(Q' \oplus P_1) < C(Q' \oplus P_2)$
for all paths Q', P_1, P_2 *with* $\omega(Q') = \alpha(P_1) = \alpha(P_2)$ *and* $\omega(P_1) = \omega(P_2)$.

Moreover, let C *be* =-*preserving in the following sense:*

(iii) $C(P_1) = C(P_2) \implies C(Q' \oplus P_1) = C(Q' \oplus P_2)$
for all paths Q', P_1, P_2 *with* $\omega(Q') = \alpha(P_1) = \alpha(P_2)$ *and* $\omega(P_1) = \omega(P_2)$.

Then $C(Y''_{i,l}) = \chi_{i,l}$ *for all* $i, l \in \mathcal{V}$.

c) *If* C *satisfies all conditions (i), (ii), (iii) then* $\boxtimes$ *is weakly associative.*

(R e m a r k : Recall the terminology of Example 2.47. Property (i) is equivalent to <-preservation with regard to $\mathbb{L}'_1 \cap \mathbb{L}'_2 \cap \mathbb{L}''_1$; property (ii) is equivalent to <-preservation with regard to $\mathbb{L}'_1 \cap \mathbb{L}'_2 \cap \mathbb{L}''_2$.)

P r o o f o f a) : Let $i, l \in \mathcal{V}$. The definition of $\boxtimes$ implies that there exists a k^+ such that the following are true:

$$X''_{i,l} = X'_{i,k^+} \oplus W_{k^+,l}, \quad C(X''_{i,l}) = \min_{k \in \mathcal{V}} C(X'_{i,k} \oplus W_{k,l}). \tag{4.147}$$

Moreover, the definition of $\boxtimes$ implies that a j^+ exists such that the following are true:

$$X'_{i,k^+} = U_{i,j^+} \oplus V_{j^+,k^+}, \quad C(X'_{i,k^+}) = \min_{j \in \mathcal{V}} C(U_{i,j} \oplus V_{j,k^+}). \tag{4.148}$$

An immediate consequence of (4.147) and (4.148) is that

$$X''_{i,l} = U_{i,j^+} \oplus V_{j^+,k^+} \oplus W_{k^+,l}. \tag{4.149}$$

The definition of $\chi_{i,l}$ implies the existence of j^*, k^* such that

$$\chi_{i,l} = C(U_{i,j^*} \oplus V_{j^*,k^*} \oplus W_{k^*,l}). \tag{4.150}$$

Moreover, there exists a $\tilde{j}$ such that

$$X'_{i,k^*} = U_{i,\tilde{j}} \oplus V_{\tilde{j},k^*}, \quad C(X'_{i,k^*}) = \min_{j \in \mathcal{V}} C(U_{i,j} \oplus V_{j,k^*}). \tag{4.151}$$

Fact (4.149) implies that $C(X''_{i,l})$ is one of the candidates when computing $\chi_{i,l}$; consequently, $\chi_{i,l} \leq C(X''_{i,l})$.

We next show that $C(X''_{i,l}) \leq \chi_{i,l}$. For this purpose, we prove the following assertion:

$$C(X'_{i,k*}) = C(U_{i,j*} \oplus V_{j*,k*}).$$ \hfill (4.152)

An immediate consequence of (4.151) is that $C(X'_{i,k*}) \leq C(U_{i,j*} \oplus V_{j*,k*})$. If $C(X'_{i,k*}) < C(U_{i,j*} \oplus V_{j*,k*})$ then the following inequality were valid:

$$C(U_{i,\tilde{j}} \oplus V_{\tilde{j},k*} \oplus W_{k*,l}) \overset{(4.151)}{=} C(X'_{i,k*} \oplus W_{k*,l}) \overset{(i)}{<} C(U_{i,j*} \oplus V_{j*,k*} \oplus W_{k*,l}),$$

and that is a contradiction to the minimality of $C(U_{i,j*} \oplus V_{j*,k*} \oplus W_{k*,l})$.

Now we have shown fact (4.152); this and the $=$-preservation of C imply relationship $(*)$ of the following inequality:

$$C(X''_{i,l}) \overset{(4.147)}{\leq} C\left(X'_{i,k*} \oplus W_{k*,l}\right) \overset{(*)}{=} C(U_{i,j*} \oplus V_{j*,k*} \oplus W_{k*,l}) = \chi_{i,l}.$$

Consequently, $C\left(X''_{i,l}\right) \leq \chi_{i,l}$; this and the above relationship $\chi_{i,l} \leq C(X''_{i,l})$ imply that $C(X''_{i,l}) = \chi_{i,l}$.

P r o o f o f b) : This proof is quite similar to the one of part a). Let $i,l \in V$. The definition of $\boxtimes$ implies that there exist $j^+ \in V$ and $k^+ \in V$ such that the following assertions are true:

$$Y''_{i,l} = U_{i,j+} \oplus Y'_{j+,l}, \qquad C(Y''_{i,l}) = \min_{j \in V} C(U_{i,j} \oplus Y'_{j,l}),$$ \hfill (4.153)

$$Y'_{j+,l} = V_{j+,k+} \oplus W_{k+,l}, \qquad C(Y'_{j+,l}) = \min_{k \in V} C(V_{j+,k} \oplus W_{k,l}).$$ \hfill (4.154)

An immediate consequence of these facts is the following equation:

$$X''_{i,l} = U_{i,j+} \oplus V_{j+,k+} \oplus W_{k+,l+}.$$ \hfill (4.155)

We define j^* and k^* in the same way as in part a). Then there exists a $\tilde{k}$ such that

$$Y'_{j*,l} = V_{j*,\tilde{k}} \oplus W_{\tilde{k},l}, \qquad C(Y'_{j*,l}) = \min_{k \in V} C(V_{j*,k} \oplus W_{k,l}).$$ \hfill (4.156)

Fact (4.155) implies that $C(Y''_{i,l})$ is one of the candidates when computing $\chi_{i,l}$; consequently, $\chi_{i,l} \leq C(Y''_{i,l})$.

We next show that $C(Y''_{i,l}) \leq \chi_{i,l}$. For this purpose, we prove the following assertion:

$$C(Y'_{j*,l}) = C(V_{j*,k*} \oplus W_{k*,l}).$$ \hfill (4.157)

An immediate consequence of (4.156) is that $C(Y'_{j*,l}) \leq C(V_{j*,k*} \oplus W_{k*,l})$. If $C(Y'_{j*,l}) < C(U_{j*,k*} \oplus W_{k*,l})$ then the following inequality were valid:

$$C(U_{i,j*} \oplus V_{j*,\tilde{k}} \oplus W_{\tilde{k},l}) \overset{(4.156)}{=} C(U_{i,j*} \oplus Y'_{j*,l}) \overset{(ii)}{<} C(U_{i,j*} \oplus V_{j*,k*} \oplus W_{k*,l}),$$

and that is a contradiction to the minimality of $C(U_{i,j*} \oplus V_{j*,k*} \oplus W_{k*,l})$.

Now we have shown fact (4.157); this and (iii) imply relationship $(*)$ in the following inequality:

$$C(Y''_{i,l}) \overset{(4.153)}{\leq} C(U_{i,j*} \oplus Y'_{j*,l}) \overset{(*)}{=} C(U_{i,j*} \oplus V_{j*,k*} \oplus W_{k*,l}) = \chi_{i,l}.$$

Consequently, $C(Y''_{i,l}) \leq \chi_{i,l}$; this and the above assertion $\chi_{i,l} \leq C(Y''_{i,l})$ imply that $C(Y''_{i,l}) = \chi_{i,l}$.

P r o o f   o f   c) : If all assumptions (i), (ii), (iii) are given then part a) and part b) imply that $C(X_{i,l}'') = \chi_{i,l} = C(Y_{i,l}'')$ for all $i, l \in V$. That means that $\boxtimes$ is weakly associative.   □

Remark 4.159 The conditions (i), (ii), (iii) in Theorem 4.158 are satisfied if C is additive. That means that $C = \phi \circ o \circ p$ where $p(v) = 0$ for all $v \in V$ and where all functions $\phi_{(i,j)}$ are of the form $\xi \mapsto \xi + a_{i,j}$
It remains to the reader to find further classes of functions $\phi_{(i,j)}$ such that the resulting path function $C = \phi \circ o \circ p_0$ has the properties (i), (ii), (iii).   □

The following result is similar to 2.23. Here, however, we assume that (4.118) be true and not that that $\odot$ have a neutral element.

Lemma 4.160 *Given a digraph $\mathcal{G} = (V, \mathcal{R})$ with $V = \{1, \ldots, n\}$ and $\mathcal{R} \subseteq V \times V$. Let $C : \mathcal{P}(\mathcal{G}) \to \mathbf{R}$ be easily computable of type B with the help of the operation $\odot$ and $h : \mathcal{R} \to \mathbb{R}$. We assume that the operation $\odot$ is associative and that (4.118) is true. Then $C(P \oplus Q) = C(P) \odot C(Q)$ for all paths $P, Q \in \mathcal{P}(\mathcal{G})$.*

P r o o f : Let $\ell(Q) = 0$; then $Q = [j]$ where $j := \omega(P)$. We obtain the following equation, in which (*) follows from $C([j, j']) = h(j, j')$ for all arcs (j, j').

$$C(P \oplus Q) = C(P) \overset{(4.118)}{=} C(P \oplus [j, j]) = C(P) \odot h(j, j) \overset{(*)}{=} C(P) \odot C([j, j]) =$$
$$C(P) \odot C([j] \oplus (j, j) \oplus [j]) \overset{(4.118)}{=} C(P) \odot C([j]) = C(P) \odot C(Q).$$

The proof in the case $\ell(Q) > 0$ is almost the same as in Theorem 2.23; we must only replace the path function H by the function C.   □

4.4.2   Fast versions of the previous matrix algorithms

Here, we develop a matrix algorithm that is faster than Algorithm 12; the idea is to organize the matrix multiplication better. Roughly speaking, we replace the multiplications with a constant matrix in (4.129), (4.130), and (4.135) by a successive squaring of matrices.

In this subsection, we shall assume that all functions $\phi_{(i,j)}$ are total, i.e. $\mathrm{def}(\phi_{(i,j)}) = \mathbb{R}$. Moreover, we shall make the following assumption:

The weak associative law is valid for matrix multiplication in (4.129).
The associative law is valid for the matrix multiplication in (4.135)   (4.158)
as far as C is easily computable of type B.

First, we transform the formulas (4.127), (4.129), and (4.131) into formulas with successively squared matrices.

Introductory Remark 4.161 In (4.127) and (4.129), we constructed paths of the form $P(i, k) \oplus [k, j]$. In (4.159) – (4.161), however, we shall construct paths of the form $\widetilde{P}(i, k) \oplus \widetilde{P}(k, j)$. The multiplication with a constant matrix in (4.129) and (4.131) will be replaced by squaring matrices in (4.160) and in (4.161).

We now describe the details. Equation (4.159) shows the recursive definition of paths, equation (4.160) is the resulting recursion formula for path matrices, and equation (4.161) is a compact version of (4.160).

$$(\forall\, i, j, t)\quad \widetilde{P}_1(i,j) := [i,j],\quad \widetilde{P}_{t+1}(i,j) := \overset{n}{\underset{k=1}{\boxplus}}\ \widetilde{P}_t(i,k) \boxtimes \widetilde{P}_t(k,j), \qquad (4.159)$$

$$\begin{pmatrix} \widetilde{P}_{t+1}(1,1) \cdots \widetilde{P}_{t+1}(1,n) \\ \vdots \qquad\quad \vdots \\ \widetilde{P}_{t+1}(n,1) \cdots \widetilde{P}_{t+1}(n,n) \end{pmatrix} = \begin{pmatrix} \widetilde{P}_t(1,1) \cdots \widetilde{P}_t(1,n) \\ \vdots \qquad\quad \vdots \\ \widetilde{P}_t(n,1) \cdots \widetilde{P}_t(n,n) \end{pmatrix} \boxtimes \begin{pmatrix} \widetilde{P}_t(1,1) \cdots \widetilde{P}_t(1,n) \\ \vdots \qquad\quad \vdots \\ \widetilde{P}_t(n,1) \cdots \widetilde{P}_t(n,n) \end{pmatrix}. \qquad (4.160)$$

$$\left(\widetilde{P}_{t+1}(i,j)\right) = \left(\widetilde{P}_t(i,k)\right) \boxtimes \left(\widetilde{P}_t(k,j)\right). \qquad (4.161)$$

In this way, we obtain a sequence of matrices $\widetilde{\mathbf{A}}_1, \widetilde{\mathbf{A}}_2, \widetilde{\mathbf{A}}_3, \ldots$ where $\widetilde{\mathbf{A}}_1 = ([i,j])$ and $\widetilde{\mathbf{A}}_{t+1} = \widetilde{\mathbf{A}}_t \boxtimes \widetilde{\mathbf{A}}_t$ for all t.

Assumption (4.158) implies that

$$(\forall\, t \in \mathbb{N})\quad \widetilde{\mathbf{A}}_t \cong_C \mathbf{A}_{2^t-1} \qquad (4.162)$$

where $\mathbf{A}_1, \mathbf{A}_2, \mathbf{A}_3, \mathbf{A}_4 \ldots$ are the matrices defined in Theorem 4.152.   □

Next, we transform the formulas (4.128), (4.130) and (4.132) into formulas with successively squared matrices. In particular, we shall obtain the matrices $\left(\chi_{1+2^{t-1}}(i,j)\right) = \left(C(P_{1+2^{t-1}}(i,j))\right)$ without computing the path matrices $\mathbf{A}_{1+2^{t-1}} = \mathbf{A} \boxtimes \widetilde{\mathbf{A}}_t$.

The problem is that objects of different types are multiplied in the formulas (4.130) and (4.132); for example, in (4.130), the matrix $\left(\chi_t(i,j)\right)$ of real numbers is multiplied with the matrix $\left(\phi_{(i,j)}\right)$ of real functions. So, the matrix multiplication in (4.130) and (4.132) cannot be used to multiply a matrix with itself.

The idea to avoid this problem is the following: We shall introduce a matrix multiplication $\Phi^{(1)} \boxtimes\boxtimes \Phi^{(2)}$ for matrices $\Phi^{(1)}, \Phi^{(2)}$ of real functions (e.g., $\Phi^{(1)} = \Phi^{(2)} = \left(\phi_{(i,j)}\right)$). This multiplication will have the property that the following is true for all path matrices $\mathbf{X}$, for all function matrices $\Phi^{(1)}, \Phi^{(2)}$, and for the matrix multiplication $\boxtimes$ in formula (4.130):

$$\left(\mathbf{X} \boxtimes \Phi^{(1)}\right) \boxtimes \Phi^{(2)} = \mathbf{X} \boxtimes \left(\Phi^{(1)} \boxtimes\boxtimes \Phi^{(2)}\right). \qquad (4.163)$$

Then a successive application of (4.163) implies that the following is true for all T and for all $\Phi^{(1)}, \ldots, \Phi^{(T)}$:

$$\begin{aligned} \left(\ldots \left(\left(\mathbf{X} \boxtimes \Phi^{(1)}\right) \boxtimes \Phi^{(2)}\right) \boxtimes \ldots\right) \boxtimes \Phi^{(T)} &= \\ \left(\ldots \left(\mathbf{X} \boxtimes \left(\Phi^{(1)} \boxtimes\boxtimes \Phi^{(2)}\right)\right) \boxtimes \ldots\right) \boxtimes \Phi^{(T)} &= \ldots \\ = \mathbf{X} \boxtimes \left(\ldots \left(\Phi^{(1)} \boxtimes\boxtimes \Phi^{(2)}\right) \boxtimes \boxtimes \ldots \boxtimes\boxtimes \Phi^{(T)}\right). & \end{aligned} \qquad (4.164)$$

If $\Phi^{(1)} = \ldots \Phi^{(T)} = \left(\phi_{(i,j)}\right)$ we shall replace the successive application of $\boxtimes\boxtimes$ by a successive squaring operation of matrices.

Next, we describe the matrix multiplication $\boxtimes\boxtimes$ in detail. For this purpose, we define the binary operations $\boxplus\boxplus$ and $\boxtimes\boxtimes$ as follows: For all $f, g : \mathbb{R} \to \mathbb{R}$ and for all $x \in \mathbb{R}$, let

$$(f \boxplus g)(x) := \min(f(x), g(x)), \qquad (f \boxtimes \boxtimes g)(x) := g \circ f(x) = (x^f)^g. \quad (4.165)$$

We recursively define the following real functions $(\phi_{(i,j)})_t : \mathbb{R} \to \mathbb{R}$:

$$(\forall\, i,j,t) \quad (\phi_{(i,j)})_1 := \phi_{(i,j)}, \quad (\phi_{(i,j)})_{t+1} := \underset{k=1}{\overset{n}{\boxplus}} \left((\phi_{(i,k)})_t \boxtimes \boxtimes (\phi_{(k,j)})_1 \right) \quad (4.166)$$

for all $t \in \mathbb{N}$. Then we define the matrix multiplication $\boxtimes\boxtimes$ such that the following formula means the application of (4.166) to each pair (i,j).

$$\begin{pmatrix} (\phi_{(1,1)})_{t+1} & \cdots & (\phi_{(1,n)})_{t+1} \\ \vdots & & \vdots \\ (\phi_{(n,1)})_{t+1} & \cdots & (\phi_{(n,n)})_{t+1} \end{pmatrix} = \begin{pmatrix} (\phi_{(1,1)})_t & \cdots & (\phi_{(1,n)})_t \\ \vdots & & \vdots \\ (\phi_{(n,1)})_t & \cdots & (\phi_{(n,n)})_t \end{pmatrix} \boxtimes\boxtimes \begin{pmatrix} (\phi_{(1,1)})_1 & \cdots & (\phi_{(1,n)})_1 \\ \vdots & & \vdots \\ (\phi_{(n,1)})_1 & \cdots & (\phi_{(n,n)})_1 \end{pmatrix}.$$

A more compact version of this recursion formula is the following:

$$\left((\phi_{(i,j)})_{t+1} \right) = \left((\phi_{(i,j)})_t \right) \boxtimes\boxtimes \left((\phi_{(i,j)})_1 \right). \quad (4.167)$$

The definitions of the matrix multiplications in the equations (4.128), (4.130) and in (4.165), (4.167) imply that (4.163) (and consequently, (4.164)) is true.

We define $\Phi_t := \left((\phi_{(i,j)})_t \right)$ for all $t \in \mathbb{N}$.

We make the following assumption, which might be a consequence of (4.158):

$$\text{The associative law is true for the matrix multiplication } \boxtimes\boxtimes. \quad (4.168)$$

Then we may abbreviate the t-fold multiplication of a function matrix Ψ with itself as Ψ^t; in particular, the following are true for all $t \in \mathbb{N}$:

$$\Phi_1^t = \Phi_t, \quad (\chi_{t+1}(i,j)) \overset{(4.132),(4.164)}{=} (\chi_1(i,j)) \boxtimes \Phi_1^t = (\chi_1(i,j)) \boxtimes \Phi_t. \quad (4.169)$$

In the following remark, we replace the formulas (4.128), (4.130), and (4.132) by formulas with successively squared matrices; we use the matrix multiplication $\boxtimes\boxtimes$.

Introductory Remark 4.162   We will replace the computation of the matrices $\Phi^1, \Phi^2, \Phi^3, \ldots$ by a sucessive squaring procedure. For this purpose, we define the real functions $\overline{\phi}_{(i,j)}$ with the following formula, which is similar to (4.166):

$$(\forall\, i,j,t) \quad (\overline{\phi}_{(i,j)})_1 := \phi_{(i,j)}, \quad (\overline{\phi}_{(i,j)})_{t+1} := \underset{k=1}{\overset{n}{\boxplus}} \left((\overline{\phi}_{(i,k)})_t \boxtimes\boxtimes (\overline{\phi}_{(k,j)})_t \right) \quad (4.170)$$

for all $t \in \mathbb{N}$. Let $\boxtimes\boxtimes$ be the same matrix multiplication as in (4.167). Then we can write the application of (4.170) to each pair (i,j) as follows:

$$\begin{pmatrix} (\overline{\phi}_{(1,1)})_{t+1} & \cdots & (\overline{\phi}_{(1,n)})_{t+1} \\ \vdots & & \vdots \\ (\overline{\phi}_{(n,1)})_{t+1} & \cdots & (\overline{\phi}_{(n,n)})_{t+1} \end{pmatrix} = \begin{pmatrix} (\overline{\phi}_{(1,1)})_t & \cdots & (\overline{\phi}_{(1,n)})_t \\ \vdots & & \vdots \\ (\overline{\phi}_{(n,1)})_t & \cdots & (\overline{\phi}_{(n,n)})_t \end{pmatrix} \boxtimes\boxtimes \begin{pmatrix} (\overline{\phi}_{(1,1)})_t & \cdots & (\overline{\phi}_{(1,n)})_t \\ \vdots & & \vdots \\ (\overline{\phi}_{(n,1)})_t & \cdots & (\overline{\phi}_{(n,n)})_t \end{pmatrix}.$$

A more compact version of this recursion formula is the following:

$$\left((\overline{\phi}_{(i,j)})_{t+1} \right) = \left((\overline{\phi}_{(i,j)})_t \right) \boxtimes\boxtimes \left((\overline{\phi}_{(i,j)})_t \right). \quad (4.171)$$

We define the tth matrix in (4.171) as $\overline{\Phi}_t$ $(t = 1,2,3,\ldots)$. Then

$$(\forall\, t \in \mathbb{N}) \quad \overline{\Phi}_1 = (\phi_{(i,j)}), \quad \overline{\Phi}_{t+1} = \overline{\Phi}_t \boxtimes\boxtimes \overline{\Phi}_t. \quad (4.172)$$

Then we define the cost matrices $\overline{\mathbf{X}}_t := \big(\overline{\chi}_t(i,j)\big)$ $(t = 1,2,3,\ldots)$ as follows; the matrix multiplication $\boxtimes$ is the same as in (4.130).

$$\overline{\mathbf{X}}_t := \big(\overline{\chi}_t(i,j)\big) := \big(C([i,j])\big) \boxtimes \big((\phi_{(i,j)})_t\big) = \big(C([i,j])\big) \boxtimes \overline{\Phi}_t. \tag{4.173}$$

Let $\overline{P}_t(i,j) := P_{1+2^{t-1}}(i,j)$ for all nodes i,j. Then $\overline{\chi}_t(i,j) = C(\overline{P}_t(i,j))$ for all nodes i,j; this is a consequence of the following equation:

$$\overline{\mathbf{X}}_t = \big(\overline{\chi}_t(i,j)\big) \overset{(4.173)}{=} \big(C([i,j])\big) \boxtimes \overline{\Phi}_t \overset{(4.168),(4.172)}{=} \big(C([i,j])\big) \boxtimes \overline{\Phi}_1^{2^{t-1}} \overset{(4.172)}{=}$$

$$\big(C([i,j])\big) \boxtimes \big(\phi_{(i,j)}\big)^{2^{t-1}} \overset{(4.169)}{=} \big(\chi_{1+2^{t-1}}(i,j)\big) = \big(C(P_{1+2^{t-1}}(i,j))\big) = \big(C(\overline{P}_t(i,j))\big).$$

It follows that the essential operations to compute $\overline{\mathbf{X}}_t = \big(\chi_{1+2^{t-1}}(i,j)\big)$ are computing the squares of $\overline{\Phi}_1, \overline{\Phi}_2, \ldots, \overline{\Phi}_{t-1}$, and that means that the matrix multiplication $\boxtimes\boxtimes$ is applied $(t-1)$ times. This matrix multiplication, however, is very complicated; therefore, it is not clear whether in this case, successive squaring is actually faster than a 2^{t-1}-fold application of formula (4.130). $\qquad\square$

Next, we transform the formulas (4.134), (4.135) and (4.136) into formulas with recursive squaring operations.

Introductory Remark 4.163 Let C be easily computable of type B with the help of the operation $\odot$ and the path function h. We recall assumption (4.158).

We now define the numbers $\widetilde{\chi}_t(i,j)$ recursively; we shall show that each value $\widetilde{\chi}_t(i,j)$ is equal to $C(\widetilde{P}_t(i,j))$. (These paths have been defined in (4.159).) The operation $\boxplus$ and $\boxtimes$ are the same as in (4.133).

$$(\forall i,j,t) \quad \widetilde{\chi}_1(i,j) := C([i,j]); \quad \widetilde{\chi}_{t+1}(i,j) := \overset{n}{\underset{k=1}{\boxplus}} \widetilde{\chi}_t(i,k) \boxtimes \widetilde{\chi}_t(k,j). \tag{4.174}$$

Then the following assertion is true for all t; the matrix multiplication in (4.175) and (4.176) is the same as in (4.135).

$$\begin{pmatrix} \widetilde{\chi}_{t+1}(1,1) \cdots \widetilde{\chi}_{t+1}(1,n) \\ \vdots \qquad\qquad \vdots \\ \widetilde{\chi}_{t+1}(n,1) \cdots \widetilde{\chi}_{t+1}(n,n) \end{pmatrix} = \begin{pmatrix} \widetilde{\chi}_t(1,1) \cdots \widetilde{\chi}_t(1,n) \\ \vdots \qquad\qquad \vdots \\ \widetilde{\chi}_t(n,1) \cdots \widetilde{\chi}_t(n,n) \end{pmatrix} \boxtimes \begin{pmatrix} \widetilde{\chi}_t(1,1) \cdots \widetilde{\chi}_t(1,n) \\ \vdots \qquad\qquad \vdots \\ \widetilde{\chi}_t(n,1) \cdots \widetilde{\chi}_t(n,n) \end{pmatrix}, \tag{4.175}$$

$$\big(\widetilde{\chi}_{t+1}(i,j)\big) = \big(\widetilde{\chi}_t(i,j)\big) \boxtimes \big(\widetilde{\chi}_t(i,j)\big). \tag{4.176}$$

We define $\widetilde{\mathbf{X}}_t := \big(\widetilde{\chi}_t(i,j)\big)$ for each t. Then the following is true for all t:

$$\widetilde{\mathbf{X}}_t = \big(\widetilde{\chi}_t(i,j)\big) \overset{(4.158)}{=} \big(\chi_{2^{t-1}}(i,j)\big) = \big(C(P_{2^{t-1}}(i,j))\big) \overset{(4.162)}{=} \big(C(\widetilde{P}_t(i,j))\big). \tag{4.177}$$

$\qquad\square$

We now describe a fast matrix algorithm; it is based on Remarks 4.161 – 4.163.

Algorithm 13

1. (* *Initialization* *)

   $\widetilde{\mathbf{A}} := \big([i,j]\big);\ \ \mathbf{X} := \widetilde{\mathbf{X}} := \big(C([i,j])\big);\ \ \ \Phi := \overline{\Phi} := \big(\phi_{(i,j)}\big);\ \ \overline{\mathbf{X}} := \mathbf{X} \boxtimes \overline{\Phi}.$

   $\tau := 0;$
   (* *The variable* τ *is used to count the iterations.* *)

   $b := \text{false};$
   (* *The variable* b *is used to check whether all entries of the cost matrix* $\mathbf{X}$ *or* $\widetilde{\mathbf{X}}$ *are minimal costs.* *)

   If C is easily computable of type B **then** $bb := \text{true}$ **else** $bb := \text{false}$.

2. **While** COND **do**

   2.1. Let $\widetilde{\mathbf{A}} := \widetilde{\mathbf{A}} \boxtimes \widetilde{\mathbf{A}}$ (where $"\boxtimes"$ means the matrix multiplication in (4.129) and (4.160)).

   2.2. **If not** bb **then**
      let $\overline{\Phi} := \overline{\Phi} \boxtimes \boxtimes \overline{\Phi}$ (where $"\boxtimes \boxtimes"$ means the multiplication in (4.167) and in (4.172));
      let $\overline{\mathbf{X}} := \overline{\mathbf{X}} \boxtimes \Phi$ (where $"\boxtimes"$ means the matrix multiplication in (4.130)).

   2.3. **If** bb **then**
      let $\widetilde{\mathbf{X}} := \widetilde{\mathbf{X}} \boxtimes \widetilde{\mathbf{X}}$ (where $"\boxtimes"$ means the multiplication in (4.135) and in (4.176)).

   2.4. **If** bb is false **then** (**if** $\overline{\mathbf{X}} = \overline{\mathbf{X}} \boxtimes \Phi$ **then** $b := \text{true}$ **else** $b := \text{false}$);

      If bb is true **then** (**if** $\widetilde{\mathbf{X}} = \widetilde{\mathbf{X}} \boxtimes \mathbf{X}$ **then** $b := \text{true}$ **else** $b := \text{false}$);

   2.5. $\tau := \tau + 1;$

3. Perform the output procedure WRITE.

The following theorems 4.164 – 4.166 say that Algorithm 13 works correctly; we shall use Theorems 4.152 – 4.154 to show 4.164 – 4.166. Here, we confine ourselves to the case that C is easily computable of type B (e.g, C is additive). Similar results can be shown if C is not easily computable of type B; fact (4.162) and the equation at the end of Remark 4.162 will be helpful in this case.

The next result is similar to 4.152.

Theorem 4.164 *Let* $\mathcal{G}$ *be the same digraph as in* 4.149. *Let* $C : \mathcal{P}(s) \to \mathbb{R}$ *be a path function. We assume that* C *is easily computable of type B with the operation* $\odot$ *and the arc function* h. *(Then each restriction* $C|_{\mathcal{P}(i)}$ *is automatically* $=$*-preserving.)*

We assume that $\odot$ *is associative and that* (4.158) *is true.*

Moreover, we assume that C *has the weak Bellman property of type 0, and we make the following assumption:*

   (i) *For all nodes* i, j *there exists a* C*-minimal* i-j*-connection.*

We define all matrices in the current theorem in the same way as the matrices in 4.152 *and in* 4.161 – 4.163. *Let* $L(x) := \lceil \log_2(x) + 1 \rceil$ *for all* $x > 0$.

Then all paths $\widetilde{P}_{L(n)}(i,j)$ *in* $\widetilde{\mathbf{A}}_{L(n)}$ *are* C*-optimal, and the costs* $\widetilde{\chi}_{L(n)}(i,j)$ *in the matrix* $\widetilde{\mathbf{X}}_{L(n)}$ *are minimal.*

That means that Algorithm 13 works correctly if we specify COND *and* WRITE *as follows:*

   • COND $:= (\tau < L(n) - 1)$,   • WRITE $:= (\text{Write } \widetilde{\mathbf{A}} \text{ and } \widetilde{\mathbf{X}})$.

P r o o f : Fact (4.162) implies that $\widetilde{\mathbf{A}}_{L(n)} \cong_C \mathbf{A}_{n'}$ for $n' := 2^{L(n)-1} \geq n$. Theorem 4.152 says that all paths $P_{n'}(i,j)$ in the matrix $\mathbf{A}_{n'}$ are C-optimal. Consequently, all paths $\widetilde{P}_{L(n)}(i,j)$ in $\widetilde{\mathbf{A}}_{L(n)}$ are optimal, too, because $C(\widetilde{P}_{L(n)}(i,j)) = C(P_{n'}(i,j))$ for all i,j.

Remark 4.163 implies that $\widetilde{\chi}_{L(n)}(i,j) = C(\widetilde{P}_{L(n)}(i,j))$ for all i and j; hence, $\widetilde{\chi}_{L(n)}(i,j)$ is the minimal cost value for all i-j-paths.   □

The next result is similar to Theorem 4.153.

Theorem 4.165 *Let $\mathcal{G} = (\mathcal{V}, \mathcal{V} \times \mathcal{V})$ be the complete digraph with $\mathcal{V} = \{1,2,\ldots,n\}$. Let $C : \mathcal{P}(\mathcal{G}) \to \mathbb{R}$ be easily computable of type B with the help of the operation $\odot$ and the arc function h. We assume that $\odot$ is associative and that this operation is strictly monotone in the following sense:*

$$(\forall\, x_1, x_2 \in \mathbb{R}\,,\ r \in \mathcal{R})\quad x_1 < x_2 \implies x_1 \odot h(r) < x_2 \odot h(r).^{44}$$

We define all matrices in the current theorem in the same way as the matrices in 4.152 and in 4.161 – 4.163. Moreover, let $L(x) := \lceil \log_2(x) + 1 \rceil$ for all $x > 0$.

Then the following assertions are true:

a) *Let $t = 1,2,3,\ldots$. Then the path $\widetilde{P}_t(i,j)$ in the matrix $\widetilde{\mathbf{A}}_t$ has the minimum C-value among all i-j-paths of a length $\leq 2^{t-1}$.*
   Moreover, the value $C(\widetilde{P}_t(i,j))$ is equal to the entry $\widetilde{\chi}_t(i,j)$ in the matrix $\widetilde{\mathbf{X}}_t$.

b) *If $\widetilde{\mathbf{X}}_{L(n)} \boxtimes \mathbf{X} = \widetilde{\mathbf{X}}_{L(n)}$ then all paths $\widetilde{P}_{L(n)}(i,j)$ in $\widetilde{\mathbf{A}}_{L(n)}$ are C-optimal, and the costs $\widetilde{\chi}_{L(n)}(i,j)$ in $\widetilde{\mathbf{X}}_{L(n)}$ are minimal.*

c) *If $\widetilde{\mathbf{X}}_{L(n)} \boxtimes \mathbf{X} \neq \widetilde{\mathbf{X}}_{L(n)}$ then it is not possible to connect each pair (i,j) by a C-minimal i-j-path.*

d) *It follows that Algorithm 13 can decide whether or not each pair (i,j) can be connected by a C-optimal i-j-path; if YES then Algorithm 13 will output an optimal i-j-path and its cost for each pair (i,j).*
   For this purpose, we specify COND and WRITE as follows:

   - COND $:= (\tau < L(n) - 1)$.

   - WRITE $:= \left(\begin{array}{l} \text{If } b = \text{true} \quad \text{then write } \mathbf{A} \text{ and } \mathbf{X} \\ \text{else write :} \quad \text{"Not all pairs } (i,j) \text{ have} \\ \qquad\qquad\quad\ a\ C\text{-minimal } i\text{-}j\text{-path."} \end{array} \right).$

e) *Algorithm 13 requires $\Theta(n^3 \cdot \log n)$ time.*

P r o o f o f a) : Fact (4.162) implies that $\widetilde{\mathbf{A}}_{L(n)} \cong_C \mathbf{A}_{n'}$ for $n' := 2^{L(n)-1} \geq n$. Theorem 4.153 a) implies that each path $\widetilde{P}_t(i,j) = P_{2^{t-1}}(i,j)$ is C-minimal among all i-j-connections of a length $\leq 2^{t-1}$. Moreover, $C(\widetilde{P}_t(i,j)) \overset{(4.177)}{=} \widetilde{\chi}_t(i,j)$ for all i,j.

P r o o f o f b) : Let $n' := 2^{L(n)-1}$; then $n' \geq n$. The definition of the matrices $\widetilde{\mathbf{X}}_t$ and $\mathbf{X}_t$ implies that $\widetilde{\mathbf{X}}_{L(n)} = (\widetilde{\chi}_{L(n)}(i,j)) \overset{(4.177)}{=} (\chi_{n'}(i,j)) = \mathbf{X}_{n'}$. Consequently,

---

[44]Then C is automatically =-preserving and <-preserving.

if $\widetilde{\mathbf{X}}_{L(n)} \boxtimes \mathbf{X} = \widetilde{\mathbf{X}}_{L(n)}$ then $\mathbf{X}_{n'} = \mathbf{X}_{n'} \boxtimes \mathbf{X} = \mathbf{X}_{n'+1}$. Then Theorem 4.153 b) implies
that all paths $\widetilde{P}_{L(n)}(i,j) = P_{n'}(i,j)$ in $\widetilde{\mathbf{A}}_{L(n)} \overset{(4.162)}{\cong_C} \mathbf{A}_{n'}$ are C-minimal.

A further consequence of (4.177) is that $\widetilde{\chi}_{L(n)}(i,j) = C(\widetilde{P}_{L(n)}(i,j))$ for all i,j; hence,
$\widetilde{\chi}_{L(n)}(i,j)$ is the minimal cost for all i-j-paths.

P r o o f   o f   c) :   Let $\widetilde{\mathbf{X}}_{L(n)} \boxtimes \mathbf{X} \neq \widetilde{\mathbf{X}}_{L(n)}$, and let again $n' := 2^{L(n)-1}$. Then
$\mathbf{X}_{n'} \boxtimes \mathbf{X} \neq \mathbf{X}_{n'}$ because $\widetilde{\mathbf{X}}_{L(n)} = \mathbf{X}_{2^{L(n)-1}} = \mathbf{X}_{n'}$. Then 4.153 c) says that there is no
C-minimal i-j-connection for some pair (i,j).

P r o o f   o f   d) :   This assertion follows immediately from b) and c).

P r o o f   o f   e) :   This assertion follows from the fact that $L(n)$ formal matrix
multiplications are executed, and each of them takes $O(n^3)$ time.   □

The next result is analogous to 4.154.

Theorem 4.166 *Let* $\mathcal{G} = (\mathcal{V}, \mathcal{V} \times \mathcal{V})$ *where* $\mathcal{V} = \{1,\dots,n\}$. *Let* $C : \mathcal{P}(\mathcal{G}) \to \mathbb{R}$ *be
easily computable of type B with the help of the operation* $\odot$ *and the arc function* h.
We assume that C *is order preserving. Also, we assume that* $\odot$ *is associative and that
(4.158) is true. We define all matrices in the current theorem in the same way as the
matrices in 4.152 and in 4.161 – 4.163.*

Then the following assertions are true:

   a) *Let* $t = 1,2,3,\dots$. *Then the path* $\widetilde{P}_t(i,j)$ *in the matrix* $\widetilde{\mathbf{A}}_t$ *has the minimum* C-*value among all* i-j-*paths of a length* $\leq 2^{t-1}$.
   Moreover, the value $C(\widetilde{P}_t(i,j))$ *is equal to the entry* $\widetilde{\chi}_t(i,j)$ *in the matrix* $\widetilde{\mathbf{X}}_t$.

   b) *If* $\widetilde{\mathbf{X}}_t = \widetilde{\mathbf{X}}_t \boxtimes \mathbf{X}$ *for some* t *then all paths* $P_t(i,j)$ *in* $\widetilde{\mathbf{A}}_t$ *are* C-*optimal, and the
   costs* $\widetilde{\chi}_t(i,j)$ *in* $\mathbf{X}_t$ *are minimal.*

   c) *It follows that Algorithm 13 can find an optimal* i-j-*path and its cost for each pair
   (i,j) *if a* C-*minimal* i-j-*path exists for all* (i,j).
   For this purpose, we define COND *and* WRITE *as follows:*

      • COND $:= (b = \text{false})$,   • WRITE $:= (\text{Write } \widetilde{\mathbf{A}} \text{ and } \widetilde{\mathbf{X}})$.

   If not each pair (i,j) *can be connected by a* C-*minimal path then Algorithm 13
   will never terminate.*

P r o o f   o f   a) :   Equation (4.162) implies that $\widetilde{\mathbf{A}}_{L(n)} \cong_C \mathbf{A}_{n'}$ for $n' := 2^{L(n)-1} \geq n$.
Theorem 4.154 a) implies that each path $\widetilde{P}_t(i,j) = P_{2^{t-1}}(i,j)$ is C-optimal among all
i-j-connections of a length $\leq 2^{t-1}$. Moreover, $C(\widetilde{P}_t(i,j)) \overset{(4.177)}{=} \widetilde{\chi}_t(i,j)$ for all i,j.

P r o o f   o f   b) :   The definition of the matrices $\widetilde{\mathbf{X}}_\tau$ and $\mathbf{X}_\tau$ implies that $\widetilde{\mathbf{X}}_t =$
$\left(\widetilde{\chi}_t(i,j)\right) \overset{(4.177)}{=} \left(\chi_{2^{t-1}}(i,j)\right) = \mathbf{X}_{2^{t-1}}$. Consequently, if $\widetilde{\mathbf{X}}_t \boxtimes \mathbf{X} = \widetilde{\mathbf{X}}_t$ then
$\mathbf{X}_{2^{t-1}} = \mathbf{X}_{2^{t-1}} \boxtimes \mathbf{X} = \mathbf{X}_{2^{t-1}+1}$. Then 4.154 implies that all paths $\widetilde{P}_t(i,j) = P_{2^{t-1}}(i,j)$
in $\widetilde{\mathbf{A}}_t \overset{(4.162)}{\cong_C} \mathbf{A}_{2^{t-1}}$ are C-minimal.

It follows from (4.177) that $\widetilde{\chi}_t(i,j) = C(\widetilde{P}_{2^t-1}(i,j))$ for all i,j; hence, $\widetilde{\chi}_t(i,j)$ is the minimal cost for all i-j-paths.

P r o o f   o f   c) : This assertion follows immediately from a) and b).   □

Remark 4.167 We conjecture that Theorem 4.165 b) remains true if we replace the condition $\widetilde{\mathbf{X}}_{L(n)} \boxtimes \mathbf{X} = \widetilde{\mathbf{X}}_{L(n)}$ by $\widetilde{\mathbf{X}}_{L(n)+1} = \widetilde{\mathbf{X}}_{L(n)}$. Moreover, we conjecture that Theorem 4.165 c) remains true if we replace the condition $\widetilde{\mathbf{X}}_{L(n)} \boxtimes \mathbf{X} \neq \widetilde{\mathbf{X}}_{L(n)}$ by $\widetilde{\mathbf{X}}_{L(n)+1} \neq \widetilde{\mathbf{X}}_{L(n)}$. If these conjectures are true, we need not compute the matrices $\mathbf{X} \boxtimes \mathbf{X}$ in step 2.4 of Algorithm 13; it is sufficient to compare $\widetilde{\mathbf{X}}_t$ with $\widetilde{\mathbf{X}}_{t+1}$ for all t.

Also, we conjecture that Theorem 4.166 b) remains true if we replace the condition $\widetilde{\mathbf{X}} = \widetilde{\mathbf{X}} \boxtimes \mathbf{X}$ in step 2.4 of Algorithm 13 by the condition $\widetilde{\mathbf{X}} = \widetilde{\mathbf{X}} \boxtimes \widetilde{\mathbf{X}}$, i.e. $\widetilde{\mathbf{X}}_t = \widetilde{\mathbf{X}}_{t+1}$ □

4.4.3   Floyd's fast search strategies

Here we describe Floyd's matrix algorithms. They solve the problem 10 a) and b) in $O(n^3)$ time; that means that the time complexity given in 4.165 is improved by the factor $\log(n)$.

Floyd algorithms are described in [Chri75], [PaSt82], and in many other graph theoretic books. The algorithm in [Pol92] is based on Floyd's ideas.

We describe two versions of Floyd's search strategy. The first constructs optimal paths, and the second computes minimal costs. Both algorithms process path matrices $(P(i,j))$ or cost matrices $(\chi(i,j))$.

For all nodes $k = 1, \ldots, n$, Floyd's optimal path algorithm replaces the current path $P(i,j)$ by $P(i,k) \oplus P(k,j)$ if the detour via k is cheaper than $P(i,j)$. We now give the algorithm in detail.

Algorithm 14 (Floyd's method, first version)

1. Let $P_0(i,j) := [i,j]$ for all $i,j = 1, \ldots, n$.

2. **For** $k := 1$ **to** n **do**

   For $i := 1$ **to** n **do**

   For $j := 1$ **to** n **do**
   If $C\left(P_{k-1}(i,k) \oplus P_{k-1}(k,j)\right) < C\left(P_{k-1}(i,j)\right)$
   then $P_k(i,j) := P_{k-1}(i,k) \oplus P_{k-1}(k,j)$
   else $P_k(i,j) := P_{k-1}(i,j)$.

We next describe the second version of Floyd's method. Here we assume that C is easily computable of type B with an arc function h and an associative operation $\odot$. The following search method is similar to the Triple Algorithm to find the transitive reflexive closure of a digraph (see also Remark 4.148).

Algorithm 15 (Floyd's method, second version)

    1. Let $\chi_0(i,j) := h(i,j) = C([i,j])$ for all $i,j = 1,\ldots,n$.

    2. **For** $k := 1$ **to** n **do**

        For $i := 1$ **to** n **do**

            For $j := 1$ **to** n **do**

                $\chi_k(i,j) := \min\left(\chi_{k-1}(i,j), \chi_{k-1}(i,k) \odot \chi_{k-1}(k,j)\right)$

We describe the behaviour of Algorithm 14.

Theorem 4.168 *Given the digraph $\mathcal{G} = (\mathcal{V},\mathcal{R})$ with $\mathcal{V} = \{1,\ldots,n\}$ and $\mathcal{R} = \mathcal{V} \times \mathcal{V}$. Let $C : \mathcal{P}(\mathcal{G}) \to \mathbb{R}$ be a cost function with property (4.118). Moreover, we make the following assumptions:*

    (i) *C has the strong prefix and the strong suffix oriented Bellman property of type 0.*
    (ii) *C satisfies conditions (i) and (ii) of Lemma 2.27.*
    (iii) *$\mathcal{G}$ has no C-negative cycles.*

Then the path $P_n(i,j)$ generated by Algorithm 14 is a C-minimal i-j-connection for each pair (i,j).

(This implies that the first version of Floyd's method finds optimal paths in $O(n^3)$ time.)

P r o o f : First, we show the following fact:

    For each pair (i,j), there exists a C-minimal i-j-path $P^*(i,j)$, which is acyclic.   (4.178)

Proof of (4.178): Assumption (iii) and Lemma 2.132 imply that each i-j-path Q can be replaced by an acyclic i-j-path P with $C(P) \leq C(Q)$. Let $P^*(i,j)$ be a C-minimal path among all finitely many acyclic i-j-paths; then $P^*(i,j)$ is C-minimal among all i-j-paths.

Let $W \subseteq \mathcal{V}$, and let $P = [v_0, v_1, \ldots, v_k]$ be a path in $\mathcal{G}$. We say that P is a W-*path* if all of its internal nodes lie in W; more precisely, we require that $v_\kappa \in W$ for all $\kappa \notin \{0,k\}$. Moreover, we say that P is an (i,W,j)-*path* if P is a W-path from $v_0 = i$ to $v_k = j$.

We next show the following fact:

    Let $k \in \{1,\ldots,n\}$, and let $W := \{1,\ldots,k\}$. Let $i,j \in \mathcal{V}$. We assume
    that a C-minimal i-j-path P^* exists, which is also an acyclic W-path.   (4.179)
    Then the path $P_k(i,j)$ generated by Algorithm 14 is C-minimal as well.

Proof of (4.179): The proof is an induction on k. Let $k = 1$, and let P^* be the $(i,\{1\},j)$-path described in (4.179). We show that the following is true:

$$C(P^*) = C([i,j]) \quad \text{or} \quad C(P^*) = C([i,1,j]).$$ (4.180)

If $\ell(P^*) = 0$ then i equals j and $P^* = [i]$. Consequently, $C(P^*) = C([i]) \overset{(4.118)}{=} C([i,i]) = C([i,j])$, and assertion (4.180) is true in this case.
If $\ell(P^*) = 1$ then $P^* = [i,j]$, and $i \neq j$ because P^* is acyclic. Consequently, $C(P^*) = C([i,j])$, and assertion (4.180) is true in this case.

If $\ell(P^*) = 2$ then $P^* = [i, 1, j]$, and $1, i, j$ are pairwise distinct because P^* is acyclic. Consequently, $C(P^*) = C([i, 1, j])$, and assertion (4.180) is true in this case. It is not possible that $\ell(P^*) > 2$ because P^* may only visit each of the nodes $1, i, j$ once. Consequently, (4.180) is true in any case.

Then the path $P_1(i, j)$ constructed by Algorithm 14 has the following property:
$$C(P_1(i, j)) \;=\; \min\{C([i, j]), \, C([i, 1, j])\} \;\overset{(4.180)}{\leq}\; C(P^*).$$
This and the minimality of P^* imply that $P_1(i, j)$ is minimal as well.

Let $k > 1$. We assume that fact (4.179) is true for $k - 1$.
Let P^* be a W-path. If P^* is even a $(W \backslash \{k\})$-path then $C(P_k(i, j)) \leq C(P_{k-1}(i, j)) \overset{(*)}{\leq} C(P^*)$ where $(*)$ follows from the induction hypothesis.

If, however, P^* is not a $(W \backslash \{k\})$-path then P^* visits k. Hence, there exist an i-k-path $\widetilde{P}$ and a k-j-path $\widetilde{Q}$ such that $P^* = \widetilde{P} \oplus \widetilde{Q}$. Then $\widetilde{P}$ and $\widetilde{Q}$ are $(W \backslash \{k\})$-paths because otherwise, P^* would visit k more than once. Moreover, $\widetilde{P}$ and $\widetilde{Q}$ are C-minimal paths by assumption (i). Consequently, the induction hypothesis implies that $C(P_{k-1}(i, k)) = C(\widetilde{P})$ and $C(P_{k-1}(k, j)) = C(\widetilde{Q})$. This and assumption (ii) have the consequence that we may apply Lemma 2.27 to $X_1 := P_{k-1}(i, k)$, $X_2 := \widetilde{P}$, $Y_1 := P_{k-1}(k, j)$, and $Y_2 := \widetilde{Q}$; thus, we obtain the following assertion:
$$C(P_{k-1}(i, k) \oplus P_{k-1}(k, j)) = C(\widetilde{P} \oplus \widetilde{Q}) = C(P^*). \tag{4.181}$$
Algorithm 14 constructs a path $P_k(i, j)$ with the following property:
$$C(P_k(i, j)) \;=\; \min\{C(P_{k-1}(i, j)), \, C(P_{k-1}(i, k) \oplus P_{k-1}(k, j))\} \;\overset{(4.181)}{\leq}\; C(P^*).$$
Consequently, $P_k(i, j)$ is C-minimal, and (4.179) is true for k.

Let $k = n$. Then $W = V$. Consequently, the path $P^*(i, j)$ described in (4.178) is an acyclic and C-minimal i-j-path. This and (4.179) imply that $P_n(i, j)$ is a C-minimal i-j-path.
$\qquad\qquad\qquad\qquad\qquad\qquad\qquad\qquad\qquad\qquad\qquad\qquad\qquad\qquad\qquad\quad\square$

Next, we describe the behaviour of Algorithm 15.

Theorem 4.169 *Given the digraph $\mathcal{G} = (V, V \times V)$ and $V = \{1, \ldots, n\}$. Let C be easily computable of type B with an arc function $h : \mathcal{R} \to \mathbb{R}$ and an associative operation $\odot$. We assume that $P_k(i, j)$ are the paths generated by Algorithm 14 and that $\chi_k(i, j)$ are the C-values computed by Algorithm 15. Then the following are true for all i, j:*

a) *$\chi_k(i, j) = C(P_k(i, j))$ for all $k = 1, \ldots, n$.*

b) *If C satisfies the conditions in Theorem 4.168 then the quantity $\chi_n(i, j)$ equals the minimal cost of all i-j-connections.*

(That means that the second version of Floyd's method computes the minimal cost of all i-j-paths in $O(n^3)$ time.)

P r o o f o f a) : Algorithm 14 generates paths $P_k(i, j)$ with the following property:
$$(\forall\, i, j, k) \;\; C(P_k(i, j)) = \min\{C(P_{k-1}(i, j)), \, C(P_{k-1}(i, k) \oplus P_{k-1}(k, j))\}. \tag{4.182}$$
Algorithm 15 generates numbers $\chi_k(i, j)$ with the following property:

$$(\forall\, i,j,k)\ \chi_k(i,j)\ =\ \min\{\chi_{k-1}(i,j)\,,\,\chi_{k-1}(i,k)\odot\chi_{k-1}(k,j)\}\,. \tag{4.183}$$

Result 4.160 implies the following assertion:

$$(\forall\, P,Q)\ C(P\oplus Q)\ =\ C(P)\odot C(Q)\,. \tag{4.184}$$

Now, we show our result by an induction of k. Let $k=0$ then $\chi_0(i,j)=C([i,j])$ by step 1 of Algorithm 15, and $[i,j]=P_0(i,j)$ by step 1 of Algorithm 14. Consequently, $\chi_0(i,j)=C([i,j])=C(P_0(i,j))$.

Let $k>1$. We assume that the result is true for $k-1$. Then we obtain the following equation, in which $(*)$ follows from the induction hypothesis.

$$\chi_k(i,j)\ \overset{(4.183)}{=}\ \min\{\chi_{k-1}(i,j)\,,\,\chi_{k-1}(i,k)\odot\chi_{k-1}(k,j)\}\ \overset{(*)}{=}$$
$$\min\{C(P_{k-1}(i,j))\,,\,C(P_{k-1}(i,k))\odot C(P_{k-1}(k,j))\}\ \overset{(4.184)}{=}$$
$$\min\{C(P_{k-1}(i,j))\,,\,C(P_{k-1}(i,k)\oplus P_{k-1}(k,j))\}\ \overset{(4.182)}{=}\ C(P_k(i,j))\,.$$

Consequently, $\chi_k(i,j)=C(P_k(i,j))$ for all i,j.

P r o o f o f b) : Part a) implies that $\chi_n(i,j)=C(P_n(i,j))$ for all i,j, and Theorem 4.168 implies that $P_n(i,j)$ is a C-minimal i-j-connection. Consequently, $\chi_n(i,j)$ is the minimal cost of all i-j-paths. $\qquad\Box$

4.4.4   Complexity theoretical results

We give upper and lower bounds to the time complexity of the Problem 10. In 4.153, 4.165, and 4.169, we investigated the time complexity of particular matrix algorithms. We now give the complexity of several further methods solving Problem 10.

Remark 4.170 Given the digraph $\mathcal{G}=(\mathcal{V},\mathcal{V}\times\mathcal{V})$ with $\mathcal{V}=\{1,\ldots,n\}$. Let C be a path function. We assume that there exist no C-negative cycles in $\mathcal{G}$ so that a C-minimal i-j-path $P^*(i,j)$ exists for each pair (i,j). We define X^* as the matrix consisting of all minimal costs; more precisely, $X^* := \big(C(P^*(i,j))\big)\,.$

The following parts of this remark are arranged such that the conditions about C become more and more restrictive.

a) Let C be prefix monotone (but not necessarily additive). Then the following algorithm will solve Problem 10 a) in $O(n^3)$ time:

> **For** $i:=1$ **to** n **do**
>> **begin**
>> Apply DIJKSTRA-BF* to $\mathcal{G}$ with start node i.
>> (Let DIJKSTRA-BF* run until OPEN is empty.)
>> **end.**

It follows from 4.141 b) that for any start node i, DIJKSTRA-BF* will find an optimal i-j-path to every node j; moreover, each $i=1,\ldots,n$ causes $O(n^2)$ time. Consequently, the above algorithm will solve Problem 10 a) in $O(n^3)$ time.

b) Let $C=SUM_h$ be additive, and let $h(r)\geq 0$ for all arcs r; then C is prefix monotone. Then Problem 10 b) can be solved in less than $O(n^3)$ time. In [Fred76], a matrix algorithm $\mathcal{A}_n$ with complexity $O(n^{2.5})$ is given for any fixed number n of nodes; moreover, Fredman describes an algorithm with a running time of $O\left(n^3 \sqrt[3]{\frac{\log \log n}{\log n}}\right)$; this time is shorter than $O(n^3)$.

This result has been improved in [Taka91]; Takaoka's matrix algorithm computes X^* in $O\left(n^3 \sqrt{\frac{\log \log n}{\log n}}\right)$ time units; hence, Fredman's time bound has been improved by the factor $\sqrt[6]{\frac{\log \log n}{\log n}}$.

c) Let $C = SUM_h$, and let $h(r) \in \mathbb{N} \cup \{0\}$ for all r.

In this case, the matrix X^* can be computed in a time of $O(n^{2.52})$. An algorithm with this time complexity is given in [Roma80]; Romani's algorithm uses fast matrix multiplication.

We compare the results of Romani and Takaoka. The advantage of Romani's algorithm is that it works in $O(n^{2.52})$ time units, which is faster than the $O\left(n^3 \sqrt{\frac{\log \log n}{\log n}}\right)$ time units of Takaoka's algorithm. On the other hand, Romani assumes that $h(r)$ is an integer number for each r whereas Takaoka does not make this assumption.

d) Let $C = SUM_h$, and let $h(r) \in \mathbb{N} \cup \{0\}$ for all r.

All algorithms mentioned in part c) of this remark have the following property: When computing $z := x + y$, $z := x - y$, $z := x \cdot y$ or $z := x/y$ for integer numbers x, y, then they consume one time unit. In contrast to this, the author of [Pan81] counts the number of bit operations that executed when computing z; using this measure of complexity, he studies the time complexity of algorithms that compute X^*.

e) Let $C = SUM_h$, and let $h(r) \in \mathbb{N}$ for all arcs r.

In [Wata81], an algorithm is given that constructs a C-minimal i-j-path for each pair (i, j) of nodes. The time complexity of this algorithm is $O(n^{2.81})$. Watanabe assumes that particular functions like $x \mapsto 2^x$ or $x \mapsto \lfloor x \rfloor$ can be evaluated in constant time.

f) Let $\mathcal{G}$ be an undirected graph, and let $C = SUM_h$ where $h(r) := 1$ for all edges r. Then Problem 10 b) can be solved in $O(n^{2376} \cdot \log n)$ time. This follows from [Seid92].

Many optimal path algorithms mentioned in this remark employ fast matrix multiplication. We conjecture that most of these algorithms can be made faster by using efficient methods of matrix multiplication. Fast methods for matrix multiplication are presented in [CoWi87], [CoWi90], [Ling91], and in many other sources. □

The paper [Mahr80] gives further upper bounds on the complexity of algorithms. The difference between Mahr's results and the ones in 4.170 is the following: Almost all complexity bounds in 4.170 are of the form $O(T(n))$ where T is a function and n is the number of the nodes of $\mathcal{G}$.[45] Mahr, however, describes the complexity of optimal path algorithms with further parameters, for example the number m of all arcs in $\mathcal{G}$.[46]

---

[45] The only exceptions are the results of [Pan81] about the bit complexity of optimal path algorithms.
[46] Mahr does not a priori assume that $\mathcal{R} = \mathcal{V} \times \mathcal{V}$.

A similar result about the complexity of Problem 10 is shown in [KKPh93].

The next remark is about lower complexity bounds.

Remark 4.171 Given the terminology of Remark 4.170.

It has been shown in [SpiP73] and in [GrYY80] that $\Omega(n^2)$ is a lower bound of the complexity of computing the matrix X^*. This bound is not surprising because any algorithm solving Problem 10 should check each of the $\Omega(n^2)$ arcs of the given digraph. Nevertheless, it seems that up to now, a sharper lower bound is not given in literature.

It has been shown in [Mahr80] that every optimal path algorithm $\mathcal{A}$ requires $\Omega(n^3)$ time if $\mathcal{A}$ belongs to a particular class $\mathcal{C}$. The algorithms $\mathcal{A} \in \mathcal{C}$ are called "label independent"; that means that in particular situations, $\mathcal{A}$ must not use any information about the path function C.

Let m be the number of arcs of the given digraph. The authors of [KKPh93] show that $\Omega(mn)$ is a lower bound for the complexity of a particular class of algorithms solving Problem 10. $\qquad\square$

4.4.5   Further solutions of All-Pairs-Optimal-Path problems

The papers [GoSV78], [PanR87], [PanR89], [Mang90], [Rote90], [Arl91] [Ausi91], and [LenT91] describe algorithms to solve Problem 10 a) or b). Most of these algorithms use well-known matrix transformations of Linear Algebra and transfer them into the situation of path matrices or cost matrices; for example, a modified Gauß-Jordan elimination is used in [Rote90] to compute minimal path costs. Problem 10 b) has been solved in [PanR87], [PanR89], [Rote90], and in [LenT91] even if the path function C is of the form $C : \mathcal{P}(\mathcal{G}) \rightarrow (\mathbf{R}, \preceq)$ where $\mathbf{R} \neq \mathbb{R}$ need not even be a field; in this case, the properties of path algebras or other algebraic structures are used to minimize the costs of paths. The paper [Ausi91] introduces a data structure to update the system of optimal paths very fast if the given digraph is modified.

The next remark is about the quickest path problem.

Remark 4.172 Given the same notations as in Problem 8. The papers [ChHu92] and [LeeP91] introduce algorithms that solve the problem to find quickest v-w-paths for all pairs (v, w). $\qquad\square$

Optimal path algorithms for undirected networks are studied in [Gab83], [Bern84], [Wai90], [Bard91], and in [Seid92]; the cost function C in [Bard91] has been described in Remark 4.144, and the paper [Seid92] was quoted in Remark 4.170.

There are many sources about Problem 10 in special graphs (for example planar graphs); we shall quote several of them in Section 4.5. Typical results about almost optimal i-j-paths for all pairs (i, j) will be described in Section 4.8.

4.5 Path problems and related questions in special graphs or in other special situations

This section is mainly about the search for optimal paths in graphs G where G has special properties. These properties often result from the given practical situation symbolized by the graph G.

In many cases, the additional properties of G make the search of optimal paths easier. Therefore, the current situation is inverse to the one in Section 4.10 where the search for paths is difficult; in the current section, the given graph has additional properties whereas in Section 4.10, the unknown optimal paths must have additional properties.

4.5.1 Path problems arising from Geometry

Here, we consider optimal path problems that are closely related to geometric problems. In these cases, we often have special cost functions C basing on distance functions in metric spaces.

The simplest case is that all nodes of the given graph G are elements of the space $\mathbb{R}^k$ with $k \geq 2$. Then each arc (x,y) or each edge $\{x,y\}$ of G is marked with a value $h((x,y)) := d(x,y)$ or $h(\{x,y\}) := d(x,y)$ where d is a fixed distance function for $\mathbb{R}^k$. The following measures of distance are used very often:

- the *Euclidean distance*; it is defined as $d(a,b) := \sqrt{\sum_{i=1}^{k}(a_i - b_i)^2}$ for all a,b;
- the *Manhattan distance*; it is defined as $d(a,b) := \sum_{i=1}^{k}|a_i - b_i|$ for all a,b;
- the *link distance*; it is defined as $d(a,b) := \left\{ \begin{smallmatrix} 0 & \text{if} & a=b \\ 1 & \text{if} & a \neq b \end{smallmatrix} \right\}$ for all a,b;

  this implies that each line segment (= link) $\overline{a,b}$ has the length 1 if $a \neq b$.

If $C := SUM_h$ then the cost $C(P)$ of a path $P = [v_0, v_1, \ldots, v_l]$ can be directly computed from the nodes $v_0, \ldots, v_l$; it is not necessary to record the values $h(r)$, $r \in \mathcal{R}$. Moreover, the following triangle inequalities are valid:

$(\forall\, x,y,z \in V)\ h((x,z)) \leq h((x,y)) + h((y,z))$ in the directed case,
$(\forall\, x,y,z \in V)\ h(\{x,z\}) \leq h(\{x,y\}) + h(\{y,z\})$ in the undirected case.

These properties of C make the search for C-optimal paths very easy. — We shall write $C = SUM_d$ instead of $C = SUM_h$ if h is based on a distance measure d.

First, we consider the following problem, which is a special case of Problem 1.

Problem 11 *Given a directed or an undirected finite graph G with $V \subseteq \mathbb{R}^k$ and $C = SUM_d$ where d is the Euclidean distance. Let $s \in V$ and $\Gamma \subseteq V$.*
Find a C-minimal s-Γ-path

Solutions of this problem are given in [GoBa78], [SeVi86], and [Bey93]. The authors of [GoBa78] and [Bey93] have developed new versions of well-known shortest path algorithms; experimental results show that all these algorithms are very fast. In [SeVi86], a shortest path algorithm is presented; if the nodes s and γ are given, the search strategy finds a C-minimal s-γ-path in an average time $O(n)$.

A modification of Problem 11 is solved in [Kan94]. The set $\mathcal{V}$ of nodes is a subset of $\{0, \pm1, \pm2, \pm3, \ldots\}^k$; if $r \in \mathcal{R}$ then $h(r)$ is defined as the Manhattan distance between the endpoints of r. The algorithm in [Kan94] finds an optimal s-γ-path in $O(|\mathcal{V}| + |\mathcal{R}|)$ time.

A further class of geometric graphs consists of hypercubes and their modifications (see Definition 3.97). All these graphs have the property that $\mathcal{V} \subseteq \{0,1\}^k$. The paper [ZKC90] is about finding shortest paths in twisted hypercubes.

We next consider the problem of finding shortest paths in geometric scenes. Solving this problem is very important for motion planning of robots. In contrast to Problem 11, we are searching for optimal paths in infinite graphs.

Problem 12 *Let $X \subseteq \mathbb{R}^k$ where $k \geq 2$. Let $\mathcal{G}_X = (\mathcal{V}_X, \mathcal{R}_X)$ be the undirected graph with $\mathcal{V}_X := X$ and $\mathcal{R}_X := \big\{ \{x,y\} \,\big|\, \text{The line segment } \overline{x,y} \text{ is a subset of } X \big\}$. Given the points $s, \gamma \in \mathcal{V}_X = X$. Let $C = SUM_d$ be a path function where d is the Euclidean, the Manhattan, or the link distance. Find a C-minimal s-γ-path.*

This problem is studied in [LeeP81], [LeeP84], [PrMi], [ChRa91], [CoGi92], and in [StR94]. All these sources are about the following special situation: The dimension k equals 2. The set X is equal to $X := \mathbb{R}^k \backslash (\Pi_1 \cup \ldots \cup \Pi_j)$ where $\Pi_1, \ldots, \Pi_j$ are rectangles or other polygons. The distance d is defined as the Euclidean distance. In this case, Problem 12 means to find a very short path from s to γ that avoids the forbidden polygons $\Pi_1, \ldots, \Pi_j$.
Coffman and Gilbert [CoGi92] and Mitchell [Mi92] solve almost the same problem; the only difference is that Coffman, Gilbert and Mitchell use the Manhattan distance and not the Euclidean distance to describe the lengths of paths.

In [PrMi], a parallel search algorithm is given; this algorithm is to find Euclidean shortest paths for a robot moving in a plane with obstacles. Praßler and Milios create a network $\mathcal{G}'$ of processors; this network has the structure of a grid and is embedded into the plane. Praßler's and Milios' ideas are very plausible, but the authors do not prove that their algorithm works correctly.

A modification of Problem 12 is solved in [KeW93]. The authors describe the search for an optimal pair of paths (P_1, P_2) in $\mathcal{G}_X$; both paths P_1, P_2 must not intersect each other, and the sum of their Euclidean lengths must be minimal.

A further modification of 12 is solved in [Gew91]; the author of the paper give an algorithm that finds a so-called monotone s-γ-path visiting the maximal number of obstacles.

A three dimensional version of the shortest path problem in a scene of obstacles is considered in [Akm86]. In this case, $k = 3$, and the sets $\Pi_1, \ldots, \Pi_j$ are three-dimensional polyhedra.

The next problem is to find a shortest s-γ-path inside a polygon. The boundaries of the polygon are parallel to the axes of the Cartesian coordinate system, and the same must be true for the edges of the path. The problem is similar to the problems 11 and 12; the only difference is that another graph is given:

Problem 13 *Let $X \subseteq \mathbb{R}^k$ where $d \geq 2$. Let $\mathcal{G}'_X = (\mathcal{V}'_X, \mathcal{R}'_X)$ be the undirected graph with $\mathcal{V}'_X := X$ and*

$$\mathcal{R}'_X := \{ \{x, y\} \mid \text{The line segment } \overline{x,y} \text{ is a subset of } X, \text{ and } \overline{x,y} \text{ is axis parallel} \}.$$

Given the points $s, \gamma \in V_X = X$. Let $C = SUM_d$ be a path function where d is the Euclidean, the Manhattan, or the link distance.

Find a C-minimal s-γ-path.

This problem is solved in [deBerg] and [Schu93] if the following situation is given: $k = 2$, X is a polygon whose boundaries are axis-parallel, d is the link distance. In [Schu93], the same problem is also solved if d is the Manhattan distance.

The article [GuiH89] presents a fast algorithm that computes the Euclidean length of the shortest x-y-path inside a polygon X. The paper [CaJo95] describes the fast construction of optimal paths inside a polygon; these paths must satisfy a side constraint about visibility. The paper [ChNt91] presents an algorithm to compute shortest paths on the boundary of a polygon.

Next, we consider the following problem, which often appears in motion planning as well as Problem 12. Roughly speaking, we want to move a circular disk from s to γ without colliding with particular points; the disk shall be as large as possible. We are searching for a path along which the disk of maximal size must be moved.

Problem 14 *Let $S \subseteq \mathbb{R}^2$ be a finite set of points. Moreover, let $G = (V, \mathcal{R})$ be a finite and undirected graph with $V \subseteq \mathbb{R}^2 \backslash S$. We assume that each edge $r \in \mathcal{R}$ is represented by a continuous curve $c_r \subseteq \mathbb{R}^2 \backslash S$ between x and y. Let $s, \gamma \in V$.*

Find an s-γ-path P^* and a radius r^* with the following properties:

- *A circular disk Δ of radius r^* can be moved along P^* from s to γ without collisions with S. More formally, if $P^* = r_1 \oplus \ldots \oplus r_l$ then no point of S enters the open kernel of Δ when moving its centre along the curves $c_{r_1}, \ldots, c_{r_l}$.*

- *The radius r^* is maximal; more precisely, $r' \leq r^*$ for all disks Δ' with radius r' that can be moved along some s-γ-path $P' \in \mathcal{P}(G)$ without collisions with S.*

This problem is equivalent to the search for a path of maximal capacity. More precisely, we define the capacity $\overline{\beta}(r)$ for each edge $r = \{x, y\}$ as follows: $\overline{\beta}(r) := \min\{\|u - v\|_2 \mid u \in c_r, \, v \in S\}$ where $\|\xi\|_2$ is the Euclidean norm of $\xi \in \mathbb{R}^2$; the capacity $\overline{\beta}(r)$ is equal to the maximal radius of a disk that can be moved along c_r without collisions. For each path $P = r_1 \oplus \ldots \oplus r_l$, we define the capacity of P as $\overline{\beta}(P) := \min\{\overline{\beta}(r_1), \ldots, \overline{\beta}(r_l)\}$. Then Problem 14 means to find a s-γ-path P^* of maximal capacity $\overline{\beta}(P^*)$.

As mentioned in Remark 4.7 a), Problem 14 can be solved with the help the maximum spanning tree of G where each edge r of G has the weight $\overline{\beta}(r)$. The path P^* is the only acyclic s-γ-path in this tree.

The problem of moving objects (for example circular disks or line segments) without collisions has been studied in [MaPl87], [Rohn91], [IRWY93], and in many further sources.

Geometric path problems may also appear in pixel scenarios. For example, given a rectangle R. We assume that R is divided into m rows and n columns of squares. Each square an be described by an ordered pair (μ, ν) where $1 \leq \mu \leq m$ and $1 \leq \nu \leq n$. A *binary image* Φ is defined as a colouring of each square with the colours WHITE or BLACK; more formally, a binary image is a function

$$\Phi : \{1, \ldots, m\} \times \{1, \ldots, n\} \to \{WHITE, BLACK\}\,.$$

There exist several definitions of paths and of distances in binary images. The typical optimal path problems have the following structure:

Problem 15 *Given a binary image Φ and two squares (μ, ν) and (μ', ν'). Find an optimal path from (μ, ν) to (μ', ν') in Φ or compute the cost of such an optimal path.*

This and similar problems are solved in [Shne81], [HüKW82] and other sources. The authors of [HüKW82] and of [Shne81] reduce Problem 15 to optimal-path-problems in a graph.

Now, we consider location problems; they can be formulated as follows:

Problem 16 *Given a set of objects (for example, circles, line segments, words). Find optimal places for these objects in $\mathbb{R}^k$.*

In [Ichi85], optimal places for facilities in the plane $\mathbb{R}^2$ must be found. Ichimori solves this problem by searching for shortest paths in a graph. The problem in [Vanth91] is to design tables. That means that words and line segments must be optimally arranged. Vathienen reduces this problem to the search for maximal paths in graphs.

A further geometric problem in relationship to optimal paths is to find shortest paths preserving graphs. These undirected graphs have the property that the shortest paths between particular points do not become longer when these paths must use the edges of $\mathcal{G}$. We now define these graphs exactly:

Definition 4.173 Given a set $X \subseteq \mathbb{R}^2$ and N points $t^{(1)}, \ldots, t^{(N)} \in X$. Let $\mathcal{G}_X$ be the same graph as in Problem 12. Let d be the Manhattan distance, and let $C := SUM_d$.

A *shortest paths preserving graph* (*spp graph*) is a finite, undirected graph $\mathcal{G} = (\mathcal{V}, \mathcal{R})$ with the following properties:

- $\{t^{(1)}, \ldots, t^{(N)}\} \subseteq \mathcal{V} \subseteq \mathbb{R}^2$.
- If $\{x, y\} \in \mathcal{R}$ then the line segment $\overline{x, y}$ is horizontal or vertical; moreover, $\overline{x, y} \subseteq X$.[47]

---

[47] It follows that $\mathcal{G}$ is a subgraph of $\mathcal{G}_X$.

- If $i, j \in \{1, \ldots, n\}$ then the minimal C-value of all $t^{(i)}$-$t^{(j)}$-paths in $\mathcal{G}$ equals the minimal C-value of all $t^{(i)}$-$t^{(j)}$-paths that may use all points of X; that means, a C-minimal $t^{(i)}$-$t^{(j)}$-path in $\mathcal{G}$ is also a C-minimal $t^{(i)}$-$t^{(j)}$-path in the greater graph $\mathcal{G}_X$.

$\square$

Shortest path preserving graphs are useful for VLSI-design. If we place wires at the edges of such a graph then all $t^{(i)}$-$t^{(j)}$-connections are as short as possible. The Manhattan distance is an appropriate distance measure in VLSI-design because all wires must be vertical or horizontal.

We now formulate the problem of finding spp graphs.

Problem 17 *Given a set $X \subseteq \mathbb{R}^2$ and the points $t^{(1)}, \ldots, t^{(N)} \in X$.*
Find an spp graph $\mathcal{G}$ for these points $t^{(1)}, \ldots, t^{(N)}$.

This problem is solved in [Widm91] in the following situation: There are polygonal obstacles $\Pi_1, \ldots, \Pi_j$ whose boundaries are axis-parallel; each admissible path must avoid the open kernels $\overset{\circ}{\Pi}_i$ of the polygons Π_i, $i = 1, \ldots, j$; that means that we are searching for an spp graph inside $X := \mathbb{R}^2 \setminus \left(\overset{\circ}{\Pi}_1 \cup \ldots \cup \overset{\circ}{\Pi}_j \right)$. Widmayer's algorithm solves this problem in $O(N \log N)$ time units.

The spp problem has almost the inverse structure of the usual optimal path problems. This is shown by the following comparison:

Structure of the usual extremal path problems:
The graph $\mathcal{G}$ is given, the properties of the extremal paths in $\mathcal{G}$ are unknown.

Structure of the spp problem:
The properties of the extremal paths of $\mathcal{G}$ are given, the graph $\mathcal{G}$ is unknown.

We now consider the following problem:

Problem 18 *Given a finite, undirected, connected graph $\mathcal{G} = (\mathcal{V}, \mathcal{R})$ and two natural numbers σ, ζ. Find two sets $S, Z \subseteq \mathcal{V}$ with $|S| = \sigma$ and $|Z| = \zeta$ such that the maximal distance between any elements $s \in S$ and $z \in Z$ is minimized.*

Problem 18 includes a typical minimaximin problem because the quantity

$$\min_{\substack{|S| = \sigma, \\ |Z| = \zeta}} \ \max_{\substack{s \in S, \\ z \in Z}} \ \min_{P \in \mathcal{P}(s,z)} \ \ell(P)$$

must be computed. — In [KS94], Problem 18 is solved for hypercubes and many other grids of high dimension.

4.5.2   Extremal paths in planar graphs

In this subsection we quote several results of literature about the search for optimal paths in planar graphs. Finding optimal paths in planar graphs is often less time-consuming than solving this problem in arbitrary graphs. One of the main reasons is that a planar and parallel free graph with n nodes can only have $O(n)$ arcs or edges; if $\mathcal{G}$ is not planar, then $\mathcal{G}$ may have $\Theta(n^2)$ arcs or edges.

The current subsection is a good link between the previous one and the next one. The reason is that planarity is a geometric property and a structural property at the same time. A graph is planar if and only if it can be embedded into the plane without crossings of edges; therefore, planar graphs have a special geometric property like the graphs in Subsection 4.5.1. On the other hand, a graph is planar if and only if it does not contain a complete graph $\mathcal{K}_5$ or $\mathcal{K}_{3,3}$. Therefore, planar graphs have a special structural property like the graphs in Subsection 4.5.3.

One of the earliest solution of optimal path problems in planar graphs is given in [FoFu56], page 404. The search for optimal paths is reduced to the search for an optimal flow in the so-called dual graph. Roughly speaking, the dual graph is constructed as follows: The nodes of the dual graph $\mathcal{G}'$ are the faces of the given planar graph $\mathcal{G}$; two nodes f_1, f_2 of the dual graph are connected with each other if f_1 and f_2 have a common boundary in $\mathcal{G}$.

The well-known paper of Frederickson [Fred87] describes several optimal path algorithms for undirected planar graphs with an additive and prefix monotone path function C. One of these algorithms can find C-optimal paths from a single source s to each node v and takes $O(n\sqrt{\log n})$ time; the complexity of this problem would be $\Theta(n^2)$ if $\mathcal{G}$ were not planar. The authors of [KRRS95] have given an linear time algorithm for planar graphs with an additive, prefix monotone cost function C.

The algorithms in [FeMS91] work as fast as the ones in [Fred87]. Moreover, Feuerstein's and Marchetti-Spaccamela's algorithms are very fast in the dynamic case: If a system of optimal s-v-paths is given and the weight $h(r)$ of one edge r is changed then the modified system of optimal paths is computed in $O(n\sqrt{\log \log n})$ time. More recent results about finding extremal paths in dynamic planar graphs are given in [DjPZ94].

In [MehS83], an algorithm for Problem 9 with additive cost measure C is given; C need not be prefix monotone The algorithm works in $O(n\sqrt{n}\log n)$ time. Hence, Mehlhorn's and Schmidt's algorithm is slower than Frederickson's one. On the other hand, Mehlhorn's and Schmidt's algorithm is more universal than Frederickson's because C need not be prefix monotone. Also, Mehlhorn's and Schmidt's search strategy may even be applied to particular non-planar graphs; this will be described in the Subsection 4.5.3 in this book. A more recent algorithm for additive cost measures with negative weights has been given in [KRRS95].

Problem 10 for planar graphs is solved in [Fred87] if the cost measure C is additive and prefix monotone; Frederickson's algorithm works in $O(n^2)$ time; this is better than Takaoka's time bound $O\left(n^3\sqrt{\frac{\log \log n}{\log n}}\right)$ for non-planar graphs (see [Taka91] and 4.170 b)). A further algorithm for problem 10 is presented in [Fred91]; it works correctly if C is additive and $\mathcal{G}$ does not contain C-negative cycles.

A parallel shortest path algorithm for planar, undirected networks can be found in [Fred90].

The authors of [BiMo90] present polynomial time algorithms to compute a embeddings for planar graphs such that particular measures of distance are minimized.

4.5.3   Path problems in graphs with a special structure or with a special application

This subsection is about path problems in graphs with special structural properties or in graphs symbolizing a special mathematical or practical situation.

We start with papers about optimal paths in graphs with a **special structure**.

The graphs in [MiRo80] are complete trees, and the distances between their leaves are computed.

The paper [ChZa95] is about searching for shortest paths in digraphs with a small treewidth.

The paper[HöWa95] describes the search for optimal paths in so-called periodic graphs.

The sources [MehS83], [PanR87], and [PanR89], give optimal path algorithms for graphs that have separators with special properties. (The definition of separators can be found in [MehS83] and in [PanR87], [PanR89].) In particular, Mehlhorn's and Schmidt's algorithm can be applied to many graphs that are not planar; this fact has been mentioned in Subsection 4.5.2.

The paper [Lawl89] describes shortest path algorithms for particular graphs that are recursively constructed; the algorithms use dynamic programming and branch-and-bound techniques. — The graphs in [Pros81] result likewise from a recursive process.

The papers [AChL93] and [Ravi92] give solutions to optimal path problems in interval graphs. An *interval graph* is an undirected graph $\mathcal{G} = (\mathcal{V}, \mathcal{R})$ with the following properties: Each node v represents an interval $I_v \subseteq \mathbb{R}$; if $\{v, w\} \in \mathcal{R}$ for two nodes v, w then $\{v, w\} \in \mathcal{R} \iff I_v \cap I_w \neq \emptyset$.

Shortest paths in so-called diagonal mesh networks are computed in [TaPa92].

The search for Hamiltonian paths in tournaments is investigated in [BaN90] and in [MaTu90]. The authors of [BJMa94] describe the search for a cycle through specified vertices in a tournament.

The paper [ShWN92] is about the search for approximate shortest paths in so-called level graphs.

The following papers are about optimal paths in graphs symbolizing a **special mathematical or practial problem**.

The graph in [Ran81] describes the behaviour of a finite automaton, and the graph in [Nije79] symbolizes a number theoretic problem. The paper [Hi92] is about finding shortest paths in graphs that describe a generalized Tower of Hanoi problem.

4.6 Parallel search algorithms for optimal paths

This section is about parallel search algorithms for optimal paths in graphs.

A fair comparison between the time complexities of parallel algorithms is often very difficult or even impossible. The reason is that there exist many architectures of (abstract) parallel computers. Therefore, we may compare the complexity of two algorithms only if we make sure that both algorithms use the same parallel computer.

4.6.1 Using computers without common memory

Here, we consider the search for optimal paths in graphs by parallel computers whose processing elements do not have a common memory. There are two version of processor networks: The first version is a network that is independent of the graph $\mathcal{G}$ in which optimal paths are searched for. The second version is a network that is isomorphic to $\mathcal{G}$.

We start with the first version. In [Pri81], a VLSI-network of n processors is described. This network finds a C-minimal s-v-path for each node v in a given acyclic digraph $\mathcal{G}$. The path function C is additive, negative weights of arcs are admissible. C.C.Price's algorithm is a parallel version of the search strategy FORD-BELLMAN 2 in Subsection 4.3.2. FORD-BELLMAN 2 and similar sequential Ford-Bellman algorithms require $\Theta(n^2)$ steps of computation; the parallel algorithm in [Pri81], however, works in $O(n)$ time and uses n processors.

We make two remarks about VLSI-algorithms for optimal paths:
- a) Probably, C.C.Price's algorithm can also be applied to a digraph $\mathcal{G}$ with cycles if $\mathcal{G}$ has a Bellman-Ford-enumeration.
- b) VLSI-networks can be used to find optimal v-w-paths for each pair (v, w) if the underlying digraph $\mathcal{G}$ is strongly connected. This has been demonstrated in the papers [LeKa72] and [GKTh79].
  It is easy to see that VLSI-networks are very good when solving Problem 10; this problem can be reduced to modified matrix multiplications, and these operations can easily be carried out with VLSI-networks.

A very fast algorithm for Problem 10 is given in [GhBh86]. The computer is a SIMD-network with $\Theta(n^3)$ processors. (This makes it obvious that the network is not isomorphic to the given digraph $\mathcal{G}$.) The time of computation is $O(\log d(\mathcal{G}) \cdot \log n)$ where $d(\mathcal{G})$ is the diameter of $\mathcal{G}$. The cost function C must be additive, and all arcs must be labelled with positive weights.

Next, we describe parallel algorithms for processor networks with the following property: The network is a copy of the graph $\mathcal{G}$ in which we are searching for the optimal path. The nodes of $\mathcal{G}$ are represented by the processing elements of the computer, and the edges or arcs of $\mathcal{G}$ are represented by directed connections from one processor to another one.

The networks of processors in [Chau87] and [ChMi82] are undirected versions of the given digraph $\mathcal{G}$; the parallel computer is organized as a MIMD-network. (MIMD = multiple data multiple instruction). Both papers [Chau87] and [ChMi82] give solutions of Problem 9 a).

The algorithm on page 220 – 222 in [Chau87] is easy to understand and works in $O(d(\mathcal{G}))$ time. The digraph $\mathcal{G}$ must be acyclic; moreover, C must be additive, and all arcs must have positive weights.

The algorithm in [ChMi82] is somewhat more complicated than the one in [Chau87]; moreover, Chandy and Misra do not estimate the time of computation. On the other hand, Chany's and Misra's algorithm has the following advantages: The digraph $\mathcal{G}$ may contain cycles, and a slight modification of the original algorithm can find C-minimal paths if C is not additive but order preserving; this modification is described in Section 6.2 of [ChMi82].

The paper [Hald93] gives a distributed algorithm to solve Problem 10 in undirected networks with an additive path function C.

Single source shortest path trees in planar, undirected networks can be constructed by the distributed algorithm in [Fred90].

The parallel computer of Praßler and Milios in [PrMi] is structured as a grid $\mathcal{G}'$, which represents and approximates the continuous and undirected graph $\mathcal{G}$. Praßler's and Milios' network is a SIMD-network. (SIMD = single instruction multiple data.) Further details about the paper [PrMi] were given on page 325 of this book.

4.6.2   Using computers with a common memory

This subsection is about the search for optimal paths with the help of computers whose processing elements have a common memory.

The algorithm OTO_{par} in [MoPa88] performs a parallel bidirectional search for an optimal s-γ-path. The search from s in direction to γ is carried out by a process $PROC_s$, and the search from γ in direction to s is done by a process $PROC_\gamma$; Both $PROC_s$ and $PROC_\gamma$ are similar to Algorithm A*.

The two processes run asynchronically. $PROC_s$ writes its intermediate s-v-paths into the common memory and reads the v-γ-paths found by $PROC_\gamma$; this process writes its v-γ-paths into the common memory and reads the s-v-paths found by $PROC_s$. In particular cases, $PROC_s$ and $PROC_\gamma$ may merge an s-v-path and a v-γ-path to an s-γ-path.

The papers [PanR87] and [PanR89] describe parallel algorithms for the problems 9 b) and 10 b). Pan and Reif use several versions of PRAM's (parallel random access machines) as models of computing. The algorithms make use of separator properties of the given graph $\mathcal{G}$.

The solution of Problem 10 a) with a CREW-PRAM is described in [MTM94]. (A CREW-PRAM is a parallel random access machine with concurrent reading and exclusive writing in the common memory.) The authors of [MTM94] achieve a time of computation in $O(n^3/p)$ where n is the number of nodes of the graph and p is the number of processors.

The articles [And84] and [AnMa87] describe the search for maximal paths with the help of CREW-PRAM's. (Here, a *maximal path* is an acyclic path P with the property that each path of the form $P \oplus r$ with $r \in \mathcal{R}$ contains a substantial cycle.) The authors of [AnMa87] study the search for lexicographically minimal candidates among all maximal paths.

The search for Hamiltonian cycles in tournaments is described in [BaN90]; the model of computation is a CRCW-PRAM (CRCW = concurrent read, concurrent write).

The paper [Frie87] is addressed to parallel search algorithms for Hamiltonian cycles in random graphs. The model of computation is an EREW-PRAM. (An EREW-PRAM is a parallel random access machine with exclusive reading and exclusive writing in the shared memory.)

4.7 Searching for optimal paths in random graphs

This section is about the search for optimal paths in graphs whose nodes, arcs, edges, or cost functions are generated by a random process. This makes the difference between this section and the other parts of the book, which are mainly about optimal paths in deterministically chosen graphs.

We do not give experimental results in this section; all quoted results about average complexities of algorithms are obtained with the help of mathematical proofs and not with the help of experiments.

We assume that only the graphs and the cost functions but not the algorithms are chosen by a random process; once the input graph has been randomly generated, the optimal path algorithm works deterministically.

When writing random variables as $f : A \to B$, we assume that σ-algebras are given for A and B with the property that f is measurable.

Now we consider the following problem; we shall see that it is closely related to standard optimal path problems.

Problem 19 *Given a finite, simple digraph $\mathcal{G} = (\mathcal{V}, \mathcal{R})$, a start node $s \in \mathcal{V}$, and a set $\Gamma \subseteq \mathcal{V}$ of goal nodes. Let Ω be the set of all arc functions $h : \mathcal{R} \to [0, \infty)$, and let $\mathbf{p}$ be a probability distribution of the elements $h \in \Omega$. For each path $P \in \mathcal{P}(s)$, we define the random variable $C_P : \Omega \to \mathbb{R}$ as follows:*

If $P = [s]$ then $C_P(h) := 0$ for all $h \in \Omega$.
If $P = r_1 \oplus \ldots \oplus r_k$ then $C_P(h) := h(r_1) + \ldots + h(r_k)$ for all $h \in \Omega$.
Moreover, let $c : \mathbb{R} \to \mathbb{R}$ be a monotonically increasing function, and let
$$C^{(h)}(P) := c \circ C_P(h) = c(C_P(h))$$
be the cost of P (depending on h, which has been randomly chosen).

We define the path function $C : \mathcal{P}(s) \to \mathbb{R}$ such that $C(P)$ is the expected cost of P where $h \in \Omega$. More formally,

$$(\forall P \in \mathcal{P}(s)) \quad C(P) := E_{h \in \Omega}\left(C^{(h)}(P)\right) = E_{h \in \Omega}(c \circ C_P(h)) = \int_\Omega c(C_P(h)) \cdot \mathbf{p}(dh).$$

Find a C-minimal path P^* from s to Γ.

This problem is almost the same as Problem 1. In particular, the cost function C is given because C can be computed from $\mathbf{p}$ and c. Therefore, Problem 19 means to find a C-minimal path where C is a given cost function. The only difference to Problem 19 is that C is computed as an expected cost value.
The path function C is not always additive because the function c may make C be non-additive.

A modified version of Problem 19 has been formulated in [EiMS85]. The main differences between the problem in [EiMS85] and Problem 19 are the following: The authors of [EiMS85] use a monotonically decreasing function u instead of an monotonically increasing function c; they define an utility function $U^{(h)}(P) := u(C_P(h))$ instead of a cost function $C^{(h)}(P) := c(C_P(h))$; they maximize the expected utility $U(P) := E_{h \in \Omega}(u \circ C_P)$ instead of minimizing the expected cost $C(P)$.
The following transformations are helpful to understand [EiMS85].

Remark 4.174 Given the terminology of Problem 19.

Moreover, let $u : \mathbb{R} \to \mathbb{R}$ be a monotonically decreasing function, and let

$$U^{(h)}(P) := u \circ C_P(h) = u(C_P(h))$$

be the utility of P (depending on h, which is randomly chosen).
We define the path function $U : \mathcal{P}(s) \to \mathbb{R}$ such that $U(P)$ is the expected utility of P where $h \in \Omega$. More formally,

$$(\forall P \in \mathcal{P}(s)) \quad U(P) := E_{h \in \Omega}\left(U^{(h)}(P)\right) = E_{h \in \Omega}(u \circ C_P(h)) = \int_\Omega u(C_P(h)) \cdot \mathbf{p}(dh).$$

Next, we transform the integrals in the definition of $C(P)$ and $U(P)$.

For this purpose, we define $\mathbf{p}_{C_P}$ as the probability distribution on $\mathbb{R}$ that is induced by C_P. That means that $\mathbf{p}_{C_P}(X) := \mathbf{p}(\{h \in \Omega \,|\, C_P(h) \in X\})$ for all measurable subsets X of $\mathbb{R}$.
Let then $f_P : \mathbb{R} \to \mathbb{R}$ be the probability distribution function of $\mathbf{p}_{C_P}$; that means that

$$(\forall\, x < y) \quad \mathbf{p}\big(x \leq C_P(h) \leq y\big) = \mathbf{p}_{C_P}([x, y]) = \int_x^y w \cdot f_P(w) \cdot dw.$$

Then we obtain two equations about $C(P)$ and $U(P)$; the correctness of these equations is proven in [Bau74] and many other books on Probability Theory.

$$C(P) := \int_\Omega c(C_P(h)) \cdot \mathbf{p}(dh) = \int_{\mathbb{R}} c(w) \cdot \mathbf{p}_{C_P}(dw) = \int_{\mathbb{R}} c(w) \cdot f_P(w) \cdot dw.$$

$$U(P) := \int_\Omega u(C_P(h)) \cdot \mathbf{p}(dh) = \int_{\mathbb{R}} u(w) \cdot \mathbf{p}_{C_P}(dw) = \int_{\mathbb{R}} u(w) \cdot f_P(w) \cdot dw. \quad \square$$

The next problem is similar to Problem 19 and arises from [DeWK79] and [Trie].

Problem 20 *Given the same notations as in Problem 19. We assume that Γ has a single element γ and that the function c is continuous.*
Let the random variable $M : \Omega \to [0, \infty)$ be defined such that $M(h)$ is the the infimum of all path costs $C_P(h)$ where $P \in \mathcal{P}(s, \gamma)$; more formally, $M(h) := \inf_{P \in \mathcal{P}(s,\gamma)} C_P(h)$ for all $h \in \Omega$.
Compute the expectation of the infimum costs $c \circ M$; that means, compute the value $E_{h \in \Omega}(c(M(h))) = \int_{h \in \Omega} c(M(h)) \cdot \mathbf{p}(dh)$.

Remark 4.175 Given the terminology of Problems 19 and 20. Then the following equation is valid where $(*)$ follows from the the assumption that c is monotonically increasing and continuous. For all $h \in \Omega$,

$$c(M(h)) = c \left(\inf_{P \in \mathcal{P}(s,\gamma)} C_P(h) \right) \stackrel{(*)}{=} \inf_{P \in \mathcal{P}(s,\gamma)} c(C_P(h)) = \inf_{P \in \mathcal{P}(s,\gamma)} C^{(h)}(P). \quad (4.185)$$

An immediate consequence of this equation is the following representation of the expectation that must be computed in Problem 20:

$$E_{h \in \Omega}(c(M(h))) = \int_{h \in \Omega} c(M(h)) \cdot \mathbf{p}(dh) = \int_{h \in \Omega} c \left(\inf_{P \in \mathcal{P}(s,\gamma)} C_P(h) \right) \cdot \mathbf{p}(dh) =$$

$$\int_{h \in \Omega} \left(\inf_{P \in \mathcal{P}(s,\gamma)} c(C_P(h)) \right) \cdot \mathbf{p}(dh) = \int_{h \in \Omega} \left(\inf_{P \in \mathcal{P}(s,\gamma)} C^{(h)}(P) \right) \cdot \mathbf{p}(dh). \qquad \square$$

At first sight, there is no substantial difference between the problems 19 and 20. In the first case, we compute the infimum of all expectations of $c \circ C_P(h) = C^{(h)}(P)$, and in the second case, we compute the expectation of all infima of $c \circ C_P(h) = C^{(h)}(P)$. In reality, however, the order of expectation and infimum must not be changed. This is shown in the following result.

Theorem 4.176 *Given the terminology of Problems 19 and 20; in particular, let c be monotinically increasing and continuous. Let P^* be a C-minimal s-γ-path. Then the following assertions are true:*

a) $E_{h \in \Omega}(c(M(h))) = E_{h \in \Omega} \left(\inf_{P \in \mathcal{P}(s,\gamma)} C^{(h)}(P) \right)$, *and*
$\inf_{P \in \mathcal{P}(s,\gamma)} \left(E_{h \in \Omega} (C^{(h)}(P)) \right) = C(P^*) = E_{h \in \Omega}(c(C_{P^*}(h)))$.

b) $E_{h \in \Omega}(c(M(h))) \leq E_{h \in \Omega}(c(C_{P^*}(h))) = C(P^*)$.

c) $E_{h \in \Omega} \left(\inf_{P \in \mathcal{P}(s,\gamma)} C^{(h)}(P) \right) \leq \inf_{P \in \mathcal{P}(s,\gamma)} \left(E_{h \in \Omega} (C^{(h)}(P)) \right)$.

d) *It may occur that* $E_{h \in \Omega}(c(M(h))) < E_{h \in \Omega}(c(C_{P^*}(h)) = C(P^*)$ *and*
$E_{h \in \Omega} \left(\inf_{P \in \mathcal{P}(s,\gamma)} C^{(h)}(P) \right) < \inf_{P \in \mathcal{P}(s,\gamma)} \left(E_{h \in \Omega} (C^{(h)}(P)) \right)$.

P r o o f o f a) : The first claim follows from fact (4.185).

The second claim is a consequence of the next equation; relationship $(*)$ follows from the definition of C, relationship $(**)$ follows from the minimality of P^*, and relationship $(***)$ follows from the definition of C and $C^{(h)}$.

$$\inf_{P \in \mathcal{P}(s,\gamma)} \left(E_{h \in \Omega}\left(C^{(h)}(P)\right)\right) \overset{(*)}{=} \inf_{P \in \mathcal{P}(s,\gamma)} C(P) \overset{(**)}{=} C(P^*) \overset{(***)}{=} E_{h \in \Omega}(c(C_{P^*}(h))).$$

P r o o f o f b) : If $h \in \Omega$ then $C_{P^*}(h)$ is one of the candidates for the infimum in the definition of $M(h)$; consequently, $M(h) = \inf_{P \in \mathcal{P}(s,\gamma)} C_P(h) \leq C_{P^*}(h)$. This and the monotonicity of c imply that $c(M(h)) \leq c(C_{P^*}(h))$ for all $h \in \Omega$. This fact and the monotonicity of the expectation are used in relation $(*)$ of the following inequality:

$$E_{h \in \Omega}(c(M(h))) = \int_\Omega c(M(h)) \cdot \mathbf{p}(dh) \overset{(*)}{\leq} \int_\Omega c(C_{P^*}(h)) \cdot \mathbf{p}(dh) = E_{h \in \Omega}(c(C_{P^*}(h))) \overset{a)}{=} C(P^*).$$

P r o o f o f c) : The assertion follows immediately from the parts a) and b) of this theorem.

P r o o f o f d) : Let $\mathcal{G} = (\mathcal{V}, R)$ be the digraph with $\mathcal{V} := \{s, x, \gamma\}$ and $\mathcal{R} := \{(s,x), (x,\gamma), (s,\gamma)\}$. Then the only s-γ-paths in $\mathcal{G}$ are $P_0 := [s, x, \gamma]$ and $P_1 := [s, \gamma]$.

We define the probability distribution $\mathbf{p}$ such that each triple $(h(s,x), h(x,\gamma), h(s,\gamma)) \in \{0,1\}^3$ has the probability $1/8$. Let c be the identity on $\mathbb{R}$.

We now compute $E_{h \in \Omega}(c(M(h))) = E_{h \in \Omega}(M(h))$. For all $h \in \Omega$, the inequality $M(h) \leq C_{P_1}(h) = h(s,\gamma) \leq 1$ is valid; consequently, $M(h) \in \{0,1\}$. The minimum $M(h)$ of $C_{P_0}(h)$ and $C_{P_1}(h)$ is equal to 1 if and only if h satisfies each of the following conditions (α) and (β):

(α) $h(s,x) = 1$ or $h(x,\gamma) = 1$. (β) $h(s,\gamma) = 1$.

These two assertions are true if and only if

$$(h(s,x), h(x,\gamma), h(s,\gamma)) \in \{(0,1,1), (1,0,1), (1,1,1)\}.$$

Therefore, $M(h) = 1$ with probability $3/8$, and $M(h) = 0$ with probability $5/8$. Consequently, $E_{h \in \Omega}(c(M(h))) = E_{h \in \Omega}(M(h)) = 3/8$.

Now, we compute $C(P^*) = E_{h \in \Omega}(c(C_{P^*}(h))) = E_{h \in \Omega}(C_{P^*}(h))$. The definition of P^* and the equation $\mathcal{P}(s,\gamma) = \{P_0, P_1\}$ imply that

$$C(P^*) = \min\{C(P) \mid P \in \mathcal{P}(s,\gamma)\} = \min\{C(P_0), C(P_1)\}. \tag{4.186}$$

Moreover, the definitions of C and c have the following consequence:

$$(\forall i \in \{0,1\})\ C(P_i) = E_{h \in \Omega}(c(C_{P_i}(h))) = E_{h \in \Omega}(C_{P_i}(h)). \tag{4.187}$$

We compute $C(P_0)$. For this purpose, we observe that $C_{P_0}(h) = h(s,x) + h(s,\gamma)$ for all h. This implies that

$$C_{P_0}(h) = \begin{cases} 0 & \text{if } (h(s,x), h(x,\gamma), h(s,\gamma)) \in \{(0,0,0), (0,0,1)\}, \\ 1 & \text{if } (h(s,x), h(x,\gamma), h(s,\gamma)) \in \{(1,0,0), (0,1,0), (1,0,1), (0,1,1)\}, \\ 2 & \text{if } (h(s,x), h(x,\gamma), h(s,\gamma)) \in \{(1,1,0), (1,1,1)\}. \end{cases}$$

Consequently, $C(P_0) \overset{(4.187)}{=} E_{h \in \Omega}(C_{P_0}(h)) = 0 \cdot \dfrac{1}{4} + 1 \cdot \dfrac{1}{2} + 2 \cdot \dfrac{1}{4} = 1.$

Next, we compute $C(P_1)$. Let $h \in \Omega$; then $C_{P_1}(h) = 0$ if $h(s, \gamma) = 0$, and $C_{P_1}(h) = 1$ if $h(s, \gamma) = 1$. So, $C_{P_1}(h)$ attains each value 0 and 1 with probability $1/2$. Therefore,

$$C(P_1) \overset{(4.187)}{=} E_{h \in \Omega}(C_{P_1}(h)) = 0 \cdot \frac{1}{2} + 1 \cdot \frac{1}{2} = \frac{1}{2}.$$

We have seen that $C(P_0) = 1$ and $C(P_1) = \frac{1}{2}$. This and (4.186) imply that $P^* = P_1$. Consequently, $C(P^*) = E_{h \in \Omega}(C_{P^*}(h)) = \frac{1}{2}$. Moreover, $E_{h \in \Omega}(c(M(h))) = \frac{3}{8} < \frac{1}{2} = E_{h \in \Omega}(C_{P^*}(h)) = C(P^*)$; that means that the first part of assertion d) has been proven.

The relation $E_{h \in \Omega}(c(M(h))) < E_{h \in \Omega}(c(C_{P^*}(h))$ is used in the following inequality, which proves the second part of assertion d):

$$E_{h \in \Omega} \left(\inf_{P \in \mathcal{P}(s, \gamma)} C^{(h)}(P) \right) \overset{a)}{=} E_{h \in \Omega}(c(M(h))) <$$

$$E_{h \in \Omega}(c(C_{P^*}(h)) \overset{a)}{=} \inf_{P \in \mathcal{P}(s, \gamma)} \left(E_{h \in \Omega}\left(C^{(h)}(P)\right)\right). \qquad \square$$

A more complicated version of Problem 20 is solved in [Kulk86], [HaSh91] and in [CoKu93] where not only the expectation but even the exact distribution of optimal path lengths are computed.

Next, we describe a further modification of Problem 19.

Problem 21 *Let **A** be a class of search strategies for paths in digraphs. The input of each algorithm $\mathcal{A} \in \mathbf{A}$ is a graph $\mathcal{G}$, a start node s, a set Γ of goal nodes, and a cost function C. The function C (and possibly $\mathcal{G}$, s, or Γ) is generated by a random process.*
We assume that each algorithm $\mathcal{A} \in \mathbf{A}$ outputs an s-Γ-path $P^*(\mathcal{A}, C)$ if the cost measure C is given; we do not assume that this path $P^*(\mathcal{A}, C)$ is always C-minimal. Let $\Omega(\mathcal{A})$ be the set of all path functions C for which $P^*(\mathcal{A}, C)$ is actually C-minimal; more formally,

$$\Omega(\mathcal{A}) := \{C \in \Omega \mid (\forall P \in \mathcal{P}(s, \Gamma)) \; C(P^*(\mathcal{A}, C)) \leq C(P)\}.$$

Find a search strategy $\mathcal{A}^* \in \mathbf{A}$ with the following property:
The probability that $\mathcal{A}$ finds a C-optimal s-γ-path is maximal; that means that $\mathbf{p}(\Omega(\mathcal{A}^*)) \geq \mathbf{p}(\Omega(\mathcal{A}))$ for all $\mathcal{A} \in \mathbf{A}$.

It is not clear whether this problem has been formulated in literature. But it is almost sure that most problems of optimal stopping can be described in this way. For example, the secretary problem can be considered as a special case of Problem 21. The secretary problem has been investigated in [Lor79], [Lor81], [SmDe75], and many other sources. It can be formulated as follows:

Problem 22 (Secretary problem)
Suppose that n secretaries $S_1, \ldots, S_n$ are applying for a job. For all ν, let r_ν be the rank of S_ν, i.e., S_ν is the r_ν-best candidate. We assume that the probability of each permutation $(r_1, \ldots, r_n)$ equals $1/n!$.
The personnel manager interviews the secretaries S_1, S_2, S_3, and so on. After interviewing S_ν, $\nu < n$, he may engage S_ν without considering $S_{\nu+1}, S_{\nu+2}, \ldots, S_n$; or he may reject S_ν and interview $S_{\nu+1}$; in this case, he must not reconsider S_ν later. If the personnel manager has rejected S_{n-1} he must give the job to S_n.
Let $\mathbf{D}$ be the set of all admissible decision policies of the personnel manager.

Find a strategy $\mathcal{D}^* \in \mathbf{D}$ such that the personnel manager will find the actually best secretary with maximum likelihood.

We symbolize the secretary problem with the help of the digraph $\mathcal{G} = (\mathcal{V}, \mathcal{R})$ with

$$\mathcal{V} := \{x_1, x_2, x_3, \ldots, x_n, \ S_1, \ldots, S_n\} \text{ and}$$
$$\mathcal{R} := \{(x_i, x_{i+1}) \,|\, i = 1, \ldots, n-1\} \cup \{(x_i, S_i) \,|\, i = 1, \ldots, n\}$$

Let $s := x_1$ and $\Gamma := \{S_1, \ldots S_n\}$.

For all $\nu \in \{1, \ldots, n\}$, we represent the decision for S_ν by the path $P_\nu := [s = x_1, x_2, \ldots, x_\nu, S_\nu]$. We define the cost $C(P_\nu) := r_\nu$.[48] Then each decision strategy $\mathcal{D} \in \mathbf{D}$ for a secretary corresponds to a search strategy $\mathcal{A}$ for an s-Γ-path in $\mathcal{G}$. Consequently, maximizing the probability of finding the best secretary means to maximize the probability of finding the C-minimal s-Γ-path. That means that Problem 22 is indeed a special case of the graph theoretic problem 21.

A further important problem is to compute minimal path costs in a short average time; when estimating this time complexity we choose the input graph for an algorithm by a random process and compute the expected running time. Here, we focus on the following version of Problem 10 b):

Problem 23 *Given the digraph $\mathcal{G} = (\mathcal{V}, \mathcal{R})$ with $\mathcal{R} = (\mathcal{V} \times \mathcal{V}) \setminus \{(v, v) \,|\, v \in \mathcal{V}\}$.*
Let $\mathbf{p}$ be a probability measure on $[0, \infty)$. We assume that for each arc r, the weight $h(r)$ is randomly chosen according to $\mathbf{p}$; that means that for all $r \in \mathcal{R}$ and for all measurable sets $X \subseteq [0, \infty)$, the probability that $h(r) \in X$ is equal to $\mathbf{p}(X)$. Moreover, we assume that the values $h(r)$, $r \in \mathcal{R}$, are stochastically independent. Let $C^{(h)} := SUM_h$ for all arc functions $h : \mathcal{R} \to [0, \infty)$.

If h is given, find the minimal $C^{(h)}$-value of all v-w-paths for each pair $(v, w) \in \mathcal{V} \times \mathcal{V}$. Moreover, use a search strategy with a good average time complexity.

One of the fastest algorithm for this problem is given in [MoTa87]; Moffat's and Takaoka's search strategy works in average time $O(n^2 \log n)$. The stochastic independence of the values $h(r)$, $r \in \mathcal{R}$, may even be relaxed to the so-called end-point independence, which is described in [MoTa87]. The algorithms in [MoTa87] have often a better average time behaviour than the ones in [TaMo80], [Blon83], and [FrGr85].

---

[48] Recall that the ranks r_ν are randomly chosen; so, the same is true for C.

By the way, the average time bound $O(n^2 \log n)$ in [MoTa87] is better than the worstcase time bound $O\left(n^3 \cdot \sqrt{\frac{\log n}{\log \log n}}\right)$ in [Taka91] (see Remark 4.170).

Most of the search strategies in [TaMo80], [Blon83], [FrGr85], and in [MoTa87] are based on the following principles:

- *Decomposing the problem into n single source problems:*
  For each start node $v \in \mathcal{V}$, the $C^{(h)}$-values of the best v-w paths ($w \in \mathcal{V}$) are computed.

- *Using fast data structures:*
  The algorithms use data structures that can be updated very fast when going on from one start node v to the next one.

- *Ignoring irrelevant arcs:*
  Solving a single source path problem with Dijkstra's algorithm would take $\Theta(n^2)$ time for each start node v; the reason is that this algorithm processes each arc $r \in \mathcal{R}$. The algorithms in in [Blon83], [FrGr85] and [MoTa87], however, may ignore particular arcs that do certainly not belong to an optimal path. Therefore, the running time of these algorithms depends on the random path function $C^{(h)}$, and it is very likely that this time is substantially shorter than $\Theta(n^2)$ for the given start node v. Thus, the average running time can be drastically reduced. For example, Moffat and Takaoka achieve $O(n \log n)$ time for the single source problem (see [MoTa87], page 1030).

In [LuRa89], the following version of Problem 23 is solved: Let $s, \gamma \in \mathcal{V}$, and let the values $h(r) \geq 0$ be independently chosen from an exponential distribution; find an optimal s-γ-path. The authors of the paper introduce a bidirectional search strategy with a short expected time of computation.

In general, if the average time of an algorithm is better than its worstcase time, the algorithm must be adaptive; that means that the running time depends on the input.

Next, we describe several further path problems and solutions in the context of random graphs.

In [DeWK79], the average costs of optimal paths in random graphs are investigated; moreover, an average case analysis of a particular shortest path algorithm is given; it turns out that this search strategy has a very good average time complexity.

In [SeVi86], an algorithm is given that only requires $O(n)$ average time to compute an optimal s-γ-path for a given pair (s, γ) of vertices; Sedgewick and Vitter assume that the nodes of the graph are randomly chosen in a Euclidean space, and the weights $h(r)$, $r \in \mathcal{R}$ are equal to the Euclidean distances between the endpoints of r.

The papers [KaPe83], [Pe83], and [BagS88] are about the following situation: The graph $\mathcal{G}$ is a tree with root s and with a set Γ of goal nodes; the cardinality of Γ is very great. The problem is to find an optimal s-Γ-path. In [KaPe83], the weights of the arcs are randomly chosen in the set $\{0, 1\}$, and in [Pe83], [BagS88], the heuristic function is found by a random process. All these articles analyse the average case complexity of optimal path algorithms. — A further article about paths in random trees is [Dav90].

In [NoMM78], the behaviour of the Dijkstra-Algorithm is studied; in particular, Noshita, Masuda, and Machida compute the probability $p(v, t)$ of the event that the current s-v-path $P(v)$ is redirected in the t-th iteration (see [NoMM78], page 241).

Hamiltonian cycles in random graphs are considered in [Frie87].

The paper [SpTs86] estimates the average time of finding cycles with negative weights in random graphs.

The papers [Bal87], [Bal90], and [Bal93] are about searching for an unknown (goal) point $x \in \mathbb{R}$; this point x is generated by a random mechanism. Let $T(x)$ be the time that a given search strategy needs to find x. The author of [Bal87], [Bal90], [Bal93] develops search strategies for which the expected time $E_{x \in \mathbb{R}}(T(x))$ is very short. The search problem in these three papers is similar to the following optimal path problem in digraphs:

Problem 24 *Let $\mathcal{G} = (\mathcal{V}, \mathcal{R})$ be a digraph with a start node s and a fixed path function $C : \mathcal{P}(s) \to \mathbb{R}$. We assume that a random process generates a goal set $\Gamma \subseteq \mathcal{V}$ (or a single goal node $\gamma \in \mathcal{V}$). For each algorithm $\mathcal{A}$ and each set $\Gamma \subseteq \mathcal{V}$, let $T(\mathcal{A}, \Gamma)$ be the time consumed by $\mathcal{A}$ to find a C-minimal s-Γ-path[49].*
Find a search strategy $\mathcal{A}^*$ with a minimal average time; that means that the the expected time $E_{\Gamma \subseteq \mathcal{V}}(T(\mathcal{A}, \Gamma))$ shall be minimal if $\mathcal{A} = \mathcal{A}^*$.

The problem in [EaY90] is essentially the following:

Problem 25 *Given a network of cells and a number l. Suppose that a target uses a path $Q = [j_1, \ldots, j_l]$, and searcher uses a path $P = [i_1, \ldots, i_k]$; then we say that the searcher finds the target if $i_\lambda = j_\lambda$ for some λ. Let Q be generated by a random process.*
Determine a path P^* such that the searcher finds the target with maximal likelihood if he uses the path P^*.

4.8   Searching for almost optimal paths in graphs

In this section, we describe several papers about the search for almost optimal paths in graphs. We consider a path P as "almost optimal" if it has one of the following properties:

- The cost of P is a good approximation to the minimal cost.

- The path P is the second best candidate or the third best candidate, and so on.

---

[49]We only consider search strategies that do not know the set Γ in advance; however, the search strategy can decide whether $v \in \Gamma$ when processing a node v.

4.8.1 Searching for paths with approximately minimal costs

This subsection is about the following problem:

Problem 26 *Given a directed or undirected graph G, two nodes s,γ, a path function C, and an $\varepsilon > 0$. We assume that there exists a C-minimal s-γ-path P^*.*

Find an s-γ-path P^+ with one of the properties A), B), or with a similar property.

A) *The absolute deviation is smaller than ε, i.e.,* $0 \leq C(P) - C(P^*) < \varepsilon$.

B) *The relative deviation is smaller than ε, i.e.,* $0 \leq \dfrac{C(P) - C(P^*)}{C(P^*)} < \varepsilon$.

This problem is obviously easier than the search for a really optimal path P^*. Solutions to this problem are given in [ItRo78], [BoLi81], and [Kova88]; the search strategy in [BoLi81] is a matrix algorithm. The article [CoGi92] describes the search for a short path in a scene of rectangles if only local information is available.

The search for almost optimal paths in so-called level graphs is described in [ShWN92].

An approximately optimal path can often be found very fast by applying Algorithm A with a non-admissible heuristic function; this situation is studied in [DePe85], page 512, and in [CGPS89]; the authors of [DePe85] use the suboptimal path to find a really optimal path in a further stage of their algorithm.

A complicated version of Problem 26 is to find all s-γ-paths with property A) or all s-γ-paths with property B). This kind of search problem is even more difficult than the search for one really optimal candidate because finding all approximately optimal paths includes finding all optimal ones. In [Pete83], a matrix algorithm is given that solves this kind of problems.

4.8.2 Searching for the K best paths

In this subsection, we consider the search for the second best path, the third best path, and so on.

Problem 27 *Given a directed or undirected graph $G = (V, \mathcal{R})$ with n nodes and m arcs or edges. Let s, γ be two nodes in G. Let $\mathcal{P} := \mathcal{P}(s,\gamma)$ or $\mathcal{P} := \mathcal{P}^+(s,\gamma)$ where $\mathcal{P}^+(s,\gamma)$ denotes the set of all acyclic s-γ-paths. Let $C = SUM_h$ be additive where $h(r) \geq 0$ for all $r \in \mathcal{R}$.*

Find the K paths $P_1^*, \ldots, P_K^* \in \mathcal{P}$ with the lowest C-values; or find the Kth best path P_K^*.

Next, we show that the formulation of Problem 27 is not clear if the restriction $C\big|_{\mathcal{P}(s,\gamma)}$ is not injective.

Example 4.177 Let $\mathcal{P}(s,\gamma) = \{P_1, \ldots, P_4\}$, and let $C(P_1) = 0$, $C(P_2) = C(P_3) = 2$, $C(P_4) = 3$.

Then the second best path is not uniquely determined because there is a tie between $C(P_2)$ and $C(P_3)$.

The third best path is not uniquely determined, too, because three paths may be considered as the third best candidate: the path P_2 because $C(P_1) \le C(P_3) \le C(P_2)$; the path the path P_3 because $C(P_1) \le C(P_2) \le C(P_3)$; the path P_4 because P_4 has the third best C-value.

☐

All these different interpretations of Problem 27 can be avoided if assuming that $C|_{\mathcal{P}(s,\gamma)}$ is injective.[50]

The solution of Problem 27 may depend on whether we are searching for the Kth best of all s-γ-paths (i.e. $\mathcal{P} = \mathcal{P}(s,\gamma)$) or for the Kth best of all acyclic s-γ-paths (i.e. $\mathcal{P} = \mathcal{P}^+(s,\gamma)$).

If $K = 1$ then the search for a Kth best path in $\mathcal{P}^+(s,\gamma)$ automatically yields a Kth best path in $\mathcal{P}(s,\gamma)$; the reason is that C is prefix monotone so that each C-minimal acyclic path is also a C-minimal path.

If, however, $K = 2$ then the Kth best s-γ-path cannot be found by searching for the Kth best acyclic s-γ-path. For example, the following situation may occur: The path P_κ, $\kappa = 1, 2, 3$, is the κth best s-γ-path; P_1 and P_3 are acyclic but P_2 is not acyclic. Then the second best s-γ-path is P_2, but the second best acyclic s-γ-path is P_3.

First, we consider the case that $\mathcal{P} = \mathcal{P}(s,\gamma)$.

In [Fox78], an algorithm is given to find the K best s-γ-paths in an undirected graph $\mathcal{G}$ in $O(rn + Kn \log r)$ time where r is the maximum degree of all nodes. If $r \approx n$ then the time of computation is $O(n^2 + Kn \log n)$.

The paper [SkGo89] gives solutions of Problem 27 for undirected and for directed graphs.

In [Rote90], page 183 – 184, a matrix algorithm is given to find the K minimal cost values of all s-γ-paths. Rote also considers the situation that $C|_{\mathcal{P}(s,\gamma)}$ is not injective. For example, if $K = 3$ and if Rote's algorithm is applied to Example 4.177 then the algorithm will output $C(P_1) = 0$ as the best C-value, $C(P_2) = 2$ as the second best C-value, and $C(P_4) = 3$ as the third best C-value.

In [EvMi92], the following problem is solved: Let s be a start node; find the K best s-v-path for each node v.

Next, we consider the case that $\mathcal{P} = \mathcal{P}^+(s,\gamma)$.

This case is considered in the book [Chri75]. In [KaIM78], an algorithm is given that finds the K best s-γ-paths in an undirected graph in $O(K \cdot n^2)$ time. This complexity is reduced to $O(\min(K \cdot n^2, K \cdot m \log n))$ in [KaIM82].

In [SkGo89], an algorithm is given that finds the K best acyclic s-γ-paths in undirected and in directed graphs.

---

[50] We do not say that this assumption has been made in the following sources of literature.

The paper [Per86] describes several experimental results about algorithms that find the K best acyclic s-γ-paths in directed graphs.

In [MTZC81], an algorithm is given that works in $O(n \log^2 n)$ time and finds the path of the Kth highest (and not of the Kth lowest) cost in an undirected tree.

The Kth quickest acyclic path is constructed in [Chen94]; the quickest path problem was formulated as Problem 8 on page 266.

In [SuTa84], a problem is solved that is similar to Problem 27 for $K = 2$:

Problem 28 *Given a digraph $\mathcal{G}$, two nodes s, γ, and an additive cost measure $C = SUM_h$ with $h(r) \geq 0$ for all arcs r.*

Find two s-γ-paths P^* and Q^* with the following properties:

- *P^* and Q^* are arc disjoint; that means that both paths have no common arcs.*
- *$C(P^*) + C(Q^*)$ is the minimum of all sums $C(P) + C(Q)$ where P and Q are arc disjoint s-γ-paths.*

If the best s-γ-path P_1^* and the second best s-γ-path P_2^* are arc disjoint then the solution of Problem 27 with $K = 2$ yields a solution of Problem 28. If, however, P_1^* and P_2^* have an arc in common then it is possible that $C(P^*) + C(Q^*)$ is minimal but neither P^* nor Q^* is a C-minimal s-γ-path.
The algorithm in [SuTa84] finds an optimal pair of paths P^* and Q^* in $O(n^2 \log n)$ time.

A modification of Problem 28 is solved in [LMCSL]. The problem is to find two arc disjoint s-γ-paths P and Q such that $\max(C(P), C(Q))$ is minimized. The following problem is studied in [WuMa92]: Given an undirected graph $\mathcal{G} = (\mathcal{V}, \mathcal{R})$. Find pairwise disjoint sets $\{u_1, v_1\}, \ldots, \{u_k, v_k\}$ and pairwise edge-disjoint u_κ-v_κ-paths P_κ $(\kappa = 1, \ldots, k)$ such that $\bigcup_{\kappa=1}^{k} \{u_\kappa, v_\kappa\} = \mathcal{V}$ and the maximal cost of all paths P_κ is minimized.

General versions of Problem 28 are considered in [LMS92] for directed and for undirected graphs; the authors study the search for $k \geq 2$ disjoint paths such that the total costs of these paths are minimized.

The complexity of finding edge disjoint paths with given start and end nodes is investigated in [Fra90]; the underlying graph is undirected.

The paper [deKo91] is about the search for k edge-disjoint Hamiltonian tours with minimal total cost.

The paper [KeW93] is about searching for an optimal pair of paths in a geometric scene.

All sources of literature in this section were about the case that the cost measure C is additive. It seems to be an interesting problem to find search algorithms for almost C-optimal paths where C is not additive.

4.9 Paths with extremal values $C_1(P)/C_2(P)$ where C_1, C_2 have a simple structure

This section is mainly about the following modification of Problem 9 on page 267:

Problem 29 *Given a digraph $\mathcal{G} = (\mathcal{V}, \mathcal{R})$ and a node $s \in \mathcal{G}$; let $\mathcal{P}(s,v) \neq \emptyset$ for all $v \in \mathcal{V}$. Let $C : \mathcal{P}(\mathcal{G}) \to \mathbb{R}$ be a path function with the following structure: There exist two additive path functions C_1 and C_2 such that $C(P) = \dfrac{C_1(P)}{C_2(P)}$ for all paths P of length > 0.*

Find a C-minimal or a C-maximal s-v-path for each node v.

One of the most interesting special case is that $C_1 = SUM_{h_1}$ and $C_2 = \ell$. Then $C(r_1 \oplus \cdots \oplus r_l) = \big(h_1(r_1) + \ldots + h_1(r_l)\big)/l$ for all paths $P = r_1 \oplus \ldots \oplus r_l$; consequently, $C(P)$ is the average h_1-value of all arcs along P.

It has been shown in Remark 2.20 that $C = C_1/C_2$ is not always $=$-preserving even if $C_2 = \ell$. Consequently, C is not always order preserving. Therefore, the previous search strategies do not find C-minimal paths.

Several problems of Approximation Theory are equivalent to the search for $\left(\dfrac{C_1}{C_2}\right)$-extremal cycles. This will be shown in Subsection 4.9.3.

We assume that $C_1 = SUM_{h_1}$ and $C_2 = SUM_{h_2}$ where $h_2(r) > 0$ for all arcs r. The definition $C := C_1/C_2$ means that $C(P) := C_1(P)/C_2(P)$ if $\ell(P) > 0$ and $C(P) := 0$ if $\ell(P) = 0$. We shall often use the following notation:

$$a_{uv} := a(u,v) := h_1(u,v) \quad \text{and} \quad b_{uv} := b(u,v) := h_2(u,v) \quad ((u,v) \in \mathcal{R}).$$

4.9.1 (C_1/C_2)-extremal paths in acyclic digraphs

Now, we concentrate on Problem 29 in the special case that $\mathcal{G}$ is acyclic. We will show in Remark 4.185 that cycles in $\mathcal{G}$ may cause problems.

A solution of Problem 29 is given in [WiKoCe] for $C_2 := \ell$. For each length $l = 1, \ldots, n-1$ and for each node $v \in \mathcal{V}$, an s-v-path $P_l(v)$ of length l is constructed such that $P_l(v)$ is C_1-minimal among all s-v-paths of length l; the paths $P_l(v)$ are found with an algorithm that is similar to FORD-BELLMAN 2. When all paths $P_l(v)$ have been constructed then $P_{l(v)}(v)$ is chosen such that $C_1\big(P_{l(v)}(v)\big)/l(v)$ is minimal among all candidates $C_1(P_l(v))/l$ with $l = 1, 2, \ldots, n-1$. Then $P(v) := P_{l(v)}(v)$ is a C-optimal s-v-path. — By the way, a similar search strategy is used in [Karp78] to find C-minimal cycles; we shall describe Karp's algorithm in more detail in Subsection 4.9.2.

A more general case is that all weights $h_2(r)$ are rational numbers. In this situation, we can use Algorithm 16 on page 345; this algorithm is a modification of the one in [WiKoCe].

Algorithm 16

1. Choose a number $\Delta \in \mathbb{N}$ such that $\Delta \cdot h_2(r)$ is an integer number for all $r \in \mathcal{R}$. Let $\widetilde{h}_i(r) := \Delta \cdot h_i(r)$ for all $i = 1, 2$ and $r \in \mathcal{R}$. Moreover, let $\widetilde{C}_i(P) := \Delta \cdot C_i(P)$ for all $i = 1, 2$ and all paths P. Then $C = C_1/C_2 = \widetilde{C}_1/\widetilde{C}_2$ and $\widetilde{C}_i = SUM_{\widetilde{h}_i}$, $i = 1, 2$.

$\big(*$ All values $\widetilde{C}_2(P)$, $P \in \mathcal{P}(s)$, are natural numbers. Moreover, $\mathcal{P}(s)$ is finite because $\mathcal{G}$ is acyclic. Therefore, there exists a number L such that $\widetilde{C}_2(P) \in \{0, \ldots, L\}$ for all $P \in \mathcal{P}(s)$. $*\big)$

2. $\big(*$ An $(L+1)$-tuple $A(v) := (P_0(v), \ldots, P_L(v))$ of paths is generated for each node v; each path $P_l(v)$ is constructed as a $\widetilde{C}_1$-minimal s-v-path among all candidates P with $\widetilde{C}_2(P) = l$; if no s-v-path P with $\widetilde{C}_2(P) = l$ exists then $P_l(v) := \bot$.
The tuples $A(v)$, $v \in \mathcal{V}$, are constructed with a dynamic programming method similar to FORD-BELLMAN 2; in particular, each tuple $A\big(v^{(j)}\big)$, $j > 2$, is constructed with the help of the components of $A\big(v^{(1)}\big), \ldots, A\big(v^{(j-1)}\big)$. $*\big)$

Let $s = v^{(1)}, \ldots, v^{(n)}$ be a topological ordering of $\mathcal{G}$.
Construct the tuples $A\big(v^{(1)}\big), \ldots, A\big(v^{(n)}\big)$ as follows:
Define $A(s) = A\big(v^{(1)}\big) := ([s], \bot, \ldots, \bot)$.
For $j := 2$ **to** n **do**

2.1 $(*$ The sets $X_l^{(j)}$, $l = 0, \ldots, L$ are defined; $X_l^{(j)}$ is the set of all s-$v^{(j)}$-paths P with $\widetilde{C}_2(P) = l$. $*)$
For $l := 0$ **to** L **do**
$$X_l^{(j)} := \left\{ P_{v'}\big(v^{(i)}\big) \oplus \big(v^{(i)}, v^{(j)}\big) \;\middle|\; \begin{array}{l} \big(v^{(i)}, v^{(j)}\big) \in \mathcal{R}, \quad l' \in \{0, \ldots, L\}, \\ P_{l'}\big(v^{(i)}\big) \neq \bot, \quad l' + \widetilde{h}_2\big(v^{(i)}, v^{(j)}\big) = l \end{array} \right\}.$$

2.2. $(*$ The entries $P_1\big(v^{(j)}\big), \ldots, P_L\big(v^{(j)}\big)$ of $A\big(v^{(j)}\big)$ are generated. $*)$
If $X_l^{(j)}$ is empty **then** $P_l\big(v^{(j)}\big) := \bot$.
If $X_l^{(j)}$ is not empty **then** choose $P_l\big(v^{(j)}\big) \in X_l^{(j)}$ such that
$$\widetilde{C}_1\big(P_l(v^{(j)})\big) = \min\left\{\widetilde{C}_1(P) \;\middle|\; P \in X_l^{(j)}\right\}.$$

3. After constructing all tuples $A(v)$, $v \in \mathcal{V}$, choose a number $l(v)$ for each v such that
$$\frac{\widetilde{C}_1\big(P_{l(v)}(v)\big)}{l(v)} = \min\left\{\frac{\widetilde{C}_1(P_l(v))}{l} \;\middle|\; l = 1, \ldots, L, \; P_l(v) \neq \bot\right\}. \tag{4.188}$$

Let then $P(v) := P_{l(v)}(v)$.

By the way, the relationship $l = \widetilde{C}_2(P_l(v))$ for all l and v implies that we may replace (4.188) by (4.189) or by (4.190):
$$\frac{\widetilde{C}_1\big(P_{l(v)}(v)\big)}{\widetilde{C}_2\big(P_{l(v)}(v)\big)} = \min\left\{\frac{\widetilde{C}_1(P_l(v))}{\widetilde{C}_2(P_l(v))} \;\middle|\; l = 1, \ldots, L, \; P_l(v) \neq \bot\right\}, \tag{4.189}$$
$$C\big(P_{l(v)}\big) = \min\{C(P_l(v)) \mid l = 1, \ldots, L, \; P_l(v) \neq \bot\}. \tag{4.190}$$

Algorithm 16 requires much time if L is very great. Next, we introduce another algorithm whose complexity is independent of C_2. For this purpose, we prove the next result. If λ is a given real and $\lambda^*(v)$ is the minimal C-value of all s-v-paths then the following result says how to find out whether $\lambda < \lambda^*(v)$, $\lambda = \lambda^*(v)$ or $l > \lambda^*(v)$.

Theorem 4.178 *Given a finite, acyclic digraph $\mathcal{G} = (\mathcal{V}, \mathcal{R})$. Let $s, v \in \mathcal{V}$ with $s \neq v$. We assume that $\mathcal{P}(s, v) \neq \emptyset$.*

Let $C_1 = SUM_{h_1}$ and $C_2 = SUM_{h_2}$ be two additive cost measures with $h_2(r) > 0$ for all $r \in \mathcal{R}$. Let $C := C_1/C_2$.

Let $P^*(v)$ be a C-minimal s-v-path. (This path exists because $\mathcal{G}$ is finite and acyclic and $\mathcal{P}(s, v) \neq \emptyset$.) Let $\lambda^*(v) := C(P^*(v))$.

For each $\lambda \in \mathbb{R}$, we define the path function $\widehat{C}_\lambda := C_1 - \lambda \cdot C_2$. (Then $\widehat{C}_\lambda = SUM_{\widehat{h}_\lambda}$ whith $\widehat{h}_\lambda := h_1 - \lambda \cdot h_2$.)

For each λ, let $\widehat{P}^*_\lambda$ be a $\widehat{C}_\lambda$-minimal s-v-path.

Then the following assertions are true for all $\lambda \in \mathbb{R}$:

   a) *If $\widehat{C}_\lambda(\widehat{P}^*_\lambda) > 0$ then $\lambda < \lambda^*(v)$.*

   b) *If $\widehat{C}_\lambda(\widehat{P}^*_\lambda) = 0$ then $\lambda = \lambda^*(v)$.*

   c) *If $\widehat{C}_\lambda(\widehat{P}^*_\lambda) < 0$ then $\lambda > \lambda^*(v)$.*

   d) *If $\widehat{C}_\lambda(\widehat{P}^*_\lambda) = 0$ then $\widehat{P}_\lambda$ is a C-minimal s-v-path (and not only a $\widehat{C}_\lambda$-minimal s-v-path).*

P r o o f   o f   a) : Let $\widehat{C}_\lambda(\widehat{P}^*_\lambda) > 0$. Then we obtain the following inequality, in which relationship $(*)$ follows from the $\widehat{C}_\lambda$-minimality of $\widehat{P}^*_\lambda$ and relationship $(**)$ follows from the assumption that $\widehat{C}_\lambda(\widehat{P}^*_\lambda) > 0$.

$$C_1(P^*(v)) - \lambda C_2(P^*(v)) \;=\; \widehat{C}_\lambda(P^*(v)) \;\overset{(*)}{\geq}\; \widehat{C}_\lambda(\widehat{P}^*_\lambda) \;\overset{(**)}{>}\; 0.$$

Consequently $C_1(P^*(v)) - \lambda C_2(P^*(v)) > 0$. Dividing this inequality by $C_2(P^*(v)) > 0$, we see that $\lambda^*(v) - \lambda = C(P^*(v)) - \lambda = \dfrac{C_1(P^*(v)) - \lambda C_2(P^*(v))}{C_2(P^*(v))} > 0$. This fact implies that $\lambda < \lambda^*(v)$.

P r o o f   o f   b) : The first part of this proof is almost the same as the proof to a). Let $\widehat{C}_\lambda(\widehat{P}^*_\lambda) = 0$. Then we obtain the following inequality, in which relationship $(*)$ follows from the $\widehat{C}_\lambda$-minimality of $\widehat{P}^*_\lambda$ and relationship $(**)$ follows from the assumption that $\widehat{C}_\lambda(\widehat{P}^*_\lambda) = 0$.

$$C_1(P^*(v)) - \lambda C_2(P^*(v)) \;=\; \widehat{C}_\lambda(P^*(v)) \;\overset{(*)}{\geq}\; \widehat{C}_\lambda(\widehat{P}^*_\lambda) \;\overset{(**)}{=}\; 0.$$

Consequently $C_1(P^*(v)) - \lambda C_2(P^*(v)) \geq 0$. Dividing this inequality by $C_2(P^*(v)) > 0$, we see that $\lambda^*(v) - \lambda = C(P^*(v)) - \lambda = \dfrac{C_1(P^*(v)) - \lambda C_2(P^*(v))}{C_2(P^*(v))} \geq 0$. This implies that $\lambda \leq \lambda^*(v)$.

We now show that $\lambda \geq \lambda^*(v)$. By the C-minimality of $P^*(v)$, we obtain the following relationship:

$$\lambda^*(v) \;=\; C(P^*(v)) \;\leq\; C(\widehat{P}^*_\lambda) \;=\; \frac{C_1(\widehat{P}^*_\lambda)}{C_2(\widehat{P}^*_\lambda)}. \qquad (4.191)$$

Moreover, $C_1(\widehat{P}^*_\lambda) - \lambda \cdot C_2(\widehat{P}^*_\lambda) = \widehat{C}_\lambda(\widehat{P}^*_\lambda) = 0$ by our assumption. Dividing this relationship by $C_2(\widehat{P}^*_\lambda) > 0$, we see that $C(\widehat{P}^*_\lambda) - \lambda = \dfrac{C_1(\widehat{P}^*_\lambda) - \lambda \cdot C_2(\widehat{P}^*_\lambda)}{C_2(\widehat{P}^*_\lambda)} = 0$. Hence,
$\lambda = C(\widehat{P}^*_\lambda) \overset{(4.191)}{\geq} \lambda^*(v)$. Consequently, $\lambda \geq \lambda^*(v)$. This and the fact that $\lambda \leq \lambda^*(v)$
imply that $\lambda = \lambda^*(v)$.

P r o o f   o f   c) : Our argumentation is almost the same as in the last part of the
proof to b). Let $\widehat{C}_\lambda(\widehat{P}^*_\lambda) < 0$. Then fact (4.191) is true also in this case. Moreover,
$C_1(\widehat{P}^*_\lambda) - \lambda \cdot C_2(\widehat{P}^*_\lambda) = \widehat{C}_\lambda(\widehat{P}^*_\lambda) < 0$ by our assumption. Dividing this relation-
ship by $C_2(\widehat{P}^*_\lambda) > 0$, we obtain $C(\widehat{P}^*_\lambda) - \lambda = \dfrac{C_1(\widehat{P}^*_\lambda) - \lambda \cdot C_2(\widehat{P}^*_\lambda)}{C_2(\widehat{P}^*_\lambda)} < 0$. Hence,
$\lambda > C(\widehat{P}^*_\lambda) \overset{(4.191)}{\geq} \lambda^*(v)$. Consequently, $\lambda > \lambda^*(v)$.

P r o o f   o f   d) : Let $\widehat{C}_\lambda(\widehat{P}^*_\lambda) = 0$. By part b), $\lambda = \lambda^*(v)$. Consequently,
$C_1(\widehat{P}^*_\lambda) - \lambda^*(v) \cdot C_2(\widehat{P}^*_\lambda) = C_1(\widehat{P}^*_\lambda) - \lambda \cdot C_2(\widehat{P}^*_\lambda) = \widehat{C}_\lambda(\widehat{P}^*_\lambda) = 0$. This implies
that $C_1(\widehat{P}^*_\lambda) = \lambda^*(v) \cdot C_2(\widehat{P}^*_\lambda)$ so that $C(\widehat{P}^*_\lambda) = C_1(\widehat{P}^*_\lambda)\big/ C_2(\widehat{P}^*_\lambda) = \lambda^*(v)$.
Hence, $C(\widehat{P}_\lambda) = \lambda^*(v) = C(P^*(v))$ so that $\widehat{P}^*_\lambda$ is C-minimal as well.   □

Next, we use Theorem 4.178 to find C-minimal paths. For this purpose, we define the
function $g : \mathbb{R} \to \mathbb{R}$ as $g(\lambda) := \widehat{C}_\lambda(\widehat{P}^*_\lambda)$. Then each value $g(\lambda)$ can be computed
by constructing a $\widehat{C}_\lambda$-minimal path $\widehat{P}_\lambda$. It follows from Theorem 4.178 that $\lambda^*(v)$ is
the only zero of g. Consequently, the following search strategy will find a C-optimal
s-v-path:

Algorithm 17 Compute a zero λ_0 of g and a $\widehat{C}_{\lambda_0}$-minimal s-v-path $\widehat{P}_{\lambda_0}$.

Then $\lambda_0 = \lambda^*(v)$, and the path $\widehat{P}_{\lambda_0} = \widehat{P}_{\lambda^*(v)}$ is a C-minimal s-v-connection by Theorem 4.178 d).

The remaining problem is how to find the zero of g. This problem is solved in the next
remark.

Remark 4.179 The Newton method is very good to find the zero of g; in this special
case, finitely many iteration steps are sufficient to find the exact value $\lambda^*(v)$.

We give reasons for this behaviour of the Newton method. Our argumentation is similar to the one in [vGoS84], Theorem 4 and Remark 3.
The set $\mathcal{P}(s,v)$ is finite because $\mathcal{G}$ is acyclic. Let $g_P(\lambda) := \widehat{C}_\lambda(P) = C_1(P) -$
$C_2(P) \cdot \lambda$ for all paths P in the finite set $\mathcal{P}(s,v)$ and for all $\lambda \in \mathbb{R}$. Then each
function g_P is a linear polynomial in λ. Moreover, for all λ the following is true:

$$g(\lambda) \;=\; \widehat{C}_\lambda(\widehat{P}^*_\lambda) \;=\; \min_{P \in \mathcal{P}(s,v)} \widehat{C}_\lambda(P) \;=\; \min_{P \in \mathcal{P}(s,v)} g_P(\lambda). \qquad (4.192)$$

That means that g is the minimum of finitely many linear functions. Therefore, g is a
concave and piecewise linear function.

Consequently, g looks like a linear function in a small neighbourhood to the left of $\lambda^*(v)$, and the same is true in a small neighbourhood to the right of $\lambda^*(v)$. More precisely, there exist $\varepsilon > 0$ and two s-v-paths X, Y such that

$$\begin{aligned} g(\lambda) &= g_X(\lambda) \quad \text{if} \quad \lambda^*(v) - \varepsilon \le \lambda \le \lambda^*(v), \\ g(\lambda) &= g_Y(\lambda) \quad \text{if} \quad \lambda^*(v) \le \lambda \le \lambda^*(v) + \varepsilon. \end{aligned} \tag{4.193}$$

(The case $X = Y$ means that the graph of g is is equal to a line segment in a neighbourhood of $\lambda^*(v)$; the case $X \ne Y$ means that two different line segments of the graph of g meet each other at the point $(\lambda^*(v), g(\lambda^*(v)))$.)

The graphs of g_X and g_Y are tangents to g at each point $(\lambda, g(\lambda))$ with $\lambda \in [\lambda^*(v) - \varepsilon, \lambda^*(v)]$ and $\lambda \in [\lambda^*(v), \lambda^*(v) + \varepsilon]$, resp. Therefore, the Newton method follows the tangent g_X or g_Y when the approximate value $\overline{\lambda}$ is sufficiently close to $\lambda^*(v)$; in this case, the Newton method will find the exact value $\lambda^*(v)$ as a zero of g_X or g_Y.

Now we describe how the slope of g can be computed. Let $\overline{\lambda} \in \mathbb{R}$ be an approximation of $\lambda^*(v)$. We make the following assumption:

$$\text{The left slope of } g \text{ at } \overline{\lambda} \text{ is equal to the right slope of } g \text{ at } \overline{\lambda}. \tag{4.194}$$

(Otherwise we replace the approximation $\overline{\lambda}$ by an approximation $\overline{\lambda} + \delta$ where δ is very small.) Let $\widehat{P}^*_{\overline{\lambda}}$ be a $(\widehat{C}_{\overline{\lambda}})$-minimal s-v-path. Then

$$g\left(\overline{\lambda}\right) \overset{(4.192)}{=} \widehat{C}_{\overline{\lambda}}\left(\widehat{P}^*_{\overline{\lambda}}\right) = C_1\left(\widehat{P}^*_{\overline{\lambda}}\right) - C_2\left(\widehat{P}^*_{\overline{\lambda}}\right) \cdot \overline{\lambda}.$$

This and (4.194) imply that g has the same slope at $\overline{\lambda}$ as the function $\lambda \mapsto g_{\widehat{P}^*_{\overline{\lambda}}}(\lambda) = C_1\left(\widehat{P}^*_{\overline{\lambda}}\right) - C_2\left(\widehat{P}^*_{\overline{\lambda}}\right) \cdot \lambda$. Hence, the slope of g at $\overline{\lambda}$ is equal to $-C_2\left(\widehat{P}^*_{\overline{\lambda}}\right)$. $\square$

We have seen two methods to find C-optimal s-v-paths, namely Algorithms 16 and 17. Next, we show that these search strategies are closely related to a construction in Analysis.

Remark 4.180 Let $h_2(r)$ be natural numbers for all r. Then $\widetilde{C}_1 = C_1$ and $\widetilde{C}_2 = C_2$ in the description of Algorithm 16.

This algorithm solves several subproblems of the following type:

Problem 30   *Minimize* $C_1(X)$ *subject to* $C_2(X) = l$.

Algorithm 17 solves a problem of the following type:

Problem 31 *Find a path* X^* *and a number* λ *with the following properties:*

$$C_1(X^*) - \lambda \cdot C_2(X^*) \text{ is minimal among all candidates } C_1(X) - \lambda \cdot C_2(X). \tag{4.195}$$

$$C_1(X^*) - \lambda \cdot C_2(X^*) = 0. \tag{4.196}$$

Problem 30 is similar to an optimization problem with side constraints in Analysis, and Problem 31 is similar to the use of *Lagrange multipliers*. More precisely, let $c_1, c_2 : \mathbb{R}^2 \to \mathbb{R}$ and $l \in \mathbb{R}$. We formulate the following problem:

Problem 32 *Minimize $c_1(X)$ subject to $c_2(X) = l$.*

Using Lagrange multipliers, we obtain the following formulation of Problem 32.

Problem 33 *Compute an $X^* \in \mathbb{R}^2$ and a number λ such that*

$$c_1(X^*) - \lambda \cdot c_2(X^*) \text{ is minimal among all candidates } c_1(X) - \lambda \cdot c_2(X). \quad (4.197)$$

$$c_2(X^*) = l. \quad (4.198)$$

The interesting observation is that (4.195) and (4.197) have the same structure. $\square$

Next, we consider the case that $C = C_1/C_2$ where C_1 or C_2 are not additive. It will be shown in 4.181 – 4.183 that in many cases, we can find two paths P^+, P^{++} and a bound M such that $M \leq C(P^+)$ and $C(P^{++}) \leq M$.
Now, we describe a situation in which a path with bounded C-value exists.

Theorem 4.181 *Let $\mathcal{G} = (\mathcal{V}, \mathcal{R})$ be a finite, acyclic digraph. Let s be a node with the property that an s-v-path exists for all nodes v. Let Γ be the set of all nodes with outdegree $= 0$.*

Let C_1, H, C_2, $C : \mathcal{P}(s, \Gamma) \to \mathbb{R}$ be four path functions with the following properties: C_1 is an arbitrary path function. The function H is defined as $H(P) := \prod_{\kappa=0}^{k-1} \deg^+(v_\kappa)$ for all s-Γ-paths $P = [s = v_0, v_1, \ldots, v_k]$; in particular, $H([s]) := 1$. The function C_2 is defined as $C_2(P) := \frac{1}{H(P)}$, and the function C is defined as $C(P) := C_1(P)/C_2(P) = C_1(P) \cdot H(P)$ for all $P \in \mathcal{P}(s, \Gamma)$.

Let $M := \sum_{P \in \mathcal{P}(s,\Gamma)} C_1(P)$.

Then the following assertions are true:

a) There exists an s-Γ-path $P^+ = [s = v_0, v_1, \ldots, v_k]$ with the following property:

$$C_1(P^+) \geq \frac{1}{H(P^+)} \cdot \sum_{P \in \mathcal{P}(s,\Gamma)} C_1(P) \left(= \frac{M}{\deg^+(v_0) \cdot \deg^+(v_1) \cdot \ldots \cdot \deg^+(v_{k-1})} \right). \quad (4.199)$$

b) This path P^+ has the following property: $C(P^+) \geq M$.

c) Each C-maximal s-Γ-path P^* has the property that $C(P^*) \geq M$.

P r o o f o f a) : Let $\mathcal{G}^s$ the s-expansion of $\mathcal{G}$ as defined on page 17. Moreover, let $\mathcal{B}$ be the set of all leaves of $\mathcal{G}^s$. The definition of Γ implies that every leaf P of $\mathcal{G}^s$ is an s-Γ-path in $\mathcal{G}$ and every s-Γ-path in $\mathcal{G}$ is a leaf of $\mathcal{G}^s$. Consequently, $\mathcal{B} = \mathcal{P}(s, \Gamma)$.

Let $P \in \mathcal{P}(s)$. Then P is a node of $\mathcal{G}^s$. We define $\widetilde{\deg}^+(P)$ as the outdegree of the node P in $\mathcal{G}^s$. The following relationship is obvious:

$$(\forall P \in \mathcal{P}(s)) \quad \widetilde{\deg}^+(P) = \deg^+(\omega(P)). \quad (4.200)$$

We define $w(P) := C_1(P)$ for each leaf $P \in \mathcal{B} = \mathcal{P}(s, \Gamma)$. Moreover, we define
$w(\mathcal{G}^s) := \sum\limits_{P \in \mathcal{B}} w(P) \; \left(= \sum_{P \in \mathcal{P}(s,\Gamma)} C_1(P) = M \right)$.

Then Lemma G in [Hu88][51] yields a $\mathcal{G}^s$-path $\widetilde{P}^+ := \left[P_0 = [s], P_1, \ldots, P_k \right]$ with the following properties:

$$P_k \text{ is a leaf of } \mathcal{G}^s; \text{ that means that } P_k \in \mathcal{B} = \mathcal{P}(s,\Gamma). \qquad (4.201)$$

$$w(P_k) \;\geq\; \frac{w(\mathcal{G}^s)}{\widetilde{\deg}^{+}(P_0) \cdot \widetilde{\deg}^{+}(P_1) \cdot \ldots \cdot \widetilde{\deg}^{+}(P_{k-1})}. \qquad (4.202)$$

Let $P^+ := P_k$. Then $\prod\limits_{\kappa=0}^{k-1} \widetilde{\deg}^{+}(P_\kappa) \overset{(4.200)}{=} \prod\limits_{\kappa=0}^{k-1} \deg^{+}(w(P_\kappa)) = H(P_k) = H(P^+)$.
This and the fact that $w(\mathcal{G}^s) = \sum_{P \in \mathcal{P}(s,\Gamma)} C_1(P)$ are used in relationship $(*)$ of the following inequality:

$$C_1(P^+) = w(P^+) = w(P_k) \overset{(4.202)}{\geq} \frac{w(\mathcal{G}^s)}{\prod_{\kappa=0}^{k-1} \widetilde{\deg}^{+}(P_\kappa)} \overset{(*)}{=} \frac{\sum_{P \in \mathcal{P}(s,\Gamma)} C_1(P)}{H(P^+)}.$$

So, $C_1(P^+) \geq \frac{1}{H(P^+)} \cdot \sum\limits_{P \in \mathcal{P}(s,\Gamma)} C_1(P)$; consequently, P^+ has property (4.199).

P r o o f   o f   b) :   The assertion follows from the fact that
$$C(P^+) = C_1(P^+) \cdot H(P^+) \overset{(4.199)}{\geq} \sum\limits_{P \in \mathcal{P}(s,\Gamma)} C_1(P) = M.$$

P r o o f   o f   c) :   The assertion is a trivial consequence of part b). $\square$

Remark 4.182 In 4.181 a), we may replace the symbol $"\geq"$ by $"\leq"$. More precisely, there exists an s-Γ-path s-Γ-path $P_+ = [s = z_0, z_1, \ldots, z_l]$ such that

$$C_1(P_+) \;\leq\; \frac{1}{H(P_+)} \cdot \sum\limits_{P \in \mathcal{P}(s,\Gamma)} C_1(P) \; \left(= \frac{M}{\deg^{+}(z_0) \cdot \deg^{+}(z_1) \cdot \ldots \cdot \deg^{+}(z_{k-1})} \right). \qquad (4.203)$$

To see this we define $\check{C}_1(P) := -C_1(P)$ for all $P \in \mathcal{P}(s,\Gamma)$. Then Theorem 4.181 says that a path P_+ exists with $\check{C}_1(P_+) \geq \frac{1}{H(P_+)} \cdot \sum\limits_{P \in \mathcal{P}(s,\Gamma)} \check{C}_1(P)$; hence, $C_1(P_+) =$

$$-\check{C}_1(P_+) \;\leq\; \frac{1}{H(P_+)} \cdot \sum\limits_{P \in \mathcal{P}(s,\Gamma)} -\check{C}_1(P) = \frac{1}{H(P_+)} \cdot \sum\limits_{P \in \mathcal{P}(s,\Gamma)} C_1(P).$$

Fact (4.203) implies that $C(P_+) \leq M$ where M is the number defined in Theorem 4.181. — It is trivial that $C(P_*) \leq M$ for all C-minimal s-Γ-paths P_*. $\square$

Next, we describe how to find a path P^+ with property (4.199). The construction of a path P_+ with property (4.203) is almost the same as the construction of P^+.

---

[51]It has been assumed in Lemma G of [Hu88] that $w(P) \geq 0$ for all $P \in \mathcal{B}$. This assumption, however, may be omitted.

Remark 4.183 We describe an algorithm to find a path with property (4.199); this algorithm resembles the one in [Hu88], page 408 – 409. We use the terminology of Theorem 4.181. Moreover, we define the path function $\overline{C}_1 : \mathcal{P}(s) \to \mathbb{R}$ as follows: For all paths $P \in \mathcal{P}(s)$, let $\overline{C}_1(P) := \sum\limits_{\substack{\alpha(Q) = \omega(P) \\ \omega(Q) \in \Gamma}} C_1(P \oplus Q)$.

By the definition of Γ in 4.181, the following is true:
$$\left(\forall \, P \in \mathcal{P}(s, \Gamma) \right) \quad \overline{C}_1(P) \;=\; C_1(P). \tag{4.204}$$

Moreover, we define $\overline{w}(P) := \overline{C}_1(P)$ for each node P of $\mathcal{G}^s$. By the definition of Γ and by (4.204), the following is true:
$$(\forall \, P \in \mathcal{B}) \quad \overline{w}(P) \;=\; \overline{C}_1(P) \;\overset{(4.204)}{=}\; C_1(P) \;=\; w(P).$$

Then we construct the $\mathcal{G}^s$-path $\widetilde{P}^+ = [P_0, \dots, P_k]$ in the proof of 4.181 a) as follows:
$P_0 := [s] = $ root of $\mathcal{G}^s$; $\kappa := 0$;

While P_κ is not a leaf of $\mathcal{G}^s$ **do**
     begin
        Choose $P_{\kappa+1} \;=\; P_\kappa \oplus r'$ such that
$$\overline{C}_1(P_{\kappa+1}) \;=\; \max\{\overline{C}_1(P_\kappa \oplus r) \,|\, r \in \mathcal{R} \,,\; \alpha(r) = \omega(P_\kappa)\}. \tag{4.205}$$
        $\kappa := \kappa + 1$;
     end;

When $\widetilde{P}^+$ has been found the path P^+ is defined as $P^+ := P_k = \omega(\widetilde{P}^+)$; then P^+ has property (4.199); this follows from the proof of Lemma G in [Hu88], page 409.

The construction of the path $\widetilde{P}^+$ in $\mathcal{G}^s$ is an example for depth first search. If a prefix $\widetilde{P}_\kappa^+ := [P_0, \dots, P_\kappa]$ of $\widetilde{P}^+$ is given then the next prefix $P_{\kappa+1}$ is constructed as the locally best extension of $\widetilde{P}_\kappa^+$; more precisely, $\widetilde{P}_{\kappa+1}$ is constructed as a $\mathcal{G}^s$-path of the form $\widetilde{P}^+ \oplus (P_\kappa, P)$ with maximal value $\overline{C}_1(P) = \overline{w}(P)$.

A further algorithm to find a path P^+ has been described in [Huck92], Remark 4.101 b). The algorithm is a modified Ford-Bellman strategy; its disadvantage is that it is slower than the above algorithm.

As mentioned above, the construction of a path P_+ with property (4.203) is almost the same as the one of P^+; we must only replace the maximization in (4.205) by a minimization.   ☐

Remark 4.184 Let $C = C_1/C_2$ where C_1, C_2 need not be additive.
In general, the search for C-minimal paths seems to be easy if C_1 and C_2 have particular Bellman properties or order preserving properties. For example, Algorithm 16 seems to work correctly if $C_2(P) > 0$ for all paths P and if C_1 has the Bellman property of type B1); this property was described in 2.111 b); in the current situation, let $H := C_1$ and $H_* := C_2$.   ☐

Next, we return to the case of additive path functions C_1, C_2. We give reasons why we have always assumed that $\mathcal{G}$ is acyclic.

Remark 4.185 If G has cycles then we cannot guarantee that a C-minimal s-v-path exists; this is even the case if $h_1(r) > 0$ for all $r \in \mathcal{R}$ and $C_2 = \ell$.

For example, let $G = (\mathcal{V}, \mathcal{R})$ with $\mathcal{V} = \{s, x, y, z\}$ and $\mathcal{R} = \{(s, z), (z, x), (x, y), (y, z)\}$. We define $X := [z, x, y]$ and X^k as the k-fold concatenation of X. Let $h(s, z) := 2$, and let $h(r) := 1$ for all other arcs. We define $C_1 := SUM_h$, $C_2 := \ell$, and $C := C_1/C_2$. Then the following is true for all $k \in \mathbb{N} \cup \{0\}$:

$$\frac{C_1([s, z] \oplus X^k)}{C_2([s, z] \oplus X^k)} = \frac{2 + 3k}{1 + 3k} = 1 + \frac{1}{1 + 3k}.$$

Consequently, the infimum of all values $C(P)$, $P \in \mathcal{P}(s, z)$, is equal to 1; but this infimum is not achieved for any s-z-path P. Hence, there exists no C-minimal s-v_5-path. □

4.9.2 (C_1/C_2)-extremal cycles

Here, we focus on the following problem:

Problem 34 *Let $G = (\mathcal{V}, \mathcal{R})$ be a finite digraph. Let $C = C_1/C_2$ where $C_i = SUM_{h_i}$ $(i = 1, 2)$ and $h_2(r) > 0$ for all $r \in \mathcal{R}$.*
Find a substantial cycle P for which $C(P)$ is minimal or maximal.

Recall that a substantial cycle is defined as a cycle of a length ≥ 1. Note that the candidate paths P may have different start nodes.

The following lemma shows that we always can find cycles with extremal C-values. The reason is that node injective cycles are the only relevant candidates.

Lemma 4.186 *Given a substantial cycle $P = X' \oplus Y \oplus X''$; we assume that Y and $X' \oplus X''$ are substantial cycles, too.*
Then $\min\{C(X' \oplus X''), C(Y)\} \leq C(P) \leq \max\{C(X' \oplus X''), C(Y)\}.$
(That means that we may replace the complicated cycle P by one of the simpler competitors $X' \oplus X''$ or Y.)

P r o o f : We use the following well-known fact about fractions:

$$(\forall a_1, b_1 \in \mathbb{R})(\forall a_2, b_2 > 0) \quad \min\left\{\frac{a_1}{a_2}, \frac{b_1}{b_2}\right\} \leq \frac{a_1 + b_1}{a_2 + b_2} \leq \max\left\{\frac{a_1}{a_2}, \frac{b_1}{b_2}\right\}. \quad (4.206)$$

Let $a_i := C_i(X' \oplus X'')$ and $b_i := C_i(Y)$, $i = 1, 2$. Then $a_2, b_2 > 0$. This and (4.206) imply that

$$\min\left\{\frac{C_1(X' \oplus X'')}{C_2(X' \oplus X'')}, \frac{C_1(Y)}{C_2(Y)}\right\} \leq \frac{C_1(X' \oplus X'') + C_1(Y)}{C_2(X' \oplus X'') + C_2(Y)} \leq \max\left\{\frac{C_1(X' \oplus X'')}{C_2(X' \oplus X'')}, \frac{C_1(Y)}{C_2(Y)}\right\}.$$

This fact and the relationships $C_i(P) = C_i(X' \oplus X'') + C_i(Y')$ $(i = 1, 2)$ imply that indeed $\min\{C(X' \oplus X''), C(Y)\} \leq C(P) \leq \max\{C(X' \oplus X''), C(Y)\}$. □

Let $C_2 = \ell$. The next algorithm finds the minimal C-value of all substantial cycles (but not the minimal cycles themselves). The algorithm has been presented in [Karp78], and it is similar to Algorithm 16 in this section. Karp assumes that G is strongly connected; if not then Algorithm 18 is applied to each strongly connected component of G.

Algorithm 18

1. Choose an arbitrary start node s in $\mathcal{G}$.

2. **For** each node $v \in \mathcal{V}$ and **for** $l := 0$ to n **do**

    If an s-v-path of length l exists **then** compute the value

    $$F_l(v) \ := \ \min\{C_1(P) \,|\, P \in \mathcal{P}(s,v),\ \ell(P) = l\}\,;$$

    else let $F_l(v) := \infty$.

3. Compute the quantity $\lambda^* := \displaystyle\min_{v \in \mathcal{V}} \left(\max_{0 \leq k \leq n-1} \left[\frac{F_n(v) - F_k(v)}{n-k} \right] \right).$

It has been shown in [Karp78] that this value λ^* is the minimal C-value of all substantial cycles in $\mathcal{G}$; as mentioned above, $\mathcal{G}$ is assumed to be strongly connected.

Next, we introduce Algorithm 19, which will output a C-minimal cycle (and not only compute its cost) if C_2 is an arbitrary additive path function with $h_2(r) > 0$ for all arcs. Algorithm 19 is similar to Algorithm 17. The following result will be helpful to understand Algorithm 19.

Theorem 4.187 (*see Theorem 4.178*)
Given a finite digraph $\mathcal{G} = (\mathcal{V}, \mathcal{R})$. Let $v \in \mathcal{V}$ be an arbitrary node. We assume that a substantial v-v-path exists.
Let $C_1 = \mathrm{SUM}_{h_1}$ and $C_2 = \mathrm{SUM}_{h_2}$ be two additive cost measures with $h_2(r) > 0$ for all $r \in \mathcal{R}$. Let $C := C_1/C_2$.
Let $P^*(v)$ be a C-minimal candidate among all substantial cycles P with $\alpha(P) = \omega(P) = v$. (The existence of $P^*(v)$ follows from Lemma 4.186, which says that only the finitely many node injective cycles are relevant candidates.) Let $\lambda^*(v) := C(P^*(v))$.
For each $\lambda \in \mathbb{R}$, we define the path function $\widehat{C}_\lambda := C_1 - \lambda \cdot C_2$. (Then $\widehat{C}_\lambda = \mathrm{SUM}_{\widehat{h}}$ with $\widehat{h}_\lambda := h_1 - \lambda \cdot h_2$.)
For each λ, let $\widehat{P}^*_\lambda$ be a cycle with minimal $\widehat{C}_\lambda$-value among all candidates P with $\alpha(P) = \omega(P) = v$ and $\ell(P) \in \{1, \ldots, n\}$.
Then the following assertions are true for all $\lambda \in \mathbb{R}$:

a) *If $\widehat{C}_\lambda(\widehat{P}^*_\lambda) > 0$ then $\lambda < \lambda^*(v)$.*

b) *If $\widehat{C}_\lambda(\widehat{P}^*_\lambda) = 0$ then $\lambda = \lambda^*(v)$.*

c) *If $\widehat{C}_\lambda(\widehat{P}^*_\lambda) < 0$ then $\lambda > \lambda^*(v)$.*

d) *If $\widehat{C}_\lambda(\widehat{P}^*_\lambda) = 0$ then $\widehat{P}_\lambda$ is C-minimal among all substantial v-v-paths.*

P r o o f : The proof is almost the same as the one of 4.178. The only difference is that all paths $P^*(v)$ and $\widehat{P}^*_\lambda$ are substantial v-v-paths and not s-v-paths. ☐

We define $g : \mathbb{R} \to \mathbb{R}$ as $g(\lambda) := \widehat{C}_\lambda(\widehat{P}^*_\lambda)$ for all λ. By Theorem 4.187 d), the following algorithm will find a C-optimal cycle:

Algorithm 19

For all nodes v **do**

Find a zero λ_0 of g and a $(\widehat{C}_{\lambda_0})$-minimal substantial v-v-cycle $\widehat{P}_{\lambda_0}$.

Let $P^*(v) := \widehat{P}_{\lambda_0}$.

Choose P^* as a C-minimal candidate of all paths $P^*(v)$, $v \in \mathcal{V}$.

We now study two details of Algorithm 19.

Remark 4.188 a) The computation of the values $g(\lambda)$ includes the search for $\widehat{C}_\lambda$-minimal v-v-cycles $\widehat{P}_\lambda^*$ in $\mathcal{G}$ of a length in $\{1, 2, \ldots, n\}$; in particular, we must avoid candidates of length 0. For this end, we generate a copy v' of v and search for an optimal v'-v-path. Since $v' \neq v$, all candidates are forced to use at least one arc. More precisely, we define the graph $\mathcal{G}' = (\mathcal{V}', \mathcal{R}')$ with $\mathcal{V}' := \mathcal{V} \cup \{v'\}$ and $\mathcal{R}' := \mathcal{R} \cup \{(v', w) \,|\, (v, w) \in \mathcal{R}\}$. We extend the arc functions h_1 and h_2 to the functions $h_1', h_2' : \mathcal{R}' \to \mathbb{R}$ such $h_i'(v', w) := h_i'(v, w)$ for all $i = 1, 2$ and for all arcs $(v, w) \in \mathcal{R}$. Let $C_i' := SUM_{h_i'}$ $(i = 1, 2)$ and $\widehat{C}_\lambda' := C_1' - \lambda \cdot C_2'$ for all $\lambda \in \mathbb{R}$.

Then $\widehat{P}_\lambda^*$ can be found as follows: Search for a $\widehat{C}_\lambda'$-minimal v'-v-path $\widehat{Q}_\lambda^*$ in $\mathcal{G}'$ with a length $\leq n$. (Theorem 4.121 says that this can be done with n iterations of FORD-BELLMAN 1.) The path $\widehat{Q}_\lambda^*$ has at least the length 1 because $v' \neq v$. If (v', w_1) is the first arc of $\widehat{Q}_\lambda^*$ then the path $\widehat{P}_\lambda^*$ is generated by replacing (v', w_1) by (v, w_1).

b) The zero $\lambda^*(v)$ of g can be found by Newton's method as well as in 4.179. To see this we define $\mathcal{P}_n^\circ(v)$ as the finite set of all cycles X in $\mathcal{G}$ with $\alpha(X) = \omega(X) = v$ and with $\ell(X) \in \{1, \ldots, n\}$. Then we define $g_X(\lambda) := \widehat{C}_\lambda(X) = C_1(X) - C_2(X) \cdot \lambda$ for all cycles $X \in \mathcal{P}_n^\circ(v)$ and all $\lambda \in \mathbb{R}$.

The further argumentation is almost the same as the one in 4.179: $g(\lambda)$ is the minimum of the finitely many linear functions $g_X(\lambda)$, $X \in \mathcal{P}_n^\circ(v)$. Consequently, g is piecewise linear and concave. This implies that the Newton method will fine the zero $\lambda^*(v)$ of g in finitely many steps. $\qquad \Box$

Remark 4.189

The relationships between (C_1/C_2)-optimal cycles and $\widehat{C}_\lambda$-minimal paths are also studied in [YTO91].

An interesting open question is how to find C-minimal cycles if C_1 or C_2 is not additive (see also 4.184) This problem seems to be difficult because the proof of Theorem 4.186 has been based on the additivity of C_1 and C_2. Consequently, if C_1 or C_2 is not additive, then Theorem 4.186 is false of very difficult to prove. Therefore, it is possible that very complicated cycles are C-optimal. $\qquad \Box$

4.9.3 Relationships between approximation theoretical problems and (C_1/C_2)-extremal cycles

The basic problem in this subsection is to approximate an arc function with a potential difference.

Problem 35 *Given a finite digraph $\mathcal{G} = (\mathcal{V}, \mathcal{R})$ and a number $\lambda \in \mathbb{R}$. For each arc $(u, v) \in \mathcal{R}$, let $a(u, v)$ be a real number.*
Find a potential difference $\delta(u, v) = f(v) - f(u)$ $((u, v) \in \mathcal{R})$ such that

$$(\forall\ (u, v) \in \mathcal{R}) \quad a(u, v) - \delta(u, v) = a(u, v) + f(u) - f(v) \leq \lambda. \qquad (4.207)$$

Moreover, compute the minimal λ with the property that (4.207) can be solved.

Let $C_1 := SUM_a$ and $C := C_1/\ell$. We shall later see that the following number are equal to each other:

- the minimal λ^* for which this problem can be solved,
- the maximal C-value q^* of all substantial cycles.

To show that the differences $a(r) - \delta(r)$ are closely related to the C-values of cycles we define $D := SUM_\delta$. Then $D(P) = 0$ for all cycles P. Let $P = r_1 \oplus \ldots \oplus r_k$. Then

$$\frac{1}{k} \cdot \sum_{\kappa=1}^{k} \left(a(r_\kappa) - \delta(r_\kappa) \right) = \frac{C_1(P) - D(P)}{\ell(P)} = \frac{C_1(P)}{\ell(P)} = C(P).$$

That means that the average of all differences $a(r_\kappa) - \delta(r_\kappa)$ is equal to $C(P)$. A similar fact will be used in the proof of Theorem 4.191.

We shall use the following notation: Let $f : \mathcal{V} \to \mathbb{R}$ be a node function; then we define the potential difference Δf as $\Delta f(u, v) := f(v) - f(u)$ for all arcs $(u, v) \in \mathcal{R}$.

We now formulate several modifications of Problem 35.

Problem 36 *Given the situation of Problem 35. Let $b : \mathcal{R} \to \mathbb{R}$ be a further arc functions with $b(u, v) > 0$ for all arcs (u, v).*
Find a node function f such that the following is true for $\delta := \Delta f$:

$$(\forall\ (u, v) \in \mathcal{R}) \quad a(u, v) - \delta(u, v) = a(u, v) + f(u) - f(v) \leq \lambda\, b(u, v). \qquad (4.208)$$

Moreover, compute the minimal number λ for which (4.208) can be solved.

Problem 36 is a generalization of Problem 35; this can be seen by defining $b(u, v) := 1$ for all $(u, v) \in \mathcal{R}$.

We now formulate the following modification of Problem 36:

Problem 37 *Given the situation of Problem 35. Let $b : \mathcal{R} \to \mathbb{R}$ be a further arc functions with $b(u, v) > 0$ for all arcs (u, v).*
Find two node functions f and g such that

$$(\forall\ (u, v) \in \mathcal{R}) \quad |a(u, v) - f(u) - g(v)| \leq \lambda\, b(u, v). \qquad (4.209)$$

Moreover, find the minimal number λ for which (4.209) can be solved.

The following problem a generalization of Problem 36; this can be seen by defining the functions $\varphi_{(u,v)}$ in Problem 38 as $\varphi_{(u,v)}(\zeta, \xi) := a(u,v) + \zeta - \xi$ for all (u,v) and for all ζ, ξ.

Problem 38 *Given a function* $\varphi_{(u,v)} : \mathbb{R}^2 \to \mathbb{R}$ *for each arc* $(u,v) \in \mathcal{R}$; *we assume that each* $\varphi_{(u,v)}$ *is monotonically increasing in the first argument and strictly decreasing in the second.*

Find a node function $f : \mathcal{V} \to \mathbb{R}$ *such that*

$$(\forall\, (u,v) \in \mathcal{R}) \quad \varphi_{(u,v)}(f(u), \lambda) \;\leq\; f(v). \tag{4.210}$$

Moreover, compute the minimal number λ *for which* (4.210) *can be solved.*

4.9.3.1   Results and remarks in connection with Problem 36

Next, we show how to solve Problem 36, which is a generalized version of Problem 35. For this purpose, we define a property of potential differences.

Definition 4.190 Given a digraph $\mathcal{G} = (\mathcal{V}, \mathcal{R})$. Let s be a node such that an s-v-path exists for each node v. Given the arc functions $h_1 = a$ and $h_2 = b$ with $h_2(u,v) > 0$ for all arcs (u,v). Let $C_i := SUM_{h_i}$, $i = 1, 2$. Given a node function $f : \mathcal{V} \to \mathbb{R}$ and the potential difference $\delta(u,v) := f(v) - f(u)$ for all $(u,v) \in \mathcal{R}$. Let $\lambda \in \mathbb{R}$.

Then δ is called λ-*admissible* if δ is a solution of Problem 36; that means, δ is λ-admissible if and only if

$$(\forall\, (u,v) \in \mathcal{R}) \quad a(u,v) - \delta(u,v) \;=\; a(u,v) + f(u) - f(v) \;\leq\; \lambda \cdot b(u,v). \qquad \square$$

We now show that the existence of λ-admissible potential differences is closely related to the costs of substantial cycles.

Theorem 4.191 *Given the same situation as in Definition* 4.190. *In addition, we assume that* $\mathcal{G}$ *is finite. Let* $C := C_1/C_2$. *We define the following subsets* $\mathcal{L}$, $\mathcal{Q} \subseteq \mathbb{R}$ *and the following quantities* λ^*, q^*:

$$\mathcal{L} := \{\lambda \in \mathbb{R} \mid \text{There exists a } \lambda\text{-admissible potential difference } \delta\}\,, \quad \lambda^* := \inf \mathcal{L},$$

$$\mathcal{Q} := \{C(P) = C_1(P)/C_2(P) \mid P \text{ is a substantial cycle in } \mathcal{G}\}\,, \quad q^* := \sup \mathcal{Q}.$$

(Then $\mathcal{L} \neq \emptyset$; to see this we choose an arbitrary potential difference δ'; then δ' is admissible for a sufficiently great number λ' because $\mathcal{G}$ is finite. It follows that $\lambda^* < \infty$. Moreover, $q^* < \infty$ because Lemma 4.186 says that only the finitely many node injective cycles of $\mathcal{G}$ must be considered when computing the supremum q^*. — It follows that λ^*, $q^* \in \mathbb{R} \cup \{-\infty\}$.)

Then the following assertions are true:

a) *If* δ *is* λ_1-*admissible and if* $\lambda_2 \geq \lambda_1$ *then* δ *is also* λ_2-*admissible.*

b) $\lambda^* = q^*$.

c) $\mathcal{L} = [\lambda^*, \infty)$ *if* $\lambda^* > -\infty$, *and* $\mathcal{L} = \mathbb{R}$ *if* $\lambda^* = -\infty$.

P r o o f   o f   a) :   We have assumed that $b(u,v) > 0$ for all arcs; consequently, $\lambda_1 \cdot b(u,v) \le \lambda_2 \cdot b(u,v)$. This and the λ_1 admissibility of δ imply that

$$(\forall\ (u,v) \in \mathcal{R})\ \ a(u,v) - \delta(u,v)\ \le\ \lambda_1 \cdot b(u,v)\ \le\ \lambda_2 \cdot b(u,v)\,.$$

Consequently, δ is λ_2-admissible.

P r o o f   o f   b) :   *Case 1:* $Q \ne \emptyset$.

Then $q^* \ne -\infty$.

Now we show that $\lambda^* \ge q^*$. Our argumentation is similar to the one immediately after the description of Problem 35. Moreover, the same idea proof can be used when a function F is approximated by a function G whose sum or integral is zero along circular paths or closed curves; an example will be given in the proof of Theorem 5.55.

Let $q \in Q$. Then there exists a substantial cycle $P = r_1 \oplus \cdots \oplus r_k$ in $\mathcal{G}$ such that $C(P) = C_1(P)/C_2(P) = q$. That means that

$$\sum_{\kappa=1}^{k} a(r_\kappa)\ =\ C_1(P)\ =\ q \cdot C_2(P)\ =\ q \cdot \sum_{\kappa=1}^{k} b(r_\kappa)\,. \tag{4.211}$$

Let $\lambda \in \mathcal{L}$. Then there exists a λ-admissible potential difference δ. Moreover,

$$\sum_{\kappa=1}^{k} \delta(r_\kappa) = 0 \tag{4.212}$$

because P is a cycle and δ is a potential difference.

Then the following inequality is valid, in which $(*)$ follows from the λ-admissibility of δ:

$$q \cdot C_2(P)\ =\ q \cdot \sum_{\kappa=1}^{k} b(r_\kappa) \overset{(4.211)}{=} \sum_{\kappa=1}^{k} a(r_\kappa) \overset{(4.212)}{=}$$

$$\sum_{\kappa=1}^{k} \Big(a(r_\kappa) - \delta(r_\kappa) \Big) \overset{(*)}{\le} \sum_{\kappa=1}^{k} \lambda \cdot b(r_\kappa)\ =\ \lambda \cdot C_2(P)\,.$$

Consequently, $q \cdot C_2(P) \le \lambda \cdot C_2(P)$; this and $C_2(P) > 0$ imply that $\lambda \le q$. This relationship is true for every $\lambda \in \mathcal{L} \overset{\text{see above}}{\ne} \emptyset$ and every $q \in Q \overset{\text{Case 1}}{\ne} \emptyset$. Consequently, $q^* \le \lambda^*$. In particular, $\lambda^* > -\infty$, i.e. $\lambda^* \in \mathbb{R}$.

Next, we show that $\lambda^* \le q^*$. For this purpose, we give an explicit construction of a λ-admissible potential difference δ if $\lambda \ge q^*$.

Let $\lambda \ge q^*$. In order to construct δ we observe that $\lambda \ge C(X) = C_1(X)/C_2(X)$ for all substantial cycles. This and the fact that $C_2(X) > 0$ imply that

$$C_1(X) - \lambda \cdot C_2(X) \le 0 \quad \text{for all substantial cycles } X. \tag{4.213}$$

Let $\widehat{C}_\lambda(P) := C_1(P) - \lambda C_2(P)$ for all paths P in $\mathcal{G}$. Then $\widehat{C}_\lambda = SUM_{\widehat{h}_\lambda}$ where $\widehat{h}_\lambda$ is defined as $\widehat{h}_\lambda(r) := h_1(r) - \lambda h_2(r)$ for all $r \in \mathcal{R}$.

For all nodes v there exists a $\widehat{C}_\lambda$-maximal s-v-path $P^*(v)$. This can be seen as follows: By (4.213), there exist no substantial cycles X with a positive $\widehat{C}_\lambda$-value. Therefore, the only relevant candidates are acyclic paths. Since $\mathcal{G}$ is finite, there exist only finitely many acyclic s-v-paths. Let $P^*(v)$ be a $\widehat{C}_\lambda$-maximal candidate among all acyclic s-v-paths; then $P^*(v)$ has maximal $\widehat{C}_\lambda$-value among all s-v-paths.

We define $f(v) := \widehat{C}_\lambda(P^*(v))$ for all nodes v. Let $(u,v) \in \mathcal{R}$. Then we obtain inequality (4.214), in which $(*)$ follows from the additivity of $\widehat{C}_\lambda$ and $(**)$ follows from the maximality of $P^*(v)$.

$$f(u) + \widehat{C}_\lambda([u,v]) = \widehat{C}_\lambda(P^*(u)) + \widehat{C}_\lambda([u,v]) \overset{(*)}{=}$$
$$\widehat{C}_\lambda(P^*(u) \oplus (u,v)) \overset{(**)}{\leq} \widehat{C}_\lambda(P^*(v)) = f(v). \tag{4.214}$$

Consequently, $f(u) + \widehat{C}_\lambda([u,v]) \leq f(v)$; this is used in $(*)$ of the following inequality.

$$f(u) + a(u,v) - \lambda\, b(u,v) = f(u) + \widehat{h}_\lambda(u,v) = f(u) + \widehat{C}_\lambda([u,v]) \overset{(*)}{\leq} f(v).$$

That means that $a(u,v) + f(u) - f(v) \leq \lambda\, b(u,v)$ for all arcs $(u,v) \in \mathcal{R}$. Therefore, $\delta := \Delta f$ is λ-admissible.

Consequently, $\lambda \in \mathcal{L}$ for each $\lambda \geq q^*$. This implies that

$$[q^*, \infty) \subseteq \mathcal{L}. \tag{4.215}$$

Therefore, $q^* = \inf [q^*, \infty)) \geq \inf(\mathcal{L}) = \lambda^*$.

Case 2: $\mathcal{Q} = \emptyset$.

Then $\mathcal{G}$ is acyclic. Hence, there exists a topological ordering $f_0 : \mathbb{R} \to \mathbb{N}$; let $\delta_0 := \Delta f_0$. Then $\delta_0(u,v) = f_0(v) - f_0(u) \geq 1$ for all $(u,v) \in \mathcal{R}$. So, if $\lambda \in \mathbb{R}$ then (4.208) can be made true by choosing $\delta := c \cdot \delta_0$ where $c \in \mathbb{R}$ is sufficiently great. Consequently, $\lambda \in \mathcal{L}$ for all $\lambda \in \mathbb{R}$; this implies that $\lambda^* = -\infty$. Moreover, $q^* = -\infty$ because $\mathcal{Q} \overset{\text{Case 2}}{=} \infty$.

P r o o f   o f   c) : First, let $\lambda^* > -\infty$. It follows from part a) that $\mathcal{L}$ is equal to one of the intervals $[\lambda^*, \infty)$ or (λ^*, ∞). Part b) and (4.215) imply that $\lambda^* = q^* \overset{(4.215)}{\in} \mathcal{L}$. Therefore, $\mathcal{L} = [\lambda^*, \infty)$.

Next, let $\lambda^* = -\infty$. Then part a) implies that $\mathcal{L} = \mathbb{R}$. $\square$

In the following remarks 4.192 – 4.194, we shall discuss Theorem 4.191.

Remark 4.192 We give answers to the question whether assertion 4.191 b) is true for infinite digraphs. For this purpose, we define $\widehat{C}_\lambda := C_1 - \lambda \cdot C_2$ for all $\lambda \in \mathbb{R}$.

Then Theorem 4.191 b) is true for all infinite digraphs $\mathcal{G}$ with $-\infty < \lambda^* < \infty$, with $-\infty < q^* < \infty$, and with the following property:

There exists a node s such that for all $v \in \mathcal{V}$ and all $\lambda \geq q^*$ the following is true:
If no $\widehat{C}_\lambda$-positive cycles in $\mathcal{G}$ exist then there exists a $\widehat{C}_\lambda$-maximal s-v-connection. $\tag{4.216}$

Unfortunately, there exist infinite digraphs $\mathcal{G}$ that do not have property (4.216). For example, let $\mathcal{G} := (\mathcal{V}, \mathcal{R})$ with $\mathcal{V} := \{x, z\} \cup \{y_0, y_1, y_2\} \cup \{u_1, u_2, u_3, \ldots\}$ and $\mathcal{R} := \{(x, y_0), (y_0, y_1), (y_1, y_2), (y_2, y_0)\} \cup \{(x, u_i), (u_i, z) | i \in \mathbb{N}\}$. We define $X := [y_0, y_1, y_2, y_0]$ and $P_i := [x, u_i, z]$ for all $i \in \mathbb{N}$.
Let $h_1(u_i, z) := i$ for all $i \in \mathbb{R}$, and let $h_1(r) := 1$ for all other $r \in \mathcal{R}$. We define

$C_1 := SUM_{h_1}$, $C_2 := \ell$, and $C := C_1/C_2$.

We show that $\mathcal{G}$ does not have property (4.216). First, we observe that $q^* = 1$ because

$$\mathcal{Q} = \{C(X^k)| k \in \mathbb{N}\} = \{C_1(X^k)/C_2(X^k)| k \in \mathbb{N}\} = \left\{ \left. \frac{3k}{3k} \right| k \in \mathbb{N} \right\} = \{1\}.$$

Let $\overline{\lambda} := 2$; then $\overline{\lambda} > q^*$. Then $\mathcal{G}$ does not have any $(\widehat{C}_{\overline{\lambda}})$–positive cycles; to see this we note all cycles X^k of $\mathcal{G}$ have the following $\widehat{C}_{\overline{\lambda}}$–values:

$$\widehat{C}_{\overline{\lambda}}(X^k) = k \cdot (C_1(X) - \overline{\lambda} \cdot C_2(X)) = k \cdot (3 - 2 \cdot 3) = -3 \cdot k < 0.$$

On the other hand, no node $s \in \{V\}$ satisfies the condition in (4.216). All nodes $s \neq x$ have the property that not all $v \in V$ can be reached by an s-v-connection. Let $s := x$. Then there does not exist any $(\widehat{C}_{\overline{\lambda}})$-maximal s-z-path because the following is true for all paths P_i:

$$\widehat{C}_{\overline{\lambda}}(P_i) = C_1(P_i) - \overline{\lambda} \cdot C_2(P_i) = (i+1) - 2 \cdot 2 = i - 4.$$

So, we have found an $\overline{\lambda} > q^*$ with the property that the absence of $(\widehat{C}_{\overline{\lambda}})$–positive cycles does not imply the existence of a $\widehat{C}_{\overline{\lambda}}$–maximal s-v-path. Consequently, $\mathcal{G}$ does not have property (4.216).

It is an open question whether λ^* is always equal to q^* if the underlying digraph does not have property (4.216). $\qquad \square$

Remark 4.193 In [vGol87], page 4, the values $f(v) = \widehat{C}_\lambda(P^*(v))$ are computed with an algorithm that is similar to FORD-BELLMAN 1'; there are only two substantial differences:

(i) FORD-BELLMAN 1' computes the costs of minimal paths, the algorithm in [vGol87], however, computes the costs of $\widehat{C}_\lambda$-maximal paths.

(ii) FORD-BELLMAN 1' processes only the predecessors u of v with $u \in S$. The algorithm in [vGol87] processes all predecessors.

We next give the recursion formula for currently maximal $\widehat{C}_\lambda$-values. Let $\lambda \in \mathbb{R}$; then $\widehat{C}_\lambda$ is defined as $C_1 - \lambda \cdot C_2$. Let $\chi_k(v)$ be the value $\chi(v)$ that has been computed immediately after the $(k-1)$st iteration of FORD-BELLMAN 1' with the above modifications (i), (ii). Then the following recursion formula is valid for all nodes $v \in V$:

$$\chi_k(v) := \sup\{\chi_{k-1}(v, \lambda), \chi_{k-1}(u, \lambda) + a(u,v) - \lambda b(u,v)\,|\,(u,v) \in \mathcal{R}\}. \quad (4.217)$$

Defining $f_k(v, \lambda) := \chi_k(v)$, we obtain the following recursion formula, which is almost the same as formula (2.1) in [vGol87].

$$f_k(v, \lambda) := \sup \{ f_{k-1}(v, \lambda), f_{k-1}(u, \lambda) + a(u,v) - \lambda b(u,v)\,|\,(u,v) \in \mathcal{R} \}. \quad (4.218) \quad \square$$

Remark 4.194 The search for λ^* is also the solution of a linear program. To see this we assume that $V = \{w_1, \ldots, w_n\}$; we write x_i instead of $f(w_i)$. Then the search for a minimal λ can be formulated as the following linear program in the unknown variables $x_1, \ldots, x_n, \lambda$:

Minimize λ *subject to* $x_i - x_j - b(w_i, w_j) \cdot \lambda \leq -a(w_i, w_j)$ *for all* $(w_i, w_j) \in \mathcal{R}$.

That means that λ^* can be found with the simplex method, with the ellipsoid method or with another method to solve linear programs.

The equation $\lambda^* = q^*$ in 4.191 implies that Linear Programming can also be used to find the maximal (or the minimal) (C_1/C_2)-value of cycles; these methods to compute q^* are studied in [Pra91] and in [YJ91]. $\qquad\qquad\qquad\qquad\qquad\qquad\qquad\qquad\qquad\Box$

Next, we consider further problems that are closely related to Problem 35.

The following problem is to approximate a function $a(x, y)$ by a function of the form $\delta(x, y) = f(x) - f(y)$.

Problem 39 *Given a set* X, *which may be finite or infinite. Given a function* $a : X \times X \to \mathbb{R}$ *and a real number* $\lambda \in \mathbb{R}$.

Find a function $f : X \to \mathbb{R}$ *such that for* $\delta(x, y) := f(x) - f(y)$ *the following is true:*

$$(\forall x, y \in X) \quad a(x, y) - \delta(x, y) \;=\; a(x, y) + f(y) - f(x) \;\leq\; \lambda \qquad (4.219)$$

Moreover, compute the minimal number λ^* *with the property that at least one function* f *satisfies (4.219).*

If X is finite then the digraph $\mathcal{G}_X := (X, X \times X)$ is finite, too, and Problem 39 is the same as Problem 35 for the special digraph $\mathcal{G}_X$.

The following problem for matrices is studied in [vGoS84].

Problem 40 (Symmetric Scaling Problem)

Given an $n \times n$-*matrix* $\mathbf{A} = \big(A_{i,j}\big)$ *and a real number* $\Lambda > 0$.

Find a diagonal matrix $\mathbf{X}$ *with entries* $X_1, \dots, X_n \neq 0$ *such that for* $\mathbf{B} := \big(B_{i,j}\big) := \mathbf{X}\,\mathbf{A}\,\mathbf{X}^{-1}$ *the following is true:*

$$(\forall\, i, j = 1, \dots, n) \quad |B_{i,j}| \;=\; |X_i\, A_{i,j}\, X_j^{-1}| \;\leq\; \Lambda. \qquad (4.220)$$

Such a product $\mathbf{X}\,\mathbf{A}\,\mathbf{X}^{-1}$ is called a *symmetric scaling* of the matrix $\mathbf{A}$.

Next, we reduce the Symmetric Scaling Problem to Problem 35. For this purpose, we define the digraph $\mathcal{G} = (\mathcal{V}, \mathcal{R})$ with $\mathcal{V} := \{1, \dots, n\}$ and $\mathcal{R} := \{(i, j)\,|\,A_{i,j} \neq 0\}$. For all $(i, j) \in \mathcal{R}$, we define $a(i, j) := \ln\big(|A_{i,j}|\big)$ where "ln" denotes the natural logarithm. Let $\lambda := \ln \Lambda$. Then we formulate the following problem:

Problem 41 *Compute real numbers* $\xi(1), \dots, \xi(n)$ *such that*

$$(\forall\, (i, j) \in \mathcal{R}) \quad \xi(i) + a(i, j) - \xi(j) \;\leq\; \lambda. \qquad (4.221)$$

Then the systems (4.220) and (4.221) of inequalities are equivalent; if $X_1, \dots, X_n$ satisfy all conditions in (4.220) then the numbers $\xi(i) := \ln(|X_i|)$, $i = 1, \dots, n$, are a solution of (4.221); conversely, if $\xi(1), \dots, \xi(n)$ satisfy all conditions in (4.221) then the numbers $X_i := e^{\xi(i)}$, $i = 1, \dots, n$, are a solution of (4.220).

It is obvious that the system (4.221) of inequalities is equivalent to condition (4.207) in Problem 35.

Further details about finding symmetric scalings are given in [vGoS84].

In Remarks 4.195 and 4.196, we describe two problems with a remote relationship to Problem 36.

Remark 4.195 The following problem is studied in [Chre84]:

> **Problem 42** *Given a finite digraph $\mathcal{G} = (\mathcal{V}, \mathcal{R})$ and two additive path functions C_1, C_2. Let s, γ be two nodes of $\mathcal{G}$, and let K be a fixed value.*
> *Find the minimal value $C_1(P)$ of all paths $P \in \mathcal{P}(s, \gamma)$ with $C_2(P) = K$.*

There are two relationships between [Chre84] and the current section of this book:

- Chrètienne optimizes the values $C_1(P)$ of paths with a fixed value $C_2(P)$. The same idea is used in Algorithms 16 and 18.
- Chrètienne studies optimal cycles with regard to the cost measure C_1/C_2. □

Remark 4.196 In [BHSh90] and [BHHS94], an optimization problem about arc functions is formulated, and this problem is reduced to the search for an optimal node function; the resulting problem is similar to Problem 35.

To formulate the optimization problem in [BHSh90] and [BHHS94] we make the following assumptions:

Given a finite and acyclic digraph $\mathcal{G} = (\mathcal{V}, \mathcal{R})$. Let γ be a node in $\mathcal{G}$ such that there exists a v-γ-path for each node v. Let a be an arc function, and let $C = SUM_a$.

Let $\mathcal{H}$ be the set of all arc functions $h : \mathcal{R} \to [0, \infty)$ such that $H := SUM_h$ has the following property: $C(P) + H(P) = C(Q) + H(Q)$ for all paths P, Q with the same start node $\alpha(P) = \alpha(Q)$ and with the fixed end node $\omega(P) = \omega(Q) = \gamma$.

We formulate the original problem in [BHSh90] and [BHHS94] as follows:

> **Problem 43**   *Minimize* $\Phi(h) := \sum_{(u,v) \in \mathcal{R}} h(u,v)$ *subject to* $h \in \mathcal{H}$.

In [BHSh90], Problem 43 is reduced to the search for an optimal node function. The idea is that $C(P) + H(P)$ only depends on $v := \alpha(P)$ and not on P itself. We can therefore replace the search for H by a search for the node function $g(v) := C(P) + H(P)$. Thus, we obtain the following problem:

> **Problem 44**   *Minimize* $\Psi(g) := \sum_{(u,v) \in \mathcal{R}} (g(u) - g(v) - a(u,v))$
>
> *subject to* $g(u) - g(v) - a(u,v) \geq 0$ *for all* $(u,v) \in \mathcal{R}$.

The problems 43 and 44 are equivalent. This can be seen as follows:

Given a function $h \in \mathcal{H}$ (see Problem 43). Then let $g_h(v) := C(P_v) + H(P_v)$ where P_v is an arbitrary v-γ-path. Then $g_h(u) - g_h(v) = C((u,v) \oplus P_v) + H((u,v) \oplus P_v) - C(P_v) - H(P_v) = a(u,v) + h(u,v)$ for all $(u,v) \in \mathcal{R}$. Consequently, $0 \leq h(u,v) = g_h(u) - g_h(v) - a(u,v)$ for all arcs; this implies that g_h satisfies the side constraint of Problem 44. Moreover, $\Psi(g_h) = \Phi(h)$.

Conversely, let g satisfy the side constraint of Problem 44. Then we define $h_g(u,v) := g(u) - g(v) - a(u,v)$ for all arcs (u,v), and we define $H_g := SUM_{h_g}$. It is easy to see that $h_g \in \mathcal{H}$ and that $\Phi(h_g) = \Psi(g)$.

To show the similarity between the problems 43, 44 and Problem 35 we replace the node function g by $-f$. Then $-\big(g(u) - g(v) - a(u,v)\big) = a(u,v) - g(u) + g(v) = a(u,v) + f(u) - f(v)$ for all arcs (u,v). So, the problems 43 and 44 are equivalent to

Problem 45   *Maximize* $\sum\limits_{(u,v)\in \mathcal{R}} \big(a(u,v) + f(u) - f(v)\big)$

subject to: $a(u,v) + f(u) - f(v) \;\leq\; 0 \text{ for all } (u,v) \in \mathcal{R}.$

The common properties of the problems 35 and 45 are that they mean a search for a potential difference $\delta(u,v) = f(v) - f(u)$. Moreover, condition (4.207) is similar to the side constraint in Problem 45. The main differences between the problems 35 and 45 are the following: Problem 35 means to minimize the maximum of all values $\big(a(u,v) + f(u) - f(v)\big)$ where $(u,v) \in \mathcal{R}$; Problem 45, however, is to maximize the sum of all values $\big(a(u,v) + f(u) - f(v)\big)$ where $(u,v) \in \mathcal{R}$. Moreover, the underlying digraph in Problem 35 may have cycles, the digraph in Problem 45 must be acyclic. A further difference between the problems 35 and 45 is the way how to solve them. In 4.191, Problem 35 is reduced to the search for (C_1/C_2)-maximal cycles where $C_1 = SUM_h$ and $C_2 = \ell$. Problem 45, however, is solved by searching for optimal flows; this is described in [BHSh90].   □

4.9.3.2   Results and remarks in connection with Problem 37

Now we consider Problem 37. This problem arises from [vGol82] and from [vGol87]. In the next remark, we compare the problems 36 and 37.

Remark 4.197 First, Problem 37 is equivalent to the search for two functions $\overline{f}, \overline{g}$ with the following property:[52]

$$(\forall\, (u,v) \in \mathcal{R}) \quad |a(u,v) + \overline{f}(u) - \overline{g}(v)| \;\leq\; \lambda\, b(u,v). \qquad (4.222)$$

We compare condition (4.222) with (4.208). On the one hand, condition (4.222) is not so hard as (4.208) because we may choose two functions $\overline{f}, \overline{g}$ to satisfy (4.222) and only one function f to satisfy (4.208). On the other hand, condition (4.222) is harder than (4.208) because we must bound the absolute values $|a(u,v) + \overline{f}(u) - \overline{g}(v)|$ in (4.222) whereas we must only bound the values $a(u,v) + f(u) - f(v)$ themselves in (4.208). So, we cannot say that condition (4.222) and the equivalent condition (4.209) is less or more restricted than condition (4.208).   □

Next, we reduce Problem 37 to the search for a potential difference in a somewhat complicated digraph; our methods are similar to the ones in [vGoS84], page 124 – 125:

Remark 4.198 Given the notations of Problem 37. Then (4.209) is equivalent to $((4.223) \wedge (4.224))$ where (4.223) and (4.224) mean the following conditions:

$$a(x,y) - f(x) - g(y) \;\leq\; \lambda\, b(x,y), \quad (x,y) \in \mathcal{R}, \qquad (4.223)$$
$$-a(x,y) + f(x) + g(y) \;\leq\; \lambda\, b(x,y), \quad (x,y) \in \mathcal{R}. \qquad (4.224)$$

We now construct a bipartite digraph $\widetilde{\mathcal{G}} := (\widetilde{\mathcal{V}}, \widetilde{\mathcal{R}})$ in the following way: We generate a copy u' for each node u of $\mathcal{G}$; each arc (u,v) of $\mathcal{G}$ is represented by the two arcs (u,v') and (v',u) in $\widetilde{\mathcal{G}}$. More formally, we define $\widetilde{\mathcal{V}} := \mathcal{V} \cup \{u' \,|\, u \in \mathcal{V}\}$ and

---

[52]The equivalence of (4.209) and (4.222) can be seen by identifying $\overline{f}$ with $-f$ and $\overline{g}$ with g.

$\widetilde{\mathcal{R}} := \{(u,v'),(v',u) \mid (u,v) \in \mathcal{R}\}.$ — Moreover, we define for all $(x,y) \in \mathcal{R}$:

$c(y',x) := a(x,y), \quad c(x,y') := -a(x,y), \qquad d(y',x) := b(x,y), \quad d(x,y') := b(x,y).$

Then we formulate the following problem:

Problem 46 *Find a node function $\widetilde{f} : \widetilde{\mathcal{V}} \to \mathbb{R}$ with the following properties:*

$$c(y',x) + \widetilde{f}(y') - \widetilde{f}(x) \leq \lambda d(y',x), \quad (y',x) \in \widetilde{\mathcal{R}}, \qquad (4.225)$$
$$c(x,y') + \widetilde{f}(x) - \widetilde{f}(y') \leq \lambda d(x,y'), \quad (x,y') \in \widetilde{\mathcal{R}}. \qquad (4.226)$$

Then the solution of (4.225) $\wedge$ (4.226) is equivalent to the solution of (4.223) $\wedge$ (4.224). To see this we assume that $f,g : \mathcal{V} \to \mathbb{R}$ satisfy conditions (4.223) and (4.224); in this case, we define $\widetilde{f}(x) := f(x) \ (x \in \mathcal{V})$ and $\widetilde{f}(y') := -g(y) \ (y \in \mathcal{V})$; then the node function $\widetilde{f} : \widetilde{\mathcal{V}} \to \mathbb{R}$ satisfies (4.225) and (4.226).
Conversely, let $\widetilde{f} : \widetilde{\mathcal{V}} \to \mathbb{R}$ satisfy (4.225) and (4.226); in this case, we define $f(x) := \widetilde{f}(x)$ for all $x \in \mathcal{V}$ and $g(y) := -\widetilde{f}(y')$ for all $y \in \mathcal{V}$; then the node functions $f,g : \mathcal{V} \to \mathbb{R}$ satisfy both conditions (4.223) and (4.224).

Consequently, solving ((4.225) $\wedge$ (4.226)) is equivalent to solving ((4.223) $\wedge$ (4.224)), and this is equivalent to satisfying condition (4.209). □

Now we describe how to find the minimal λ^* for which Problem 37 can be solved.

Remark 4.199 In this remark, we use the notations of Remark 4.198.

In particular, let c and d be the arc functions defined in 4.198. Then we define the following additive path functions for $\widetilde{\mathcal{G}}$: $\widetilde{C}_1 := SUM_c$ and $\widetilde{C}_2 := SUM_d$. Moreover, let $\widetilde{C} := \widetilde{C}_1/\widetilde{C}_2$, and let $\widetilde{q}^*$ be the supremum of all values $\widetilde{C}(P)$ where P is a substantial cycle in $\widetilde{\mathcal{G}}$.

It follows from Theorem 4.191 that $\widetilde{q}^*$ is equal to the minimal value λ^* for which conditions (4.225) and (4.226) in 4.198 can be satisfied. Moreover, ((4.225) $\wedge$ (4.226)) is equivalent to condition (4.209). Therefore, $\widetilde{q}^*$ is also equal to the minimal number λ^* for which (4.209) of Problem 37 can be solved.

Next, we give several representations and descriptions of $\widetilde{q}^* = \lambda^*$. This helps to understand particular constructions, definitions and formulas in [vGol80] – [vGol87].

- *Representing $\widetilde{\mathcal{G}}$-cycles in $\mathcal{G}$*

Now we describe how cycles P in $\widetilde{\mathcal{G}}$ can be represented in the original digraph $\mathcal{G}$. Roughly speaking, we must alternate between using an arc in the given direction and in the opposite one. Our intention is to make the definitions of paths in [vGol80], page 46 and in [vGol82], page 76 more comprehensible.

First, we show that we may restrict our attention to circles $P \in \mathcal{P}(\widetilde{\mathcal{G}})$ with $\alpha(P) \in \widetilde{\mathcal{V}}\backslash\mathcal{V}$. Let $P' = r_1 \oplus r_2 \oplus \ldots \oplus r_k$ be a substantial cycle in $\widetilde{\mathcal{G}}$ with $\alpha(P') \in \mathcal{V}$. Then the cycle $P := r_2 \oplus \ldots \oplus r_k \oplus r_1$ has the property that $\alpha(P) \in \widetilde{\mathcal{V}}\backslash\mathcal{V}$; moreover,

$\widetilde{C}_i(P) = \widetilde{C}_i(P')$, $i = 1, 2$, and in consequence, $\widetilde{C}(P) = \widetilde{C}_1(P)/\widetilde{C}_2(P) = \widetilde{C}_1(P')/\widetilde{C}_2(P') = \widetilde{C}(P')$. So, $\widetilde{q}^*$ is also the supremum of all values $C(P)$ where P is a substantial cycle in $\widetilde{\mathcal{G}}$ with $\alpha(P) \in \widetilde{\mathcal{V}}\backslash\mathcal{V}$. Let $\widetilde{\mathcal{P}}_0$ denote the set of these cycles. Then each cycle $P \in \widetilde{\mathcal{P}}_0$ has the following form:

$$
\begin{aligned}
P &= [w_1', v_1, w_2', v_2, w_3', v_3, \ldots, w_k', v_k, w_{k+1}' := w_1'] \\
&= (w_1', v_1) \oplus (v_1, w_2') \oplus (w_2', v_2) \oplus \ldots \oplus (w_k', v_k) \oplus (v_k, w_{k+1}') \\
&\quad \text{with } v_\rho \in \mathcal{V}, \ w_\rho \in \mathcal{V}, \ (v_\rho, w_\rho) \in \mathcal{R}, \ (v_\rho, w_{\rho+1}) \in \mathcal{R} \\
&\quad \text{for all } \rho = 1, \ldots, k.
\end{aligned}
\tag{4.227}
$$

We represent P as the sequence Q_P of all original nodes in P:
$$
Q_P := (w_1, v_1, w_2, v_2, \ldots, w_k, v_k, w_{k+1} = w_1). \tag{4.228}
$$
Obviously, Q uses the arc $(v_1, w_1) \in \mathcal{R}$ in its opposite direction, the next arc $(v_1, w_2) \in \mathcal{R}$ in its given direction, the next arc $(v_2, w_2) \in \mathcal{R}$ in its opposite direction and so on. This behaviour of Q_P is shown in the following sketch:

$$
Q_P : w_1 \leftarrow v_1 \rightarrow w_2 \leftarrow v_2 \rightarrow w_3 \leftarrow \ldots \rightarrow w_{k-1} \leftarrow v_{k-1} \rightarrow w_k \leftarrow v_k \rightarrow w_{k+1} = w_1.
$$

That means that Q_P has the same structure as the closed paths in [vGol80], page 46 and in [vGol82], page 76.

We define $\mathbf{Q}$ as the set of all sequences Q_P with $P \in \widetilde{\mathcal{P}}_0$.

It is easy to reconstruct $P \in \widetilde{\mathcal{P}}_0$ if $Q_P \in \mathbf{Q}$ is given.

• *Expressing $\widetilde{C}_1(P)$ and $\widetilde{C}_2(P)$ with the help of the given arc functions a, b*

For all cycles $P \in \widetilde{\mathcal{P}}_0$, we express $\widetilde{C}_1(P)$ and $\widetilde{C}_2(P)$ in terms of the arc functions a, b, which have been given in the original digraph $\mathcal{G}$.

Let P have the form described in (4.227). By the definition of c and d, we obtain the following equations:

$$
\widetilde{C}_1(P) = \sum_{\rho=1}^{k}\left[c(w_\rho', v_\rho) + c(v_\rho, w_{\rho+1}')\right] = \sum_{\rho=1}^{k}[a(v_\rho, w_\rho) - a(v_\rho, w_{\rho+1})], \tag{4.229}
$$
$$
\widetilde{C}_2(P) = \sum_{\rho=1}^{k}\left[d(w_\rho', v_\rho) + d(v_\rho, w_{\rho+1}')\right] = \sum_{\rho=1}^{k}[b(v_\rho, w_\rho) + b(v_\rho, w_{\rho+1})]. \tag{4.230}
$$

That means that can express $\widetilde{C}_1(P)$ and $\widetilde{C}_2(P)$ with the arc functions a, b.

• *Defining pseudolengths of the sequences Q_P*

We define the functions $\mathbf{C}_1, \mathbf{C}_2 : \mathbf{Q} \to \mathbb{R}$ as follows: Let $\mathbf{C}_i(Q_P) := \widetilde{C}_i(P)$ for all $Q_P \in \mathbf{Q}$ and $i = 1, 2$. In [vGol82], page 76, the values $\mathbf{C}_1(Q_P)$ and $\mathbf{C}_2(Q_P)$ are called *pseudolengths* of Q_P. The following formula shows how to compute these values.

$$
\mathbf{C}_1(Q_P) \overset{(4.228),(4.229)}{=\!=\!=\!=\!=} \sum_{\rho=1}^{k}[a(v_\rho, w_\rho) - a(v_\rho, w_{\rho+1})],
$$
$$
\mathbf{C}_2(Q_P) \overset{(4.228),(4.230)}{=\!=\!=\!=\!=} \sum_{\rho=1}^{k}[b(v_\rho, w_\rho) + b(v_\rho, w_{\rho+1})].
$$

So, $\mathbf{C}_1(Q_P)$ is computed by adding each value $a(v_\kappa, w_\kappa)$ and by subtracting each value $a(v_\kappa, w_{\kappa+1})$; $\mathbf{C}_2(Q_P)$ is computed by adding all values $b(v_\kappa, w_\kappa)$ and $b(v_\kappa, w_{\kappa+1})$. The computation of $\mathbf{C}_1(Q_P)$ is shown in the following sketch:

$$Q_P \; : \quad w_1 \xleftarrow{+a(v_1,w_1)} v_1 \xrightarrow{-a(v_1,w_2)} w_2 \xleftarrow{+a(v_2,w_2)} v_2 \xrightarrow{-a(v_2,w_3)} w_3 \xleftarrow{+a(v_3,w_3)} \cdots\cdots$$

$$\cdots \xleftarrow{+a(v_{k-1},w_{k-1})} v_{k-1} \xrightarrow{-a(v_{k-1},w_k)} w_k \xleftarrow{+a(v_k,w_k)} v_k \xrightarrow{-a(v_k,w_{k+1})} w_{k+1} \; = \; w_1 \,.$$

- *Expressing $\lambda^* = \tilde{q}^*$ with absolute values of pseudolengths*

We show an equation that is analogous to assertion (7.4) in [vGol82]:

$$\lambda^* \; = \; \tilde{q}^* \; = \; \sup \left\{ \frac{|C_1(Q)|}{C_2(Q)} \,\Big|\, Q \in \mathbf{Q} \right\} . \tag{4.231}$$

For this purpose, we observe that for each arc $r \in \tilde{\mathcal{R}}$, the reverse arc $-r := (\omega(r), \alpha(r))$ is also in $\tilde{\mathcal{R}}$. Consequently, if $P = r_1 \oplus \cdots \oplus r_k \in \tilde{\mathcal{P}}_0$ then the reverse path $-P := (-r_k) \oplus (-r_{k-1}) \oplus \cdots \oplus (-r_1)$ is in $\tilde{\mathcal{P}}$ as well. It is $\tilde{C}_1(-P) = -\tilde{C}_1(P)$ and $\tilde{C}_2(-P) = \tilde{C}_2(P)$ because $c(-r) = -c(r)$ and $d(-r) = d(r)$ for all arcs r in $\tilde{\mathcal{G}}$. This implies that $\tilde{C}(-P) = (-\tilde{C}_1(P))/(\tilde{C}_2(P)) = -\tilde{C}(P)$ for all P in $\tilde{\mathcal{P}}_0$. Consequently, if $\tilde{q} = \tilde{C}(P)$ is a candidate for $\tilde{q}^*$ then $-\tilde{q} = -\tilde{C}(P) = \tilde{C}(-P)$ is a candidate, too. Consequently, $\tilde{q}^*$ is also the supremum of all absolute values $|\tilde{C}(P)| = |\tilde{C}_1(P)|/\tilde{C}_2(P)$ $= |C_1(Q_P)|/C_2(Q_P)$ where $P \in \tilde{\mathcal{P}}_0$. This implies fact (4.231).

By the way, if $b(r) = 1$ for all $r \in \mathcal{R}$ then $d(\tilde{r}) = 1$ for all arcs $\tilde{r} \in \mathcal{R}$ so that $\tilde{C}_2 = \ell$. In this case, the candidates for $\tilde{q}^*$ can be written in a simpler form:

$$|\tilde{C}(P)| \; = \; \frac{|\tilde{C}_1(P)|}{\ell(P)} \; = \; \frac{\left| \sum_{\rho=1}^{k} [+a(v_\rho, w_\rho) - a(v_\rho, w_{\rho+1})] \right|}{2\,k} . \qquad \Box$$

In the next remark, we describe the computation of the functions $\tilde{f}$, f, and g in Remark 4.198 with the help of a Ford-Bellman strategy; in particular, we derive two formulas (4.237) and (4.238) that are almost the same as the recursion formulas (7.6) and (7.7) in [vGol82].

Remark 4.200 We use the notations of Remark 4.198.

If λ is given, the function $\tilde{f}$ can be computed with the modified version of FORD-BELLMAN 1' that was described in Remark 4.193. We transform the recursion formula (4.218) into the current situation. Thus, we obtain the following formula, which is valid for all nodes $v \in \tilde{\mathcal{V}}$.

$$\tilde{f}_k(v, \lambda) \; := \; \sup\{\tilde{f}_{k-1}(v, \lambda)\,,\, \tilde{f}_{k-1}(u, \lambda) + c(u,v) - \lambda\, d(u,v) \,|\, (u,v) \in \tilde{\mathcal{R}}\} \,.$$

Changing the order of the summands, we derive the following equivalent formula:

$$\tilde{f}_k(v, \lambda) := \sup\{\tilde{f}_{k-1}(v, \lambda)\,,\, c(u,v) - \lambda\, d(u,v) + \tilde{f}_{k-1}(u, \lambda) \,|\, (u,v) \in \tilde{\mathcal{R}}\} \,.$$

We formulate this formula for the nodes $v = x \in \mathcal{V}$ and the nodes $v = y'$ with $y \in \mathcal{V}$; thus, we obtain the following equations:

$$\tilde{f}_k(x, \lambda) := \sup\left\{\tilde{f}_{k-1}(x, \lambda)\,,\, c(y', x) - \lambda\, d(y', x) + \tilde{f}_{k-1}(y', \lambda) \,\Big|\, (y', x) \in \tilde{\mathcal{R}}\right\} \; (x \in \mathcal{V}), \tag{4.232}$$

$$\tilde{f}_k(y', \lambda) := \sup\left\{\tilde{f}_{k-1}(y', \lambda)\,,\, c(x, y') - \lambda\, d(x, y') + \tilde{f}_{k-1}(x, \lambda) \,\Big|\, (x, y') \in \tilde{\mathcal{R}}\right\} \; (y \in \mathcal{V}). \tag{4.233}$$

Moreover, (4.233) is equivalent to

$$-\widetilde{f}_k(y',\lambda) := \inf\left\{-\widetilde{f}_{k-1}(y',\lambda), -c(x,y') + \lambda\, d(x,y') - \widetilde{f}_{k-1}(x,\lambda)\,\middle|\,(x,y')\in\widetilde{\mathcal{R}}\right\}\quad (y\in\mathcal{V}). \qquad (4.234)$$

We use these equations to derive recursion formulas for the solutions f,g of the original problem 37. At the end of Remark 4.198, we have seen that $f(x) := \widetilde{f}(x)$ $(x\in\mathcal{V})$ and $g(y) := -\widetilde{f}(y')$ $(y\in\mathcal{V})$ solve Problem 37 if $\widetilde{f}$ satisfies $((4.225)\wedge(4.226))$. Therefore, we can therefore find all values $f(x)$ and $g(y)$ by computing the intermediate values $f_k(x,\lambda) := \widetilde{f}_k(x,\lambda)$ and $g_k(y,\lambda) := -\widetilde{f}_k(y',\lambda)$ for all $x,y\in\mathcal{V}$ and all $k = 1,\ldots,n$. That means that we transform the formulas (4.232) and (4.234) as follows:
We replace

- $\widetilde{f}_k(x,\lambda)$ by $f_k(x,\lambda)$,
- $\widetilde{f}_k(y',\lambda)$ by $-g_k(y,\lambda)$,
- $c(y',x)$ by $a(x,y)$,
- $d(y',x)$ by $b(x,y)$,

- $\widetilde{f}_{k-1}(x,\lambda)$ by $f_{k-1}(x,\lambda)$,
- $\widetilde{f}_{k-1}(y',\lambda)$ by $-g_{k-1}(y,\lambda)$,
- $c(x,y')$ by $-a(x,y)$,
- $d(x,y')$ by $b(x,y)$.

Thus, we obtain (4.235) and (4.236), respectively.

$$f_k(x,\lambda) := \sup\left\{f_{k-1}(x,\lambda),\, a(x,y) - \lambda\, b(x,y) - g_{k-1}(y,\lambda)\,\middle|\,(x,y)\in\mathcal{R}\right\}\quad (x\in\mathcal{V}), \qquad (4.235)$$

$$g_k(y,\lambda) := \inf\left\{g_{k-1}(y,\lambda),\, a(x,y) + \lambda\, b(x,y) - f_{k-1}(x,\lambda)\,\middle|\,(x,y)\in\mathcal{R}\right\}\quad (y\in\mathcal{V}). \qquad (4.236)$$

We may assume that the kth iteration of (4.235) has been been finished before the kth iteration of (4.236) has begun. In this case, we may replace the values $f_{k-1}(x,\lambda)$ in (4.235) by $f_k(x,\lambda)$. Thus, we obtain the following formulas:

$$f_k(x,\lambda) := \sup\left\{f_{k-1}(x,\lambda),\, a(x,y) - \lambda\, b(x,y) - g_{k-1}(y,\lambda)\,\middle|\,(x,y)\in\mathcal{R}\right\}\quad (x\in\mathcal{V}), \qquad (4.237)$$

$$g_k(y,\lambda) := \inf\left\{g_{k-1}(y,\lambda),\, a(x,y) + \lambda\, b(x,y) - f_k(x,\lambda)\,\middle|\,(x,y)\in\mathcal{R}\right\}\quad (y\in\mathcal{V}). \qquad (4.238)$$

These recursion formulas are almost the same as the equations (7.6) and (7.7) in [vGol82]. $\qquad\square$

Next, we discuss several problems that are closely related to Problem 37.

First, we consider the following problem; a function $a(x,y)$ must be approximated by a function of the form $f(x) + f(y)$.

Problem 47 (*see also Problem 39*)
Given two sets X,Y, two functions $a,b : X\times Y\to\mathbb{R}$, and a real number $\lambda\in\mathbb{R}$.
Find two functions $f : X\to\mathbb{R}$ and $g : Y\to\mathbb{R}$ such that the following is true:

$$(\forall\,x\in X,\ y\in Y)\ \ |a(x,y) - (f(x) + g(y))| \ \leq\ \lambda b(x,y).$$

Moreover, compute the minimal number λ^* with the property that at least pair (f,g) satisfies the above condition.

If X and Y are finite then Problem 47 is almost the same as Problem 37 for the digraph $\mathcal{G}_{X,Y} := (\mathcal{V},\mathcal{R})$ with $\mathcal{V} := X\cup Y$ and $\mathcal{R} := X\times Y$.[53]

The above approximation problem 47 solved in [vGol82].

---

[53]There is, however, the following difference between Problem 37 and in Problem 47: The node functions f and g in Problem 47 must only be defined for all $x\in X$ and $y\in Y$, respectively. The

A modification of the above approximation problem is that X and Y are compact subsets of $\mathbb{R}$, the given function a is continuous and the unknown functions f, g are required to be continuous, too. This modified problem is the same as Problem B in [vGol82]; the solution of this Problem B is described in Theorem 7.1 of [vGol82].

Next, we discuss a matrix problem closely related to Problem 37. For this purpose, we introduce the following notation: Let $\mathbf{U} = (U_{i,j})$ be an $n \times n$-matrix. Let $\mathbf{T}$ and $\mathbf{V}$ be two diagonal matrices with entries $T_1, \ldots, T_n \neq 0$ and $V_1, \ldots, V_n \neq 0$, respectively. Let $\mathbf{W} := (W_{i,j}) := \mathbf{T} \cdot \mathbf{U} \cdot \mathbf{V}$. Then we define the following numbers:

$$W(\mathbf{T}, \mathbf{U}, \mathbf{V}) := \sup\{|W_{i,j}| \,|\, W_{i,j} \neq 0\}, \quad w(\mathbf{T}, \mathbf{U}, \mathbf{V}) := \inf\{|W_{i,j}| \,|\, W_{i,j} \neq 0\}.$$

Note that $W_{i,j} = T_i \cdot U_{i,j} \cdot V_j$ for all i, j. Therefore, the following equations are valid:

$$W(\mathbf{T}, \mathbf{U}, \mathbf{V}) = \sup\{|T_i \cdot U_{i,j} \cdot V_j| \,|\, U_{i,j} \neq 0\}, \quad w(\mathbf{T}, \mathbf{U}, \mathbf{V}) = \inf\{|T_i \cdot U_{i,j} \cdot V_j| \,|\, U_{i,j} \neq 0\}.$$

Now we formulate the following matrix problem, which is studied in [vGol80].

Problem 48 (Asymmetric Scaling Problem)

Given an $n \times n$-matrix $\mathbf{A} = (A_{i,j})$ and a real number $\Lambda > 0$.

Compute a diagonal matrix $\mathbf{T}$ with entries $T_1, \ldots, T_n \neq 0$ and a diagonal matrix $\mathbf{Z}$ with entries $Z_1, \ldots, Z_n \neq 0$ such that such that

$$\frac{W(\mathbf{T}, \mathbf{A}, \mathbf{Z})}{w(\mathbf{T}, \mathbf{A}, \mathbf{Z})} \leq \Lambda. \tag{4.239}$$

This scaling problem can be normalized as follows:

Problem 49

Given an $n \times n$-matrix $\mathbf{A} = (A_{i,j})$ and a real number $\Lambda > 0$.

Compute a diagonal matrix $\mathbf{X}$ with entries $X_1, \ldots, X_n \neq 0$ and a diagonal matrix $\mathbf{Y}$ with entries $Y_1, \ldots, Y_n \neq 0$, respectively, such that

$$W(\mathbf{X}, \mathbf{A}, \mathbf{Y}) \leq \Lambda \quad \text{and} \quad w(\mathbf{X}, \mathbf{A}, \mathbf{Y}) \geq 1. \tag{4.240}$$

The problems 48 and 49 are equivalent. If the matrices $\mathbf{T}$ and $\mathbf{Z}$ have property (4.239) then the matrices $\mathbf{X} := \mathbf{T}$ and $\mathbf{Y} := \mathbf{Z}/w(\mathbf{T}, \mathbf{A}, \mathbf{Z})$ have property (4.240). Conversely, if the matrices $\mathbf{X}$ and $\mathbf{Y}$ have property (4.240) then the matrices $\mathbf{T} := \mathbf{X}$ and $\mathbf{Z} := \mathbf{Y}$ have property (4.239).

Next, we reduce Problem 49 to a graph theoretic one that resembles Problem 37. For this purpose, we define the digraph $\mathcal{G}_0 = (\mathcal{V}_0, \mathcal{R}_0)$ with $\mathcal{V}_0 := \{1, \ldots, n\}$ and $\mathcal{R}_0 := \{(i,j) \,|\, A_{i,j} \neq 0\}$. For each $(i,j) \in \mathcal{R}$, we define $a(i,j) := a_{i,j} := \ln(|A_{i,j}|)$ where "ln" denotes the natural logarithm. Let $\lambda := \frac{1}{2} \ln \Lambda$. Then we formulate the following problem:

---

node functions f and g in Problem 37, however, must be defined for all $v \in \mathcal{V} = X \cup Y$. But this difference is not substantial for the following reason: If $f : X \to \mathbb{R}$ and $g : Y \to \mathbb{R}$ solve Problem 47 then each extension of f and g to $\mathcal{V}$ is a solution of Problem 37. (Note that $(x,y) \in \mathcal{R}$ only if $x \in X$, $y \in Y$.) Conversely, if $f, g : X \cup Y \to \mathbb{R}$ solve Problem 37 then the restrictions $f|_X$ and $g|_Y$ solve Problem 47.

Problem 50 *Compute real numbers $x_1, \ldots, x_n$ and $y_1, \ldots, y_n$ such that*

$$(\forall\, (i,j) \in \mathcal{R}) \quad 0 \;\leq\; a_{i,j} - x_i - y_j \;\leq\; 2\lambda. \tag{4.241}$$

Then the problems 49 and 50 are equivalent; if the diagonal matrices $\mathbf{X}$ and $\mathbf{Y}$ satisfy condition (4.240) then the numbers $x_i := -\ln(|X_i|)$ and $y_i := -\ln(|Y_i|)$, $i = 1, \ldots, n$, are a solution of (4.241); conversely, if $x_1, \ldots, x_n$ and $y_1, \ldots, y_n$ satisfy all conditions in (4.241) then the diagonal matrices $\mathbf{X}$ and $\mathbf{Y}$ with entries $X_i := e^{-x_i}$ and $Y_i := e^{-y_i}$ are a solution of (4.240).

Solving all inequalities in (4.241) is equivalent to solving Problem 37 for $\mathcal{G}_0$. The reason is that condition (4.241) is equivalent to the following condition:

$$(\forall\, (i,j) \in \mathcal{R}) \quad \left| a(i,j) - \underbrace{(x_i + \lambda)}_{=:f(i)} - \underbrace{y_j}_{=:g(j)} \right| = \left| a_{i,j} - (x_i + \lambda) - y_j \right| \;\leq\; \lambda. \tag{4.242}$$

Consequently, condition (4.242) is equivalent to condition (4.209) in Problem 37; we must only replace $x_i + \lambda$ by $f(i)$ and y_i by $g(i)$ in condition (4.242).

We now have reduced the Asymmetric Scaling Problem to Problem 37. For this purpose, we have shown that the following problems are equivalent:

- finding two matrices $\mathbf{T}$ and $\mathbf{Z}$ such that assertion (4.239) is true,
- finding two matrices $\mathbf{X}$ and $\mathbf{Y}$ such that assertion (4.240) is true,
- finding real numbers $x_1, \ldots, x_n$ and $y_1, \ldots, y_n$ such that assertion (4.241) and the equivalent assertion (4.242) are true,
- finding two node functions $f, g : \mathcal{V}_0 \to \mathbb{R}$ such that assertion (4.209) in Problem 37 is true.

Next, we give two recursion formulas to compute the values $x_1, \ldots, x_n$ and $y_1, \ldots, y_n$. For this end, we recall the formulas (4.237) and (4.238), which help to find a solution of Problem 37. Let $M := 2\lambda \, (= \ln \Lambda)$. For all nodes i, j and for all k, we define

$$x_i(k, M) := f_k(i, \lambda) - \lambda = f_k(i, M/2) - M/2, \quad y_j(k, M) := g_k(j, \lambda) = g_k(j, M/2). \tag{4.243}$$

(We use the names $x_i(k, M)$ and $y_j(k, M)$ to make our terminology equal to the one in [vGol80].)

Let $b(i,j) := 1$ for all nodes i, j. Then the following equations are valid if $f_k(i, \lambda)$ and $g_k(j, \lambda)$ are generated according to (4.237) and (4.238):

$$(\forall\, i \in \mathcal{V}_0) \quad x_i(k, M) \stackrel{(4.243)}{=} f_k(i, \lambda) - \lambda \stackrel{(4.237)}{=}$$

$$\sup\left\{ \left(f_{k-1}(i, \lambda) - \lambda \right), a(i,j) - \lambda \cdot \underbrace{b(i,j)}_{=1} - \lambda - g_{k-1}(j, \lambda) \,\middle|\, (i,j) \in \mathcal{R}_0 \right\}$$

$$\stackrel{2\lambda = M, \,(4.243)}{=\!=\!=\!=} \sup\left\{ x_i(k-1, M), a_{i,j} - y_j(k-1, M) - M \,\middle|\, (i,j) \in \mathcal{R}_0 \right\}.$$

$$(\forall\, j \in \mathcal{V}_0) \quad y_j(k, M) \stackrel{(4.243)}{=} g_k(j, \lambda) \stackrel{(4.238)}{=}$$

$$\inf\left\{ g_{k-1}(j, \lambda), a(i,j) + \lambda \cdot \underbrace{b(i,j)}_{=1} - f_k(i, \lambda) \,\middle|\, (i,j) \in \mathcal{R}_0 \right\}$$

$$\stackrel{(4.243)}{=} \inf\left\{ y_j(k-1, M), a_{i,j} - x_i(k, M) \,\middle|\, (i,j) \in \mathcal{R}_0 \right\}.$$

Similar formulas can be obtained by combining (3) with (4) and (5) with (6) in [vGol80].

Moreover, the quantities $f(i)$ and $g(j)$ in (4.242) play a similar rôle as the notations x_i^* and y_j^*, respectively; these notations are defined in [vGol80], formula (8).

4.9.3.3 Results and Remarks in connection with Problem 38

Theorem 4.115 of [Huck92] gives estimations of the minimal number λ for which condition (4.210) can be solved; the result in [Huck92] is a generalization of Theorem 4.191 a) and b) in this book. Perhaps, Theorem 4.115 of [Huck92] will also be proven in a second edition of this book.

4.10 Path problems with several path functions or with side constraints

This section is about the following problems:

- Given several cost functions $C_1, \ldots, C_k$. Find a path P such that all values $C_1(P), \ldots, C_k(P)$ are very small.
- Given a cost function C. Find a C-minimal path (or its C-value) among all candidates satisfying a particular side constraint (e.g., see Problem 42).

The second of these problems can be solved with Algorithm 16; this algorithm searches for a $\widetilde{C}_1$-minimal s-v-path $P_l^*(v)$ among all candidates with the fixed value $\widetilde{C}_2(P^*(v)) = l$.

The presence of several path functions or of side constraints often makes the search for optimal paths difficult. Many problems are so hard that they probably cannot deterministically be solved in polynomial time; these problems are closely related to Section 4.34.

4.10.1 Path problems with several path functions

We now concentrate on the following type of problems:

Problem 51 *Given a digraph* $\mathcal{G} = (\mathcal{V}, \mathcal{R})$. *Let* $\mathcal{P}$ *be a set of paths, and let* $C_1, C_2 : \mathcal{P} \to \mathbb{R}$ *be two path functions.*
Find a path $P^* \in \mathcal{P}$ *with small* C_1- *and* C_2-*values.*

In most cases, there does not exist a C_1-minimal path that is C_2-minimal at the same time. For this reason, there exist many criteria for the optimality of paths. We give five examples.

- *Pareto-Minimality:* We define $(a,b) \preceq (c,d) :\iff (a \le c \lor b \le d)$ for all real numbers a, b, c, d. A path P^* is *Pareto-minimal* among all paths $P \in \mathcal{P}$ if and only if the pair $(C_1(P^*), C_2(P^*))$ is $\preceq$-minimal among all pairs $(C_1(P), C_2(P))$, $P \in \mathcal{P}$. That means that no path $P \in \mathcal{P}$ exists with $C_1(P) < C_1(P^*)$ and $C_2(P) < C_2(P^*)$.

• *Minimality with regard to domination:* This version of minimality is similar to Parato-minimality. We say that P^* is *minimal with regard to domination* if no path P exists with the following properties:
$$C_1(P) \le C_1(P^*), \; C_2(P) \le C_2(P^*), \text{ and } (C_1(P), C_2(P)) \ne (C_1(P^*), C_2(P^*)).$$

• *Lexicographic Minimality:* We define the relation "$\le$" as follows: Let
$$(a, b) \le (c, d) \; :\Longleftrightarrow \; \big(a < c \lor (a = c \land b \le d)\big) \text{ for all real numbers } a, b, c, d.$$
Then a path P^* is called *lexicographically minimal with respect to* C_1, C_2 if and only if $(C_1(P^*), C_2(P^*)) \preceq (C_1(P), C_2(P))$ for all paths $P \in \mathcal{P}$.

• *Extremal position in a set of points:* We consider the pairs $(C_1(P), C_2(P))$, $(P \in \mathcal{P})$ as points in a two-dimensional Cartesian coordinate system. Let $\mathcal{C}$ be the convex hull of these points.
Then P^* is considered as a very good or a very bad path if $(C_1(P^*), C_2(P^*))$ is an extremal point (= vertex) of $\mathcal{C}$.

(That means that we have a path problem in geometric context. But there is a substantial difference between this problem and the ones in Section 4.5.1. Roughly speaking, in Section 4.5.1, we reduced geometric problems to path problems in graphs; here, however, we reduce path problems in graphs to geometric problems.)

• *Optimality with respect to a new real-valued cost function:* A new cost function $C : \mathcal{P} \to \mathbb{R}$ is defined with the help of C_1 and C_2; then we search for a C-minimal path P^*. Typical examples for C are

   – convex combinations $C(P) := \alpha \cdot C_1(P) + (1 - \alpha)C_2(P)$ where $(0 < \alpha < 1)$,

   – the maximum norm $C(P) := \|(C_1(P), C_2(P))\|_{\max} = \max(|C_1(P)|, |C_2(P)|)$.

Most of these optimality criteria are studied in [Hen85]. The paper [EmPe89] is addressed to the search for Pareto-optimal paths. The Best-First Search algorithm MOA* in [StWh91] finds optimal paths with regard to domination. In [GuLe87], the search for a C-minimal path is described where C is computed from two order preserving functions C_1, C_2.

The search for optimal paths with regard to the above optimality criteria is often a hard problem. For example, let $C(P) := \max(|C_1(P)|, |C_2(P)|)$ for all paths P where C_1 and C_2 are additive and prefix monotone. Then [Vaĭn85] says that finding a C-minimal s-γ-path is $\mathcal{NP}$-hard.

Remark 4.201
   a) In most cases, the following search problems have different solutions $P^* \ne Q^*$:
      a1) Find a path $P^* \in \mathcal{P}$ with lexicographically minimal value $(C_1(P^*), C_2(P^*))$.
      a2) Find a path $Q^* \in \mathcal{P}$ with lexicographically minimal value $(C_2(P^*), C_1(P^*))$.
     Both problems, however, are equivalent in the following situation:
     Let $\mathcal{P}_1, \mathcal{P}_2 \subseteq \mathcal{P}(\mathcal{G})$ with $\mathcal{P}_1 \cap \mathcal{P}_2 \ne \emptyset$; we define $C_i(P) := 0$ for all $i = 1, 2$ and all $P \in \mathcal{P}_i$; we define $C_i(P) := 1$ for all $i = 1, 2$ and all $P \in \mathcal{P} \backslash \mathcal{P}_i$. Then both problems a1), a2) have the same set of solutions, and this set is $\mathcal{P}_1 \cap \mathcal{P}_2$.

b) A natural extension of Problem 51 is the following: Given more than two path functions $C_1, \ldots, C_j : \mathcal{P} \to \mathbb{R}$. Find a path P^* with small values $C_1(P^*), \ldots, C_j(P^*)$.
All above optimality criteria can be formulated if $j > 2$:
Pareto minimality, lexicographic minimality, extremal position in a set of points, and optimality with respect to a new real-valued cost function. □

A further path problem with two path functions is the following:

Problem 52 *Given a digraph* $\mathcal{G} = (\mathcal{V}, \mathcal{R})$. *Let* $\mathcal{P}$ *be a set of paths, and let* $C_1, C_2 : \mathcal{P} \to \mathbb{R}$ *be two path functions. Let* $b \in \mathbb{R}$.
Find a path C_1-*minimal or a* C_1-*maximal path* $P^* \in \mathcal{P}$ *among all candidates* P *with* $C_2(P) \leq b$.

For example, the traveller's problem in [HiLa92] and [AgTo93] can be reduced to Problem 52. The original problem in [HiLa92] and [AgTo93] is to compute the maximum distance a traveller can cover if the following constraints are given: The traveller must stop at an inn each night and pay the cost of that inn out of his limited pocket; moreover, the travelling distance per day is limited, too. The reduction to Problem 52 is the following: Let $\mathcal{G} := (\mathcal{V}, \mathcal{R})$ where $\mathcal{V}$ contains all inns and the traveller's starting point s; we define $\mathcal{R}$ as the set of all pairs (u, v) with the property that the traveller can move from u to v within one day. Let $h_1(u, v)$ be equal to the distance between u and v, and let $h_2(u, v)$ be equal to the room rate of the hotel v. Let $C_i := SUM_{h_i}$ $(i = 1, 2)$, and let b equal the traveller's budget. If the traveller uses a path $P \in \mathcal{P}(s)$ then $C_1(P)$ is the length of the traveller's trip, and $C_2(P)$ is the traveller's cost for overnight stays. Consequently, the traveller's problem is to find a C_1-maximal path P^* such that $C_2(P^*) \leq b$.

A modification of Problem 52 is Problem 57; this problem is solved in [AnNa78].

The authors of [HöWa95] describe the search for a C_1-minimal path P with the side constraint that $C_2(P)$ must have a fixed value; the values of C_2 are vectors. An easier version is Problem 42, which has been studied in [Chre84].

4.10.2 Path problems with rational side constraints

Now we study path problems with so-called *rational* side constraints; these side constraints are introduced in the following definition; the rough idea is that a finite automaton can test whether a rational side constraint is satisfied.

Definition 4.202 Let Σ be a finite alphabet, and let Σ^* be the set of all words over Σ. We assume that each arc or edge r of the given graph $\mathcal{G}$ is marked with a symbol $w(r) \in \Sigma$. For all paths $P = r_1 \oplus \ldots \oplus r_k$ in $\mathcal{G}$, we define the word $w(P) := w(r_1) \ldots w(r_k)$.[54]
Then a *rational* side constraint about P is a condition of the form $w(P) \in L$ where $L \subseteq \Sigma^*$ is a rational (= regular) language; that means that L can be decided by a finite automaton $\mathcal{A}$. □

---

[54] If the length of P is 0 then $w(P)$ is defined as the empty word ε.

Next, we formulate the optimal path problem with rational side constraints:

Problem 53 *Given a finite directed or undirected graph $\mathcal{G}$, a set $\mathcal{P}$ of paths and a path function $C : \mathcal{P} \to \mathbb{R}$.*
Given a finite alphabet Σ and a rational ($=$ regular) language $L \subseteq \Sigma$. We assume that each arc or edge r of $\mathcal{G}$ is marked with a symbol $w(r) \in \Sigma$; for all paths $P \in \mathcal{P}(\mathcal{G})$, the word $w(P)$ is defined as above.
Find a C-minimal path P^* among all candidates $P \in \mathcal{P}$ with $w(P) \in L$.

Many types of side constraints can be formulated as rational ones. Here are several examples:

(1) *Side constraints of the form $j' \leq \ell(P) \leq j''$ where j', j'' are two constants*
In this case, mark each arc or edge r of $\mathcal{G}$ with the same symbol $w(r) := a$; the language $\{a^l \mid j' \leq l \leq j''\}$ is rational.

(2) *Side constraints of the form $\ell(P) \equiv j \,(\mathrm{mod}\,K)$ or $\ell(P) \not\equiv j \,(\mathrm{mod}\,K)$ where k and K are two constants*
In this case, mark each arc or edge r of $\mathcal{G}$ with the same symbol $w(r) := a$; the languages $\{a^l \mid l \equiv j \,(\mathrm{mod}\,K)\}$ and $\{a^l \mid l \not\equiv j \,(\mathrm{mod}\,K)\}$ are rational.

(3) *Side constraints about the nodes, the arcs or the edges that must be visited or avoided*
Typical examples of such side constraints are the following conditions that must be satisfied by all candidate paths P:
- P must visit a particular node v.
- P must not use the arc or edge r.
- P must start with a particular node s, and P must end with a node in a particular set Γ.
- P must be a cycle; that means that $\omega(P) = \alpha(P)$.
- P must use the arc or edge r if it visits the node v.
- P must be node injective.
- P must start with the node s, visit all nodes $v \neq s$ exactly once, and at last return to s. (This is the Traveling Salesman Problem.)
- P must visit exactly one of the nodes v and w.
- P may visit the node w only if P has visited the node v before.

In these or similar cases, we define the alphabet $\Sigma := R \times V \times V$. If r is an arc, we define $w(r) := (r, \alpha(r), \omega(r))$; if r is an edge with the endpoints x_1, x_2, we define $w(r) := (r, x_1, x_2)$. Consequently, the word $w(P)$ gives the exact information about the nodes, arcs, and edges used by P. We define L such that $w(P) \in L$ is equivalent to the assertion that P satisfies the side constraint.

Problem 52, too, has a rational side constraint if $C_2 = SUM_{h_2}$ is additive, $h_2(r) \in \mathbb{N} \cup \{0\}$ for all arcs, and $b \in \mathbb{N} \cup \{0\}$. The finite automaton has the states $\{0, 1, \ldots, b, b+1\}$. A state $\beta \in \{0, 1, \ldots, b\}$ means that $C_2(P) = \beta$, and the state $\beta = b+1$ means that $C_2(P) > b$.

The solution of Problem 53 is described in [Rom88]: Let $\mathcal{A}$ be a finite automaton that decides L; let Φ be the set of all states of $\mathcal{A}$, and let Φ_L be the set of all accepting states. Then a directed graph $\mathcal{G}_\mathcal{A}$ is defined with the set $\mathcal{V}_\mathcal{A} := \mathcal{V} \times \Phi$ of nodes. Each arc in $\mathcal{G}_\mathcal{A}$ from a node (v, z_1) to a node (w, z_2) means the following: The arc $r := (v, w)$ or the edge $r := \{v, w\}$ belongs to $\mathcal{G}$, and the automaton $\mathcal{A}$ moves from the state z_1 to z_2 if reading the symbol $w(r)$. It follows that each path P in $\mathcal{G}$ can be represented as a path $P_\mathcal{A}$ in $\mathcal{G}_\mathcal{A}$; if P ends with v and if $w(P)$ causes the state $z \in \Phi$ then $P_\mathcal{A}$ will end with the pair $(v, z) \in \mathcal{V}_\mathcal{A}$. Consequently, the search for an optimal path $P^* \in \mathcal{P}$ with a rational side constraint is reduced to the search for an optimal path $P_\mathcal{A}^* \in \mathcal{P}_\mathcal{A} := \{P_\mathcal{A} \mid P \in \mathcal{P}, \omega(P_\mathcal{A}) \in \mathcal{V} \times \Phi_L\}$; that means that the rational side constraint about $P_\mathcal{A}^*$ is replaced by a side constraint about the last node of $P_\mathcal{A}^*$; more precisely, we require $\omega(P_\mathcal{A}^*)$ to be in a set Γ of goal nodes where $\Gamma := \mathcal{V} \times \Phi_L$.

The construction of $\mathcal{G}_\mathcal{A}$ is a simple and universal method to solve Problem 53. The slight disadvantage of this method is that it may cause an exponential time of computation. For example, if the Traveling Salesman Problem in a digraph is given then we interpret the set Z of all nodes visited by P as a state $Z \in \Phi$; the worst case is that (almost) each subset of $\mathcal{V}$ is a state of $\mathcal{A}$ so that Φ has the exponential cardinality $\approx 2^{|\mathcal{V}|}$; then $\mathcal{G}_\mathcal{A}$ has very many nodes, and the optimal path problem in this digraph has exponential complexity.

Next, we quote several papers about special cases of rational constraints.

The search for shortest cycles of even length is described in [Mon85]; the condition to have even length is of type (2). The path problems in [Kum70], [Das81], and [LaNo83] have side constraints of type (3). There are several sets $\mathcal{V}_1, \ldots, \mathcal{V}_k$ of nodes; all candidate paths must visit at least one node $w_\kappa \in \mathcal{V}_\kappa$ for all $\kappa = 1, \ldots, k$; the candidate paths may visit these sets $\mathcal{V}_\kappa$ in arbitrary order, and they may enter them more than once. In [LaNo83], each set $\mathcal{V}_1, \ldots, \mathcal{V}_k$ has exactly one element.
Similar problems are studied in [Lomo88]; a (not necessarily extremal) cycle must be found such that this cycle visits several specified nodes and edges of an undirected graph.

The candidate paths P in [IhmN82] and in [IhmN84] must likewise satisfy side constraints of type (3). A set $\mathcal{Q}$ of forbidden paths is given, and the admissible candidates P must not have any path $Q \in \mathcal{Q}$ as infixes. These constraints make path problems $\mathcal{NP}$-hard if the set $\mathcal{Q}$ has a great cardinality. That is the reason why we shall again consider the papers [IhmN82] and [IhmN84] in Section 4.11.

Searching for a Hamiltonian cycle is also a problem of type (3) because every Hamiltonian cycle P^* has the maximal length among all node injective candidates P with $\alpha(P) = \omega(P)$. Algorithms to search for Hamiltonian cycles are given in [Chri75], [BaN90], [CSK90], and in [Koc90].

Another side constraint of type (3) is considered in [AnMa87]. The paper is about the search for a C-minimal path in undirected graphs where C is a lexicographic cost function; the side constraint is that P must be *maximal*; that means that P itself does not visit any node more than once, and each path $P \oplus r$ does visit some node more than once. The maximality of a path P is a rational constraint. To see this we construct a finite automaton $\mathcal{A}$ that works as follows: If P is a path of a length $\leq n$ then $\mathcal{A}$ will reach a state $(\ell(P), \mathcal{V}_P, N_P)$ where $\ell(P)$ is the length of P, $\mathcal{V}_P$ is the set of all nodes visited by P, and N_P is the set of all neighbours of $\omega(P)$ in the graph $\mathcal{G}$. The automaton $\mathcal{A}$ accepts the state $(\ell(P), \mathcal{V}_P, N_P)$ if and only if the following two assertions are true:

- $|\mathcal{V}_P| = \ell(P) + 1$. (That means that all nodes of P are pairwise distinct.)
- $N_P \subseteq \mathcal{V}_P$. (That means that each path $P \oplus r$ must visit some node of P twice.)

An interesting detail of [AnMa87] is the path function C, which is similar to the lexicographic function $\mathcal{W}_h$ in Definition 2.159. [55]

4.10.3 Further path problems with side constraints

The article [San94] is about finding optimal paths if each node and each arc of the given digraph may only be used during a limited period of time.

The paper [CaJo95] is about searching for optimal paths in polygons; these paths must satisfy a condition about visibility.

We leave it to the reader to check whether the side constraints in [San94] or in [CaJo95] are rational.

4.10.4 General remarks about path problems with side constraints

Here we formulate several general remarks about path problems with side constraints.

Remark 4.203 Given a finite graph $\mathcal{G}$, a set $\mathcal{P}$ of paths, and a cost function $C : \mathcal{P} \to \mathbb{R}$.

a) Almost every search problem for C_1-minimal paths with side constraints can be formulated as a search problem for a lexicographically (C_2, C_1)-minimal path where C_2 is a further path function. To see this we define $C_2 : \mathcal{P} \to \mathbb{R}$ as follows:

$$(\forall P \in \mathcal{P}) \quad C_2(P) := \left\{ \begin{matrix} 0 \\ 1 \end{matrix} \right\} \iff P \left\{ \begin{matrix} \text{satisfies} \\ \text{does not satisfy} \end{matrix} \right\} \text{ the side constraint.}$$

Let $\preceq$ denote the lexicographic order on $\mathbb{R}^2$. We define $C : \mathcal{P} \to (\mathbb{R}^2, \preceq)$ as follows: For all paths $P \in \mathcal{P}$, let $C(P) := (C_2(P), C_1(P))$.

Suppose that at least one path $P \in \mathcal{P}$ satisfies the side constraint. Then we can formulate the optimal path problem with that side constraint as follows: We are searching for a path P^* such that

- P^* itself satisfies the side constraint. That means that $C_2(P^*)$ is equal to 0, and this is equivalent to the C_2-minimality of P^*.
- $C_1(P^*) \leq C_1(P)$ for all paths P with $C_2(P) = 0$.

Obviously, finding a path P^* with these two properties is equivalent to finding a C-minimal path P^*.

---

[55] These lexicographic functions have nothing to do with the lexicographic order of pairs $(C_1(P), C_2(P))$ with path functions $C_1, C_2 : \mathcal{P} \to \mathbb{R}$.

b) Part a) of this remark has the consequence that the following classes **P1** and **P2** contain equivalent problems:

> **P1** := the class of all search problems for C_1-minimal paths with arbitrary satisfiable side constraints; C_1 is an arbitrary real-valued path function.

> **P2** := the class of all search problems for (C_2, C_1)-minimal paths with respect to the lexicographic order on $\mathbb{R}^2$; C_2 and C_1 are arbitrary real-valued path functions.

We give reasons why **P1** and **P2** are equivalent.

If the problem **P** belongs to **P1** then it can be formulated as a problem $\mathbf{P'} \in \mathbf{P2}$; that has been shown in part a) of this remark. On the other hand, if **P** belongs to **P2** then **P** is the search for a (C_2, C_1)-minimal path; this is the same as searching for a C_1-minimal path with the side constraint that all candidates must be C_2-minimal. Hence, **P** belongs to **P1**. $\qquad\square$

Remark 4.204 Optimal path problems without side constraints become easier if the underlying cost functions are order preserving or have a Bellman property. Probably, the same is true for optimal path problems with side constraints. More precisely, given two path functions $C_1, C_2 : \mathcal{P}(\mathcal{G}) \to \mathbb{R}$, and let $C(P) := (C_2(P), C_1(P))$. Then searching for a lexicographically C-minimal path is the same as searching for a C_1-minimal path among all candidates with minimal value $C_2(P)$; this has been shown in Remark 4.203. An algorithmic search for C-minimal paths would be easy if C had an order preserving property of a Bellman property. The following remarks are about Bellman and order preserving principles of C, C_1, C_2 and about actual or probable relationships between these properties.

a) It is easy to show that the following is true: If C_2 is $<$- and $=$-preserving and if C_1 is order preserving then C is order preserving.

   We conjecture that similar assertions are true about other order preserving principles and about particular Bellman properties.

b) In 2.111, we defined the Bellman principle of type B1. We conjecture that this Bellman principle (with $H := C_1$, $H_* := C_2$ or $H := C_2$, $H_* := C_1$) is equivalent to a Bellman property of C.

   The inverse question is the following: Let C satisfy a particular Bellman condition (for example, the strong prefix oriented Bellman condition of type 0); is this condition equivalent to any Bellman property of C_1 and C_2?

c) The following definition is given in [DaAJ91].

   A side constraint **SC** about paths is called *monotonically negative* if for all paths P and all infixes $Q \subseteq P$, the following is true: If P satisfies **SC** then Q satisfies **SC**, too.

   This property of **SC** is equivalent to the assertion that the following path function $C_2 : \mathcal{P}(\mathcal{G}) \to \mathbb{R}$ has the strong infix oriented Bellman property of type 0:

$$(\forall\, P \in \mathcal{P}(\mathcal{G}))\quad C_2(P) := \begin{cases} 0 & \text{if} \quad P \text{ satisfies } \mathbf{SC}, \\ 1 & \text{if} \quad P \text{ does not satisfy } \mathbf{SC}. \end{cases}$$

   (This path function is essentially the same as in Remark 4.203 a).)

At the end of this remark, we recall the order preserving principles in Definition 2.46 and the Bellman principle of type A) in 2.111. These properties are probably useful if the candidate paths have different start nodes and different end nodes. $\square$

4.11   Hard problems

This section is mainly about hard search problems in connection with optimal paths in graphs; almost all problems in this section are so difficult that a deterministic solution in polynomial time has not yet been found.

This section is closely related to Section 4.10 because many hard search problems contain side constraints. Moreover, this section is also related to Section 4.12 because many search problems in that section are hard.

We shall consider three classes of problems:
1. Path problems whose hardness is caused by the underlying cost function,
2. Path problems whose hardness is caused by side constraints (for example the Traveling Salesman Problem),
3. Further hard problems in connection with extremal paths.

4.11.1   Path problems whose hardness is caused by the underlying cost function

In this subsection, we study the usual search problem for C-minimal paths in digraphs; the cost function C, however, has special properties making the search problem very difficult.

The following result shows that the search for C-minimal paths may take an arbitrarily long time even if $C = C_0$ or $C = (C_0)_{\max}$ where C_0 is order preserving.

Theorem 4.205 *Let $\mathcal{A}$ be the class of all algorithms $\mathbf{A}$ with the following properties:*
- *The algorithm $\mathbf{A}$ processes a digraph $\mathcal{G}$, two nodes s, γ of $\mathcal{G}$ and a path function C.*
- *The algorithm $\mathbf{A}$ outputs an s-γ-path P^+. (This path P^+ may be C-minimal but it need not be.)*
- *The algorithm $\mathbf{A}$ requires at least $\ell(P^+)$ time units.*
  (This assumption is reasonable because writing P^+ causes $\ell(P^+)$ steps.)

Then for all $n \geq 5$ and for all $M > n$, there exist
- *a digraph $\mathcal{G} = (\mathcal{V}, \mathcal{R})$ with n nodes and with two nodes $s, \gamma \in \mathcal{V}$,*
- *an order preserving path function $C_0 : \mathcal{P}(s) \to \mathbb{R}$*
such that the following is true for all algorithms $\mathbf{A} \in \mathcal{A}$ and for all path functions $C \in \{C_0, (C_0)_{\max}\}$:
If the path P^+ output by $\mathbf{A}$ is C-minimal then $\mathbf{A}$ consumes at least M time units.

P r o o f : We construct a digraph $\mathcal{G}$ with very long C-minimal paths; our example is similar to the one in 4.124.

More precisely, we define the digraph $\mathcal{G} = (\mathcal{V}, \mathcal{R})$ with $\mathcal{V} := \{s, x_1, \ldots, x_{n-2}, \gamma\}$ and

$$\mathcal{R} := \{(s, x_1)\} \cup \{(x_1, x_2), (x_2, x_3), \ldots (x_{n-3}, x_{n-2}), (x_{n-2}, x_1)\} \cup \{(x_1, \gamma)\}.$$

Moreover, we define the cycle $X := [x_1, x_2, \ldots, x_{n-2}, x_1]$.

Let $M \geq n$. We define a function $\phi_r : \mathbb{R} \to \mathbb{R}$ for each arc $r \in \mathcal{R}$; for all $\xi \in \mathbb{R}$, let
$\phi_r(\xi) := \xi + M + 1$ if $r = (s, x_1)$, $\phi_r(\xi) := \max(1, \xi - 1)$ if $r = (x_1, x_2)$,
$\phi_r(\xi) := \xi + M^2 + 1$ if $r = (x_1, \gamma)$ $\phi_r(\xi) := \xi$ for all other $r \in \mathcal{R}$.

Let $p(v) := 0$ for all nodes v. We define the path function $C_0 : \mathcal{P} \to \mathbb{R}$ as $C_0 := \phi_{\infty} p$.

All functions ϕ_r are monotonically increasing; consequently, C_0 is order preserving.

Moreover, C_0 has the following property:

$$C_0(P) = (C_0)_{\max}(P) \text{ for all } s\text{-}\gamma\text{-paths } P. \tag{4.244}$$

To see this we assume that P be an s-γ-path and that Q be a prefix of P with $Q \neq P$. Then $C_0(Q) \leq M + 1 < M^2 + 1 \leq C_0(P)$. Consequently, $C_0(P)$ is the maximum of all values $C_0(Q)$ with $Q \leq P$. Therefore, $(C_0)_{\max}(P) = C_0(P)$.

We next show the following fact:

> Let $C \in \{C_0, (C_0)_{\max}\}$, and let P^* be a C-minimal s-γ-path.
> Then $P^* = (s, x_1) \oplus X^{\nu} \oplus (x_1, \gamma)$ where $\nu \geq M$.   (4.245)
> Consequently, $\ell(P^*) \geq M$.

By fact (4.244), it is sufficient to show assertion (4.245) for $C = C_0$. Let P^* be a C_0-minimal s-γ-path. Then $P^* = (s, x_1) \oplus X^{\nu} \oplus (x_1, \gamma)$, and we must show that $\nu \geq M$. For this purpose, we observe that $C_0([s, x_1]) = M + 1$. This value can only be reduced by moving along the cycle X; the minimum $C_0((s, x_1) \oplus X^{\nu}) = 1$ is only achieved if $\nu \geq M$. So, the minimal value $C_0((s, x_1 \oplus X^{\nu} \oplus (x_1, \gamma)) = M^2 + 2$ is only realized if $\nu \geq M$.

Consequently, if an algorithm $\mathbf{A} \in \mathcal{A}$ finds and writes a C-minimal path P^+ in $\mathcal{G}$ then the time consumed by $\mathbf{A}$ is at least $\ell(P^+) \geq \ell(P^*) \overset{(4.245)}{\geq} M$.   □

Remark 4.206 It is not clear whether the following special case of Problem 5 (see page 179) can be solved in in polynomial time (with respect to the number of nodes of the input graph):

> **Problem 54** *Given an acyclic digraph* $\mathcal{G} = (\mathcal{V}, \mathcal{R})$ *with* n *nodes. Let* $s, \gamma \in \mathcal{V}$; *we assume that an* s-v-*path exists for each node* $v \in \mathcal{V}$. *Let* $C : \mathcal{P}(s) \to \mathbb{R}$ *is a path function with the following property: There exists an order preserving path function* C_0 *such that* $C = (C_0)_{\max}$. *Find a* C-*minimal* s-γ-*path.*

An example for exponential running time of Algorithm A* was given in Theorem 4.66.

Problem 54 can be modified as follows: We use the cost function $C := C_0$ instead of $C = (C_0)_{\max}$. Then the search problem for C-minimal s-γ-paths can be solved in polynomial time. This follows from Theorem 4.130 among others. $\qquad\square$

Next, we describe another type of path functions C that often make the search for C-minimal paths very hard. Let $C(P) := \Psi(C_1(P), \ldots, C_j(P))$ for all paths P where $C_1, \ldots, C_j$ are j order preserving path functions and $\Psi : \mathbb{R}^j \to \mathbb{R}$. Then the search for C-minimal paths is often $\mathcal{NP}$-hard.

A simple example is the following: All functions $C_1, \ldots, C_j$ are additive and prefix monotone, and that Ψ is the maximum norm; that means that

$$C(P) = \Psi(C_1(P), \ldots, C_j(P)) = \|(C_1(P), \ldots, C_j(P))\|_{\max} = \max_{i=1,\ldots,j} |C_j(P)|.$$

It is shown in [Vaĭn85] that finding a C-minimal s-γ-path for a given pair (s, γ) of nodes is $\mathcal{NP}$-complete.

The next remarks help to understand why the search for C-minimal paths is so complicated. We show that C does not always have the structural properties that make the search for C-optimal paths easy.

Remark 4.207 The following example shows that $C := \|C_1, C_2\|_{\max}$ does not always have the weak prefix oriented Bellman property.

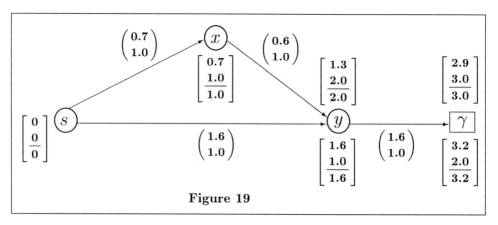

Figure 19

Let G be the digraph shown in Figure 19. Each arc r is marked with a column $\binom{h_1(r)}{h_2(r)}$. Let $C_i := SUM_{h_i}$, $i = 1, 2$. Obviously, C_1 is prefix monotone, and $C_2 = \ell$. Let $C(P) := \max\left(|C_1(P)|, |C_2(P)|\right)$.

We define $\mathcal{C}(P) := \begin{bmatrix} C_1(P) \\ C_2(P) \\ \overline{C(P)} \end{bmatrix}$ for all paths P. The column at s in Figure 19 equals $\mathcal{C}([s])$; the column at x is $\mathcal{C}([s, x])$; the upper column at y is $\mathcal{C}([s, x, y])$, and the lower column at y is $\mathcal{C}([s, y])$; the upper column at γ is $\mathcal{C}([s, x, y, \gamma])$, and the lower column at γ is $\mathcal{C}([s, y, \gamma])$.

The only C-optimal s-γ-path is $[s, x, y, \gamma]$ because $C([s, x, y, \gamma]) = 3 < 3.2 = C([s, y, \gamma])$. But it is not true that all prefixes of $[s, x, y, \gamma]$ are C-minimal; the prefix $[s, x, y]$ has the property that $C([s, x, y]) = 2 > 1.6 = C([s, y])$.

Consequently, there does not exist any C-minimal s-γ-connection with exclusively C-minimal prefixes. Therefore, C does not have the weak prefix oriented Bellman property of type 0. $\qquad\square$

Remark 4.208 The function $C(P) = \max\left(|C_1(P)|, |C_2(P)|\right)$ is not always $=$-preserving even if C_1, C_2 are additive. It is easy to construct a counterexample such that $C_1 = SUM_{h_1}$, $C_2 = SUM_{h_2}$, $h_1(r) = 1$, $h_2(r) = 4$, and

$$\begin{array}{llll}
C_1(P) = 5, & C_1(P') = 5 & C_1(P \oplus r) = 6, & C_1(P' \oplus r) = 6, \\
C_2(P) = 1, & C_2(P') = 3, & C_2(P \oplus r) = 5, & C_2(P' \oplus r) = 7, \\
C(P) = 5, & C(P') = 5, & C(P \oplus r) = 6, & C(P' \oplus r) = 7.
\end{array}$$

Then $C(P) = 5 = C(P')$ but $C(P \oplus r) = 6 \neq 7 = C(P' \oplus r)$. $\qquad\square$

Remark 4.209 Given the arc functions h_1, h_2, and let $C_1 = SUM_{h_1}$, $C_2 = SUM_{h_2}$, and $C(P) := \|C_1(P), C_2(P)\|_{\max}$ for all $P \in \mathcal{P}(\mathcal{G})$. Then the search for C-minimal cycles may require arbitrarily many steps if h_1 attains negative values; the reason is that C-minimal paths may be very long.

To see this we recall the digraph $\mathcal{G} = (V, \mathcal{R})$ in 4.124. Let $M \geq 5$ be an arbitrary number; we define C_1 and C_2 such the only C-minimal s-γ-path has a length $\geq M$.

For this purpose, we define the following arc functions $h_1, h_2 : \mathcal{R} \to \mathbb{R}$:

$$\begin{array}{llll}
h_1(s, x_1) := 2M, & h_1(x_1, x_2) := -1, & h_1(r) := 0 & \text{for all other arcs } r \in \mathcal{R}, \\
h_2(s, x_1) := 0, & h_2(x_1, x_2) := 1, & h_2(r) := 0 & \text{for all other arcs } r \in \mathcal{R}.
\end{array}$$

Moreover, we define the cycle $X := [x_1, x_2, x_3, x_1]$. For all $\nu = 1, 2, 3, \ldots$, let $P_\nu := (s, x_1) \oplus X^\nu \oplus (x_1, z)$. Then $C_1(P_\nu) = 2M - \nu$, $C_2(P_\nu) = \nu$, and $C(P_\nu) = \max(|2M - \nu|, |\nu|)$ for all ν. This value is minimal if $\nu = M$ and $C(P_\nu) = C(P_M) = M$. The length of this C-minimal s-z path P_M is greater than M.

Let $\mathcal{A}$ be the class algorithms described in 4.205; then each search strategy $\mathbf{A} \in \mathcal{A}$ will require at least M steps to find and to write the optimal path P_M. $\qquad\square$

Searching for a path with minimal value $\|(C_1(P), \ldots, C_j(P))\|_{\max}$ is a good method to solve the following problem:

Problem 55 *Given a digraph $\mathcal{G} = (V, \mathcal{R})$ with n nodes. Given the path functions $C_1, \ldots, C_j$ such that $C_1(P), \ldots, C_j(P) \geq 0$ for all paths P in $\mathcal{G}$. Let $\mathbf{c} \geq 0$. Find a path $P \in \mathcal{P}$ such that*

$$C_1(P) \leq \mathbf{c}, \ldots, C_j(P) \leq \mathbf{c}. \tag{4.246}$$

Moreover, find the minimum $\mathbf{c}^*$ of all numbers $\mathbf{c}$ for which the inequalities in (4.246) have a solution $P \in \mathcal{P}$.

This problem can be solved as follows: Let $C(P) := \|(C_1(P), \ldots, C_j(P))\|_{\max}$ for all paths $P \in \mathcal{P}$. Then the following is true for all $\mathbf{c} \in \mathbb{R}$ and all $P \in \mathcal{P}$:

$$C(P) = \|(C_1(P), \ldots, C_j(P))\|_{\max} \leq \mathbf{c} \iff C_1(P), \ldots, C_j(P) \leq \mathbf{c} \qquad (4.247)$$

Consequently, solving the inequalities in (4.246) is equivalent to finding a path P with $C(P) \leq \mathbf{c}$. Moreover, finding $\mathbf{c}^*$ is equivalent to finding a C-minimal path $P^* \in \mathcal{P}$ because $\mathbf{c}^* = C(P^*)$.

If $C_1, \ldots, C_j$ are additive then the above result of Vaĭnshteĭn[56] says that finding a C-minimal path P^* is $\mathcal{NP}$-complete.

The following special case of Problem 55 is studied in Example 3 and in Subsection 8.1 of [LT91] and of [LTh91].

Problem 56 *Given a digraph $\mathcal{G} = (\mathcal{V}, \mathcal{R})$ with n nodes. Let $\tau : \mathcal{R} \to \{1, \ldots, j\}$ be an arc function. Let $h_1, \ldots, h_j : \mathcal{R} \to \{0, 1\}$ be defined as follows:*

$$(\forall i = 1, \ldots, j, \ r \in \mathcal{R}) \ \big(h_i(r) := 1 \ \text{if} \ \tau(r) = i, \quad h_i(r) := 0 \ \text{if} \ \tau(r) \neq i\big).$$

Let $C_i := \mathrm{SUM}_{h_i}, \ i = 1, \ldots, j$.[57]
Find a path P such that (4.246) is true.

Solving Problem 56 means to find a path P with the following property: For all $i = 1, \ldots, j$, the path P uses at most $\mathbf{c}$ arcs r with $\tau(r) = i$.

Let $C(P) := \max(C_1(P), \ldots, C_j(P))$. Lemma 6 of [LT91] says that the search for a C-minimal s-γ-path is $\mathcal{NP}$-complete; the proof is rather simple.

Next, we formulate the following formal generalization of Problem 55:

Problem 57 *Given a digraph $\mathcal{G} = (\mathcal{V}, \mathcal{R})$ with n nodes. Let $\mathbf{c}_1, \ldots, \mathbf{c}_j \geq 0$. Find a path $P \in \mathcal{P}$ such that*

$$C_1(P) \leq \mathbf{c}_1, \ldots, C_j(P) \leq \mathbf{c}_j. \qquad (4.248)$$

We reduce this problem to Problem 57. The first step is to formulate Problem 57 such that all upper bounds are unequal to zero; our idea is to replace each condition $C_i(P) \leq \mathbf{c}_i$ by $C_i(P) + 1 \leq \mathbf{c}_i + 1$ if $\mathbf{c}_i = 0$. More precisely, let

$$C_i'(P) := \begin{cases} C_i(P) & \text{if } \mathbf{c}_i > 0, \\ C_i(P) + 1 & \text{if } \mathbf{c}_i = 0. \end{cases}, \quad \mathbf{c}_i' := \begin{cases} \mathbf{c}_i & \text{if } \mathbf{c}_i > 0, \\ \mathbf{c}_i + 1 & \text{if } \mathbf{c}_i = 0 \end{cases}, \quad C_i''(P) := \frac{C_i'(P)}{\mathbf{c}_i'}$$

for all paths $P \in \mathcal{P}$ and all $i = 1, \ldots, j$. Then Problem 57 is equivalent to searching for a path $P \in \mathcal{P}$ such that $C_i'(P) \leq \mathbf{c}_i'$ for all $i = 1, \ldots, j$, and this is equivalent to searching for a path P such that

$$C_1''(P) \leq 1, \ldots, C_j''(P) \leq 1. \qquad (4.249)$$

Consequently, Problem 57 can be reduced to Problem 55 with $\mathbf{c} := 1$.

Let $C(P) := \|(C_1''(P), \ldots, C_j''(P))\|_{\max}$ for all paths $P \in \mathcal{P}$. Then solving (4.248) is equivalent to finding a path P with $C(P) \leq 1$. Moreover, let $P^* \in \mathcal{P}$ be a C-minimal path. If $C(P^*) \leq 1$ then P^* is a solution of (4.249) and of (4.248); if, however, $C(P^*) > 1$ then neither (4.249) nor (4.248) is solvable.

---

[56] See page 378 of this book.
[57] Then $C_i(P)$ says how often P uses an arc with $\tau(r) = i$.

The author of [Jaf84] gives approximate solutions of Problem 57 in the case that $j = 2$ and the functions C_1, C_2 are additive with arc functions $h_1, h_2 : \mathcal{R} \to \mathbb{N} \cup \{0\}$, respectively.

The authors of [AnNa78] give an algorithm that solves a modified version of the problems 52 and 57; more precisely, the algorithm finds a C-minimal s-γ-path P^* such that $C_1'(P^*) \leq \mathbf{c}_1'$, $\ldots$, $C_j'(P^*) \leq \mathbf{c}_j'$. The functions $C, C_1', \ldots, C_j'$ are additive.

Next, we consider another problem. It is mainly a hard version of a very easy problem: Find a C-minimal s-γ-path where C is an additive path function such that all arcs have non-negative, integer weights.

Problem 58 (M-PATH)
Given an acyclic digraph $\mathcal{G} = (\mathcal{V}, \mathcal{R})$ with n nodes. Let $s, \gamma \in \mathcal{V}$. Let $C = SUM_h$ where $h : \mathbb{R} \to \mathbb{N} \cup \{0\}$. Moreover, let $M \in \mathbb{N} \cup \{0\}$.
Decide whether an s-γ-path P with $C(P) = M$ exists.

The problem M-PATH itself does not belong to the current subsection because M-PATH is not an optimal path problem and its hardness is not caused by the cost function. But M-PATH makes it possible to describe the complexity of the following problem, which does belong to this subsection:

Problem 59 *Given an acyclic digraph $\mathcal{G} = (\mathcal{V}, R)$ with n nodes. Let $s, \gamma \in \mathcal{V}$. Let $C = SUM_h$ where $h : \mathbb{R} \to \mathbb{N} \cup \{0\}$. Moreover, let $M \in \mathbb{N} \cup \{0\}$, and let $C'(P) := |C(P) - M|$ for all paths $P \in \mathcal{P}(\mathcal{G})$.*
Find a C'-minimal s-γ-path.

The next result is about the complexity of the problems 58 and 59.

Theorem 4.210 *The problems 58 and 59 are $\mathcal{NP}$-complete.*

P r o o f : First, we consider M-PATH. This problem is in $\mathcal{NP}$ because it can be solved by a nondeterministic Turing machine that starts with s, guesses the arcs of P and tests whether $C(P) = M$. If YES then the Turing machine writes P, if NO then the Turing machine gives the information that no path P with $C(P) = M$ exists. Since $\mathcal{G}$ is acyclic, the Turing machine does not generate any path with more than $(n-1)$ arcs; consequently, the machine works in $O(n)$ time.

Next, we show that M-PATH is $\mathcal{NP}$-hard. For this purpose, we reduce the problem PARTITION to M-PATH. According to [GaJo79], the problem PARTITION is $\mathcal{NP}$-complete.

Problem 60 (PARTITION)
Given a finite set $A = \{a_1, \ldots, a_n\}$ and a weight function $\sigma : A \to \mathbb{N}$.
Decide whether a subset $B \subseteq A$ exists such that $\sum_{a \in B} \sigma(a) = \sum_{a \in A \setminus B} \sigma(a)$.

We reduce this problem to M-PATH in polynomial time. For this purpose, we define the acyclic digraph $\mathcal{G} = (\mathcal{V}, \mathcal{R})$ as follows: Let $\mathcal{V} := \{s\} \cup \{a_1, \ldots, a_n\} \cup \{\gamma\}$, and let $a_0 := s$, $a_{n+1} := \gamma$; let $\mathcal{R} := \{(a_i, a_j) \mid i < j\}$. For each arc $r = (a_i, a_j) \in \mathcal{R}$, we define $h(r) := \sigma(a_j)$, if $j \leq n$ and $h(r) = 0$ if $j = n+1$; we define the path function $C := SUM_h$. Then the following is true for all subsets $X = \{a_{i_1}, \ldots, a_{i_k}\} \subseteq A$ with $i_1 < \ldots < i_k$:

$$\sum_{a \in X} \sigma(a) \; = \; \sum_{\kappa=1}^{k} \sigma(a_{i_\kappa}) \; = \; C([\, s = a_0\, , a_{i_1}\, , \ldots, \, a_{i_k}\, , a_{n+1} = \gamma\,]) \,. \qquad (4.250)$$

Then we define $M := \frac{1}{2} \cdot \sum_{a \in A} \sigma(a)$. If M is not an integer number then PARTITION is not solvable. Otherwise, PARTITION has a solution $B \subseteq A$ if and only if there exists a path $P = [\, s, b_1, \ldots, b_k, \gamma \,]$ in $\mathcal{G}$ with $C(P) = M$. To see this we assume that B be a solution to PARTITION; then define the path P such that P visits all elements of B and no element of $A \backslash B$; then $C(P) = M$ by fact (4.250). On the other hand, let P be an s-γ-path in $\mathcal{G}$ with $C(P) = M$; then we define $B := \{b \in P \mid b \neq s, \, b \neq \gamma\}$; this set B is a solution of PARTITION.

The $\mathcal{NP}$-completeness of Problem 59 follows immediately from the $\mathcal{NP}$-completeness of M-PATH. $\square$

Remark 4.211   a) Several search strategies for paths with a given value $C(P) = M$ are described in the article [Mon85].

b) We conjecture that a similar reduction as in 4.210 can be used to show the $\mathcal{NP}$-completeness of many other problems about paths in digraphs. The main idea of these reductions is the following: We choose an $\mathcal{NP}$-complete problem **P** about the search for a particular subset $B \subseteq A$; we then replace the search for B with the search for a path P that represents the set B. $\square$

4.11.2   The Traveling Salesman Problem and other hard path problems with side constraints

Here, we discuss optimal path problems whose hardness is caused by side constraints. A prominent example for a hard path problem with side constraints is the Traveling Salesman Problem (see Problem 2 on page 175). Further path problems with side constraints are studied in [AnNa78].

We concentrate on hard problems whose side constraints describe the nodes, the arcs or the edges of the admissible candidate paths P. In particular, we consider the Traveling Salesman Problem and the search for paths in so-called valve graphs.

4.11.2.1   The Traveling Salesman Problem (TSP)

One of the most well-known problems about optimal paths is the Traveling Salesman Problem. It is well-known that this problem is $\mathcal{NP}$-complete.

Next, we describe several typical sources of literature about the TSP and its variants. There exist so many further papers about this problem that it is possible to write a thick book about the TSP alone.

Overviews about the TSP are given in [PaRa83], in [LaLe85], and in [MSS89b]. Variants of the TSP are described in [Ho91].

Approximative methods to solve the TSP are described in the papers [Mart79], [YUOK83], [MSS89a], and in [JThR94].

The paper [JogSuh] presents a parallel algorithm to solves the TSP.

The TSP can be considered as problem in Linear Programming. This point of view is studied in [SmMT90] and in [BPu91]. In particular, the paper [SmMT90] gives lower bounds for the cost of an optimal TSP tour.

Next, we discuss *complicated or extended versions* of the TSP.

The problem in [Frie83] and in [SLB90], [deKo91] is that each of $k \geq 1$ traveling salesmen must make a tour such that sum of the costs of all tours is minimal.

The salesman in [KaHA85] must first visit a pickup customer to buy a particular product, and he must later visit the delivery customer to whom he sells this product.

The authors of [LMW93] assume that the given cities are grouped as clusters (for example regions). The problem is to find an optimal tour that visits at least one city in each cluster.

A further complicated version of the TSP is the Traveling Purchaser Problem in [KuKa88].

Next, we describe *restricted versions or easy variants* of the Traveling Salesman Problem.

If the values $h(\{v, w\})$ are interpreted as the distances between the cities v and w then h and C satisfy the following triangle inequality: For all $u, v, w \in V$,

$$C(\{u, w\}) = h(\{u, w\}) \leq h(\{u, v\}) + h(\{v, w\}) = C(\{u, v\} \oplus \{v, w\}). \qquad (4.251)$$

Then we formulate the following special case of the TSP

Problem 61 (Triangle-TSP)
Given the same situation as in Problem 2. Moreover, let (4.251) be true.
Find a C-minimal node injective cycle P^* among all candidates visiting each node of the graph $\mathcal{G}$.

Triangle-TSP is $\mathcal{NP}$-complete as well as the original TSP. In [GaJo79], several algorithms are given that find an almost optimal solution to Triangle-TSP in polynomial time; more precisely, there exist a constant $c > 1$ with the following property: If the path P^+ is found by the particular search strategy and if if P^* is an actually optimal path then $C(P^+) \leq c \cdot C(P^*)$.

Triangle-TSP can be solved in polynomial time $O(n^N)$ if N is a fixed number and the n cities lie on N parallel lines in the plane; this result is shown in [Rote92].

Another special situation of Triangle-TSP is studied in [StSn89]. The set $\mathcal{V}$ of nodes is a subset of the d-dimensional hypercube $[0, 1]^d \subseteq \mathbb{R}^d$, and each value $h(\{v, w\})$, $v, w \in V$, is defined as the Euclidean distance between v and w. Steele and Snyder estimate the length of an optimal tour of the travelling salesman.

The authors of [PaYa93] describe approximate solutions for following special case of the TSP: All edges r have a weight $h(r) \in \{1, 2\}$.

A further easy version of TSP is given in [TaNS80]. The underlying graph $\mathcal{G}$ need not be complete, and the cost $C(P)$ is defined as the number of arcs of the path P. The problem is to find a cost minimal cycle P that visits all nodes of $\mathcal{G}$; here, the cycle P need not be node injective. The authors of [TaNS80] give a deterministic algorithm that solves this problem in polynomial time.

The paper [Warr91] is also about easy versions of the Traveling Salesman Problem.

Further modifications of the TSP are described in [LaMa90], [NaRi93].

The TSP can also be formulated for the complete digraph $\mathcal{G} = (\mathcal{V}, \mathcal{R})$ with $\mathcal{R} := \mathcal{V} \times \mathcal{V} \setminus \{(v, v) \,|\, v \in \mathcal{R}\}$. The problem is to find a C-minimal Hamiltonian cycle where $C = SUM_h$. The TSP in digraphs is called *symmetric* if $h(u, v) = h(v, u)$ for all arcs (u, v), and it is called *asymmetric* otherwise; obviously, the symmetric TSP is equivalent to the TSP in undirected graphs. The TSP in digraphs is studied in [MaSa91].

4.11.2.2   The Chinese Postman Problem (CPP)

A further $\mathcal{NP}$-complete path problem with side constraints is the Chinese Postman Problem (CPP) for half directed graphs; this problem is similar to the TSP.

Problem 62 (Chinese Postman Problem)
Given a finite half directed graph $\mathcal{G}$ with a set $\mathcal{R}_1$ of arcs and a set $\mathcal{R}_2$ of edges. Let $h(r) = 1$ for all $r \in \mathcal{R}_1 \cup \mathcal{R}_2$, and let $C := SUM_h$; then $C = \ell$.
Find a C-minimal path in $\mathcal{G}$ that uses each arc and each edge at least once.

The arcs and edges of $\mathcal{G}$ may be interpreted as streets; a postman must deliver mail in every street, and he searches for a route of minimal length.

Details about the CPP are given in [GaJo79]; one of the results in Garey's and Johnson's book is that CPP can be solved in polynomial time if $\mathcal{R}_1 = \emptyset$ or $\mathcal{R}_2 = \emptyset$, i.e., $\mathcal{G}$ is a completely undirected or a completely directed graph. If, however, $\mathcal{G}$ has both arcs and edges then the Chinese Postman Problem is $\mathcal{NP}$-complete.

Further sources about the Chinese Postman Problem are [Pa76], [Mini79], [Bru80], and [Vich82].

4.11.2.3   Searching for paths whose lengths are even, odd, or fixed

The papers [Karm84] and [LaPP84] are about $\mathcal{NP}$-complete variants of the problem to find paths of even lengths in directed graphs; in [LaPP84], the hardness of the problem is caused by the side constraint that all admissible paths must be node injective.
If, however, the given graph is undirected then a node injective s-γ-path of even length can be found in polynomial time; this follows from [LaPP84].

Three $\mathcal{N}\mathcal{P}$-complete problems about paths in undirected graphs are given in [Bien92]; for example, deciding the existence of an s-γ path with odd length is $\mathcal{N}\mathcal{P}$-complete; this path need not be node injective.

Another $\mathcal{N}\mathcal{P}$-complete problem is to decide whether a vertex set of an undirected graph can be partitioned into vertex disjoint paths of a length $k \geq 2$; a fast randomized algorithm to solve this problem is given in [Stou93].

4.11.2.4 Searching for acyclic paths with maximal C-value

Given the following problem:

Problem 63 (ACYCLIC MAXPATH)
Given a digraph $\mathcal{G} = (\mathcal{V}, \mathcal{R})$ with n nodes. Let $s, \gamma \in \mathcal{V}$, let $h : \mathcal{R} \to \mathbb{N} \cup \{0\}$, and let $C := SUM_h$.
Find a C-maximal s-γ-path P^* among all acyclic s-γ-paths P.[58]

One of the statements in [GaJo79] is that this maximal path problem is $\mathcal{N}\mathcal{P}$-complete. Now, we describe two polynomial variants of the problem.

Remark 4.212

a) The problem to find a C-minimal acyclic s-γ-path can be solved in polynomial time. The first step of the solution is searching for a C-minimal s-γ-path P^+ without side constraints. The second step is removing all substantial cycles of P^+; this procedure does not influence the C-minimality of the resulting acyclic path P^* because $C(X) \geq 0$ for all removed cycles X.
This method cannot be used to find a C-maximal acyclic s-γ-path because the search for a C-maximal s-γ-path is often unsuccessful. For example, given a graph $\mathcal{G}$ with an s-γ-path $P \oplus X \oplus Q$ where X is a cycle with $C(X) > 0$. Then $C(P \oplus X^n \oplus Q) = C(P) + C(Q) + \nu \cdot C(X)$ for all ν, and this quantity can be arbitrarily great. Therefore, there exists no C-maximal s-γ-path in $\mathcal{G}$. Hence, the search for a C-maximal s-γ-path cannot be used to solve ACYCLIC MAXPATH.

b) The hardness of ACYCLIC MAXPATH is caused by the side constraint that all candidate paths P must be acyclic. If this side constraint is not given then we have the following problem, which can be solved in polynomial time.

> **Problem 64** *Given the same situation as in Problem 63.*
> *Decide whether there exists a C-maximal s-γ-path P^*; if yes then find such a path.*

This problem can be solved with Algorithm 20, which is given immediately after the current remark. We shall prove the correctness and the polynomial complexity of Algorithm 20 in Theorem 4.213. □

---

[58] Here, the path function C is interpreted as a utility function.

Algorithm 20

For all arcs $r = (u, v)$ with $h(r) > 0$, check whether r has the following three properties:

$$\text{There exists an } s\text{-}u\text{-path } U_r. \tag{4.252}$$
$$\text{There exists an } v\text{-}u\text{-path } V_r. \tag{4.253}$$
$$\text{There exists an } v\text{-}\gamma\text{-path } W_r. \tag{4.254}$$

C a s e 1 : **If** there exists an arc r with properties (4.252) – (4.254) **then** write: "There exists no C-maximal s-γ-path."

C a s e 2 : **If** there exists no arc r with properties (4.252) – (4.254) **then** search for a C-maximal s-γ-path P^+ among all candidates of a length $\leq n - 1$.[59] Output the path P^+.

Now we show the correctness and the polynomial complexity of this algorithm.

Theorem 4.213

  a) *Algorithm 20 can decide whether C-maximal s-γ-path exists; if yes then Algorithm 20 will find such a path.*

  b) *The running time of Algorithm 20 is polynomial.*

P r o o f o f a) : To show the correctness of Algorithm 20 we prove the equivalence of the following assertions:

There exists an arc r with $h(r) > 0$ and with properties (4.252) – (4.254). $\quad$ (4.255)

There exists an s-γ-path with a positive cycle. More formally, there exists an s-γ-path $P \oplus X \oplus Q$ such that X is a cycle with $C(X) > 0$. $\quad$ (4.256)

Proof of (4.255) $\Rightarrow$ (4.256) : If (4.255) is true then there exist the paths U_r, V_r, W_r as described in (4.252) – (4.254). Let $P := U_r$, $X := r \oplus V_r$, $Q := r \oplus W_r$. Then $P \oplus X \oplus Q$ is an s-γ-path, and X is a cycle with $\alpha(X) = \omega(X) = u$ and $C(X) \geq h(r) > 0$.

Proof of (4.256) $\Rightarrow$ (4.255) : Let (4.256) be true. The fact that $C(X) > 0$ implies that there exists an arc $r = (u, v)$ on X with $h(r) > 0$. This arc r has the properties (4.252) – (4.254). To see this we define the paths X' and X'' such that $X = X' \oplus (u, v) \oplus X''$; then $\omega(X'') = \alpha(X')$ because X is a cycle. Let $U_r := P \oplus X'$, $V_r := X'' \oplus X'$, and $W_r := X'' \oplus Q$. Then U_r is an s-u-path, V_r is a v-u-path and W_r is a v-γ-path. Hence, r has the properties (4.252) – (4.254).

Now we show that Algorithm 20 works correctly. If Case 1 is given then (4.256) is true, and the paths $P \oplus X^\nu \oplus Q$, $\nu \in \mathbb{N}$, have arbitrarily great C-values; hence, there exists no C-maximal s-γ-path.

If Case 2 is given then (4.255) is false, and consequently, assertion (4.256) is false as well. Therefore, a C-maximal path can be found by comparing all acyclic s-γ-paths with each other, and all these candidates are at most of the length $n - 1$.

---

[59]This can be done in polynomial time by applying FORD-BELLMAN 1 with cost function $(-C)$; it follows from 4.121 that n iterations of FORD-BELLMAN 1 are sufficient to generate a $(-C)$-minimal s-γ-path among all candidates of a length $\leq n - 1$.

P r o o f o f b) : Algorithm 20 decides the existence of x-y-paths for particular nodes $x, y \in V$. If Case 2 is given then Algorithm 20 will search for a C-maximal path, and this can be done by n iterations of FORD-BELLMAN 1. Hence, Algorithm 20 does not require more than polynomial time.
□

By the way, Algorithm 20 cannot be used to solve ACYCLIC MAXPATH; the reason is that the algorithm does not find a C-maximal acyclic path if Case 1 is given.

4.11.2.5 Finding optimal paths if particular subpaths are forbidden

Here, we consider the following situation: Given a set Q of paths in a graph $\mathcal{G}$. We say that a path P is *legal* if it has no infix $Q \subseteq P$ with $Q \in \mathcal{Q}$; all other paths P are forbidden.

The results in [IhmN82] and [IhmN84] imply that the following problem is $\mathcal{NP}$-hard:

Problem 65 *Given a digraph $\mathcal{G} = (V, \mathcal{R})$ with n nodes. Let $s, \gamma \in V$. Given the functions $h : \mathcal{R} \to \mathbb{N} \cup \{0\}$ and $C := SUM_h$. Given a set $\mathcal{Q} \subseteq \mathcal{P}(\mathcal{G})$ of paths. Find a C-minimal legal s-γ-path.*

It has been shown in [IhmN84] that this problem is $\mathcal{NP}$-hard even if $\mathcal{Q}$ has the following structure: $\mathcal{Q} = \bigcup_{\kappa=1}^{k} \mathcal{P}(v_\kappa, w_\kappa)$ where $v_1, w_1, \ldots, v_k, w_k$ are nodes of $\mathcal{G}$; in this case, P must not use the node w_κ later than v_κ ($\kappa = 1, \ldots, k$).

4.11.2.6 Path problems in graphs with valves

When searching for a path in a graph, we usually make the following assumption: Each node v may be entered by an arbitrary arc $(u, v) \in \mathcal{R}$, and v may be left by an arbitrary arc $(v, w) \in \mathbb{R}$. Here, however, we consider the case that a valve is installed in v. It dependes on the current valve adjustment whether or not an outgoing arc (v, w) may be used immediately after an incoming arc (u, v). E.g., if the valve adjustment in Figure 20 a) is given then the node y_1 may only be left via (y_1, x_2) if y_1 has been entered via (s, y_1); moreover, the arc (y_1, x_3) must not be used because it is not connected to (x_1, y_1). The valve adjustments in Figure 20 b) and d) permit to use (y_2, x_6) immediately after using (x_5, y_2), and moving along (y_2, x_4) is forbidden.

Directed networks with valves are of practical relevance. Typical examples are networks of pipes in breweries or refineries. We assume that pumps are installed in the pipes so that the stream of liquid can use them only in one direction. The valves in the network can open or interrupt particular connections between pipes. This practical situation was studied in [Br90]. The author of that thesis developed a program system for the control of a machine produced by a German company; this machine automatically adjusts valves in pipe systems of breweries and refineries.

Another example for the practical relevance of paths in directed networks with valves is the situation of a computer network; the switches connecting particular pairs of ports may be interpreted as valves.

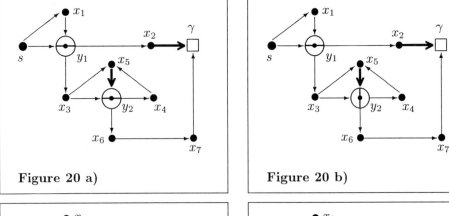

Figure 20 a)   **Figure 20 b)**

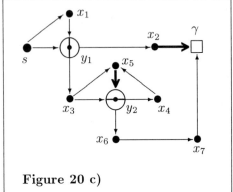   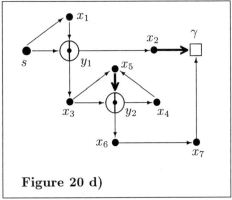

Figure 20 c)   **Figure 20 d)**

Let $\mathcal{G} = (\mathcal{V}, \mathcal{R})$ be a graph with valves, and let $\widehat{\mathcal{V}} \subseteq \mathcal{V}$ be the set of all nodes with valves. Let $v, w \in \mathcal{V}\backslash\widehat{\mathcal{V}}$, and let P be a v-w-path. We say that a path P in $\mathcal{G}$ is *admissible* if we can adjust the valves such that each arc of P is connected with the next arc used by P; otherwise, P is called *forbidden*. For example, let $\mathcal{G}$ be the graph in Figure 20 a) – d); then $\widehat{\mathcal{V}} = \{y_1, y_2\}$. The path $[s, y_1, x_2, \gamma]$ is admissible because the valve adjustments in Figure 20 a) and b) connect (s, y_1) to (y_1, x_2). The path $[s, y_1, x_3]$ is forbidden because no valve adjustment connects (s, y_1) to (y_1, x_3). The path $[x_3, y_2, x_4, x_5, y_2, x_6]$ is forbidden, too, because no valve adjustment simultaneously connects the arcs (x_3, y_2), (y_2, x_4) and the arcs (x_5, y_2), (y_2, x_6).

Of course, one can construct valves with another structure than the valves in Figure 20. For example, the valve in Figure 21 a) – c) has three entrances and four exits; the arc r_1 can be connected to r_1', r_3', the arc r_2 can be connected to r_2', r_3', and the arc r_3 can be connected to r_3'. The *type*[60] of a valve describes its structure; for example, the valve y_2 in Figure 20 are of the same type as y_1; moreover, y_1 and y_2 are of another type than the valve in Figure 21 a) – c).

---

[60]The type of a valve is exactly defined in [Hu93] and in [Huc97].

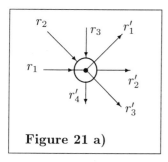

Figure 21 a)

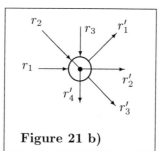

Figure 21 b)

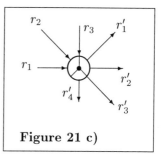

Figure 21 c)

Next, we formulate three problems concerning graphs with valves.

Problem 66 (Existence of admissible paths)
Let $\mathcal{G} = (\mathcal{V}, \mathcal{R})$ be a graph. Let $\widehat{\mathcal{V}} \subseteq \mathcal{V}$ be the set of all valve nodes in $\mathcal{G}$, and let $s, \gamma \in \mathcal{V} \backslash \widehat{\mathcal{V}}$.
Decide whether or not an admissible s-γ-path exists.[61]

Problem 67 (Optimal admissible paths)
Let $\mathcal{G} = (\mathcal{V}, \mathcal{R})$ be a graph. Let $\widehat{\mathcal{V}} \subseteq \mathcal{V}$ be the set of all valve nodes in $\mathcal{G}$, and let $s, \gamma \in \mathcal{V} \backslash \widehat{\mathcal{V}}$. We assume that at least one admissible s-γ-path exists. Let $C : \mathcal{P}(s) \to \mathbb{R}$ be a path function.
Find a C-minimal admissible s-γ-path.

Problem 68 (Existence of valve adjustments interrupting all s-γ-paths)
Let $\mathcal{G} = (\mathcal{V}, \mathcal{R})$ be a graph. Let $\widehat{\mathcal{V}} \subseteq \mathcal{V}$ be the set of all valve nodes in $\mathcal{G}$, and let $s, \gamma \in \mathcal{V} \backslash \widehat{\mathcal{V}}$.
Decide whether a valve adjustment exists that interrupts all s-γ-paths.[62]

First, we focus on **Problem 66**. For example, there exists an admissible s-γ-path in Figure 20 because the valve adjustment in Figure 20 a) permits the path $[s, y_1, x_2, \gamma]$. It is shown in [Hu93] that Problem 66 is $\mathcal{NP}$-complete if all valves in $\mathcal{G}$ have the same structure as the valves y_1, y_2 in Figure 20.
Moreover, a class $\mathbb{T}_0$ of valve types is defined in [Hu93], and this class has the following property:
   If we assume that the type of each valve node $v \in \widehat{\mathcal{V}}$ belongs to $\mathbb{T}_0$
   then Problem 66 can be solved in polynomial time.
   If, however, we do not make this assumption then Problem 66 is $\mathcal{NP}$-complete.

Next, we focus on **Problem 67**. When solving this problem, we must not ignore the valves. For example, assume that $h(r) = 1$ and $h(r) = 100$, respectively, for all thin and all thick arcs in Figure 20 a) – d). If ignoring the valves in y_1 and y_2, we shall find the C-minimal path $P' := [s, y_1, x_3, y_2, x_6, x_7, \gamma]$ with $C(P') = 6$; this path, however, is not admissible. The best admissible path is $P'' := [s, y_1, x_2, \gamma]$ with $C(P'') = 102$.

---

[61]That means: Can we adjust the valves of $\mathcal{G}$ such that *at least one* connection from s to γ exists?
[62]That means: Can we adjust the valves of $\mathcal{G}$ such that *no* connection from s to γ exists?

A heuristic approach to Problem 67 can be found in [Br90]. It is shown in [Hu92] that this problem is $\mathcal{NP}$-complete if all valves have the same structure as y_1, y_2 in Figure 20 a) – d). The same result follows from [Hu93] because one can reduce Problem 66 to 67. This reduction works as follows. Let $\mathcal{G} = (\mathcal{V}, \mathcal{R})$ be an instance of Problem 66; if $(s, \gamma) \in \mathcal{R}$ then $[s, \gamma]$ solves Problem 66. Otherwise define $\widetilde{\mathcal{G}} := (\mathcal{V}, \widetilde{\mathcal{R}})$ with $\widetilde{\mathcal{R}} := \mathcal{R} \cup \{(s, \gamma)\}$. Let $h(r) := 1$ for each arc $r \in \mathcal{R}$, let $h(s, \gamma) := |\mathcal{R}|^2 + 2$, and let $C := SUM_h$. Let P^* be a C-minimal s-γ-path in $\widetilde{\mathcal{G}}$. (Such a path can be found by solving Problem 67.) Then $\mathcal{G}$ has an admissible s-γ-path if and only if $P^* \neq [s, \gamma]$. [63]

Problem 67 is also $\mathcal{NP}$-complete if $\mathcal{G}$ is *undirected* and all valves have the structure of y_1, y_2 in Figure 20. This fact has been shown in [Hu92], Theorem 4.2.

Next, we focus on **Problem 68**. For example, the valve adjustment in Figure 20c) does not allow any s-γ-path.
It is shown in [Huc93] and in [Huc97] that Problem 68 is $\mathcal{NP}$-complete if all valves in $\mathcal{G}$ have the same structure as the valves y_1, y_2 in Figure 20.
Moreover, a class $\mathbf{T}^\bullet$ of valve types is defined in [Huc93], [Huc97], and this class has the following property:

> If we assume that the type of each valve node $v \in \widehat{\mathcal{V}}$ belongs to $\mathbf{T}^\bullet$ then Problem 68 can be solved in polynomial time.
> If, however, we do not make this assumption then Problem 68 is $\mathcal{NP}$-complete.

4.11.3   Further hard problems

The following two problems are considered in [Li93]:

Problem 69 *Given an undirected graph $\mathcal{G}$ and a constant d.*
Add a minimal number of edges to $\mathcal{G}$ such that the resulting graph has a diameter $\leq d$.

Problem 70 *Given an undirected graph $\mathcal{G}$ and a constant k.*
Add k or fewer edges to $\mathcal{G}$ such that the resulting graph has a minimal diameter.

---

[63]This follows from the fact that $\mathcal{G}$ contains an admissible s-γ-path of a length $\leq |\mathcal{R}|^2 + 1 = C([s, \gamma]) - 1$ if $\mathcal{G}$ contains an admissible s-γ-path at all.
To see this we assume that $P = [s = v_0, v_1, \ldots, v_k = \gamma]$ be an admissible path in $\mathcal{G}$ with $\ell(P) = k \geq |\mathcal{R}|^2 + 2$. The path P contains $(k-1)$ pairs $\big((v_{i-1}, v_i), (v_i, v_{i+1})\big)$, $i = 1, \ldots, k-1$, and it is $k - 1 \geq |\mathcal{R}|^2 + 1$. The pidgeon hole principle implies that one of these pairs of arcs appears twice in P. Hence, there exist $i < j$ such that $(v_{i-1}, v_i) = (v_{j-1}, v_j)$ and $(v_i, v_{i+1}) = (v_j, v_{j+1})$. We delete the cycle $[v_i, v_{i+1}, \ldots, v_j = v_i]$ in P and obtain the shorter path $P' := \big[s = v_0, v_1 \ldots, v_i = v_j, v_{j+1}, \ldots, v_k = \gamma\big]$; this path is admissible even if $v_i = v_j$ is a valve node; the reason is that both P and P' use the same arcs when entering and when leaving $v_i = v_j$.
Hence, we can replace each admissible s-γ-path P in $\mathcal{G}$ by a shorter one if $\ell(P) \geq |\mathcal{R}|^2 + 2$; so, the shortest admissible s-γ-path has a length $\leq |\mathcal{R}|^2 + 1$ if such a path exists.
By the way, the existence of an admissible s-γ-path P implies even the existence of a *node injective* admissible s-γ-path P' if each valve v has the following property: Any adjustment of v can only connect one incoming arc with outgoing arcs. This is the case for all valves in Figure 20 , in Figure 21, and in the sources [Hu92], [Hu93], [Huc93], [Huc97]. In this situation, the shortest admissible path has a length $\leq n - 1$.

The authors of [Li93] show that the first problem is $\mathcal{N}\mathcal{P}$-hard even in the special case that $d = 2$.

The geodetic number of a graph is defined in [HLTs93]; the authors show that computing this number is $\mathcal{N}\mathcal{P}$-complete.

A further $\mathcal{N}\mathcal{P}$-complete problem is the following:

Problem 71
Given an undirected graph $\mathcal{G}$, a number $k \geq 2$, and $2\,k$ nodes $u_1, v_1, \ldots, u_k, v_k$.
Find k pairwise edge disjoint paths $P_1, \ldots, P_k$ such that $\alpha(P_\kappa) = u_\kappa$ and $\omega(P_\kappa) = v_\kappa$
for all κ.

The paper [Fra90] describes tractable modifications of this problem. A modified version of Problem 71 has been studied in [Schr94]; the paper is about pairwise node disjoint paths in digraphs.

The paper [LMS92] is about finding disjoint paths such that the sum of their costs is minimal; the underlying network may be directed or undirected, and the word "disjoint" may be interpreted as "node disjoint", "arc disjoint", or "edge disjoint". The problem is $\mathcal{N}\mathcal{P}$-complete in any case. (By the way, the search for cost minimal arc disjoint paths in digraphs was formulated as Problem 28.)

Further hard search problems will be discussed in Section 4.12.

4.12 Searching for optimal graphs and for other discrete objects similar to paths in graphs

This section is about discrete search problems similar to the search for optimal paths in graphs; in particular, we consider the situation that an optimal graph must be found.

Many search problems in this section are problems of Artificial Intelligence. We shall often quote the book [KaKu88]; further problems and results about search in Artificial Intelligence can be found in [Nils82] and [GlGr89].

This section is organized as follows: Subsections 4.12.1 and 4.12.2 are about the search for optimal graphs, and Subsection 4.12.3 is about further problems in a remote connection with extremal paths.

4.12.1 Searching for optimal subgraphs in a given graph

Many problems of Combinatorial Optimization or of Artificial Intelligence can be reduced to the search for a subgraph $\mathcal{G}'$ of a given graph $\mathcal{G}$. In many cases, the graph $\mathcal{G}'$ must satisfy additional conditions; for example, $\mathcal{G}'$ must be isomorphic to a given graph, or $\mathcal{G}'$ must have particular nodes or arcs. The cost or the utility of a subgraph $\mathcal{G}'$ is often measured by a real number $C(\mathcal{G}')$; the problem is to find a subgraph $\mathcal{G}'$ of $\mathcal{G}$ with the minimal or the maximal value $C(\mathcal{G}')$.

4.12.1.1   Well-known problems

Probably, the most well-known search problem for optimal subgraphs is the search for minimum spanning trees in undirected graphs. More details about this problem were given in Section 4.1 (see Problem 4). A modification of this task is the search for a spanning tree of minimal diameter; this problem is solved in [HaTa95].

Another modification of the Minimum Spanning Tree Problem is the following:

Problem 72 (Steiner tree problem)
Given an undirected graph $G = (V, R)$ and a subset $V' \subseteq V$.
Find a cost minimal connected subgraph of G that contains all vertices of V'.

Surveys over this problem are given in [Win87] and [HwRi92]. The sources [Cie91], [EvMi92], [Cie95] give further details about the Steiner tree problem. The paper [WiSm92] describes heuristic solutions of Problem 72; one of these solutions is based on shortest paths.

4.12.1.2   Optimal subgraphs and optimal paths

Here we describe a relationship between the search for optimal paths in graphs and the search for optimal subgraphs.

Remark 4.214 Let G be a digraph, and let $\mathcal{P} \subseteq P(G)$.

a) If all paths in $\mathcal{P}$ are acyclic then the search for an optimal path is a special case of the search for an optimal subgraph; more precisely, we may replace the search for an optimal path $P \in \mathcal{P}$ with the search for an optimal subgraph G'_P consisting of all nodes and arcs of P. The reason is that all paths $P \in \mathcal{P}$ are acyclic so that they can be identified by the subgraph G'_P.

b) If, however, not all paths in $\mathcal{P}$ are node injective then the search for an optimal path $P \in \mathcal{P}$ cannot always be replaced with the search for a good subgraph G'_P. For example, let X be a substantial cycle; given the candidates $P_\nu = Y' \oplus X^\nu \oplus Y'' \in \mathcal{P}$ where $\nu \in \mathbb{N}$; then $G'_{P_1} = G'_{P_2} = G'_{P_3} = \ldots$. Therefore, the search for an optimal path among $P_1, P_2, P_3, \ldots$ cannot be reduced to the search for an optimal subgraph $G'_{P_1}, G'_{P_2}, G'_{P_3}, \ldots$.   □

Further relationships between optimal subgraphs and paths in graphs are described in [NgPa86]; Nguyen and Pallotino define hyperpaths, which are subgraphs with particular properties; the authors study the search for an optimal hyperpaths.

4.12.1.3   Optimal subgraphs in AND/OR graphs

AND/OR graphs are often constructed in Artificial Intelligence. This type of graph is used to describe the following situation: A hard problem ξ and several easier problems $\xi_1, \ldots, \xi_k$ are given, and one of the following assertions ($\vee$) or ($\wedge$) is true:

    ($\wedge$) Solving problem ξ means to solve ξ_1 *and* ξ_2 *and* ... *and* ξ_k.
    ($\vee$) Solving problem ξ means to solve ξ_1 *or* ξ_2 *or* ... *or* ξ_k.

Next, we give an exact definition of AND/OR graphs; our definition is mainly the same as the one in [KuNK88]. Moreover, we define a measure of utility for solution trees; this utility function is often used in literature. Most objects in Definition 4.215 are illustrated in Figure 22.

Definition 4.215 An *AND/OR graph* is an acyclic digraph $\mathcal{G} = (V, \mathcal{R})$ with the following properties:

There exists a node $s \in V$ of indegree $\deg^-(s) = 0$; moreover, $\deg^-(v) > 0$ for all $v \neq s$. There exists an s-v-path in $\mathcal{G}$ for each $v \in V$. The set V is the disjoint union of three sets V_{and}, V_{or}, and V_{end}. The node s is an element of $V_{and} \cup V_{or}$, and V_{end} is equal to the set of all nodes v with outdegree $\deg^+(v) = 0$.

The elements of V_{and}, V_{or}, and V_{end}, respectively, are called *AND-nodes*, *OR-nodes* and *END-nodes*. If $\mathcal{G}$ is a tree with root s then $\mathcal{G}$ is called an *AND-OR-tree*.

An AND/OR graph $\mathcal{G}$ is illustrated in Figure 22. The nodes $v \in V_{and}$ are drawn as cycles, and all arcs $r \in \mathcal{R}$ with $\alpha(r) = v$ are connected with a circular arc. The nodes $v \in V_{or}$ are drawn as white squares, and the nodes $v \in V_{end}$ are drawn as black squares.

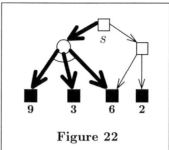

Figure 22

Let $\mathcal{G}' = (V', \mathcal{R}')$ be a subgraph of $\mathcal{G}$ with the property that $\mathcal{G}'$ is a rooted tree in s. Then $\mathcal{G}'$ is called a *solution tree* of $\mathcal{G}$ if the following is true for all nodes $v \in V'$:

If v is an OR-node then $\mathcal{R}'$ contains exactly one arc $r \in \mathcal{R}$ with $\alpha(r) = v$.
If v is an AND-node then $\mathcal{R}'$ contains all arcs $r \in \mathcal{R}$ with $\alpha(r) = v$.   (4.257)

For example, the subgraph in Figure 22 with the bold arcs is a solution tree of $\mathcal{G}$.

We interpret any solution tree $\mathcal{G}'$ as a solution of the problem s. Assertion (4.257) guarantees that each subproblem $\xi \in V'$ is solved correctly.

Let $p : V_{end} \rightarrow \mathbb{R}$; the value $p(v)$ is interpreted as the utility of v. For example, each end node v in Figure 22 is marked with $p(v)$.

For each solution tree $\mathcal{G}'$, we define the *minimum profit* $f_p(\mathcal{G}')$ of $\mathcal{G}'$ as the minimal value $p(v)$ of all leaves v of $\mathcal{G}'$. For example, $f_p(\mathcal{G}') = 3$ in Figure 22.
The *maximin profit* $F_p(\mathcal{G})$ of $\mathcal{G}$ is defined as the maximum of all values $f_p(\mathcal{G}')$ where $\mathcal{G}'$ is a solution tree of $\mathcal{G}$. $\square$

The maximin profit of an AND/OR tree $\mathcal{G}$ is an important quantity in Game Theory. Roughly speaking, $F_p(\mathcal{G})$ can be interpreted as the the maximum profit that player A can achieve if his adversary B always uses his best strategy.

Next, we shall quote several sources in connection with the following problem:

Problem 73 *Given a finite AND/OR tree $\mathcal{G} = (V, \mathcal{R})$ with root s. Let $p : V_{end} \rightarrow \mathbb{R}$.*
a) *Find an optimal (i.e. an f_p-maximal) solution tree $\mathcal{G}^* \subseteq \mathcal{G}$.*
b) *Compute the maximin profit $F_p(\mathcal{G})$.*

A simple algorithm to solve this problem is described in Definition 2.4 of [Sto79] and in the proof of Theorem 1 in the same paper. Stockman's algorithms have been slightly modified in [Huck92], Theorem 4.129.

All these algorithms often execute more computations than actually needed to solve Problem 73. The algorithm *alpha-beta pruning* avoids many unnessecary computations. Alpha-beta pruning has been studied in [Sto79], [Pe82], [Tar83], [KuNK88], and in many other sources.

A further method to compute $F_p(\mathcal{G})$ very fast is the use of heuristic functions; roughly speaking, these heuristic functions help to decide whether or not a computation step is necessary. Heuristic functions are used by the algorithm B* in [Berl79], by the algorithm SSS* in [Sto79], [McGl83], [RoPe88], by the algorithm AO* in [NKK84], [MaBa85], by the best first search algorithm GEN-AO* in [ChGh92], and by the algorithm REV* in [Chak94], which is similar to Dijkstra's algorithm.

The algorithm in [Riv88] is based on replacing the functions $\min(x, y)$ and $\max(x, y)$ by approximations $\phi_{\min}(x, y)$ and $\phi_{\max}(x, y)$, respectively; the functions $\phi_{\min}$ and $\phi_{\max}$ have continuous partial derivatives.

Further algorithms to solve Problem 73 are described in [Nils82] and in [KuNK88].

We next describe relationships between extremal subgraphs of AND/OR-trees and extremal paths in these trees.

Remark 4.216 Let $\mathcal{G} = (\mathcal{V}, \mathcal{R})$ be an AND/OR tree with root s. Let Γ be the set of all leaves of $\mathcal{G}$, and let $p : \Gamma \to \mathbb{R}$. For each node $v \in \mathcal{V}$, let $P(v)$ be the unique s-v-path in $\mathcal{G}$. For all paths P in $\mathcal{G}$, let $\mathcal{G}'_P$ be the subgraph of $\mathcal{G}$ that consists of all nodes and arcs of P. (This graph has been introduced in Remerk 4.214.) Let $C(P) := p(\omega(P))$ for all s-Γ-paths P. We define **G** as the set of all solution trees of $\mathcal{G}$.

There are two relationships between extremal subgraphs and extremal paths in $\mathcal{G}$.

a) The digraph $\mathcal{G}$ is acyclic. So, a C-minimal s-Γ-path P can be found by searching for an extremal subgraph $\mathcal{G}'_P$.

b) Problem 73 a) can be formulated as a problem concerning minimal paths. To see this we assume that $\mathcal{G}' \in$ **G** be an arbitrary solution tree, and let $\Gamma' \subseteq \Gamma$ be the set of all leaves of $\mathcal{G}$. Then the following is true for all leaves $v' \in \Gamma'$:
$$p(v') = f_p(\mathcal{G}') \text{ if and only if } P(v') \text{ is a } C\text{-minimal } s\text{-}\Gamma'\text{-path in } \mathcal{G}'.$$

Consequently, we can formulate Problem 73 a) as follows:

Problem 74 *Given an AND/OR tree $\mathcal{G}$, a node function $p : \mathcal{V} \to \mathbb{R}$, and the path function $C(P) := p(\omega(P))$. Let **G** be the set of all solution trees of $\mathcal{G}$. For all digraphs $\mathcal{G}' \in$ **G**, let $P^*_{\mathcal{G}'}$ be a C-minimal s-Γ'-path in $\mathcal{G}'$.*

*Find a graph $\mathcal{G}^* \in$ **G** such that $C\left(P^*_{\mathcal{G}^*}\right)$ is maximal among all values $C\left(P^*_{\mathcal{G}'}\right)$ where $\mathcal{G}' \in$ **G**.*

This problem has a similar structure as Problem 17 in Section 4.5 because the

search for an spp graph is also the search for a graph such that its extremal
paths have particular properties.
□

The construction of solution trees in AND/OR
graphs may cause problems. For example, the
AND/OR graph $\mathcal{G}$ in Figure 23 does not contain a solution tree at all; if $\mathcal{G}'$ were a solution
tree then $\mathcal{G}'$ would contain all arcs with start
nodes s, a, and b. Consequently, $\mathcal{G}' = \mathcal{G}$, and
this graph is not a rooted tree. This prob-
lem can be avoided by not requiring that $\mathcal{G}'$ is
a tree; that means, we define solution graphs
and search for these graphs and not for solution trees.

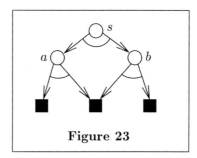

Figure 23

Further problems arising from solution trees have been described in Remark 4.131 and
4.132 of [Huck92].

4.12.1.4   The Constraint-Satisfaction Problem

The Constraint-Satisfaction Problem (CSP) is another problem that can be reduced
to the search for a (possibly optimal) subgraph; this will be shown immediately after
Definition 4.218.
An overview over the CSP can be found in [Mese89]. We formulate a very simple
version of this problem.

Problem 75 (Binary Constraint Satisfaction Problem, Binary CSP)
Given a finite and nonempty set $X = \{x_1, \ldots, x_n\}$.
Let $l_1, \ldots, l_n \geq 0$, and let $W_\nu := \{0, \ldots, l_\nu\}$ for all $\nu = 1, \ldots, n$.
Given a relation $C_{i,j} \subseteq W_i \times W_j$ for each pair (i, j) with $1 \leq i < j \leq n$.
Find a function $f : X \to W_1 \cup \ldots \cup W_n$ with the following properties (i) *and* (ii):

   (i) $(\forall\, i = 1, \ldots, n)\ f(x_i) \in W_i,$ (ii) $(\forall\, i < j)\ (f(x_i), f(x_j)) \in C_{i,j}.$

The relations $C_{i,j}$ are called the *constraints* of the problem. If the pairs $(f(x_i), f(x_j))$
need not satisfy any substantial condition then we define $C_{i,j} := W_i \times W_j$. We call
$C_{i,j}$ *substantial* if $C_{i,j}$ is a proper subset of $W_i \times W_j$.

The next remark is about the *general* constraint-satisfaction problem; the constraints
in this problem may be J-ary relations with $J \geq 2$.

Remark 4.217

The general CSP is formulated with the help of J-ary relations C where $2 \leq J \leq n$.
A constraint $C \subseteq W_{i_1} \times \ldots \times W_{i_J}$ means that the J-tuple $\left(f(x_{i_1}), \ldots, f(x_{i_J})\right)$ must
be an element of C.

It is mentioned in [Mese89], page 4, that each CSP with J-ary relations can be transformed into an equivalent CSP with exclusively binary relations.

The general Constraint Satisfaction Problem is $\mathcal{NP}$-complete. This has been shown in [Mack77] and on page 297 of [Huck92]; the proof in [Huck92] is a reduction of 3SAT to the CSP with ternary relations (i.e. $J = 3$).   □

We transform the CSP with binary relations into a subgraph problem. This is the purpose of the next definition, which is strongly influenced by [Mese89].

Definition 4.218 Given an instant **P** of Problem 75 with the binary relations $C_{i,j}$ as constraints.

We define the *constraint graph* of the problem **P** as the digraph $\mathcal{G} = (\mathcal{V}, \mathcal{R})$ with the nodes $x_1, \ldots, x_n$ and an arc (x_i, x_j) for each substantial constraint $C_{i,j}$; more formally, we define $\mathcal{V} := X = \{x_1, \ldots, x_n\}$ and $\mathcal{R} := \{(x_i, x_j) \mid i < j, \; C_{i,j} \neq W_i \times W_j\}$.

The *extended constraint graph* $\widetilde{\mathcal{G}} = (\widetilde{\mathcal{V}}, \widetilde{\mathcal{G}})$ describes the search for appropriate values $f(x_1), \ldots, f(x_n)$. The nodes of $\widetilde{\mathcal{G}}$ are pairs (x_i, λ) and symbolize the case that $f(x_i) = \lambda$. The constraints are represented by arcs. More formally, $\widetilde{\mathcal{V}}$ and $\widetilde{\mathcal{R}}$ are defined as follows:

$$\widetilde{\mathcal{V}} := \bigcup_{\nu=1}^{n} \widetilde{\mathcal{V}}_\nu \text{ where } \widetilde{\mathcal{V}}_\nu := \{(x_\nu, \lambda) \mid \lambda \in W_\nu\}, \; \nu = 1, \ldots, n.$$

$$\widetilde{\mathcal{R}} := \{((x_i, \lambda), (x_j, \mu)) \mid i < j, \; (\lambda, \mu) \in C_{i,j}, \; C_{i,j} \neq W_i \times W_j\}.$$ □

The problem **P** is equivalent to the search for a subgraph $\mathcal{G}' = (\mathcal{V}', \mathcal{R}')$ of $\widetilde{\mathcal{G}}$ with the following properties:

> $\mathcal{V}' = \{v^{(1)}, \ldots, v^{(n)}\}$ with $v_i^{(\nu)} \in \widetilde{\mathcal{V}}_\nu$, $\nu = 1, \ldots, n$; that means that $\mathcal{V}'$ contains exactly one pair $v^{(\nu)} = (x_\nu, \lambda_\nu)$ for each $\nu = 1, \ldots, n$.   (4.258)

> $\mathcal{G}'$ is isomorphic to the constraint graph $\mathcal{G}$ in the following way: The nodes $v^{(\nu)} = (x_\nu, \lambda_\nu) \in \widetilde{\mathcal{V}}$ and $x_\nu \in \mathcal{V}$ ($\nu = 1, \ldots, n$) are isomorphic images of each other.   (4.259)

We give reasons why the problem **P** is equivalent to the search for a graph $\mathcal{G}'$.
Let $f : X \to W_1 \cup \ldots \cup W_n$ be a solution of **P**. We then define $\lambda_\nu := f(x_\nu)$ for all ν. Then $v^{(\nu)} = (x_\nu, \lambda_\nu) = (x_\nu, f(x_\nu))$ for all ν. This and the fact that f is a solution of **P** imply that $\mathcal{G}'$ has properties (4.258) and (4.259).
Conversely, let $\mathcal{G}'$ be a digraph for which (4.258) and (4.259) are true. We define $f(x_\nu) := \lambda_\nu$ for all ν. Then $v^{(\nu)} = (x_\nu, \lambda_\nu) = (x_\nu, f(x_\nu))$ for all ν. By assertion (4.259) and by the definition of the sets $\mathcal{R}'$ and $\widetilde{\mathcal{R}}$, the function f satisfies all constraints $C_{i,j}$.

A natural variant of the CSP is to find *optimal* functions that satisfy all constraints. We only must define a cost $C(f)$ for all functions $f : X \to W_1 \cup \ldots \cup W_n$ or a cost $C'(\mathcal{G}')$ for all subgraphs $\mathcal{G}'$ of the extended constraint graph $\widetilde{\mathcal{G}}$. Then the modified CSP is to find an admissible function f with minimal value $C(f)$; this is equivalent to find a C'-minimal subgraph $\mathcal{G}'$ with properties (4.258) and (4.259).

More details about the Constraint Satisfaction Problem can be found in [Freud85], [Dav87], [DePe88], [Mese89], and in many other sources. A geometric CSP is solved in [Sapo83]; the graph defined on page 112 of that paper is similar to the extended constraint graph in Definition 4.218.

4.12.2   Construction of optimal graphs that are not a part of a given supergraph

Many problems of Artificial Intelligence can be reduced to the search for an optimal graph $\mathcal{G}$; this graph $\mathcal{G}$ is an element of a given class of graphs, but $\mathcal{G}$ is not a subgraph of a given graph. For reasons of brevity we confine ourselves to the search for optimal rooted trees.

An easy problem is the construction of binary search trees with minimal expected time of searching. More precisely, several real numbers $x_1, \ldots, x_n$ are given and each element x_ν is sought with probability p_ν. The problem is to arrange the x_n's in a binary search tree such that the expectation of the searching time is minimal. This problem was described in Section 2.4; more details can be found in [Nolt82], in [Aign88], and in many other sources. Another source about optimal search trees is [LeLS84].

A similar problem is solved in [Larm87]. Several objects $x_1, \ldots, x_n$ are given, and each object has a weight $q_\nu > 0$. The problem is to arrange the x_n's in a height-restricted binary search tree such that $\sum_{\nu=1}^{n} q_\nu \cdot \mathbf{h}(x_\nu)$ is minimal where $\mathbf{h}(x_\nu)$ denotes the height of x_ν in the tree.

The paper [Sik88] is about the construction of trees with the property that particular path lengths are minimized.

A further special search problem for optimal trees is solved in [Hwa80].

The definitions and problems in [Helm88] are very general. Helman assumes that a universe $\mathcal{U}$ of objects and a subset $X \subseteq \mathcal{U}$ of elementary objects are given; all elements $u \in \mathcal{U}$ are called *conjuncts*. Moreover, Helman assumes that a binary operation $\odot : \mathcal{U} \times \mathcal{U} \to \mathcal{U}$ is given; this operation need not be associative.

The construction of a conjunct $u \in \mathcal{U}$ from elementary objects can be represented by a binary tree.

To see this we assume that $\mathcal{G}$ is a binary tree with the property that every internal node has exactly two sons. An example is shown in Figure 24. Let s be the root of $\mathcal{G}$. We assume that the internal nodes of $\mathcal{G}$ are marked with $"\odot"$, and the leaves v are marked with an element $g(v) \in X$. We define $g(v) := g(v') \odot g(v'')$ for any internal node v with sons v' and v''. Then we interpret $g(s)$ as the object represented by $\mathcal{G}$. E.g., the tree in Figure 24 represents the conjunct $\big((a \odot b) \odot (a \odot c)\big) \odot (a \odot c)$.

It is easy to see that each object $u \in \mathcal{U}$ can be represented by a tree if u can be computed from elementary objects $x \in X$ by repeatedly applying $\odot$. This implies that the construction optimal conjuncts can often be reduced to the construction of optimal binary computation trees.

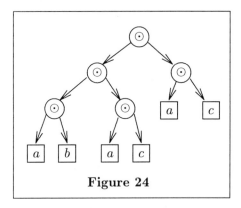

Figure 24

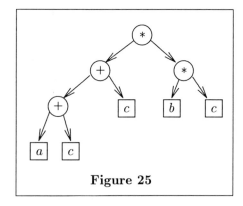

Figure 25

Next, we see that constructing optimal computation trees can often be reduced to the search for optimal paths in graphs. For this purpose, we first describe how to represent a tree as a path in a graph. Let $\mathcal{G} = (\mathcal{V}, \mathcal{R})$ be a rooted binary tree with the property that all internal nodes have two children. Let s be the root of $\mathcal{G}$. Let $\rho(\mathcal{G}')$ be the root of $\mathcal{G}'$ if $\mathcal{G}'$ is a subtree of $\mathcal{G}$; in particular, $\rho(\mathcal{G}') = s$ for $\mathcal{G}' = \mathcal{G}$. We assume that two disjoint finite alphabets Σ', Σ'' are given; let $\Sigma := \Sigma' \cup \Sigma''$. Moreover, we assume that each internal node v [each leaf v] is marked with a symbol $m(v) \in \Sigma'$ [$m(v) \in \Sigma''$]. An example is given in Figure 25 where $\Sigma' = \{+, *\}$, $\Sigma'' = \{a, b, c\}$, and $\Sigma = \{a, b, c, +, *\}$.

We define the word $F(\mathcal{G})$ over $\Sigma \cup \{(,)\}$ as follows: If the root s is the only node of $\mathcal{G}$ then we define $F(\mathcal{G}) := m(s)$. Otherwise let $\mathcal{G}_L$ and $\mathcal{G}_R$ be the left and the right subtree of $\mathcal{G}$, respectively. Then we define $F(\mathcal{G}) := \big(F(\mathcal{G}_L)\, m(\rho(\mathcal{G}))\, F(\mathcal{G}_R) \big)$, and we apply this formula recursively to $\mathcal{G}_L$ and $\mathcal{G}_R$ until reaching a leaf. For example, $F(\mathcal{G}) = \big(\big((a + c) + c \big) * (b * c) \big)$ for the graph $\mathcal{G}$ in Figure 25.

It is easy represent the word $F(\mathcal{G})$ (and consequently the tree $\mathcal{G}$ itself) by a path in the graph $\widehat{\mathcal{G}} := (\widehat{\mathcal{V}}, \widehat{\mathcal{R}})$ with $\widehat{\mathcal{V}} := \Sigma \cup \{(,)\}$ and $\widehat{\mathcal{R}} := \widehat{\mathcal{V}} \times \widehat{\mathcal{V}}$; if $F(\mathcal{G}) = z_1 \ldots z_k$ with $z_1, \ldots, z_k \in \Sigma \cup \{(,)\}$, we represent $F(\mathcal{G})$ by the path $P_{\mathcal{G}} := [z_1, \ldots, z_k]$.

Let $\mathcal{P}'$ be the set of all paths P in $\widehat{\mathcal{G}}$ with the property that $P = P_{\mathcal{G}}$ for some tree $\mathcal{G}$. Then the search for an optimal tree $\mathcal{G}$ can be replaced by the search for an optimal path $P \in \mathcal{P}'$.

The set $\mathcal{P}'$ is not closed under concatenation of paths because not each compound path $Q_1 \oplus Q_2$ with $Q_1, Q_2 \in \mathcal{P}'$ represents a tree. On the other hand, the set $\mathcal{P}'$ is closed under the operations $\boxplus_y$, which are now defined, and which represent the construction of a new tree if the left subtree, the right subtree, and a root s with $m(s) = y \in \Sigma'$ are given. More precisely, we define $Q_1 \boxplus_y Q_2$ as the path constructed by inserting a new start node "(", a new end node ")", and the symbol $y \in \Sigma'$ between Q_1 and Q_2; that means, if $Q_1 = [x_1, \ldots, x_j]$, $Q_2 = [z_1, \ldots, z_k]$ and $y \in \Sigma'$ then we define

$$
\begin{aligned}
Q_1 \boxplus_y Q_2 &:= \big[(, x_1, \ldots, x_j, y, z_1, \ldots, z_k,) \big] \\
&= \big[(, \alpha(Q_1)] \oplus Q_1 \oplus [\omega(Q_1), y] \oplus [y, \omega(Q_1)] \oplus Q_2 \oplus [\omega(Q_2),) \big].
\end{aligned}
$$

We now show that $\mathcal{P}'$ is closed under $\boxplus_y$. Let $Q_1, Q_2 \in \mathcal{P}'$. Then there exist two trees $\mathcal{G}_L$ and $\mathcal{G}_R$ with $Q_1 = P_{\mathcal{G}_L}$ and $Q_2 = P_{\mathcal{G}_R}$. Then $Q_1 \boxplus_y Q_2 = P_{\mathcal{G}}$ for the tree $\mathcal{G}$ that is constructed as follows: Generate a root s, mark s with y, and generate a left and a right arc starting from s; append $\mathcal{G}_L$ at the left arc and $\mathcal{G}_R$ at the right one.

It is easy to see that these operations $\boxplus_y$ are not associative.

If χ is a cost measure for trees then we define the path function $C_\chi : \mathcal{P}' \to \mathbb{R}$ such that $C_\chi(P_\mathcal{G}) := \chi(\mathcal{G})$ for all trees $\mathcal{G}$; this function C_χ is a reasonable cost measure because the tree $\mathcal{G}$ and the $P_\mathcal{G}$ should have the same cost. Many interesting questions arise from the algebraic structure of C_χ; for example, how can $C_\chi(Q_1 \boxplus_y Q_2)$ be computed from $C_\chi(Q_1)$, $C_\chi(Q_2)$ and y?

Probably, the algebraic structure of C_χ and the missing associativity of $\boxplus_y$ make the problem of finding C_χ-optimal path $P \in \mathcal{P}'$ very difficult; that means that the equivalent problem of finding χ-minimal trees is difficult, too.

4.12.3 Further problems in connection with extremal paths in graphs

Besides the problems in Subsections 4.12.1 and 4.12.2), there are many further problems in a remote connection with extremal paths in graphs.

One of these problems is to get information about a complete set $\mathcal{P}$ of paths (and not only about the extremal elements of $\mathcal{P}$). In [Ord87], an algorithm is given that finds all paths in a network. The paper [MaDe76] gives an overview over algorithms that enumerate all cycles of a graph, and in [Johns77], two algorithms are given to determine all *simple* cycles of a graph. The paper [Tarj81] is about generating regular expressions $p(x, y)$ that represent all paths from node x to a node y.

In [RoPe84], the following problem is studied: Given two nodes s, γ and a number k, find the maximal number of pairwise node disjoint s-γ-paths of length $\leq k$.
Several other results about these paths have been quoted in Theorems 3.11, 3.42 of this book. The difference, however, between those results and the ones of Ronen and Perl is the following: In 3.11 and 3.42, the maximal number of node disjoint s-γ-paths is described; in [RoPe84], however, an algorithm is given to compute or approximate this maximal number.

A fast algorithm to find node disjoint paths is given in [Shi80].

The problem in [Hort87] is likewise similar to the search for optimal paths in graphs; Horton develops an algorithm to find an optimal cycle basis of an undirected graph. There exist many other papers about cycle bases. The cycles are not defined as paths P with $\alpha(P) = \omega(P)$; instead of this, a cycle is a set S of edges such that the following is true for each node v: The number of all edges in S with endpoint v is even. The paper [Hart93] investigates objects that are similar to cycle bases.

Another problem is the search for a path with a given property; this property is not necessarily the optimality of P. The paper [SpTs86] is about the search for C-negative cycles where C is an additive path function; these cycles need not be minimal, but they are important when searching for C-minimal paths. The article [RiLi85] describes the search for cycles of a given length; a similar problem is M-PATH in Theorem 4.210. The algorithm in [Trusz81] finds diagonal free cycles of a length greater than three. (The definition of diagonal free cycles was given in Theorem 3.91.)

The article in [Blum86] is about finding a proof that a given graph has a Hamiltonian cycle; to find such a proof is a problem of Artificial Intelligence in the context of extremal paths in graphs.

Chapter 5

Relationships between paths in graphs and continuous curves

The previous sections were about paths in graphs, which are discrete objects. We now compare (optimal) paths to (optimal) continuous curves; we shall give reasons why curves in a plane or in a space may be considered as continuous versions of paths in graphs.

Most curves appearing in this chapter are of the form $\{(\xi, f(\xi)) \mid a \leq \xi \leq b\}$ where $f : [a, b] \to \mathbb{R}$ is a differentiable function and $0 \leq a < b \leq 1$. We shall define the cost of such a curve as a value $L(f) \in \mathbb{R}$; an example is $L(f) := \int_a^b f(\xi) \mathrm{d}\xi$. So, a functional L is an analogic object to a path function.

The problem of finding optimal curves is similar to finding optimal paths in graphs. The search for optimal curves is the central problem of *Variational Calculus*.

Also, we shall define structural properties of cost measures for continuous curves; these properties are similar to additivity, monotonicity, order preservation, and Bellman properties of path functions. Moreover, we discuss the problem of finding optimal curves if the cost measure L has a particular structural property. Furthermore, we shall prove a continuous version of Theorem 4.191 b).

This chapter is structured a follows: A detailed description of the analogies between paths in graphs and continuous curves is given in Section 5.1. Sections 5.2 – 5.8 are about particular properties of cost measures for curves; in most cases, we describe the special properties earlier than the more general ones. Section 5.9 describes a translation of Theorem 4.191 b) into the setting of continuous curves.

Our main intention is to describe examples of definitions and of (optimization) problems that can be transferred from the world of graphs into the world of continuous curves. We do, however, not intend to give a systematic or detailed description of

continuous optimization problems and their solutions. In particular, we often formulate problems but we do not give their complete solution. More details about solving
continuous optimization problems are given in every book about Variational Calculus,
for example in [Akh62], [GeFo63], [Bre83], and [Leit86].

5.1 Continuous curves: basic definitions, elemen- tary facts, and relationships to paths in graphs

In this section, we introduce the basic terminology in the context of continuous curves.
Moreover, we describe analogies between curves and paths in a digraph.

We start with technical conventions.

Remark/Definition 5.1   a) Let $F : \mathbb{R}^k \dashrightarrow \mathbb{R}$. We define $\partial_i F$ as the partial
derivative of F with respect to the ith argument.
The advantage of this notation is that it is independent of the letters we use.
For example, let $F(u,v) := u^3 v^2$ for all $u, v \in \mathbb{R}^2$. Then $\partial_1 F(x,y) = 3x^2 y^2$, and
$\partial_1 F(y,x) = 3y^3 x^2$ for all $x, y \in \mathbb{R}$.
The notation $\partial_{i_1,\ldots,i_k} F$ means the iterated partial derivative with respect to the
i_1th, i_2th, ... , i_kth argument; more formally, $\partial_{i_1,\ldots,i_k} F := \partial_{i_k} \partial_{i_{k-1}} \cdots \partial_{i_1} F$.

b) We shall often omit arguments of functions. E.g., we shall write $\int_0^x X(\xi, f, f') \, d\xi$
instead of $\int_0^x X(\xi, f(\xi), f'(\xi)) \, d\xi$. $\Box$

Next, we define Lipschitz functions.

Definition 5.2 Let $X \subseteq \mathbb{R}^k$, let $f : X \to \mathbb{R}$, and let $1 \leq \kappa \leq k$. The function f is
called a *Lipschitz function with respect to the κth argument* if there exists an $c \geq 0$
such that

$$|f(x_1, \ldots., x_{\kappa-1}, x'_\kappa, x_{\kappa+1}, \ldots., x_k) - f(x_1, \ldots, x_{\kappa-1}, x''_\kappa, x_{\kappa+1}, \ldots., x_k)| < c \cdot |x'_\kappa - x''_\kappa|$$

for all $(x_1, \ldots, x_{\kappa-1}, x'_\kappa, x_{\kappa+1}, \ldots, x_k) \in X$ and $(x_1, \ldots, x_{\kappa-1}, x''_\kappa, x_{\kappa+1}, \ldots, x_k) \in X$.
We call c a *Lipschitz constant*. $\Box$

Next, we define several notations in the context of differentiable functions.

Definition 5.3   a) Let $0 \leq a \leq b \leq 1$. If $a < b$ then let $\mathbb{F}[a,b]$ denote the set of all
differentiable functions $f : [a,b] \to \mathbb{R}$ with continuous derivatives; in particular, $f'(a)$ is defined as the right derivative at a, and $f'(b)$ is defined as the left
derivative at b.   If $a = b$ then $\mathbb{F}[a,b]$ is the set of all functions f
from $[a,a] = \{a\}$ to $\mathbb{R}$; the derivative $f'(a)$ is not defined.
Let $\mathbb{F} := \bigcup_{0 \leq a \leq b \leq 1} \mathbb{F}[a,b]$, and let $\mathbb{F}^+ := \bigcup_{0 \leq a < b \leq 1} \mathbb{F}[a,b]$.

b) Let $f \in \mathbb{F}$ and $\mathrm{def}(f) = [a,b]$. Then we define $\mathsf{a}(f) := a$ and $\mathsf{b}(f) := b$.

c) Let $f \in \mathbb{F}$, $a = \mathsf{a}(f)$, $b = \mathsf{b}(f)$, and $a \leq x \leq b$. Then we define $f|_x := f|_{[a,x]}$.
The restrictions $f|_x$ are similar to prefixes of paths in graphs.

d) A subset $\widetilde{\mathbb{F}} \subseteq \mathbb{F}$ is called *prefix closed* if it is closed under restrictions $f|_x$; that means that $f \in \widetilde{\mathbb{F}} \Rightarrow f|_x \in \widetilde{\mathbb{F}}$ for all $f \in \mathbb{F}$ and all $x \in \mathrm{def}(f)$.

e) Any partial function $L : \mathbb{F} \dashrightarrow \mathbb{R}$ is called a *functional*. If L is given, we define $\overline{L}(f,x) := L\left(f|_x\right)$ for all f and x with $\mathsf{a}(f) \le x \le \mathsf{b}(f)$ and $f|_x \in \mathrm{def}(L)$.
For example, let $L(f) := \int_a^b f(\xi)\,\mathrm{d}\xi$ for all a, b and all $f \in \mathbb{F}[a, b]$. Then L is a functional, and $\overline{L}(f, x) = \int_a^x f(\xi)\,\mathrm{d}\xi$ for all $x \in [a, b]$.
Another example of a functional is $L(f) := \int_a^b F\big(\xi, f(\xi), f'(\xi)\big)\,\mathrm{d}\xi$ where $F : \mathbb{R}^3 \to \mathbb{R}$ is a given function; then $\overline{L}(f, x) = \int_a^x F\big(\xi, f(\xi), f'(\xi)\big)\,\mathrm{d}\xi$ for all f and all x.

f) Let $0 \le a \le b \le 1$. A functional $l : \mathbb{F}[a, b] \to \mathbb{R}$ is called *linear* if the following is true for all $f, g \in \mathbb{F}[a, b]$ and all $c_1, c_2 \in \mathbb{R}$: $l(c_1 \cdot f + c_2 \cdot g) = c_1 \cdot l(f) + c_2 \cdot l(g)$.
For example, $l(f) := \int_a^b f(\xi)\,\mathrm{d}\xi$ is a linear functional. $\qquad\Box$

Next, we define a digraph that is often helpful when comparing paths in graphs to continuous functions.

Definition/Remark 5.4

a) We define the digraph $\mathcal{G}_0 = (\mathcal{V}_0, \mathcal{R}_0)$ as follows:
$$\mathcal{V}_0 := [0, 1] \times \mathbb{R}, \quad \mathcal{R}_0 := \left\{ \big((x_1, y_1), (x_2, y_2)\big) \mid 0 \le x_1 < x_2 \le 1 \right\}.$$

b) Let $f : [a, b] \to \mathbb{R}$, and let $a \le x \le x + \delta \le b$. Then we define
$$r_\delta^f(x) := \big((x, f(x)), (x + \delta, f(x + \delta))\big). \tag{5.1}$$

c) Let $P = [(x_0, y_0), (x_1, y_1), \dots, (x_k, y_k)]$ be a path in $\mathcal{G}_0$; then we define $\mathcal{X}(P) := \{x_0, \dots, x_k\}$.
Let $x_0 \le x$, and let $x_j = \max\{x_\kappa | x_\kappa \le x\}$. Then $P|_x := [(x_0, y_0), \dots, (x_j, y_j)]$.

d) Given a path function $H : \mathcal{P}(\mathcal{G}_0) \dashrightarrow \mathbb{R}$, we define $\overline{H}(P, x) := H\left(P|_x\right)$ for all P and x for which the path $P|_x$ and the H-value $H\left(P|_x\right)$ exist.

e) Let $f : [a, b] \to \mathbb{R}$, and let $P \in \mathcal{P}(\mathcal{G}_0)$. We say that the path P *approximates* f if $P = [(x_0, y_0), \dots, (x_k, y_k)]$ has the properties that $x_0 = a$, $x_k = b$, and $y_\kappa = f(x_\kappa)$ for all $\kappa = 0, \dots, k$; in this case, we write $P \approx f$.
An example is shown in Figure 26. $\qquad\Box$

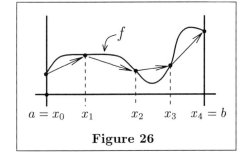

Figure 26

Next, we describe analogies between graph theoretic and continuous objects.

Remark 5.5 Let $f \in \mathbb{F}$, $a = \mathsf{a}(f)$, $b = \mathsf{b}(f)$. Moreover, let $P \in \mathcal{P}(\mathcal{G}_0)$ approximate f, and let $x \in \mathcal{X}(P)$.

a) We consider f as a continuous version of P. If $a < b$ and $a = b$, respectively, then f is analogous to a path P with $\ell(P) > 0$ and $\ell(P) = 0$.

b) We consider the restriction $f|_x$ as a continuous version of the prefix $P|_x$. It is easy to show that even $f|_x$ approximates $P|_x$ if $x \in \mathcal{X}(P)$.

c) Next, we see why we may consider the triple $r^f(x) := (x, f(x), f'(x))$ as an infinitesimal object in analogy to an arc $r_\delta^f(x)$. For this purpose, we define

$$\mathcal{R}_1 := \{(x, y, \rho, \delta) \mid \delta > 0, \ 0 \le x < x + \delta \le 1, \ y, \rho \in \mathbb{R}\}.$$

Let $\mathbf{t} : \mathcal{R}_0 \to \mathcal{R}_1$ be defined as $\mathbf{t}\big((x_1, y_1), (x_2, y_2)\big) := \left(x_1, y_1, \frac{y_2 - y_1}{x_2 - x_1}, x_2 - x_1\right)$ for all arcs of $\mathcal{R}_0$. Then $\mathbf{t}$ is bijective, and $\mathbf{t}^{-1}(x, y, \rho, \delta) = \big((x, y), (x + \delta, y + \rho \cdot \delta)\big)$ for all $(x, y, \rho, \delta) \in \mathcal{R}_1$.

In particular, each arc $r_\delta^f(x)$ is uniquely determined by $\mathbf{t}\big(r_\delta^f(x)\big)$ $\overset{(5.1)}{=} \left(x, f(x), \frac{f(x+\delta) - f(x)}{\delta}, \delta\right)$. If δ tends to zero, we obtain an infinitesimal version of $\mathbf{t}(r_\delta^f(x))$, namely $(x, f(x), f'(x), 0)$. The fourth component of this quadruple is always equal to 0; so, we may omit it and obtain the triple $r^f(x) = (x, f(x), f'(x))$.

d) We have motivated why $r^f(x) = (x, f(x), f'(x))$ is analogous to an arc $r := r_\delta^f(x)$. The same argumentation implies that $(x, f(x), f'(x))$ is also analogous to $r' := \big((x - \delta, f(x - \delta)), (x, f(x))\big)$. Consequently, the same object $(x, f(x), f'(x))$ is considered as an infinitesimal version of the two different objects r and r'. We leave it to the reader to solve this philosophic problem. □

Next, we define several analytic terms in connection to functionals.

Definition 5.6 Let $L : \mathbb{F} \dashrightarrow \mathbb{R}$ and $\|\bullet\| : \mathbb{F}^+ \to \mathbb{R}$ be two functionals.

a) We call $\|\bullet\|$ a *norm* if each restriction $\big(\|\bullet\|\big)|_{\mathbb{F}[a,b]}$ is a norm for $\mathbb{F}[a, b]$ in the usual sense. For example, let

$$\|f\|_1 := \int_a^b |f(\xi)|\, d\xi, \qquad \|f\|_{max} := \max\big\{|f(\xi)| \mid a \le \xi \le b\big\}, \qquad \text{and}$$

$$\|f\|_* := |f(a)| + \int_a^b |f'(\xi)|\, d\xi = |f(a)| + \|f'\|_1$$

for all $f \in \mathbb{F}^+$ and for $a := a(f)$, $b := b(f)$. Then $\|\bullet\|_1$, $\|\bullet\|_{max}$, and $\|\bullet\|_*$ are norms.

b) Let $\|\bullet\|$ be a norm. Then L is called $\|\bullet\|$-*continuous* if for all $f \in \operatorname{def}(L) \cap \mathbb{F}^+$ and all $\varepsilon > 0$, there exists a $\delta > 0$ such that the following is true for all $g \in \operatorname{def}(L)$ with $\operatorname{def}(g) = \operatorname{def}(f)$: $\|g - f\| \le \delta \implies |L(g) - L(f)| < \varepsilon$.

c) Let $a < b$, and let $\operatorname{def}(L) \supseteq \mathbb{F}[a, b]$; moreover, let $f \in \mathbb{F}[a, b]$. Given a set $\mathbf{D} \subseteq \mathbb{F}[a, b]$ of "directions" and a linear, functional $l : \mathbb{F}[a, b] \to \mathbb{R}$.

Then we call l the $\mathbf{D}$-*differential of L at f* if a function $s : \mathbb{F}[a, b] \times \mathbb{F}[a, b] \times \mathbb{R} \to \mathbb{R}$ exists with the following properties:

$$(\forall \varepsilon \ne 0)\,(\forall \eta \in \mathbf{D}) \quad L(f + \varepsilon\,\eta) = L(f) + \varepsilon \cdot l(\eta) + s(f, \eta, \varepsilon).$$

$s(f, \eta, \varepsilon) \in o(\varepsilon)$; that means that $\lim_{\varepsilon \to 0} s(f, \eta, \varepsilon)/\varepsilon = 0$ for all f and η.

The term **D**-differential is closely related to the terms Gâteaux differential, Gâteaux derivative, Fréchet differential, Fréchet derivative.

d) Let $a < b$, $f \in \mathbb{F}[a,b]$ and $x \in [a,b]$. We say that $\overline{L}$ is *continuous* and *differentiable* at (f,x), respectively, if the real function $\xi \mapsto \overline{L}(f,\xi)$ is continuous and differentiable at $\xi = x$; if $\overline{L}$ is differentiable at (f,x) then $\overline{L}'(f,x)$ will denote the derivative of $\overline{L}(f,x)$ with respect to x. □

Example 5.7 We describe the **D**-differential of a special functional L, which is very important in Variational Calculus.

Let $0 \le a < b \le 1$. Given a constant $c > 0$, and let
$$\mathbf{D} := \{\eta \in \mathbb{F}[a,b] \mid \|\eta\|_{\max} \le 1,\ \|\eta'\|_{\max} \le c\}\ .$$
Let $F : \mathbb{R}^3 \to \mathbb{R}$; we assume that the partial derivatives $\partial_1 F$, $\partial_2 F$ and $\partial_3 F$ exist and are continuous. For all $f \in \mathbb{F}[a,b]$ and all $\eta \in \mathbf{D}$, we define
$$L(f) := \int_a^b F\big(\xi, f(\xi), f'(\xi)\big)\mathrm{d}\xi\ ,\quad, \tag{5.2}$$
$$l(\eta) := \int_a^b \Big[\partial_2 F\big(\xi, f(\xi), f'(\xi)\big)\cdot\eta(\xi) + \partial_3 F\big(\xi, f(\xi), f'(\xi)\big)\cdot\eta'(\xi)\Big]\mathrm{d}\xi\ . \tag{5.3}$$

Then l is the **D**-differential of L at f. This is a consequence of the following equation where $o(\varepsilon)$ means a summand $s(f,\eta,\varepsilon)$ with $\lim\limits_{\varepsilon\to 0} s(f,\eta,\varepsilon)/\varepsilon = 0$.

$$(\forall\, f,\eta)\quad L(f+\varepsilon\cdot\eta) = \int_a^b F\big(\xi, f+\varepsilon\cdot\eta, f'+\varepsilon\cdot\eta'\big)\mathrm{d}\xi\ =$$
$$\int_a^b \Big[F(\xi,f,f') + \varepsilon\,\partial_2 F(\xi,f,f')\cdot\eta(\xi) + \varepsilon\,\partial_3 F(\xi,f,f')\cdot\eta' + o(\varepsilon)\Big]\mathrm{d}\xi \tag{5.4}$$
$$\overset{\eta\in\mathbf{D}}{=} L(f) + l(\eta)\cdot\varepsilon + o(\varepsilon)\ . \qquad\qquad □$$

The next lemma establishes a relationship between the two norms $\|\bullet\|_*$ and $\|\bullet\|_{\max}$.

Lemma 5.8 *Let $0 \le a < b \le 1$. Then $\|f\|_{\max} \le \|f\|_*$ for all $f \in \mathbb{F}[a,b]$.*

(By the way, a consequence of this result is that each functional L is $\|\bullet\|_*$-continuous as far as L is $\|\bullet\|_{\max}$-continuous; this follows from the assertion $\|f-g\|_* \le \delta \Rightarrow \|f-g\|_{\max} \le \delta$, which is true for all $f,g \in \mathbb{F}[a,b]$ and $\delta > 0$.)

P r o o f : Let $f \in \mathbb{F}[a,b]$. Then the following is true for all $x \in [a,b]$:
$$|f(x)| = \Big|f(a) + \int_a^x f'(\xi)\mathrm{d}\xi\Big| \le |f(a)| + \Big|\int_a^x f'(\xi)\mathrm{d}\xi\Big| \le |f(a)| + \int_a^b |f'(\xi)|\mathrm{d}\xi = \|f\|_*\ .$$
Hence, $|f(x)| \le \|f\|_*$ for all $x \in [a,b]$ so that $\|f\|_{\max} \le \|f\|_*$. □

Next, we define minima and maxima of functionals.

Definition 5.9 Let $L : \mathbb{F} \dashrightarrow \mathbb{R}$ be a functional, and let $f \in \mathbb{F}[a,b] \cap \mathrm{def}(L)$.

Then we call f a *strong relative minimum* for L if there exists an $\varepsilon > 0$ such that $L(f) \le L(g)$ for all $g \in \mathbb{F}[a,b] \cap \mathrm{def}(L)$ with the following properties (i) – (iii):

(i) $g(a) = f(a)$,   (ii) $g(b) = f(b)$,   (iii) $\|f - g\|_{\max} \leq \varepsilon$.

We say that f is a *weak relative minimum* for L if $L(f) \leq L(g)$ is only true for all candidates g with the properties (i) – (iii) and the additional property

$$\text{(iv)} \quad \|f' - g'\|_{\max} \leq \varepsilon.$$

A relative minimum is always meant to be a *strong* relative minimum.

Strong and weak relative maxima are defined in analogy to strong and weak relative minima, respectively.   □

5.2 Additive functionals

Here, we introduce functionals that are similar to additive path functions. It turns out that solving the most well-known problem of Varational Calculus is the same as finding the minimal value $\overline{L}(f, x)$ where L is an additive functional.

Our first idea is the following: A path function H is additive if an arc function h exists such that $H(P) = \sum_{\kappa=1}^{k} h(r_\kappa)$ for all paths $P = r_1 \oplus \cdots \oplus r_k$. Since integrals are analogous to sums and since triples $(\xi, f(\xi), f'(\xi))$ are analogous to arcs in graphs, we shall call a functional L additive if an "arc function" F exists such that $L(f) = \int_a^b F(\xi, f(\xi), f'(\xi))$ for all $f \in \mathbb{F}[a, b]$.

Next, we give a more detailed motivation of this definition of additivity; we shall later derive other properties of functionals in the same way as in the following remark. Note that the conclusions in Remark 5.10 are meant to be heuristic and not mathematically exact.

Introductory Remark 5.10 If a path function H is additive, there exists an arc function h such that $H(P \oplus r) = H(P) + h(r)$ for all paths P and all arcs r.

In particular, let $P \in \mathcal{P}(\mathcal{G}_0)$ approximate a function $f \in \mathbb{F}[a, b]$ where $a < b$. Let $r := r_\delta^f(x)$ be an arc of P. Then $H\left(P|_{x+\delta}\right) = H\left(P|_x \oplus r_\delta^f(x)\right) = H\left(P|_x\right) + h\left(r_\delta^f(x)\right)$ so that

$$H\left(P|_{x+\delta}\right) = H\left(P|_x\right) + h\left(r_\delta^f(x)\right). \tag{5.5}$$

We now transform (5.5) into an equation with continuous objects. For this purpose, we replace the prefixes $P|_x$ and $P|_{x+\delta}$ by the restrictions $f|_x$ and $f|_{x+\delta}$, respectively. As motivated in 5.5 c), we may replace the arc $r_\delta^f(x)$ by the triple $(x, f(x), f'(x))$. Moreover, we replace the path function H by the functional L and the arc function h by an "arc function" $F : \mathbb{R}^3 \to \mathbb{R}$. Thus, we obtain

$$L\left(f|_{x+\delta}\right) = L\left(f|_x\right) + F(x, f(x), f'(x)). \tag{5.6}$$

In order to approximate the *infinitesimal* addition of $F(x, f(x), f'(x))$ we replace (5.6) by an equation in which the addition of $F(x, f(x), f'(x))$ is executed only δ times. We thus obtain the following equation:

$$L\left(f|_{x+\delta}\right) = L\left(f|_x\right) + \delta \cdot F(x, f(x), f'(x)). \tag{5.7}$$

We now transform the left and the right side of (5.7) as follows: First, we subtract $L\left(f\big|_x\right)$; next, we divide by δ; then we replace $L\left(f\big|_x\right)$ and $L\left(f\big|_{x+\delta}\right)$ by $\overline{L}(f,x)$ and $\overline{L}(f, x+\delta)$, respectively. Thus, we generate the following equation:

$$\frac{\overline{L}(f,x+\delta)-\overline{L}(f,x)}{\delta} \;=\; F\big(x, f(x), f'(x)\big). \tag{5.8}$$

If δ tends to zero we obtain the following version of (5.8):

$$\overline{L}'(f,x) \;=\; F\big(x, f(x), f'(x)\big). \tag{5.9}$$

If we assume that $\overline{L}(f,a) \;=\; 0$ then (5.9) implies that $L(f) \;=\; \overline{L}(f,b) \;=\; \int_a^b F\big(x, f(x), f'(x)\big)$ where F is a given function. □

We have described the analogies between additive path functions and functionals of the form $L(f) = \int_a^b F(x, f(x), f'(x))$. This is a motivation of the following definition of additivity.

Definition 5.11 Let $\widetilde{\mathbb{F}} \subseteq \mathbb{F}$; we do not assume that $\widetilde{\mathbb{F}}$ is prefix closed. A functional $L : \widetilde{\mathbb{F}} \to \mathbb{R}$ is called *additive of type 1* if there exists a continuous function $F : [0,1] \times \mathbb{R}^2 \to \mathbb{R}$ such that for all $a \le b$ and for all functions $f \in \mathbb{F}[a,b] \cap \widetilde{\mathbb{F}}$,

$$L(f) \;=\; \int_a^b F\big(\xi, f(\xi), f'(\xi)\big)\, d\xi .$$

If F is given we define the following additive functional $INT_F : \mathbb{F} \to \mathbb{R}$:

$$(\forall\, a, b)\ (\forall\, f \in \mathbb{F}[a,b])\quad INT_F(f) \;:=\; \int_a^b F\big(\xi, f(\xi), f'(\xi)\big)\, d\xi . \qquad \square$$

Remark 5.12 As mentioned above, the transformations in 5.10 have been based on analogies and not on mathematically exact conclusions; in particular, L has not been computed as a limes of H-values. This, however, is possible if there are particular relationships between F and h.

For example, let $0 \le a < b \le 1$ and $f \in \mathbb{F}[a,b]$. Moreover, let $L = INT_F$ and $H = SUM_h$ where the following be true for all x and δ:

$$h\big(r_\delta^f(x)\big) = F\big(x, f(x), f'(x)\big) \tag{5.10}$$

Let $\delta_N := \frac{b-a}{N}$ for all $N \in \mathbb{N}$; then $b = a + N \cdot \delta_N$. Let $P_N \in \mathcal{P}(\mathcal{G}_0)$ be the following path: $P_N := \big[(a, f(a)),\, (a + \delta_N, f(a + \delta_N)),\, (a + 2\,\delta_N, f(a + 2\,\delta_N)),\, \dots,\, (b, f(b))\big]$. Then the following is true:

$$P_N \;=\; r_{\delta_N}^f(a) \oplus r_{\delta_N}^f(a + \delta_N) \oplus r_{\delta_N}^f(a + 2 \cdot \delta_N) \oplus \cdots \oplus r_{\delta_N}^f(a + (N-1)\cdot \delta_N). \tag{5.11}$$

Consequently,

$$L(f) \;=\; \int_a^b F\big(\xi, f(\xi), f'(\xi)\big)\, d\xi \;=$$
$$\lim_{N\to\infty} \delta_N \cdot \sum_{\nu=0}^{N-1} F\big(a + \nu \cdot \delta_N,\, f(a + \nu \cdot \delta_N),\, f'(a + \nu \cdot \delta_N)\big) \overset{(5.10)}{=}$$

$$\lim_{N\to\infty} \delta_N \cdot \sum_{\nu=0}^{N-1} h\big(r_{\delta_N}^f(a + \nu \cdot \delta_N)\big) \overset{(5.11)}{=} \lim_{N\to\infty} \delta_N \cdot H(P_N) .$$

That means that $L(f)$ can be computed as a limes of H-values. □

The following example is about finding a function f such that $INT_F(f)$ is minimal or maximal; this is the classical problem of Variational Calculus.

Example 5.13 Let $0 \le a < b \le 1$, and let $F : \mathbb{R}^3 \to \mathbb{R}$. The classical problem of Variational Calculus is to search for a function $f \in \mathbb{F}[a, b]$ such that $INT_F(f) = \int_a^b F(\xi, f(\xi), f'(\xi)) \, d\xi$ has its minimal or the maximal value. The side constraints are $f(a) = y_0$ and $f(b) = y_1$ where $y_0, y_1 \in \mathbb{R}$ are given.

We outline the solution of this problem. We define $\mathbf{D}$ in the same way as in 5.7, and we define $L := INT_F$. Then we compute the $\mathbf{D}$-differential $l(\eta)$ at f and x; this $\mathbf{D}$-differential has been given in equation (5.3). The following assertion is an immediate consequence of fact (5.4).

$$(\forall \, \eta \in \mathbf{D})(\forall \, \varepsilon \neq 0) \qquad \frac{\overline{L}(f + \varepsilon \cdot \eta, x) - \overline{L}(f, x)}{\varepsilon} \;=\; l(\eta) + \frac{o(\varepsilon)}{\varepsilon}. \tag{5.12}$$

Let $\eta(a) = \eta(b) = 0$. Then $\overline{L}(f, x)$ is extremal among all candidates $\overline{L}(f + \varepsilon \cdot \eta, x)$ where $\varepsilon \in \mathbb{R}$. Therefore, $\lim\limits_{\varepsilon \to 0} \frac{\overline{L}(f + \varepsilon \cdot \eta, x) - \overline{L}(f, x)}{\varepsilon} \;=\; 0$. This and (5.12) imply that $l(\eta) = 0$. Consequently,

$$\int\limits_a^b \big[\, \partial_2 F(\xi, f(\xi), f'(\xi)) \cdot \eta(\xi) + \partial_3 F(\xi, f(\xi), f'(\xi)) \cdot \eta'(\xi) \,\big] \, d\xi \;=\; 0 \,.$$

By partial integration of the second summand, we obtain the following equation:

$$\int\limits_a^b \big[\, \partial_2 F(\xi, f, f') \cdot \eta(\xi) - \tfrac{d}{d\xi} \partial_3 F(\xi, f, f') \cdot \eta(\xi) \,\big] \, d\xi \;+\; \partial_3 F(\xi, f, f') \cdot \eta(\xi) \Big|_a^b \;=\; 0 \,.$$

The second summand is equal to zero because $\eta(a) = \eta(b) = 0$. Consequently,

$$\int\limits_a^b \big[\, \partial_2 F(\xi, f, f') \cdot \eta(\xi) - \tfrac{d}{d\xi} \partial_3 F(\xi, f, f') \cdot \eta(\xi) \,\big] \, d\xi \;=\; 0 \,, \tag{5.13}$$

and this equation is valid for all $\eta \in \mathbf{D}$ with $\eta(a) = \eta(b) = 0$. Fact (5.13) is used to derive the Euler-Lagrange differential equation, namely

$$\partial_2 F(\xi, f, f') - \partial_{31} F(\xi, f, f') - \partial_{32} F(\xi, f, f') \cdot f' - \partial_{33} F(\xi, f, f') \cdot f'' \;=\; 0. \tag{5.14} \qquad \square$$

Next, we introduce another version of additivity of a functional.

Introductory Remark 5.14 Let $0 \le a' < b' \le 1$, let P_1, P_2 be two paths in $\mathcal{G}_0$ with $a', b' \in \mathcal{X}(P_1) \cap \mathcal{X}(P_2)$, and let $Q_i \subseteq P_i$ such that $P_i|_{b'} = P_i|_{a'} \oplus Q_i$, $i = 1, 2$. If H is an additive path function for $\mathcal{P}(\mathcal{G}_0)$ then the following is true:

$$Q_1 = Q_2 \;\Rightarrow\; H\left(P_1|_{b'}\right) - H\left(P_1|_{a'}\right) \;=\; H\left(P_2|_{b'}\right) - H\left(P_2|_{a'}\right) . \tag{5.15}$$

Let $f_1, f_2 \in \mathbb{F}$ with $[a', b'] \subseteq \mathrm{def}(f_1) \cap \mathrm{def}(f_2)$. We consider f_i as analogous object to P_i and $f_i|_{[a', b']}$ as an analogous object to Q_i, $i = 1, 2$. Thus, we transform (5.15) into the following equation about a functional L:

$$f_1|_{[a', b']} = f_2|_{[a', b']} \;\Rightarrow\; L\left(f_1|_{b'}\right) - L\left(f_1|_{a'}\right) \;=\; L\left(f_2|_{b'}\right) - L\left(f_2|_{a'}\right) .$$

This equation is used in Definition 5.15 to define another version of additivity. $\qquad \square$

Definition 5.15 We say that a functional $L : \mathbb{F} \to \mathbb{R}$ is *additive of type 2* if it satisfies the following conditions:

- $\overline{L}(f, \mathsf{a}(f)) = 0$ for all $f \in \mathbb{F}$.

- $f_1|_{[a',b']} = f_2|_{[a',b']} \implies L\left(f_1|_{b'}\right) - L\left(f_1|_{a'}\right) = L\left(f_2|_{b'}\right) - L\left(f_2|_{a'}\right)$
  for all a', b' and all $f_1, f_2 \in \mathbb{F}$ with $[a', b'] \subseteq \mathrm{def}(f_1) \cap \mathrm{def}(f_2)$.

The second condition is equivalent to

$$f_1\big|_{[a',b']} = f_2\big|_{[a',b']} \implies \overline{L}(f_1, b') - \overline{L}(f_1, a') = \overline{L}(f_2, b') - \overline{L}(f_2, a').$$ $\square$

Remark 5.16 The functional $L = INT_F$ is additive of type 2. To see this we assume
that $f_1, f_2 \in \mathbb{F}$ and $0 \le a' \le b' \le 1$ such that

$$f_1\big|_{[a',b']} = f_2\big|_{[a',b']}. \tag{5.16}$$

Let $a_i := \mathsf{a}(f_i)$, $i = 1, 2$; then $\overline{L}(f_i, b') - \overline{L}(f_i, a') = \int_{a_i}^{b'} F(\xi, f_i, f_i')\mathrm{d}\xi - \int_{a_i}^{a'} F(\xi, f_i, f_i')\mathrm{d}\xi$
$= \int_{a'}^{b'} F(\xi, f_i, f_i')\mathrm{d}\xi$ for $i = 1, 2$. Consequently,

$$\overline{L}(f_1, b') - \overline{L}(f_1, a') = \int_{a'}^{b'} F(\xi, f_1, f_1')\,\mathrm{d}\xi \stackrel{(5.16)}{=} \int_{a'}^{b'} F(\xi, f_2, f_2')\,\mathrm{d}\xi = \overline{L}(f_2, b') - \overline{L}(f_2, a').$$

We conjecture that the reverse conclusion is also true:

Given an functional L with the following properties: $\overline{L}(f, \mathsf{a}(f)) = 0$ for all f, L is additive
of type 2, and L is "sufficiently smooth". Then an F exists with $L = INT_F$.

We give reasons for this conjecture, and we leave it to the reader to transform our
argumentation into an exact proof.
When showing the existence of F with $L = INT_F$, the main step is to show that the
derivative $\overline{L}'(f, x)$ only depends on the triple $(x, f(x), f'(x))$ if $\mathsf{a}(f) < x < \mathsf{b}(f)$. That
means that

$$(\alpha) \ (f(x) = g(x), \ f'(x) = g'(x)) \quad \text{implies} \quad (\beta) \ \frac{\mathrm{d}}{\mathrm{d}x}\overline{L}(f, x) = \frac{\mathrm{d}}{\mathrm{d}x}\overline{L}(g, x) \tag{5.17}$$

for each x with

$$\mathsf{a}(f) < x < \mathsf{b}(f) \quad \text{and} \quad \mathsf{a}(g) < x < \mathsf{b}(g). \tag{5.18}$$

To prove $(\alpha) \Rightarrow (\beta)$ in (5.17) we assume that $(5.17)(\alpha)$ is true for some x with property (5.18). Let $a_0 := \max(\mathsf{a}(f), \mathsf{a}(g))$ and $b_0 := \min(\mathsf{b}(f), \mathsf{b}(g))$; then $[a_0, b_0] \subseteq$
$\mathrm{def}(f) \cap \mathrm{def}(g)$, and (5.18) implies that $a_0 < x < b_0$. We define the following function
$q : [a_0, b_0] \to \mathbb{R}$:

$$(\forall x \in \mathbb{R}) \ q(\xi) := \begin{cases} g(\xi), & a_0 \le \xi \le x, \\ f(\xi), & x \le \xi \le b_0. \end{cases}$$

Then $q \in \mathbb{F}$ because $f'(x) = g'(x)$ by $(5.17)(\alpha)$ so that q is differentiable at x.
Let $\delta > 0$ such that $x + \delta < b_0$. The type-2-additivity of L and the equation
$q|_{[x,x+\delta]} = f|_{[x,x+\delta]}$ imply that

$$\overline{L}(f, x + \delta) - \overline{L}(f, x) = \overline{L}(q, x + \delta) - \overline{L}(q, x). \tag{5.19}$$

Now we draw a heuristic conclusion. The functions q and g have the same values on
$[a_0, x]$ by the definition of q; moreover, $f'(x) \stackrel{(5.17)(\alpha)}{=} g'(x) = q'(x)$; therefore, it is ob-
vious that the restriction $q|_{[x,x+\delta]} = f|_{[x,x+\delta]}$ is a "good" approximation of $g|_{[x,x+\delta]}$.

Therefore, the following conclusion is reasonable if L is sufficiently smooth.
$$\overline{L}(q, x + \delta) - \overline{L}(q, x) = \overline{L}(g, x + \delta) - \overline{L}(g, x) + o(\delta). \tag{5.20}$$
Then we obtain the following equation:
$$\frac{\overline{L}(f,x+\delta)-\overline{L}(f,x)}{\delta} \overset{(5.19)}{=} \frac{\overline{L}(q,x+\delta)-\overline{L}(q,x)}{\delta} \overset{(5.20)}{=} \frac{\overline{L}(g,x+\delta)-\overline{L}(g,x)}{\delta} + \frac{o(\delta)}{\delta}. \tag{5.21}$$
Let δ tend to 0. Then (5.17)(β) follows from (5.21). $\qquad\square$

Remark 5.17 Any functional L is additive of type 2 if and only if L has the following properties:
- $\overline{L}(f, \mathsf{a}(f)) = 0$ for all $f \in \mathbb{F}$.
- For all $a', b' \in [0,1]$ with $a' \leq b'$ and all functions $g \in \mathbb{F}[a', b']$, there is a real number σ_g such that the following is true: If $f \in \mathbb{F}$, $a', b' \in \mathrm{def}(f)$ and $f|_{[a',b']} = g$ then
$$\overline{L}(f, b') = \overline{L}(f, a') + \sigma_g.$$
A similar result is true for path functions: A path function H is additive if and only if it has the following properties:
- $H([\alpha(P)]) = 0$ for all paths P.
- For each path Q, there exists a real number σ_Q such that the following is true: If P is a path of the form $P = P' \oplus Q \oplus P''$ then
$$H(P' \oplus Q) = H(P') + \sigma_Q. \qquad\square$$

5.3 Recursively linear functionals

In this section, we introduce recursive linearity of path functions. This property is more general than additivity and not so general as order preservation. Later, we shall define recursively linear functionals.

Definition 5.18 Let $H : \mathcal{P}(\mathcal{G}) \to \mathbb{R}$ be a path function with $H(P) = 0$ for all paths P with length 0. We call H *recursively linear* if the following is true: For each arc r, there exists a linear function $\phi_r(\zeta) := h_1(r) \cdot \zeta + h_2(r)$ with $h_1(r) > 0$ such that
$$H(P) = \phi_{r_k} \circ \cdots \circ \phi_{r_1}(H([\alpha(P)])) \quad \text{for all paths } P = r_1 \oplus \ldots \oplus r_k. \qquad\square$$

Remark 5.19 The class of all recursively linear path functions contains the class all additive ones. Let $H = SUM_h$; then H is generated with the linear functions $\phi_r(\zeta) := 1 \cdot \zeta + h(r)$; therefore, H is recursively linear. $\qquad\square$

Remark 5.20 Recursively linear functions are a good model of financial planning. We interpret each arc as a financial transaction; moreover, we interpret ζ and $\phi_r(\zeta)$ as the wealth before and after the transaction, respectively. For example, let $\phi_r(\zeta) := 1.03\,\zeta + 3000$ for some arc r; then r symbolizes the transactions of a saver who gets 3% interest and pays \$ 3000,– into his account. $\qquad\square$

We shall transfer the definition of recursively linear path functions into the continuous situation. In particular, we shall replace a discrete sequence of linear functions ϕ_r by a continuous stream of linear functions; for this purpose, we imagine that each linear function is applied δ times where δ tends to zero. The following result helps to define the δ-fold iteration of a linear function if δ is not an integer number.

Theorem 5.21 *Given a linear function* $\phi(\zeta) := A \cdot \zeta + B$ $(\zeta \in \mathbb{R})$ *with* $A > 0$*. Let*

$$\phi^l(\zeta) \quad := \quad A^l \cdot \zeta + \frac{A^l - 1}{A - 1} \cdot B \tag{5.22}$$

for all $l \in \mathbb{R}$ *and* $\zeta \in \mathbb{R}$*; in particular, we define* $\frac{A^l - 1}{A - 1} := l$ *for* $A = 1$*.*

Then the following are true:

    a) *If* $l \in \mathbb{N} \cup \{0\}$ *then the function* ϕ^l *is the* l*-fold iteration of* ϕ*; that means that*
$$\phi^l(\zeta) \;=\; \underbrace{\phi \circ \cdots \circ \phi(\zeta)}_{l \text{ times}} \text{ for all } \zeta \text{ and all } l.$$
    In particular, $f^1 = f$*, and* f^0 *is the identity.*

    b) *Let* $l_0, l_1 \in \mathbb{R}$*; then* $\phi^{l_0} \circ \phi^{l_1} = \phi^{l_0 + l_1}$*.*

    c) *Let* $l_0, l_1 \in \mathbb{R}$*. Then* $\zeta \mapsto \phi^{l_0}(\zeta)$ *is a linear function so that* $\left(\phi^{l_0}\right)^{l_1}$ *is well-defined; moreover,* $\left(\phi^{l_0}\right)^{l_1} = \phi^{l_0 \cdot l_1}$*.*

    d) *Let* $A = 1$*; the our definition of* $\frac{A^l - 1}{A - 1}$ *implies that*
$$\phi(\zeta) = \zeta + B, \text{ and } \phi^l(\zeta) = \zeta + l \cdot B \text{ for all } l, \zeta \in \mathbb{R}.$$

P r o o f : The proof is straightforward.   □

Theorem 5.21 implies that it is reasonable to consider ϕ^l as the l-fold iteration of ϕ even if $l \notin \mathbb{N} \cup \{0\}$.

Next, we introduce recursively linear[1] functionals; for this purpose, we translate the recursive linearity of path functions into a similar property of functionals; the conclusions in the following remark are rather heuristic than mathematically exact.

Introductory Remark 5.22 (*see also Remark* 5.10)
If a path function H is recursively linear, each arc r is marked with a linear function $\phi_r(\zeta) = h_1(r) \cdot \zeta + h_2(r)$ such that $H(P \oplus r) = \phi_r(H(P)) = h_1(r) \cdot H(P) + h_2(r)$ for all paths P and all arcs r; moreover, $h_1(r) > 0$ for all r.

In particular, let $P \in \mathcal{P}(\mathcal{G}_0)$ approximate a function $f \in \mathbb{F}[a, b]$ where $a < b$. Let $r := r_\delta^f(x)$ be an arc of P. Then $H\left(P|_{x+\delta}\right) = H\left(P|_x \oplus r_\delta^f(x)\right) = \phi_{r_\delta^f(x)}\left(H\left(P|_x\right)\right) = h_1\left(r_\delta^f(x)\right) \cdot H\left(P|_x\right) + h_2\left(r_\delta^f(x)\right)$ so that
$$H\left(P|_{x+\delta}\right) = \phi_{r_\delta^f(x)}\left(H\left(P|_x\right)\right) = h_1\left(r_\delta^f(x)\right) \cdot H\left(P|_x\right) + h_2\left(r_\delta^f(x)\right). \tag{5.23}$$
We now transform this formula into an equation with continuous objects. For this

---

[1]The property "recursively linear" has nothing to do with linearity of functionals, which was introduced in Definition 5.3 f) on page 402.

purpose, we replace the prefixes $P|_x$ and $P|_{x+\delta}$ by the restrictions $f|_x$ and $f|_{x+\delta}$, respectively. As motivated in 5.5 c), we replace the arc $r_\delta^f(x)$ by the triple $r^f(x) = \big(x, f(x), f'(x)\big)$. Moreover, we replace the path function H by the functional L and the arc functions h_1, h_2 by the "arc functions" $A, B : \mathbb{R}^3 \to \mathbb{R}$, respectively. In particular, the linear function $\phi_{r_\delta^f(x)}(\zeta) = h_1\big(r_\delta^f(x)\big)\cdot\zeta + h_2\big(r_\delta^f(x)\big)$ is replaced by the linear function $\phi_{r^f(x)}(\zeta) := A\big(r_\delta^f(x)\big)\cdot\zeta + B\big(r_\delta^f(x)\big)$. Thus, we obtain

$$L\left(f|_{x+\delta}\right) = \phi_{r^f(x)}\left(L\left(f|_x\right)\right)$$
$$= A\big(x, f(x), f'(x)\big)\cdot L\left(f|_x\right) + B\big(x, f(x), f'(x)\big). \tag{5.24}$$

In order to approximate the *infinitesimal* application of $\phi_{r^f(x)}$ we replace (5.24) by an equation in which $\phi_{r^f(x)}$ is applied only δ times. That means that we apply $\big(\phi_{r^f(x)}\big)^\delta(\zeta) \overset{(5.22)}{=} A\big(x, f(x), f'(x)\big)^\delta\cdot\zeta + \frac{A(x,f(x),f'(x))^\delta - 1}{A(x,f(x),f'(x))-1}\cdot B\big(x, f(x), f'(x)\big)$. Thus, we obtain the following equation:

$$L\left(f|_{x+\delta}\right) = \big(\phi_{r^f(x)}\big)^\delta\left(L\left(f|_x\right)\right)$$
$$= A\big(x, f(x), f'(x)\big)^\delta\cdot L\left(f|_x\right) + \frac{A(x,f(x),f'(x))^\delta - 1}{A(x,f(x),f'(x))-1}\cdot B\big(x, f(x), f'(x)\big). \tag{5.25}$$

We now transform the left and the right side of (5.25) as follows: First, we subtract $L\left(f|_x\right)$; next, we divide by δ; then we replace $L\left(f|_x\right)$ and $L\left(f|_{x+\delta}\right)$ by $\overline{L}(f,x)$ and $\overline{L}(f, x + \delta)$, respectively; moreover, we omit the arguments of A and B. Thus, we generate the following equation:

$$\frac{\overline{L}(f, x + \delta) - \overline{L}(f, x)}{\delta} = \frac{A^\delta - 1}{\delta}\cdot\overline{L}(f, x) + \frac{B}{A - 1}\cdot\frac{A^\delta - 1}{\delta}. \tag{5.26}$$

Now we let δ tend to zero and observe that $\lim\limits_{\delta\to 0}\left(\frac{A(r^f(x))^\delta - 1}{\delta}\right) = \ln\big(A(r^f(x))\big)$. Thus, we obtain the following version of (5.26):

$$\overline{L}'(f, x) = \overline{L}(f, x)\cdot\ln\big(A\big(x, f(x), f'(x)\big)\big) + \frac{\ln\left(A(x,f(x),f'(x))\right)\cdot B(x,f(x),f'(x))}{A(x,f(x),f'(x)) - 1}, \tag{5.27}$$

i.e. $L'(f, x) = \overline{L}(f, x)\cdot\ln(A) + \frac{\ln(A)\cdot B}{A - 1}$. $\qquad\square$

Requiring that $\overline{L}(f, \mathsf{a}(f)) = 0$ for all functions f, we define recursively linear functionals as follows:

Definition/Remark 5.23 Let $\widetilde{\mathbb{F}} \subseteq \mathbb{F}$; we do not assume $\widetilde{\mathbb{F}}$ to be prefix closed.

A functional $L : \widetilde{\mathbb{F}}$ is called *recursively linear* if two continuous functions $A, B : [0, 1] \times \mathbb{R}^2 \to \mathbb{R}$ exist such that the following are true for all $f \in \widetilde{\mathbb{F}}$ and for $a := \mathsf{a}(f)$, $b := \mathsf{b}(f)$

- $A\big(x, f(x), f'(x)\big) > 0$ for all $x \in [a, b]$.

- If $a < b$ then a function $y : [a, b] \to \mathbb{R}$ exists such that $\overline{L}(f, \xi) = L\left(f|_{[a,\xi]}\right) = y(\xi)$ for all $\xi \in [a, b]$ with $f|_{[a,\xi]} \in \widetilde{\mathbb{F}}$;[2] this y solves the following initial value problem:

$$(\forall x \in [a, b]) \ y'(x) = y(x) \cdot \ln A(x, f(x), f'(x)) + \frac{\ln\left(A(x, f(x), f'(x))\right) \cdot B(x, f(x), f'(x))}{A(x, f(x), f'(x)) - 1}, \quad [3]$$

$$y(a) = 0.$$

In particular, $L(f) = \overline{L}(f, b) = y(b)$.

- If $a = b$ then $L(f) = 0$.

We have not explicitly required that the above initial value problem has exactly one solution y; in reality, however, y is always unique. $\square$

Remark 5.24 The differential equation in (5.27) is linear; we can therefore describe $\overline{L}(f, x)$ explicitly. Let $a := \mathsf{a}(f)$ and $b = \mathsf{b}(f)$, and let f be recursively linear. Then the following is true fo all $x \in [a, b]$ with $f|_x \in \widetilde{\mathbb{F}}$:

$$\overline{L}(f, x) = $$

$$e^{\int\limits_a^x \ln\left(A(\xi, f(\xi), f'(\xi))\right) d\xi} \cdot \left[\int\limits_a^x \frac{\ln\left(A(\xi, f(\xi), f'(\xi))\right) \cdot B(\xi, f(\xi), f'(\xi))}{A(\xi, f(\xi), f'(\xi)) - 1} \cdot e^{-\int\limits_a^\xi \ln\left(A(\zeta, f(\zeta), f'(\zeta))\right) d\zeta} \, d\xi \right].$$

This equation can be used to compute $L(f)$ because $L(f) = \overline{L}(f, b)$. $\square$

Example 5.25 We consider several special cases of A and B. Let $0 \le a < b \le 1$ and let L be defined on a set $\widetilde{\mathbb{F}} \subseteq \mathbb{F}^+$.

- $A\big(x, f(x), f'(x)\big) = 1$ and $B\big(x, f(x), f'(x)\big) = F\big(x, f(x), f'(x)\big)$ for all $f \in \mathbb{F}[a, b]$ and all $x \in [a, b]$.[4]

  Then the following is true for all $f \in \mathbb{F}[a, b] \cap \widetilde{\mathbb{F}}$ and all x with $f|_x \in \widetilde{\mathbb{F}}$:

$$\overline{L}(f, x) = \int\limits_a^x F(\xi, f(\xi), f'(\xi)) \, d\xi.$$

  Consequently, additivity of type 1 is a special version of recursive linearity.

- $A\big(x, f(x), f'(x)\big) = 1$ and $B\big(x, f(x), f'(x)\big) = f(x)$ for all $f \in \mathbb{F}[a, b]$ and all $x \in [a, b]$.

  Then $\overline{L}(f, x)$ is the integral of f; that means that the following

---

[2] If $\widetilde{\mathbb{F}}$ is not prefix closed it may occur that $f|_\xi \notin \widetilde{\mathbb{F}}$ for some $\xi \in [a, b]$. In this case, $y(\xi)$ is defined but $\overline{L}(f, \xi)$ is not defined, and the equation $\overline{L}(f, \xi) = y(\xi)$ is not valid.

[3] We define $\frac{\ln(A)}{A-1} := 1$ if $A = 1$.

[4] This case is given if $A(u, v, w) = 1$ for all u, v, w.

is true for all $f \in \mathbb{F}[a,b] \cap \widetilde{\mathbb{F}}$ and all x with $f|_x \in \widetilde{\mathbb{F}}$:

$$\overline{L}(f,x) = \int_a^x f(\xi)\,d\xi.$$

- $A\big(x, f(x), f'(x)\big) = 2$ and $B\big(x, f(x), f'(x)\big) = f(x)$
  for all $f \in \mathbb{F}[a,b]$ and all $x \in [a,b]$.

  Then the following is true for all $f \in \mathbb{F}[a,b] \cap \widetilde{\mathbb{F}}$ and all x with $f|_x \in \widetilde{\mathbb{F}}$:

$$\overline{L}(f,x) = 2^x \cdot \int_a^x f(\xi) \cdot (\ln 2) \cdot 2^{-\xi}\,d\xi.$$

- $A\big(x, f(x), f'(x)\big) = e$ and $B\big(x, f(x), f'(x)\big) = f(x)$
  for all $f \in \mathbb{F}[a,b]$ and all $x \in [a,b]$.

  Then the following is true for all $f \in \mathbb{F}[a,b] \cap \widetilde{\mathbb{F}}$ and all x with $f|_x \in \widetilde{\mathbb{F}}$:

$$\overline{L}(f,x) = e^x \cdot \int_a^x f(\xi) \cdot \left(\frac{1}{e-1}\right) \cdot e^{-\xi}\,d\xi.$$

- There exists $g \in \mathbb{F}$ such that $A(x, f(x), f'(x)) = g(x) > 0$ for all $f \in \mathbb{F}[a,b]$ and all $x \in [a,b]$; [5] moreover, $B(x, f(x), f'(x)) = f(x)$ for all f and x.

  Then the following is true for all $f \in \mathbb{F}[a,b] \cap \widetilde{\mathbb{F}}$ and all x with $f|_x \in \widetilde{\mathbb{F}}$:

$$\overline{L}(f,x) = e^{\int_a^x \ln(g(\xi))\,d\xi} \cdot \left(\int_a^x f(\xi) \cdot \frac{\ln(g(\xi))}{g(\xi)-1} \cdot e^{-\int_a^\xi \ln(g(\zeta))\,d\zeta}\,d\xi \right).$$

We call $\overline{L}(f,x)$ the g-integral of f and write $\overline{L}(f,x) =: \displaystyle\int_{g\,a}^{x} f(\xi)\,d\xi.$

If $g(x) = 1$ for all x then $\displaystyle\int_{g\,a}^{x} f(\xi)\,d\xi = \int_a^x f(\xi)\,d\xi.$

- $A\big(x, f(x), f'(x)\big) = f(x) > 0$ for all $f \in \mathbb{F}[a,b] \cap \widetilde{\mathbb{F}}$ and all $x \in [a,b]$; [6] moreover, there exists $g \in \mathbb{F}$ such that $B\big(x, f(x), f'(x)\big) = g(x)$ for all f and x.

  Then the following is true for all $f \in \mathbb{F}[a,b] \cap \widetilde{\mathbb{F}}$ and all x with $f|_x \in \widetilde{\mathbb{F}}$:

$$\overline{L}(f,x) = \int_{f\,a}^{x} g(\xi)\,d\xi.$$

Hence, the f-integral of a fixed function g is also a recursively linear functional.

---

[5] For example, if $A(u, v, w) = u^2 + 1$ for all (u, v, w) then $g(x) = x^2 + 1$.

[6] The assumption that $f(x) > 0$ for all $f \in \mathbb{F}[a,b] \cap \widetilde{\mathbb{F}}$ means that $\mathbb{F}[a,b] \cap \widetilde{\mathbb{F}}$ is a subset of $\{f \in \mathbb{F}[a,b] \mid (\forall x \in [a,b])\ f(x) > 0\}$.

- $A(x, f(x), f'(x)) > 0$ and $B(x, f(x), f'(x)) = 0$ for all $f \in \mathbb{F}[a, b]$ and all $x \in [a, b]$.

This case is analogous to the graph theoretic situation that each arc r is assigned a function $\phi_r(\xi) := h_1(r) \cdot \xi$ with $h_1(r) > 0$. Consequently, the resulting path function H is multiplicative; in this case, we should require that $H(P) = 1$ and not $H(P) = 0$ if the length of P is zero.

If we require that $\overline{L}(f, a) = 1$, we obtain the following equation:

$$\overline{L}(f, x) = e^{\int\limits_0^x A(\xi, f(\xi), f'(\xi)) \, d\xi} \qquad . \qquad \Box$$

Remark 5.26

a) Let $L : \tilde{\mathcal{R}} \to \mathbb{R}$ be a recursively linear functional where $\tilde{\mathbb{F}} \subseteq \mathbb{F}$ is prefix closed. Let $f \in \tilde{\mathbb{F}}$, $a = \mathsf{a}(f)$, $b = \mathsf{b}(f)$, and $y(x) := \overline{L}(f, x)$ for all $x \in [a, b]$.

Then $y(a) = \overline{L}(f, a) = 0$, and y solves the differential equations (5.27). Consequently, y is a solution of the following initial value problem:

$$y'(x) = y(x) \cdot \ln(A(x, f(x), f'(x))) + \frac{B(x, f(x), f'(x)) \cdot \ln(A(x, f(x), f'(x)))}{A(x, f(x), f'(x)) - 1}, \qquad (5.28)$$

$$y(a) = 0 \, .$$

Hence, there exist functions $U, V : [a, b] \times \mathbb{R}^2$ such that $y(x)$ is a solution of the following initial value problem:

$$y' = U(x, f, f') \cdot y + V(x, f, f'), \quad y(a) = 0 \, . \qquad (5.29)$$

This can be seen as follows: For all triples (u, v, w), let

$$U(u, v, w) := \ln(A(u, v, w)), \quad V(u, v, w) := \frac{\ln(A(u, v, w)) \cdot B(u, v, w)}{A(u, v, w) - 1}$$

b) Conversely, let (5.29) be valid for all functions $y(x) = \overline{L}(f, x)$ with $f \in \tilde{\mathbb{F}}$; then L is recursively linear. This can be seen by defining $A := e^U$ and $B := \dfrac{V \cdot (e^U - 1)}{U}$ (with $B(u, v, w) := V(u, v, w)$ if $U(u, v, w) = 0$).

That means that the initial value problems (5.28) and (5.29) are equivalent.

c) The class of recursively linear functionals can be extended as follows: We assume that the current linear function depends on x, $f(x)$, $f'(x)$, and on $\overline{L}(f, x) = y(x)$; then $\overline{L}(f, x) = y(x)$ is generated by infinitesimal applications of linear functions

$$\zeta \mapsto A(x, f(x), f'(x), y(x)) \cdot \zeta + B(x, f(x), f'(x), y(x)) \, .$$

Then $y(x) = \overline{L}(f, x)$ is a solution of the following initial value problem, which is analogous to (5.28):

$$y'(x) = y(x) \cdot \ln(A(x, f, f', y)) + \frac{\ln(A(x, f, f', y)) \cdot B(x, f, f', y)}{A(x, f, f', y) - 1}, \qquad y(a) = 0 \, . \qquad (5.30)$$

The solution of this differential equation, however, is very difficult. $\qquad \Box$

We now define recursive linearity of type 2; our definition is analogous to Remark 5.17.

Definition 5.27 We say that $L : \mathbb{F} \to \mathbb{R}$ is *recursively linear of type 2* if L has the following properties:

- $\overline{L}(f, a(f)) = 0$ for all $f \in \mathbb{F}$.
- For all $a', b' \in [0, 1]$ with $a' \le b'$ and all functions $g \in \mathbb{F}[a', b']$, there are two real numbers $\eta_g > 0$ and σ_g such that the function $\lambda_g(\zeta) := \eta_g \cdot \zeta + \sigma_g$ $(\zeta \in \mathbb{R})$ has the following property: If $f \in \mathbb{F}$, $a', b' \in \mathrm{def}(f)$ and $f|_{[a',b']} = g$ then

$$\overline{L}(f, b') = \eta_g \cdot \overline{L}(f, a') + \sigma_g = \lambda_g(\overline{L}(f, a')).$$

$\square$

5.4 Generalizations of recursive linearity

We have seen that recursively linear functionals are analogic to recursively linear path functions. Here, however, we try to find functionals that are analogous to path functions of the form $H = \phi \circ \circ p$ where the functions $\phi_r(\zeta)$ need not be linear with respect to ζ. Our first concept is analogous to the one in 5.22.

Introductory Remark 5.28 Let H be a path function with $H = \phi \circ \circ p$ where p is a node function and $\phi_r : \mathbb{R} \to \mathbb{R}$ for each arc of the given digraph. Then $H(P \oplus r) = \phi_r(H(P))$ for all paths P and all arcs r.

In particular, let $P \in \mathcal{P}(\mathcal{G}_0)$ approximate a function $f \in \mathbb{F}[a, b]$ where $a < b$. Let $r := r_\delta^f(x)$ be an arc of P. Then $H\left(P|_{x+\delta}\right) = H\left(P|_x \oplus r_\delta^f(x)\right) = \phi_{r_\delta^f(x)}\left(H\left(P|_x\right)\right)$ so that

$$H\left(P|_{x+\delta}\right) = \phi_{r_\delta^f(x)}\left(H\left(P|_x\right)\right) . \tag{5.31}$$

We now transform (5.31) into an equation with continuous objects. For this purpose, we replace the prefixes $P|_x$ and $P|_{x+\delta}$ by the restrictions $f|_x$ and $f|_{x+\delta}$, respectively. As motivated in 5.5 c), we replace the arc $r_\delta^f(x)$ by the triple $r^f(x) = (x, f(x), f'(x))$. Moreover, we replace the path function H by the functional L. In particular, the function $\phi_{r_\delta^f(x)}$ is replaced by the function $\phi_{r^f(x)}$. Thus, we obtain

$$L\left(f|_{x+\delta}\right) = \phi_{r^f(x)}\left(L\left(f|_x\right)\right) . \tag{5.32}$$

In order to approximate the *infinitesimal* application of $\phi_{r^f(x)}$ we replace (5.32) by an equation in which $\phi_{r^f(x)}$ is applied only δ times; i.e., we apply the δ-fold iteration $\left(\phi_{r^f(x)}\right)^\delta$ if such an iteration is defined. Thus, we obtain the following equation:

$$L\left(f|_{x+\delta}\right) = \left(\phi_{r^f(x)}\right)^\delta\left(L\left(f|_x\right)\right) \tag{5.33}$$

We now transform the left and the right side of (5.33) as follows: First, we subtract $L\left(f|_x\right)$; next, we divide by δ; then we replace $L\left(f|_x\right)$ and $L\left(f|_{x+\delta}\right)$ by $\overline{L}(f, x)$ and $\overline{L}(f, x + \delta)$, respectively. Thus, we generate the following equation:

$$\frac{\overline{L}(f, x+\delta) - \overline{L}(f,x)}{\delta} = \frac{\left(\phi_{rf(x)}\right)^{\delta}(\overline{L}(f,x)) - \overline{L}(f,x)}{\delta} \tag{5.34}$$

The last (and often very difficult) step is to transform this formula into a differential equation for $y(x) := \overline{L}(f, x) = L\left(f|_x\right)$. □

Next, we comment on the ideas in Remark 5.28.

Remark 5.29 When generating formula (5.33), we have replaced $\phi_{rf(x)}$ by the δ-fold iteration of this function. The advantage of this idea is that the δ-fold iteration is a reasonable approximation of the infinitesimal application of $\phi_{rf(x)}$.

Now we see how to define such iterations. In general, if ϑ is a real function and $\delta = 1/N$ then the δ-fold iteration ϑ^{δ} should have the property that $\underbrace{\vartheta^{\delta} \circ \cdots \circ \vartheta^{\delta}}_{N \text{ times}}$ is equal to ϑ itself.

In many special cases, there exists a reasonable definition of the δ-fold iteration of a function $\vartheta := \phi_{rf(x)}$ even if ϑ is not linear. In particular, this is the case if there exists a function Θ with $\Theta(t + 1) = \vartheta(\Theta(t))$ for all $t \in X$ where $X \subseteq \mathbb{R}$; then we define $\vartheta^{\delta}(\Theta(\xi)) := \Theta(\xi + \delta)$ for all ξ. Here are three examples of ϑ and Θ.

$$\begin{array}{lll}
\vartheta(\xi) := \xi^a \ (\text{with } a > 0) & : & \Theta(t) := e^{a^t}, \\
\vartheta(\xi) := 2\xi^2 - 1 & : & \Theta(t) := \cos(2^t), \\
\vartheta(\xi) := \xi^2 - 2 & : & \Theta(t) := 2\cosh(t).
\end{array}$$

On the other hand, it is often difficult or impossible to define the δ-fold iteration of a given function $\vartheta = \phi_{rf(x)}$ in a reasonable way. For example, it is a hard problem to define $\vartheta^{1/2}$ if $\vartheta(\xi) := -\xi$ or $\vartheta(\xi) = \xi^2 + 1$. Moreover, the δ-fold iteration is not always unique; if $\theta(\xi) = \xi$ then $\vartheta^{1/2} \circ \vartheta^{1/2} = \vartheta$ is true for $\vartheta^{1/2}(\xi) := \xi$ and for $\vartheta^{1/2}(\xi) := -\xi$.

Moreover, there is no general method to transform equation (5.34) into an analogous differential equation. □

Next, we approximate the infinitesimal application of $\phi_{rf(x)}$ in another way. Our idea can be applied to all real functions $\phi_{rf(x)}$, and we shall obtain a general differential equation for $\overline{L}$.

Remark 5.30 In Remark 5.28, we have approximated the infinitesimal application of $\vartheta := \phi_{rf(x)}$ by a δ-fold iteration of this function. Here, however, we write $\vartheta(\zeta) = \zeta + [\vartheta(\zeta) - \zeta]$, and we simulate the infinitesimal application of ϑ by adding $[\vartheta(\zeta) - \zeta]$ only δ times to ζ; that means, we compute $\zeta + \delta \cdot [\vartheta(\zeta) - \zeta]$ instead of the δ-fold iteration ϑ^{δ}.

Next, we describe this idea in detail. We use again equations (5.31) and (5.32). Then we replace the δ-fold iteration of $\phi_{rf(x)}$ by the function $\zeta \mapsto \zeta + \delta \cdot [\phi_{rf(x)}(\zeta) - \zeta]$. Thus, we obtain the following equation:

$$L\left(f|_{x+\delta}\right) = L\left(f|_x\right) + \delta \cdot \left[\phi_{rf(x)}\left(L\left(f|_x\right)\right) - L\left(f|_x\right)\right]. \tag{5.35}$$

We now transform the left and the right side of (5.35) as follows: First, we subtract

$L\left(f|_x\right)$; next, we divide by δ; then we replace $L\left(f|_x\right)$ and $L\left(f|_{x+\delta}\right)$ by $\overline{L}(f,x)$ and $\overline{L}(f,x+\delta)$, respectively. Thus, we generate the following equation:

$$\frac{\overline{L}(f,x+\delta) - \overline{L}(f,x)}{\delta} \;=\; \phi_{r^f(x)}\big(\overline{L}(f,x)\big) \,-\, \overline{L}(f,x). \tag{5.36}$$

We now let δ tend to zero and obtain:

$$\overline{L}'(f,x) \;=\; \phi_{r^f(x)}\big(\overline{L}(f,x)\big) - \overline{L}(f,x) \;=\; \phi_{(x,f(x),f'(x))}\big(\overline{L}(f,x)\big) - \overline{L}(f,x). \tag{5.37}$$

To make (5.37) more comprehensible we define $U(u,v,w,\zeta) := \phi_{(u,v,w)}(\zeta)$ for all (u,v,w,ζ). Let $f \in \mathrm{def}(L)$, and let $y(x) := \overline{L}(f,x)$, $\mathsf{a}(f) \le x \le \mathsf{b}(f)$. Then (5.37) is equivalent to the differential equation $y' = U(x,f,f',y) - y$.   □

Remark 5.31 Given the recursively linear case. Then there exist two functions A, B such that $\phi_{(x,f(x),f'(x))}(\zeta) = A(x,f(x),f'(x)) \cdot \zeta + B(x,f(x),f'(x))$ for all x and all ζ. Then (5.37) is of the following form:

$$L'(f,x) \;=\; [A(x,f(x),f'(x)) - 1] \cdot \overline{L}(f,x) + B(x,f(x),f'(x)). \tag{5.38}$$

This differential equation is almost always different from (5.27).

If, however, $A(u,v,w) = 1$ for all u,v,w then (5.38) is equivalent to (5.27). The reason is that $\phi_{(u,v,w)}(\zeta) = \zeta + B(u,v,w)$ for all u,v,w,ζ; consequently,

$$\left(\phi_{r^f(x)}\right)^{\delta}(\zeta) \;=\; \zeta + \delta \cdot B\left(r^f(\zeta)\right) \;=\; \zeta + \delta \cdot \left[\phi_{r^f(x)}(\zeta) - \zeta\right]$$

for all ζ. That means that the ideas in 5.22 and in 5.30 yield the same result.   □

We now consider the recursive application of quadratic functions.

Remark 5.32 Let $\widetilde{\mathbb{F}} \subseteq \mathbb{F}$ be prefix closed, and let $L : \widetilde{\mathbb{F}} \to \mathbb{R}$. Given three functions $A, B, C : \mathbb{R}^3 \to \mathbb{R}$. For all u,v,w,ξ, we define

$$\phi_{(u,v,w)}(\zeta) \;:=\; A(u,v,w) \cdot \zeta^2 + B(u,v,w) \cdot \zeta + C(u,v,w)$$

Then 5.28 and 5.30 provide two methods to simulate the recursively infinitesimal application of the functions $\phi_{r^f(x)}$, $\mathsf{a}(f) \le x \le \mathsf{b}(f)$.

The method described in 5.28 can only be applied if a δ-fold application of $\phi_{r^f(x)}$ is defined.

Remark 5.30 yields the following quadratic differential equation:

$$\overline{L}'(f,x) \;=\; A(x,f,f') \cdot \overline{L}(f,x)^2 + [B(x,f,f') - 1] \cdot \overline{L}(f,x) + C(x,f,f').$$

Next, we discuss a further idea to define functionals with the help of quadratic functions. In analogy to Definition 5.27, we say that L is *quadratic of type 2* if L has the following properties:

- $\overline{L}(f,\mathsf{a}(f)) = 0$ for all $f \in \widetilde{\mathbb{F}}$.
- For all $a',b' \in [0,1]$ with $a' \le b'$ and all $g \in \mathbb{F}[a',b']$, there are three real numbers $\beta_g, \eta_g, \sigma_g$ such that the function $\lambda_g(\zeta) := \beta_g \cdot \zeta^2 + \eta_g \cdot \xi + \sigma_g$ ($\zeta \in \mathbb{R}$) has the following property: If $f \in \widetilde{\mathbb{F}}$, $a',b' \in \mathrm{def}(f)$ and $f|_{[a',b']} = g$ then
$$\overline{L}(f,b') \;=\; \beta_g \cdot \overline{L}(f,a')^2 + \eta_g \cdot \overline{L}(f,a') + \sigma_g \;=\; \lambda_g\big(\overline{L}(f,a')\big).$$

It is, however, doubtful whether this class of functionals is greater than the class of all recursively linear functionals of type 2. The reason is that the class of all quadratic functions is not closed under composition, and this causes problems.

For example, let $g : [0.4, 0.6] \to \mathbb{R}$, and let $g_1 := g|_{[0.4, 0.5]}$ and $g_2 := g|_{[0.5, 0.6]}$. We assume that $\lambda_{g_1}(\xi) = \lambda_{g_2}(\xi) = \xi^2$. Let $f|_{[0.4, 0.6]} = g$. Then $\overline{L}(f, 0.6) = \lambda_{g_2}(\overline{L}(f, 0.5)) = \lambda_{g_2}\left(\lambda_{g_1}(\overline{L}(f, 0.4))\right) = \overline{L}(f, 0.4)^4$. On the other hand, $\overline{L}(f, 0, 6) = \lambda_g(\overline{L}(f, 0.4))$ where λ_g is a quadratic function. Consequently, $\overline{L}(f, 0.4)^4 = \overline{L}(f, 0.6) = \lambda_g(\overline{L}(f, 0.4))$ so that $y^4 = \lambda_g(y)$ for $y := \overline{L}(f, 0.4)$. This equation about y does not have more than four solutions because λ_g is a quadratic function. So, $y = \overline{L}(f, 0.4)$ can attain at most four values, and this is a drastical restriction about the definition of $\overline{L}$. — This problem has been caused by the fact that λ_{g_1}, λ_{g_2} are quadratic functions but $\lambda_{g_2} \circ \lambda_{g_1}$ is not. $\qquad\square$

5.5   Functionals arising from initial value problems

Here, we consider functionals L with the property that $\overline{L}$ satisfies one of the following initial value problems:

Implicit: $\overline{L}(f, x) = y(x)$ with $X(x, f, f', y, y') = 0$, $y(a(f)) = 0$.
Explicit: $\overline{L}(f, x) = y(x)$ with $y' = W(x, f, f', y)$, $y(a(f)) = 0$.

For example, recursively linear functionals can be described in this way; this follows from Remark 5.26 a). We now give a precise definition of functionals that are generated with the help of initial value problems.

Definition/Remark 5.33 (*see Definition/Remark 5.23*)

Let $\widetilde{\mathbb{F}} \subseteq \mathbb{F}$; we do not assume $\widetilde{\mathbb{F}}$ to be prefix closed.

a) Let $X : [0, 1] \times \mathbb{R}^4 \to \mathbb{R}$ be continuous. A functional $L : \widetilde{\mathbb{F}} \to \mathbb{R}$ is called an X-*functional* if the following are true for all $f \in \widetilde{\mathbb{F}}$ and for $a := \mathsf{a}(f)$, $b := \mathsf{b}(f)$:

 • If $a < b$ then an $y : [a, b] \to \mathbb{R}$ exists such $\overline{L}(f, \xi) = L\left(f|_{[a,\xi]}\right) = y(\xi)$ for all $\xi \in [a, b]$ with $f|_{[a,\xi]} \in \widetilde{\mathbb{F}}$;[7] this y solves the following initial value problem:

$$(\forall\, x \in [a, b])\ \ X(x, f(x), f'(x), y(x), y'(x)) = 0, \quad y(a) = 0 \qquad (5.39)$$

 In particular, $L(f) = \overline{L}(f, b) = y(b)$.
 We write $f \searrow y$ if y is used to define $L(f)$.

 • If $a = b$ then $L(f) = 0$.

b) Let $W : [0, 1] \times \mathbb{R}^3 \to \mathbb{R}$ be continuous. A functional $L : \widetilde{\mathbb{F}} \to \mathbb{R}$ is called a W-*functional* if the following are true for all $f \in \widetilde{\mathbb{F}}$ and for $a := \mathsf{a}(f)$, $b := \mathsf{b}(f)$:

---

[7] If $\widetilde{\mathbb{F}}$ is not prefix closed it may occur that $f|_\xi \notin \widetilde{\mathbb{F}}$ for some $\xi \in [a, b]$. In this case, $y(\xi)$ is defined but $\overline{L}(f, \xi)$ is not defined, and the equation $\overline{L}(f, \xi) = y(\xi)$ is not valid.

- If $a < b$ then an $y : [a,b] \to \mathbb{R}$ exists such $\overline{L}(f,\xi) = L\left(f|_{[a,\xi]}\right) = y(\xi)$ for all $\xi \in [a,b]$ with $f|_{[a,\xi]} \in \widetilde{\mathbb{F}};$[8] this y solves the following initial value problem:

$$(\forall\, x \in [a,b])\ y'(x) = W(x, f(x), f'(x), y(x)), \quad y(a) = 0 \qquad (5.40)$$

In particular, $L(f) = \overline{L}(f,b) = y(b)$.
We write $f \searrow y$ if y is used to define $L(f)$.

- If $a = b$ then $L(f) = 0$.

c) In general, it may occur that (5.39) and (5.40) have more than one solution; in this case, we choose one solution y and let $f \searrow y$.

d) Our definition has the following consequence: Given two functions $f_1, f_2 \in \widetilde{\mathbb{F}} \cap \mathbb{F}[a,b]$ with the same restriction $f_1|_{[a,x]} = f_2|_{[a,x]} \in \widetilde{F}$; let $f_i \searrow y_i$, $i = 1,2$; then $y_1(x) = y_2(x)$ even if the given problem (5.39) or (5.40) has several solutions for $f = f_1|_{[a,x]} = f_2|_{[a,x]}$. The equality of $y_1(x)$ and $y_2(x)$ follows from the equation $y_1(x) = L\left(f_1|_x\right) = L\left(f_2|_x\right) = y_2(x)$.
If $\widetilde{\mathbb{F}}$ is even prefix closed and $f_1|_{[a,x]} = f_2|_{[a,x]}$ then $f_1|_{[a,\xi]} = f_2|_{[a,\xi]} \in \widetilde{\mathbb{F}}$ for all $\xi \in [a,x]$; then $y_1(\xi) = L\left(f_1|_\xi\right) = L\left(f_2|_\xi\right) = y_2(\xi)$ for all ξ so that $y_1|_{[a,x]} = y_2|_{[a,x]}$.

e) Of course, each explicit initial value problem can be formulated as an implicit one; if W is given then $W(x,f,f',y) - y' = 0$, $y(a) = 0$ is an implicit initial value problem.

f) If an initial value problem (5.39) or (5.40) is given we define
$$\check{\mathbb{F}} := \{f \in \mathbb{F} \,|\, \text{The given initial value problem has a solution } y : [a(f), b(f)] \to \mathbb{R}\}.$$
Then $\check{\mathbb{F}}$ is automatically prefix closed. It follows from our definition of X- and of W-functionals that they are automatically defined on a subset $\widetilde{\mathbb{F}} \subseteq \check{\mathbb{F}}$. $\square$

We give several results about extremal functionals L solving initial value problems. In particular, we shall study the problem of finding a function f such that $L(f)$ is extremal.

The next result says that a somewhat generalized version of (5.40) has exactly one solution if W satisfies a Lipschitz condition.

Lemma 5.34 *Let $W : [0,1] \times \mathbb{R}^3 \to \mathbb{R}$ be a Lipschitz function with respect to the fourth argument. Let $f \in \mathbb{F}^+$, let $a := a(f)$, $b := b(f)$, and let $\overline{x} \in [a,b]$, $\overline{y} \in \mathbb{R}$.*
Then the following initial value problem has exactly one solution.

$$(\forall\, x \in [a,b])\ y'(x) = W(x,f(x),f'(x),y(x)), \quad y(\overline{x}) = \overline{y} \qquad (5.41)$$

P r o o f : We define $W_f(u,v) := W(u,f,f',v)$ for all $u \in [a,b]$ and $v \in \mathbb{R}$. Then (5.41) is equivalent to the following equation:

---

[8]If $\widetilde{\mathbb{F}}$ is not prefix closed it may occur that $f|_\xi \notin \widetilde{\mathbb{F}}$ for some $\xi \in [a,b]$. In this case, $y(\xi)$ is defined but $\overline{L}(f,\xi)$ is not defined, and the equation $\overline{L}(f,\xi) = y(\xi)$ is not valid.

$$(\forall\, x \in [a,b]) \quad y'(x) = W_f(x, y(x)), \quad y(\overline{x}) = \overline{y}. \tag{5.42}$$

Moreover, W_f is a Lipschitz function with respect to the second variable. The results I and II in [Wal72], paragraph 6, imply that (5.42) has exactly one solution on $[a,b]$.[9] (Those results should be applied to $[a, \overline{x}]$ and to $[\overline{x}, b]$ if $a < \overline{x} < b$.) Consequently, (5.41) has exactly one solution, too. □

Next, we give an auxiliary result about upper bounds for solutions of differential inequalities. This result will be used in Theorem 5.36.

Lemma 5.35 *Let* $0 \le a < b \le 1$. *Let* $c \ge 0$ *and* $k : [a,b] \to [0, \infty)$. *Let* $u \in \mathbb{F}$ *with the following property:*

$$(\forall\, x \in [a,b]) \quad |u'(x)| \le c \cdot |u(x)| + k(x). \tag{5.43}$$

Then $|u(x)| \le e^{cx} \cdot \left(|u(a)| + \left| \int\limits_a^x k(\xi)\, e^{-c\xi} d\xi \right| \right)$.

P r o o f : The assertion follows from Lemma 3.1 in [KnKa74]. We translate Knobloch's and Kappel's proof into our special situation.

Let $v_\varepsilon(x) := \sqrt{u(t)^2 + \varepsilon}$ for all $x \in [a,b]$ and $\varepsilon > 0$. Then

$$v_\varepsilon(x) > |u(x)|, \quad \text{and consequently,} \quad |v_\varepsilon(x)| = v_\varepsilon(x). \tag{5.44}$$

Moreover, $v_\varepsilon(x)$ is differentiable for all x and ε, and the equation $v_\varepsilon(x)^2 = u(x)^2 + \varepsilon$ implies that $v_\varepsilon(x) \cdot v_\varepsilon'(x) = u'(x) \cdot u(x)$. Consequently,

$$v_\varepsilon(x) \cdot |v_\varepsilon'(x)| \overset{(5.44)}{=} |v_\varepsilon(x) \cdot v_\varepsilon'(x)| = |u(x)| \cdot |u'(x)| \overset{(5.44),(5.43)}{\le}$$

$$v_\varepsilon(x) \cdot (c \cdot |u(x)| + k(x)) \overset{(5.44)}{\le} v_\varepsilon(x) \cdot (c \cdot v_\varepsilon(x) + k(x)).$$

Dividing this inequality by $v_\varepsilon(x) \overset{(5.44)}{>} 0$ yields the inequality $|v_\varepsilon'(x)| \le c \cdot v_\varepsilon(x) + k(x))$; this and $v_\varepsilon'(x) \le |v_\varepsilon'(x)|$ imply the following assertion:

$$(\forall\, \varepsilon, x) \quad v_\varepsilon'(x) \le c \cdot v_\varepsilon(x) + k(x). \tag{5.45}$$

This inequality is similar to (5.43) but (5.45) does not contain absolute values.

We now employ the so-called variation of constants, which works as follows: Given the differential equation $y' = c \cdot y + k$. Then the equation $\widetilde{y}' = c \cdot \widetilde{y}$ has the solution $\widetilde{y} = e^{cx}$; then the equation $y' = c \cdot y + k$ is transformed into a differential equation about $z := y/\widetilde{y} = y \cdot e^{-cx}$. Here, we apply the same idea; we define $w_\varepsilon(x) := v_\varepsilon(x) \cdot e^{-cx}$, and we shall use (5.45) to estimate $w_\varepsilon(x)$.

The fact that $c, x \ge 0$ implies that

$$(\forall\, \varepsilon, x) \quad v_\varepsilon(x) = e^{cx} \cdot w_\varepsilon(x) \ge w_\varepsilon(x) \tag{5.46}$$

---

[9]Similar results, for example the theorem of Picard-Lindelöf, can be found in [KnKa74], [Heus89], [Amm90] and in many other books about differential equations.

Moreover, $v'_\varepsilon(x) = \frac{d}{dx}\left(e^{cx} \cdot w_\varepsilon(x)\right) = c \cdot e^{cx} \cdot w_\varepsilon(x) + e^{cx} \cdot w'_\varepsilon(x) = c \cdot v_\varepsilon(x) + e^{cx} \cdot w'_\varepsilon(x)$.

$$(5.45)$$

Consequently, $c \cdot v_\varepsilon(x) + e^{cx} \cdot w'_\varepsilon(x) = v'_\varepsilon(x) \overset{(5.45)}{\leq} c \cdot v_\varepsilon(x) + k(x)$ so that the following is true:

$$(\forall\, \varepsilon, x) \quad w'_\varepsilon(x) \cdot e^{cx} \; \leq \; k(x). \qquad (5.47)$$

Consequently $w'_\varepsilon(\xi) \leq k(\xi) \cdot e^{-c\xi}$ for all ξ and ε. This implies that the following is true for all x and ε: $w_\varepsilon(x) - w_\varepsilon(a) = \int_a^x w'_\varepsilon(\xi)\, d\xi \leq \int_a^x k(\xi) \cdot e^{-c\xi}\, d\xi$ so that

$$w_\varepsilon(x) \; \leq \; w_\varepsilon(a) + \int_a^x k(\xi) \cdot e^{-c\xi}\, d\xi \overset{(5.46)}{\leq} v_\varepsilon(a) + \int_a^x k(\xi) \cdot e^{-c\xi}\, d\xi.$$

We multiply this inequality with e^{cx} and obtain

$$v_\varepsilon(x) \; \leq \; e^{cx} \cdot \left(v_\varepsilon(a) + \int_a^x k(\xi) \cdot e^{-c\xi}\, d\xi\right).$$

Moreover, $\lim_{\varepsilon \to 0} v_\varepsilon(a) = |u(a)|$ and $\lim_{\varepsilon \to 0} v_\varepsilon(x) = |u(x)|$. Consequently, $|u(x)| \leq e^{cx} \cdot \left(|u(a)| + \int_a^x k(\xi) \cdot e^{-c\xi}\, d\xi\right)$, and this is the desired assertion. $\qquad\square$

The following result says that L is continuous if $\overline{L}(f, x)$ is a solution of (5.40); this result will be quoted in Corollary 5.48.

Theorem 5.36 (Continuity of L)
Let $W : [0,1] \times \mathbb{R}^3 \to \mathbb{R}$ be continuous. We assume that $\partial_1 W$ and $\partial_2 W$ exist and are continuous. Moreover, we assume that W is a Lipschitz function with respect to the two last arguments; more precisely, we assume that there exist two constants c_1, c_2 such that for all $t, u, v, v_1, v_2, w, w_1, w_2$,

$$|W(t, u, v_1, w) - W(t, u, v_2, w)| \leq c_1 \cdot |v_1 - v_2|, \quad \text{and}$$
$$|W(t, u, v, w_1) - W(t, u, v, w_2)| \leq c_2 \cdot |w_1 - w_2|.$$

Let $\widetilde{\mathbb{F}} \subseteq \mathbb{F}^+$ be prefix closed, and let $L : \widetilde{\mathbb{F}} \to \mathbb{R}$ be a W-functional.

Then L is $\|\bullet\|_*$-continuous.

P r o o f : We prove our result by deriving the following local Lipschitz property of $\overline{L}$: If f is given then there exists a constant $c(f)$ such that the following is true for all $g \in \widetilde{\mathbb{F}}$ with $\|f - g\|_* \leq 1$:

$$|\overline{L}(f, x) - \overline{L}(g, x)| \; \leq \; c(f) \cdot \|f - g\|_*. \qquad (5.48)$$

For this purpose, we consider two arbitrary functions $f, g \in \widetilde{\mathbb{F}} \cap \mathbb{F}^+$ with $\mathrm{def}(f) = \mathrm{def}(g)$ and $\|f - g\|_* \leq 1$. Let $a := \mathsf{a}(f) = \mathsf{a}(g)$ and $b := \mathsf{b}(f) = \mathsf{b}(g)$; then $a < b$ because $f, g \in \mathbb{F}^+$.

Let $\eta := g - f$; then $g = f + \eta$.

Let $f \searrow y_0$ and $g \searrow y_1$. (The existence of y_0 and y_1 follows from Lemma 5.34.) Then

$$y_0' = W(x, f, f', y_0), \quad y_1' = W(x, f + \eta, f' + \eta', y_1), \quad y_0(a) = y_1(a) = 0.$$

Consequently, the following assertion is true for all $x \in [a, b]$.

$$|y_0'(x) - y_1'(x)| = |W(x, f, f', y_0) - W(x, f + \eta, f' + \eta', y_1)| =$$

$$\left| \begin{array}{c} W(x, f, f', y_0) - W(x, f + \eta, f', y_0) + W(x, f + \eta, f', y_0) - W(x, f + \eta, f' + \eta', y_0) \\ + W(x, f + \eta, f' + \eta', y_0) - W(x, f + \eta, f' + \eta', y_1) \end{array} \right|$$

$$\leq \underbrace{\left| W\left(x, f + \eta, f', y_0\right) - W\left(x, f, f', y_0\right) \right|}_{=: \text{ I}} + \underbrace{\left| W\left(x, f + \eta, f' + \eta', y_0\right) - W\left(x, f + \eta, f', y_0\right) \right|}_{=: \text{ II}} \qquad (5.49)$$

$$+ \underbrace{\left| W\left(x, f + \eta, f' + \eta', y_0\right) - W\left(x, f + \eta, f' + \eta', y_1\right) \right|}_{=: \text{ III}} .$$

By the mean value theorem for derivatives, there exists a $\zeta_x \in [0, 1]$ with the following property:
$$\text{I} = |\partial_2 W(x, f(x) + \zeta_x \cdot \eta(x), f'(x), y_0(x))| \cdot |\eta(x)| . \qquad (5.50)$$

The following assertion is a consequence of Lemma 5.8 and of $\|\eta\|_* \leq 1$:
$$|f(x) + \zeta_x \cdot \eta(x)| \leq \|f\|_{\max} + \|\eta\|_{\max} \leq \|f\|_{\max} + \|\eta\|_* \leq \|f\|_{\max} + 1 . \qquad (5.51)$$

Therefore, the argument of $\partial_2 W$ in equation (5.50) is an element of the following compact set:

$$\mathbf{S} := \left\{ (t, u, v, w) \in \mathbb{R}^4 \mid t \in [0, 1], \ |u| \overset{(5.50)}{\leq} \|f\|_{\max} + 1, \ |v| \leq \|f'\|_{\max}, \ |w| \leq \|y_0\|_{\max} \right\} .$$

Then $\partial_2 W$ is bounded on $\mathbf{S}$ because we have assumed $\partial_2 W$ to be continuous. Hence, there exists a real number c_0 such that
$$\text{I} \leq c_0 \cdot |\eta(x)| . \qquad (5.52)$$

This number depends only on f and y_0 because only these two parameters appear in the definition of $\mathbf{S}$; moreover, Lemma 5.34 implies that y_0 is uniquely determined if f is given; consequently,
$$c_0 = c_0(f) . \qquad (5.53)$$

The Lipschitz property of W implies the following assertion where c_1, c_2 are the above Lipschitz constants.
$$\text{II} \leq c_1 \cdot |\eta'(x)| , \qquad \text{III} \leq c_2 \cdot |y_0(x) - y_1(x)| . \qquad (5.54)$$

A consequence of (5.49), (5.52), and (5.54) is the following fact:

$$(\forall x \in [a, b]) \quad |y_0'(x) - y_1'(x)| \leq c_2 \cdot |y_0(x) - y_1(x)| + c_1 \cdot |\eta'(x)| + c_0 \cdot |\eta(x)| . \qquad (5.55)$$

Let $u(x) := y_0(x) - y_1(x)$, $c := c_2$ and $k(x) := c_1 \cdot |\eta'(x)| + c_0 \cdot |\eta(x)|$ for all $x \in [a, b]$. Then Lemma 5.35 says that fact (5.55) has the following consequence:

$$(\forall x \in [a, b]) \quad |y_0(x) - y_1(x)| \leq e^{c_2 x} \cdot \left(|y_0(a) - y_1(a)| + \left| \int_a^x k(\xi) e^{-c_2 \xi} \mathrm{d}\xi \right| \right) . \qquad (5.56)$$

We next estimate several terms in (5.56). If $x \in [a, b] (\subseteq [0, 1])$ then $c_2 \cdot x \leq c_2$ so that
$$e^{c_2 x} \leq e^{c_2} . \qquad (5.57)$$

Moreover,

$$|y_0(a) - y_1(a)| = |0 - 0| = 0.\tag{5.58}$$

We estimate the integral in (5.56). The definition of $\|\bullet\|_*$ and Lemma 5.8 imply that

$$\|\eta'\|_1 \leq \|\eta\|_* \quad\text{and}\quad \|\eta\|_{\max} \leq \|\eta\|_*.\tag{5.59}$$

Then we obtain the following inequality; all expressions in curled brackets are elements of $[0,1]$; relationship $(*)$ follows from the fact that $0 \leq a \leq b \leq 1$ so that $b - a \leq 1$.

$$\left|\int_a^x k(\xi)\, e^{-c_2\xi}\mathrm{d}\xi\right| = \int_a^x \left(c_1 \cdot |\eta'(\xi)| + c_0 \cdot |\eta(\xi)|\right) \cdot \left\{e^{-c_2\xi}\right\}\mathrm{d}\xi \leq$$

$$c_1 \int_a^b |\eta'(\xi)| \cdot \left\{e^{-c_2\xi}\right\}\mathrm{d}\xi + c_0 \int_a^b |\eta(\xi)| \cdot \left\{e^{-c_2\xi}\right\}\mathrm{d}\xi \leq\tag{5.60}$$

$$c_1 \cdot \|\eta'\|_1 + c_0 \cdot (b-a) \cdot \|\eta\|_{\max} \overset{(*)}{\leq} c_1 \cdot \|\eta'\|_1 + c_0 \cdot \|\eta\|_{\max} \overset{(5.59)}{\leq} (c_1 + c_0) \cdot \|\eta\|_*$$

$$= (c_1 + c_0) \cdot \|f - g\|_* \overset{(5.53)}{=} (c_1 + c_0(f)) \cdot \|f - g\|_*.$$

We rewrite inequality (5.56) with the help of (5.57), (5.58), and (5.60).

$$(\forall\, x \in [a,b]) \quad |y_0(x) - y_1(x)| \leq e^{c_2} \cdot (c_1 + c_0(f)) \cdot \|f - g\|_*.\tag{5.61}$$

Let $c(f) := e^{c_2} \cdot (c_1 + c_0(f))$. Then $|L(f) - L(g)| = |\overline{L}(f,b) - \overline{L}(g,b)| = |y_0(b) - y_1(b)| \overset{(5.61)}{\leq} c(f) \cdot \|f - g\|_*$. This proves assertion (5.48).
The $\|\bullet\|_*$-continuity of L is an immediate consequence of fact (5.48). $\qquad\square$

Next, we describe the optimization of L if $x \mapsto \overline{L}(f,x)$ is a solution of an initial value problem. We start with the simpler case that (5.40) is given.

Problem 76 *Given a function $W : [0,1] \times \mathbb{R}^3 \to \mathbb{R}$.*
Let $0 \leq a < b \leq 1$. We assume that (5.40) has a solution $y : [a,b] \to \mathbb{R}$ for all functions $f \in \mathbb{F}[a,b]$; we do not assume that this solution is unique.
Let $L : \widetilde{\mathbb{F}} \to \mathbb{R}$ be a W-functional where $\widetilde{\mathbb{F}} \supseteq \mathbb{F}[a,b]$ is prefix closed. Let $y_0, y_1 \in \mathbb{R}$. Find a function f such that $L(f)$ is minimal among all candidates $L(g)$ with $g \in \mathbb{F}[a,b]$, $g(a) = y_0$, and and $g(b) = y_1$.

We now give several examples in which (5.40) can be easily solved.

Example 5.37 Given the terminolgy of Problem 76.
• *Let $W(x_1, x_2, x_3, x_4) = F(x_1, x_2, x_3)$ for all $x_1, \ldots, x_4 \in \mathbb{R}$ where $F : \mathbb{R}^3 \to \mathbb{R}$.*
Then (5.40) has the following form:
$$y' = F(x, f, f'), \quad y(a) = 0.\tag{5.62}$$

Hence, $y(x) = \int_a^x F(\xi, f(\xi), f'(\xi))\,\mathrm{d}\xi$ for all x. That means that Problem 76 can be formulated as follows:

Find f such that $f(a) = y_0$, $f(b) = y_1$, and

$$L(f) = \overline{L}(f, b) = y(b) = \int_a^b F(\xi, f(\xi), f'(\xi)) \, d\xi \text{ is extremal.}$$

This is the classical problem of Variational Calculus, which has now turned out to be a special case of Problem 76.

This special version of Problem 76 may be degenerate; we give two examples.

Let $W(x_1, x_2, x_3, x_4) := F(x_1, x_2, x_3) := x_2$ for all x_1, x_2, x_3, x_4. Then $L(f) = \overline{L}(f, b) = y(b) = \int_a^b f(\xi) \, d\xi$, and this integral has now lower or upper bound even if $f(a) = y_0$ and $f(b) = y_1$.

Let $W(x_1, x_2, x_3, x_4) := F(x_1, x_2, x_3) := x_3$ for all x_1, x_2, x_3, x_4. Then $L(f) = \overline{L}(f, b) = y(b) = \int_a^b f'(\xi) d\xi = f(b) - f(a) = y_1 - y_0$ for all candidates f; that means that each candidate $f \in \mathbb{F}[a, b]$ with $f(a) = y_0$ and $f(b) = y_1$ solves Problem 76 in this special situation.

- *Let $W(x_1, x_2, x_3, x_4) = F_1(x_1, x_2, x_3) \cdot x_4 + F_2(x_1, x_2, x_3)$ for all $x_1, \ldots, x_4 \in \mathbb{R}$ where $F_1, F_2 : \mathbb{R}^3 \to \mathbb{R}$.*

Then (5.40) is a linear differential equation:

$$y' = F_1(x, f, f') \cdot y + F_2(x, f, f'), \quad y(a) = 0. \tag{5.63}$$

Hence, $y(x) = \exp\left[\int_a^x F_1(\xi, f(\xi), f'(\xi)) \, d\xi\right] \cdot \int_a^x \left(\frac{F_2(\xi, f(\xi), f'(\xi))}{\exp\left[\int_a^\xi F_1(\zeta, f(\zeta), f'(\zeta)) \, d\zeta\right]}\right) \cdot d\xi.$

That means that Problem 76 can be formulated as follows:

Find f such that $f(a) = y_0$, $f(b) = y_1$, and

$$L(f) = \overline{L}(f, b) = y(b) = \exp\left[\int_a^b F_1(\xi, f(\xi), f'(\xi)) \, d\xi\right] \cdot \int_a^b \left(\frac{F_2(\xi, f(\xi), f'(\xi))}{\exp\left[\int_a^\xi F_1(\zeta, f(\zeta), f'(\zeta))\right] d\zeta}\right) \cdot d\xi$$

is extremal.

This special version of Problem 76 can be degenerate as well; we give two examples.

Let $F_1(x_1, x_2, x_3) := F_2(x_1, x_2, x_3) := x_2$ for all x_1, x_2, x_3. Then (5.63) is of the form $y' = f \cdot y + f$ so that $\overline{L}(f, x) = y(x) = \exp\left[\int_a^x f(\xi) \, d\xi\right] - 1$ for all $x \in [a, b]$. There is no f such that $L(f) = \overline{L}(f, b) = \exp\left[\int_a^b f(\xi) \, d\xi\right] - 1$ is minimal or maximal.

Let $F_1(x_1, x_2, x_3) := F_2(x_1, x_2, x_3) := x_3$ for all x_1, x_2, x_3. Then (5.63) is of the form $y' = f' \cdot y + f'$ so that $\overline{L}(f, b) = y(b) = e^{f(b) - f(a)} - 1 = e^{y_1 - y_0} - 1$ for all f. That means that each candidate $f \in \mathbb{F}[a, b]$ with $f(a) = y_0$ and $f(b) = y_1$ solves Problem 76 in this special situation. $\qquad\qquad \square$

It is often difficult or impossible to find an explicit solution of (5.40). In this case, Problem 76 can be (partially) solved with methods of Variational Calculus; we shall give more details in Remark 5.38. Moreover, the special techniques of Optimal Control seem to be useful when solving Problem 76.

Remark 5.38 Given the situation of Problem 76. If $f \in \mathbb{F}[a,b]$ then $\overline{L}(f,x) = y(x) = \int_a^x y'(\xi)d\xi$ for all x. Then Problem 76 is equivalent to the search for f and for y such that

$$(\alpha) \quad \int_a^b y'(\xi)d\xi \overset{!}{=} \text{extremum} \quad \text{subject to} \quad (\beta) \quad W(\xi, f(\xi), f'(\xi), y(\xi)) - y'(\xi) = 0. \quad (5.64)$$

This is an optimization problem with a differential equation as a side constraint. To solve this problem we introduce a Lagrange multiplier $\lambda \in \mathbb{F}[a,b]$. Then the problem is to find f, y and λ such that $(5.64)(\beta)$ is true and the following integral attains an extremal value:

$$\int_a^b \left[y'(\xi) + \lambda(\xi) \cdot \left(W(\xi, f(\xi), f'(\xi), y(\xi)) - y'(\xi) \right) \right] d\xi =$$

$$\int_a^b \left[(1 - \lambda(\xi)) \cdot y'(\xi) + \lambda(\xi) \cdot W(\xi, f(\xi), f'(\xi), y(\xi)) \right] d\xi.$$

Let $H(x, u, v, w, z) := (1-\lambda(x)) \cdot z + \lambda(x) \cdot W(x, u, v, w)$ for all $x \in [a,b]$ and $u, v, w \in \mathbb{R}$. Then we must find three functions f, y and λ such that $(5.64)(\beta)$ is true and the integral $\int_a^b H(\xi, f(\xi), f'(\xi), y(\xi), y'(\xi)) d\xi$ is extremal. This problem can be reduced to three differential equations; the first and the second are Euler-Lagrange differential equations, and the third is the same as $(5.64)(\beta)$.

$$\partial_2 H(\xi, f, f', y, y') = \frac{d}{d\xi} \partial_3 H(\xi, f, f', y, y') \quad (5.65)$$

$$\partial_4 H(\xi, f, f', y, y') = \frac{d}{d\xi} \partial_5 H(\xi, f, f', y, y') \quad (5.66)$$

$$W(\xi, f(\xi), f'(\xi), y(\xi)) - y'(\xi) = 0 \quad (5.67)$$

It is possible to eliminate λ. This has been done in [Huck92], [10] page 321 – 322, and it will perhaps be done in a second edition of this book. Thus, the equations (5.65) – (5.67) can be transformed into the following equations with the unknown functions f and y.

$$\partial_{31} W(\xi, f, f', y) + \partial_{32} W(\xi, f, f', y) + \partial_{33} W(\xi, f, f', y) = \\ \partial_2 W(\xi, f, f', y) - \partial_{34} W(\xi, f, f', y) + \partial_4 W(\xi, f, f', y) \cdot \partial_3 W(\xi, f, f', y) \quad (5.68)$$

$$W(\xi, f, f', y) = y'. \quad (5.69) \quad \square$$

Next, we describe the search for an optimal f if L is an X-functional.

Problem 77 *Given a function* $X : [0,1] \times \mathbb{R}^4 \to \mathbb{R}$.
Let $0 \le a < b \le 1$. *We assume that (5.39) has a solution* $y : [a,b] \to \mathbb{R}$ *for all* $f \in \mathbb{F}[a,b]$; *we do not assume that this solution is unique.*
Let $L : \widetilde{\mathbb{F}} \to \mathbb{R}$ *be a* X-functional where $\widetilde{\mathbb{F}} \supseteq \mathbb{F}[a,b]$ *is prefix closed. Let* $y_0, y_1 \in \mathbb{R}$.

---

[10]The description in [Huck92] only concerns the special case that $a = 0$, $b = 1$; but λ can be eliminated in the same way if $0 \le a < b \le 1$.
The equation $\lambda(\xi) = c \cdot e^{\int [\partial_4 W(\tau_{f,y}(\xi))]d\xi}$ on page 322 of [Huck92] should be replaced by the equation $\lambda(x) = c \cdot e^{\int_a^x [\partial_4 W(\tau_{f,y}(\xi))]d\xi}$ (where possibly $a = 0$).
Equation (5.68) follows by dividing the last equation in [Huck92], page 322, by λ.

Find a function f such that $L(f)$ is minimal among all candidates $L(g)$ with $g \in \mathbb{F}[a,b]$, $g(a) = y_0$, and and $g(b) = y_1$.

Now we describe how this problem can be (partially) solved:

Remark 5.39 Problem 77, too, is an optimization problem with a differential equation as a side constraint. We are searching for two functions y and f such that

$$(\alpha) \ \int_a^b y'(\xi)d\xi \overset{!}{=} \text{extremum} \quad subject \ to \quad (\beta) \ X(\xi, f(\xi), f'(\xi), y(\xi), y'(\xi)) = 0. \quad (5.70)$$

This problem, too, can be solved with a Lagrange multiplier. We must find f, y and λ such that condition $(5.70)(\beta)$ is satisfied and the following integral is extremal:

$$\int_a^b \left[y'(\xi) + \lambda(\xi) \cdot X(\xi, f, f', y, y') \right] d\xi.$$

We define $H(x, u, v, w, z) := z + \lambda(x) \cdot X(x, u, v, w, z)$ for all $x \in [a, b]$, $u, v, w, z \in \mathbb{R}$. Then we must find f, y and λ such that $(5.70)(\beta)$ is true and

$$\int_a^b H\big(\xi, f(\xi), f'(\xi), y(\xi), y'(\xi)\big) \, d\xi$$

is extremal. This optimization problem can be reduced to the Euler-Lagrange differential equations (5.71), (5.72). Equation (5.73) is almost the same as $(5.70)(\beta)$.

$$\partial_2 H(\xi, f, f', y, y') = \frac{d}{d\xi} \partial_3 H(\xi, f, f', y, y') \quad (5.71)$$

$$\partial_4 H(\xi, f, f', y, y') = \frac{d}{d\xi} \partial_5 H(\xi, f, f', y, y'), \quad (5.72)$$

$$X\big(\xi, f(\xi), f'(\xi), y(\xi), y'(\xi)\big) = 0. \quad (5.73)$$

The term $\frac{d}{d\xi} \partial_5 H(\xi, f, f', y, y')$ is much more complicated than the one in Remark 5.39. Nevertheless, one can eliminate λ also in this case. This has been done in [Huck92], page 323 – 324, and it will perhaps be done in a second edition of this book. $\square$

5.6   Order preservation of functionals

In this section, we define principles of order preservation for functionals; our definitions are analogous to the definition of order preservation of path functions. Theorem 5.42 and Corollary 5.43 imply that in many cases, L is $<$- and $\le$-preserving if L defined with the help of (5.40).

Next, we define order preservation of functionals. For this purpose, we recall the definition of $\widehat{\Re}$-preservation of path functions if $\widehat{\Re} \in \{<, \le, =\}$.

> Given two paths $P_1, P_2 \in \mathcal{P}$ with $\alpha(P_1) = \alpha(P_2)$ and $\omega(P_1) = \omega(P_2)$;
> let Q be a common continuation of P_1 and P_2.
> Then $H(P_1) \ \widehat{\Re} \ H(P_2)$ implies that $H(P_1 \oplus Q) \ \widehat{\Re} \ H(P_2 \oplus Q)$.

We define a similar properties of functionals.

Definition/Remark 5.40 Let $\widetilde{\mathbb{F}} \subseteq \mathbb{F}$ be prefix closed, and let $L : \widetilde{\mathbb{F}} \to \mathbb{R}$ be a functional. Let $\widehat{\mathfrak{R}} \in \{<, \leq, =\}$.

Then L is called $\widehat{\mathfrak{R}}$-*preserving* if the following is true for all $0 \leq a \leq b \leq 1$, for all $f_1, f_2 \in \widetilde{\mathbb{F}} \cap \mathbb{F}[a, b]$, and for all $x_0 \in [a, b]$:

If $f_1(a) = f_2(a)$ [11] and $f_1|_{[x_0, b]} = f_2|_{[x_0, b]}$ [12]

then $\overline{L}(f_1, x_0) \, \widehat{\mathfrak{R}} \, \overline{L}(f_2, x_0) \implies L(f_1) = \overline{L}(f_1, b) \, \widehat{\mathfrak{R}} \, \overline{L}(f_2, b) = L(f_2)$.

L is called *order preserving* if L is $\leq$-preserving. $\qquad\qquad\qquad\qquad\qquad\qquad$ □

Remark 5.41 Let $\widetilde{\mathbb{F}} \subseteq \mathbb{F}$ be prefix closed, and let $L : \widetilde{\mathbb{F}} \to \mathbb{R}$.

a) If L is additive of type 1 then L is $\leq$-preserving and $<$-preserving.

b) Not all functionals are order preserving; this follows from 5.51. $\qquad\qquad$ □

Next, we show that L is $<$-preserving if L is a W-functional. In the proof, we shall use a result about the unique solution of differential equations.

Theorem 5.42 Let $W : [0, 1] \times \mathbb{R}^3 \to \mathbb{R}$ be continuous; we assume that W is a Lipschitz function with respect to the fourth argument.
Let $\widetilde{\mathbb{F}} \subseteq \mathbb{F}$ be prefix closed, and let $L : \widetilde{\mathbb{F}} \to \mathbb{R}$ be a W-functional. Then L is $<$-preserving.

P r o o f : The idea is the following: Let $\overline{L}(f_1, x_0) < \overline{L}(f_2, x_0)$, and suppose that $\overline{L}(f_1, b) \geq \overline{L}(f_2, b)$. Then there exists an $\bar{x} \in [x_0, b]$ with $\overline{L}(f_1, \bar{x}) = \overline{L}(f_2, \bar{x})$; moreover, the functions $x \mapsto \overline{L}(f_i, x)$, $i = 1, 2$, solve the same differential equation (namely (5.77)). Then Lemma 5.34 implies that $\overline{L}(f_1, x) = \overline{L}(f_2, x)$ for all $x \in [x_0, b]$, which is a contradiction to the assumption $\overline{L}(f_1, x_0) < \overline{L}(f_2, x_0)$.

We now give the details. Let $0 \leq a \leq b \leq 1$, let $x_0 \in [a, b]$, and let $f_1, f_2 \in \mathbb{F}[a, b] \cap \widetilde{\mathbb{F}}$ such that the following three assertions $(\alpha) - (\gamma)$ are true:

$$(\alpha) \;\; f_1(a) = f_2(a), \quad (\beta) \;\; f_1|_{[x_0, b]} = f_2|_{[x_0, b]}, \quad (\gamma) \;\; \overline{L}(f_1, x_0) < \overline{L}(f_2, x_0). \quad (5.74)$$

We must show that $L(f_1) < L(f_2)$.

If $x_0 = b$ then $L(f_1) = \overline{L}(f_1, b) = \overline{L}(f_1, x_0) \overset{(5.74)(\gamma)}{<} \overline{L}(f_2, x_0) = \overline{L}(f_2, b) = L(f_2)$.

The other case is that $a \leq x_0 < b$. Let $y_1, y_2 : [a, b] \to \mathbb{R}$ such that $f_i \searrow y_i$, $i = 1, 2$. Then y_1 and y_2 have the following properties:

$$(\forall \, i = 1, 2, \; x \in [a, b]) \quad y_i'(x) = W(x, f_i(x), f_i'(x), y_i(x)), \qquad (5.75)$$

$$y_1(x_0) \overset{(5.74)(\gamma)}{<} y_2(x_0) \qquad\qquad\qquad\qquad (5.76)$$

---

[11] That means that the "prefixes" $f_1|_{[a, x_0]}$ and $f_2|_{[a, x_0]}$ have the same "start node" $(a, f_1(a)) = (a, f_2(a))$.

[12] That means that the functions $f_1|_{[a, x_0]}$ and $f_2|_{[a, x_0]}$ have a common continuation $f_1|_{[x_0, b]} = f_2|_{[x_0, b]}$ = from the "node" $(x_0, f_1(x_0)) = (x_0, f_2(x_0))$ to the "node" $(b, f_1(b)) = (b, f_2(b))$.

Let $\widetilde{y}_i := y_i|_{[x_0,b]}$, $i = 1, 2$. Then

$$(\forall\, i = 1, 2,\ x \in [x_0, b])\quad \widetilde{y}_i'(x) \overset{(5.75),(5.74)(\beta)}{=\!=\!=\!=\!=\!=} W(x, f_1(x), f_1'(x), \widetilde{y}_i(x)), \qquad (5.77)$$

$$\widetilde{y}_1(x_0) \overset{(5.76)}{<} \widetilde{y}_2(x_0) \qquad (5.78)$$

We now show that $\widetilde{y}_1(b) < \widetilde{y}_2(b)$. If not then (5.78) implies that an $\overline{x} \in (x_0, b]$ exists with $\widetilde{y}_1(\overline{x}) = \widetilde{y}_2(\overline{x})$. Then Lemma 5.34 says that there exists exactly one solution of (5.77) through the point $(\overline{x}, \widetilde{y}_1(\overline{x}))=(\overline{x}, \widetilde{y}_2(\overline{x}))$. That would mean that $\widetilde{y}_1 = \widetilde{y}_2$, which is a contradiction to (5.78).

We have shown that $\widetilde{y}_1(b) < \widetilde{y}_2(b)$. Moreover, $L(f_i) = \overline{L}(f_i, b) = y_i(b) = \widetilde{y}_i(b)$ for $i = 1, 2$; consequently, $L(f_1) < L(f_2)$.

That means $L(f_1) < L(f_2)$ if (5.74) (α) – (γ) are true. Hence, L is $<$-preserving. $\qquad \square$

Corollary 5.43 *Let $\widetilde{\mathbb{F}}$ be prefix closed, and let $L : \widetilde{\mathbb{F}} \to \mathbb{R}$ be the solution of (5.40). We assume that W is a Lipschitz function with respect to the fourth variable. Then L is $=$-preserving and order preserving.*

P r o o f : First, we show that L is $=$-preserving. For this purpose, we define the functions $y_1, y_2, \widetilde{y}_1, \widetilde{y}_2$ in the same way as in 5.42. Let $\overline{L}(f, x_0) = \overline{L}(g, x_0)$. Then $\widetilde{y}_1(x_0) = \widetilde{y}_2(x_0) =: \overline{y}$. This and fact (5.77) in the proof of 5.42 imply that each function $\widetilde{y} \in \{\widetilde{y}_1, \widetilde{y}_2\}$ solves the initial value problem

$$\widetilde{y}' = W(x, f(x), f'(x), \widetilde{y}(x)), \quad \widetilde{y}(x_0) = \overline{y}.$$

The solution of this problem is unique by the Lipschitz property of W and by Theorem 5.34; consequently, $\widetilde{y}_1 = \widetilde{y}_2$. Hence, $\overline{L}(f, b) = \widetilde{y}_1(b) = \widetilde{y}_2(b) = \overline{L}(g, b)$.

The order preservation of L follows from the $=$-preservation and the $<$-preservation. $\qquad \square$

We now describe the idea to define monotone functionals in analogy to monotone path functions.

Remark 5.44 Perhaps, it is sensible to define monotone functionals in analogy to monotone path functions (see Definition 2.24); that means that L would be called monotone if L were the result from a recursive and infinitesimal application of monotone functions $\phi_{r^f(x)} : \mathbb{R} \to \mathbb{R}$ (see 5.28 and 5.30).

The monotonicity of $\phi_{r^f(x)}$ is equivalent to the property that $\frac{\mathrm{d}}{\mathrm{d}\xi}\phi_{r^f(x)}(\xi) \geq 0$ for all ξ. Perhaps, this condition can be used to define or to characterize monotonicity of functionals.

A further question is whether a result like 2.28 is true for functionals. $\qquad \square$

5.7   Bellman properties of functionals

In this section, we define and investigate Bellman principles for functionals. We show that $<$-preservation of functionals implies their Bellman property. This result is similar to 2.95.

Definition 5.45   Let $\widetilde{\mathbb{F}} \subseteq \mathbb{R}$ be prefix closed, and let $L : \widetilde{\mathbb{F}} \to \mathbb{R}$ be a functional.

We say that L has the *strong prefix oriented Bellman property of type 0 with respect to strong [weak] minima* if the following is true for all $f^* \in \widetilde{\mathbb{F}}$ and all $x_0 \in [a, b]$:

> If f^* is a strong [weak] minimum of L then the "prefix" $f|_{x_0}$ is a strong [weak] minimum of L.
>
> $\square$

Next, we show that $<$-preservation implies the above Bellman property with regard to strong and to weak minima.

Theorem 5.46 *Let $L : \mathbb{F} \to \mathbb{R}$ be a functional with the following properties:*

$\qquad$ (i) *L is $\|\bullet\|_*$-continuous.*[13] $\qquad$ (ii) *L is $<$-preserving.*

Then L has the strong prefix oriented Bellman property of type 0 with respect to strong and with respect to weak minima.

P r o o f : Several parts of this proof have been written in square brackets. These parts are only relevant in the case of weak minima; if, however, considering strong minima then the text in the square brackets must be ignored.

Let $0 \le a \le b \le 1$ and $x_0 \in [a, b]$. Moreover, let $f \in \mathbb{F}[a, b]$ be a strong [weak] relative minimum of L. We must show that then $f|_{x_0}$ is a strong [weak] relative minimum of L.

This is trivial if $x_0 = b$. We therefore assume that $a \le x_0 < b$.

Suppose that $f|_{x_0}$ is not a strong [weak] relative minimum of L. Then for all $\varepsilon > 0$, there exists a $g \in \mathbb{F}[a, b]$ such that the following are true:

$$(\alpha) \;\; g(x_0) = f(x_0), \quad (\beta) \;\; \|f - g\|_{\max} \le \varepsilon/3 \;\; [(\gamma) \;\; \|f' - g'\|_{\max} \le \varepsilon/3]. \quad (5.79)$$

$$\overline{L}(g, x_0) < \overline{L}(f, x_0); \quad \text{that means that } \Delta := \overline{L}(f, x_0) - \overline{L}(g, x_0) > 0. \quad (5.80)$$

Then we define a function $h \in \mathbb{F}[a, b]$ with the help of f and g; this function will have the property $L(h) < L(f)$, and that is a contradiction to the minimality of $L(f)$.

We now describe the construction of h in detail. A simple but false idea is to define h as follows:

$$(\forall \, x \in [a, b]) \;\; h(x) \; := \; \begin{cases} g(x), & a \le x \le x_0, \\ f(x), & x_0 \le x \le b. \end{cases} \quad (5.81)$$

---

[13]It was mentioned in Lemma 5.8 that this condition is somewhat weaker than the $\| \bullet \|_{\max}$-continuity of L.

Then (5.80) and the $<$-preservation of L would immediately imply that $L(h) < L(f)$, which is a contradiction to the minimality of $L(f)$. The disadvantage of h, however, is that h is not differentiable at x_0 if $f'(x_0) \neq g'(x_0)$; then $h \notin \mathbb{F}[a, b]$, and $L(h)$ is not defined at all.

To avoid this disadvantage we construct the function h as follows:

First, we choose an auxiliary function η. Next, we use η to define functions η_λ and ψ_λ, $0 < \lambda < 1$. After this, we choose a sufficiently small λ_0 and define $\widetilde{g} := g + \psi_{\lambda_0}$. Then $\widetilde{g}'(x_0) = f'(x_0)$. Consequently, when replacing the given function g by $\widetilde{g}$ in definition (5.81), the resulting function h is differentiable at x_0.

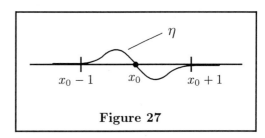

Figure 27

We now give the details of the construction. First, we choose a function $\eta : \mathbb{R} \to \mathbb{R}$ with the following properties (see Figure 27):

$$(\alpha)\ \eta(x_0) = 0, \qquad (\beta)\ \eta'(x_0) = f'(x_0) - g'(x_0), \qquad (5.82)$$

$$(\alpha)\ \|\eta\|_{\max} \leq 1, \qquad (\beta)\ \|\eta'\|_{\max} \leq |f'(x_0) - g'(x_0)|, \qquad (5.83)$$

$$\eta(x) = 0 \text{ for all } x \text{ with } |x - x_0| \geq 1. \qquad (5.84)$$

Next, we define $\eta_\lambda(x) := \lambda \cdot \eta\left(x_0 + \frac{1}{\lambda} \cdot (x - x_0)\right)$ for all $\lambda \in (0, 1)$ and all $x \in \mathbb{R}$; moreover, let $\psi_\lambda := \eta_\lambda|_{[a,b]}$ for all λ. Then $\psi_\lambda \in \mathbb{F}[a, b]$, and $\psi_\lambda'(x) = \eta_\lambda'(x) = \eta'\left(x_0 + \frac{1}{\lambda} \cdot (x - x_0)\right)$ so that the following are true for all λ and all $x \in [a, b]$:

$$(\alpha)\ \psi_\lambda(x_0) = \eta_\lambda(x_0) \overset{(5.82)(\alpha)}{=} 0, \quad (\beta)\ \psi_\lambda'(x_0) = \eta_\lambda'(x_0) \overset{(5.82)(\beta)}{=} f'(x_0) - g'(x_0), \qquad (5.85)$$

$$(\alpha)\ |\psi_\lambda(x)| = |\eta_\lambda(x)| \overset{(5.83)(\alpha)}{\leq} \lambda, \quad (\beta)\ |\psi_\lambda'(x)| = |\eta_\lambda'(x)| \overset{(5.83)(\beta)}{\leq} |f'(x_0) - g'(x_0)|, \quad (5.86)$$

$$\psi_\lambda(x) = \eta_\lambda(x) = 0, \text{ if } |x - x_0| \geq \lambda, \qquad (5.87)$$

Moreover, $\lim\limits_{\lambda \to 0} \|\psi_\lambda\|_* = 0$.

This is a consequence of the following inequality, which is valid for all $\lambda \in (0, 1)$:

$$\|\psi_\lambda\|_* = |\psi_\lambda(a)| + \int_a^b |\psi_\lambda'(\xi)|\, d\xi = |\eta_\lambda(a)| + \int_a^b |\eta_\lambda'(\xi)|\, d\xi \overset{(5.86)(\alpha),\ (5.87)}{\leq}$$

$$\lambda + \int_{x_0 - \lambda}^{x_0 + \lambda} |\eta_\lambda'(\xi)|\, d\xi \overset{(5.86)(\beta)}{\leq} \lambda + 2\lambda \cdot |f'(x_0) - g'(x_0)|.$$

We have assumed that L be $\|\bullet\|_*$-continuous. That means that for all $\widetilde{\varepsilon} > 0$, there exists a $\widetilde{\delta} = \widetilde{\delta}(\widetilde{\varepsilon})$ such that the following is true for all functions $\widetilde{\psi} \in \mathbb{F}[a, b]$:

$$\|\widetilde{\psi}\|_* \leq \widetilde{\delta} \ \Rightarrow\ |\overline{L}(g + \widetilde{\psi}, x_0) - \overline{L}(g, x_0)| \leq \widetilde{\varepsilon}. \qquad (5.88)$$

Let $\varepsilon > 0$, and let $\widetilde{\varepsilon} := \min(\varepsilon, \Delta/2)$. The fact that $\lim\limits_{\lambda \to 0} \|\psi_\lambda\|_* = 0$ (see above) implies that there exists a $\lambda_0 \leq \widetilde{\varepsilon}/3$ such that $\|\psi_{\lambda_0}\|_* \leq \widetilde{\delta}(\widetilde{\varepsilon})$. Applying (5.88) to $\widetilde{\psi} := \psi_{\lambda_0}$,

we see the following:
$$|\overline{L}(g+\psi_{\lambda_0},x_0)-\overline{L}(g,x_0)| \le \widetilde{\varepsilon} \le \frac{\Delta}{2}. \tag{5.89}$$

Let $\widetilde{g} := g+\psi_{\lambda_0}$; then the following is true: $\overline{L}(f,x_0)-\overline{L}(\widetilde{g},x_0) = \overline{L}(f,x_0)-\overline{L}(g,x_0)-$
$(\overline{L}(\widetilde{g},x_0)-\overline{L}(g,x_0)) \ge \overline{L}(f,x_0)-\overline{L}(g,x_0)-|\overline{L}(\widetilde{g},x_0)-\overline{L}(g,x_0)| \overset{(5.80),(5.89)}{\ge} \Delta-\frac{\Delta}{2} > 0$.
Consequently,
$$\overline{L}(\widetilde{g},x_0) < \overline{L}(f,x_0); \tag{5.90}$$
that means that $\widetilde{g}|_{[a,x_0]}$ is "better" than $f|_{[a,x_0]}$. Moreover, $|f(x)-\widetilde{g}(x)| \le$
$|f(x)-g(x)|+|\psi_{\lambda_0}(x)| \overset{(5.79)(\beta),(5.86)(\alpha)}{\le} \frac{\varepsilon}{3}+\lambda_0 \le \frac{\varepsilon}{3}+\frac{\widetilde{\varepsilon}}{3} \le \frac{2\varepsilon}{3}$ for all $x \in [a,b]$ so
that
$$\|f-\widetilde{g}\|_{\max} \le \frac{2}{3}\varepsilon. \tag{5.91}$$

$$\left[\begin{array}{l} \text{Moreover, } |f'(x)-\widetilde{g}'(x)| \le |f'(x)-g'(x)|+|\psi'_{\lambda_0}(x)| \overset{(5.79)(\gamma),(5.86)(\beta)}{\le} \\ \varepsilon/3+|f'(x_0)-g'(x_0)| \overset{(5.79)(\gamma)}{\le} \varepsilon/3+\varepsilon/3 \text{ so that} \\ \qquad\qquad \|f'-\widetilde{g}'\|_{\max} \le \frac{2}{3}\varepsilon. \tag{5.92} \end{array}\right]$$

Also, $\widetilde{g}$ has the following properties:
$$\begin{aligned} \widetilde{g}(x_0) &= g(x_0)+\psi_{\lambda_0}(x_0) \overset{(5.79)(\alpha),(5.85)(\alpha)}{=\!=\!=} f(x_0), \\ \widetilde{g}'(x_0) &= g'(x_0)+\psi'_{\lambda_0}(x_0) \overset{(5.85)(\beta)}{=\!=\!=} f'(x_0). \end{aligned} \tag{5.93}$$

Therefore, the following function h is differentiable at x_0.
$$(\forall\, x \in [a,b]) \ \ h(x) := \left\{ \begin{array}{ll} \widetilde{g}(x), & \text{if} \quad a \le x \le x_0, \\ f(x), & \text{if} \quad x_0 \le x \le b. \end{array} \right\}$$

If $x \le x_0$ then $|f(x)-h(x)| \overset{(5.91)}{\le} \frac{2}{3}\varepsilon$; if $x > x_0$ then $|f(x)-h(x)|=0$. Consequently,
$$(\forall\, x \in [a,b]) \ |f(x)-h(x)| \le \frac{2}{3}\varepsilon. \tag{5.94}$$

$$\left[\begin{array}{l} \text{If } x \le x_0 \text{ then } |f'(x)-h'(x)| \overset{(5.92)}{\le} \frac{2}{3}\varepsilon; \text{ if } x > x_0 \text{ then } |f'(x)-h'(x)|=0. \\ \text{Consequently,} \\ \qquad\qquad (\forall\, x \in [a,b]) \ |f'(x)-h'(x)| \le \frac{2}{3}\varepsilon. \tag{5.95} \end{array}\right]$$

Furthermore, $h|_{[a,x_0]} = \widetilde{g}|_{[a,x_0]}$ implies that $\overline{L}(h,x_0)=\overline{L}(\widetilde{g},x_0)$. Therefore,
$$\overline{L}(h,x_0) = \overline{L}(\widetilde{g},x_0) \overset{(5.90)}{<} \overline{L}(f,x_0). \tag{5.96}$$
We have assumed that L is $<$-preserving. Moreover, $h|_{[x_0,b]}=f|_{[x_0,b]}$. Consequently,
$$L(h) \overset{(5.96)}{<} L(f). \tag{5.97}$$

We now have seen the following: For all $\varepsilon > 0$, there exists a function $h \in \mathbb{F}[a, b]$ with the following properties: $\|f - h\|_{\max} \overset{(5.94)}{\leq} \frac{2}{3}\varepsilon$, $\left[\|f' - h'\|_{\max} \overset{(5.95)}{\leq} \frac{2}{3}\varepsilon\right]$, and $L(h) < L(f)$. That is a contradiction to the assumption that f is a strong [weak] relative minimum of L. This contradiction has been caused by the assumption that f is not a strong [weak] minimum of L. Consequently, that assumption is false, and the assertion of the theorem is true. $\qquad\square$

Remark 5.47 It is easier to prove Theorem 5.46 if L is defined for all functions f where f is piecewise differentiable. In this case, we may use the function h defined in (5.81).

On the other hand, there are two reasons why the given theorem 5.46 is at least as good as the proposed modification:

1. The assumption that f is a strong minimum of L is weaker in the original theorem 5.46 than in the modified version. In Theorem 5.46, $\overline{L}(f, x_1) \leq \overline{L}(g, x_1)$ must only be true for all differentiable candidates g; in modified version, $\overline{L}(f, x_1) \leq \overline{L}(g, x_1)$ would have to be true for all piecewise differentiable candidates.

2. The original theorem 5.46 can be formulated for weak minima. This, however, is difficult or impossible when admitting piecewise differentiable functions. The reason is the condition $\|f'(x) - g'(x)\|_{\max} \leq \varepsilon$ in Definition 5.9; this condition cannot be formulated if the piecewise differentiable function f or g is not differentiable at x. $\quad\square$

Corollary 5.48 *Let $L : \mathbb{F} \to \mathbb{R}$ be a W-functional where $W[0, 1] \times \mathbb{R}^3 \to \mathbb{R}$ satifies all conditions in Theorem 5.36.*

Then L has the strong prefix oriented Bellman property of type 0 with respect to strong and to weak maxima.

P r o o f : L is $\|\bullet\|_*$-continuous by Theorem 5.36, and L is $<$-preserving by Theorem 5.42. The desired Bellman properties of L follow from Theorem 5.46. $\quad\square$

Next, we describe another aspect of functionals with Bellman properties.

Remark 5.49 Bellman properties can be used to derive Euler-Lagrange differential equations of Variational Calculus.

More precisely, let $\overline{L}(f, x) := \int_a^x F(\xi, f(\xi), f'(\xi)) \, d\xi$ $(f \in \mathbb{F}[a, b])$ where F is a given function. The problem of Variational Calculus is to find a relative strong minimum of L among all candidates f with given values $f(a)$ and $f(b)$.

The functional L has the strong prefix oriented Bellman property of type 0 with regard to strong minima; this property can be used to derive a Euler-Lagrange differential equation about f. More details are given in [BeDr62], [BeKa65], [Bell67], and in [Huck92], Remark 5.36, which is a very careful derivation of the Euler-Langrange differential equations. $\quad\square$

5.8   Functionals with many relative extrema

In this section, we desribe functionals L with the following property: There exist many functions f such that L attains an extremal value. We obtain these functionals L by defining $L(f) := \psi(L_0(f))$ where $\psi : \mathbb{R} \to \mathbb{R}$ and L_0 is another functional.

Example 5.50 We define $L_0 : \mathbb{F} \to \mathbb{R}$ as the arc length; more precisely, let $f \in \mathbb{F}$, $a = \mathsf{a}(f)$, $b = \mathsf{b}(f)$; then $L_0(f,x) := \int_a^b \sqrt{1 + f'(\xi)^2}\, d\xi$.

Let $\psi(u) := u^2 - 2u$ for all $u \in \mathbb{R}$; then we define

$$L(f) := \psi \circ L_0(f) = L_0(f)^2 - 2\, L_0(f) = \left[\int_a^b \sqrt{1 + f'(\xi)^2} d\xi \right]^2 - 2 \int_a^b \sqrt{1 + f'(\xi)^2} d\xi.$$

It is clear that $\psi(u)$ is minimal if and only if $u = 1$. Consequently, $L(f)$ is minimal if and only if $L_0(f) = 1$.

Let $a < b$, and let $y_0, y_1 \in \mathbb{R}$. If the Euclidean distance between the points (a, y_0) and (b, y_1) is smaller than 1 then there exist infinitely many functions $f \in \mathbb{F}[a,b]$ with $f(a) = y_0$, $f(b) = y_1$ and $L_0(f) = 1$; all these functions f are strong relative minima of L; therefore, L has infinitely many strong relative minima f with $f(a) = y_0$ and $f(b) = y_1$.

The functional L has infinitely many minima; nevertheless, the problem to minimize $L(f)$ can be formulated as a formal problem of Variational calculus with a differential equation as a side constraint. This has been done in [Huck92], page 327 – 328 in two different ways.   ☐

Remark 5.51 Let L_0 and L be the same functionals as in 5.50; then L is not order preserving.

To see this we define $a := 0$, $b := 1$, and we choose two functions $h_1, h_2 : [0, 0.5] \to \mathbb{R}$ with the following properties:

$$h_1(0) = h_2(0) = 0, \qquad h_1(0.5) = h_2(0.5) = 0.5, \qquad h_1'(0.5) = h_2'(0.5) = 0, \quad \text{and}$$
$$c_1 := \overline{L}_0(h_1, 0.5) = 0.8, \quad c_2 := \overline{L}_0(h_2, 0.5) = 1.$$

Then we define $f_1, f_2 \in \mathbb{F}[0,1]$ as follows:

$$f_1(x) := \begin{cases} h_1(x) & \text{if } 0 \le x \le 0.5, \\ 0.5 & \text{if } 0.5 \le x \le 1, \end{cases} \qquad f_2(x) := \begin{cases} h_2(x) & \text{if } 0 \le x \le 0.5, \\ 0.5 & \text{if } 0.5 \le x \le 1. \end{cases}$$

for all $x \in [0,1]$. Then f_1 and f_2 have the following properties:

$$(\alpha)\ \ f_i(0) = h_i(0) = 0,\ i = 1, 2, \quad (\beta)\ \ f_1(0.5) = h_1(0.5) = h_2(0.5) = f_2(0.5), \tag{5.98}$$
$$(\gamma)\ \ (f_1)\big|_{[0.5,1]} = (f_2)\big|_{[0.5,1]}.$$

Moreover, the arc length $C_1 := L_0(f_1)$ equals 1.3, and $C_2 := L_0(f_2)$ equals 1.5. Hence,

$$\overline{L}(f_1, 0.5) = c_1^2 - 2\, c_1 = -0.96, \qquad \overline{L}(f_2, 0.5) = c_2^2 - 2\, c_2 = -1,$$
$$L(f_1) = C_1^2 - 2\, C_1 = -0.91, \qquad L(f_2) = C_2^2 - 2\, C_2 = -0.75.$$

This implies that $\overline{L}(f_1, 0.5) > \overline{L}(f_2, 0.5)$ and $L(f_1) < L(f_2)$ although (5.98)(α) – (γ) are true. Hence, L is not order preserving.   ☐

The next result is about sufficient conditions for the fact that L has infinitely many strong relative minima. We leave it to the reader to check whether the functions L_0, ψ, and L in Example 5.50 satisfy these conditions.

Theorem 5.52 *Given a functional $L_0 : \mathbb{F} \to \mathbb{R}$ and a differentiable function $\psi :$ $\mathbb{R} \to \mathbb{R}$ such that ψ' is continuous; we define $L := \psi \circ L_0$.*
Let $s_0 \in \mathbb{R}$ be a relative minimum of ψ. Let $0 \leq a < b \leq 1$, and let $f \in \mathbb{F}[a,b]$ such that $f(a) = 0$ and $L_0(f) = s_0$; let $\overline{y} := f(b)$.
Let $\eta_1, \eta_2 \in \mathbb{F}[a,b]$ such that $\eta_1(a) = \eta_2(a) = 0$ and $\eta_1(b) = \eta_2(b) = 0$.

Let $h(\tau, \zeta) := L_0(f + \tau \cdot \eta_1 + \zeta \cdot \eta_2)$ for all $(\tau, \zeta) \in \mathbb{R}^2$. We make the following assumptions:

   (i) *There exists a norm $\|\bullet\|$ such that L_0 is continuous with respect to $\|\bullet\|$.*
   (ii) *The functions η_1, η_2 are linearly independent; that means that there is no constant c such that $\eta_1 = c \cdot \eta_2$ or $\eta_2 = c \cdot \eta_1$.*
   (iii) *There exists an $\varepsilon' > 0$ such that $\partial_1 h(\tau, \zeta)$ and $\partial_2 h(\tau, \zeta)$ exist and are continuous for all $\tau, \zeta \in [-\varepsilon', +\varepsilon']$.*
   (iv) *$\partial_1 h(0,0) \neq 0 \neq \partial_2 h(0,0)$. (This implies that f is not a relative minimum of L_0.)*

Then there exist an $\varepsilon > 0$ and a differentiable function $\chi : [-\varepsilon, \varepsilon] \to [-\varepsilon', \varepsilon']$ with the following properties:

   a) *$\chi(0) = 0$.*
   b) *If $\tau \in [-\varepsilon, +\varepsilon]$ then the function $f_\tau(x) := f(x) + \tau \cdot \eta_1(x) + \chi(\tau) \cdot \eta_2(x)$ $(x \in [a,b])$ is a relative strong minimum of L.*
   (That means that $L(f_\tau)$ is a minimal among all candidates $L(g)$ for which $g(a) = 0$, $g(b) = \overline{y}$, and $\|g - f_\tau\|$ is sufficiently small; in particular, $f_0 = f$ is a relative strong minimum of L)

P r o o f : By (iii) and (iv), we may apply the implicit function theorem to the equation $h(\tau, \zeta) - h(0,0) = 0$. Consequently, there exist an $\varepsilon \in (0, \varepsilon']$ and a function $\chi : [-\varepsilon, +\varepsilon] \to [-\varepsilon', +\varepsilon']$ such that $\chi(0) = 0$ and $h(\tau, \chi(\tau)) - h(0,0) = 0$ for all $\tau \in [-\varepsilon, +\varepsilon]$. It follows that $L_0(f_\tau) = h(\tau, \chi(\tau)) = h(0,0) = L_0(f) = s_0$ for all τ so that $L_0(f_\tau) = s_0$ for all $\tau \in [-\varepsilon, +\varepsilon]$. This implies that $L(f_\tau) = \psi(L_0(f_\tau)) = \psi(s_0)$ for all τ so that

$$(\forall \tau \in [-\varepsilon, +\varepsilon]) \quad L(f_\tau) = \psi(s_0). \tag{5.99}$$

Next, we show that each f_τ with $\tau \in [-\varepsilon, +\varepsilon]$ is a strong relative minimum of L. For this purpose, we choose ε_0 such that $\psi(s_0)$ is an absolute minimum of $\psi|_{[s_0 - \varepsilon_0, s_0 + \varepsilon_0]}$.

By (i), there exists a $\delta_0 > 0$ with the following property: If $g \in \mathbb{F}[a,b]$ and $\|g - f_\tau\| \leq \delta_0$ then $|L_0(g) - L_0(f_\tau)| \leq \varepsilon_0$; this and (5.99) imply that $L_0(g) \in [s_0 - \varepsilon_0, s_0 + \varepsilon_0]$ if $\|g - f_\tau\| \leq \delta_0$. Consequently, $\overline{L}(g, x_0) = \psi(L_0(g)) \geq \psi(s_0) = \psi(L_0(f_\tau)) = L(f_\tau)$ so that $L(g) \geq L(f_\tau)$. Therefore, f_τ is a strong relative minimum of L.

Moreover, the functions f_τ are pairwise distinct. To see this we suppose that $\tau' \neq \tau''$ but $f_{\tau'} = f_{\tau''}$. Then the definition of $f_{\tau'}$ and $f_{\tau''}$ would imply that $f + \tau' \cdot \eta_1 + \chi(\tau') \cdot \eta_2 = f + \tau'' \cdot \eta_1 + \chi(\tau'') \cdot \eta_2$ so that $\eta_1 = \frac{\chi(\tau'') - \chi(\tau')}{\tau' - \tau''} \cdot \eta_2$; that is a contradiction to assumption (ii).

So, there exist infinitely many functions f_τ with $\tau \in [-\varepsilon, +\varepsilon]$; each of these functions is a strong relative minimum of L as seen above. ∎

5.9   Approximation theoretical results

In this section, we transform the approximation theoretic result 4.191 b) into the setting of continuous curves.

The problem in 4.191 b) was the following: Given an arbitrary arc function h; approximate h by a potential difference δ. That problem is closely related to C-minimal cycles where $C(P)$ was defined as the quotient of $C_1(P) := SUM_h(P)$ and the length $\ell(P)$. When proving 4.191 b), we used fact (4.212), i.e., $SUM_\delta(P) = 0$ for all potential differences δ and all cycles P.

A continuous version of the above problem is the following: Given an arbitrary pair (f, g) of continuous function $f, g : \mathbb{R}^2 \dashrightarrow \mathbb{R}$; approximate (f, g) by an integrable pair $(\tilde{f}, \tilde{g})$ of functions. (A pair $(\tilde{f}, \tilde{g})$ is called *integrable* if a function $H : \mathbb{R} \dashrightarrow \mathbb{R}$ exists such that $\tilde{f} = \partial_1 H$ and $\tilde{g} = \partial_2 H$.) We give several results about the best approximation. There are many analogies between objects and ideas in 4.191 b) and the ones in the current situation. For example, one of the results in the proof of 5.55 is that $\int_Q (\partial_1 H \, dx + \partial_2 H \, dy) = 0$ for all integrable pairs $(\partial_1 H, \partial_2 H)$ and all closed curves Q; this fact is analogous to (4.212).

In the previous sections of this chapter, we considered functions $f \in \mathbb{F}$ as continuous analogies to paths in digraphs; moreover, we used arbitrary functionals as cost measures. Here, however, we consider arbitrary continuous curves as analogies to paths in digraphs, and we use curve integrals as cost measures. We now give a list of analogic objects.

Graph theoretic objects	Continuous objects
Paths P	Curves $Q \subseteq \mathbb{R}^2$
Cycles	Closed curves
$C_1(P) := SUM_h(P) =$ sum of all h-values along P	$C_1(Q) := \int_Q (f \, dx + g \, dy) =$ curve integral over (f, g) along Q
$\ell(P) :=$ length of P	$|Q| :=$ arc length of Q

We now introduce the following notation:

Definition 5.53 Let $G \subseteq \mathbb{R}^2$ be a simply connected, bounded, open set with $(0, 0) \in G$. Then we define $D_1(G)$ as the set of all functions $H : G \to \mathbb{R}$ for which $\partial_1 H$, $\partial_2 H$ exist and are continuous. ∎

Next, we show an auxiliary result, which is well-known in Analysis.

Lemma 5.54 *Given a simply connected, bounded, open set $G \subseteq \mathbb{R}^2$ with the property that $(0, 0) \in G$. Given two functions $a, b \in D_1(G)$. Let $Q \subseteq G$ be a continuous, rectifiable curve, and let $M := \max_{(x,y) \in Q} \sqrt{a(x, y)^2 + b(x, y)^2}$.*

Then $\left| \int_Q a(\sigma,\tau)\,d\sigma + b(\sigma,\tau)\,d\tau \right| \le M \cdot |Q|$.

P r o o f : We assume that Q is represented by the parameter functions $u, v : [0,1] \to G$. We then obtain the following fact, in which $(*)$ is a consequence of the Cauchy-Schwartz inequality.

$$\left| \int_Q a(\sigma,\tau)\,d\sigma + b(\sigma,\tau)\,d\tau \right| = \left| \int_0^1 [a(u(\zeta),v(\zeta)) \cdot u'(\zeta) + b(u(\zeta),v(\zeta)) \cdot v'(\zeta)] \cdot d\zeta \right|$$

$$\le \int_0^1 \sqrt{[a(u(\zeta),v(\zeta)) \cdot u'(\zeta) + b(u(\zeta),v(\zeta)) \cdot v'(\zeta)]^2}\,d\zeta \overset{(*)}{\le}$$

$$\int_0^1 \sqrt{a(u(\zeta),v(\zeta))^2 + b(u(\zeta),v(\zeta))^2} \cdot \sqrt{u'(\zeta)^2 + v'(\zeta)^2}\,d\zeta \le \int_0^1 M \cdot \sqrt{u'(\zeta)^2 + v'(\zeta)^2}\,d\zeta$$

$$= M \cdot |Q|. \qquad \square$$

Next, we show the following result, which is analogous to 4.191 b).

Theorem 5.55 *Given a simply connected, bounded, open set $G \subseteq \mathbb{R}^2$ with $(0,0) \in G$. Given two functions $f, g \in D_1(G)$.*

For each $H \in D_1(G)$, we define $\Lambda(H)$ as the supremum of all Euclidean distances between any points $(f(x,y), g(x,y))$ and $(\partial_1 H(x,y), \partial_2 H(x,y))$ with $(x,y) \in G$; more formally, $\Lambda(H) := \sup_{(x,y) \in G} \sqrt{(f(x,y) - \partial_1 H(x,y))^2 + (g(x,y) - \partial_2 H(x,y))^2}$.[14]

We define $\mathcal{L}$, $\mathcal{Q}$ and λ^* and q^* in analogy to 4.191 b):

$$\mathcal{L} := \{\Lambda(H) \mid H \in D_1(G)\}, \qquad\qquad \lambda^* := \inf(\mathcal{L}),$$

$$\mathcal{Q} := \left\{ \frac{1}{|Q|} \cdot \left| \int_Q f(\sigma,\tau)d\sigma + g(\sigma,\tau)d\tau \right| \;\middle|\; \begin{matrix} Q \subseteq G \text{ is a closed,} \\ \text{rectifiable curve} \end{matrix} \right\}, \qquad q^* := \sup(\mathcal{Q}).$$

Then $\lambda^* \ge q^*$.[15]

P r o o f : The argumentation is similar to the one in 4.191 b).

Let $H \in D_1(G)$, and let Q be a closed, rectifiable curve. Then $\int_Q \partial_1 H \cdot dx + \partial_2 H \cdot dy = 0$. This is used in relationship $(*)$ of the following assertion; relationship $(**)$ follows from Lemma 5.54.

$$\frac{1}{|Q|} \cdot \left| \int_Q f(\sigma,\tau)d\sigma + g(\sigma,\tau)d\tau \right| \overset{(*)}{=}$$

$$\frac{1}{|Q|} \cdot \left| \int_Q [f(\sigma,\tau) - \partial_1 H(\sigma,\tau)] \cdot d\sigma + [g(\sigma,\tau) - \partial_2 H(\sigma,\tau)]\,d\tau \right| \overset{(**)}{\le}$$

$$\frac{1}{|Q|} \cdot \max_{(x,y) \in Q} \sqrt{[f(x,y) - \partial_1 H(x,y)]^2 + [g(x,y) - \partial_2 H(x,y)]^2} \cdot |Q| \le \Lambda(H).$$

Hence, $q^* = \sup_Q \left(\frac{1}{|Q|} \cdot \left| \int_Q f(\sigma,\tau)d\sigma + g(\sigma,\tau)d\tau \right| \right) \le \inf_{H \in D_1(G)} \Lambda(H) = \lambda^*$. $\qquad \square$

---

[14] Our intention is to minimize $\Lambda(H)$.

[15] That means: When approximating (f,g) by a pair $(\partial_1 H, \partial_2 H)$ we cannot achieve a better precision than q^*.

The next result is about an upper bound of λ^*.

Theorem 5.56 *Given a simply connected, bounded, and open set $G \subseteq \mathbb{R}^2$ with $(0,0) \in G$. Given $f, g \in D_1(G)$. We define $\Lambda(H)$, λ^* and σ^* in the same way as in 5.55. Let $\mu > 0$, and let $\widehat{C}(K) := \int_K f(u,v)du + g(u,v)dv - \mu \cdot |K|$ for each rectifiable curve $K \subseteq G$.[16] Moreover, we define the function $H_0 : G \to \mathbb{R} \cup \{\infty\}$ as follows:*

$$(\forall (x,y) \in G) \quad H_0(x,y) := \sup\{\widehat{C}(K) \,|\, K \subseteq G \text{ is a rectifiable curve from } (0,0) \text{ to } (x,y)\}.$$

We make the following assumptions:

(i) *For each pair $(x,y) \in G$, there is a curve $K^{(x,y)} \subset G$ from $(0,0)$ to (x,y) such that $H_0(x,y) = \widehat{C}\left(K^{(x,y)}\right)$ and $|H_0(x,y)| < \infty$.[17]*

(ii) *Each curve $K^{(x,y)}$ is given by two parameter functions $u, v : [0,1] \to \mathbb{R}$; more precisely, we assume that $u(0) = 0$, $v(0) = 0$, $u(1) = x$, $v(1) = y$, and*

$$K^{(x,y)} := \{(u(\xi), v(\xi)) \,|\, \xi \in [0,1]\}.$$

(iii) *The partial derivatives $\partial_1 H_0$ and $\partial_2 H_0$ exist and are continuous; that means that $H_0 \in D_1(G)$.*

Then $\lambda^* \leq \mu$.

P r o o f : In the proof of 4.191 b), we used the maxima of $\widehat{C}$ to define potentials. Here, we apply a similar idea. We use $\widehat{C}(K^{(x,y)}) = H_0(x,y)$ to define an an approximation $(\partial_1 H_0, \partial_2 H_0)$ of (f, g).

With the help of u and v, we obtain the following assertion for all $(x,y) \in G$.

$$H_0(x,y) = \int_0^1 \left[f(u(\xi), v(\xi)) \cdot u'(\xi) + g(u(\xi), v(\xi)) \cdot v'(\xi) - \mu \cdot \sqrt{u'(\xi)^2 + v'(\xi)^2} \right] d\xi.$$

Let $(x,y) \in G$, and let $\varepsilon > 0$ such that $(x+h', x+h'') \in G$ for all $h', h'' \in [-\varepsilon, +\varepsilon]$. Let $h_1, h_2 \in [-\varepsilon, +\varepsilon]$, and let S be the directed line segment from (x,y) to $(x+h_1, y+h_2)$; then $S \subseteq G$. We define $\widetilde{K}^{(x+h_1, y+h_2)}$ as the curve generated by first moving along $K^{(x,y)}$ and then moving along S. Then

$$\widehat{C}\left(\widetilde{K}^{(x+h_1,y+h_2)}\right) \leq \widehat{C}\left(K^{(x+h_1,y+h_2)}\right). \tag{5.100}$$

by the maximality of $\widehat{C}\left(K^{(x+h_1,y+h_2)}\right)$. (Here, we have used the same argumentation as in (**) of fact (4.214).)

By assumption (iii), we may write $\widehat{C}(K^{(x+h_1,y+h_2)})$ as follows:

$$\begin{aligned} \widehat{C}\left(K^{(x+h_1,y+h_2)}\right) &= H_0(x + h_1, y + h_2) \\ &= H_0(x,y) + h_1 \cdot \partial_1 H_0(x,y) + h_2 \cdot \partial_2 H_0(x,y) + o(\varepsilon). \end{aligned} \tag{5.101}$$

(The term $o(\varepsilon)$ represents a summand $s(\varepsilon)$ with $\lim\limits_{\varepsilon \to 0} s(\varepsilon)/\varepsilon = 0$.)

---

[16] This definition is similar to the one of $\widehat{C}_\lambda$ in 4.191 b).
[17] This implies that the supremum $H_0(x,y)$ is a maximum.

Next, we show the following fact:

$$\int_S f(\sigma,\tau)\,d\sigma + g(\sigma,\tau)\,d\tau = h_1 \cdot f(x,y) + h_2 \cdot g(x,y) + o(\varepsilon). \qquad (5.102)$$

For this purpose, we observe that $S = \{(x + h_1\xi, y + h_2\xi) \,|\, 0 \leq \xi \leq 1\}$. This is used in relationship $(*)$ of the next equation; fact $(**)$ follows from the Taylor expansions of f and g, which exist because $f, g \in D_1(G)$. It is

$$\int_S f(\sigma,\tau)d\sigma + g(\sigma,\tau)d\tau \overset{(*)}{=} \int_0^1 [f(x + h_1\xi, y + h_2\xi) \cdot h_1 + g(x + h_1\xi, y + h_2\xi) \cdot h_2]\,d\xi$$

$$\overset{(**)}{=} \int_0^1 \begin{bmatrix} (f(x,y) + h_1 \cdot \xi \cdot \partial_1 f(x,y) + h_2 \cdot \xi \cdot \partial_2 f(x,y) + o(\varepsilon^2)) \cdot h_1 + \\ (g(x,y) + h_1 \cdot \xi \cdot \partial_1 g(x,y) + h_2 \cdot \xi \cdot \partial_2 g(x,y) + o(\varepsilon^2)) \cdot h_2 \end{bmatrix} =$$

$$\underline{h_1 \cdot f(x,y)} + \tfrac{1}{2}h_1^2 \cdot \partial_1 f(x,y) + \tfrac{1}{2}h_1 h_2 \cdot \partial_2 f(x,y) + h_1 \cdot o(\varepsilon^2) +$$

$$\underline{h_2 \cdot g(x,y)} + \tfrac{1}{2}h_1 h_2 \cdot \partial_1 g(x,y) + \tfrac{1}{2}h_2^2 \cdot \partial_2 g(x,y) + h_2 \cdot o(\varepsilon^2).$$

Then (5.102) follows from the fact that all summands in the last term are $o(\varepsilon)$ up to the underlined ones. (Note that $|h_1|, |h_2| \leq \varepsilon$.)

Assertion (5.102) is used in the following equation:

$$\widehat{C}\left(\widetilde{K}^{(x+h_1, y+h_2)}\right) = \widehat{C}(K^{(x,y)}) + \widehat{C}(S) =$$

$$H_0(x,y) + \int_S f(\sigma,\tau)\,d\sigma + g(\sigma,\tau)\,d\tau - \mu \cdot |S| \overset{(5.102)}{=}$$

$$H_0(x,y) + h_1 f(x,y) + h_2 g(x,y) + o(\varepsilon) - \mu\sqrt{h_1^2 + h_2^2}.$$

Consequently, the following is true:

$$\widehat{C}(\widetilde{K}^{(x+h_1, y+h_2)}) = H_0(x,y) + h_1 f(x,y) + h_2 g(x,y) - \mu \cdot \sqrt{h_1^2 + h_2^2} + o(\varepsilon). \qquad (5.103)$$

Now we obtain the following fact with the help of (5.100) – (5.103):

$$H_0(x,y) + h_1 \cdot f(x,y) + h_2 \cdot g(x,y) - \sqrt{h_1^2 + h_2^2} + o(\varepsilon) \overset{(5.103)}{=} \widehat{C}\left(\widetilde{K}^{(x+h_1, y+h_2)}\right) \overset{(5.100)}{\leq}$$

$$\widehat{C}\left(K^{(x+h_1, y+h_2)}\right) \overset{(5.101)}{=} H_0(x,y) + h_1 \cdot \partial_1 H_0(x,y) + h_2 \cdot \partial_2 H_0(x,y) + o(\varepsilon).$$

Consequently,

$$h_1 \cdot (f(x,y) - \partial_1 H_0(x,y)) + h_2 \cdot (g(x,y) - \partial_2 H_0(x,y)) \leq \mu \cdot \sqrt{h_1^2 + h_2^2} + o(\varepsilon). \qquad (5.104)$$

Next, we show the following fact:

$$\sqrt{[f(x,y) - \partial_1 H_0(x,y)]^2 + [g(x,y) - \partial_2 H_0(x,y)]^2} \leq \mu + \tfrac{o(\varepsilon)}{\varepsilon}. \qquad (5.105)$$

This fact trivial if $[f(x,y) - \partial_1 H_0(x,y)]^2 + [g(x,y) - \partial_2 H_0(x,y)]^2 = 0$. If, however, $[f(x,y) - \partial_1 H_0(x,y)]^2 + [g(x,y) - \partial_2 H_0(x,y)]^2 > 0$ we define the positive number

$$\delta := \frac{\varepsilon}{\sqrt{[f(x,y) - \partial_1 H_0(x,y)]^2 + [g(x,y) - \partial_2 H_0(x,y)]^2}}.$$

Then we specify h_1 and h_2 as follows:

$$h_1 := \delta \cdot [f(x,y) - \partial_1 H_0(x,y)], \qquad h_2 := \delta \cdot [g(x,y) - \partial_2 H_0(x,y)].$$

The numbers h_1 and h_2 are elements of $[-\varepsilon, +\varepsilon]$ by the definition of δ. So, we may apply (5.104) to these special numbers h_1, h_2 and obtain the following inequality:

$$\delta \cdot [f(x,y) - \partial_1 H_0(x,y)]^2 + \delta \cdot [g(x,y) - \partial_2 H_0(x,y)]^2 \leq$$

$$\mu \cdot \delta \cdot \sqrt{[f(x,y) - \partial_1 H_0(x,y)]^2 + [g(x,y) - \partial_2 H_0(x,y)]^2} + o(\varepsilon).$$

We divide this inequality by $\delta \cdot \sqrt{[f(x,y) - \partial_1 H_0(x,y)]^2 + [g(x,y) - \partial_2 H_0(x,y)]^2} =$ $\varepsilon > 0$ and obtain assertion (5.105) also in the case that $[f(x,y) - \partial_1 H_0(x,y)]^2 + [g(x,y) - \partial_2 H_0(x,y)]^2 > 0$.

If ε tends to 0 then the summand $o(\varepsilon)/\varepsilon$ in (5.105) tends likewise to zero. Consequently, $\sqrt{[f(x,y) - \partial_1 H_0(x,y)]^2 + [g(x,y) - \partial_2 H_0(x,y)]^2} \leq \mu$ for all $(x,y) \in G$. This and the definitions of λ^* and $\Lambda(H_0)$ imply that $\lambda^* \leq \Lambda(H_0) \leq \mu$.   □

We now describe the relationships between the constructions in 5.56 and a prominent problem of Variational Calculus.

Remark 5.57 Given the situation of Theorem 5.56. Then the curves $K^{(x,y)}$ solve the following problem of Variational Calculus, which is called the isoperimetric problem.

    Given $a > 0$.

    Maximize $\int_K f \, d\sigma + g \, d\tau$ where K is a curve from $(0,0)$ to (x,y) of length $|K| = a$.

We can easily see that $K^{(x,y)}$ solves the isoperimetric problem. Let $a := |K^{(x,y)}|$. If K and K' are two curves of length a then $\int_K f \, d\sigma + g \, d\tau \geq \int_{K'} f \, d\sigma + g \, d\tau$ if and only if $\widehat{C}(K) \geq \widehat{C}(K')$. This and the maximality of $\widehat{C}\left(K^{(x,y)}\right)$ imply that $\int_{K^{(x,y)}} f \, d\sigma + g \, d\tau$ is maximal among all candidates $\int_K f \, d\sigma + g \, d\tau$ with $|K| = a = |K^{(x,y)}|$.

By the way, the isoperimetric problem is often reduced to the search for K and μ such that $\int_K f \, d\sigma + g \, d\tau - \mu \cdot |K|$ is maximal among all candidates K with $|K| = a$. That means that $\widehat{C}(K)$ must be maximized where μ is unknown and $|K|$ is given. μ is a Lagrange multiplier.   □

In the next example, we estimate the quantity λ^* from Theorem 5.55 in a concrete situation; also, we describe the curves $K^{(x,y)}$; the situation of the following example is illustrated in Figure 28.

Example 5.58 Let G be equal to the open disk $B(0,1)$ around $(0,0)$ with radius 1. Let λ^* be the same quantity as in Theorem 5.55. Let $f(\sigma, \tau) := \frac{1}{2}\tau$, $g(\sigma, \tau) := -\frac{1}{2}\sigma$ for all $(\sigma, \tau) \in G$. Moreover, let $C_1(Q) := \int_Q f(\sigma, \tau) \, d\sigma + g(\sigma, \tau) \, d\tau$ for all rectifiable curves Q. The integral $C_1(Q)$ has the following properties:

> If Q is a simple, closed curve then $-C_1(Q)$ or $C_1(Q)$ is the area enclosed by Q, depending on whether we are moving clockwise or counterclockwise along Q.   (5.106)
>
> Moreover, if Q is a simple curve from $(0,0)$ to (x,y) and Q lies on the left side of the directed line S from $(0,0)$ to (x,y) then $C_1(Q)$ is the area bounded by Q and S. (An example is the hatched area in Figure 28.)

Now we estimate the quantity q^* from Theorem 5.55. For this purpose, we define

$$\widetilde{Q} := \left\{ \frac{1}{|Q|} \cdot \left| \int_Q f(\sigma,\tau)\mathrm{d}\sigma + g(\sigma,\tau)\mathrm{d}\tau \right| \;\middle|\; \begin{array}{l} Q \subseteq G \text{ is a simple, closed,} \\ \text{rectifiable curve} \end{array} \right\}, \quad \widetilde{q}^* := \sup(\widetilde{Q}).$$

Then $q^* \geq \widetilde{q}^*$ because $\widetilde{Q}$ is a subset of the set Q in Theorem 5.55. We now show that $\widetilde{q}^* = 0.5$.

Let $Q \subseteq G$ be a simple, closed curve. Then (5.106) implies that $|C_1(Q)|$ is the area enclosed by Q. Consequently,

$$|C_1(Q)| \text{ is smaller than the area of the region } G \text{ itself.}^{18} \tag{5.107}$$

It is well-known that a cycle has the smallest circumference among all curves with the same enclosed area. Hence, there exists a cycle Q' with

$$(\alpha) \quad |C_1(Q')| = |C_1(Q)| \qquad \text{and} \qquad (\beta) \quad \frac{|C_1(Q')|}{|Q'|} \geq \frac{|C_1(Q)|}{|Q|}. \tag{5.108}$$

Fact (5.107) and fact (5.108)(α) imply that Q' can be chosen as a cycle lying in G. This and (5.108)(β) imply that the supremum $\widetilde{q}^*$ can be achieved by only considering circular lines Q'. The quotient $|C_1(Q')|/|Q'|$ is equal to the half of the radius of Q'. The radii of all cycles $Q' \subseteq G$ have the supremum 1; consequently, the supremum of all values $|C_1(Q')|/|Q'|$ is equal to 0.5. This implies that $\widetilde{q}^* = 0.5$.

We have seen that $q^* \geq \widetilde{q}^* \geq 0.5$. Theorem 5.55 says that $\lambda^* \geq q^* \geq 0.5$. That means: If approximating (f,g) by a pair $(\partial_1 H, \partial_2 H)$, we cannot achieve a better precision than 0.5.

Next, we describe the curves $K^{(x,y)}$ in Theorem 5.56. Let $\mu > 0$, and let $(x,y) \in G \backslash \{(0,0)\}$ be a point with $\sqrt{x^2+y^2} \leq \mu$; suppose that we are searching for the maximal value of $\widehat{C}(K) = \int_K \left(\frac{1}{2}\tau \cdot \mathrm{d}\sigma - \frac{1}{2}\sigma\,\mathrm{d}\tau \right) - \mu \cdot |K|$ where K is a curve from $(0,0)$ to (x,y); then $\widehat{C}(K)$ is maximal if K is the arc of a cycle with diameter μ.[19]
In particular, let $\mu = 1$; then $\sqrt{x^2+y^2} \leq \mu$ for all $(x,y) \in G$ so that we can reach $(x,y) \in G$ by an arc of a cycle with diameter 1. This situation is illustrated in Figure 28.

Theorem 5.56 implies that $\lambda^* \leq \mu = 1$. Moreover, $\lambda^* \geq 0.5$ as seen above. Consequently, $0.5 \leq \lambda^* \leq 1$. □

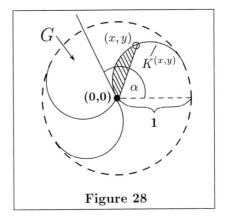

Figure 28

---

[18] This assertion may be false if Q is not simple; for example, if Q results from moving N-times along a closed curve K then $|C_1(Q)|$ may become arbitrarily great.

[19] We give reasons why K is a circular arc. Suppose that we want to maximize the area $C_1(K) = \int_K \left(\frac{1}{2}\tau \cdot \mathrm{d}\sigma - \frac{1}{2}\sigma\,\mathrm{d}\tau \right)$ where K is a curve from $(0,0)$ to (x,y) with a given length $|K| = a$. It is

Now we describe how the constructions in Example 5.58 can be modified.

Remark 5.59 Given the situation of Example 5.58.

a) Let $H_1(\sigma, \tau) := \frac{1}{2}\sigma \cdot \tau$ for all $(\sigma, \tau) \in G$. Then $\Lambda(H_1) = 1 = \Lambda(H_0)$. This follows from the fact that $f(\sigma, \tau) - \partial_1 H_1(\sigma, \tau) = 0$ and $g(\sigma, \tau) - \partial_2 H_1(\sigma, \tau) = -\sigma$ for all $(\sigma, \tau) \in G$.

The advantage of H_1 is that it has a simpler structure than H_0. The advantage of H_0 is that it has been generated with a general method.

b) Let $\mu < 1$. Then it is doubtful whether the curves $K^{(x,y)}$ exist. In particular, the construction of circular arcs from $(0,0)$ to (x, y) fails if the distance from $(0,0)$ to (x, y) is greater than μ. □

Next, we discuss the case that the number μ in 5.56 is chosen too small.

Remark 5.60 An immediate consequence of Theorems 5.55 and 5.56 is that $q^* \leq \lambda^* \leq \Lambda(H_0) \leq \mu$. Consequently, Theorem 5.56 cannot be proven for $\mu < q^*$.

If we try to show 5.56 for $\mu < q^*$, the following problem arises: The curves $K^{(x,y)}$ do not exist for all (x, y) because there exist "positive cycles" with respect to $\widehat{C}$. More precisely, if $\mu < q^*$ then a closed, rectifiable curve Q exists such that

$$\mu < \frac{1}{|Q|} \cdot \left| \int_Q f(\sigma, \tau)d\sigma + g(\sigma, \tau)d\tau \right|. \tag{5.109}$$

To see this we assume that $\int_Q f(\sigma, \tau)d\sigma + g(\sigma, \tau)d\tau \geq 0$. (Otherwise we move along Q in the opposite direction.) Therefore, $\mu \overset{(5.109)}{<} \frac{1}{|Q|} \cdot \int_Q f(\sigma, \tau)d\sigma + g(\sigma, \tau)d\tau$ so that $\widehat{C}(Q) = \int_Q f(\sigma, \tau)d\sigma + g(\sigma, \tau)d\tau - \mu \cdot |Q| > 0$. That means that $\widehat{C}(Q) > 0$ and that the curve Q is a "positive cycle".

Let $(x, y) \in Q$. Let $K \subseteq G$ be a curve from $(0,0)$ to (x, y). If $m \in \mathbb{N}$ we define K_m as the curve generated by first moving along the curve K and then moving m times along the closed curve Q. Then $\widehat{C}(K_m) = \widehat{C}(K) + m \cdot \widehat{C}(Q)$. This and the fact that $\widehat{C}(Q) > 0$ imply that there is no upper bound on $\{\widehat{C}(K_m) \,|\, m \in \mathbb{N}\}$ so that no maximal value $\widehat{C}\left(K^{(x,y)}\right)$ exists. □

Now we give reasons for the existence of the curves $K^{(x,y)}$ and the function H_0 in Theorem 5.56; moreover, we describe how to compute the curves $K^{(x,y)}$; more details can be found in [Huck92], page 342 – 343.

Remark 5.61 The idea is to cover the region G with curves L_α such that each $K^{(x,y)}$ is an initial part of a curve L_α. Each curve L_α starts at $(0,0)$, and the angle between the x-axis and the tangent to L_α at $(0,0)$ is equal to α. The simplest example is that L_α is equal to the ray $R_\alpha := \{(\zeta \cdot \cos \alpha, \zeta \cdot \sin \alpha) \,|\, 0 \leq \zeta < \infty\}$. Another example is the situation in 5.58 where the curves L_α should be chosen as the semicircles

---

well-known that a circular arc is the best candidate. Using μ as a Lagrange multiplyier, we transform the maximization of $C_1(K)$ into a maximization of $\widehat{C}(K)$; so, it is plausible that $\widehat{C}(K)$ is maximal if K is a circular arc.

shown in Figure 28.

We describe this idea more formally. Each line L_α is parametrized with two functions $U_\alpha, V_\alpha : [0, \infty) \to \mathbb{R}$, i.e., $L_\alpha = \{(U_\alpha(\xi), V_\alpha(\xi)) \mid \xi \in [0, \infty)\}$. We assume that for each $(x, y) \in G$, there exist ζ and α such that

$K^{(x,y)}$ is equal to the initial part $\{(U_\alpha(\xi), V_\alpha(\xi)) \mid \xi \in [0, \zeta]\}$ of L_α; in particular, $(0,0) = (U_\alpha(0), V_\alpha(0))$, and $(x, y) = (U_\alpha(\zeta), V_\alpha(\zeta))$. $\qquad$ (5.110)

The condition that α is the angle between the x-axis and the tangent to L_α at $(0,0)$ can be formulated as follows:

$$U'_\alpha(0) = \cos(\alpha), \qquad V'_\alpha(0) = \sin(\alpha). \qquad (5.111)$$

Next, we describe how the functions U_α, V_α can be found. Let $F(x, u_0, v_0, u_1, v_1) := f(u_0, v_0) \cdot u_1 + g(u_0, v_0) \cdot v_1 - \sqrt{u_1^2 + v_1^2}$ for all $(x, u_0, v_0, u_1, v_1) \in \mathbb{R}^5$ with $(u_0, v_0) \in G$. It follows from (5.110) and the maximality of $K^{(x,y)}$ that U_α and V_α solve the problem to maximize the integral

$$\int_0^\zeta \left(f(U_\alpha(\xi), V_\alpha(\xi)) \cdot U'_\alpha(\xi) + g(U_\alpha(\xi), V_\alpha(\xi)) \cdot V'_\alpha(\xi) - \mu \cdot \sqrt{U'_\alpha(\xi)^2 + V'_\alpha(\xi)^2} \right) d\xi =$$

$$\int_0^\zeta F(\xi, U_\alpha(\xi), V_\alpha(\xi), U'_\alpha(\xi), V'_\alpha(\xi)) d\xi.$$

The resulting Euler-Lagrange differential equations are given in (5.112); we have abbreviated $\partial_i F$ and $\partial_i \partial_j F$ by F_i and F_{ij}, respectively; moreover, the arguments of all functions have been omitted.

$$\begin{aligned} F_2 - F_{41} - F_{42} \cdot U'_\alpha - F_{43} \cdot V'_\alpha - F_{44} \cdot U''_\alpha - F_{45} \cdot V''_\alpha &= 0 \\ F_3 - F_{51} - F_{52} \cdot U'_\alpha - F_{53} \cdot V'_\alpha - F_{54} \cdot U''_\alpha - F_{55} \cdot V''_\alpha &= 0 \end{aligned} \qquad (5.112)$$

We may consider the equations (5.112) as linear equations with the unknown variables U''_α and V''_α. When solving this system of equations, we obtain two functions X and Y such that $U''_\alpha(\xi) = X(\xi, U_\alpha(\xi), U'_\alpha(\xi), V_\alpha(\xi), V'_\alpha(\xi))$ and $V''_\alpha = Y(\xi, U_\alpha(\xi), U'_\alpha(\xi), V_\alpha(\xi), V'_\alpha(\xi))$ for all $\xi \in [0, \infty)$. Defining $\widehat{U}_\alpha := U'_\alpha$ and $\widehat{V}_\alpha := V_\alpha$, we obtain the following initial value problem:

$$\begin{aligned} U'_\alpha(\xi) &= \widehat{U}_\alpha(\xi) & U_\alpha(0) &\overset{(5.110)}{=} 0 \\ \widehat{U}'_\alpha(\xi) &= X(\xi, U_\alpha(\xi), U'_\alpha(\xi), V_\alpha(\xi), V'_\alpha(\xi)) & \widehat{U}_\alpha(0) &\overset{(5.111)}{=} \cos \alpha \\ V'_\alpha(\xi) &= \widehat{V}_\alpha(\xi) & V_\alpha(0) &\overset{(5.110)}{=} 0 \\ \widehat{V}'_\alpha(\xi) &= Y(\xi, U_\alpha(\xi), U'_\alpha(\xi), V_\alpha(\xi), V'_\alpha(\xi)) & \widehat{V}_\alpha(0) &\overset{(5.111)}{=} \sin \alpha \end{aligned} \qquad (5.113)$$

Solving this initial value problem yields often the desired curves L_α and $K^{(x,y)}$.

Let H_0 be defined in the same way as in Theorem 5.56; we now give reasons why H_0 has the following property:

For all $(x, y) \in G$, the partial derivatives $\partial_1 H_0(x, y)$ and $\partial_2 H_0(x, y)$ exist and are continuous. $\qquad$ (5.114)

Let $U(\xi, \alpha) := U_\alpha(\xi)$ and $V(\xi, \alpha) := V_\alpha(\xi)$ for all ξ and α. We may often conclude that the solutions $U_\alpha(\xi) = U(\xi, \alpha)$ and $V_\alpha(\xi) = V(\xi, \alpha)$ of (5.113) are partially differentiable functions of the parameter α and that all functions $\frac{\partial U}{\partial \xi}$, $\frac{\partial U}{\partial \alpha}$, $\frac{\partial V}{\partial \xi}$, $\frac{\partial V}{\partial \alpha}$ are continuous.[20]

Suppose that the determinant of the matrix $\mathbf{D}(\zeta, \alpha) := \begin{pmatrix} \partial_1 U(\zeta, \alpha) & \partial_2 U(\zeta, \alpha) \\ \partial_1 V(\zeta, \alpha) & \partial_2 V(\zeta, \alpha) \end{pmatrix}$ is different from 0 if $\big(U(\zeta, \alpha), V(\zeta, \alpha)\big) \neq (0, 0)$.[21] Moreover, suppose that

$$\partial_1 H_0(x, y) \text{ and } \partial_2 H_0(x, y) \text{ exist and are continuous at } (x, y) = (0, 0). \qquad (5.115)$$

Let $(x, y) \in G$. If $(x, y) = (0, 0)$ then (5.114) follows from (5.115).

Let $(x, y) \in G \backslash \{(0, 0)\}$. We choose ζ and α such that (5.110) is true. Then

$$\big(U(\zeta, \alpha), V(\zeta, \alpha)\big) = \big(U_\alpha(\zeta), V_\alpha(\zeta)\big) = (x, y) \neq (0, 0) \qquad (5.116)$$

We have assumed that $det(\mathbf{D}(\zeta, \alpha)) \neq 0$. So, the pair (U, V) of functions can be locally inverted. More precisely, there exist a neighbourhood $N(x, y) \subseteq G$ and two functions $Z, A : N(x, y) \to \mathbb{R}$ with the following properties:

$$Z(x, y) = \zeta, \quad A(x, y) = \alpha,$$
$$(\forall (\tilde{x}, \tilde{y}) \in N(x, y)) \quad \big(U(Z(\tilde{x}, \tilde{y}), A(\tilde{x}, \tilde{y})) = \tilde{x} \text{ and } V(Z(\tilde{x}, \tilde{y}), A(\tilde{x}, \tilde{y},)) = \tilde{y}\big);$$

moreover, all derivatives $\partial_1 Z$, $\partial_2 Z$, $\partial_1 A$, $\partial_2 A$ exist and are continuous. Then we can write $H_0(\tilde{x}, \tilde{y})$ as follows:

$$H_0(\tilde{x}, \tilde{y}) = \int_{K^{(\tilde{x}, \tilde{y})}} f \, d\sigma + g \, d\tau - \mu \cdot \left| K^{(\tilde{x}, \tilde{y})} \right| \qquad \overset{(5.110)}{=}$$

$$\int_0^{Z(\tilde{x}, \tilde{y})} \left[\begin{array}{l} f\big(U(\xi, A(\tilde{x}, \tilde{y})), V(\xi, A(\tilde{x}, \tilde{y}))\big) \cdot \partial_1 U(\xi, A(\tilde{x}, \tilde{y})) + \\ g\big(U(\xi, A(\tilde{x}, \tilde{y})), V(\xi, A(\tilde{x}, \tilde{y}))\big) \cdot \partial_1 V(\xi, A(\tilde{x}, \tilde{y})) - \\ \mu \cdot \sqrt{[\partial_1 U(\xi, A(\tilde{x}, \tilde{y}))]^2 + [\partial_1 V(\xi, A(\tilde{x}, \tilde{y}))]^2} \end{array} \right] \cdot d\xi.$$

This equation and the continuity of $\partial_1 Z$, $\partial_2 Z$, $\partial_1 A$, $\partial_2 A$ imply that $H_0|_{N(x, y)}$ is partially differentiable with continuous derivatives. Consequently, $\partial_1 H_0$ and $\partial_2 H_0$ exist and are continuous at (x, y).

Hence, we have shown the partial differentiability of H_0 if the following conditions are satisfied: The curves $K^{(x, y)}$ are initial parts of particular lines L_α; the parameter functions U_α, V_α and the derivatives $\hat{U}_\alpha$, $\hat{V}_\alpha$ solve the initial value problem (5.113); the solutions $U_\alpha(\xi) = U(\xi, \alpha)$ and $V_\alpha(\xi) = V(\xi, \alpha)$ of (5.113) have the property that $\frac{\partial U}{\partial \xi}$, $\frac{\partial U}{\partial \alpha}$, $\frac{\partial V}{\partial \xi}$, $\frac{\partial V}{\partial \alpha}$ exist and are continuous; each determinant $det(\mathbf{D}(\zeta, \alpha))$ with $(\zeta, \alpha) \neq (0, 0)$ is unequal to 0; moreover, (5.115) is true. $\qquad \Box$

---

[20]The correctness of this conclusion can be proven with results saying that the solutions of an initial value problem are differentiable functions of the parameters appearing in this problem; such results are given in [KnKa74], page 129 – 131, and in many other books about differential equations.

[21]The determinant of $\mathbf{D}(\zeta, \alpha)$ may be equal to zero if $(U_\alpha(\zeta), V_\alpha(\zeta)) = (U(\zeta, \alpha), V(\zeta, \alpha)) = (0, 0)$ For example, the coordinate functions of the ray R_α can be choosen as $U(\zeta, \alpha) := \zeta \cdot \cos \alpha$ and $V(\zeta, \alpha) := \zeta \cdot \sin \alpha$. Consequently, $\mathbf{D}(\zeta, \alpha) = \begin{pmatrix} \cos \alpha & -\zeta \cdot \sin \alpha \\ \sin \alpha & \zeta \cdot \cos \alpha \end{pmatrix}$ for all $\zeta \geq 0$ and $\alpha \in \mathbb{R}$. If $\zeta = 0$ then the determinant of this matrix is equal to zero and $(U(\zeta, \alpha), V(\zeta, \alpha)) = (0, 0)$.

Next, we describe how the problem of approximating a pair (f, g) of functions can be normalized.

Remark 5.62 The problem of approximating a given pair (f, g) by an integrable pair $(\partial_1 H, \partial_2 H)$ can often be solved by approximating a pair $(\tilde{f}, \tilde{g})$ with $\tilde{g} \equiv 0$.

To see this we assume that G have the following property:[22]

The whole line segment $\overline{(x, 0), (x, y)}$ lies in G for each $(x, y) \in G$.   (5.117)

Then the integral $\partial_2^{-1} h(x, y) := \int_0^y h(x, \xi) \, d\xi$ is well-defined for all $(x, y) \in G$ and for all continuous functions $h : G \to \mathbb{R}$. It is clear that $\partial_2 \partial_2^{-1} h = h$ for all h.

Given $f, g : G \to \mathbb{R}$. Then we define $\tilde{f} := f - \partial_1 \partial_2^{-1} g$ and $\tilde{g} :\equiv 0$. Then the assertions (5.118) and (5.119) imply that approximating (f, g) is equivalent to approximating $(\tilde{f}, \tilde{g}) = (\tilde{f}, 0)$.

> For each function $H \in D_1(G)$, there exists an $\tilde{H} \in D_1(G)$ such that $\tilde{f} - \partial_1 \tilde{H} = f - \partial_1 H$ and $\tilde{g} - \partial_2 \tilde{H} = -\partial_2 \tilde{H} = g - \partial_2 H$.   (5.118)
> The function $\tilde{H} := H - \partial_2^{-1} g$ has the desired properties.

> For each function $\tilde{H} \in D_1(G)$, there exists an $H \in D_1(G)$ such that $f - \partial_1 H = \tilde{f} - \partial_1 \tilde{H}$ and $g - \partial_2 H = -\partial_2 \tilde{H} = \tilde{g} - \partial_2 \tilde{H}$.   (5.119)
> The function $H := \tilde{H} + \partial_2^{-1} g$ has the desired properties.

For example, let $f(x, y) := \frac{1}{2} y$ and $g(x, y) := -\frac{1}{2} x$ for all $(x, y) \in G$. Then $\partial_2^{-1} g(x, y) = -\frac{1}{2} xy$ and $\partial_1 \partial_2^{-1} g(x, y) = -\frac{1}{2} y$ so that $\tilde{f}(x, y) = y$.   □

Remark 5.63 Given the terminology of Theorem 5.56. Let $(x, y) \in G$.

The curve $K^{(x,y)}$ can often be parametrized with $u(\xi) := \xi$ and a function $v(\xi)$; then

$$K^{(x,y)} = \{(\xi, v(\xi)) \mid 0 \le \xi \le x\}.$$

The search for v is very easy if the following situation is given:

(α)   The function f only depends on y; that means that there exists a function φ with $f(x, y) = \varphi(y)$ for all $(x, y) \in G$.

(β) $g(x, y) = 0$ for all $(x, y) \in G$.

(If f and g do not have these properties, replace (f, g) by the functions $(\tilde{f}, \tilde{g})$ in 5.62; then (β) is automatically true, and it may be that $\tilde{f}$ has property (α).)

In this case, the search for $K^{(x,y)}$ is equivalent to the search for a function v such that

$$\int_0^x \left[f(\xi, v(\xi)) - \mu \cdot \sqrt{1 + (v'(\xi))^2} \right] d\xi = \int_0^x \left[\varphi(v(\xi)) - \mu \cdot \sqrt{1 + (v'(\xi))^2} \right] d\xi$$

is maximal. The Euler-Lagrange differential equation of this problem is

---

[22]Property (5.117) has nothing to do with convexity. For example, if G is the open kernel of the triangle with the corners $(-1, 1)$, $(0, -1)$, $(1, 1)$ then G is convex but G does not have property (5.117); if, however, $G = \{(x, y) \mid -1 < x < 1, \ -1 < y < x^2 + 1\}$ then G has property (5.117) but G is not convex.

$$\varphi'(v(\xi)) \;=\; \frac{\mu\,v''(\xi)}{\left(\sqrt{1+(v'(\xi))^2}\,\right)^3}\,.$$

We multiply this equation with v' and obtain: $\varphi'(v(\xi))\cdot v'(\xi) \;=\; \dfrac{\mu\,v''(\xi)\cdot v'(\xi)}{\left(\sqrt{1+(v'(\xi))^2}\,\right)^3}\,.$

Integrating both sides, we obtain $\varphi(v(\xi)) \;=\; \dfrac{-\mu}{\sqrt{1+(v'(\xi))^2}} + c$ where $c \in \mathbb{R}$ is a constant. Consequently,

$$v'(\xi) \;=\; \pm\sqrt{\left[\frac{\mu}{c-\varphi(v(\xi))}\right]^2 - 1}\,.$$

This differential equation, however, is of the form $v'(\xi) = \psi(v(\xi))$ where ψ is a real function; this type of differential equations can be easily solved.   □

Bibliography

[AChL93] M. ATALLAH, D. Z. CHEN, D. T. LEE, *An optimal algorithm for shortest paths on weighted interval and circular-arc graphs, with applications.* In: T. Lengauer (ed.), Algorithms (ESA '93) (Bad Honnef, 1993). Lecture Notes in Comput. Sci., Vol. 726, Springer, Berlin, Heidelberg, New York, 1993, 13 – 24, MR 95c:68090.

[AgTo93] A. AGGARWAL, T. TOKUYAMA, *An improved algorithm for the traveler's problem.* In: Algorithms and computation, (Hong Kong, 1993). Lecture Notes in Comput. Sci., Vol. 762, Springer, Berlin, Heidelberg, New York, 1993, 476 – 485, MR 95g:90023.

[AH92] F. AURENHAMMER, J. HAGAUER, *Computing equivalence classes among the edges of a graph with applications.* Discrete Math. **109** (1992), 3 – 12, MR 93k:05167.

[Aign88] M. AIGNER, *Combinatorial Search.* Teubner, Stuttgart, Wiley, Chichester, New York, Brisbane, Toronto, Singapore, 1988.

[AjKS81] M. AJTAI, J. KOMLÓS, E. SZEMERÉDY, *The longest path in a random graph.* Combinatorica **1** (1981), no. 1, 1 – 12.

[Akh62] N. I. AKHIEZER, *The Valculus of Variations.* Blaisdell Publ. Comp., New York, London, 1962.

[Akm86] V. AKMAN, *Unobstructed shortest paths in polyhedral environments.* Lecture Notes in Comput. Sci., Vol. 251 (1986).

[AlHM87] B. ALSPACH, K. HEINRICH, B. MOHAR, *A note on Hamilton cycles in block-intersection graphs.* In: Finite geometries and combinatorial designs (Lincoln, NE, 1987). Contemp. Math., Vol. 111, Amer. Math. Soc., Providence, RI, 1990, 1 – 4, MR 91h:05079.

[Alon86] N. ALON, *The longest cycle of a graph with a large minimal degree.* J. Graph Theory **10** (1986), 123 – 127.

[AlPo89] P. ALLES, SV. POLJAK, *Long induced paths and cycles in Kneser graphs.* Graphs Combin. **5** (1989), no. 4, 303 – 306, MR 90m:05076.

[Alsp80] B. ALSPACH, *The search for long paths and cycles in vertex-transitive graphs and digraphs.* Combinatorial Mathematics, VIII (Geelong, 1980). Lecture Notes in Math., Vol. 884, 1981, 14 – 22, MR 83b:05080.

[Alth90] I. ALTHÖFER, *Average distances in undirected graphs and the removal of vertices.* J. Combin. Theory Ser. B **48** (1990), no. 1, 140 – 142, MR 91c:05071.

[AltS92] I. ALTHÖFER, T. SILLKE, *An "average distance" inequality for large subsets of the cube.* J. Combin. Theory Ser. B **56** (1992), no. 2, 296 – 301, MR 93i:05057.

[Amm90] H. AMANN, *Ordinary Differential Equations. An Introduction to Nonlinear Analysis.* Walter de Gruyter, Berlin, New York, 1990.

[AmMa90] D. AMAR, Y. MANOUSSAKIS, *Cycles and paths of many lengths in bipartite digraphs.* J. Combin. Theory Ser. B **50** (1990), no. 2, 254 – 264, MR 92c:05087.

[AMOT90] R. K. AHUJA, K. MEHLHORN, J. B. ORLIN, R. E. TARJAN, *Faster algorithms for the shortest path problem.* J. Assoc. Comput. Mach. **37** (1990), 219 – 223.

[AnAN83]   Y. P. ANEJA, V. AGGARWAL, K. P. K. NAIR, *Shortest chain subject to side constraints.*
            Networks **13** (1983), 295 – 302.

[AnB93]   F. ANNEXSTEIN, M. BAUMSCHLAG, *On the diameter and bisector size of Cayley graphs.*
            Math. Systems Theory **26** (1993), no. 3, 271 – 291, MR 94c:05036.

[And84]   R. ANDERSON, *A parallel algorithm for the maximal path problem.* Combinatorica **7**
            (1984), no. 4, 315 – 326.

[AndM87]   R. ANDERSON, E. MAYR *Parallelism and the maximal path problem.* Inform. Process.
            Lett. **24** (1987), 121 – 126.

[AnMa87]   R. ANDERSON, E. W. MAYR, *Parallelism and the maximal path problem.* Inform. Process.
            Lett. **24** (1987), 121 – 126.

[AnNa78]   Y. P. ANEJA, K. P. K. NAIR, *The constraint shortest path problem.* Naval Res. Logist.
            Quart. **25** (1978), 549 – 555.

[Arl91]   W. C. ARLINGHAUS, *Shortest path problems.* In: Applications of discrete mathematics,
            McGraw-Hill, New York, 1991, 322 – 331, MR 94j:90049.

[Ausi91]   G. AUSIELLO, G. F. ITALIANO, A. MARCHETTI-SPACCAMELA, U. NANNI, *Incremental al-
            gorithms for minimal length paths.* J. Algorithms **12** (1991), no. 4, 615 – 638,
            MR 92g:68025.

[Ayel81]   J. AYEL, *Degrees and longest paths in bipartite digraphs.* In: Combinatorial mathematics
            (Marseille-Luminy, 1981). North-Holland Math. Stud., Vol. 75, North-Holland, Amster-
            dam, New York, 1983, 33 – 38, MR 87h:05097.

[Ayel83]   J. AYEL, *Degrees and longest paths in bipartite digraphs.* Ann. Discrete Math. **17** (1983),
            33 - 38.

[Bab79]   L. BABAI, *Long cycles in vertex transitive graphs.* J. Graph Theory **3** (1979), no. 3,
            301 – 304.

[BagS88]   A. BAGCHI, A. K. SEN, *Average-case analysis of heuristic search in tree-like networks.*
            In: L. Kanal, V. Kumar (eds.), Search in Artificial Intelligence, Springer, Berlin, Heidel-
            berg, New York, 1988, 131 – 165.

[Bal87]   Z. BALKHI, *The generalized linear search problem, existence of optimal search paths.*
            J. Oper. Res. Soc. Japan **30** (1987), no. 4, 399 – 421.

[Bal90]   Z. BALKHI, *On the minimality of search paths under absolutely continuous univariate
            distributions.* Z. Oper. Res. **34** (1990), no. 1, 43 – 58, MR 91c:90062.

[Bal93]   Z. BALKHI, *More on the minimality of search paths under continuous univariate distribu-
            tions.* Belg. J. Oper. Res. Statist. Comput. Sci. **33** (1993), no. 3, 3 – 12, MR 94m:90060.

[BaMa83]   A. BAGCHI, A. MAHANTI, *Search algorithms under different kinds of heuristics – a com-
            parative study.* J. Assoc. Comput. Mach. **30** (1983), no. 1, 1 – 21.

[BaMa85]   A. BAGCHI, A. MAHANTI, *Three approaches to heuristic search in networks.* J. Assoc.
            Comput. Mach. **32** (1985), no. 1, 1 – 27.

[BaN90]   A. BAR-NOY, J. NAOR, *Sorting, minimal feedback sets, and Hamilton paths in tourna-
            ments.* SIAM J. Discrete Math. **3** (1990), no. 1, 7 – 20, MR 91d:68007.

[Bard91]   V. A. BADADYM, *Minimax paths and minimal spanning trees.* Kibernetika (Kiev) **1990**,
            no. 2, 122, MR 91k:90058.

[Bau74]   H. BAUER, *Wahrscheinlichkeitstheorie und Grundzüge der Maßtheorie.* 2nd edition, Wal-
            ter de Gruyter, Berlin, New York, 1974.

[BB94]   D. BAUER, H. J. BROERSMA, J. VAN DEN HEUVEL, H. J. VELDMAN, *On hamiltonian prop-
            erties of 2-tough graphs.* J. Graph Theory **18** (1994), no. 6, 543 – 563.

[BBVL89]   D. BAUER, H. J. BROERSMA, H. J. VELDMAN, R. LI, *A generalization of a result of
            Häggkvist and Nicoghossian.* J. Combin. Theory Ser. B **47** (1989), no. 2, 237 – 243,
            MR 91a:05066.

[BCDJ93]   N. BRAND, ST. CURRAN, S. DAS, T. JACOB, *Probability of diameter two for Steinhaus graphs.* Discrete Appl. Math. **41** (1993), no. 2, 165 – 171, MR 93m:05130.

[BCS93]   P. BEDROSSIAN, G. CHEN, R. H. SCHELP, *A generalization on Fan's condition for Hamiltonicity, pancyclicity, and Hamiltonian connectedness.* Discrete Math. **115** (1993), 39 – 50, MR 94e:05173.

[BeaP90]   L. B. BEASLEY, N. J. PULLMAN, *Linear operators strongly preserving digraphs whose maximum cycle length is small.* Linear and Multilinear Algebra **28** (1990), no. 1 – 2, 111 – 117, MR 91m:05090.

[BeBo81]   J. C. BERMOND, B. BOLLOBÁS, *The diameter of graphs — a survey.* In: Proceedings of the twelfth Southeastern conference on combinatorics, graph theory, and computing, Vol. I (Baton Rouge, La., 1981). Congr. Numer. **32** (1981), 3 – 27, MR 84e:05061.

[BeDQ82]   J. C. BERMOND, C. DELORME, J. J. QUISQUATER, *Tables of large graphs with given degree and diameter.* Inform. Process. Lett. **15** (1982), no. 1, 10 – 13, MR 84a:05038.

[BeDr62]   R. E. BELLMAN, S. E. DREYFUS, *Applied Dynamic Programming.* Princeton University Press, Princeton, New Jersey, 1962.

[Behz73]   M. BEHZAD, *Minimal 2-regular digraphs with given girth.* J. Math. Soc. Japan **25** (1973), 1 – 6 , MR **46**#8889.

[BeKa65]   R. BELLMAN, R. KALABA, *Dynamic programming and modern control theory.* Academic Press, New York, London, 1965.

[Bell58]   R. BELLMAN, *On a routing problem.* Quart. Appl. Math. **16** (1958), 87 – 90.

[Bell67]   R. BELLMAN, *Dynamische Programmierung und selbstanpassende Regelprozesse.* Oldenbourg, München, Wien, 1967.

[Benh84]   A. BENHOCINE, *On the existence of a specified cycle in digraphs with constraints on degrees.* J. Graph Theory **8** (1984), 101 – 107.

[Berl79]   H. BERLINER, *The B* tree search algorithm: a best-first proof procedure.* Artificial Intelligence **12** (1979), 23 – 40.

[Bern84]   R. BERNSTEIN, *Shortest paths in undirected graph with negative edges.* In: Proceedings of the fifteenth Southeastern conference on combinatorics, graph theory, and computing (Boca Rouge, La., 1984). Congr. Numer. **43** (1984), 117 – 126, MR 86j:05085.

[BeTh81]   J. C. BERMOND, C. THOMASSEN, *Cycles in digraphs – a survey.* J. Graph Theory **5** (1981), 1 – 43.

[Bey93]   O. BEYERSDORFF, *Kürzeste Wege in Graphen.* "Jugend forscht" (a German contest for young researchers), 1993.

[BGHS81]   J. C. BERMOND, A. GERMA, M. C. HEYDEMANN, D. SOTTEAU, *Longest paths in digraphs.* Combinatorica **1** (1981), no. 4, 337 – 341, MR 83d:05048.

[BHBA91]   R. BEIVIDE, E. HERRADA, J. L. BACÁZAR, A. ARRUABARRENA, *Optimal distance networks of low degree for parallel computers.* IEEE Trans. Comput. **40** (1991), no. 10, 1109 – 1124, MR 92i:68013.

[BHHS94]   E. BOROS, P. L. HAMMER, M. E. HARTMAN, R. SHAMIR, *Balancing problems in acyclic networks.* Discrete Appl. Math. **49** (1994), 77 – 93, MR 95j:68077.

[BHSh90]   E. BOROS, P. L. HAMMER, R. SHAMIR, *Balancing rooted data flow graphs.* RUTCOR Research Report # 28-90, 1990.

[Bien92]   D. BIENSTOCK, *On the complexity of testing for odd holes and induced odd paths.* Discrete Math. **90** (1992), no. 1, 85 – 92, MR 92m:68040a.
            Corrigendum: "On the complexity of testing for odd holes and induced odd paths". Discrete Math. **102** (1992), no. 1, 109, MR 92m:68040b.

[BiMo90]   D. BIENSTOCK, CL. L. MONMA, *On the complexity of embedding planar graphs to minimize certain distance measures.* Algorithmica **5** (1990), no. 1, 93 – 109, MR 91a:68119.

[BJMa94]   J. BANG-JENSEN, Y. MANOUSSAKIS, *Cycles through k vertices in bipartite tournaments.* Combinatorica **14** (1994), no. 2, 243 – 246, MR 95f:05068.

[BJS89]   D. BAUER, H. A. JUNG, E. F. SCHMEICHEL, *On 2-connected graphs with small circumference.* In: Graphs, designs and combinatorial geometries (Catania, 1989). J. Combin. Inform. System Sci. **15** (1990), no. 1 – 4, 20 – 28, MR 92g:05117.

[BLC91]   D. BAUER, L. L. LASSER, G. CHEN, *A degree condition for Hamiltonian cycles in t-tough graphs with t > 1.* In: Advances in graph theory. Vishwa, Gulbarga, 1991, 19 – 32, MR 94a:05132.

[BlG93]   A. BLUM, R. J. GOULD, *Generalized degree sums and Hamiltonian graphs.* Ars Combinatoria **35–A** (1993), 35 – 54.

[BlKQ81]   G. S. BLOOM, J. W. KENNEDY, L. V. QUINTAS *Some problems concerning distance and path degree sequences.* In: Graph theory (1981). Lecture Notes in Math., Vol. 1018, Springer, Berlin, New York, 1983, 179 – 190, MR 85j:05029.

[Blon83]   P. A. BLONIARZ, *A shortest-path algorithm with expected time $O(n^2 \log n \log^* n)$.* SIAM J. Comput. **12** (1983), no. 3, 588 – 600, MR 84i:68060.

[Blum86]   M. BLUM, *How to prove a theorem so no one else can claim it.* In: Proceedings of the International Congress of Mathematicians, Vol. 1, 2 (Berkeley, Calif. 1986). Amer. Math. Soc. Providence, RI, 1987, 1444 – 1451, MR 91h:68141.

[BoEx82]   ST. M. BOYLES, G. EXOO, *A counterexample to a conjecture on paths of bounded length.* J. Graph Theory **6** (1982), no. 2, 205 – 209, MR 84a:05040.

[BoFa85]   J. A. BONDY, G. FAN, *Optimal paths and cycles in weighted graphs.* Graph Theory in Memory of G. A. Dirac (Sandbjerg, 1985). Ann. Discrete Math. **41**, North–Holland, Amsterdam, New York, 1989, 53 – 69, MR 90c:05123.

[BoFF90]   B. BOLLOBÁS, T. I. FENNER, A. M. FRIEZE, *Hamilton cycles in random graphs of minimal degree at least k.* In: A tribute to Paul Erdős. Cambridge Univ. Press, Cambridge, 1990, 59 – 95.

[BoHa81]   B. BOLLOBÁS, F. HARARY, *The trail number of a graph.* In: Graph theory (Cambridge, 1981). North-Holland Math. Stud., Vol. 62, North-Holland, Amsterdam, New York, 1982, 51 – 60, MR 83k:05066.

[BoHä90]   B. BOLLABÁS, R. HÄGGKVIST, *Powers of Hamilton cycles in tournaments.* J. Combin. Theory Ser. B **50** (1990), no. 2, 309 – 318, MR 91j:05050.

[BoLi81]   K. S. BOOTH, R. J. LIPTON, *Computing extremal and approximate distances in graphs having unit cost edges.* Acta Inform. **15** (1981), 319 – 328.

[BoLo85]   J. A. BONDY, L. LOVÁSZ, *Lengths of cycles in Halin graphs.* J. Graph Theory **9** (1985), no. 3, 397 – 410, MR 87a:05092.

[Bon80]   J. A. BONDY, *Longest paths and cycles in graphs of high degree.* Research Report CORR 80–16, University of Waterloo, Waterloo, Ontario, 1980.

[Bou93]   A. BOUCHET, *Compatible Euler tours and supplementary Eulerian vectors.* European J. Combin. **14** (1993), no. 6, 513 – 520, MR 94k:05127.

[BPu91]   S. C. BOYD, W. R. PULLEYBLANK, *Optimizing over the subtour polytope of the travelling salesman problem.* Math. Programming **49** (1990/91), no. 2 (Ser. A), 153 – 187, MR 92b:90147.

[Br90]   U. BRAUN, *Graphentheoretische Modellierung und Lösungsverfahren bei der rechnergestützten Wegeplanung in der Prozeßautomatisierung.* Diploma thesis, Würzburg University, Würzburg, 1990.

[Bra91]   A. BRANDSTÄDT, *Short disjoint cycles in cubic bridgeless graphs.* In: G. Schmidt, R. Berghammer (eds.), Proceedings of the 17th International Workshop on Graph-Theoretic Concepts in Computer Science (WG '91) (Fischbachau, Bayern, 1991). Lecture Notes in Comput. Sci., Vol. 570, Springer, Berlin, Heidelberg, New York, 1992, 239 – 249.

[Bre83]   U. BRECHTKEN-MANDERSCHEID, *Einführung in die Variationsrechnung*. Wissenschaft-
            liche Buchgesellschaft Darmstadt, Darmstadt, 1983.

[Broe93]   H. J. BROERSMA, J. VAN DEN HEUVEL, H. A. JUNG, H. J. VELDMAN, *Long paths and cycles
            in tough graphs*. Graphs Combin. **9** (1993), no. 1, 3 – 17, MR 94e:05157

[Bru80]   P. BRUCKER *The Chinese postman problem for mixed graphs*. In: Proceedings of the 6th
            Workshop on Graph-Theoretic Concepts in Computer Science (WG '80) (Bad Honnef,
            1980). Lecture Notes in Comput. Sci., Vol. 100, 354 – 367.

[BSV88]   D. BAUER, E. SCHMEICHEL, H. J. VELDMAN, *Some recent results on long cycles in tough
            graphs*. In: Y. Alavi, G. Chartrand, O. R. Oellermann, A. Schwenk (eds.), Graph Theory,
            Combinatorics and Applications; Proceedings of the 6th Quadrennial International Con-
            ference on the Theory and Applications of Graphs (Western Michigan University, 1988).
            Wiley, New York, 1991, 113 – 123, MR 93b:05002.

[BuSu81]   F. BUCKLEY, L. SUPERVILLE, *Distance distributions and mean distance problems*. In:
            Proceedings of the 3rd Caribbean Conference on Combinatorics and Computing
            (Bridgetown, 1981). Univ. West Indies, Cave Hill Campus, Bridgetown, Barbados, 1981,
            67 – 76, MR 83h:05054.

[BuT94]   D. BURTON, PH. L. TOINT, *On the use of an inverse shortest paths algorithm for recov-
            ering linearly correlated costs*. Math. Programming **63** (1994), no. 1, Ser. A, 1 – 22,
            94k:90127.

[BvdH90]   H. J. BROERSMA, J. VAN DEN HEUVEL, H. J. VELDMAN, *Long cycles, degree sums and
            neighborhood unions*. Graph Theory (Niedzica Castle, 1990). Discrete Math. **121** (1993),
            25 – 35, MR 94f:05089.

[BvdH93]   H. J. BROERSMA, J. VAN DEN HEUVEL, H. J. VELDMAN, *A generalization of Ore's theorem
            involving neighborhood unions*. Discrete Math. **122** (1993), 37 – 49, MR 94i:05056.

[BVMS90]   D. BAUER, H. J. VELDMAN, A. MORGANA, E. F. SCHMEICHEL, *Long cycles in graphs with
            large degree sums*. Discrete Math. **79** (1989/90), no. 1, 59 – 70.

[CaJo95]   S. CARLSSON, H. JONSSON, *Computing a shortest watchman path in a simple polygon in
            polynomial-time*. In: S G. Akl, F. Dehne, J. R. Sack, N. Santoro (eds.), Proceedings of the
            4th International Workshop on Algorithms and Data Structures (WADS '95) (Kingston,
            Cananda, 1995). Lecture Notes in Comput. Sci., Vol. 955, Springer, Berlin, Heidelberg,
            New York, 122 – 134.

[CamW94]   H. CAMERON, D. WOOD, *Maximal path length of binary trees*. Discrete Appl. Math. **55**
            (1994), no. 1, 15 – 35, MR 95h:05046.

[CaVi91]   L. CACCETTA, K. VIJAYAN, *Long cycles in subgraphs with prescribed minimum degree*.
            Discrete Math. **97** (1991), 69 – 81, MR 93g:05078.

[CGPS89]   P. P. CHAKRABARTI, S. GHOSE, A. PANDEY, S. C. DE SARKAR, *Increasing search effi-
            ciency using multiple heuristics*. Inform. Process. Lett. **30** (1989), 33 – 36.

[Chak94]   P. P. CHAKRABARTI, *Algorithms for searching explicit AND/OR graphs and their appli-
            cations to problem reduction search*. Artificial Intelligence **65** (1994), no. 2, 329 – 345,
            MR 94m:68150.

[Chau87]   P. CHAUDHURI, *Algorithms for some graph problems on a distributed computational
            model*. Information Sciences **43** (1987), 205 – 228.

[ChCh90]   Y. L. CHEN, Y. H. CHIN, *The quickest path problem*. Comput. Oper. Res. **17** (1990), 153
            – 161.

[Ched93]   F. B. CHEDID, R. B. CHEDID, *A new variation on hypercubes with smaller diameter*. In-
            form. Process. Lett. **46** (1993), no. 6, 275 – 280, MR 94f:68041.

[Chen91]   CHEN, *One sufficient condition for Hamiltonian graphs*. J. Graph Theory **14** (1990),
            no. 4, 501 – 508, MR 91h:05080.

[Chen93]   G. CHEN, *Hamiltonian graphs involving neighborhood intersections*. Discrete Math. **112**
            (1993), 253 – 257, MR 94a:05133.

[Chen94]   Y. L. CHEN, *Finding the k quickest simple paths in a network.* Inform. Process. Lett. **50** (1994), no. 2, 89 – 92, MR 95b:05188.

[ChGh89]   P. P. CHAKRABARTI, S. GHOSE, A. PANDEY, S. C. DE SARKAR, *Increasing search efficiency using multiple heuristics.* Inform. Process. Lett. **30** (1989), 33 – 36.

[ChGh92]   P. P. CHAKRABARTI, S. GHOSE, *A general best first search algorithm in AND/OR graphs.* J. Algorithms **13** (1992), no. 2, 177 – 187, MR 93i:68084.

[ChHM93]   D. CHILLAG, M. HERZOG, A. MANN, *On the diameter of a graph related to conjugacy classes of groups.* Bull. London Math. Soc. **25** (1993), no. 3, 255 – 262, MR 94a:20038.

[ChHu92]   G. H. CHEN, Y. C. HUNG, *On the quickest path problem.* Inform. Process. Lett. **46** (1992), 125 – 128, MR 94b90023.

[ChiH92]   K. B. CHILAKAMARRI, P. HAMBURGER, *A note on packing paths into complete bipartite graphs.* Bull. Inst. Combin. Appl. **4** (1992), 32 – 34, MR 93h:05123.

[ChJa93]   G. CHARTRAND, E. JARRET, *Transformations, distance, and distance graphs.* In: Proceedings of the 22nd Manitoba Conference on Numerical Mathematics and Computing (Winnipeg, MB, 1992). Congr. Numer. **92** (1993), 225 – 241, MR 95b:05066.

[ChJT93]   G. CHARTRAND, G. L. JOHNS, S. T. TIAN, *Detour distance in graphs.* In: Quo vadis, graph theory? Ann. Discrete Math. **55** (1993), 127 – 136, MR 94a:05066.

[ChMi82]   K. M. CHANDY, J. MISRA, *Distributed computation on graphs: shortest path algorithms.* Communications of the ACM **25** (1982), no. 11, 833 – 837.

[ChNt91]   W. P. CHIN, S. NTAFOS, *Shortest watchman routes in simple polygons.* Discrete Comput. Geom. **6** (1991), no. 1, 9 – 31, MR 91g:68159.

[ChOe89]   G. CHARTRAND, O. R. OELLERMAN, M. SCHULTZ, *Distance: a graphical tour.* In: Proceedings of graph theory, combinatorics, algorithms, and applications (San Francisco, CA, 1989). SIAM, Philadelphia, PA, 1991, 441 – 458, MR 92i:05088.

[ChOeTZ]   G. CHARTRAND, O. R. OELLERMANN, S. L. TIAN, H. B. ZOU, *Steiner distance in graphs.* Časopis Pěst. Mat. **114** (1989), no. 4, 399 – 410, MR 91a:05040.

[ChPS86]   N. S. CHAUDHARI, D. B. PHATAK, N. L. SARDA *On characterizing a clique by paths in a digraph.* J. Combin. Inform. System Sci. **11** (1986), no. 2–4, 129 – 137, MR 89j:05039.

[ChRa91]   Y. M. CHEN, PR. RAMANAN, *Euclidean shortest paths in the presence of obstacles.* Networks **21** (1991), no. 3, 257 – 265, MR 92e:68191.

[Chre84]   P. CHRÉTIENNE, *Chemins extrémeaux d'un graphe doublement valué.* RAIRO Rech. Opér. **18** (1984), no. 3, 221 – 245, MR 86g:05055.

[Chri75]   N. CHRISTOFIDES, *Graph theory. An algorithmic approach.* Academic Press, New York, London, San Francisco, 1975.

[ChS94]   G. CHEN, R. H. SCHELP, *Hamilton graphs with neighborhood intersections.* J. Graph Theory **18** (1994), no. 5, 497 – 513.

[ChTi90]   G. CHARTRAND, S. L. TIAN, *Maximum distance in digraphs.* Graph theory, combinatorics, algorithms, and applications (San Francisco, CA, 1989), 525 – 538, SIAM, Philadelphia, PA, 1991.

[Chung92]   F. R. K. CHUNG, *Graphs with small diameter after edge deletion.* Discrete Appl. Math. **37/38** (1992), 73 – 94, MR 93k:05084.

[Chva73]   V. CHVÁTAL, *Tough graphs and hamiltonian circuits.* Discrete Math. **5** (1973), 215 – 228.

[ChZa95]   S. CHAUDHURI, CH. D. ZAROLIAGIS, *Shortest path queries in digraphs of small treewidth.* In: S G. Akl, F. Dehne, J. R. Sack, N. Santoro (eds.), Proceedings of the 4th International Workshop on Algorithms and Data Structures (WADS '95) (Kingston, Cananda, 1995). Lecture Notes in Comput. Sci., Vol. 955, Springer, Berlin, Heidelberg, New York, 244 – 255.

[Cie91]   D. CIESLIK, *The 1-Steiner-minimal-tree problem in Minkowski-Spaces*. Optimization **22** (1991), no. 2, 291 – 296.

[Cie95]   D. CIESLIK, *Graph-theoretical methods to approximate Steiner minimal trees in Banach-Minkowski planes*. In: Proceedings of ISORA '95 (Beijing, 1995). Lecture Notes in Operations Research, Vol. 1, Springer, Berlin, Heidelberg, New York, 213 – 220.

[CIJS90]   P. CATLIN, IQBALUNNISA, T. N. JANAKIRAMAN, N. SRINIVASAN, *Hamilton cycles and closed trails in iterated line graphs*. J. Graph Theory **14** (1990), no. 3, 347 – 364, MR 91f:05077.

[CJT92]   G. CHARTRAND, G. L. JOHNS, S. L. TIAN, *Directed distance in digraphs: centers and peripheries*. In: Proceedings of the 23rd Southeastern International Conference on Combinatorics, Graph Theory, and Computing (Boca Raton, FL, 1992). Congr. Numer. **89** (1992), 89 – 95, MR 93j:05051

[CoGi92]   E. G. COFFMAN JR., E. N. GILBERT, *Paths through a maze of rectangles*. Networks **22** (1992), 349 – 367.

[Coh94]   E. COHEN, *Polylog time and near-linear work approximation scheme for undirected shortest paths*. In: Proceedings of the 26th ACM Symposium on Theory of Computing (STOC '94) (Montreal, Quebec, Canada, 1994), 16 – 26.

[CoKu93]   G. A. COREA, V. G. KULKARNI, *Shortest paths in stochastic networks with arc lengths having discrete distributions*. Networks **23** (1993), 175 – 183, MR 94b:90027.

[CoMi80]   E. J. COCKAYNE, D. J. MILLER, *Optimum addition of an edge to a path*. IEEE Trans. Circuits and Systems **27** (1980), no. 7, 649 – 651, MR 81f:68074.

[CoWi87]   D. COPPERSMITH, S. WINOGRAD, *Matrix multiplication via arithmetic progressions*. In: Proceedings of the 19th ACM Symposium on Theory of Computing (STOC '87), (New York, 1987), 1 – 6.

[CoWi90]   D. COPPERSMITH, S. WINOGRAD, *Matrix multiplication via arithmetic progressions*. J. Symb. Comput. **9** (1990), no. 3, 251 – 280, MR 91i:68058.

[Cri91]   D. CRIPPA, *A special case of the dynamization problem for least cost paths*. Inf. Process. Lett. **39** (1991), no. 6, 297 – 302, MR 92g:68112.

[CSK90]   M. CHROBAK, T. SZYMACHA, A. KRAWCZYK, *A data structure useful for finding Hamiltonian cycles*. Theoret. Comput. Sci. **71** (1990), no. 3, 419 – 424, MR 91e:68026.

[CuRu89]   J. CULBERSON, P. RUDNICKI, *A fast algorithm for construction trees from distance matrices*. Inform. Process. Lett. **30** (1989), 215 – 220.

[DaAJ91]   S. DAR, R. AGRAWAL, H. V. JAGADISH, *Optimization of generalized transitive closure queries*. In: Proceedings of the Seventh International Conference on Data Engineering (Kobe, Japan, 1991). IEEE Computer Society Press, Los Alamitos, California, Washington, Brussels, Tokyo, 345 – 354.

[Dan60]   G. B. DANTZIG, *On the shortest route through a network*. Management Sci. **6** (1960), 187 – 190.

[Dang93]   K. Q. DANG, *The circumferences of graphs*. J. Northeast Univ. Tech. **14** (1993), no. 1, 84 – 87, MR 94f:05091.

[Darb82]   S. KH. DARBINYAN, *Estimation of lengths of cycles and paths in regular digraphs*. Tanulmányok—MTA Számítástech. Automat. Kutató Int. Budapest no. 135 (1982), 131 – 144, MR 84b:05053.

[Das81]   S. DAS, *On the shortest route through sets of specified nodes*. Journ. Math. Phy. Sci. **15**, no. 5 (1981), 423 – 434, MR 83b:90049.

[Dav87]   E. DAVIS, *Constraint propagation with interval graphs*. Artificial Intelligence **32** (1987), no. 3, 282 – 331, MR 88g:68094.

[Dav90]   H. W. DAVIS, *Cost-Error relationship in A* tree-searching*. J. Assoc. Comput. Mach. **37** (1990), no. 2, 195 – 199.

[deBerg]   M. DE BERG, *On rectilinear link distance*. Utrecht University, Technical Report.

[Deim85]   K. DEIMER, *A new upper bound for the length of snakes*. Combinatorica **5** (1985), no. 2, 109 – 120, MR 87d:05104.

[deKo91]   J. B. J. M. DE KORT, *Lower bounds for symmetric k-peripatetic salesmen problems*. Optimization **22** (1991), no. 1, 113 – 122, MR 91m:90118.

[deMo94]   O. DE MOOR, *Categories, relations and dynamic programming*. Math. Structures Comput. Sci. **4** (1994), no. 1, 33 – 69, MR 94k:90140.

[DeoP84]   N. DEO, C. PANG, *Shortest-path algorithms: taxonomy and annotation*. Networks **14** (1984), 275 – 323.

[DePe85]   R. DECHTER, J. PEARL, *Generalized best-first search strategies and the optimality of A**. J. Assoc. Comput. Mach. **32** (1985), 505 – 536.

[DePe88]   R. DECHTER, J. PEARL, *Network-based heuristics for constraint satisfaction problems*. Artificial Intelligence **34** (1988), 1 – 38.

[DesP94]   A. DE SANTIS, G. PERSIANO, *Tight upper and lower bounds on the path length of binary trees*. SIAM J. Comput. **23** (1994), no. 1, 12 – 23, MR 95e:68036.

[DeWK79]   H. K. DEWITT, M. M. KRIEGER, *Expected length of shortest paths and algorithm behavior*. In: Proceedings of the tenth Southeastern conference on combinatorics, graph theory, and computing (Florida Atlantic Univ., Boca Raton, Fla., 1979). Congr. Numer., XXIII – XXIV, Utilitas Math., Winnipeg, Man., 1979, 367 – 380, MR 81d:68056.

[Di90]   M. B. DILLENCOURT, *An upper bound on the shortness exponent of 1-tough, maximal planar graphs*. Discrete Math. **90** (1991), no. 1, 93 – 97, MR 92g:05120.

[DiHa94]   M. J. DINNEEN, P. R. HAFNER, *New results for the degree/diameter problem*. Networks **24** (1994), no. 7, 359 – 367, MR 95h:05141.

[Dijk59]   E. W. DIJKSTRA, *A Note on two problems in connection with graphs*. Numer. Math. **1** (1959), 269 – 271.

[Dir52]   G. A. DIRAC, *Some theorems on abstract graphs*. Proc. London Math. Soc. **2** (1952), 69 – 81.

[Dir59]   G. A. DIRAC, *Paths and circuits in graphs: Extreme cases*. Acta Math. Acad. Sci. Hung. **10** (1959), 357 – 362.

[Dir78]   G. A. DIRAC, *Hamilton circuits and long circuits*. Ann. Discrete Math. **3** (1978), 75 – 92.

[DjPZ94]   H. N. DJIDJEV, G. E. PANTZIOU, C. D. ZAROLIAGIS, *Fast algorithms for maintaining shortest paths in outerplanar and planar digraphs*. In: H. Reichel (ed.), Proceedings of the 10th Conference on Fundamentals of Computation Theory (FCT '95) (Dresden, 1995). Lecture Notes in Comput. Sci., Vol. 965, Springer, Berlin, Heidelberg, New York, 191 – 200.

[Dong88]   W. QU. DONG, J. Y. SHAO, CH. F. DONG, *On the exponents of primitive digraphs with the shortest elementary circuit length s*. Linear Algebra Appl. **104** (1988), 1 – 27.

[DoSk85]   A. A. DOBRYNIN, V. A. SKOROBOGATOV, *Properties of chains in graphs and isotopy* (Russian). Vychisl. Sistemy no. **112** (1985), 33 - 45, 122, MR 88h:05055.

[DoSk91]   A. A. DOBRYNIN, V. A. SKOROBOGATOV, *Metric invariants of subgraphs of molecular graphs* (Russian). Vychisl. Sistemy no. 140, Mat. Metody v Khim. Inform. (1991), 3 – 62, 207. MR 94f:92016.

[EaY90]   J. N. EAGLE, J. R. YEE, *An optimal branch-and-bound procedure for the constrained path, moving target search problem*. Oper. Res. **38** (1990), no. 1, 110 – 114, MR 91a:90097.

[Ega93]   Y. EGAWA, *Edge-disjoint Hamiltonian cycles in graphs of Ore type*. SUT J. Math. **29** (1993), no. 1, 15 – 50, MR 94i:05058.

[EgGL91]   Y. EGAWA, R. GLAS, S. C. LOCKE, *Cycles and paths through specified vertices in k-connected graphs*. J. Combin. Theory Ser. B **52** (1991), no. 1, 20 – 29, MR 92i:05132.

[EgMi89]   Y. EGAWA, T. MIYMOTO, *The longest cycles in a graph G with minimum degree at least $|G|/k$*. J. Combin. Theory Ser. B, Vol. **46** (1989), no. 3, 356 - 362, MR 90d:05139.

[EiMS85]   A. EIGER, P. B. MIRCHANDANI, H. SOROUSH, *Path preferences and optimal path in probabilistic networks*. Transportation Science **19** (1985), no. 1, 75 - 83.

[EmPe89]   V. A. EMELICHEV, V. A. PEREPELITSA, *Some algorithmic problems of multicriterial optimization on graphs*. Zh. Vychisl. Mat. i Mat. Fiz. **29** (1989), no. 2, 171 - 183, 317, MR 90b:90131.

[EnJS77]   R. C. ENTRINGER, D. E. JACKSON, P. J. SLATER, *Geodetic connectivity of graphs*. IEEE Trans. on Circuits and Systems **24** (1977), 460 - 463.

[EnMK82]   R. ENTRINGER, SH. MACKENDRICK, *Longest paths in locally Hamiltonian graphs*. In: Proceedings of the thirteenth Southeastern conference on combinatorics, graph theory, and computing (Boca Raton, Fla., 1982). Congr. Numer. **35** (1982), 275 - 281, MR 85g:05088.

[Eno84]   H. ENOMOTO, *Long paths and large cycles in finite graphs*. J. Graph Theory **8** (1984), 287 - 301, MR 85h:05060.

[EnSa84]   H. ENOMOTO, A. SAITO, *Disjoint shortest paths in graphs*. Combinatorica **4** (1984), no. 4, 275 - 279, MR 86d:05078.

[Entr81]   R. ENTRINGER, *Girth of cubic graphs with annular symmetry*. In: The Theory and Applications of Graphs (Kalamazoo, Mich., 1980). Wiley, New York, 1981, 317 - 329, MR 83d:05054.

[ErGa59]   P. ERDŐS, T. GALLAI, *On maximal paths and circuits in graphs*. Acta Math. Acad. Sci. Hung. **10** (1959), 337 - 356.

[EvMi92]   J. R. EVANS, E. MINIEKA, *Optimization algorithms for networks and graphs*. 2nd edition, M. Dekker, New York, 1992.

[Exoo83]   G. EXOO, *On line disjoint paths of bounded length*. Discrete Math. **44** (1983), no. 3, 317 - 318, MR 84i:05074.

[Fan84]   G. FAN, *New sufficient conditions for cycles in a graph*. J. Combin. Theory Ser. B **37** (1984), no. 3, 221 - 227, MR 86c:05083.

[Fan85]   G. FAN, *Longest cycles in regular graphs*. J. Combin. Theory Ser. B **39**, 325 - 345 (1985).

[Fan91]   G. H. FAN, *Long cycles and the codiameter of a graph. I.* J. Combin. Theory Ser. B **49** (1990), no. 2, 151 - 180, MR 91j:05063.

[FaSc78]   R. J. FAUDREE, R. H. SCHELP, *Various length paths in graphs*. In: Theory and applications of graphs (Proc. Internat. Conf. Western Mich. Univ., Kalamazoo, Mich., 1976). Lecture Notes in Math., Vol. 672, Springer, Berlin, 1978, 160 - 173, MR 80m:05071.

[Fau93]   R. J. FAUDREE, *Some strong variations of connectivity*. In: Combinatorics, Paul Erdős is eighty, Vol. 1, Bolyai Soc. Math. Stud., János Bolyai Math. Soc., (Budapest, 1993), 125 - 144, MR 94i:05052.

[FeMS91]   E. FEUERSTEIN, A. MARCHETTI-SPACCAMELA, *Dynamic algorithms for shortest paths in planar graphs*. In: G. Schmidt, R. Berghammer (eds.), Proceedings of the 17th International Workshop on Graph-Theoretic Concepts in Computer Science (WG '91) (Fischbachau, Bayern, 1991). Lecture Notes in Comput. Sci., Vol. 570, Springer, Berlin, Heidelberg, New York, 1992, 187 - 197.

[FeMy92]   R. FELDMANN, P. MYSLIEWIETZ, *The shuffle exchange network has a Hamiltonian path*. In: Mathematical Foundations of Computer Science (MFCS '92) (Prague, 1992). Lecture Notes in Comput. Sci., Vol. 629, Springer, Berlin, Heidelberg, New York, 1992, 246 - 254, MR 94k:68008.

[Feng88]   T. FENG, *A short proof of a theorem about the circumference of a graph*. J. Combin. Theory Ser. B **45** (1988), no. 3, 373 - 375, MR 90b:05074.

[FGJL91]   R. J. FAUDREE, R. J. GOULD, M. S. JACOBSON, L. LESNIAK, *Neighborhood unions and highly Hamiltonian graphs*. Ars Combin. **31** (1991), 139 – 148, MR 92d:05093.

[FGJL92]   R. J. FAUDREE, R. J. GOULD, M. S. JACOBSON, L. LESNIAK, *Neighborhood unions and a generalization of Dirac's theorem*. Discrete Math. **105** (1992), 61 – 71, MR 92d:05093.

[FGOR]   PH. FLAJOLET, ZH. CH. GAO, A. ODLYZKO, B. RICHMOND, *The distribution of heights of binary trees and other simple trees*. Combin. Probab. Comput. **2** (1993), no. 2, 145 – 156, MR 94k:05061.

[Filk92]   TH. FILK, *Equivalence of massive propagator distance and mathematical distance of graphs*. Modern Phys. Lett. A **7** (1992), no. 28, 2637 – 2645, MR 94e:81336.

[Fl62]   R. FLOYD, *Algorithm 97: Shortest path*. Communications of the ACM **5** (1962), 345.

[FoFr85]   I. FOURNIER, P. FRAISSE, *On a conjecture of Bondy*. J. Combin. Theory Ser. B **39** (1985), no. 1, 17 – 26, MR 87a:05094.

[FoFu56]   L. R. FORD, D. R. FULKERSON, *Maximal flow through a network*. Canad. J. Math. **8** (1956), 399 – 404.

[Ford46]   L. R. FORD JR., *Network flow theory*. Rand Corporation Report P-923 (1946).

[FOSJ87]   R. J. FAUDREE, E. T. ORDMAN, R. H. SCHELP, M. S. JACOBSON, ZS. TUZA, *Menger's theorem and short paths*. J. Combin. Math. Combin. Comput. **2** (1987), 235 – 253, MR 89c:05050.

[Four82]   I. FOURNIER, *Longest cycles in 2-connected graphs of independence number α*. In: Cycles in graphs (Burnaby, B.C., 1982). North Holland Math. Stud., Vol. 115, North-Holland, Amsterdam, New York, 1985, 201 – 204, MR 87f:05098.

[Fox78]   B. L. FOX, *Data structures and computer science techniques in operations research*. Oper. Res. **26** (1978), 686 – 717.

[Fra90]   A. FRANK, Packing paths in planar graphs. Combinatorica **10** (1990), no. 4, 325 – 331, MR 92m:05068.

[Frai86]   P. FRAISSE, *Circuits including a given set of vertices*. J. Graph Theory **10** (1986), 553 – 557.

[Fred76]   M. L. FREDMAN, *New bounds on the complexity of the shortest path problem*. SIAM J. Comput. **5** (1976), no. 1.

[Fred87]   G. N. FREDERICKSON *Fast algorithms for shortest paths in planar graphs with applications*. SIAM J. Comput. **16** (1987), no. 6, 1004 – 1022.

[Fred90]   G. N. FREDERICKSON, *A shortest path algorithm for a planar network*. Inform. and Comput. **86** (1990), no. 2, 140 – 159, MR 91j:68050.

[Fred91]   G. N. FREDERICKSON, *Planar graph decomposition and all pairs shortest paths*. J. Assoc. Comput. Mach. **38** (1991), no. 1, 162 – 204, MR 91m:68144.

[Freud85] E. C. FREUDER, *A sufficient condition for backtrack-bounded search*. J. Assoc. Comput. Mach. **32** (1985), no. 4, 755 - 761.

[FrGr85]   A. M. FRIEZE, G. R. GRIMMETT, *The shortest-path problem for graphs with random arc-lengths*. Discr. Appl. Math. **10** (1985), 55 – 77, MR 86g:05084.

[Frie83]   A. M. FRIEZE, *An extension of Christofides heuristic to the k-person travelling salesman problem*. Discrete Appl. Math. **6** (1983), no. 1, 79 – 83, MR 84f:90035.

[Frie87]   A. M. FRIEZE, *Parallel algorithms for finding Hamilton cycles in random graphs*. Inform. Process. Lett. **25** (1987), no. 2, 111 – 117, MR 88m:68019.

[Frie92]   A. FRIEZE, C. McDIARMID, BR. REED, *On a conjecture of Bondy and Fan*. Ars Combin. **33** (1992), 329 – 336, MR 93g:05079.

[FrIm89]   H. DE FRAYSSEIX, H. IMAI, *Notes on oriented depth-first search and longest paths*. Inform. Process. Lett. **31** (1989), 53 – 56.

[Fur92] Z. FÜREDI, *The maximum number of edges in a minimal graph of diameter 2.* J. Graph Theory **16** (1992), no. 1, 81 – 98, MR 92j:05067.

[Gab83] H. N. GABOW, *An efficient reduction technique for degree-constrained subgraph and bidirected network flow problems.* In: Proceedings of the 15th ACM Symposium on Theory of Computing (STOC '83) (Boston, Mass., 1983), 448 – 456.

[GaCh91] R. F. GALEEVA, N. I. CHERNOV, *Conditions for the strict exponential growth of the number of cycles in countable directed graphs.* Uspekhi Mat. Nauk **46** (1991), no. 4 (280), 143 – 144; translation in Russian Math. Surveys **46** (1991), no. 4, 172 – 172; MR 92k:58215.

[GaJo79] M. R. GAREY, D. S. JOHNSON *Computers and Intractability. A Guide to the Theory of NP-Completeness.* Freeman, New York, 1979.

[GaPa83] G. GALLO, ST. PALLOTINO, *Shortest path methods: a unifying approach.* In: Netflow at Pisa (Pisa, 1983). Math. Programming Stud. **26** (1986), 38 – 64, MR 87h:90083.

[GaPa88] G. GALLO, S. PALLOTINO, *Shortest path algorithms.* Ann. Oper. Res. **13** (1988), no. 1 – 4, 3 – 79.

[GeFo63] I. M. GELFAND, S. V. FORMIN, *Calculus of Variations.* Prentice-Hall, Englewood Cliffs, New Jersey, 1963.

[Gern86] D. GERNERT, *The cycle-length sequence of a graph and some applications.* In: XI Symposium on operations research (Darmstadt, 1986). Methods of Oper. Res. **57**, Athenäum/Hain/Hanstein, Königstein/Taunus, 1987, 121 – 130, MR 88j:05026.

[Gew91] L. P. GEWALI, *Planning monotone paths to visit a set of obstacles.* Congr. Numer. **82** (1991), 49 – 56.

[GhBh86] R. K. GHOSH, G. P. BHATTACHARJEE, *Parallel algorithm for shortest paths.* Proc. IEE-E **133** (1986), no. 2, 87 – 93, MR 87k:68071.

[GKPS85] F. GLOVER, D. D. KLINGMAN, N. V. PHILLIPS, R. F. SCHNEIDER, *New polynomial shortest path algorithms and their computational attributes.* Management Science **31** (1985), no. 9, 1106 – 1128.

[GKTh79] L. J. GUIBAS, H. T. KUNG, C. D. THOMPSON, *Direct VLSI Implementation of Combinatorial Algorithms.* In: Caltech Conference on VLSI (1979), 509 – 525.

[GlGr89] F. GLOVER, H. J. GREENBERG, *New approaches for heuristic search: a bilateral linkage with artificial intelligence.* European J. Oper. Res. **39** (1989), no. 2, 119 – 130, MR 90c:90059.

[GMO76] H. N. GABOW, S. N. MAHESWARI, L. J. OSTERWEIL, *On two problems in the generation of program test paths.* IEEE Trans. Software Engrg. SE-2 (1976), 227 – 231.

[GoBa78] B. L. GOLDEN, M. BALL, *Shortest paths with euclidean distances: an explanatory model.* Networks **8** (1978), no. 4, 297 – 314, MR 80d:90034.

[GoHK90] D. GOLDFARB, J. HAO, S. R. KAI, *Efficient shortest path simplex algorithms.* Oper. Res. **38** (1990), no. 4, 624 – 628, MR 91j:90027.

[GoHK91] D. GOLDFARB, J. HAO, S. R. KAI, *Shortest path algorithm using dynamic breadth-first search.* Networks **21** (1991), no. 1, 29 – 50, MR 91k:68162.

[Gold93] A. GOLDBERG, *Scaling algorithms for the shortest path problem.* In: Proceedings of the Fourth Annual ACM-SIAM-Symposium on Discrete Algorithms (Austin, TX, 1993). ACM, New York, 1993, 222 – 231, MR 94b:90087.

[GoMF81] D. L. GOLDSMITH, B. MANVEL, V. FABER, *A lower bound for the order of graphs in terms of the diameter and minimum degree.* J. Combin. Inform System Sci. **6** (1981), no. 4, 315 – 319, MR83h:05051.

[GoSV78] S. GOTO, A. SANGIOVANNI-VIENCENTELLI, *A new shortest path updating algorithm.* Networks **8** (1978), no. 4, 341 – 372, MR 80d:90035.

[Gou91] R. J. GOULD, *Updating the Hamiltonian problem — a survey.* J. Graph Theory **15** (1991), no. 2, 121 – 157, MR 92m:05128.

[GPSV85]   A. GYÁRFÁS, H. J. PRÖMEL, E. SZEMERÉDI, B. VOIGT, *On the sum of reciprocals of cycle lengths in sparse graphs.* Combinatorica **5** (1985), no. 1, 41 – 52.

[Grö84]   M. GRÖTSCHEL, *On intersections of longest cycles.* In: Graph Theory and Combinatorics (Cambridge, 1983). Academic Press, London, New York, 1984, 171 – 189, MR 84d:05073.

[GrYY80]   R. L. GRAHAM, A. C. YAO, F. F. YAO, *Information bounds are weak in the shortest distance problem.* J. Assoc. Comput. Mach. **27** (1980), no. 3, 428 – 444.

[Gu93]   G. H. GU, *Sufficient conditions for 3-connected graphs to have Hamiltonian properties.* J. Southeast Univ. **23** (1993), no. 6, 30 – 35, MR 95c:05081.

[GuiH89]   L. J. GUIBAS, J. HERSHBERGER, *Optimal shortest path queries in a simple polygon. Computational geometry.* J. Comput. System Sci. **39** (1989), no. 2, 126 – 152, MR 91g:68164.

[GuLe87]   N. N. GUSHCHINSKIĬ, G. M. LEVIN, *Two-level minimization of a composite function and its application to a problem of the optimization of a path in a graph.* Vestsī Akad. Navuk BSSR Ser. Fīz.-Mat. Navuk (1987), no. 2, 28 – 32, 125; translated in Soviet Comput. Syst. Sci. **29**, no. 6, 31 – 42; MR 88m:90073.

[GuPf92]   W. GUTJAHR, G. C. PFLUG, *The asymptotic distribution of leaf heights in binary trees.* Graphs and Combinatorics **8** (1992), 243 – 251.

[GYCh94]   I. GUTMAN, Y. N. YEH, J. C. CHEN, *On the sum of all distances in graphs.* Tamkang J. Math. **25** (1994), no. 1, 83 – 86, MR 95i:05053.

[GyKS84]   A. GYÁRFÁS, J. KOMLÓS, E. SZEMERÉDI, *On the distribution of cycle lengths in graphs.* J. Graph Theory **8** (1984), 441 – 462.

[GyRS84]   A. GYÁRFÁS, C. C. ROUSSEAU, R. H. SCHELP, *An extremal problem for paths in bipartite graphs.* J. Graph Theory **8** (1984), 83 – 95.

[GYu94]   R. J. GOULD, X. YU, *On hamiltonian-connected graphs.* J. Graph Theory **18** (1994), no. 8, 841 – 860.

[HaBe81]   L. H. HARPER, X. BERENGUER, *The Global Theory of Paths in Networks. I. Definitions, Examples and Limits.* Adv. in Appl. Math. **2** (1981), no. 4, 490 – 506, MR 83e:68086.

[HadS88]   R. W. HADDAD, A. A. SCHÄFFER, Recognizing Bellman-Ford-orderable Graphs. SIAM J. Discrete Math. **1** (1988), no. 4, 447 – 471, MR 89m:68111.

[Hägg93]   R. HÄGGKVIST, *Hamilton cycles in oriented graphs.* Combin. Probab. Comput. **2** (1993), no. 1, 25 – 32, MR 94k:05128.

[HaHe94]   J. H. HATTINGH, M. A. HENNING, *The ratio of the distance irredundance and domination numbers of a graph.* J. Graph Theory **18** (1994), no. 1, 1 – 9, MR 94j:05070.

[Hald93]   S. HALDAR, *An 'All Pairs Shortest Paths' distributed algorithm using $2n^2$ messages.* In: J. van Leeuwen (ed.), Proceedings of the 19th International Workshop on Graph-Theoretic Concepts in Computer Science (WG '93) (Utrecht, 1993). Lecture Notes in Comput. Sci., Vol. 790, Springer, Berlin, Heidelberg, New York, 1993, 350 – 363.

[Hans80]   P. HANSEN, *An $O(m \log D)$ algorithm for shortest paths.* Discrete Appl. Math. **2** (1980), 151 – 153, MR 81d:68083.

[Hara69]   F. HARARY, *Graph Theory.* Addison-Wesley, Reading, Mass., 1969.

[Hart93]   D. HARTVIGSEN, *Minimum path bases.* J. Algorithms **15** (1993), no. 1, 125 – 142, MR 94b:05121.

[HaS88]   R. W. HADDAD, A. A. SCHÄFFER, Recognizing Bellman-Ford-orderable Graphs. SIAM J. Discrete Math. **1** (1988), no. 4, 447 – 471, MR 89m:68111.

[HaSh91]   K. J. HAYHURST, D. R. SHIER, *A factoring approach for the stochastic shortest path problem.* Oper. Res. Lett. **10**, no. 6, 329 – 334, MR 92g:90076.

[HaTa95]   R. HASSIN, A. TAMIR, *On the minimum diameter spanning tree problem.* Inform. Process. Lett. **53** (1995), 109 – 111.

[HaWa81]   J. HARANT, J. WALTHER, *On the radius of graphs.* J. Combin. Theory Ser. B **30** (1981), 113 – 117.

[HaZe85]   R. HASSIN, E. ZEMEL, *On shortest paths in graphs with random weights.* Math. Oper. Res. **10** (1985), no. 4, 557 – 564, MR 87a:05096.

[Hck92]   U. HUCKENBECK, *On a generalization of the Bellman-Ford-Algorithm for acyclic graphs.* Würzburg University, Report no. 41, Würzburg, 1992.

[Hck93]   U. HUCKENBECK, *Dynamic programming in a generalized decision model.* International Computer Science Institute (ICSI), Technical Report TR-93-080, Berkeley, California, 1993.

[Helm86]   P. HELMAN, *The principle of optimality in the design of efficient algorithms.* J. Math. Anal. Appl. **119**, 97 – 127 (1986).

[Helm88]   P. HELMAN, *An algebra for search problems and their solutions.* In: L. N. Kanal, V. Kumar (eds.), Search in Artificial Intelligence. Springer, Berlin, Heidelberg, New York, 1988, 28 – 90.

[Hen85]   M. I. HENIG, *The shortest path problem with two objective functions.* European J. Oper. Res. **25** (1985), 281 – 291, MR 87e:90035.

[Heus89]   H. HEUSER, *Gewöhnliche Differentialgleichungen.* Teubner, Stuttgart, 1989.

[Hi92]   A. M. HINZ, *Shortest paths between regular states of the Tower of Hanoi.* Information Sci. **63** (1992), 173 – 181, MR 93b:05001.

[Hic82]   P. HÍC, *On partially directed geodetic graphs.* Math. Slovaca **32** (1982), no. 3, 255 – 262.

[HiLa92]   D. S. HIRSCHBERG, L. L. LAMORE, *The traveler's problem.* J. Algorithms **13** (1992), no. 1, 148 – 160, MR 92m:90054.

[HiNo84]   T. HILANO, K. NOMURA, *Distance degree regular graphs.* J. Combin. Theory Ser. B **37** (1984), 96 – 100, MR 85j:05036.

[HLTs93]   F. HARARY, E. LOUKAKIS, C. TSOUROS, *The geodetic number of a graph.* In: Graph-theoretic models in computer science, II (Las Crudes, NM, 1988 – 1990). Math. Comput. Modelling **17** (1993), no. 11, 89 – 95, MR 94d:05130.

[HMPT82]   F. HARARY, R. A. MELTER, U. N. PELED, I. TOMESCU, *Boolean distance for graphs.* Discrete Math. **39** (1982), no. 2, 123 – 127, MR 84f:05081.

[Ho91]   A. M. HOBBS, *Traveling salesman problem.* In: Applications of Discrete Mathematics. McGraw-Hill, New York, 1991, 263 – 287, MR 94k:90076.

[Hort87]   J. D. HORTON, *A polynomial-time algorithm to find the shortest cycle basis of a graph.* SIAM J. Comput. **16** (1987), no. 2, 358 – 365.

[HöWa95]   F. HÖFTING, E. WANKE, *Minimum cost paths in periodic graphs.* SIAM J. Comput. **24** (1995), no. 5, 1051 – 1067.

[HsLu]   D. F. HSU, T. LUCZAK, *On the k-diameter of k-regular k-connected graphs.* Discrete Math. **133** (1994), no. 1 – 3, 291 – 296, MR 95g:05041.

[Hu87]   ZH. QU. HU, *Maximal paths in connected graphs.* J. Systems Sci. Math. Sci. **7** (1987), no. 1, 35 – 39, MR 88d:05108.

[Hu88]   U. HUCKENBECK, *On path in search or decision trees which require almost worst-case time.* In: J. v. Leeuwen (ed.), Proceedings of the 14th International Workshop on Graph-Theoretic Concepts in Computer Science (WG '88) (Amsterdam, 1988). Lecture Notes in Comput. Sci., Vol. 344, Springer, Berlin, Heidelberg, New York, 1988, 406 – 423.

[Hu92]   U. HUCKENBECK, *Cost-bounded paths in networks of pipes with valves.* Würzburg University, Report no. 42, Würzburg, 1992.

[Hu93]   U. HUCKENBECK, *On paths in networks with valves.* In: P. Enjalbert, A. Finkel, K. W. Wagner (eds.), Proceedings of the 10th Symposium on Theoretical Aspects of Computer Science (STACS '93) (Würzburg, 1993). Lecture Notes in Comput. Sci., Vol. 665, Springer, Berlin, Heidelberg, New York, 90 – 99.

[Huc93]   U. HUCKENBECK, *On valve adjustments that interrupt all s-t-paths in a digraph*. International Computer Science Institute (ICSI), Technical Report TR-93-081, Berkeley, California, 1993.

[Huc97]   U. HUCKENBECK, *On valve adjustments interrupting all s-t-paths in a digraph*. To appear in: Journal of Automata, Languages and Combinatorics, probably in 1997.

[Huck92] U. HUCKENBECK, *Extremale Pfade in Graphen und verwandte Problemstellungen*. Habilitationsschrift², Würzburg University, Würzburg, 1992.

[HuDi88] M. S. HUNG, J. J. DIVOKY, *A computational study of efficient shortest path algorithms*. Comput. Oper. Res. **15** (1988), no. 6, 567 – 576.

[HüKW82] A. HÜBLER, R. KLETTE, G. WERNER, *Shortest path algorithms for graphs of restricted in-degree and out-degree*. J. Inform. Process. Cybernet. (EIK) **18** (1982), no. 3, 141 – 151.

[HuRu90] U. HUCKENBECK, D. RULAND, *A generalized best-first search method in graphs*. In: R. H. Möhring (ed.), Proceedings of the 16th International Workshop on Graph-Theoretic Concepts in Computer Science (WG '90) (Berlin, 1990). Lecture Notes in Comput. Sci., Vol. 484, Springer, Berlin, Heidelberg, New York, 1991, 41 – 60.

[Hwa80]   F. K. HWANG, *Maximum wealth trees*. Math. Oper. Res. **5** (1980), no. 4.

[HWMG94] D. HANSON, P. WANG, G. MACGILLIVRAY, *A note on minimum graphs with girth pair* $(4, 2l + 1)$. J. Graph Theory **18** (1994), no. 4, 325 – 327, MR 95f:05062.

[HwRi92] F. K. HWANG, D. S. RICHARDS, P. WINTER, *The Steiner tree problem*. Ann. Discr. Math. **53** (1992).

[Iba77]   T. IBARAKI, *The power of dominance relations in branch-and-bound algorithms*. J. Assoc. Comput. Mach. **24** (1977), no. 2, 264 – 279.

[Ichi85] T. ICHIMORI, *A shortest path approach to a multifacility minimax location problem with rectilinear distances*. J. Oper. Res. Soc. Japan **28** (1985) no. 4, MR 87c:90081.

[IhmN82] H. S. IHM, S. C. NTAFOS, *On finding legal paths in the presence of impossible paths*. In: Proceedings of the thirteenth Southeastern conference on combinatorics, graph theory, and computing (Boca Raton, Fla. 1982). Congr. Numer. **36** (1982), 311 – 323, MR 85f:05080.

[IhmN84] H. S. IHM, S. C. NTAFOS, *On legal path problems in digraphs*. Inform. Process. Lett. **18** (1984), 93 – 98.

[ImSc83] W. IMRICH, G. SCHWARZ, *Trees and length functions on groups*. Ann. Discrete Math. **17** (1983), 347 – 359.

[IRWY93] CHR. ICKING, G. ROTE, E. WELZL, CH. YAP, *Shortest paths for line segments*. In: Computational robotics: the geometric theory of manipulation, planning, and control. Algorithmica **10** (1993), no. 2 – 4, 182 – 200, MR 94m:68189.

[ItRo78] A. ITAI, M. RODEH, *Finding a minimum circuit in a graph*. SIAM J. Comput. **7** (1978), no. 4, 413 – 423, MR 80g:68069.

[Ja81]   B. JACKSON, *Long paths and cycles in oriented graphs*. J. Graph Theory **5** (1981), 145 – 157.

[Jack81] B. JACKSON, *Cycles in bipartite graphs*. J. Combin. Theory Ser. B **30** (1981), no. 3, 332 – 342.

[Jack83] B. JACKSON, *Maximal cycles in bipartite graphs*. Ann. Discrete Math. **17** (1983), 361 – 363.

[Jack91] B. JACKSON, *Neighborhood unions and Hamilton cycles*. J. Graph Theory **15** (1991), no. 4, 443 – 451, MR 92i:05141.

---

²Thesis to attain the right to give lectures at a German university.

[Jaf84]   J. M. JAFFE, *Algorithms for finding paths with multiple constraints.* Networks **14** (1984), no. 1, 95 – 116, MR 85c:68069.

[Jai92]   P. JAILLET, *Shortest path problems with node failures.* Networks **22** (1992), 585 – 605.

[JaWo90]   B. JACKSON, N. C. WORMALD, *Cycles containing matchings and pairwise compatible Euler tours.* J. Graph Theory **14** (1990), no. 1, 127 – 138, MR 91g:05087.

[JaWo93]   B. JACKSON, N. C. WORMALD, *Longest cycles in 3-connected graphs of bounded maximum degree.* In: Graphs, matrices, and designs. Lecture Notes in Pure and Appl. Math., Vol. 139, Dekker, New York, 1993, 237 – 254, MR 94g:05045.

[Jeya83]   S. JEYARATNAM *A new algorithm for finding the shortest path between a specified pair of nodes in a graph of nonnegative arcs.* European J. Oper. Res. **12** (1983), 375 – 378.

[JLee89]   G. L. JOHNS, T. CH. LEE, *S-distance in trees.* In: Computing in the 90's (Kalamazoo, MI, 1989). Lecture Notes in Comput. Sci., Vol. 507, Springer, Berlin, Heidelberg, New York, 1991, 29 – 33, MR 92j:68097.

[JogSuh]   P. JOG, J. Y. SUH, *Parallel genetic algorithms applied to the traveling salesman problem.* SIAM J. Optim. **1** (1991), no. 4, 515 – 529, MR 92i:90064.

[Johns77]   L. F. JOHNSON, *Algorithms for enumerating the simple cycles of a digraph.* Proceedings of the 6th Manitoba Conference on Numerical Mathematics (Univ. Manitoba, Winnipeg, Man., 1976). Congr. Numer., XVIII, Utilitas Math., Winnipeg, Man., 1977, 211 – 229, MR 80e:68072.

[Jor92]   L. JØRGENSEN, *Diameters of cubic graphs.* Discrete Appl. Math. **37/38** (1992), 347 – 351, MR 94b:05195.

[JThR94]   M. JÜNGER, ST. THIENES, *Provably good solutions for the Traveling Salesman Problem.* Z. Oper. Res. (ZOR) **40** (1994), 183 – 217.

[Ju94]   D. JUNGNICKEL, *Graphen, Netzwerke und Algorithmen.* 3rd edition, BI Wissenschaftsverlag, Mannheim, Wien, Zürich, 1994.

[Jung73]   H. A. JUNG, *On maximal circuits in finite graphs.* Ann. Discrete Math. **5** (1973), 215 – 228.

[Jung86]   H. A. JUNG, *Longest paths joining given vertices in a graph.* Abh. Math. Sem. Univ. Hamburg **56** (1986), 127 – 137.

[Jung90]   H. A. JUNG, *Long cycles in graphs with moderate connectivity.* In: Topics in combinatorics and graph theory (Oberwolfach, 1990), Phyisica, Heidelberg, 1990, 765 – 778, MR 91m:05117.

[KaHA85]   B. KALANTARI, A. V. HILL, S. R. ARORA, *An algorithm for the traveling salesman problem with pickup and delivery customers.* European J. Oper. Res. **22** (1985), 377 – 386.

[KaHe67]   R. M. KARP, M. HELD, *Finite-state processes and dynamic programming.* SIAM J. Appl. Math. **15**, no. 3 (1967), 693 – 718.

[KaIM78]   N. KATOH, T. IBARAKI, H. MINE, *An $O(Kn^2)$ algorithm for K shortest simple paths in an undirected graph with nonnegative arc length.* Electron. Comm. Japan **61-A**, no. 12 (1978), 1 – 8, MR 83g:68044.

[KaIM82]   N. KATOH, T. IBARAKI, H. MINE, *An efficient algorithm for K shortest simple paths.* Networks **12** (1982), 411 – 427, MR 82j:68063.

[KaKu88]   L. N. KANAL, V. KUMAR (eds.), *Search in Artificial Intelligence.* Springer, Berlin, Heidelberg, New York, 1988.

[Kan94]   K. KANCHANASUT, *A shortest-path algorithm for Manhattan graphs.* Inform. Process. Lett. **49** (1994), no. 1, 21 – 25, MR 94k:05187.

[KaPe83]   R. M. KARP, J. PEARL, *Searching for an optimal path in a tree with random costs.* Artificial Intelligence **21** (1983), 99 – 116, MR 84k:68045.

[KaPo83]   R. G. KARLSSON, P. V. POBLETE, *An $O(m \log \log D)$ algorithm for shortest paths.* Discr. Appl. Math. **6** (1983), 91 – 93, MR 84m:05039.

[Karm84]   S. B. KARMAKAR, *An algorithm for finding a circuit of even length in a directed graph.* Internat. J. Systems Sci. **15** (1984), no. 11, 1197 – 1201.

[Karp78]   R. M. KARP, *A characterization of the minimum cycle mean in a digraph.* Discrete Math. **23** (1978), 309 – 311.

[Kemp90]   R. KEMP *On the number of deepest nodes in ordered trees.* Discrete Math. **81** (1990), no. 3, 247 – 258, MR 91c:05014.

[KeW93]   H. KELLERER, G. WOEGINGER, *On the Euclidean two paths problem.* Discr. Appl. Math. **47** (1993), 165 – 173.

[KiSa80]   G. KISHI, I. SASAKI, *Intersection functions of shortest paths in a nondirected graph.* Electron. Comm. Japan **63** (1980), no. 11, 1 – 10 (1982), MR 83k:05067.

[KKPh93]   D. R. KARGER, D. KOLLER, S. J. PHILLIPS, *Finding the hidden path: time bounds for all-pairs shortest paths.* SIAM J. Comput. **22** (1993), no. 6, 1199 – 1217, MR 95j:05158.

[KlPe90]   S. KLAŽAR, M. PETKOVŠEK, *Graphs with nonempty intersection of longest paths.* Ars Combin. **19** (1990), 43 – 52, MR 91c:05110.

[KlW89]   R. KLEIN, D. WOOD, *On the path length of binary trees.* J. Assoc. Comput. Mach. **36** (1989), no. 2, 280 – 289, MR 91e:68027.

[KnKa74]   H. W. KNOBLOCH, F. KAPPEL, *Gewöhnliche Differentialgleichungen.* Teubner, Stuttgart, 1974.

[Knu82]   D. E. KNUTH, *The Art of Computer Programming. Vol. 1 – 3.* Addison-Wesley, Reading, Mass., 1982.

[Knya90]   A. V. KNYAZEV, *Dichtomous graphs with maximal girth.* Diskret. Mat. **2** (1990), no. 3, 56 – 64.

[Koc90]   W. KOCAY, *An extension of the multi-path algorithm for finding hamilton cycles.* Discrete Math. **101** (1992), 171 – 188, MR 93g:05095.

[Korf85]   R. E. KORF, *Depth first iterative deepening: an optimal admissible tree search.* Artificial Intelligence **27** (1985), 97 – 109.

[KoSW93]   A. V. KOSTOCHKA, A. A. SAPOZHENKO, K. WEBER, *Radius and diameter of random subgraphs of the hypercube.* Random Structures Algorithms **4** (1993), no. 2, 215 – 229, MR 94c:050566.

[Kova88]   M. YA. KOVALEV, *Interval ε-approximation of the optimal path in a graph.* Vestsī Akad. Navuk BSSR Ser. Fiz.-Mat. Navuk (1988), no. 2, 15 – 20, MR 89i:05178.

[Koz91]   D. C. KOZEN, *The Design and Analysis of Algorithms.* Springer, New York, Berlin, 1991.

[KrNa80]   V. KRISHNAMOORTHY, R. NANDAKUMAR, *A class of counterexamples to a conjecture on diameter critical graphs.* In: Combinatorics and graph theory (Calcutta, 1980). Lecture Notes in Math., Vol. 885, Springer, Berlin, Heidelberg, New York, 1981, 297 – 300, MR 83d:05055.

[KRRS95]   P. KLEIN, S. RAO, M. RAUCH, S. SUBRAMANIAN, *Faster shortest-path algorithms for planar graphs.* In: Zoltán Fülöp, Ferenc Gécseg (eds.), Proceedings of the 22nd International Colloquium on Automata, Languages and Programming (ICALP '95) (Szeged, 1995). Lecture Notes in Comput. Sci., Vol. 944, Springer, Berlin, Heidelberg, New York, 27 – 37.

[KrTr83]   P. KŘIVKA, N. TRINAJSTIĆ, *On the distance polynomial of a graph.* Aplikace Matematiky **28** (1983), no. 5, 357 – 363.

[KS94]   D. J. KLEITMAN, L. J. SCHULMAN, *Minimally distant sets of lattice points.* European J. Combin. **14** (1993), no. 3, 231 – 240, MR 94e:05096.

[KuKa88]   M. KUBO, H. KASUGAI, *Traveling purchaser problem.* Bull. Sci. Engrg. Res. Lab. Waseda Univ. Waseda no. **116** (1986), 7 – 12, MR 88c:90051.

[Kulk86]   V. G. KULKARNI, *Shortest paths in networks with exponentially distributed arc lengths.* Networks **16** (1986), no. 3, 255 – 274, MR 87i:90105.

[Kum70]   S. KUMAR, *Optimal path through k specified sets of nodes*. Indian J. Math. **12**, no. 1 (1970), 25 – 30.

[Kun80]   S. KUNDU, *A Dijkstra-like shortest path algorithm for certain cases of negative arc lengths*. BIT **20** (1980), 522 – 524, MR 82j:68061.

[KuNK88]   V. KUMAR, D. S. NAU, L. N. KANAL, *A general branch-and-bound formulation for AND/OR graph and game tree search*. In: L. N. Kanal, V. Kumar (eds.), Search in Artificial Intelligence. Springer, Berlin, Heidelberg, New York, 1988, 91 – 130.

[Kwa89]   J. B. H. KWA, *BS*: An admissible bidirectional staged heuristic search algorithm*. Artificial Intelligence **38** (1989), 95 – 109.

[LaLe85]   E. L. LAWLER, J. K. LENSTRA, A. H. G. RINNOOY KAN, D. B. SHMOYS (eds.), *The traveling salesman problem: A guided tour of combinatorial optimization*. Wiley, New York, 1985.

[LaMa90]   G. LAPORTE, S. MARTELLO, *The selective travelling salesman problem*. Discrete Appl. Math. **26** (1990), 193 – 207, MR 91c:90087.

[LaNo83]   G. LAPORTE, Y. NORBERT, *Finding the shortest cycle through k specified nodes*. In: Proceedings of the fourteenth Southeastern conference on combinatorics, graph theory, and computing (Boca Raton, Fla. 1983). Congr. Numer. **40** (1983), 155 – 167, MR 85h:90047.

[LaPP84]   A. S. LAPAUGH, C. H. PAPADIMITRIOU, *The even-path problem for graphs and digraphs*. Networks **14** (1984), no. 4, 507 – 513, MR 86g:05057.

[Larm87]   L. LARMORE, *Height restricted optimal binary trees*. SIAM J. Comp. **16** (1987), no. 6, 1115 – 1123.

[Lawl89]   E. L. LAWLER, *Computing shortest paths in networks derived from recurrence relations*. Topological network design (Copenhagen, 1989). Ann. Oper. Res. **33** (1991), no. 1 – 4, 363 – 377, MR 92h:90129.

[LaWo66]   E. L. LAWLER, D. E. WOOD, *Branch-and-bound methods, a survey*. Operations Research **14** (1966), 699 – 719.

[LeeP81]   D. T. LEE, F. P. PREPARATA, *Euclicdean shortest paths in the presence of parallel rectilinear barriers*. Proceedings of the 7th Conference on Graph-Theoretic Concepts in Computer Science (WG '81), Linz, 1981. Hanser, München, Wien, 1982, 303 – 313, MR 85a:68150.

[LeeP84]   D. T. LEE, F. P. PREPARATA, *Euclidean shortest paths in the presence of rectilinear barriers*. Networks **14** (1984), 393 – 410.

[LeeP91]   D. T. LEE, E. PAPADOPOULOU, *The all-pairs quickest path problem*. Inform. Process. Lett. **45** (1993), 261 – 267.

[Leit86]   G. LEITMANN, *The Calculus of Variations and Optimal Control. An introduction*. Plenum Press, New York, London, 1986.

[LeKa72]   K. N. LEVITT, W. H. KAUTZ, *Cellular arrays for the solution of graph problems*. Communications of the ACM **15** (1972), no. 9, 789 – 801.

[LeLS84]   C. LEVCOPOULOS, A. LINGAS, J. SACK, *Nearly optimal heuristics for binary search trees with geometric generalizations*. In: Th. Ottmann (ed.), Proceedings of the 14th International Colloquium on Automata Languages and Programming (ICALP '84) (Karlsruhe, 1984). Lecture Notes in Comput. Sci., Vol. 267, Springer, Berlin, Heidelberg, New York, 376 – 385.

[LenT91]   T. LENGAUER, D. THEUNE, *Unstructured path problems and the making of semirings*. In: F. Dehne, J. R. Sack, N. Santoro (eds.), Proceedings of the 2nd Workshop on Algorithms and Data Structures (WADS '91) (Ottawa, Canada, 1991). Lecture Notes in Comput. Sci., Vol. 519, Springer, Berlin, Heidelberg, New York, 189 – 200.

[Li89]   H. LI, *Edge disjoint cycles in graphs*. J. Graph Theory **13** (1989), no. 3, 313 – 322, MR 91e:05052.

[Li90a]   H. LI, *Edge-Hamiltonian property in regular 2-connected graphs.* Discrete Math. **82** (1990), no. 1, 25 – 34, MR 91f:05080.

[Li90b]   R. H. LI, *The shortest path in the 2-norm sense in a network with weighted vectors.* Shandong Kuangye Xueyuan Xuebao **9** (1990), no. 2, 188 – 192, MR 91j:90028.

[Li93]   C. L. LI, S. TH. MCCORMICK, D. SIMCHI-LEVI, *On the minimum-cardinality-bounded-diameter and the bounded-cardinality-minimum-diameter edge addition problems.* Oper. Res. Lett. **11** (1992), no. 5, 303 – 308, MR 93i:68144.

[Ling91]   A. LINGAS, *Bit complexity of matrix products.* Inform. Process. Lett. **38** (1991), no. 5, 237 – 242, MR 92e:68072.

[Liu91]   C. J. LIU, *Enumeration of Hamiltonian cycles and paths in a graph.* Proc. Amer. Math. Soc. **111** (1991), no. 1, 289 – 296, MR 91d:05056.

[LiuShi]   R. G. LIU, Y. B. SHI, *The maximum number of edges of an odd [even] cycle distributed graph.* Shanghai Shifan Daxue Xuebao Ziran Kexue Ban **21** (1992), no. 4, 21 – 23, MR 93m:05101.

[LMCSL]   C. L. LI, S. TH. MCCORMICK, D. SIMCHI-LEVI, *The complexity of finding two disjoint paths with min-max objective function.* Discrete Appl. Math. **26**, no. 1, 105 – 115, MR 91d:05061.

[LMS92]   C. L. LI, S. TH. MCCORMICK, D. SIMCHI-LEVI, *Finding disjoint paths with different path-costs: complexity and algorithms.* Networks **22** (1992), no. 7, 653 – 667, MR 93g:68056.

[LMW93]   Y. N. LIEN, E. MA, B. W. S. WAH, *Transformation of the generalized traveling-salesman problem into the standard traveling-salesman problem.* Inform. Sci. **74** (1993), no. 1 – 2, 177 – 189, MR 94j:90031.

[Loc81]   S. C. LOCKE, Doctoral thesis, University of Waterloo, Waterloo, 1981.

[Lomo88]   M. V. LOMONOSOV, *Cycles though prescribed elements in a graph.* In: Paths, flows, and VLSI-layout (Bonn, 1988). Algorithms Combin., Vol. 9, Springer, Berlin, Heidelberg, New York, 1990, 215 – 234, MR 91m:05119.

[Lor79]   T. J. LORENZEN, *Generalizing the secretary problem.* Adv. Appl. Prob. **11** (1979), 384 – 396.

[Lor81]   T. J. LORENZEN, *Optimal stopping with sampling cost: The secretary problem.* Ann. Probab. **9** (1981), 167 – 172.

[Lov85]   L. LOVÁSZ, *Computing ears and branchings in parallel.* In: IEEE Proceedings on Foundations of Computer Science 1985 (FOCS '85), 464 – 467.

[LT91]   T. LENGAUER, D. THEUNE, *Efficient algorithms for path problems with general cost criteria.* Cadlab Report 9/91, Cadlab, Paderborn, 1991.

[LTh91]   T. LENGAUER, D. THEUNE, *Efficient algorithms for path problems with general cost criteria.* In: J. L. Albert, B. Monien, M. R. Artalejo (eds.), Proceedings of the 18th International Colloquium on Automata, Languages and Programming (ICALP '91) (Madrid, 1991). Lecture Notes in Comput. Sci., Vol. 510, Springer, Berlin, Heidelberg, New York, 314 – 326.

[LTSh93]   J. LI, F. TIAN, R. Q. SHEN, *A further generalization of Jung's theorem.* Systems Sci. Math. Sci. **6** (1993), no. 1, 52 – 60, MR 95b:05113.

[LuRa89]   M. LUBY, PR. RAGDE, *A bidirectional shortest-path algorithm with good average-case behavior.* Algorithmica **4** (1989), no. 4, 551 – 567, MR 91e:68074.

[LZh94]   J. LIU, H. SH. ZHOU, *Graphs and digraphs with given girth and connectivity.* Discrete Math. **132** (1994), 387 – 390, MR 95h:05095.

[MaBa85]   A. MAHANTI, A. BAGCHI, *AND/OR graph heuristic search methods.* J. Assoc. Comput. Mach. **32** (1985), no. 1, 28 – 51.

[MaCh93]   V. V. MALYSHEV, D.È. CHERNOV, *Generalized dynamic programming. General principles.* Avtomat. i Telemekh. **1993**, no. 12, 101 – 110; translation in Automat. Remote Control **54** (1993), no. 12, part 1, 1812 – 1819 (1994); MR 94m:90120.

[Mack77]   A. K. MACKWORTH, *Consistency in Networks of Relations.* Artificial Intelligence **8** (1977), 99 – 118.

[MaDe76]   P. MATETI, N. DEO, *On algorithms for enumerating all circuits in a graph.* SIAM J. Comput. **5** (1976), no. 1, 90 – 98.

[Mahr80]   B. MAHR, *A bird's-eye view to path problems.* In: Proceedings of the 6th Workshop on Graph-Theoretic Concepts in Computer Science (WG '80) (Bad Honnef, 1980). Lecture Notes in Comput. Sci., Vol. 100, Springer, Berlin, Heidelberg, New York, 1981, 335 – 353.

[Mahr82]   B. MAHR, *Algebraic complexity of path problems.* RAIRO Inform. Théor. **16** (1982), no. 3, 263 – 292, MR 84a:68064.

[Man92]   Y. MANOUSSAKIS, *Directed Hamiltonian graphs.* J. Graph Theory **16** (1992), no. 1, 51 – 59, MR 92m:05132.

[Mang90]   R. MANGER, *New examples of the path algebra and corresponding graph-theoretic path problems.* In: VII Conference on Applied Mathematics (Osijek, 1989). Univ. Osijek, Osijek, 1990, 119 – 128, MR 91m:05095.

[MaPl87]   B. M. MAGGS, S. A. PLOTKIN, *Minimum-cost spanning tree as a path-finding problem.* Inform. Process. Lett. **26** (1987/88), 291 – 293.

[Marc83]   D. MARCU, *Note on the length of elementary cycles of a graph.* Quart. J. Math. Oxford Ser. (2) **34** (1983), no. 135, 475 – 476.

[Mart77]   A. MARTELLI, *On the complexity of admissible search algorithms.* Artificial Intelligence **8** (1977), 1 – 13.

[Mart79]   I. MARTINEC, *A combinatorial method for the solution of the travelling salesman problems.* Ekonom.-Mat. Obzor **15** (1979), no. 3, 320 – 341, MR 81a:90084.

[Maru]   DR. MARUŠIČ, *Hamiltonicity of tree-like graphs.* Discrete Math. **80** (1990), no. 2, 167 – 173, MR 91c:05118.

[MaSa91]   N. MACULAN, J. J. C. SALLES, *A lower bound for the shortest Hamiltonian path in directed graphs.* OR Spektrum **13** (1991), no. 2, 99 – 102, MR 92f:90071.

[MaTu90]   Y. MANOUSSAKIS, Z. TUZA, *Polynomial algorithms for finding cycles and paths in bipartite tournaments.* SIAM J. Discrete Math. **3** (1990), no. 4, 537 – 543, MR 91h:68080.

[McCa88]   J. E. MCCANNA, *Orientations of the n-cube with minimum diameter.* Discrete Math. **68** (1988), no. 2 – 3, 309 – 310.

[McCaWi]   J. E. MCCANNA, ST. J. WINTERS, *Multiply-sure distances in graphs.* In: Proceedings of the Twenty-Fourth Southeastern International Conference on Combinatorics, Graph Theory, and Computing (Boca Raton, FL, 1992). Congr. Numer. **97** (1993), 71 – 81. MR 94k:05074.

[McDi]   C. MCDIARMID, *Expected numbers at hitting times.* J. Graph Theory **15** (1991), no. 6, 637 – 648, 92m:05174.

[McGl83]   R. J. MCGLINN, *Is SSS* better than alpha-beta ?* Inform. Process. Lett. **16** (1983), 113 – 120.

[MehS83]   K. MEHLHORN, B. H. SCHMIDT, *A single source shortest path algorithm for graphs with separators.* In: Foundations of Computation Theory, Proceedings of the 1983 International FCT-Conference (Borgholm, Sweden, 1983). Lecture Notes in Comput. Sci., Vol. 158, Springer, Berlin, Heidelberg, New York, 302 – 309.

[Merr90]   R. MERRIS, *The distance spectrum of a tree.* J. Graph Theory **14** (1990), no. 3, 365 – 369, MR 91h:05085.

[Mese89]   P. MESEGUER, *Constraint satisfaction problems: an overview*. AICOM **2** (1989), no. 3, 3 – 17.

[MeTo81]   R. A. MELTER, I. TOMESCU, *Remarks on distances in graphs*. An. Ştiinţ. Univ. "Al. I. Cuza" Iaşi Secţ. I a Mat. (N. S.) **27** (1981), no. 2, 407 – 410, MR 83c:05081.

[Mi91]   J. G. MICHAELS, *The Chinese postman problem*. Applications of discrete mathematics, 354 – 364, MacGraw-Hill, New York, 1991, MR 94k:05200.

[Mi92]   J. S. B. MITCHELL, L_1-*shortest paths among polygonal obstacles in the plane*. Algorithmica **8** (1992), no. 1, 55 – 88, MR 94a:68133.

[MiFr88]   M. MILLER, I. FRIŠ, *Minimum diameter of diregular digraphs of degree 2*. Comput. J. **31** (1988), no. 1, 71 – 75, MR 89c:05042.

[MiFr93]   M. MILLER, I. FRIŠ, *Maximum order digraphs for diameter 2 or degree 2*. In: Graphs, matrices, and designs. Lecture Notes in Pure and Appl. Math., Vol. 139, Dekker, New York, 1993, 269 – 278, MR 94a:05110.

[Mini79]   E. MINIEKA, *The Chinese postman problem for mixed networks*. Management Sci. **25** (1979/80), no. 7, 643 – 648, MR 81a:90085.

[MiRo80]   R. E. MILLER, A. L. ROSENBERG, *On computing distances between leaves in a complete tree*. Internat. J. Comput. Math. **8** (1980), no. 4, 289 – 301, MR 82a:68126.

[Mitt64]   L. G. MITTEN, *Composition principles for synthesis of optimal multistage processes*. Oper. Res. **12** (1964), 610 – 619.

[Mon82]   B. MONIEN, *The complexity of determining a shortest cycle of even length*. In: Proceedings of the 8th Conference on Graph Theoretic Concepts in Computer Science (WG '82) (Neunkirchen, 1982). Hanser, München, Wien, 195 – 208, MR 84e:68075.

[Mon85]   B. MONIEN, *How to find long paths efficiently*. Ann. Discrete Math. **25** (1985), 239 – 254, MR 87a:05097.

[Moo57]   E. F. MOORE, *The shortest path through a maze*. Proc. Int. Symp. on the Theory of Switching (1957), Part II, The Annals of the Computation Laboratory of Havard University **30**, Harvard University Press, 1959, 285 – 292.

[MoPa88]   TH. MOHR, C. PASCHE, *A parallel shortest path algorithm*. Computing **40** (1988), 281 – 292

[Mor82]   TH. L. MORIN, *Monotonicity and the principle of optimality*. J. Math. Anal. Appl. **86** (1982), 665 – 674.

[MoTa87]   A. MOFFAT, T. TAKAOKA, *An all pairs shortest path algorithm with expected time* $O(n^2 \log n)$. SIAM J. Comput. **12** (1987), no. 6, 1023 – 1031.

[MSS89a]   I. I. MELAMED, S. I. SERGEEV, I. KH. SIGAL, *The traveling salesman problem. Approximate algorithms*. Avtomat. i Telemekh. **1989**, no. 11, 3 – 26; translation in Automat. Remote Control **50** (1989), no. 11, part 1, 1459 – 1479 (1990); MR 91b:90159.

[MSS89b]   I. I. MELAMED, S. I. SERGEEV, I. KH. SIGAL, *The traveling salesman problem. Issues in the theory*. Avtomat. i Telemekh. **1989**, no. 9, 3 – 33; translation in Automat. Remote Control **50** (1989), no. 9, part 1, 1147 – 1173 (1990); MR 91b:90160.

[MTM94]   J. MA, T. TAKAOKA, SH. H. MA, *Parallel algorithms for a class of graph-theoretic problems*. Trans. Inform. Process. Soc. Japan **35** (1994), no. 7, 1235 – 1240, MR 95h:05063.

[MTZC81]   N. MEGIDDO, A. TAMIR, E. ZEMEL, R. CHANDRASEKARAN, *An* $O(n \log^2 n)$ *algorithm for the kth longest path in a tree with applications to location problems*. SIAM J. Comput. **10**, no. 2 (1981), 328 – 337, MR 82j:68063.

[Mu92]   A. MUTHUSAMY, *Counterexample to a conjecture on Hamilton cycles*. Ars Combin. **34** (1992), 223 – 224, MR 93i:05089.

[Na94]   K. NACHTIGALL, *Time depending shortest-path problems with applications to railway networks*. European J. Oper. Res. **83** (1995), 154 – 166.

[NaRi91] D. NADDEF, G. RINALDI, *The symmetric traveling salesman polytope and its graphical relaxation: composition of valid inequalities.* Math. Programming **51** (1991), no. 3, (Ser. A), 359 – 400, MR 92i:90065.

[NaRi93] D. NADDEF, G. RINALDI, *The graphical relaxation; a new framework for the symmetric travelling salesman polytope.* Math. Programming **58** (1993), no. 1, Ser. A, 53 – 88, MR 94e:90081.

[Ne94] L. NEBESKÝ, *A characterization of the set of all shortest paths in a connected graph.* Math. Bohem. **119** (1994), no. 1, 15 – 20, MR 95i:05081.

[Neu75] K. NEUMANN, *Operations Research Verfahren. Vol. III: Graphentheorie, Netzplantechnik.* Hanser, München, Wien, 1975.

[NgPa86] S. NGUYEN, ST. PALLOTINO, *Hyperpaths and shortest hyperpaths.* In: Combinatorial optimization (Como, 1986). Lecture Notes in Math., Vol. 1403, Springer, Berlin, Heidelberg, New York, 1989, 258 – 271, MR 91c:05113.

[Nije79] A. NIJENHUIS, *A minimal-path algorithm for the "money changing problem".* Amer. Math. Monthly **86** (1979), 832 – 835, MR 82m:68070.

[Nils82] N. J. NILSSON, *Principles of Artificial Intelligence.* Springer, Berlin, Heidelberg, New York, 1982.

[Nish86] TS. NISHIMURA, *Short cycles in digraphs.* In: Proceedings of the First Japan Conference on Graph Theory and Applications (Hakone, 1986). Discrete Math. **72** (1988), no. 1 – 3, 295 – 298, MR 90a:05119.

[NKK84] D. S. NAU, V. KUMAR, L. KANAL, *General branch and bound, and its relation to A* and AO*.* Artificial Intelligence **23** (1984), no. 1, 29 – 58, MR 86e:68094.

[Nolt75] H. NOLTEMEIER, *An algorithm for the determination of longest distances in a graph.* Math. Programming **9** (1975), 350 – 357.

[Nolt76] H. NOLTEMEIER, *Graphentheorie.* Walter de Gruyter, Berlin, New York, 1976.

[Nolt82] H. NOLTEMEIER, *Informatik III. Einführung in Datenstrukturen.* Hanser, München, Wien, 1982.

[NoMM78] K. NOSHITA, E. MASUDA, H. MACHIDA, *On the expected behaviors of the Dijkstra's shortest path algorithm for complete graphs.* Inform. Process. Lett. **7** (1978), no. 5, 237 – 243, MR 80c:68022.

[Ord87] V. V. ORDIN, *The method of Markov iterations for finding paths in a nework of general form.* Avtomat. i Telemekh. **1987**, no. 1, 183 – 185, MR 88c:90056.

[OrR90] A. ORDA, R. ROM, *Shortest-path and minimum-delay algorithms in networks with time-dependent edge-length.* J. Assoc. Comput. Mach. **37** (1990), no. 3, 607 – 625, MR 91m:90063.

[OrR91] A. ORDA, R. ROM, *Minimum weight paths in time-dependent networks.* Networks **21** (1991), no. 3, 295 – 319, MR 92f:68138.

[Ota95] K. OTA, *Cycles through prescribed vertices with large degree sum.* Discrete Math. **145** (1995), 201 – 210.

[Pa76] C. H. PAPADIMITRIOU, *On the complexity of edge traversing,* J. Assoc. Comput. Mach. **23** (1976), 544 – 554.

[Pall84] ST. PALLOTINO, *Shortest-path methods: complexity, interrelations and new propositions.* Networks **14** (1984), 257 – 267.

[Pan81] V. YA. PAN, *The bit-operation complexity of matrix multiplication and of all pair shortest path problem.* Comput. Math. Appl. **7** (1981), no. 5, 431 – 438, MR 83a:68055.

[PanR87] V. PAN, J. REIF, *Fast and efficient solution of path algebra problems.* Technical Report 87-3, State University of New York at Albany (SUNY), Computer Science Department, 1987.

[PanR89]   V. PAN, J. REIF, *Fast and efficient solution of path algebra problems.* J. Comput. System
            Sci. **38** (1989), no. 3, 494 – 510, MR 91k:68077.

[PaRa83]   R. G. PARKER, R. L. RARDING, *The travelling salesman problem: an update of research.*
            Naval Res. Logist. Quart. **30** (1983), no. 1, 69 – 96, MR 84i:90060.

[PaSt82]   CH. H. PAPADIMITRIOU, K. STEIGLITZ, *Combinatorial Optimization: Algorithms and
            Complexity.* Prentice Hall, Englewood Cliffs, N.J., 1982.

[PaSu78]   J. PACH, L. SURÁNYI, *Graphs of diameter 2 and linear programming.* In: Algebraic meth-
            ods in graph theory, Vol. I, II (Szeged, 1978). Colloq. Math. Soc. János Bolyai, Vol. 25,
            North-Holland, Amsterdam, 1981, 599 – 629, MR 83d:05059.

[PaYa93]   CH. H. PAPADIMITRIOU, M. YANNAKAKIS, *The traveling salesman problem with distances
            one and two.* Math. Oper. Res. **18** (1993), no. 1, 1 – 11, MR 94k:90073.

[PBSh91]   M. J. PAUL, C. B. SHERSHIN, A. C. SHERSHIN, *Notes on sufficient conditions for a graph
            to be Hamiltonian.* Internat. J. Math. & Math. Sci. **14**, no. 4 (1991), 825 – 827,
            MR 92i:05142.

[Pe82]   J. PEARL, *The solution for the branching factor of the alpha-beta pruning algorithm and
            its optimality.* Communications of the ACM **25** (1982), no. 8, 559 –564.

[Pe83]   J. PEARL, *Knowledge versus search: a quantitative analysis of A*.* Artificial Intelligence
            20 (1983), no. 1, 1 – 13, MR 84e:68119.

[Per86]   A. PERKO, *Implementation of algorithms for K shortest loopless paths.* Networks **16**
            (1986), 149 – 160, MR 87j:05102.

[Pete83]   V. PETEANU, *ε-Minimal paths in large networks.* Matematica **25** (**48**) (1983), no. 1,
            43 – 44.

[Phil94]   A. B. PHILPOTT, *Continuous-time shortest path problems and linear programming.* SIAM
            J. Control Optim. **32** (1994), no. 2, 538 – 552, MR 94k:90136.

[Pl84]   J. PLESNÍK, *On the sum of all distances in a graph or digraph.* J. Graph Theory **8** (1984),
            1 – 21.

[PlBa89]   L. K. PLATZMAN, J. J. BARTHOLDI, *Spacefilling curves and the planar travelling salesman
            problem.* J. Assoc. Comput. Mach. **36** (1989), no. 4, 719 – 737, MR 91k:90124.

[Ples84]   J. PLESNÍK, *A construction of geodetic graphs based on pulling subgraphs homeomorphic
            to complete graphs.* J. Combin. Theory Ser. B **36** (1984), no. 3, 284 – 297, MR 86c:05082.

[Pol92]   G. G. POLAK, *On a parametric shortest path problem from primal-dual multicommodity
            network optimization.* Networks **22** (1992), 282 – 295.

[Poll83]   A. D. POLLINGTON, *There is a long path in the divisor graph.* Ars Combin. **16** (1983),
            B, 303 – 304, MR 85g:05090.

[Pomer83]   C. POMERANCE, *On the longest simple paths in divisor graphs.* In: Proceedings of
            the fourteenth Southeastern conference on combinatorics, graph theory, and comput-
            ing (Boca Raton, Fla. 1983). Congr. Numer. **40** (1983), 291 – 304.

[PPSp90]   J. PACH, R. POLLACK, J. SPENCER, *Graph distance and Euclidean distance on the grid.*
            In: Topics in combinatorics and graph theory (Oberwolfach, 1990). Physica, Heidelberg,
            1990, 555 – 559, MR 91m:05077.

[Pra91]   S. PRASSER, *Verfahren zur Bestimmung von Kreisen kleinsten mittleren Gewichts in
            einem bewerteten Graphen.* Diploma thesis, Würzburg University, Würzburg, 1991.

[PraMil]   E. PRASSLER, E. MILIOS, *Parallel distributed robot navigation in the presence of obsta-
            cles.* Research Institute for Applied Knowledge Processing (FAW) Ulm, Technical Re-
            port.

[Pri81]   C. C. PRICE, *A VLSI algorithm for shortest path through a directed acyclic graph.* In:
            Proceedings of the 11th Manitoba Conference on Numerical Mathematics and Comput-
            ing (Winnipeg, Man., 1981). Congr. Numer. **34** (1982), 363 – 371.

[PrMi]   E. PRASSLER, E. MILIOS, *Parallel distributed robot navigation in the presence of obstacles*. Research Institute for Applied Knowledge Processing (FAW) Ulm, Technical Report.

[Pros81]   A. PROSKUROWSKI, *Recursive graphs, recursive labelings and shortest paths*. SIAM J. Comput. **10** (1981), no. 2, 391 – 397, MR 82j:68065.

[PyTu93]   L. PYBER, ZS. TUZA, *Menger-type theorems with restrictions on path lengths*. Discrete Math. **120** (1993), 161 – 174, MR 94e:05171.

[Ran81]   M. V. RANCHICH, *Determination of shortest paths of a graph*. In: Systems-theoretic methods and their application in automated systems. Akad. Nauk. Ukrain. SSR, Inst. Kibernet., Kiev, 1983, 53 – 55, 99, MR 86e:05057.

[Ravi92]   R. RAVI, M. V. MARATHE, C. PANDU RAGAN, *An optimal algorithm to solve the all-pair-shortest path problem on interval graphs*. Networks **22** (1992), 21 – 35.

[RiLi85]   D. RICHARDS, A. L. LIESTMAN, *Finding cycles of a given length*. Ann. Discrete Math. **27** (1985), 249 – 256.

[Riv88]   R. L. RIVEST, *Game Tree Searching by min/max approximation*. Artificial Intelligence **34** (1988) 77 – 96.

[Rohn91]   H. ROHNERT, *Moving a disc between polygons*. Algorithmica **6** (1991), 182 – 191.

[Rom88]   J. F. ROUMEUF, *Shortest path under rational constraint*. Inform. Process. Lett. **28** (1988), 245 – 248, MR 89j:68124.

[Roma80]   F. ROMANI, *Shortest path problem is not harder than matrix multiplication*. Inform. Process. Lett. **11** (1980), no. 3, 134 – 136.

[RoPe84]   D. RONEN, Y. PERL, *Heuristics for finding a maximum number of disjoint bounded paths*. Networks **14** (1984), 531 – 544.

[RoPe88]   I. ROIZEN, J. PEARL, *A minimax algorithm better than alpha-beta? Yes and no*. Artificial Intelligence **21** (1983), no. 1 – 2, 199 – 220, MR 84k:68059.

[Rote90]   G. ROTE, *Path problems in graphs*. Comput. Suppl. **7** (1990), 155 - 189.

[Rote92]   G. ROTE, *The N-line traveling salesman problem*. Networks **22** (1992), no. 1, 91 – 108, MR 92i:90045.

[RSr93]   Y. ROUSKOV, PR. K. SRIMANI, *Fault diameter of star graphs*. Inform. Process. Lett. **48** (1993), no. 5, 243 – 251, MR 94k:68011.

[Ry90]   ZD. RYJÁČEK, *Hamiltonian circuits in N_2-locally connected $K_{1,3}$-free graphs*. J. Graph Theory **14** (1990), no. 3, 321 – 331, MR 91d:05066.

[SaChGS]   U. K. SARKAR, P. P. CHAKRABARTI, S. G. GHOSE, S. C. DE SARKAR, *Reducing reexpansions in iterative deepening search by controlling cutoff bounds*. Artificial Intelligence **50** (1991), no. 2, 207 – 221.

[San94]   N. G. F. SANCHO, *Shortest path problems with time windows on nodes and arcs*. J. Math. Anal. Appl. **186** (1994), no. 3, 643 – 648, MR 95e:90117.

[Sapo83]   M. S. SAPAROV, *The branch-and-bound method for calculating recurrent functions defined on graphs*. Engineering Cybernetics **20** (1982), no. 6, 112 – 119 (1983).

[Sav93]   C. D. SAVAGE, *Long cycles in the middle two levels of the Boolean lattice*. Ars Combin. **35** (1993), A, 97 – 108, MR 94m:05116.

[SBvL87]   A. A. SCHOONE, H. L. BODLAENDER, J. VAN LEEUWEN, *Diameter increase caused by edge deletion*. J. Graph Theory **11** (1987), no. 3, 409 – 427, MR 88k:05113.

[Scap90]   R. SCAPELLATO, *On F-geodetic graphs*. Discrete Math. **80** (1990), no. 3, 313 – 325, MR 91e:05049.

[SCGS]   U. K. SARKAR, P. P. CHAKRABARTI, S. G. GHOSE, S. C. DE SARKAR, *Effective use of memory in iterative deepening search*. Inform. Process. Lett. **42** (1992), no. 1, 47 – 52.

[Sch89]   I. SCHIERMEYER, *Neighborhood intersections and hamiltonicity*. In: Graph theory, combinatorics, algorithms, and applications (San Francisco, CA, 1989). SIAM, Philadelphia, PA, 1991, 427 – 440.

[Schr94]   A. SCHRIJVER, *Finding k disjoint paths in a directed planar graph*. SIAM J. Comput. **23** (1994), no. 4, 780 – 788, MR 95f:05070.

[Schu93]   S. SCHUIERER, *Rectilinear path queries in a simple rectilinear polygon*. In: P. Enjalbert, A. Finkel, K. W. Wagner (eds.), Proceedings of the 10th Symposium on Theoretical Aspects of Computer Science (STACS '93) (Würzburg, 1993). Lecture Notes in Comput. Sci., Vol. 665, Springer, Berlin, Heidelberg, New York, 282 – 293.

[Schw89]   A. SCHWILL, *Shortest edge disjoint paths in graphs*. In: J. Cori, B. Monien (eds.), Proceedings of the 6th Symposium on Theoretical Aspects of Computer Science (STACS '89) (Paderborn, 1989). Lecture Notes in Comput. Sci., Vol. 349, Springer, Berlin, Heidelberg, New York, 505 – 516.

[Se93]   N. SEIFTER, *On the girth of infinite graphs*. Discrete Math. **118** (1993), no. 1 – 3, 275 – 283, MR 94b:05113.

[Seid92]   R. SEIDEL, *On the all-pairs-shortest-path problem*. In: Proceedings of the 24th ACM Symposium on Theory of Computing (STOC '92) (Victoria, B.C., 1992), 745 – 749.

[SeVi86]   R. SEDGEWICK, J. S. VITTER, *Shortest paths in euclidean graphs*. Algorithmica **1** (1986), 31 – 48.

[Shen88]   Y. CH. SHEN, *The best directed Hamiltonian paths in tournaments*. Zhangzhou Shiyuan Xuebao (Ziran Kexue Ban) **1988**, no. 1, 38 – 48, MR 91c:05119.

[Shi80]   Y. SHILOACH, *A polynomial solution to the undirected two paths problem*. J. Assoc. Comput. Mach. **27**, no. 3 (1980), 445 – 456.

[Shi94]   R. H. SHI, *The average distance of trees*. Systems Sci. Math. Sci. **6** (1993), no. 1, 18 – 24, MR 94e:05098.

[Shne81]   M. SHNEIER, *Path-length distances for quadtrees*. Inform. Sci. **23** (1981), no. 1, 49 – 67, MR 82a:68130.

[ShWi81]   D. R. SHIER, C. WITZGALL, *Properties of labeling methods for determining shortest path trees*. J. Res. Nat. Bur. Standards **86** (1981), no. 3, 317 – 330, MR 82j:68067.

[ShWN92]   J. SHAPIRO, J. WAXMAN, D. NIR, *Level graphs and approximate shortest path algorithms*. Networks **22** (1992), 691 – 717.

[Sik88]   KR. SIKDAR, *Generalized t-ary trees and their path lengths with applications*. In: Combinatorial mathematics and applications (Calcutta, 1988). Sankhyā Ser. A **54** (1992), Special Issue, 443 – 459; MR 94d:05041.

[Simo87]   J. M. S. SIMOÑES-PAREIRA, *A note on convexity and submatrices of distance matrices*. Linear and Multilinear Algebra **20** (1987), no. 4, 363 – 366, MR 88f:05080.

[Simo90]   J. M. S. SIMOÑES-PAREIRA, *An algorithm and its role in the study of optimal graph realizations of distance matrices.*. Discr. Math. **79** (1990), no. 3, 299 – 312, MR 91m:05134.

[SkGo89]   C. C. SKISCIM, B. L. GOLDEN, *Solving k-shortest and constrained shortest path problems efficiently*. Network optimization and applications. Ann. Oper. Res. **20** (1989), no. 1 – 4, 249 – 282, MR 90i:90048.

[Sku87]   Z. SKUPIEŃ, *Sharp sufficient conditions for Hamiltonian cycles in tough graphs*. Combinatorics and graph theory (Warsaw, 1987). Banach Center Publ., Vol. 25, PWN, Warsaw, 1989, 163 – 175, MR 91m:05128.

[SLB90]   D. SIMCHI-LEVI, O. BERMAN, *Optimal locations and districts of two traveling salesmen on a tree*. Networks **20** (1990), no. 7, 803 – 815, MR 91k:90079.

[SmDe75]   H. M. SMITH, J. J. DEELY, *A secretary problem with finite memory*. J. Amer. Statist. Assoc. **70** (1975), 357 – 361.

[SmMT90]   T. H. C. Smith, T. W. S. Meyer, G. L. Thompson, *Lower bounds for the symmetric trav-elling salesman problem from Lagrangean relaxions.* Discrete Appl. Math. **26** (1990), 209 – 217, MR 91a:90141.

[Smy87]   W. F. Smyth, *Sharp bounds on the diameter of a graph.* Canad. Math. Bull. **30** (1987), no. 1, 72 – 74.

[Snie86]   M. Sniedovich, *A new look at Bellman's principle of optimality.* J. Optim. Theory Appl. **49** (1986), no. 1, 161 – 176.

[Šolt86]   L. Šoltés, *Orientations of graphs minimizing the radius or the diameter.* Math. Slovaca **36** (1986), 289 – 296.

[So86]   Zh. M. Song, *Long paths and cycles in bipartite oriented graphs.* J. Nanjing Inst. Tech. **2** (1986), no. 5, 1 – 5.

[Song86]   Zh. M. Song, *Long paths and cycles in oriented graphs.* J. Nanjing Inst. Tech. **16** (1986), no. 5, 102 – 108, MR 87m:05117.

[Song89]   Zh. M. Song, *Number of arcs and longest cycles in digraphs.* J. Southeast Univ. **19** (1989), no. 3, 74 – 80, MR 91a:05050.

[Song92]   Z. M. Song, *Degrees, neighborhood unions and Hamiltonian properties.* Ars Combin. **34** (1992), 205 – 211, MR 93m:05124.

[SpiP73]   P. M. Spira, A. Pan, *On finding and updating shortest paths and spanning trees.* In: Conference Record, IEEE 14th Annual Symposium on Switching and Automata Theory, 1973, 82 – 84.

[SpTs86]   P. Spirakis, A. Tsakalidis, *A very fast, practical algorithm for finding a negative cycle in a digraph.* In: L. Kott (ed.), Proceedings of the 13th International Colloquium on Automata, Languages and Programming (ICALP '86) (Rennes, France, 1986). Lecture Notes in Comput. Sci., Vol. 226, Springer, Berlin, Heidelberg, New York, 397 – 406.

[Sri87]   N. Srinivasan, *A characterization of geodetic graphs.* J. Math. Phys. Sci. **21** (1987), no. 2, 143 – 146, MR 88c:05090.

[SrOA87]   N. Srinivasan, J. Opatrný, V. S. Alagar, *Construction of geodetic and bigeodetic blocks of connectivity $k \geq 3$ and their relation to block designs.* Ars Combin. **24** (1987), 101 – 114, MR 88k:05117.

[Sti92]   S. Stifter, *Secure Path Planning.* In: M. Overmars (ed.), Workshop on Computational Geometry (CG '92). Technical Report RUU-92-10, Utrecht University, Utrecht, 1992, 7 – 8.

[Sto79]   G. C. Stockman, *A minimax algorithm better than Alpha-Beta ?* Artificial Intelligence **12** (1979), 179 – 196, MR 81d:68130.

[Stou93]   L. Stougie, *A fast randomized algorithm for partitioning a graph into paths of fixed length. – Combinatorial structures and algorithms.* Discrete Appl. Math. **42** (1993), 291 – 303, MR 94e:05238.

[StR94]   J. A. Storer, J. H. Reif, *Shortest paths in the plane with polygonal obstacles.* J. Assoc. Comp. Mach. **41** (1994), no. 5, 982 – 1012.

[StSn89]   M. Steele, T. L. Snyder, *Worst case growth rates of some classical problems of combinatorial optimization.* SIAM J. Comput. **18** (1989), no. 2, 278 – 287.

[StWh91]   Br. S. Stewart, Ch. C. White, *Multiobjective A*.* J. Assoc. Comp. Mach. **38** (1991), no. 4, 775 – 814, MR 92k:68101.

[SuTa84]   J. W. Suurballe, R. E. Tarjan, *A quick method for finding shortest pairs of disjoint paths.* Networks **14** (1984), 325 – 336, MR 86c:90046.

[Taka91]   T. Takaoka, *A new upper bound on the complexity of the all pairs shortest path problem.* In: G. Schmidt, R. Berghammer (eds.), Proceedings of the 17th International Workshop on Graph-Theoretic Concepts in Computer Science (WG '91) (Fischbachau, Bayern, 1991). Lecture Notes in Comput. Sci., Vol. 570, Springer, Berlin, Heidelberg, New York, 1992, 209 – 213.

[Taka92]   T. TAKAOKA, *A review of all pairs shortest path algorithms*. In: Combinatorial mathematics and applications (Calcutta, 1988). Sankhyā Ser. A **54** (1992), Special Issue, 475 – 501, MR 94d:05135.

[TaMo80]   T. TAKAOKA, A. MOFFAT, *An $O(n^2 \log n)$ expected time algorithm for the all shortest distance problem*. In: Proceedings of the 9th Symposium on Mathematical Foundations of Computer Science (MFCS '80) (Reisen, 1980), Lecture Notes in Comput. Sci. **88** (1980), 643 – 655, MR 82a:68087.

[TaNS80]   K. TAKAMIZAWA, T. NISHIZEKI, N. SAITO, *An algorithm for finding a short closed spanning walk in a graph*. Networks **10** (1980), no. 3, 249 – 263, MR 81k:05074.

[TaPa92]   K. W. TANG, S. A. PADUBIDRI, *Routing and diameter analysis of diagonal mesh networks*. In: Proceedings of the 1992 International Conference on Parallel Processing, Vol. I (St. Charles, IL, 1992). CRC, Boca Raton, FL, 1992, I-143 – I-150, MR 94j:68018.

[Tar83]   M. TARSI, *Optimal search on some game trees*. J. Assoc. Comput. Mach. **30** (1983), no. 3, 389 – 396, MR 84j:68045.

[Tarj81]   R. E. TARJAN, *Fast algorithms for solving path problems*. J. Assoc. Comput. Mach. **28** (1981), no. 3, 594 – 614, MR 82m:90074.

[TeoK92]   C. P. TEO, K. M. KOH, *The number of shortest cycles and the chromatic uniqueness of a graph*. J. Graph Theory **16** (1992), no. 1, 7 – 15.

[Thom87]   C. THOMASSEN, *Counterexamples to Adáms conjecture on arc reversals in directed graphs*. J. Combin. Theory Ser. B **42** (1987), no. 1, 128 – 130, MR 88c:05076.

[Tian90]   S. L. TIAN, *Sum distance in digraphs*. In: Proceedings of the Twenty-First Southeastern Conference on Combinatorics, Graph Theory, and Computing (Boca Raton, FL, 1990). Congr. Numer. **78** (1990), 179 – 192, MR 93m:05065.

[Tian93]   Y. CH. TIAN, *On the longest cycles of 1-tough graphs*. J. Northeast Univ. Tech. **13** (1993), no. 2, 187 – 192, MR 93i:05081.

[Tom94]   I. TOMESCU, *On the sum of all distances in chromatic blocks*. J. Graph Theory **18** (1994), no. 1, 83 – 102.

[Trie]   E. TRIESCH, scientific communication.

[Trusz81] M. TRUSZCZYŃSKI, *A simple algorithm for finding a cycle of length greater than three and without diagonals*. Computing **27** (1981), 89 – 91.

[Va92]   P. VACEK, *Bounds of lengths of open Hamiltonian walks*. Arch. Math. (Brno) **28** (1992), 11 – 16, MR 94b:05125.

[Vanth91] J. VANTHIENEN, *A longest path algorithm to display decision tables*. The Computer Journal **34**, no. 4, (1991), Algorithm 124, 358 – 364.

[Vaĭn85]   A. D. VAĬNSHTEĬN, *A vector problem of the shortest path in a uniform norm* (Russian). Èkonom. i Mat. Metody **21** (1985), no. 6, 1132 – 1137, MR 87f:05101.

[Ve90]   H. J. VELDMAN, *Short proofs of some Fan-type results*. Ars Combin. **29** (1990), 28 – 32.

[vGol80]   M. V. GOLITSCHEK, *An algorithm for scaling matrices and computing the minimum cycle mean in a digraph*. Numerische Mathematik **35** (1980), 45 – 55.

[vGol82]   M. V. GOLITSCHEK, *Optimal cycles in doubly weighted graphs and approximation of bivariate function by univariate ones*. Numer. Math. **39** (1982), 65 – 84.

[vGol87]   M. V. GOLITSCHEK, *The cost-to time ratio problem for large or infinite graphs*. Discr. Appl. Math. **16** (1987), 1 – 9.

[vGoS84]   M. V. GOLITSCHEK, H. SCHNEIDER, *Applications of shortest path algorithms to matrix scalings*. Numer. Math. **44** (1984), 111 – 126.

[Vich82]   S. A. VICHES, *An efficient algorithm to construct the shortest tour of a one-color connected drawing*. Avtomat. i Telemekh. (1982), no. 12, 85 – 96 (Russian); translated in Automat. Remote Control **43** (1982), no. 12, part 2, 1569 – 1579 (1983); MR 85c:68072.

[Wai90]   G. R. WAISSI, *A new $O(n^2)$ shortest chain algorithm*. Appl. Math. Comput. **37** (1990), no. 2, part II, 111 – 120, MR 91e:90126.

[Wal72]   W. WALTER, *Gewöhnliche Differentialgleichungen*. 4th edition, Springer, Berlin, Heidelberg, New York, 1990.

[Warr91]   R. H. WARREN, *Polynomially solvable traveling salesman problems*. Zastos. Mat. **21** (1991), no. 2, 283 – 287, MR 92i:90067.

[Wata81]   O. WATANABE, *A fast algorithm for finding all shortest paths*. Inform. Process. Lett. **13** (1981), no. 1, 1 – 3.

[Wei84]   A. WEISS, *Girths of bipartite sextet graphs*. Combinatorica **4** (1984), 241 – 245, MR 86c:05082.

[Wei93a]   B. WEI, *A generalization of a result of Bauer and Schmeichel*. Graphs Combin. **9** (1993), no. 5, 961 – 967, MR 95b:05139.

[Wei93b]   B. WEI, *Hamiltonian paths and Hamiltonian connectivity in graphs*. In: Graph Theory (Niedzica Castle). Discrete Math. **121** (1993), 223 – 228. MR 94i:05060.

[Whit32]   H. WHITNEY, *Congruent graphs and the connectivity of graphs*. Amer. J. Math. **54** (1932), 150 – 168.

[Widm91]   P. WIDMAYER, *On graphs preserving rectilinear shortest paths in the presence of obstacles*. Ann. Oper. Res. **33** (1991), 557 – 575, MR 92i:05134.

[WiKoCe]   S. WIMER, I. KOREN, I. CEDERBAUM, *On paths with the shortest average arc length in weighted graphs*. Discrete Appl. Math. **45** (1993), 169 – 179, MR 94g:05088.

[Win87]   P. WINTER, *Steiner Problem in Networks: A Survey*. Networks **17**, 129 – 167.

[WiSm92]   P. WINTER, M. SMITH, *Path-distance heuristics for the Steiner problem in undirected networks*. The Steiner problem. Algorithmica **7** (1992), no. 2 – 3, 309 – 327, MR 92k:05074.

[Wojc90]   E. WOJCICKA, *Hamiltonian properties of domination-critical graphs*. J. Graph Theory **14** (1990), no. 2, 205 – 215, MR 91c:05120.

[WoWo88]   A. P. WOJDA, M. WOŹNIAK, *Hamiltonian properties of balanced bipartite digraphs*. Graphs, hypergraphs and matroids, III (Kalzig, 1988). Higher College Engrg. Grünberg, Schlesien, 1989, 179 – 186, MR 91d:05067.

[WuMa92]   S. WU, U. MANBER, *Path-matching problems*. Algorithmica **8** (1992), 89 – 101.

[WuShi]   CH. X. WU, B. Y. SHI, *Graphs determined by the distribution of their cycle lengths*. Shanghai Shifan Daxue Xuebao Ziran Kexue Ban **21** (1992), no. 4, 15 – 20, MR 93m:05112.

[Xu91]   SH. J. XU, *Relationships between parameters of a graph*. Discrete Math. **89** (1991), no. 1, 65 – 68, MR 92j:05161.

[Xu92]   J. M. XU, *An inequality relating the order, maximum degree, diameter and connectivity of a strongly connected digraph*. Acta Math. Appl. Sin. **8**, no. 2 (1992), 144 – 152, MR 93m:05086.

[Yin91]   J. H. YIN, *Neighbourhood unions and Hamiltonian properties*. J. Southeast Univ. **21** (1991), no. 1, 97 – 100, MR 92g:05119.

[YJ91]   CH. E. YANG, D. Y. JIN, *A primal-dual algorithm for the minimum average weighted length circuit problem*. Networks **21** (1991), no. 7, 705 – 712, MR 92k:90115.

[YTO91]   N. E. YOUNG, R. E. TARJAN, J. B. ORLIN, *Faster parametric shortest paths and minimum-balance algorithms*. Networks **21** (1991), 205 – 221, MR 91m:68147.

[Yu94]   ZH. G. YU, Y. J. ZHU, Z. H. LIU, *Longest cycles in regular 3-connected graphs*. Systems Sci. Math. Sci. **3** (1990), no. 4, 289 – 297, MR 94c:05047.

[Yuc89]   J. L. YUCAS, *Hamiltonian cycles of binary Lyndon words*. In: Proceedings of the Nineteenth Manitoba Conference on Numerical Mathematics and Computing (Winnipeg, MB, 1989). Congr. Numer. **75** (1990), 111 – 114, MR 91f:05082.

[YUOK83]   S. YAMADA, A. UMEZU, T. OHNO, T. KASAI, *Approximation methods for finding a Hamiltonian walk with minimum cost.* Bull. Univ. Osaka Prefect. Ser. A **31** (1982), no. 1, 37 – 41 (1983), MR 84m:05047.

[Zamf92]   CH. ZAMFIRESCU, T. ZAMFIRESCU, *Hamiltonian properties of grid graphs.* SIAM J. Discrete Math. **5** (1992), no. 4, 564 – 570, MR 93m:05129.

[Zeli82]   B. ZELINKA, *Elongation in a graph.* Math. Slovaca **32** (1982), no. 3, 291 – 296.

[Zha81]   C. QU. ZHANG *Path and cycles in oriented graphs.* Kexue Tongbao (English series) **26** (1981), no. 10, 865 – 868, MR 83h:05059.

[Zha87]   M. K. ZHANG, *Longest paths and cycles in bipartite oriented graphs.* J. Graph Theory **11** (1987), no. 3, 339 – 348.

[Zhao92]   B. Z. ZHAO, *The circumference and girth of a simple graph.* J. Northeast Univ. Tech. **13** (1992), no. 3, 294 – 296, MR 93k:05090.

[Zhou92]   G. L. ZHOU, *Proof of a conjecture of S. Fajtlowicz.* Qufu Shifan Daxue Xuebao Ziran Kexue Ban **18** (1992), suppl. 14, 17; MR 93m:05057.

[ZKC90]   S. Q. ZHENG, K. H. KWON, J. CHEN, *Finding a shortest path in twisted hypercubes.* Congr. Numer. **83** (1991), 75 – 90, MR 93b:05104.

[Znam90]   S. ZNÁM, *Minimal size of graphs with diameter 2 and given maximal degree.* Ars Combin. **28** (1989), 278 – 284, MR 91b:05113.

[Znam92]   S. ZNÁM, *Minimal size of graphs with diameter 2 and given maximal degree. II.* Acta Math. Univ. Comenian. (N. S.) **61** (1992), no. 2, 209 – 217, MR 94a:05115.

Index

Remarks: 1. Algorithms with short names are listed under "Algorithm"; algorithms with long names can be found under their names themselves.
2. At the end of the subject index, there is a special list of terms beginning with a variable; these terms are ordered with regard to their constant suffixes. E.g., the term "v-w-path" can be found by searching for the suffix "-path" in this list.

Subject index

Symbol index